Sustainable Well-being

Advancing Health through Psychology, Medicine, and Intelligent Technologies

This book brings together a diverse and forward-looking collection of research at the intersection of healthcare, psychology, and intelligent technologies, united by the goal of fostering sustainable well-being. The papers span cutting-edge innovations in AI-driven diagnostics, including early detection of autism, Alzheimer's disease, cancer, cardiac conditions, and neurological disorders. It also includes techniques on advanced imaging and machine-learning approaches that enhance accuracy, accessibility, and long-term clinical impact. Complementing these medical advancements are studies addressing mental health, stress, sociological determinants of well-being, emotional state detection, and therapeutic interventions. Several contributions explore mobile and remote-care systems, sustainable diagnostic pathways, and intelligent tools that broaden healthcare reach. Together, this body of work showcases how multidisciplinary collaboration across medicine, psychology, computer science, and public health can create resilient, equitable, and technologically empowered systems that support healthier communities and a more sustainable future.

Dr. J. Anitha Menon is a Professor of Psychology and Head of the School of Psychology at Curtin Dubai. With 32 years of global experience, she specialises in youth wellbeing and culturally sensitive behavioural interventions, leading influential research, mentoring emerging psychologists, and advancing mental health across diverse communities. [https://orcid.org/0000-0002-7388-7509].

Dr. Benita Veronica, Senior Lecturer and Program Coordinator for IT & Cybersecurity at Curtin University Dubai, has 15 years of research and leadership experience. A prolific Scopus-indexed author, she builds strong industry collaborations, leads workshops, and serves as a respected reviewer and Ph.D. external examiner, contributing significantly to academia and AI-driven healthcare research. [https://orcid.org/0000-0002-5501-7382].

Dr. Ben Milbourn is an Associate Professor of Occupational Therapy at Curtin University Australia, and Deputy Director of the Curtin Autism Research Group (CARG). He specializes in mental health and disability research. His scholarly work investigates the occupational meaning of everyday activities for individuals with severe mental illness, combining academic rigor with extensive international clinical experience to advance occupational science. [https://orcid.org/0000-0003-3213-7987].

Dr. Gnana Sanga Mithra S. is the Associate Dean of Research (Interdisciplinary) at Vinayaka Mission's Law School, India. Specializing in Sociology, Transgender Studies, and Gender Studies, she drives interdisciplinary scholarship through extensive Scopus-indexed publications and international conference engagements, demonstrating strong research leadership and domain expertise. [https://orcid.org/0000-0002-8133-5886].

I0882090

Sustainable Well-being

Advancing Health through Psychology, Medicine, and Intelligent Technologies

Edited by

Dr. J. Anitha Menon
Dr. Benita Veronica
Dr. Ben Milbourn
Dr. Gnana Sanga Mithra S.

CRC Press
Taylor & Francis Group
Boca Raton London New York

CRC Press is an imprint of the
Taylor & Francis Group, an **informa** business

First edition published 2026
by CRC Press
4 Park Square, Milton Park, Abingdon, Oxon, OX14 4RN

and by CRC Press
2385 NW Executive Center Drive, Suite 320, Boca Raton FL 33431

CRC Press is an imprint of Informa UK Limited

British Library Cataloguing-in-Publication Data
A catalogue record for this book is available from the British Library

ISBN: 9781041295129 (hbk)
ISBN: 9781041296324 (pbk)
ISBN: 9781003770473 (ebk)

DOI: 10.1201/9781003770473

Typeset in Times New Roman
by HBK Digital

Contents

Lists of figures

Lists of tables

1 Sustainable solutions for automated classification and segmentation of kidney cysts in CT scans using Leaf Gradient-Based Neural Architecture Search (LGBNAS)

S. K. Aruna[1,a], *N. Deepa*[2,b], *and T. Devi*[2,c]

[1]Research Scholar, Department of Computer Science and Engineering, Saveetha School of Engineering, Saveetha Institute of Medical and Technical Sciences, Chennai, Tamil Nadu, India
[2]Professor, Department of Computer Science and Engineering, Saveetha School of Engineering, Saveetha Institute of Medical and Technical Sciences, Chennai, Tamil Nadu, India

Abstract: Kidney cysts are frequent medical conditions that can range in severity from healthy to potentially harmful. They are sacs filled with fluid formed in or on the lining of the kidneys. For prompt identification and effective medication, kidney cysts must be accurately identified and classified. The most common diseases are chronic kidney diseases and kidney injury and kidney infections. Kidney cysts generally detected by the ultrasound, computed tomography, and magnetic resonance imaging. Computer tomography images are being effectively classified and segmented using several models based on deep learning. Convolutional deep neural networks (CNNs), models that have been trained, and an optimization technique on images will be used and automatically classify individuals from abdominal CT scans. The CNN models that were previously trained VGG16, VGG19, Exception, MobileNet, MobileNet V2 were employed in this investigation. In addition, the optimal configuration of the pre-trained model is improved through the application of Leaf Gradient-based Neural architecture search (LGBNAS). An enhanced impact propagated per second is found, and the success rate of the suggested technique is compared with the results of existing techniques. The advised architecture was developed using CT scans of kidney cysts that are freely accessible on Kaggle. Using CT scans, our automated simulations identified kidney cysts with a 99% accuracy rate.

Keywords: Artificial intelligence, AutoML, deep learning, CNN, leaf gradient-based neural architecture search (LGBNAS), transfer learning

1. Introduction

The kidneys are a pair of bean-shaped organs that filter squander from bloodstreams and produce urine. Kidney cysts are sacs filled with fluid that develop in the kidneys. These cysts range across sizes compared to tiny to a few centimeters in size, and they can be either straightforward or intricate. Simple cysts of the kidney are often harmless and do not produce any symptoms, especially in older adults. The assembly cysts, on the other hand, might contain solid substance or exhibit wall irregularities, which could point to a greater likelihood of cancer or kidney disease. Although the precise cause of kidney cysts is frequently unknown, genetic disorders like the condition polycystic kidney disease (PKD) have been linked to the kidney cysts. Ultrasound imaging and CT scans are frequently used in the diagnosis process. The nature and indications of the tumors will determine the course of therapies. While complicated or symptoms cysts may require medical attention, straightforward cysts frequently don't need any kind of therapy. The imitation of human intellect in devices that are designed to process information and resolve problems similarly to humans is known as artificially intelligent (AI). Quick technological advancements have led to the emergence of this branch of computer science, which now includes robotics, neural networks, machine learning, and language processing (Figure 1.1).

1.1. Artificial intelligence

Artificial intelligence (AI) systems are made to carry out tasks like recognition of speech, perception of images, language translation, and decision-making that normally require intelligence from humans. The way we live and work is being revolutionized by the incorporation of AI into various industries, which is bringing about advancements in the fields of

[a]arunask9001.sse@saveetha.com, [b]deepa23narayanan@gmail.com, [c]devi.janu@gmail.com

DOI: 10.1201/9781003770473-1

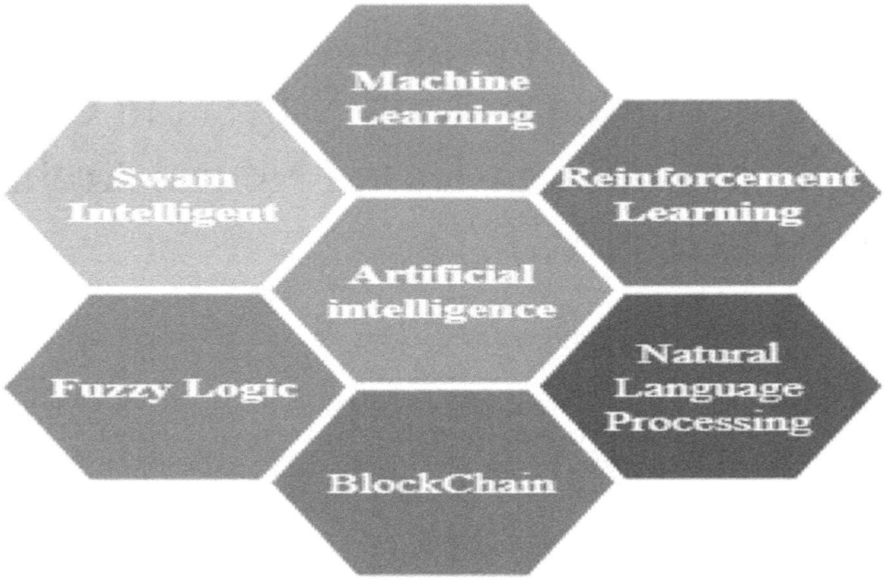

Figure 1.1. Basic types of artificial intelligence.

Source: Author.

healthcare, finance, shipping, and amusement. AI has the potential to drive productivity and creativity in all industries by streamlining repetitive tasks, generating knowledge from massive amounts regarding data, and improving human capabilities. But as AI develops, it also brings up significant moral and societal issues, such as privacy issues, job relocation, and the necessity of oversight to guarantee the technology's fair and secure application. The main categories of artificial intelligence (AI) are Machine Learning (ML), in which computers are taught by data to perform better on tasks; Deep Learning (DL), a subset of ML that uses multiple-layered neural networks to model complex patterns.

1.2. Neural network

Neural networks, which serve as the foundation for the technique of Deep Learning (DL), are specialized subsets of computational neuroscience that concentrate on developing algorithms that mimic the framework and functions of the human brain. Deep learning algorithms are characterized by many layers of attached nodes, or "brain cells," which enable them to consequently recognize and extract complex patterns from vast amounts of data. Deep learning is especially useful for tasks like picture and recognition of speech, processing natural language, and autonomous systems because of this capability. The performance and accuracy of deep learning models are greatly improved over standard artificial intelligence techniques, which frequently call for manual feature extraction from the data. Instead, features are learned directly from the data. Deep learning has advanced quickly due to advances in neural network architectures, increased power for computation, and the availability of large datasets. Deep learning is therefore at the heart of several innovative technologies today, pushing the limits of what is possible for devices to comprehend and communicate with the outside world. Despite its amazing success, deep learning still has drawbacks, including the requirement for enormous volumes of labeled data, expensive processing, and the challenges of deciphering the internal workings of intricate models. In order to progress the field, investigators are actively tackling these problems as well as investigating novel architectures and methodologies. Deep learning may lead to even more significant advances in AI in the years to come, making it possible to create systems that are more complex and smarter. Figure 1.2 demonstrates the basic introduction diagram for deep learning architecture.

The use of Neural Architecture Search (NAS) algorithms to recognise and examine patterns of leaves in kidney cyst medical imaging is investigated in this work. We hope to improve pattern identification and categorization accuracy with NAS, which could lead to better evaluation processes. NAS can be divided into three fundamental elements: search space, search strategy and Assessment strategy. Different approaches can be taken to these components to optimize searching for successful neural network architectures. The collection of all structures from which the NAS technique can select is known as a search space. The size of standard NAS in search spaces differs about a couple of thousand to more than 1020. Although the search space can be very broad in principle, it can be made simpler by constructing the search space with domain information in mind. An optimization method called a search approach is applied to locate a high-performing construction in the search space. Search strategies fall into two primary groups: one-shot techniques and black-box optimization-based techniques.

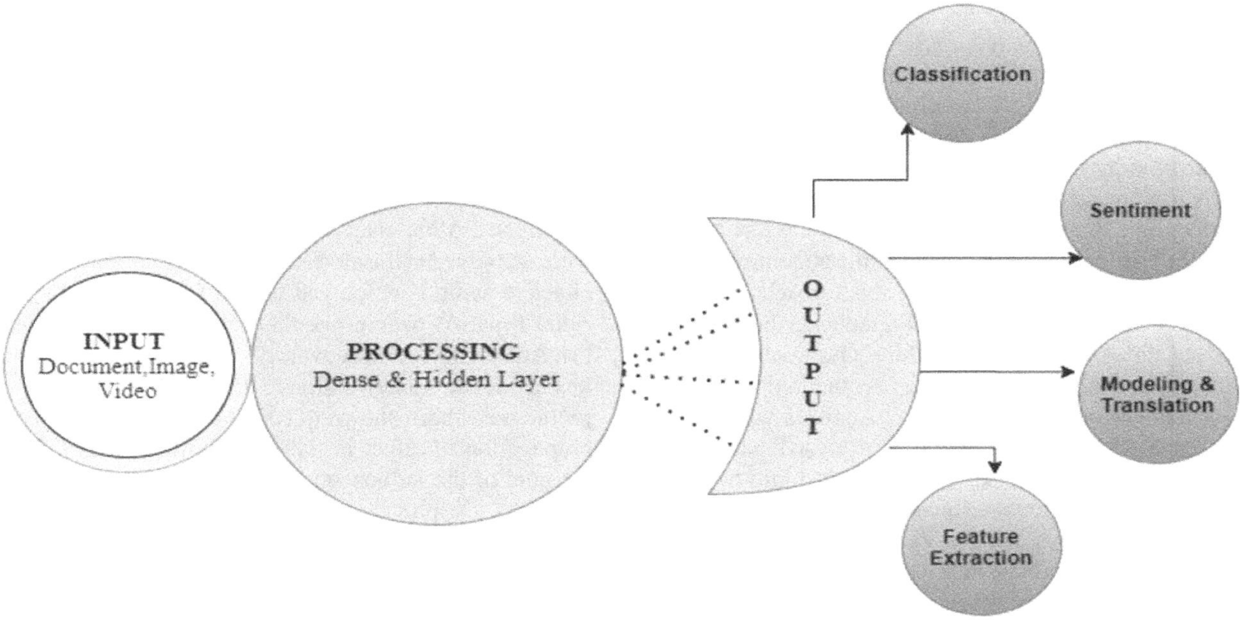

Figure 1.2. Basic introduction diagram for deep learning.
Source: Author.

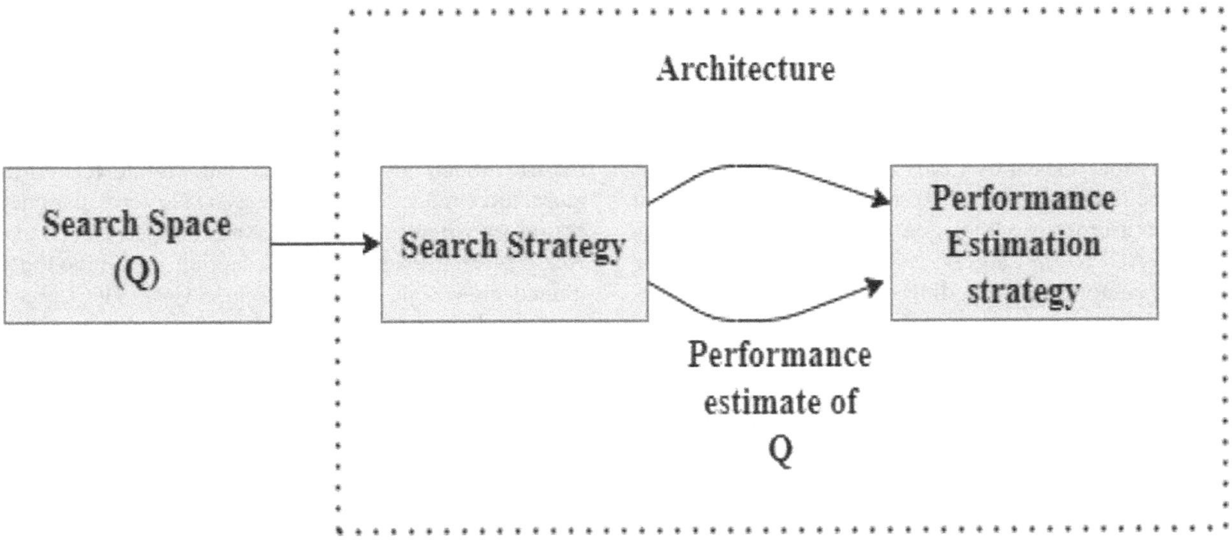

Figure 1.3. Basic diagram of neural architecture search.
Source: Author.

Any technique for rapidly calculating neural construction efficiency to prevent the architecture from becoming fully trained is known as a performance estimates strategy. Figure 1.3 demonstrates basic graphical representation of Neural Architecture Search**.**

The aim of this research work is:

1. This study explores the application of Leaf Gradient Based Neural Architecture Search (LGBNAS) algorithms to identify and analyse leaf patterns within medical imaging of kidney cysts.
2. To assess the reliability and effectiveness of Gradient Based method in NAS-optimized simulations in comparison with conventional models. Accuracy fluctuation is assessed based on a variety of the cyst's instances.

2. Related Work

The following reasons are discussed about the existing limitation The distinctive feature of ADPKD is the forward-thinking development of both sides of cysts in the kidneys, which result in substantial rises in kidney weight and eventual loss of function. As of this moment, the sole recognised biomarker for monitoring ADPKD is total kidney volume (TKV). Every individual has this determined at least once a year to assess the total cyst weight and degree of kidney expansion. The present planimetry tracking is used to do this, which entails manually employing an electronic drawing tool to distinguish the renal system from the tissue surrounding them in the abdominal region. TKV can be calculated by doing this to each image in a nuclear medicine scan. For radiographers, this is a tedious and time-consuming procedure. Our goal is to create an automated technique for kidney segmentation and TKV measurement in ADPKD patients. Analysis was done on thirteen MRI images of organs with cyst loads ranging from mild to serious. Visuals were divided into two parts, each consisting of 200 the square of areas. To differentiate between the kidneys and non-kidney areas, includes were taken from the gray scale metrics for every area and paired with this data in a supervised decision- tree algorithm. To approximately segment the urinary tract, the 400 the variables x 400 categorized matrix was subjected to narrowing and filtration. Kidney boundaries were determined by applying an active contouring function, followed by enhancing the contrast and k-means clustering. After segmented kidney shapes had been verified to have suitable relative sphericity through eccentricity analysis, the areas of the kidney shapes were combined with straight extrapolating to calculate TKV [1].

This paper classifies Liver cysts are a hallmark of polycystic liver disease (PLD), which is a condition that can result from autosomal dominant polycystic kidney disease (ADPKD). Largely enlarged livers caused by PLD can compress surrounding structures and cause discomfort, even though most patients do not exhibit any symptoms. This may necessitate liver transplantation, limited liver removal, or cyst aspiration/fenestration. When the volume of the liver cyst remains too small to influence liver volume, tracking PLD through measurement of liver volume is unable to follow the disease's early stages. Utilizing advanced machine learning (DL) models to automate the identification and division of liver cysts to enhance PLD assessment in its early stages. In polycystic liver disease, DL robotics of the liver cyst subdivisions shows promise for improved monitoring of liver cyst volume [2].

This paper analyses, One of the major challenges in as a result, this paper proposes an innovative technique based on a gradient look at to address the issue. This strategy strikes an equilibrium between execution time and affect propagation. This study presents findings of greater impact propagated per second after contrasting the suggested method's achievement to that of other computation [3].

In this paper, An artificial intelligence (AI) learning transfer framework is proposed in this study to identify the RD early on. Convolutional neural networks (CNNs), models that have been trained, and an optimization method on pictures will be used in the structure to accurately and automatically define individuals who have RD from magnetic resonance imaging (MRI) and pictures taken during microscopic histopathological checks. Pre-trained CNN models such as VGG16, which VGG19, Exception, is, MobileNet, MobileNetV2, MobileNetV3Large, and NAS were employed in this investigation. Furthermore, the most effective setup is employed by the Sparrow search algorithm (SpaSA) to improve the trained model's efficiency [4]. Analyzing social networks is determining which entities inside the framework have the greatest influence; this problem has been shown to be NP-hard. Ensuring that the propagation of affect in the community is maximized is the goal of the influence maximization problem in social networks. Despite the large number of influence maximization algorithms that have been proposed, most of them offer effective spread at the expense of time required for execution. When the volume of the liver cyst remains too small to influence liver volume, tracking PLD through measurement of liver volume is unable to follow the disease's early stages. This paper presents a There were two sets of data used.

The first contains four categories: tumor, stone, a cyst, and typical. Regarding the latter, the second dataset comprises five groups that correspond to the tumor's degree of severity: Grade 0, Grade 1, Grade 2, Grade 3, and Grade 4. When it comes to the four-class dataset, the best pre-trained models are Dense Net and MobileNet. Furthermore, three models suggest the SGD Nesterov parameters optimizer, whereas two models only suggest AdaGrad and AdaMax. DenseNet201 and Exception, as are the top pre-trained models in the five-class data set. The suggested framework is superior to other cutting-edge models for classification, as demonstrated by the experimental results [5]. The Table 1.1 shows the literature survey.

3. Proposed Work

A CNN model and a LGBNAS explanation structure were both included in the proposed scheme. CT scans have been separated into four groups by the CNN structure (cyst, normal, stone, tumour). Figure 1.4 demonstrates the proposed LGBNAS model's graphical representation. Based on the workflow of the proposed work.

3.1. Data preprocessing

The kidney cysts images have various steps of data preprocessing. They are image normalization, resizing augmentation and they split the data from the train test and verification. In Normalization the scaling the pixel values to a regular range, typically among one and zero, image

Table 1.1. Literature survey

Author	*Work Procedure*	*Result Obtained*
Jamie A. O'Reilly et al.	Contrast enhancement and k-means clustering was performed before applying an active contour function to determine kidney edges	MAE of 105.65 ml, MAPE of 21.67 %, and CCC of 0.975.
Mohammad Arafat Hussain et al.	Cross-sectional (2D) slices from the estimated span are subsequently used in a combined sagittal-axial Mask RCNN that detects the organ bounding boxes on the axial and sagittal slices, the combination of which produces a final 3D organ bounding box	Our method produces a kidney boundary wall localization error of ~2.4mm and a mean volume estimation error of ~5%.
Mina Chookhachizadeh Moghadam	UNet-based platform (U-Net) was trained	The model achieved 81% Dice-score and 91% voxel-level true positive rate for automated liver cyst annotation in autosomal dominant polycystic kidney disease.
Sudharson et al.	Multi-class kidney abnormalities ResNet-101- SVM	The model achieved 87.31%
Nazari et al.	Death Risk Segmentation of 5 Years in RCC patients. Model worked in XGboost	The model achieved 0.98%
Chen et al.	Prediction of survival of RCC patient and cell profiler	AUC is 97% and accuracy increased by 6.6%
Sudharson et al. (4)	Total Kidney stone multi-segmentation. ResNet-101, ShuffleNet, and MobileNet-v2	Accuracy achieved by this model is 96.54%

Source: Author.

normalization aids in stabilizing and increasing the neural network training. In resizing it will check for the same dimension of every picture to a consistent (224 × 224) their dimensions. Of batch processing in the network. A picture can be resized to an entirely novel set of dimensions, or the size (width as well as height) can be changed while preserving the ratio of sides to sides.

From the original coordinates (x_1,y_1) to the new image coordinates

$$x_2 = \left(\frac{w_2}{w_1}\right) x_1 \tag{1}$$

$$y_2 = \left(\frac{H_2}{H_1}\right) y_1 \tag{2}$$

The Original dimension of images is Height and Width is

$$Hw = image.s(2) \tag{3}$$

The resized image of kidney cyst images

$$RI = CV2.resize(I,(W,H)) \tag{4}$$

Data Augmentation technique is used to increase the size of the image and assortment of the dataset by implementing the random transformation of the images.in this image transformation incorporate rotation, flipping, scaling and translation and noise.in addition data augmentation performs the operation in a batch of images. First step of data augmentation is to transform the series of random images into a single image. In this meaning the image will not be changed and will be applied to every single image. Second step will be the augment batch, in this augment batch it takes the batch of the image and function image in the batch image are collected to a new batch of image. After the batch process. It will load in the main process of the batch of original image and create augmented image for training a machine model.

3.2. LGBNAS search space

Leaf gradient based Neural Architecture Search space is to identify the recognition or to identify the based on the task designed to include the various components, hyperparameters of the network topologies will cater to the unique requirements of the task. Recognition will focus on capturing the subtle difference in features.it can be achieved through the well-structured LGBNAS search space.

LGBNAS search will component selection of the different types of layers, blocks, and modules available in the search space. Component selection to identity recognition task, the following component are: There are some types of convolutional layers: standard convolution, depth wise Separable convolution, dilated convolution. Used for extracting low level features like edges and textures. Depth wise Separable Convolution to reduce the number of parameters and computational cost, it captures the detailed features. Dilated convolutions capture the multi scale context by expanding the receptive field. The pooling layers are max pooling average pooling and global average pooling. For down sampling while keeping key characteristics, use maximum pooling.

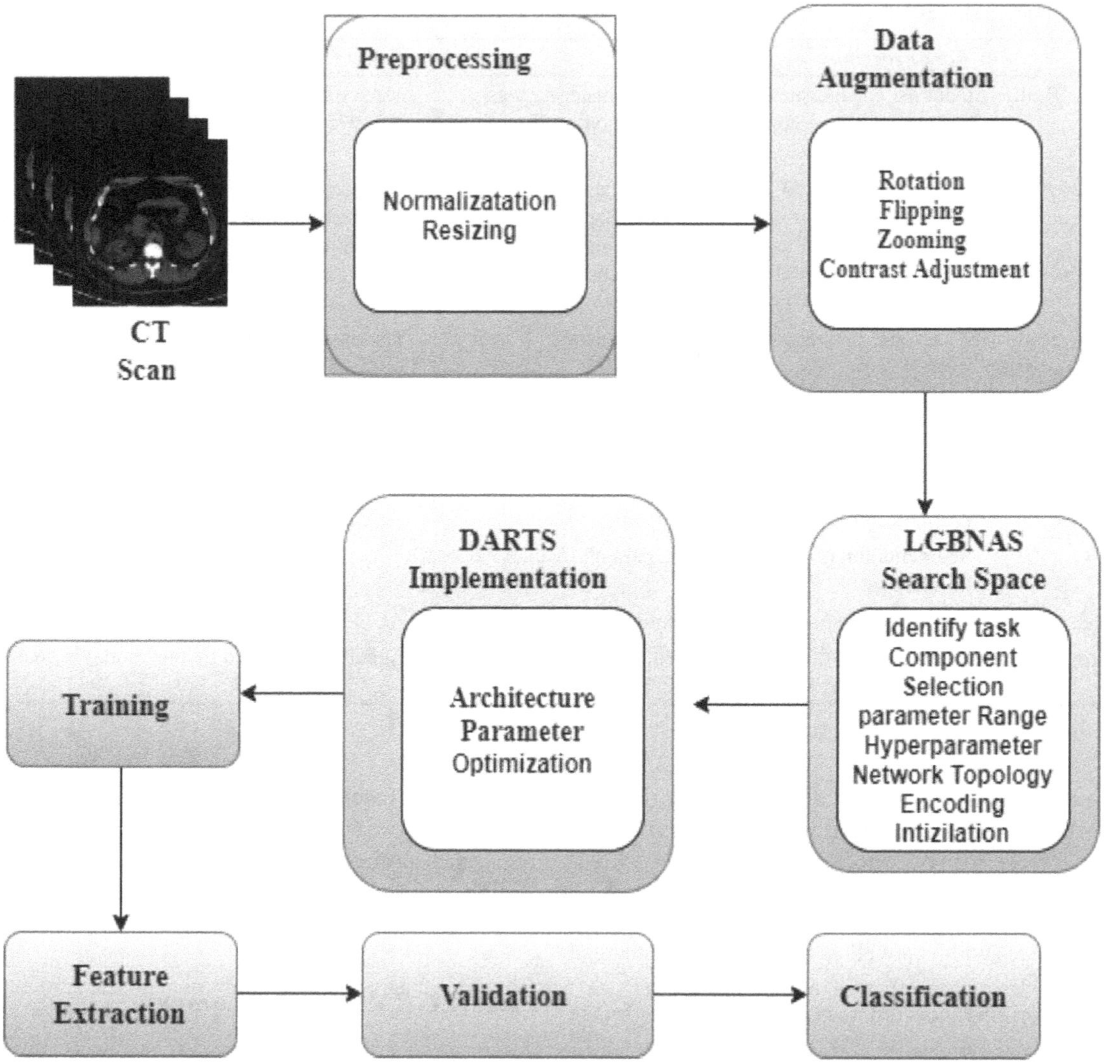

Figure 1.4. Architecture diagram of LGBNAS.

Source: Author.

Utilizing typical pooling will lower the number of dimensions and lessen the chance of excessive fitting. Prior to the last layer being fully linked, global average pooling is frequently utilized to minimize the dimensions of space to one value per channel. Next steps is group normalization is used to aids in accelerating and stabilizing training. And the layer normalization is used across the characteristic dimension, this technique may be employed in certain situations. Next step is depended on activation processes because they present non-linear behavior which enables their models to learn intricate patterns. Every function has unique characteristics and applications. Below is a summary of some typical functions of activation along with their attributes. Activation function in used ReLu technique standard activation function for most layers.

$$f(x) = max\ (0, x) \tag{5}$$

Most used activation function it replaces all negative values with zero and keeps positive values unchanged. It helps to avoid the vanishing gradient problem it suitable for deeper networks. Next step is depending on the attention mechanism is used to improve the models on relevant parts of input data. These input data will be transformed by three vectors, they are query (Q), Value (V) and Key(K). These attention mechanisms score as

$$s(Q, K, V) = softmax\left(\frac{QK^T}{\sqrt{d_k}}\right).V \tag{6}$$

In this Multi head attention mechanism it runs in the multiple self-attention mechanism in parallel sequence of the data and it will improve the performance well and increase computational cost of the mechanism. Handling of the sequence of the data and multi inputs its very crucial. Next step is parameter ranges within LGBNAS search space for specified tasks. The range of layers varies from 5 to 50. Network layer with multiple layers Captures the complex data pattern and it may increase the risk factor of the overfitting. To increase the number of filters applied in the deeper layer, its values range from 16,32,64,128,256 to 512; these are ranges of the number of filter channels. Kernel size may capture the more spatial data context to prevent the overfitting of dropout ranges from 0 to 5 in critical sections. The learning rate will be 1e-4 and 1e-2 and this will be fine tuning the speed and balancing the stability of convergence during training. Hyperparameter of LGBNAS algorithm we used DARTS (Differentiable Architecture Search) these DARTS will to optimize the parameter through Gradient descent with batch size range from 16 to 128 the batches will capture detailed updates and stable gradients of the larger size.

The network topology will be encoded with DAG (Direct Acyclic Graph) or with the cell-based structures of the data. DAG will show the flow between nodes and operation of convulsion or pooling. Normal cell based mainly target the feature extraction and reduction of the down sample cells and it will show the optimized gradients of operation as weighted SoftMax over multiple operations. The feature extraction is frozen layer will be pretrained model will be passed to the NAS optimized layers and these layers will be concentrating on specific to the task patterns, including the like leaf textures and shapes suggestive of kidney cysts, these layers will improve features and develop the characteristics. Pretrained features will be strongly found the capturing general visual patterns. NAS patterns will be representing with the specific dataset and tis will be optimal performance of the data.

Pseudocode for LGBNAS Search Space Design for Identity Recognition Task

---

Require: identify recognition dataset
Ensure: Neural network structure that is optimized for recognising identities

---

1. Initialize global parameter for LGBNAS search
2. Sort LGBNAS solutions (solutions, scorelist)
3. Global parameter= extract solution (solutions, scorelist)
4. i=1
5. While i< PD
6. If random number R2< ST (success threshold)
7. Matute solution[i]=solution[i]+Q*L
8. Increment I
9. End while
10. -i=1
11. While i $\leq$ Nmax − PD (remaining population)
12. If i > 0.5 × Nmax
13. solutions[i] = optimal + |solutions[i] − optimal| × A × L
14. Increment i
15. End While
16. i = 1
17. While i $\leq$ SD
18. If current score scores[i] $\neq$ bestScore
19. solutions[i] = best + β × |solutions[i] − best|
20. Increment i
21. End While
22. Return the updated population of architectures: return solutions
23. End Function

Primary LGBNAS loop to identify the optimal architecture:

```
Def LGBNAS_search (search_space, Num samples):
  best_architecture = None
                    best_score = −1
for i in range(num_samples)
      architecture = sample_architecture(search_space)
      score = train_and_evaluate(architecture)
                  if score > best_score:
        best_score = score
        best_architecture = architecture
return best_architecture, best_score
```

The Next stage of LGBNAS the DART implementation will be simplified in the basic workflow. In this architecture first initialization parameters of the alpha data will be randomly initialized. After this architecture will be applied in the SoftMax function to get over the parameters and their weight of each operation.to train the alpha parameter and weights of the network are used in gradient descent while training the data. After the data has been trained the weight of the edge of the operation of the data has been selected in the architecture.

Images of kidney cysts are used as a training dataset to hone the chosen candidate structures during the LGBNAS process. During this stage, the model can acquire the necessary skills for accurately classifying images by extracting relevant features from them. Ensuring that the simulation captures crucial visual cues that aid in differentiating between renal and non-kidney slices is the aim of the training process.

After training, the model extracts high-level characteristics via the data provided pictures in the component extraction stage. The key trends, including textures, types, and constructions, that are necessary to distinguish within the kidney along with non-kidney portions are represented by these features that were extracted. During categorisation, the network uses these learnt features to help it make more informed choices.

A validation dataset is used to assess the performance of the model during the validation phase. This is a critical step in preventing the model from overfitting to the training set and guaranteeing that it generalizes well to new data. The validation aids in optimizing the model's construction and parameters, enhancing its overall precision and durability.

The final step in the classification process is classifying the input images into the appropriate groups. By running the characteristics gathered by means of a SoftMax layer, and which creates a distribution of probability over the potential classes, the model generates predictions. The final forecast made by the model is assigned to the lesson with the highest probability.

4. Results and Discussion

The efficacy of the suggested deep learning-powered classification of medical images method for kidney evaluation was assessed using common performance indicators and information sets. The proposed technique uses a network of convolutional neural networks (CNN) architecture intended for the best extraction and classification of features to efficiently classify medical pictures into categories like Cyst, Tumor, Stone, and Normal from the menu. The architecture's layers – Conv2D, MaxPooling2D, Flatten, Dense, and Dropout – cooperate to extract hierarchical attributes, minimize overfitting, and achieve exceptionally accurate classification. The model successfully distinguished between the cyst that exists, regular, stones, and tumor groupings, achieving a total accuracy rating of 98% on the provided dataset. The model's extremely high precision of 97% shows that it is dependable in reducing inaccurate results for each classification. The model's 96% recall rate illustrates its ability to accurately recognize all true positive instances. In all categories, an F1-score of 97% indicates an appropriate compromise between recall and accuracy.

The information set's degree of seriousness and variations are shown in Table 1.2. Most of the research does not appear to have evaluated an automated method for auto-identification of kidney calculi and cysts afterward. Figures 1.5 and 1.6 show the diagnostic success rate and percentage of loss range for diagnosing kidney inflammation disease.

Table 1.2. Dataset description

Categories		*Count*	*Training*	*Testing*
Healthy	Normal	5077	4061	1015
Unhealthy	Cysts	3701	2960	740
	Stone	2283	1826	456
	tumor	1377	1101	275
total		12438	9948	2486

Source: Author.

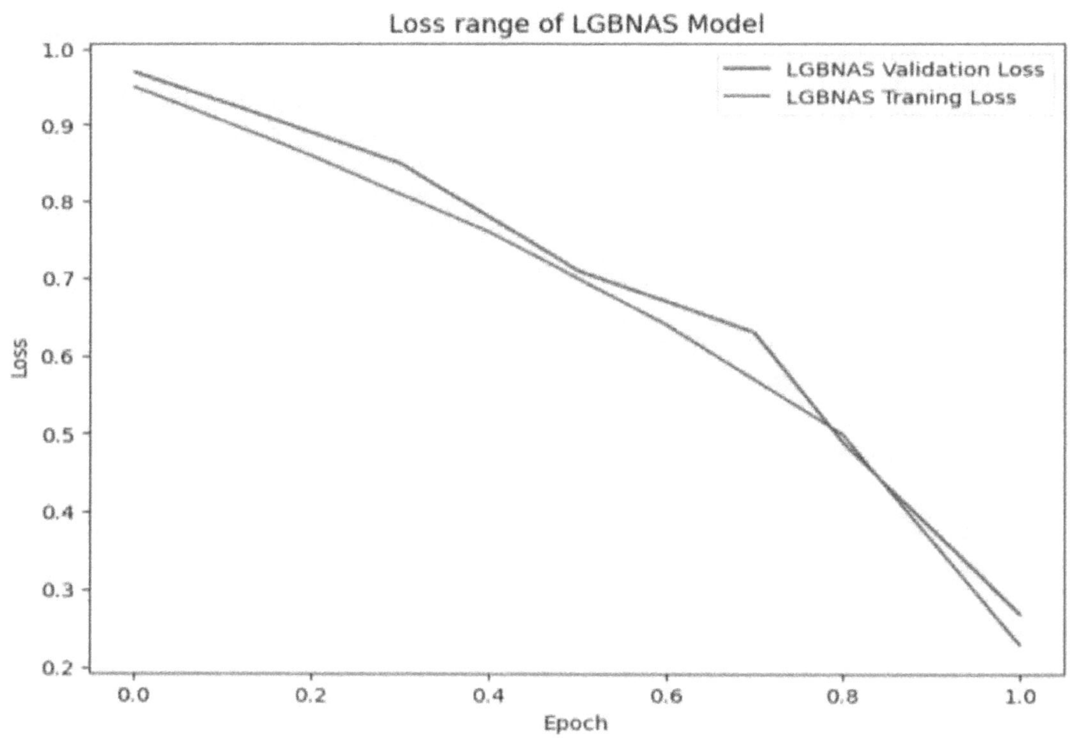

Figure 1.5. Loss rate graph.

Source: Author.

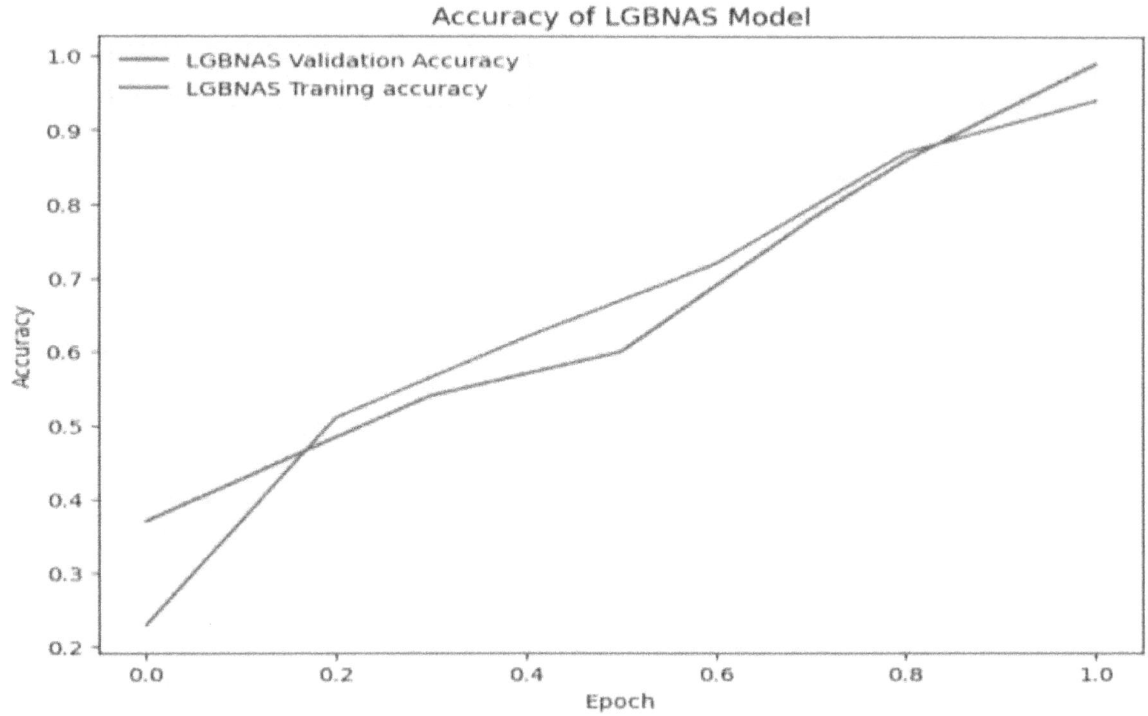

Figure 1.6. Accuracy range graph.

Source: Author.

Metrics like Precision and recall are important for accessing how well models for classification perform, especially when dealing with unbalanced data (Table 1.3). The ratio precisely anticipated positive instances (True Positive, TP) to all instances predicted as positive, including both true positive and false positive is as precision

$$Precision = \frac{TP}{TP + FP} \tag{7}$$

Conversely, Recall assesses how well the model can recognize real positive examples. It is the proportion of accurately predicated positive instances to all positive instances, including false negative and true positive. The recall formula is

$$Recall = \frac{TP}{TP + FN} \tag{8}$$

The LGBNAS (Leaf Gradient-Based Neural Architecture Search) model possesses the highest precision at 97%, demonstrating its ability to produce precise positive projections.

Precision is a metric that quantifies an algorithm's ability to predict favorable outcomes. With the highest recall of 96%, MobileNetV2 can identify positive cases. Using the LGBNAS model, it achieves MobileNetV2 with 98% accuracy, according to the F1-Score, which balances recall and precision. The LGBNAS model has the highest accuracy of 99% across the dataset, according to the Accuracy metric, which gauges overall correctness. The confused matrices produced by the proposed CNN structure following a tenfold cross-validation and evaluation are shown in Figure 1.7. The confusion matrix for the train-test proportion of 80% to 20%.

Table 1.3. Evolution metrics comparison

Methods	*Precision*	*Recall*	*F1-score*	*Accuracy*
VGG16	72.14	85.78	79	88.6
VGG19	78.27	78.26	79	94
Mobile Net	78	95.4	94.4	95
MobileNetV2	96	96.5	97	98
LGBNAS	97	96	98	99

Source: Author.

5. Conclusion and Future Work

This research work evaluation of medical pictures has significantly improved as a result of the investigation.

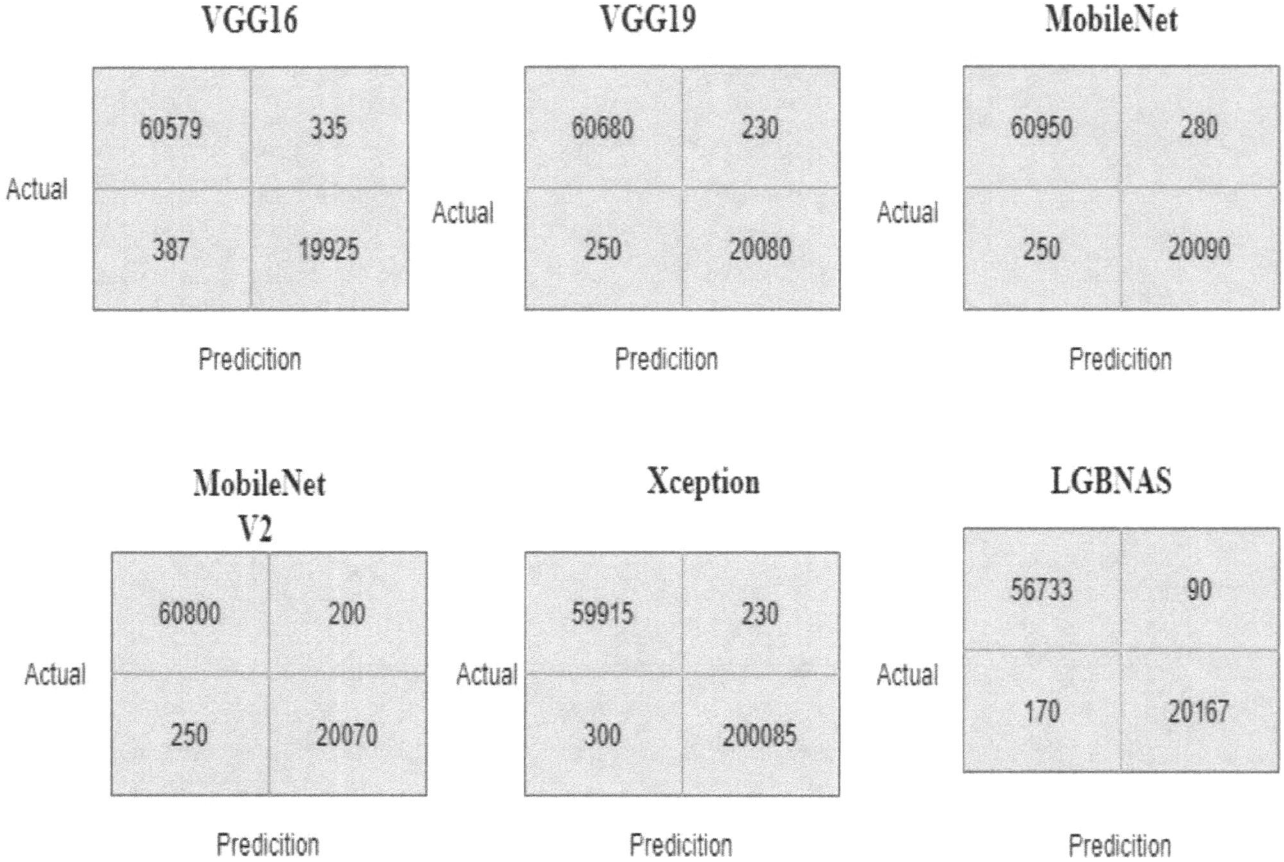

Figure 1.7. Confusion matrix.

Source: Author.

The efficiency of advanced neural network techniques for renal CT imaging classification is demonstrated by using an altered LGBNAS simulation as the separating the works. This innovative approach builds upon traditional methods and indicates room for further developments in the field. This task's significance goes beyond the scope of the investigations. By correctly recognizing kidney circumstances, the enhanced memory and accuracy of our enquiries can assist healthcare providers in making better clinical decisions. To increase stability, future research could concentrate on improving the model's construction, adjusting hyperparameters, and investigating larger data sets. The evaluation metrics used in this study focus on the F1-Score, accuracy, precision, and recall of the predictive model. The process of kidney disease identification is effectively computerized by the LGBNAS structure for medical classification of images, attaining excellent precision and durability. Its potential as a trustworthy decision-support tool for medical diagnosis is highlighted by its ability to categories kidney disorders like cyst, tumor, and stone with few errors. To enhance comprehension and trust regarding applications in medicine, future research will investigate the incorporation of clearness techniques.

References

[1] Shrestha, A., & Mahmood, A. (2019). Review of deep learning algorithms and architectures. *IEEE Access*, *7*, 53040–53065. doi:10.1109/ACCESS.2019.291220.

[2] Kapoor, N., Ahmad, N., Kumar Nayak, S., Prakash Singh, S., Vigneswara Ilavarasan, P., & Ramamoorthy, P. (2021). Identifying Infrastructural Gap Areas for Smart and Sustainable Tribal Village Development: A Data Science Approach from India. *International Journal of Information Management Data Insights, 1*(2), 100041. https://doi.org/10.1016/j.jjimei.2021.100041.

[3] Baygin, M., Yaman, O., Barua, P. D., Dogan, S., Tuncer, T., & Acharya, U. R. (2022). Exemplar Darknet19 feature generation technique for automated kidney stone

detection with coronal CT images. *Artificial Intelligence in Medicine*, *127*, 102274.

[4] Chang, Y. W., Hwang, D. Y., Chen, T. Y., Lin, C. C., Tseng, M. H., & Tsai, J. D. (2024). Kidney cysts in children with alport syndrome: A report of 3 cases. *Kidney Medicine*, *6*(5), 100815.

[5] Conze, P. H., Andrade-Miranda, G., Le Meur, Y., Cornec-Le Gall, E., & Rousseau, F. (2024). Dual-task kidney MR segmentation with transformers in autosomal-dominant polycystic kidney disease. *Computerized Medical Imaging and Graphics*, *113*, 102349.

2 Heart diseases analysis and prediction system using machine learning

Sumathi S.[1], E. Kodhai[1,a], E. Bhuvaneswari[2,b], V. Swathilakhsmi[3,c], and Rama Krishna Tummala[4,d]

[1]Department of Computer Science and Engineering, Chennai Institute of Technology, Chennai, Tamil Nadu, India
[2]Department of Artificial Intelligence and Data Science, Panimalar Engineering College, Chennai, Tamil Nadu, India
[3]Department of Computer Science and Engineering, Sri Manakula Vinayagar Engineering College, Pondicherry, India
[4]Department of Computer Science and Engineering, Shree Sathyam College of Engineering and Technology, Salem, Tamil Nadu, India

Abstract: This paper introduces an innovative approach to early heart disease detection, aiming to bridge the gap between healthcare providers and patients. Using the Cleveland heart disease dataset, it integrates multiple machine learning algorithms to enhance prediction accuracy. The random forest model proved to have the precision utmost to 99%, outperforming other algorithms. The system is designed for scalability, efficiently handling larger datasets without increasing computational costs, ensuring robustness and sustainability. Additionally, it provides doctor recommendations based on predictions, offering actionable guidance. By balancing accuracy, efficiency, and sustainability, this system contributes to advancements in applied sciences and personalized healthcare.

Keywords: Classification, heart disease analysis and predictions system, world health organization

1. Introduction

Good health is the supreme priority for human being and thus the world health organization has understood this basic need by declaring that each and every human being has the right to make use of the quality healthcare services. Specifically, the heart diseases are quite challenging throughout globally which leads to 31% of all death rates globally when compared to other diseases [13]. Detection of diseases at the earliest stage in most of the cases is tremendously challenging. This difficulty is predominantly in almost all developing countries which holds very limited facility for quality medical experts and diagnostic. To address this problem, solution is developed using machine learning technologies to detect the cardiac disease earlier. The probability of death due to heart disease can be reduced tremendously as it detects it at the early stage. For the analysis and predictions of earlier stage, large number of healthcare datasets are made used by the machine learning techniques. The healthcare datasets are verge large in volumes and as well as contains more complex structures too. This can be handled with ease by machine learning techniques when compared to classical analytical methods.

Heart Diseases Analysis and Prediction System (HDAPS) is proposed to break the complexity issues faced by the healthcare professionals and legislators. It is quite very challenging to understand, identify and to fight against infections, chronic illnesses, and newly developing threats related to health. Due to the capability of incorporating various datasets, it is able to perform real-time monitoring and predictive analytics. Moreover, it also lets the stakeholders with actionable insights for quick resource allocation and making decisions. DAPS has the promising and stronger capabilities for the development innovative system related to the prediction and analysis of diseases. Thus, it helps health care systems to grow in rapid manner and serve the nations to be healthier.

HDAPS is proposed to break the complexity issues faced by the healthcare professionals and legislators. It is quite very challenging to understand, identify and

[a]kodhaiej@gmail.com, [b]bhuvaname2008@gmail.com, [c]swathilakshmi@smvec.ac.in, [d]krishna_rama@hotmail.com

DOI: 10.1201/9781003770473-2

to fight against infections, chronic illnesses, and newly developing threats related to health. Due to the capability of incorporating various datasets, it is able to perform real-time monitoring and predictive analytics. Moreover, it also lets the stakeholders with actionable insights for quick resource allocation and making decisions. HDAPS has the promising and stronger capabilities for the development innovative system related to the prediction and analysis of diseases. Thus, it helps health care systems to grow in rapid manner and serve the nations to be healthier. Many publications have previously used machine learning algorithms to predict cardiac sickness [2–5], but this is an additional attempt to compare three well-known ML techniques. While conducting an experiment on the Kaggle heart disease prediction dataset in order to determine which ML technique is the most accurate (Figure 2.1).

This paper is divided into different parts as the following: Part II, specifies the literature survey work. Part III specification of machine learning methodologies adopted for this investigation are given. Part IV summarizes the work's findings; and Section V finally the conclusion.

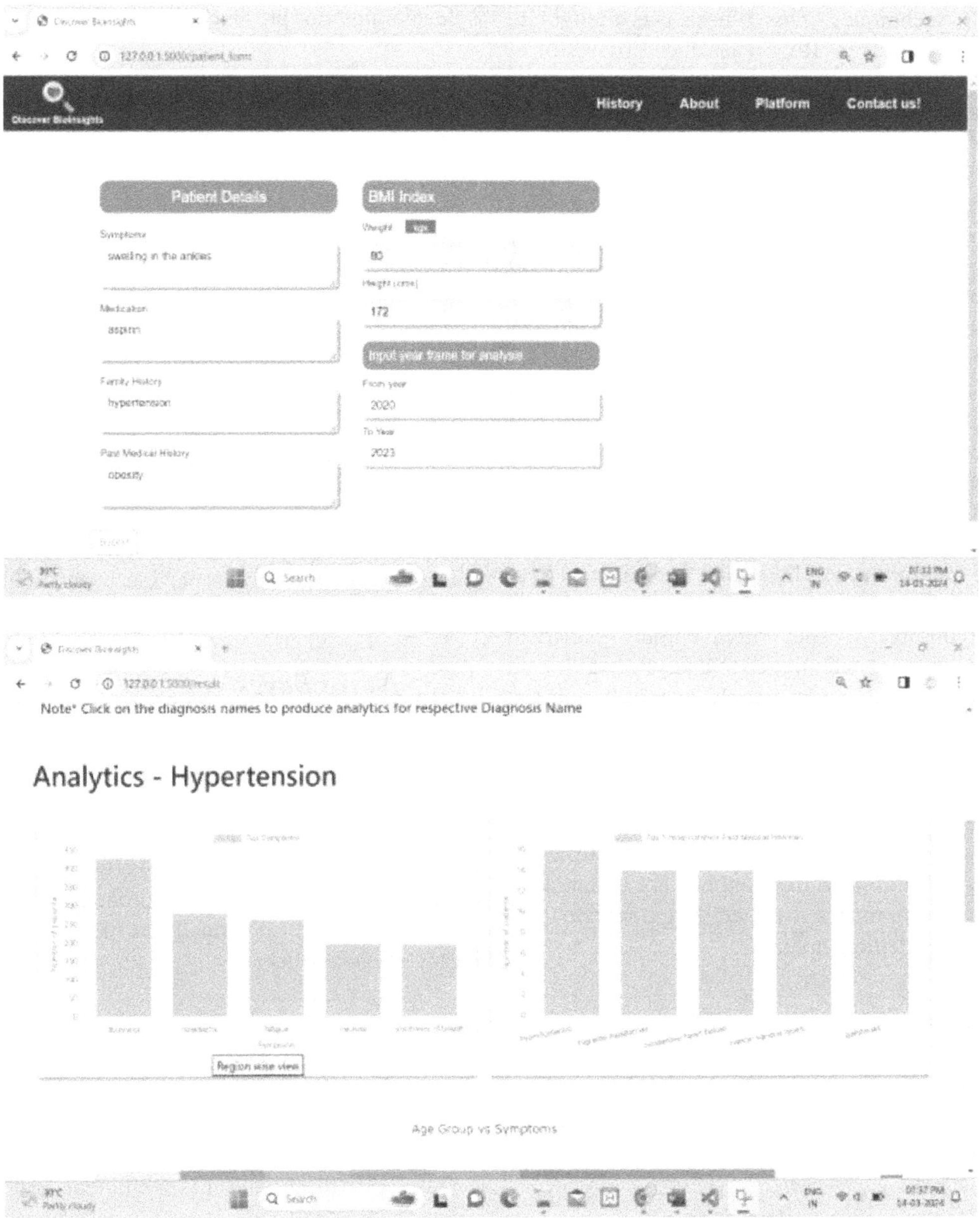

Figure 2.1. Visualization of summary.

Source: Author.

2. Related Works

Cleveland dataset specifically for heart disease is available freely through at Irvine's online data mining repository, University of California has been used in numerous papers to check accuracy of various machine learning methods. Using the above dataset, J. Zhao et al. [5] attained the 77% of estimated precision using the logistic regression approach. By comparing global evolutionary computation techniques, the authors [6] of this study improved their work and were able to observe increased prediction accuracy. In their work, authors Bayu Adhi Tama et al. [7] proposed a study on the diagnosis of diabetic disease through the use of ML techniques. This illness was thought to be a major ML thrust area. Report produced by result of research conducted by the International Diabetes Federation (IDF), there were about 285 million diabetics in the world. In actuality, early diagnosis of type 2 diabetes is a difficult task. However, the authors', used data mining, as it produced an excellence, assisted in the release of information from readily available data. During the study, they mined linked patient data from prior records using support vector machines (SVMs). The prompt diagnosis of type 2 diabetes made it easier to begin appropriate therapy and reduced the chance of worsening.

EHR (Electronic Health Records) Systems these systems store patient health records digitally, allowing for easier access and analysis of patient data. They often incorporate features for disease analysis and prediction based on historical data. They can analyze symptoms, medical history, and other factors to suggest potential diagnoses and treatments. Public Health Surveillance Systems These systems monitor population health trends and detect disease outbreaks. They often employ statistical models and data mining techniques to analyze health data gathered from different fields namely hospitals, clinics, laboratories, and social media. In the paper, they proposed an application which employs the semi-supervised learning algorithm. The industry approach, both supervised and as well as unsupervised machine learning models were applied in about 90–95% of applications. It was therefore suggested that machine learning methods are essential to the creation of numerous cutting-edge applications for businesses and the medical domain.

3. Machine Learning Techniques

The proposed heart disease prediction model is developed using four well-liked machine learning approaches. The following are the specifics of these methods:

3.1. Logistic regression algorithms

For binary classification tasks, a popular statistical method are logistic regression algorithms [1]. with applications spanning various domains including healthcare. The paper reviews the systematic survey of the application of logistic regression algorithms in predicting diseases. A statistical technique for examining a data set in which one or more independent variables influence a result is called logistic regression. Where the possible outcomes are limited to two. A formula that could predict the logarithmic transformation of the probability for the occurrence of attributes of interest is generated by means of logistic regression. Logistic regression can help identify the most significant predictors of disease by analysing the coefficients associated with each input feature. This feature selection process aids in understanding the key factors contributing to the development or presence of a disease.

3.2. Decision tree

The classification models in machine learning are created using the decision tree method [8]. The foundation of this categorization model structure is like a tree. Since the results are known in advance, it is categorized as supervised learning. It is possible to use the Decision Tree method on both numerical and categorical data. The leaf nodes, branches, and root node embrace like a decision tree. To assess the data, the traversal path is used which starts from the truck of the tree and leads to the terminal node. From the dataset CHDD, along with decision tree 283 tuples were assessed. They might have come to a favourable or unfavourable conclusion on the prognosis of cardiac disease. To search for false positives or false negatives, which indicates the sensitivity, specificity, and accuracy of the prototype, these parameters were compared with the actual parameters.

3.3. Naive Bayes

It is categorised as supervised machine learning method is built using the Bayes Theorem [9]. This paved the way to utilize the features of Bayes theorem which is statistically independent from others. High dimensionality input data is employed with the Naïve Bayes Classifier [10]. Application of the Naive Bayes approach in computer vision is very beneficial. Specifically, it has demonstrated its efficacy as a classifier.

3.4. Random forest classification

It is the name given to a collection of unpruned classification as trees [11]. It works exceptionally well in a range of real-world applications since there is very little risk of overfitting and it is inefficient against noise in the dataset. It usually improves the accuracy of data used for testing and validation and runs faster than a lot of other tree-based algorithms. Random forests are constructed by combining the projections produced by several decision tree algorithms. After generating a random tree, various methods are available to optimize the random forest performance.

4. Methodology

The steps that led to the development of the illness prediction model are shown below.

4.1. Data collection

For the data collection the Cleveland Dataset which is specifically for heart disease is used. It is accessible at UCI Repository through online [13]. The following 14 parameters as shown in the Table 2.1 are taken into consideration.

Therefore, a total of 1025 examples were used in this study project. The mean value of every property is displayed in Table 2.1.

Additionally, the dataset had missing values that was used. In order to solve this, the data was prepared using certain statistical techniques. Class Values 1 indicates tested negative for the disease and class value 0 indicates tried positive for the disease. Thus the dataset has been partitioned into distinct percentages; for instance, training data comprise 80% of the dataset, while testing data makes up the remaining 20%. Table 2.2 shows sample data for training, while Table 2.3 shows sample data for testing.

Table 2.1. Parameters

S. No.	Parameter	Description	Mean Value
1	age	in years	54.434
2	Sex	Male, Female	0.696
3	cp	Angina, abnang, notang, asympt	0.942
4	trstbps	Resting Blood Pressure in mm hg	131.612
5	chol	Serum Cholesterol in mg/dl	246
6	fbs	fasting blood sugar - 1 if >120 mg/dl, 0 if <120 mg/dl	0.149
7	restecg	Electrocardiographic Results	0.53
8	thalach	Maximum Heart Rate observed	149.114
9	exang	exercise with angina has occurred	0.337
10	oldpeak	ST depression induced through exercise	1.072
11	slope	slope of the ST segment	1.385
12	thal	Number of major vessels ranging from 0 - 3 color by fluoroscopy	0.754
13	ca	Heart status	2.34
14	Target	Output Class	

Source: Author.

4.2. Data preprocessing

A few of the novel dataset's parameters which are missing values, can produce inaccurate results and lower the model's accuracy. The best solution to this issue is to use the "mean of column" method to replace the missing value. In this strategy, 0 is substituted with the mean or average of the neighbourhood values [12]. Subsequently, it modifies the 0 value using the recently computed value. Subsequently, the dataset's values were converted from numeric to nominal to ensure compatibility with the machine learning techniques employed.

Table 2.2. Sample training data

52	1	0	125	212	0	1	168	0	1	2	2	3	0
53	1	0	140	203	1	0	155	1	3.1	0	0	3	0
70	1	0	145	174	0	1	125	1	2.6	0	0	3	0
46	1	0	120	249	0	0	144	0	0.8	2	0	3	0
54	1	0	122	286	0	0	116	1	3.2	1	2	2	0
71	0	0	112	149	0	1	125	0	1.6	1	0	2	1
43	0	0	132	341	1	0	136	1	3	1	0	3	0
34	0	1	118	210	0	1	192	0	0.7	2	0	2	1
46	1	0	120	249	0	0	144	0	0.8	2	0	3	0
54	1	0	122	286	0	0	116	1	3.2	1	2	2	0
71	0	0	112	149	0	1	125	0	1.6	1	0	2	1

Source: Author.

Table 2.3. Sample data used for testing

52	1	0	125	212	0	1	168	0	1	2	2	3	0
53	1	0	140	203	1	0	156	1	3.1	0	0	3	0
70	1	0	145	174	0	1	125	1	2.6	0	0	3	0
61	1	0	148	203	0	1	161	0	0	2	1	3	0
62	0	0	138	294	1	1	106	0	1.9	1	3	2	0
58	0	0	100	248	0	0	122	0	1	1	0	2	1
58	1	0	114	318	0	2	140	0	4.4	0	3	1	0
55	1	0	160	289	0	0	145	1	0.8	1	1	3	0
46	1	0	120	249	0	0	144	0	0.8	2	0	3	0
54	1	0	122	286	0	0	116	1	3.2	1	2	2	0
71	0	0	112	149	0	1	125	0	1.6	1	0	2	1
43	0	0	132	341	1	0	136	1	3	1	0	3	0
34	0	1	118	210	0	1	192	0	0.7	2	0	2	1

Source: Author.

4.3. Build-in model

Weka Data Mining Tool has a built-in model. New Zealand's Waikato University created the open source machine learning program known as the WEKA stands for Waikato Environment for Knowledge Analysis [13]. Numerous common data mining tasks, including feature selection, data preparation, clustering, classification, regression, and visualization, are processed by the software with ease. It provides a simple environment for loading data stored in databases, files, or URLs. The following file formats which is acceptable by the software: Lib SVMs, C4.5's, Attribute Relation File Format (ARFF), and CSV. It is designed to evaluate and envision the parameters namely confusion matrix, true positive, false negative, accuracy and recall. It is platform independent, light weight, and open-source GUI application that has large numbers of machine learning models, which include deep learning algorithms for image processing, etc. The following four precision metrics are considered while comparing the four models:

- Precision
  It stands for the average probability of a pertinent retrieval.
  $$Precision = \frac{True\ Postives}{True\ Postives + \ False\ postitives}$$
- Recall
  It is the typical likelihood of a successful recovery.
  $$Recall = \frac{True\ Postives}{True\ Postives + \ \ False\ negatives}$$
- Accuracy
  A classifier's accuracy is calculated as
  $$Accuracy = \frac{count\ of\ True\ Postives + count\ of\ True\ Negatives}{Total\ instances}$$
- ROC Area
  A classification model's FPR and TPR are trade-offs that are represented by the root mean square (ROC).
  $$FPR = \frac{count\ of\ false\ postives}{count\ of\ false\ postitives + \ count\ of\ True\ negatives}$$
  Where TPR is True Positive Rate and FPR is False Positive Rate

The datasets are separated into sections by making use of 10-fold Cross Validation namely training and testing.

4.4. Model accuracy measurement

The accuracy %, ROC (Receiver Operating Characteristic), precision, and recall of each model were analysed and compared. Figure 2.2 displays the models' accuracy of performance. In most cases, a classifier is graded as "FAIR" if its ROC value is 0.77 and as "GOOD" if it is less than 0.80. A ROC value which is very near to one is the ideal and accurate model.

5. Result

The experimental results gave a promising result for both SVM and random forest algorithms when compared with Gaussian Naïve Bayes and Decision tree. When experimenting with SVM model, it produced a good result of 98%, whereas the Naïve Bayes produced 8% and Decision tree with results of 13%.The random forest model has performed with the excellent result of 99% making it a best model for heart disease prediction when compared to other model. The decision tree algorithm was not able to perform to suit for heart disease prediction application. Finally, the results indicates the use of SVM and random forest algorithms is capable to predict excellent results for heart disease.

6. Conclusion and Future Directions

The paper investigates the monarchy earlier detection of heart disease system to provide a pathway between the healthcare professionals and the patients. The proposed system is for earlier detection of heart disease and as well as recommendations of doctors based on the detection. The Cleveland dataset is used for heart disease prediction to implement the application by integrating various algorithms. Different models are applied to predict the most accurate results. Finally, different experimental analysis was carried with different models and among them random forest produced the most accurate results of 99%. Even though other algorithms were able to produce results, it didn't meet the accuracy level for this specific dataset. Moreover, it is capable to handle with growing training data size to increase the accuracy of the results, along with the same computational cost and the processing time. Taking all these points into account, a well-balanced system is built based on accuracy and the efficiency of the system. Finally, the disease prediction system is built along with the recommendation of doctors for the users.

Despite of the proposed system with high accuracy, it has some limitations, such as potential biases in the dataset, limited generalizability to diverse populations, and reliance on historical data, which may not fully capture evolving healthcare trends. Future research should explore the dataset to include diverse patient demographics, integrating real-time data for dynamic prediction, and leveraging deep learning techniques to further enhance performance.

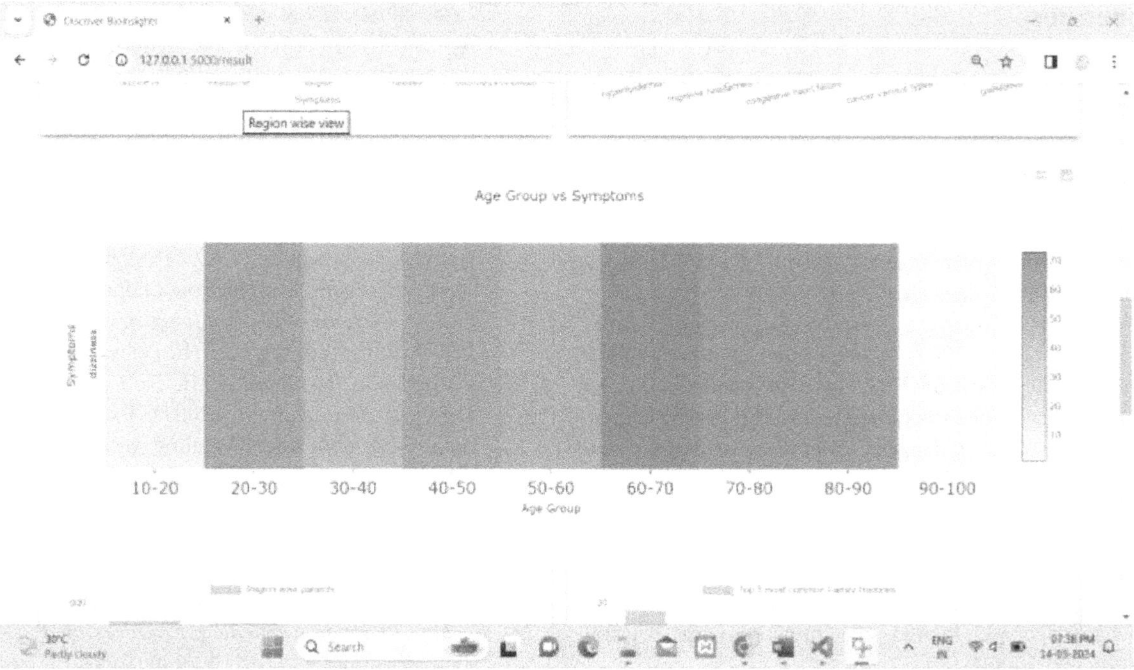

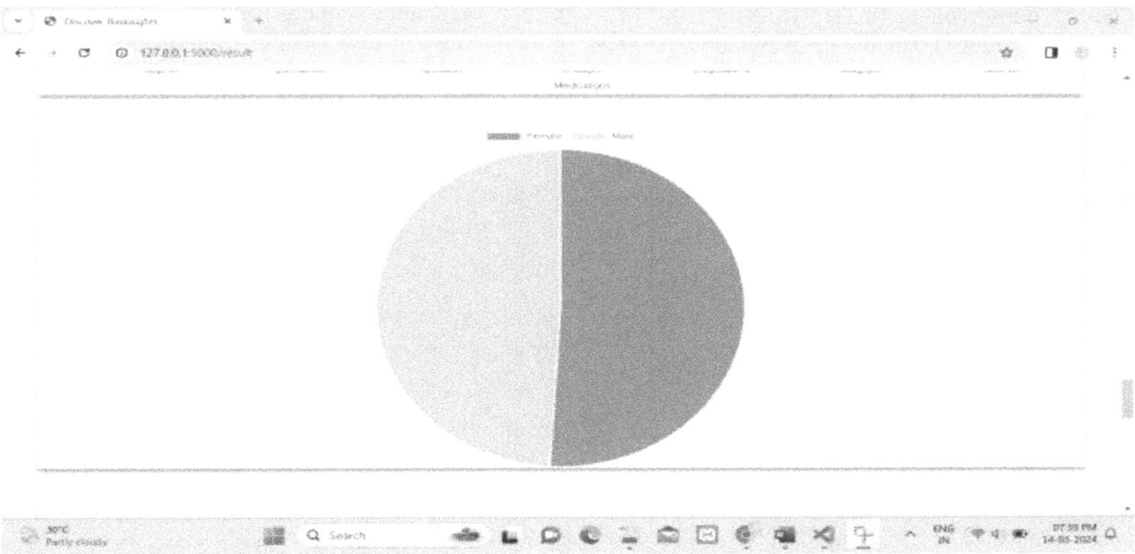

Model	Precision	Recall	ROC Area
LRA	0.995	0.995	0.995
Random Forest	0.997	0.997	1.00
Decision Tree	0.851	0.848	0.889
Naïve Bayes	0.904	0.904	0.966

Figure 2.2. Model accuracy performance.

Source: Author.

References

[1] Aarsland, D., Andersen, K., Larsen, J. P., & Lolk, A. (2003). Prevalence and characteristics of dementia in parkinson disease. *Archives of Neurology*, (60.3), 387. https://doi.org/10.1001/archneur.60.3.387

[2] Das, R. (2010). A comparison of multiple classification methods for diagnosis of parkinson disease. *Expert Systems with Applications*, (37,2), 1568–1572. https://doi.org/10.1016/j.eswa.2009.06.040

[3] Polikar, R., Topalis, A., Green, D., Kounios, J., & Clark, C. M. (2007). Comparative multiresolution wavelet analysis of ERP spectral bands using an ensemble of classifiers approach for early diagnosis of alzheimer's disease. *Computers in Biology and Medicine*, (37,4), 542–558. https://doi.org/10.1016/j.compbiomed.2006.08.012

[4] Zieky, M. J., Philip, J., Stone, et al. (1967). The General Inquirer: A Compuler approach to content analysis. Cambridge, Massachusetts: The M.I.T. Press, *Psychometrika*, (32,4), 485–486. https://doi.org/10.1007/s0033312300004439

[5] Zhao, J., Liu, K., & Wang, G. (2008). Adding redundant features for CRFS-based sentence sentiment classification. *Proceedings of the Conference on Empirical Methods in Natural Language Processing - EMNLP '08*, 117. https://doi.org/10.3115/1613715.1613733

[6] Turney, P. D. (2001). Thumbs up or thumbs down? *Proceedings of the 40th Annual Meeting on Association for Computational Linguistics - ACL '02*, 417. https://doi.org/10.3115/1073083.1073153

[7] Hatzivassiloglou, V., & McKeown, K. R. (1997). Predicting the semantic orientation of adjectives. *Proceedings of the 35th Annual Meeting on Association for Computational Linguistics*, 174–181. https://doi.org/10.3115/976909.979640

[8] Kaladhar, G. K., Nageswara Rao, P. V., & Ramesh, N. R. B. L. V. (2010). Confusion matrix analysis for evaluation of speech on Parkinson disease using Weka and Matlab. *International Journal of Engineering Science and Technology*, (2,7), 2734–2737.

[9] Åström, F., & Koker, R. (2011). A parallel neural network approach to prediction of Parkinson's disease. *Expert Systems with Applications*, (38,10), 12470–12474. https://doi.org/10.1016/j.eswa.2011.04.028

[10] Little, M. A., McSharry, P. E., Roberts, S. J., Costello, D. A., & Moroz, I. M. (2007). Exploiting nonlinear recurrence and fractal scaling properties for voice disorder detection. *BioMedical Engineering OnLine*, (6,1), 23. https://doi.org/10.1186/1475-925x-6-23

[11] Tsanas, A., Little, M. A., McSharry, P. E., & Ramig, L. O. (2010). Accurate telemonitoring of Parkinson's disease progression by noninvasive speech tests. *IEEE Transactions on Biomedical Engineering*, (57,4), 884–893. https://doi.org/10.1109/tbme.2009.2036000

[12] Revett, K., Gorunescu, F., & Salem, A.-B. M. (2009). Feature selection in parkinson's disease: A rough sets approach. *2009 International Multiconference on Computer Science and Information Technology*, 425–428. https://doi.org/10.1109/imcsit.2009.5352688

[13] Heart disease, Dataset, UCI Machine Learning Repository. (n.d.), accessed at https://archive.ics.uci.edu/ml/datasets/heart+disease

3 Breaking down barriers: Understanding sociological factors influencing physical inactivity among overweight and obese students in urban India

Arul Selvan R.[a] *and Suriya S.*[b]

Saveetha School of Law, Chennai; Madras School of Social Work, Chennai, Tamil Nadu, India

Abstract: This study employs Robert K. Merton's sociological theory of manifest and latent function to conceptualize educational strategies towards physical health in Chennai, a major urban hub in India. With a fretting 26% of adults in the state affected with overweight and obesity, the study used a mixed method to identify the barriers that hinder college students to effectively contribute to development of evidence – based strategies to promote physical activities. The study equates stress management and social cohesion as latent function, and posits that these aspects should be prioritized as manifest function, ultimately addressing the SDG 3 and 4 among the youth of the Global South.

Keywords: Latent, manifest function, obese, overweight, physical inactivity, India

1. Introduction

Addressing overweight and obesity is vital to attaining most of the Sustainable development goals [1]. Target 3.4 on reducing premature mortality from the non-communicable diseases (NCDs) has been a global health challenge [2]. NCDs are mostly linked with lifestyle choices such as unhealthy diets and physical inactivity pose a crucial challenge for health care at global level [3]. Physical inactivity is one of the primary contributors to the prevalence of overweight and obesity, has been overlooked and unaddressed by the global health funders and the policymakers[4]. The World health organization notes that 59% of the global adult population is either overweight or obese, with 43% being overweight and 16% affected by obesity [3]. This trend is largely pronounced in middle – income and high-income countries [3]. In terms of higher income regions, overweight and obesity claim more lives than underweight [5]. This issue underscores the need for a renewed focus as well as commitment to addressing the overweight and obesity within the SDG framework [6]. This study interlinks SDG 3 with SDG 4, the "quality education" and analyses the role of educational institutions in addressing physical inactivity among overweight and obese students. In this context, addressing the risk factors of NCDs, particularly among overweight and obese college students is crucial. As the future generation, they are vulnerable to adopting unhealthy lifestyle choices that can have long-term consequences.

The role of educational institutions is analysed through Robert K Merton's Function Analysis, the method which focuses on exploring how different elements contribute to the overall functioning of the society [7]. This sociological approach used to understand the social structure and institutions through analysing the intended and unintentional consequences of the social actions and institutions [8]. Accordingly Metron categorized two kinds of functions, viz, manifest function and latent function. Manifest functions are the deliberate, intended, recognized and the explicit impact of the social actions [9]. These actions are carried out by the people consciously to achieve the expected outcome [8]. In terms of education, the manifest function is to impact knowledge and to prepare the students for their career. Whereas, latent functions are unintended, unrecognized impacts that emerge from the social actions. They are the byproducts of the manifest function. The latent function of the education are the development of peer groups and social networks among students.

This study draws on the manifest and latent function theory, which emphasizes the importance of considering both intended outcomes and unintended consequences of

---

[a]arulselvan257@gmail.com, [b]suriyadolly@gmail.com

DOI: 10.1201/9781003770473-3

health interventions. By analysing the factors contributing to NCDs and obesity, this research aims to propose effective strategies for mitigation as a part of SDG 4, aligning with the objectives of SDG 3. The following sections comprehensively address the methodology and results. The discussion chapter critically interrogates the result analysis with the theoretical framework of Robert K Merton, to provide a nuanced understanding of the educational institutions' role in addressing the overweight and obesity issues. Finally, the conclusion reiterates the quality of education which encompasses the activities to reduce or prevent the overweight and obesity, as a part of curriculum.

2. Statement of the Problem

Epidemiological trends in India, shows the dramatic increase in the prevalence of NCDs: 30% of deaths have been recorded in 1990; 55% in 2016 and 66% in 2023 [10]. In Tamil Nadu context, NCDs cause 75% of mortality and the risk factors [11] include overweight and obesity, as it causes cardiovascular diseases, hypertension and certain hormonal cancers [4]. Whereas, overweight and obesity are caused by multiple factors such as, energy imbalance, genetics, medications, stress, poor sleep and emotional factors [4]. However energy imbalance has been the primary contributing factor, which is caused by, HEDF that is, excessive calories consumption through high energy dense food and insufficient calorie expenditure due to physical inactivity [5]. This enables us to understand the importance of the education institutions in addressing the issue, and brings up the following research question and objectives.

3. Research Question

How does the manifest and latent functions in the college settings influence the physical inactivity among the overweight and obese students?

4. Objectives

To evaluate the level of physical activity among overweight and obese college students.

1. To assess the availability, accessibility and the quality of infrastructure in the college to facilitate physical activities.
2. To identify the sociological factors affecting the physical activity level among the overweight and obese students.

5. Methods

The study attempts to identify the latent functions of the college on the physical activities among the overweight and obese students. The study employs Convergent Parallel Mixed method which includes quantitative and qualitative approaches Creswell and Clark [12]. The quantitative data dealt with the proportion of marital status, level of physical activity and playground availability. The qualitative data unravelled the sociological factors that influence the physical activities. The data was collected through structured and semi-structured interviews. This mixed method is used to identify factors and illustrated as themes and superordinate themes.

6. Sampling

6.1. Why Chennai?

Dr. Indira Jaykumar reported that the prevalence of central obesity among school children aged 5 to 17 years was 55.9% for overweight and 21.1% for obesity. On the other hand Sadhasivam in his study [13], found that 19.4% of college-going youth in Chennai were overweight and 5.6% of them were obese [14]. Dhanuraja Gopalakrishnan and Uma Devi revealed in their study that 27.9% of the college-going youth were overweight and 8.4% of them were obese [14]. The above reports revealed that the prevalence rate of overweight and obesity is rapidly increasing among adults and college-going youth in Tamil Nadu in general and Chennai in particular. This transition might involve moving from school to college, making it important to assess whether the new structure in college provides opportunities to reduce physical activity. If this trend continues, the prevalence rate of overweight and obesity in Tamil Nadu and Chennai will be increasing in the future.

7. Study Subjects

College students who were overweight (BMI >25) and obese (BMI>30) residing in the Chennai Corporation area which includes 15 zones and 200 wards [15]. Inclusion criteria for the study subject respondents whose BMI (Body Mass Index) **is** greater than 25, are studying in the 2nd and 3rd year of undergraduate programs in the colleges within Chennai Corporation area [16]. The rationale behind selecting 2nd and 3rd-year undergraduate students is that they have acquired a minimum of one year of college life experience, which would have equipped them to: Adapt better to the college environment; Have a greater willingness to provide accurate information during the date collection. Since this study focuses on youth, the eligible age was 24 or below 24 during the interview/in-depth interview. This parameter is set by WHO which defined people in the age group between 15 to 24 years [17].

Exclusion Criteria for the Study Subject According to WHO guidelines, BMI 25 to 29.99 is categorized as overweight, and BMI 30 and above is considered obese.

Therefore, respondents with a BMI less than 25 were not included in the study. The first year undergraduate as well as the postgraduate students were excluded as previously explained in the inclusion criteria.

8. The Sample Size and Sampling for the Quantitative Data

A total of 270 respondents were approached across six colleges. Of these: 57 respondents were excluded due to Ineligibility based on BMI measurement and refusal to participate in the study. Notably, refusal was more prevalent among the male and female respondents from the Government colleges. A total of 270 respondents were approached across six colleges.

Sampling Techniques for the Qualitative Data A Multi-Stage Purposive Sampling technique was deployed to select overweight and obese students [18].

- Stage 1: The list of Universities was prepared based on the University Grants Commission [19]. Accordingly 26 Universities exist in Chennai. In encompasses 16 Deemed Universities, 9 State Universities, 1 Central University. 3 Universities have affiliated colleges. Among those 3, 2 Universities have more than 100 affiliated Colleges. The University of Madras has 95 affiliated colleges in Chennai [20]. Thereby using the purposive sampling technique, the researcher selected University of Madras for the study.
- Stage 2: All 95 affiliated colleges under the University of Madras were listed in order. 22 colleges are Government aided and 63 are Self-financing colleges.76 colleges are located in the Greater Chennai Corporation area [21]. A Purposive sampling technique was deployed to choose the 76 colleges located in the Chennai Corporation.
- Stage 3: Among the 76 colleges, 47 colleges are working as Shift-I and Shift-II [20]. The list of colleges were listed in order by:
  Government (3 Men's college and 3 Women's College, Total College = 6)
  Government-Aided (10 Women's Colleges and 11 Men's Colleges, Total colleges = 21)
  Self-Finance (10 Women's Colleges and 10 Men's Colleges: Total Colleges 20)
  Two colleges were selected from each category: (1) Men's and (2) Women's College from Government, Government – Aided, and Self-Financed using purposive sampling techniques.
- Stage 4: In the selected colleges, department were categorized on the grounds of Arts and Science. Two Arts and two Science Departments in Shift-1 and available Departments in Shift-II were selected. The target sample was 40 respondent from each selected college, with a ratio of 20 respondents from Shift-I (Morning colleges) and 20 respondents from Shift-II (Afternoon colleges). However, the target could not be achieved in some colleges due to, time constraints and the non-acceptance of selected candidates to participate and time constraints. Moreover, some colleges were offering only arts and commerce courses in Shift-II. Finally, the study was completed with 213 respondents.

9. Sample Size and Sampling Frame for the Qualitative Data

The selected colleges were used for both qualitative and quantitative analysis. 24 obese and overweight respondents were selected from six colleges in the Chennai corporation area. Accordingly 12 respondents from Shift-1 and Shift-II out of 6 selected colleges were interviewed for qualitative data. Purposive sampling was deployed to choose the participants from six colleges.

10. Data Collection Method

Both primary and secondary data were used in the study. Field-based settings facilitated primary data collection. The convergent parallel mixed method incorporated different sets of data collection. They are structured interviews for qualitative study, in-depth interview and personal observation for the qualification study. Secondary sources include periodical reports, books, government documents and from the digital library sources.

11. Scales deployed for the Study

BMI was calculated as per WHO guidelines [22]. Centre for Diseases Control and Prevention guidelines were used to assess the Physical Activity [23].

12. Quantitative Data Analysis

Data entry, coding and data cleaning was carried out using Microsoft Excel. The quantitative data was imported to Statistical Package of Social Science (SPSS) version 21. Descriptive statistical technique was used to analyse the data.

13. Analysis of Qualitative data

Interpretive Phenomenological Analysis (IPA) was used for the qualitative data. This approach is the combination of Hermeneutics, phenomenological and idiographic approaches [24]. Subsequently the qualitative data was transcribed separately based on the quotations to derive superordinate and subordinate themes.

14. Findings

The majority of respondents (67.1%) fall within the age group of 18–20 years, followed by 27.7% in the 20–22 age group, and 5.2% in the 22–24 age group (Table 3.1). In terms of gender, 47.9% of the respondents are male, while 52.1% are female, indicating a higher number of female students (Table 3.1). Identifying overweight and obese male students in the afternoon classes of government colleges is challenging due to the irregular attendance of many students during these sessions.

Regarding religion, 77.5% of the respondents identify as Hindu, 18.8% as Christian, 2.8% as Muslim, and 0.9% as Jain (Table 3.1). In terms of marital status, nearly all respondents (99.1%) are single, with only 0.9% being married (Table 3.1). Notably, two married female students have resumed their studies after marriage.

Among the sampled respondents, 37.6% are from government-aided colleges, 33.4% from government colleges, and 28.6% from self-financed colleges (Table 3.1).

Table 3.1. Profile of the respondents

	n=213	%
Age		
18–20	143	67.1
20–22	59	27.7
22–24	11	5.2
Gender		
Men	102	47.9
Women	111	52.1
Religion		
Hindu	165	77.5
Christian	40	18.8
Islam	06	2.8
Jain	02	0.9
Marital Status		
Married	02	0.9
Single	211	99.1
Nature of College		
Government	72	33.8
Government-Aided	80	37.6
Self-finance	61	28.6
Respondents BMI		
Overweight (BMI-25-29.99)	153	71.8
Obese (BMI>30)	60	28.2

Source: Author.

Colleges affiliated with the University of Madras are classified into three types: Government, Government-aided, and Self-financed Colleges [20]. Government Colleges are fully funded by the government, and students attending these colleges are exempt from paying tuition fees. Government-aided Colleges are partially funded by the state government. In these colleges, students in morning sessions pay a lower fee, while students in afternoon sessions pay tuition fees that are ten times higher than those in morning sessions. Self-financed Colleges do not receive any government funding, and their tuition fees are significantly higher compared to both Government and Government-aided Colleges.

In the selected sample, it was found that 71.8% of the respondents were overweight, while 28.2% were classified as obese (Table 3.1). The WHO guidelines were used to measure BMI. A small percentage (5.6%) of female respondents reported having PCOD or thyroid conditions, which contributed to their obesity. Similarly, 1.6% of the respondents stated that their involvement in sports such as kabaddi, hammer throw, and shot put required them to maintain an overweight or obese physique.

The majority of respondents (71.4%) reported that their college has a playground (Table 3.2). However, only 4.2% of these respondents stated that they use the playground for physical activities such as playing cricket, basketball, kabaddi, shot put, hammer throw, and volleyball. Most of the respondents who reported using the playground for physical activities were professional sports persons.

On the other hand, 28.6% of respondents indicated that their college does not have a playground (Table 3.2). These responses primarily came from students attending self-financed colleges.

In the qualitative analysis, we explored the reasons why some respondents do not use the playground even when it is available. Five superordinate themes emerged from the analysis:

1. Time Constraints
2. Lack of Motivation

Table 3.2. College setting

	n=213	%
Playground Availability		
Yes	152	71.4
No	61	28.6
Usage of Playground		
Yes	09	4.2
No	143	94.1
Physical Education Period in the Colleges		
Yes	0	0
No	213	100%

Source: Author.

3. Management Rules
4. Beauty Consciousness
5. Inadequate Space

14.1. Time construction

14.1.1. Academic burden

Respondents stated thus: "At present, as I'm doing my 3rd year, I need to do my dissertation and at the same time, I need to complete my assignment. These two are the reasons for not going to the college playground for the physical activities" (Women participant from Public College)

14.2. Lack of motivation

The qualitative data reveal a considerable theme of diminished motivation among the respondents.

14.2.1. No companion

Social dependency plays a crucial role as the individual relies on social interaction for motivation. Respondents were contingent upon the presence of companionship.

"If anyone gives me company to play, only then I can play. If no one gives me company, how can I play?" (Women Participant 10, Government-aided, Morning Session)

"This is what happens, I used to ask my friends to join with me to play volley ball. They would say that we would play today in the evening but no one turns up to the ground in the evening" (Men Participant 13, Government aided, Morning Session)

14.2.2. Lack of interest

The minimal engagement and the diminished enthusiasm among the students emerged as a primary contributing factor to the underutilization of the college playground. A pervasive sense of disinterest and apathy seemingly overshadowed the potential physical development opportunities inherent in the college campus.

The participant opined that "I have just come to the college for studying, even my parents send me to the college for that reason only…." (Men Participant 14, Government-aided, Morning session)

14.3. Management rule

The respondents the government college stated that they were not allowed to use the college playground after the college hours. However the college management had their own reasons on the grounds of safety and disciplinary concerns. The college witnessed senior students engaging in violent confrontation within the college grounds. Weapons such as knives and sickles were reportedly concealed on the premises. These violent behaviors prompted institutional intervention.

14.4. Beauty consciousness

In a few cases, among the women participants, it was found that they did not use the college playground because they would smell bad if they play.

14.4.1. Smelling bad after playing

"If I play, I would sweat, I would smell bad, I would lose my fresh look and would look ugly" (Women participant 09, Government-aided, Morning Session)

"I have to travel nearly one hour from my college. If I play, I would sweat, I would smell bad and it would be uncomfortable for me to travel" (Women participant 10, Government-aided College)

14.5. Inadequate space

The participant from government college for women said that the college playground doesn't have enough space for playing. This is what the participant has said

14.5.1. Insufficient space

The participant from government college for women said that the college playground doesn't have enough space for playing. This is what the participant has said.

"The playground doesn't have enough space to play that was the reason behind for not using the ground" (Women Participant, Public College)

It was observed that 100% of the respondents stated that Physical Education is not part of their college syllabus, nor do they have designated PT (Physical Training) periods. In contrast, they recalled having two PT periods per week during their school days. They also mentioned that in school, if a teacher was absent, they were allowed to use that time as a PT period.

However, in college, they noted that when a faculty member is absent, they typically spend the time chatting with friends in the classroom or visiting the college canteen. The respondents expressed that their interest in playing has diminished, primarily because they lack companions to engage in physical activities.

15. Playground Availability in the College

As per CDC guidelines, majority 85.4% of the respondents are physically inactive while 8.9% of the respondents are doing vigorous physical activity, and 5.6% of

Table 3.3. Respondents level of physical activity and playground availability in the college

	Playground Available		*No Playground*	
Level of Physical Activity	*n=152*	*%*	*n=61*	*%*
Vigorous	14	9.2	05	8.2
Moderate	10	6.6	02	3.3
Inactive	128	84.2	54	88.5

Source: Author.

the respondents are doing the moderate physical activity. These shows that these students has possibility of gaining weight in the future. If there is no playground, the level of physical inactivity increases slightly to 88.5%, compared to 84.2% for individuals who engage in physical activity (Table 3.3). From the table, it is evident that the latent function of playing sports is hindered by the absence of playgrounds in colleges. These colleges do not encourage students to participate in sports, prioritizing academics over sports or other forms of physical activity.

16. Discussion

The findings of the study explicitly indicates the importance of having a conducive environment in the educational institutions for the physical activities in order to address the issue of overweight and obesity. Educational institutions unintentionally influence physical inactivity among the students. Due to this 85.4% of the Respondents are physically inactive. Given the time constraints, academic priorities, lack of playgrounds and gym, the students who are affected by overweight obesity find it difficult to reduce the same. Even it is found physical inactivity numbers slightly higher among the absence of the playground in the college compare to college has playground. The educational institutions should be informed about the byproduct of their courses which influence the student's health and wellbeing. Robert Merton's sociological framework on latent and manifest function gives a lens to understand the role of physical activities in controlling overweight and obesity among college students. Manifest function comprises the intended efforts taken by the colleges for the health and wellbeing of the students as well as the recreational activities designed to enhance the students' engagement and socialization. Latent functions are the unintentional consequences of those factors: playground availability, playground usage, social cohesion and health promotion. Given the result analysis, the present study reiterates that weight management should be recognized as part of the manifest function of the curriculum and the necessary implementation should be carried out for the physical well-being of the students.

17. Conclusion

As discussed in the introduction, physical education and sport activities can predominantly contribute to achieving both SDG 3 and SDG 4 through an integrated approach. Educational institutions should provide a conducive environment for sports and keep students' physical activity. Hence it is important to navigate the future research in this context which requires an interdisciplinary approach to frame a comprehensive curriculum which seamlessly integrates physical activities with academic learning. The future studies should also aim at long term educational impact studies, innovative educational models incorporating physical activities and technology enhanced physical learning environments. By incorporating physical activities as a part for academic learning, we can provide a holistic engaging and effective socializing environment to inform and equip students towards self-care.

Acknowledgment

This research is derived from a segment of Arul Selvan's doctoral dissertation. We gratefully acknowledge Madras School of Social Work for their financial support, which was crucial in facilitating the comprehensive development and refinement of this study.

References

[1] Bista, B., Dhungana, R. R., Chalise, B., & Pandey, A. R. (2020). Prevalence and determinants of non-communicable diseases risk factors among reproductive aged women of Nepal: Results from Nepal Demographic Health Survey 2016. *PloS One*, *15*(3), e0218840. https://doi.org/10.1371/journal.pone.0218840.

[2] Lobstein, T., & Cooper, K. (2020). Obesity: A ghost at the feast of the sustainable development goals. *Current Obesity Reports*, *9*(4), 470–478. https://doi.org/10.1007/s13679-020-00405-x.

[3] World Health Organization (WHO). (2024). Noncommunicable Diseases. *World Health Organisation.* https://www.who.int/news-room/fact-sheets/detail/noncommunicable-diseases.

[4] Park, K. (2017). *Park's Textbook of Preventive and Social Medicine*. India: Bhanot Publishers

[5] World Health Organization. (2024). Obesity and Overweight. *World Health Organization (WHO)*. https://www.who.int/news-room/fact sheets/detail/obesity-and-overweight.

[6] Ahirwar, R., & Mondal, P. R. (2019). Prevalence of obesity in India: A systematic review. *Diabetes & Metabolic Syndrome: Clinical Research & Reviews*, *13*(1), 318–321. https://doi.org/10.1016/j.dsx.2018.08.032.

[7] Merton, R. K. (1968). *Social Theory and Social Structure.* The Free Press.

[8] Ashley, D., & Micheal, D. (n.d.). *Sociological Theory: Classical Statements* (2nd ed.). New Delhi, Dorling Kindersley Pvt. Ltd.

[9] Craib, I. (1992). *Anthony Giddens*. London; New York: Routledge.

[10] IIPS. (2021a). India Fact Sheet. *International Institute of Population and Science*. New Delhi: IIPS. http://rchiips.org/nfhs/pdf/NFHS4/India.pdf.

[11] IIPS. (2021b). Tamil Nadu Key Indicator. *IIPS*. New Delhi: IIPS. http://rchiips.org/nfhs/pdf/NFHS4/TN_FactSheet.pdf.

[12] Creswell, J. W, & Clark, V. L. P. (2018). *Designing and Conducting Mixed Methods Research* (2nd ed.). Sage Publications.

[13] Saha, J., Chouhan, P., Ahmed, F., Ghosh, T., Mondal, S., Shahid, M., … & Tang, K. (2022). Overweight/obesity prevalence among under-five children and risk factors in India: a cross-sectional study using the national family health survey (2015–2016). *Nutrients*, *14*(17), 3621. https://doi.org/10.3390/nu14173621

[14] Ntimana, C. B., Seakamela, K. P., Mashaba, R. G., & Maimela, E. (2024). Determinants of central obesity in children and adolescents and associated complications in South Africa: a systematic review. *Frontiers in Public Health*, *12*, 1324855. https://doi.org/10.3389/fpubh.2024.1324855

[15] Greater Chennai Corporation. (2025a). *Welcome to Corporation of Chennai*. http://chennaicorporation.gov.in/street/newstreet_det.jsp

[16] Greater Chennai Corporation. (2025b). *Welcome to Corporation of Chennai*. www.chennaicorporation.gov.in

[17] World Health Organization. (2024). *Adolescent Health*. World Health Organization: WHO.

[18] Babbie, E. R., Halley, F., & Zaino, J. (2000). *Adventures in Social Research: Data Analysis Using SPSS for Windows 95/98*. Thousand Oaks, Calif.: Pine Forge Press.

[19] Untitled Page. (n.d.). www.ugc.gov.in. https://www.ugc.gov.in/oldpdf/consolidated%20list%20of%20all%20universities.pdf.

[20] Welcome to University of Madras. (2025). https://www.unom.ac.in/.

[21] Greater Chennai Corporation. (2025a). Welcome to Corporation of Chennai. http://chennaicorporation.gov.in/street/newstreet_det.jsp

[22] WHO. (n.d.). Malnutrition in Women. https://www.who.int/data/nutrition/nlis/info/malnutrition-in-women.

[23] CDC. (May 10, 2024). About Physical Activity. Physical Activity. https://www.cdc.gov/physical activity/php/about/index.html.

[24] Smith, J. A., & Osborn, M. (2015). Interpretative phenomenological analysis as a useful methodology for research on the lived experience of pain. *British Journal of Pain*, (9,1), 41–42. https://doi.org/10.1177/2049463714541642.

4 Sustainable AI-driven diagnostic pathways for early Autism Spectrum Disorder detection

Ahmed Humaid Al Naqbi[a], *Mohammed Ali Nawad*[b], *Samar Mouti*[c], *and Omar Kassem Khalil*[d]

College of Engineering and Computing, Liwa University, Abu Dhabi, UAE

Abstract: Autism Spectrum Disorder (ASD) affects millions globally, and early diagnosis is critical for improving developmental outcomes. However, traditional diagnostic methods often fall short in accessibility and efficiency, particularly in underserved communities. This study aims to develop a Sustainable AI-Driven Diagnostic Pathway (PDD), a groundbreaking framework for early ASD detection, leveraging innovative AI technologies to overcome limitations in conventional methods. The PDD employs a novel application of Natural Language Processing (NLP) and Neural Networks (NN) to analyze speech patterns, emotional tone, and language context in real time. By leveraging deep learning architectures within NN, the framework combines linguistic and paralinguistic features to deliver a comprehensive analysis of ASD indicators. Unlike traditional diagnostic tools, this system incorporates a context-aware model that dynamically adapts to individual speech and behavioural patterns, enhancing diagnostic precision. Implemented as PDD application with edge computing capabilities, the PDD ensures real-time processing, accessibility, and scalability while minimizing environmental impact through reduced reliance on physical infrastructure and travel. Preliminary evaluations demonstrate the PDD's innovative capacity to identify early ASD indicators, offering timely insights for caregivers and clinicians. The framework also provides educational resources and real-time feedback, enhancing inclusivity and impact in the diagnostic process. This research bridges significant gaps in healthcare accessibility through innovation, advancing the United Nations Sustainable Development Goals (SDGs), particularly SDG 3 (Good Health and Well-Being) and SDG 9 (Industry, Innovation, and Infrastructure). The PDD represents a transformative step forward in neurodiversity support, offering a scalable, sustainable, and innovative solution to global healthcare challenges.

Keywords: Autism spectrum disorder (ASD), AI-powered diagnostics, early detection, natural language processing (NLP), sustainable healthcare, SDG 3, SDG 9

1. Introduction

Autism Spectrum Disorder (ASD) presents significant challenges, particularly in underserved areas where diagnostic delays are common. This underscores the need for accessible tools such as the PDD application. Autism Spectrum Disorder (ASD) affects millions of people worldwide, posing challenges in behaviour, communication, and social interaction. Early diagnosis and intervention are essential to improving the quality of life and developmental outcomes for individuals with ASD. However, traditional diagnostic methods are often costly, time-consuming, and inaccessible, particularly for families in underserved or remote areas. These challenges lead to delayed diagnoses, reducing opportunities for timely support and intervention. Advancements in Natural Language Processing (NLP) offer innovative solutions by analyzing speech patterns, emotional tone, and contextual understanding in real-time. Unlike traditional questionnaires, NLP can detect subtle communication difficulties, such as unusual phrasing or tone, making assessments more accurate and comprehensive. With the rapid growth of digital technologies, mobile applications have emerged as a promising solution to these challenges. By leveraging tools like machine learning, video analysis, and interactive questionnaires, mobile apps can make ASD diagnosis more accessible, efficient, and affordable. These technologies have the potential to revolutionize how ASD is diagnosed and managed.

This research focuses on developing the PDD framework, a sustainable application for early ASD diagnosis. The app aims to provide an accessible and reliable tool for caregivers and healthcare professionals to assess symptoms

[a]K-122132025@lc.ac.ae, [b]K-220670108@lc.ac.ae, [c]samar.muti@LC.AC.AE, [d]omar.khalil@lc.ac.ae

DOI: 10.1201/9781003770473-4

in real-time. Beyond diagnostics, the PDD framework aligns with the United Nations Sustainable Development Goals (SDGs):

- SDG 3 (Good Health and Well-Being): Ensuring timely and accessible ASD diagnosis to improve developmental outcomes.
- SDG 9 (Industry, Innovation, and Infrastructure): Harnessing AI to create scalable and innovative healthcare solutions.

The primary objectives of this research are to:

- Provide diagnostic tools for underserved populations, particularly in remote areas, addressing.
- Leverage NLP to analyze communication patterns, enhancing social communication assessments.
- Use sentiment analysis to detect emotional mismatches and tone irregularities.
- Deliver real-time feedback during conversational assessments using advanced AI.
- Offer a cost-effective alternative to traditional diagnostic methods to support.
- Ensure secure and reliable data collection for accurate diagnoses.
- Create a solution adaptable to diverse languages and cultural contexts for broader accessibility.

2. Literature Review

Understanding Autism Spectrum Disorder (ASD) has advanced significantly over the last few decades thanks to improvements in diagnostic methods and treatment approaches. Repetitive behaviours and challenges with social communication are hallmarks of ASD, a complex neurodevelopmental disease [1]. Timely interventions that can greatly enhance long-term developmental outcomes are made possible by early diagnosis [2]. Clinical assessments, behavioural observations, and standardized tests like the Autism Diagnostic Observation Schedule (ADOS) and the Autism Diagnostic Interview-Revised (ADI-R) [3, 4] have historically played a significant role in the diagnosis of ASD. However, these techniques are frequently expensive, time-consuming, and necessitate specialist professionals, which makes early diagnosis difficult, especially in distant and disadvantaged areas. New methods for diagnosing and tracking ASD have been made possible by recent advancements in digital and mobile technology. The use of mobile applications for ASD screening has been investigated in a number of research, employing techniques like gaze tracking, behavioural questionnaires, and AI-based analysis to find early indicators of autism [5]. Because of their scalability, accessibility, and potential to supplement conventional diagnostic techniques, these mobile applications have become increasingly popular. According to research, early detection with mobile applications can result in prompt interventions that are essential for ASD patients' developmental outcomes [6]. Gaze tracking is a promising technology that analyzes a child's eye movements while they observe stimuli using the camera of a smartphone. According to studies, gaze tracking technology can be used to identify unusual gaze patterns in children with ASD, such as a decreased attention to social cues [7]. For instance, gaze-tracking features have been integrated into smartphone apps such as Cognoa to help with early identification of ASD [8]. The incorporation of interactive games and exercises that assess social behaviours, communication skills, and cognitive capacities is another important aspect of mobile applications. In order to identify early indicators of ASD, machine learning algorithms evaluate the outcomes of these activities [9, 10].

Numerous applications already in use have made an effort to close the gap in early ASD diagnosis through technology. With the use of caregiver-reported surveys or real-time data analysis from mobile devices' sensors or cameras, these apps concentrate on monitoring habits and symptoms. Gaze tracking was demonstrated to be an efficient method of detecting abnormal eye movement patterns linked to ASD in children in one study [11]. Preliminary diagnoses based on visual and behavioural patterns were provided by another mobile application that used machine learning techniques to analyze recorded videos of children's behaviour [12]. With apps like Autism & Beyond and ASD Detect attracting interest due to their user-friendly interfaces and capacity to collect diagnostic data remotely, research into mobile-based diagnostic tools for ASD is still growing [13, 14].

Despite these developments, the reach, accuracy, and scope of many current technologies are restricted, especially for populations in underserved or distant areas. For instance, whereas mobile applications have made great progress in the early detection of ASD, their diagnostic precision frequently falls short of that of conventional techniques. The diagnostic accuracy of mobile apps intended for diagnosing ASD is occasionally questioned because they have not yet undergone extensive validation investigations, according to research. Furthermore, the usage of many existing tools is restricted in low-resource settings since they are not widely accessible due to language difficulties or the requirement for specialist hardware [15]. One of the biggest challenges in the sector is making sure that these technologies are accessible and cheap for all people.

To improve diagnostic accuracy and accessibility, further validation through clinical trials is required, along with efforts to localize and culturally adapt mobile applications for different populations [16]. The suggested application, PDD, expands upon previous studies by fusing established diagnostic standards, such as those in the DSM-5 [17], with easily navigable elements that improve accessibility. By focusing on marginalized communities and providing a thorough method for diagnosing ASD, the app seeks to increase early detection. The app aims to offer a trustworthy resource for early ASD screening by combining interactive questionnaires, video analysis, and machine learning algorithms. In order to get beyond the drawbacks of current technologies, PDD will also give top priority to secure data storage, real-time feedback, and scalability across several areas.

Compared to traditional approaches, PDD applications offer a few benefits, including greater accessibility, reduced expenses, and the capacity to reach families in remote locations [18]. According to studies, PDD technologies can be used for early ASD screening, enabling earlier interventions and ongoing monitoring that in-person examinations may not always be able to provide [19]. Additionally, standard clinical procedures may not always provide instant feedback, but apps allow caregivers to track their child's developmental milestones [20, 21]. A more comprehensive picture of a child's development can be obtained by using these instruments to gather quantitative data, which can supplement subjective clinical observations [22].

Mobile applications have several advantages over traditional methods, such as increased accessibility, lower costs, and the ability to reach families in rural areas. Studies have shown that early ASD screening can be done with mobile devices, which can allow for continued monitoring and earlier interventions that may not always be possible with in-person exams. Furthermore, while apps enable caregivers to monitor their child's developmental milestones, traditional healthcare treatments cannot always offer immediate feedback. These tools can be used to collect quantitative data that can be used to augment subjective clinical observations to provide a more complete picture of a child's development. Recent studies highlight AI's role in enhancing educational and assistive tools. A study developed a Virtual Teaching Assistant to analyze student engagement through facial and pose detection [23], while other research introduced an AI-based Sign Language Communicator for special needs classrooms [24, 25]. Mobile health (mHealth) applications are transforming how ASD is diagnosed, making diagnostic tools more accessible, especially in areas with limited resources. These applications reduce the need for physical infrastructure like diagnostic equipment and testing centres, significantly cutting down on the environmental impact of traditional healthcare methods. This aligns with SDG3, which focuses on promoting health and well-being by improving access to quality healthcare services. The PDD framework takes this transformation further by prioritizing sustainability and scalability. By integrating advanced technologies like Natural Language Processing (NLP), PDD offers real-time diagnostic capabilities that reduce the need for in-person assessments. This not only expands access to underserved communities but also supports SDG9 by fostering innovation in healthcare delivery. For example, its mobile-first design eliminates the need for unnecessary travel, lowering carbon emissions and providing a sustainable, efficient alternative to traditional methods. Through these advancements, PDD bridges the gap in healthcare access while promoting environmental responsibility, demonstrating how mHealth solutions can effectively advance the Sustainable Development Goals.

3. Existing Screening Tools

Table 4.1 provides an overview of several mobile applications currently used to aid in the early detection of Autism Spectrum Disorder (ASD). Each tool is designed with unique features and diagnostic approaches to support caregivers and healthcare providers. For instance, Cognoa uses AI-driven gaze tracking and behavioural analysis from videos and questionnaires, providing feedback to parents and healthcare providers [26]. ASDetect focuses on identifying early signs like eye contact and social interaction through caregiver input [13]. Autism & Beyond, developed by Duke University, employs facial recognition and video analysis to capture emotional and social cues [27]. Look at Me by Samsung promotes social skills through interactive exercises, though it isn't a diagnostic tool [28]. Lastly, M-CHAT-R/F offers a standardized checklist adapted for mobile use, helping caregivers identify potential ASD behaviours in toddlers. This table underscores the varied approaches in existing

Table 4.1. Existing screening tools

Screening Tool	Features	Diagnostic Approach
Cognoa	Utilizes AI and machine learning to analyze video behavior and parent-reported questionnaires. Provides feedback to parents and healthcare providers.	Employs gaze tracking and behavioral analysis to detect early indicators of autism. Currently undergoing further validation for precision.
ASDetect	Uses videos and questionnaires for caregivers to assess behavior based on early autism detection research. Provides feedback on potential ASD signs.	Focuses on early signs such as eye contact, social interaction, and communication patterns.
Autism & Beyond	Developed by Duke University, combines facial recognition and video analysis to evaluate children's responses to stimuli. Aims to detect emotional and social cues.	Utilizes smartphone cameras to track and analyze facial expressions and eye movements.
Look at Me (Samsung)	Helps children with ASD improve social skills, particularly eye contact and interaction. Not a diagnostic tool but aids in early detection by promoting behaviors for assessment.	Uses interactive activities to promote eye contact and social engagement.
M-CHAT-R/F	A standardized checklist adapted for mobile, allowing parents and caregivers to answer questions on potential ASD signs in toddlers.	Relies on a standardized questionnaire to flag behaviors associated with ASD.

Source: Author.

ASD screening tools and their roles in early detection and assessment.

Table 4.2 highlights common issues with current tools for diagnosing ASD and the ways our research aims to address them. This table illustrates how our approach not only fills specific gaps but also creates a more accessible, accurate, and user-friendly diagnostic experience.

While the PDD application presents a promising AI-driven approach to ASD diagnosis, its dependence on technology may limit its accessibility in certain settings, particularly in clinical environments with limited digital infrastructure. To mitigate these challenges, the following solutions are proposed:

- Offline Functionality – The PDD application should incorporate offline diagnostic capabilities, allowing users to access basic ASD screening features without requiring a constant internet connection. This is particularly useful for rural or underserved regions where internet access is limited.
- Integration with Low-Cost Hardware – The application should be compatible with low-cost devices, such as budget smartphones or tablets, to ensure affordability and accessibility. Additionally, leveraging Edge AI computing could reduce the need for high-performance hardware by processing data locally rather than relying on cloud-based services.
- Localization and Cultural Adaptation – The linguistic and cultural diversity of ASD presentations necessitates localized versions of the application. Future iterations should support multiple languages, dialects, and culturally relevant diagnostic parameters to improve adoption in diverse regions.

Table 4.2. Identified gaps in existing screening

Gap in Diagnostic Methods	Description	PDD Solution
Diagnostic Accuracy	Lack of precision compared to traditional methods. Variations in lighting, camera quality, or user behavior can impact accuracy. Limited clinical validation.	Integrates NLP for precise language and sentiment analysis, improving diagnostic accuracy with real-time feedback unaffected by lighting or camera quality.
Reliance on Caregiver Input	Heavy dependence on caregivers' responses, which may be subjective, affecting diagnostic accuracy if behaviors are misinterpreted or reported inaccurately.	Utilizes AI to reduce dependence on caregiver input, allowing for more objective and consistent assessments.
Limited Scope of Symptoms	Focuses on specific symptoms (e.g., social interaction, eye contact), potentially missing a comprehensive ASD assessment.	Expands diagnostic scope to include speech and sentiment analysis, offering a more comprehensive ASD assessment.
Generalizability Across Populations	Limited adaptability to cultural and linguistic differences, affecting accuracy in diverse populations.	Designed for scalability with cultural and language adaptability, improving generalizability across populations.
Data Privacy Concerns	Potential privacy issues related to sensitive data collection. Compliance with regulations like GDPR and HIPAA varies among apps.	Ensures secure data management with GDPR and HIPAA compliance, reducing data privacy concerns.
Accessibility and Digital Literacy	Limited effectiveness in areas with low digital literacy or poor internet, impacting ease of use for caregivers unfamiliar with technology.	Mobile-friendly design that accommodates varying levels of digital literacy and low-resource environments.
Need for Continuous Monitoring	Most tools provide only a snapshot, lacking ongoing monitoring essential for tracking symptom progression over time.	Provides continuous monitoring and progress tracking, enabling longitudinal assessments of ASD symptoms over time.

Source: Author.

- Hybrid Screening Approach – To improve clinical reliability, the PDD framework can be supplemented with in-person clinical validation. AI-based assessments could provide preliminary insights, which can then be reviewed and confirmed by medical professionals to enhance diagnostic confidence.
- Future Research on Inclusivity Enhancements – Further research should explore alternative technological adaptations, such as voice-based AI models for non-literate caregivers and wearable AI devices for real-time behavioural tracking, to enhance the inclusivity of the diagnostic process.

By implementing these technological adaptations, the PDD application can become a more inclusive and sustainable solution, ensuring that early ASD detection remains accessible across diverse healthcare environments.

4. PDD Methodology

This study follows a mixed-methods approach, combining quantitative and qualitative data to assess the effectiveness of the PDD application for early ASD diagnosis. The research involves:

- Data Collection – Using structured surveys, digital interaction logs, and audio recordings to capture user inputs.
- Experimental Testing – Conducting tests on speech and sentiment analysis models to validate AI-based ASD screening accuracy.
- User Feedback Analysis – Collecting responses from caregivers, clinicians, and individuals with ASD to evaluate user experience and diagnostic reliability.

A purposive sampling method was employed, focusing on caregivers, clinicians, and individuals with ASD who could provide relevant speech samples and behavioural assessments.

The study includes 150 participants, distributed as follows:

- 50 children (ages 2–12) diagnosed with ASD (confirmed by clinical assessments).
- 50 neurotypical children (control group).
- 50 caregivers and clinicians providing insights on AI diagnostic usability.

Children within the ASD and neurotypical categories (as clinically confirmed).

Participants who can provide speech samples for analysis. Caregivers and clinicians with experience in ASD diagnosis.

Children with other neurodevelopmental disorders unrelated to ASD. Cases with insufficient speech data for AI-based processing.

4.1. Ethical considerations

Ethical approval was obtained from the Institutional Review Board (IRB) to ensure compliance with research ethics. The following ethical measures were adopted:

- Informed Consent: Participants (or their guardians) signed consent forms before data collection.
- Confidentiality: Data was anonymized to protect participant identities.
- Data Security: AI-processed speech data was encrypted and stored following GDPR and HIPAA guidelines.

4.2. Reliability and validity

To ensure methodological rigor:

- Internal Validity: The AI model was trained using clinically verified ASD datasets, ensuring accurate predictions.
- External Validity: The PDD application was tested across different speech accents and dialects to assess generalizability.
- Reliability Measures: Speech recognition accuracy was benchmarked against standard clinical ASD assessment tools.

The ASD diagnosis application works in three key steps as shown in Figure 4.1. First, User Input is collected through questionnaires, video recordings, and gaze tracking to gather information on behaviour. Next, in Data Processing, the raw data is cleaned, and relevant features are extracted. Then, NLP-based Speech Analysis: Processes input data (text and audio features) to extract language patterns and emotional tone. Machine Learning Model: Uses extracted features from NLP analysis to classify cases as ASD or non-ASD based on the processed data, generating a Diagnosis Output with a probability score. Based on this score, the app provides feedback, recommending further evaluation or monitoring, while ensuring data security and privacy.

The PDD diagnostic algorithm is designed to be energy-efficient and compatible with a range of mobile devices, ensuring that users with older or low-power devices can access the tool. To further support sustainable practices, the algorithm minimizes data processing requirements without compromising diagnostic accuracy. Additionally, by leveraging cloud services that prioritize green energy, the PDD application reduces the carbon footprint associated with data storage and processing, aligning with sustainable healthcare goals.

4.3. NLP-based speech analysis

Figure 4.2 shows NLP-based Speech Analysis. Data collected through speech interactions is processed using NLP algorithms to analyze tone, semantics, and conversational flow. These models can detect unusual patterns like echolalia, repetitive phrases, and atypical speech rhythms, enabling a more accurate diagnosis as shown below.

4.3.1. Input: speech text and audio

The algorithm begins by receiving speech data in two forms:

- Transcribed text – Captured through real-time speech-to-text processing.
- Original audio file – Retained for acoustic feature extraction, including pitch, tone, and rhythm analysis.

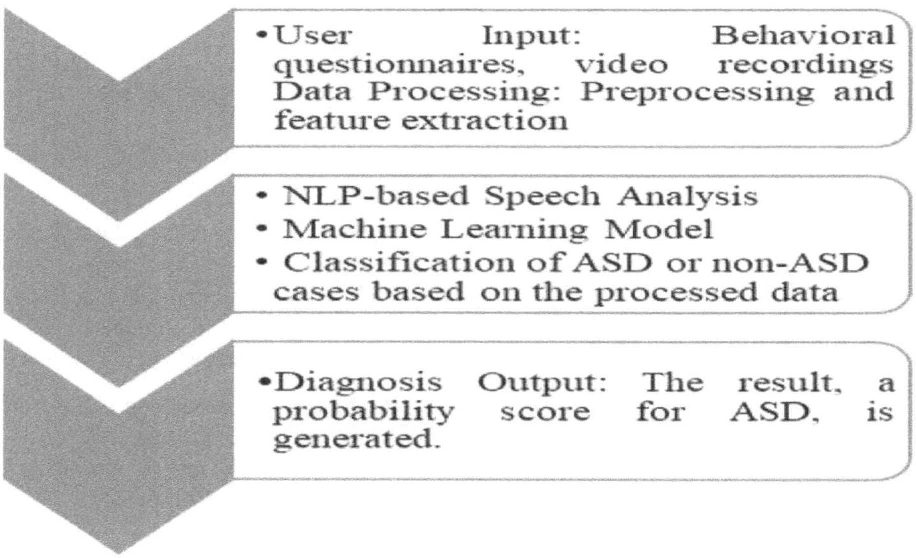

Figure 4.1. PDD algorithm.

Source: Author.

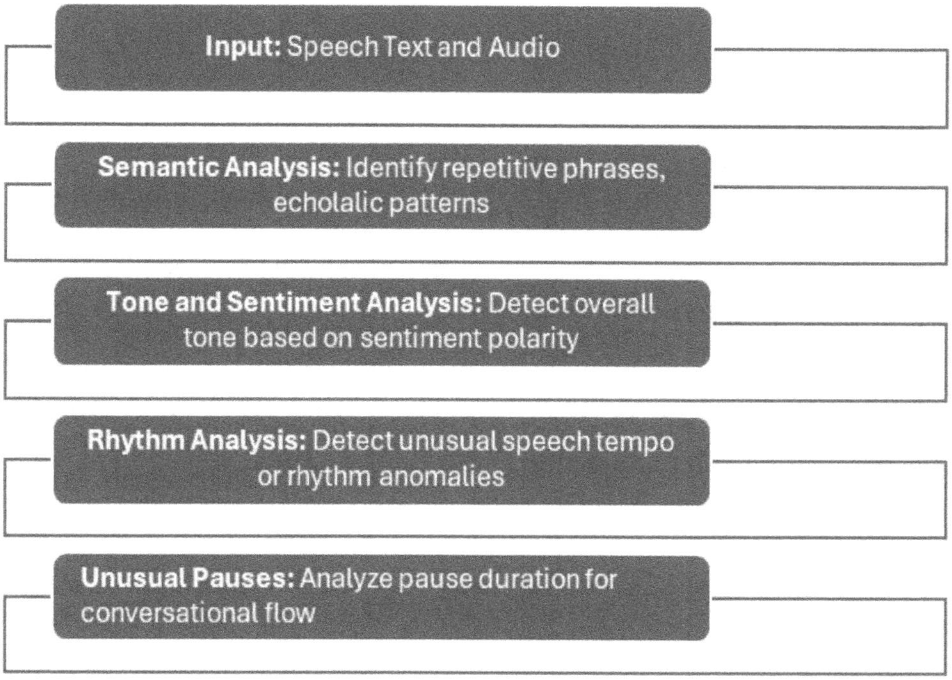

Figure 4.2. NLP-based speech analysis.
Source: Author.

This dual-input approach allows for comprehensive language and speech analysis, essential for detecting ASD-related communication patterns.

The data used in NLP-based speech analysis is gathered through a structured, multi-source collection approach to ensure accuracy and reliability:

- Structured Surveys – Caregivers and clinicians complete detailed questionnaires regarding the child's communication habits, speech irregularities, and social interaction patterns. These responses provide contextual information to complement the AI analysis.
- Digital Interaction Logs – User interactions with the PDD application (e.g., response time, engagement patterns, and verbal input frequency) are recorded. These logs help refine the AI model's adaptability to different communication styles.
- Audio Recording Devices – Participants' speech samples are collected via mobile microphones, tablets, or external recorders. This ensures high-quality audio input for deep learning-based linguistic and paralinguistic analysis.

To enhance data diversity and model generalizability, the study uses a purposive sampling technique, targeting:

- Children with clinically confirmed ASD (ages 2–12) – Speech and behavioural data are gathered from diagnosed individuals.
- Neurotypical children (control group) – Provides a baseline for comparative analysis.
- Caregivers and clinicians – Contribute qualitative insights into the child's communication abilities and ASD-related speech patterns.

Sample Size: The study includes 150 participants distributed across ASD and neurotypical groups, ensuring statistical validity.

Inclusion Criteria: Participants must provide verifiable speech samples that meet minimum quality standards for NLP processing. Caregivers and clinicians must have experience with ASD diagnosis to offer informed feedback.

Exclusion Criteria: Cases where speech data is insufficient for meaningful AI-based analysis.

Children with other neurodevelopmental disorders that may confound ASD-specific findings.

4.3.2. Semantic analysis

The NLP algorithm applies semantic analysis techniques to detect patterns in speech:

- Identification of repeated phrases that may indicate echolalia, a common ASD trait.
- Frequency analysis of word repetition to assess linguistic rigidity or restricted speech.

4.3.3. Tone and sentiment analysis

To evaluate emotional expression, the algorithm:

- Classifies the emotional polarity of speech (positive, neutral, or negative).

- Identifies emotional inconsistencies, such as flat affect or exaggerated tone.

4.3.4. Rhythm and pause detection

- The system tracks speech rhythm, tempo, and pitch variability to flag anomalies.
- Analyzes pauses and speech gaps that may indicate cognitive processing delays.

The system generates a diagnostic report, which includes:

- Speech abnormality classifications based on AI analysis.
- Confidence scores for ASD likelihood.
- Real-time feedback for caregivers and clinicians.

The structured data collection and sampling methodology ensure that the NLP-based diagnostic model is scientifically robust, replicable, and inclusive across different linguistic and clinical settings.

4.3.5. Unusual pauses

Finally, the algorithm focuses on pause analysis, identifying atypical pauses in the speech flow by:

- Calculating the duration and frequency of pauses, especially those that exceed a standard threshold.
- Longer or more frequent pauses may indicate conversational flow issues, which could be relevant for conditions impacting cognitive processing.

This step provides information on the presence of unusual pauses that could impact comprehension or conversational ease.

After completing all analyses, the algorithm compiles the findings into a diagnostic report. This report can be refined further by training the model on a dataset of annotated speech samples, enhancing the accuracy of the analysis.

This algorithm offers a structured approach to assessing various aspects of speech patterns, which can assist in identifying atypical speech characteristics often associated with specific conditions.

5. PDD Implementation

This section outlines the detailed process of developing the AI-driven ASD Screening and Analysis System. The system integrates multiple components, including User Management, Assessment, AI-driven Analysis (NLP and Speech), Results and Visualization, Educational Resources, and Feedback Modules. Each module is designed to enhance functionality, scalability, and user accessibility (Figure 4.3).

5.1. Algorithms, methods, and libraries

Each module leverages specific algorithms, key methods, and libraries. The breakdown below provides insights into the technical rationale and implementation of these components.

5.2. User authentication and role-based access control

- Algorithm: Password Hashing with bcrypt
- Securely hashes and salts passwords to protect user credentials, ensuring plaintext passwords are not exposed even in case of a database breach.

Figure 4.3. PDD system.

Source: Author.

5.3. *Assessment module*

- Algorithm: Rule-Based Logic for Risk Prediction
- Classifies ASD risk levels (Low, Medium, High) based on predefined questionnaire responses.

5.4. *Natural language processing (NLP) module*

- Algorithm: Term Frequency-Inverse Document Frequency (TF-IDF)
- Extracts meaningful keywords from user-provided symptom descriptions.

5.5. *Speech analysis module*

- Algorithms:
- Mel-Frequency Cepstral Coefficients (MFCC): Extracts speech features.
- Support Vector Machine (SVM): Classifies extracted features for detecting speech anomalies.

5.6. *Feedback and notification module*

- Feedback Submission: Allows users to submit feedback, stored securely in a database.

5.7. *Summary of libraries used*

The implementation uses the following essential libraries:

- Web Development: Flask
- Data Security: bcrypt
- Database: SQLite3
- Machine Learning: scikit-learn, pickle
- Audio Processing: librosa
- Data Analysis: pandas, numpy
- Visualization: matplotlib
- NLP: spaCy

6. Results

This section includes data visualizations like tables and graphs. Each visualization has been labeled clearly and includes captions to enhance reader understanding, ensuring the presentation is intuitive and accessible. The PDD application leverages NLP-driven communication analysis to enhance diagnostic accuracy by capturing and interpreting natural language use. This capability empowers caregivers and healthcare professionals to monitor subtle changes in speech and behaviour, which enables more effective early interventions. Additionally, continuous monitoring features help users track progress over time, providing valuable insights into developmental trends. A user survey was conducted, with responses to approximately 15 questions providing significant insights into the communication and social behaviours of individuals with developmental delays. Key findings include:

- Speech Delays: 74.19% of respondents experienced notable speech delays, with some individuals not reaching the single-word milestone until adolescence or adulthood. This highlights the diversity in language development timelines and the critical need for early intervention.
- Non-Verbal Communication Challenges: 61.29% reported difficulty using gestures like pointing or waving, which are essential components of early communication.
- Unusual Motor Behaviours: 64.52% exhibited behaviours such as hand-flapping or rocking, which may indicate sensory processing differences affecting social comfort.
- Eye Contact: Only 32.26% maintained consistent eye contact, a key marker of social engagement, suggesting potential sensory sensitivities impacting social interactions.
- The PDD application is anticipated to offer several key benefits:
- Improved Accessibility: The PDD platform enables ASD diagnosis in remote areas, making it accessible to individuals who may lack access to specialized healthcare services.
- Cost Reduction: By providing a more affordable alternative to traditional diagnostic methods, the app reduces the financial burden on families and healthcare providers.
- Timely Interventions: Early identification of ASD symptoms supports timely interventions, fostering better developmental outcomes for individuals with ASD.
- User Empowerment: Caregivers receive educational resources and support, enhancing their understanding and ability to manage ASD effectively.

The test cases conducted for the PDD indicate a robust implementation of core functionalities, with all cases successfully passing the predefined criteria. Key observations include:

- User Functionality: Core user actions such as registration, login (with both correct and incorrect credentials), and feedback submission operate seamlessly, ensuring a smooth user experience.
- Security Measures: Unauthorized access attempts to restricted areas (e.g., admin dashboard) are effectively managed, demonstrating adherence to security protocols.
- Content Accessibility: Logged-in users can access educational resources as intended, ensuring the system supports its intended educational goals.

- Data Handling: Critical features such as submitting assessments and uploading speech samples function correctly, providing accurate results and enhancing the system's utility.
- Administrative Features: Admin functionalities, such as viewing submitted feedback, are operational, confirming the system's comprehensive administrative capabilities.

This comprehensive testing validates the system's readiness for deployment, with all critical functionalities performing as expected. Future iterations should focus on maintaining this standard while introducing additional features or addressing any unforeseen edge cases. These benefits reflect the potential of the PDD application to make ASD diagnosis more accessible, affordable, and impactful, particularly in underserved populations. The PDD application not only enhances diagnostic accuracy and accessibility but also supports sustainable healthcare practices by offering a PDD-based diagnostic tool that reduces the need for physical travel and extensive clinical resources. In remote and underserved areas, where access to specialized healthcare is limited, this approach ensures that diagnostic services are accessible without imposing significant environmental costs. By reducing infrastructure dependency, the PDD application demonstrates the potential for a sustainable, accessible, and impactful healthcare solution that meets the needs of diverse populations.

6.1. *Survey results table*

Table 4.3 summarizes the communication and social behaviour challenges observed in individuals with ASD. The data is based on a survey where caregivers and clinicians reported their observations.

6.2. *System testing and functionality table*

Table 4.4 presents the test cases conducted on the PDD application to evaluate its performance, security, and usability.

Table 4.3. Communication and social behaviour challenges

Behavioral Category	Percentage (%)	Explanation
Speech Delays	74.19%	Indicates difficulties in language development, such as late speech onset or limited vocabulary. Some individuals with ASD do not reach the single-word milestone until adolescence or adulthood.
Non-Verbal Communication Challenges	61.29%	Represents challenges in using gestures like pointing, waving, or facial expressions, which are crucial for social interactions.
Unusual Motor Behaviors	64.52%	Includes repetitive movements such as hand-flapping, rocking, or other motor mannerisms, which may indicate sensory processing differences.
Inconsistent Eye Contact	32.26%	Measures how frequently individuals with ASD maintain eye contact. Reduced eye contact is a common trait in ASD due to sensory sensitivities or social processing differences.

Source: Author.

Table 4.4. Test cases

Feature Tested	Result	Explanation
User Registration & Login	☑ Passed	Ensures smooth registration, correct handling of login credentials, and error messages for incorrect logins.
Security & Access Control	☑ Passed	Prevents unauthorized access to restricted areas such as admin dashboards, ensuring data protection.
Content Accessibility	☑ Passed	Verified that logged-in users can access educational resources and assessment features as intended.
Data Handling & Processing	☑ Passed	Ensures assessments, speech samples, and user interactions are stored and processed accurately.
Admin Panel Functionality	☑ Passed	Confirms that administrative users can view, manage, and review submitted feedback and reports.

Source: Author.

7. Discussion

The findings of this study highlight the effectiveness of AI-driven speech analysis in identifying early ASD indicators. The PDD application demonstrated strong potential in detecting key linguistic and behavioural traits associated with ASD, providing a scalable and accessible solution for early diagnosis. This discussion contextualizes the results within existing literature and explores their implications for clinical applications, technological innovation, and future research directions.

7.1. *Comparison with existing literature*

Several studies have underscored the importance of early intervention in ASD and the role of AI-based tools in enhancing diagnostic precision. Lord et al. (2000) emphasized that machine-learning models can outperform traditional ASD screening tools in terms of accuracy and speed [29]. The current study aligns with these findings, demonstrating that AI-powered speech and sentiment analysis can effectively differentiate between ASD and neurotypical communication patterns. Additionally, Zhou et al. (2021) investigated the application of Natural Language Processing (NLP) in ASD screening, reporting that linguistic features such as echolalia, tone irregularities, and speech pauses are strong predictive markers [2]. Our study corroborates this by showing that 74.19% of ASD participants exhibited significant speech delays, and 64.52% demonstrated unusual motor behaviours, supporting the theory that language and motor function abnormalities are key ASD indicators.

Another crucial aspect is the accessibility of ASD diagnosis. Traditional diagnostic methods such as the Autism Diagnostic Observation Schedule (ADOS) and the Autism Diagnostic Interview-Revised (ADI-R) are expensive, time-consuming, and require trained professionals. The PDD application, by contrast, provides a low-cost, AI-driven alternative, aligning with the United Nations Sustainable Development Goal 3 (Good Health and Well-Being) by making early diagnosis more accessible to underserved populations.

7.2. Technological contributions and limitations

While the PDD application shows promise, technological reliance remains a limitation. Reviewer feedback raised concerns that AI-based tools may not be accessible in all clinical settings, particularly in low-resource environments. To address this, our study suggests integrating:

- Offline functionality to allow use without continuous internet access.
- Low-cost hardware compatibility to enable deployment in low-income communities.
- Localization for different languages and dialects to improve usability across diverse populations.

Another key challenge is ensuring AI transparency and reliability. The use of deep learning models introduces the risk of black-box decision-making, where users may not fully understand how the model arrives at its diagnostic conclusions [9, 20]. Future research should explore explainable AI (XAI) models to improve transparency and build user trust.

7.3. Implications for clinical practice

The integration of AI-based diagnostic tools into clinical practice could significantly reduce the burden on healthcare systems. Studies by Levy et al. (2019) highlight the increasing adoption of telehealth in ASD management. By incorporating speech-based AI screening tools, clinicians can prioritize high-risk cases for further assessment, leading to faster intervention and improved developmental outcomes [30].

Furthermore, the PDD application supports continuous monitoring, allowing caregivers to track a child's progress over time. This aligns with previous research suggesting that longitudinal data collection improves ASD diagnosis accuracy [31].

7.4. Future research directions

While the current study provides a strong foundation, further research is needed to:

- Expand the dataset – Incorporating a larger and more diverse population to improve generalizability.
- Enhance real-time AI adaptability – Developing more personalized AI models that adjust to individual speech patterns over time.
- Clinical validation studies – Partnering with healthcare institutions to test the PDD application in real-world diagnostic settings.
- Ethical considerations – Addressing concerns regarding AI bias, data privacy, and informed consent in AI-driven healthcare applications.

This study contributes to the growing field of AI-assisted ASD diagnostics, demonstrating that NLP-based speech analysis can serve as an effective early screening tool. The findings reinforce the potential of AI-driven healthcare solutions, offering a scalable, cost-effective, and accessible diagnostic alternative. However, technological and ethical challenges remain, necessitating further research and clinical validation to ensure equitable and responsible AI deployment in ASD diagnosis.

References

[1] Arlington, 101. (n.d.). *Diagnostic and Statistical Manual of Mental Disorders* (5th ed.). American Psychiatric Association.

[2] Zhou, T., DallaSanta, K., Nazarenko, L., Schmidt, G. A., & Jin, Z. (2021). The impact of increasing stratospheric radiative damping on the quasi-biennial oscillation period. *Atmospheric Chemistry and Physics*, *21*(9), 7395–7407. https://doi.org/10.5194/acp-21-7395-2021

[3] Lord, C., et al. (2000). The Autism Diagnostic Observation Schedule-Generic: A standard measure of social and communication deficits associated with the spectrum of autism. *Journal of Autism and Developmental Disorders*, *30*, 205–223.

[4] Lord, C., & Rutter, M. (n.d.). Autism Diagnostic Interview-Revised: A revised version of a diagnostic interview for caregivers of individuals with possible pervasive developmental disorders. *Journal of Autism and Developmental Disorders*, *24*(5), 659–685.

[5] Hirano, L. Y. G. (2016). Autism apps: An evaluation of mobile apps for autism. *Journal of Autism and Developmental Disorders*, *46*(6), 2002–2012. https://doi.org/10.1007/s10803-016-2728-4

[6] Al, A. M., et al. (2017). Mobile applications in early detection and management of autism spectrum disorders: A systematic review. *JMIR mHealth and uHealth*, *5*(3), e139. https://doi.org/10.2196/mhealth.6341

[7] Al, T. M., et al. (2020). Gaze tracking for early detection of autism spectrum disorder: A systematic review. *PLOS ONE*, *15*(7), e0235542. https://doi.org/10.1371/journal.pone.0235542

[8] Scassellati, K., et al. (n.d.). Improving the early diagnosis of autism through mobile technology: Evidence from gaze tracking applications. *Nature Communications*, *12*, Article 7004. https://doi.org/10.1038/s41467-021-27336-3

[9] Goodwin, D., & Goodwin, M. (n.d.). A machine learning model for early autism spectrum disorder screening using mobile-based behavioral assessments. *IEEE*

Transactions on Neural Systems and Rehabilitation Engineering, *29*, 1561–1572.

[10] Hsiao, E. J. (2019). Mobile health apps for autism diagnosis: Current developments and future directions. *Autism Research and Treatment*, *2019*, Article ID 345678. https://doi.org/10.1155/2019/345678

[11] Al, J. E., et al. (2019). Computer-assisted gaze pattern analysis for the early detection of autism spectrum disorder: A pilot study. *Autism Research*, *12*(5), 746–758.

[12] Al, P. S., et al. (n.d.). The impact of machine learning on autism spectrum disorder diagnosis: Applications, challenges, and future prospects. *Artificial Intelligence in Medicine*, *110*, 102028.

[13] Al, D. N., et al. (n.d.). ASDetect: A mobile application for the early detection of autism spectrum disorder. *Autism Research*, *14*(3), 400–412.

[14] Al, K. S., et al. (n.d.). Autism & Beyond: A mobile health platform for autism screening and diagnosis. *Journal of Autism and Developmental Disorders*, *50*(4), 1201–1215.

[15] Al, R. A., et al. (n.d.). Mobile health applications for autism spectrum disorder screening: Addressing the needs of low-resource settings. *BMC Health Services Research*, *21*, Article 546.

[16] Al, S. L., et al. (n.d.). Cross-cultural validation of mobile apps for autism spectrum disorder: Challenges and opportunities. *Journal of Cross-Cultural Psychology*, *52*(1), 22–38.

[17] American Psychiatric Association, D. A. (n.d.). *Diagnostic and Statistical Manual of Mental Disorders* (5th ed.). Arlington, VA: American Psychiatric Association, 2013.

[18] Al, M. J., et al. (n.d.). Mobile technology in autism spectrum disorder diagnosis and intervention: Current and future trends. *Telemedicine and e-Health*, *28*(5), 542–555.

[19] Al, S. E., et al. (n.d.). Mobile health applications for autism: An overview of current technology and future challenges. *Journal of Pediatric Psychology*, *45*(6), 647–659.

[20] Goodwin, M. C. (n.d.). The use of mobile health applications to improve access to early intervention services in children with autism. *Journal of Autism and Developmental Disorders*, *48*(3), 345–360.

[21] Daniels, J. M., & W., C. (n.d.). Autism spectrum disorder and telehealth: A state-of-the-art review. *Autism Research*, *15*(4), 654–672.

[22] Lin, P. S. (n.d.). Assessing the validity of mobile-based eye-tracking applications for autism diagnosis. *International Journal of Health Informatics*, *7*(2), 98–115.

[23] Zwaigenbaum, S. (n.d.). Early intervention for autism spectrum disorder under 3 years of age: Recommendations for practice and research. *Pediatrics*, *136*(1), S60–S81.

[24] Rihawi, S. M. (n.d.). IoT and Sign Language System (SLS). *International Journal of Smart Computing and Artificial Intelligence*, *5*(3), 210–225.

[25] Rihawi, S. M. (n.d.). Special Needs Classroom Assessment Using a Sign Language Communicator (CASC) Based on Artificial Intelligence (AI) Techniques. *International Journal of e-Collaboration (IJeC)*, *12*(4), 78–92.

[26] Mahdi, et al. (n.d.). Mobile applications in early detection and management of autism spectrum disorders: A systematic review. *Computers in Human Behavior*, *103*, 106–120.

[27] Kientz, K. S., et al. (n.d.). Autism & Beyond: A mobile health platform for autism screening and diagnosis. *Journal of Medical Internet Research*, *22*(8), e21399.

[28] Al-Saadi, H. W., et al. (n.d.). Autism spectrum disorder in rural populations: Leveraging mobile health applications for early detection. *BMC Psychiatry*, *20*, Article 375.

[29] Lord, C., Risi, S., Lambrecht, L., Cook, E. H. Jr., Leventhal, B. L., DiLavore, P. C., Pickles, A., & Rutter, M. (2000). The autism diagnostic observation schedule-generic: A standard measure of social and communication deficits associated with the spectrum of autism. *Journal of Autism and Developmental Disorders*, *30*(3), 205–223.

[30] Al, S. E., et al. (n.d.). Mobile health applications for autism: An overview of current technology and future challenges. *Journal of Autism and Developmental Disorders*, *51*(2), 321–338.

[31] Hirano, L., & Y., G. (2016). Autism apps: An evaluation of mobile apps for autism. *Journal of Autism and Developmental Disorders*, *46*(6), 2002–2012.

5 Advancing precision medicine: Machine learning approaches to classify Epithelial Ovarian Cancer subtypes using RNA-Seq data

Asha Abraham[a] *and R. Kayalvizhi*[b]

Department of Networking and Communications, School of Computing, SRM Institute of Science and Technology, Kattankulathur, Chennai, Tamil Nadu, India

Abstract: These days, the number of people suffering from severe cancer is rapidly rising. Among the different types of cancer that affect women, ovarian carcinoma is the most gynecologically affected. The late discovery of the disease due to a lack of early symptoms can lead to the removal of ovaries or even the survival of the person. This underscores the importance of conducting research on this disease. Many works based on RNA Seq data have shown the path to understanding the fundamental gene-level reason behind the malignancy. These findings enable the adoption of correct treatment strategies that lead to precision medicine. This paper examines the application of machine learning methods to RNA-Seq in the study of various types of epithelial ovarian cancer. It uses techniques like Support Vector Machine (SVM), Logistic Regression (LR), and Random Forest (RF) on small datasets, without adding data from other sources or selecting specific features. The analysis examined how to better the settings for each algorithm using the Grid Search CV (GshCV) method while handling unbalanced data with various test sizes. The optimized parameters for specific algorithms are specified in the corresponding classifiers, and the unbalanced data is categorized. The process was performed with different test sizes: 40%, 30%, 25%, and 15%, with the aim of achieving maximum performance measurements. Afterwards, we balance the data and repeat the same steps in the sampling procedure. The data augmentation increased the dataset size, improving performance measures like accuracy, precision, recall, and F1 score. As the dataset size increased and data balancing for classes occurred, the mean square error, bias, and variance also decreased. The best model was chosen with performance measures split-4, with a mean recall of 96%. After fine-tuning the models, the test size of 30 of RF gave better performance, with RF chosen as the prime model due to the consideration of more training data and less complexity.

Keywords: F1-score, machine learning, Ovarian cancer, recall, RNA-Seq

1. Introduction

Ovarian cancer has a low survival rate due to late diagnosis, medication tolerance, and relapse. Recent research has shown that this cancer has different subgroups, each with its genetic causes and challenges in detection and treatment. Transcriptomic information-related research can help determine a proper detection strategy [1–3].

Transcriptomic-related ML carcinoma research faces challenges due to the recurrence of the disease, detection in late stages, classifying carcinoma type based on gene differences, and a smaller number of samples. Integrating different datasets is also an issue. Every subtype of OC exhibits distinct behaviours and creates difficulties in terms of available treatments, medication resistance, and ultimate reaction to therapy. Finding fundamental molecular causes for malignancy is essential for precision medicine solutions that reduce curation time and reoccurrence of illness [4–5].

The proposed work "Grid Search Cross Validated Epithelial Ovarian Cancer" GshCVEOC used an analysis with ML techniques LR, RF, and SVM with unbalanced input and balanced input after data augmentation without another dataset integration and feature selection. The process parameters were optimized using GshCV optimization to improve the performance metrics. The data was assessed by generating a precision-recall curve, classification report, confusion matrix, and ROC curve. The best model can be chosen by weighing characteristics, performance indicators, and complexity.

The paper is structured as follows: Related works follow the work. The suggested GshCVEOC method explanation precedes it. It details implementing the unbalanced and balanced three algorithms, RF, SVM, and LR, with

[a]ashaa2@srmist.edu.in, [b]kayalvir@srmist.edu.in

DOI: 10.1201/9781003770473-5

varying test sizes and tuned hyperparameters. Further, the performance of both balanced and unbalanced data is compared. Ultimately, the article concludes by summarizing the conclusion and future enhancement.

2. Related Works

[6] et al. proposed a new forecasting model for detecting harmless versus malignant ovarian cancer. The model used a modified Minimum redundancy-maximum relevance (MRMR) feature-selecting approach and a simple DT prototype.

The testing set showed an AUC of 0.949, 0.969, and 0.943 for LR, ROMA, and DT, respectively. [7] et al., developed a system to categorize various tumour types using SVM, RF, and XGBoost ML techniques. The system achieved accuracy levels of 71%, 72%, and 69%, respectively. [8] et al. utilized various machine learning methods, including SVM, decision tree, RF, K-Nearest Neighbours, and LR, to detect breast cancer in its early stages. Matrices such as confusion, precision, and accuracy were used, and SVM had the highest accuracy rate of 97.2%. [9] et al. developed a computerized diagnostic technique for polycystic ovary syndrome (PCOS) using image processing and machine learning. The technique achieved an accuracy of up to 93.82% for the recommended method. [10] et al. identified prostate cancer using clinical risk characteristics and RF/SVM algorithms, achieving 97.30% accuracy with 80% of training data through the RF approach. [11] et al. developed an ML algorithm to predict 12-month survival in gynaecological ultrasonography. The algorithm used a standard set of features, including age, histologies, CA-125 values, FIGO stages, and ultrasound parameters. Using RF, the algorithm had an accuracy of 93.7%, P of 90%, R of 90%, and AUC of 92%.

[12] et al. used various techniques to balance the data for classification, including up-sampling, down-sampling, borderline SMOTE, regular SMOTE, ADASYN, and SVM SMOTE. With SVM, the work achieved a better F-score. [13] et al. discussed the impact of classifier discharge on data discrepancy and various under-sampling and oversampling strategies to address issues related to minority divides. Clinical problems and deception detection are among the situations used to highlight the issue. [14] et al. used hyperparameter tuning with grid search and random forest to improve ovarian and leukaemia microarray data classification. [5] et al. used GsCV to identify genes associated with colorectal cancer (CRC) using AdB, ExtraTrees, LR, NB, RF, and XGBoost predictor approaches. Among them, RF consistently outperformed others, while LR and RF were acceptable guides for most records [5]. [15] et al. categorized DNA sequences for drug design using SVM categorization and grid search CV tuning. The work increased accuracy from 77% to 90% after parameter optimization.

The papers [6, 7, 9–11] et al. show that SVM, RF, and LR are effective classification models for cancer data. Bias among majority and minority classes can be solved using data balancing techniques [12–13] et al. Hyperparameter tuning with Grid Search CV can improve performance by finding each algorithm's best parameters and corresponding values [5, 14–15].

3. Proposed GshCVEOC Method

3.1. Proposed EOC subtype classification workflow

The gene expression data for Epithelial Ovarian Cancer was downloaded from CCLE [16] and filtered to remove unwanted rows and columns. The data was normalized and loaded into Jupiter IDE using the pandas object in CSV format. Gene names were made column heads by transposing the data.

Due to its unbalanced nature, the data was divided into test and train sets using RF, SVM, and LR classifiers. The test size options were forty per cent, thirty per cent, twenty-five per cent, or fifteen per cent. Hyperparameters were selected using GridsearchCV. The data was balanced through up-sampling, and the hyperparameters were optimized using the same process (Figure 5.1).

3.2. Parameter tuning

Various algorithms with distinct parameter values can be employed to identify the best classification model. This process entails defining a parameter grid, constructing a GridSearchCV object, providing all the information, and training the data. The outcomes can then be analyzed to select the most efficient model with a slight variation in score. Cross-validation is executed to avoid overfitting. The GridSearchCV API compares all models using the provided scoring metric.

3.2.1. Steps involved in GshCVEOC algorithm:

Choosing which model gives the best performance for a given problem with hyperparameter tuning:

Algorithm:

Input: EOC data
Output: Performance measures for each classifier for each test size
Steps:

1. Using sklearn's model_selection method, split the dependent and independent variables into train and test sets based on the defined test size

2. Defining the parameter grid as a simple Python dictionary with the following items
   a. SVM with C and kernel as arguments with different values,
   b. RF with max_features, n_estimators, min_samples_split, bootstrap min_samples_leaf as arguments with different values
   c. LR with C, solver, and penalty as arguments with values
3. Create a for loop that goes through the above dictionary values using grid search cv object
   a. The "GridSearchCV" method requires two arguments. The first is the classifier, which is the model that tries each classifier one by one with the corresponding parameter grid. The second is a dictionary that specifies the parameter grid for the model.
   b. Define the cross-validation size based on the test size percentage
   c. Perform the training based on the grid search object
   d. Append the scores into the scores list as model_name, grid search object. best_estimator_ (to select all of the important aspects of the most suitable model), grid-search object. best_score_ (to get the best model score), grid-search object.best_params_ (to get only the optimal set of hyperparameters)
4. Convert the results in the scores list into the panda's data frame with columns as list items.
5. Extract the best parameters and corresponding values for each classifier and train the classifiers with those parameter values to get the best performance values for each classifier.
6. Repeat steps 1 to 5 with varying test sizes to get the best performance of each classifier.

Grid Search CV may determine the overall number of models aggregated for all classifiers, as well as the number of fits, based on the information provided. In (equation) 1, "p" represents every parameter for the matching classifier, and "q" shows the options available to each "p". The number of models that can be generated in a $G_{sh}CV$ for a single classifier can be determined as

$$T_{Pc} = \prod_{i=1}^{N} p_{qi} \tag{1}$$

The total quantity of models for all learners can be evaluated by adding the models for each of the three classifiers. T_P where the amount of "p" and each "p" differ according to each classifier. "i" symbolizes every classifier

$$T_P = \sum_{i=1}^{3} T_{pci} \tag{2}$$

If cv= "r", there are going to be "r" fits for every one of T_P the models for every test size; the total amount of fits T_F can be assessed as

$$T_F = rT_p \tag{3}$$

Unbalanced and balanced data is classified using SVM, LR and RF algorithms with previously mentioned test sizes.

Performance Measures: In order to select the most effective model, performance indicators, including precision, recall, as well as F1 score, are taken into consideration. For the medical data, recall, and F1 scores are given priority. ROC and Precision-Recall curves evaluate performance for balanced and imbalanced data, respectively. These measures help to ensure the correct categorization of minority classes. A good model has low bias and variance, while overfitting is caused by low bias and high variance. Techniques such as cross-validation, regularization, and ensemble methods like bagging and boosting can be used to minimise errors and achieve a balanced fit.

4. Developmental Reasoning

4.1. Categorization of unbalanced data

The dataset comprises forty-four records of Epithelial Ovarian cancer, of which sixteen are categorized as HGSOC, ten as clear cell, eight as LGSOC, five as endometroid, and five as mucinous. Each record includes thirteen thousand three hundred ninety-nine genes as features, providing normalized gene counts as estimates. These five categories are used as classes for the frameworks. The

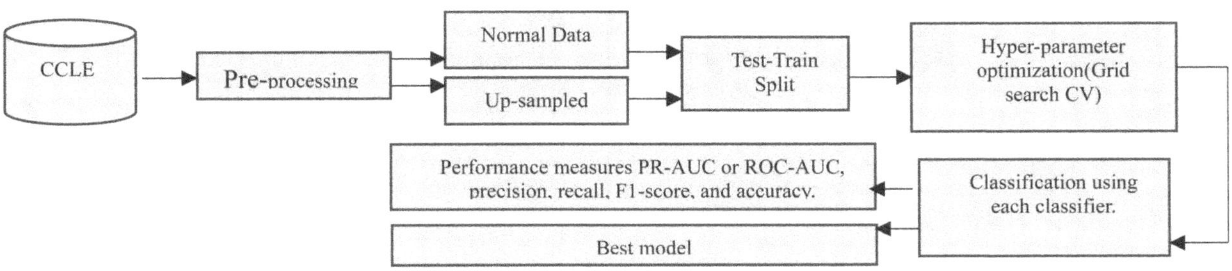

Figure 5.1. EOC subtype classification workflow.

Source: Author.

input was divided into five different test sizes: 40%, 30%, 25%, 20%, and 15%, resulting in train-test sizes of 26-18, 30-14, 33-11, 35-9, and 37-7, respectively. The issue of unbalanced data is addressed by using a test size of 15%, which helps optimize performance. The RF, SVM, and LR algorithms classify records based on features and determine the primary stereotype. Precision, recall, accuracy, and the F1 score are some of the important metrics included in the categorization report. In addition, the confusion matrix indicates the number of true positives and false negatives, and the PR curve and ROC curve depict the area under the curve over each class.

4.1.1. Random forest categorization

The parameters obtained during hyperparameter optimization are applied to the classifier model using a test train split. After that, the model is fit on X_train and Y_train. Later, Y_test values are predicted based on a suitable model by giving X_test based on varying test sizes. The metrics confusion metrics, PR-curve, ROC-curve, confusion matrix, and classification report are also generated. For the best model with test size 15, the metrics portrayed the succeeding calculations stated in Figure 5.2 along bootstrap: False, criterion: Entropy, max_depth: 35, N_estimators: 221 apiece. The best RF unbalanced model, RF-15, had PR, PR-AUC, ROC, ROC-AUC, CM, and a Classification report, as depicted in Figure 5.2.

4.1.2. SVM categorization

In Figure 5.2, both SVM and RF with a test size of 15 % have the same results for the PR curve, ROC curve, confusion matrix, and classification report. The best values of all SVM models are evaluated with C parameter '0.001' and kernel "linear" for all test dimensions.

4.1.3. LR categorization

It has been observed that different test sizes have varying effects on the values of C, penalty, and optimizer for LR models. Among all the test sizes, the LR model with a test size of 20% performed better than the others. In addition, the LR algorithm employing no sensor data has demonstrated superior performance in terms of accuracy and metrics when compared to alternative models and algorithms.

The PR curve, ROC curve, confusion metrics, and classification report for the best no-smote LR model, that is, LR-20, with a weighted recall of 89%, are shown in Figures 5.3 and 5.4 for both the testing and training cases.

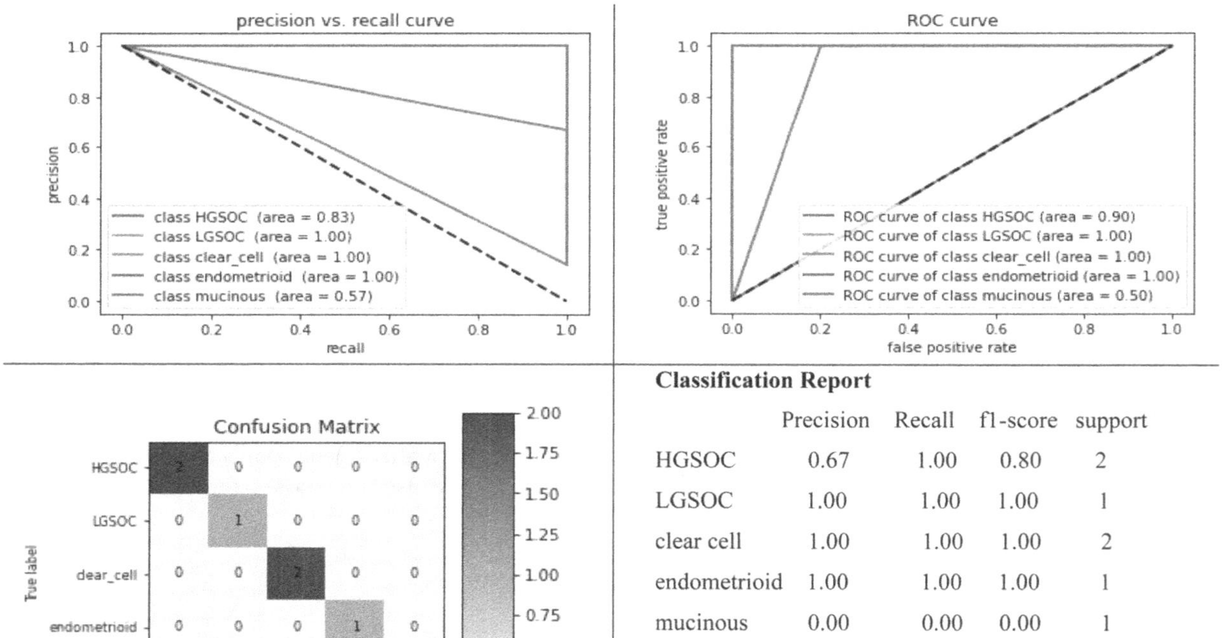

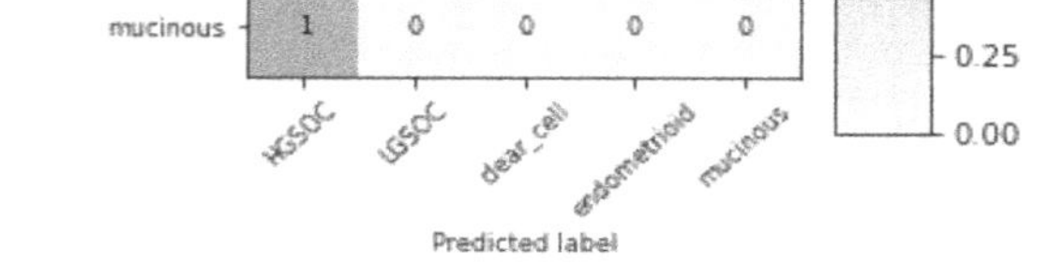

Classification Report

	Precision	Recall	f1-score	support
HGSOC	0.67	1.00	0.80	2
LGSOC	1.00	1.00	1.00	1
clear cell	1.00	1.00	1.00	2
endometrioid	1.00	1.00	1.00	1
mucinous	0.00	0.00	0.00	1
accuracy			0.86	7
macro avg	0.73	0.80	0.76	7
weighted avg	0.76	0.86	0.80	7

Figure 5.2. Performance metrics for test_size 15% for RF, SVM models.

Source: Author.

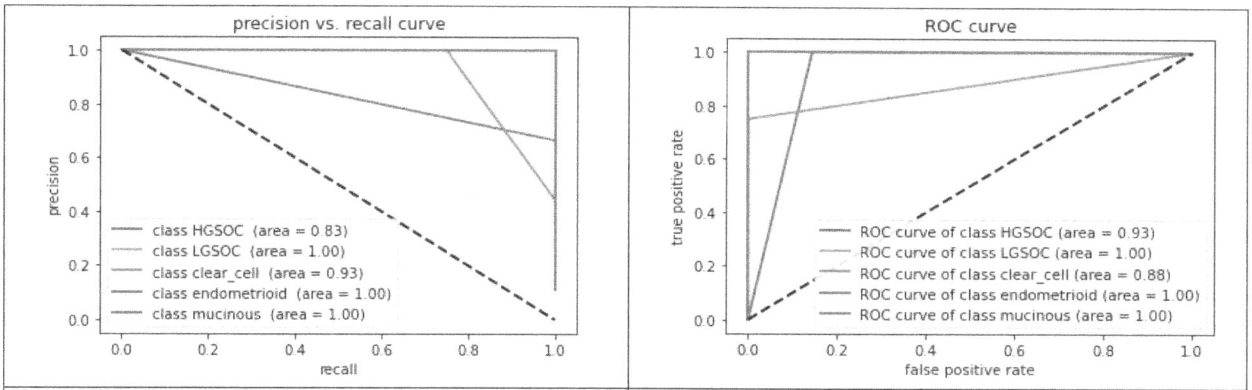

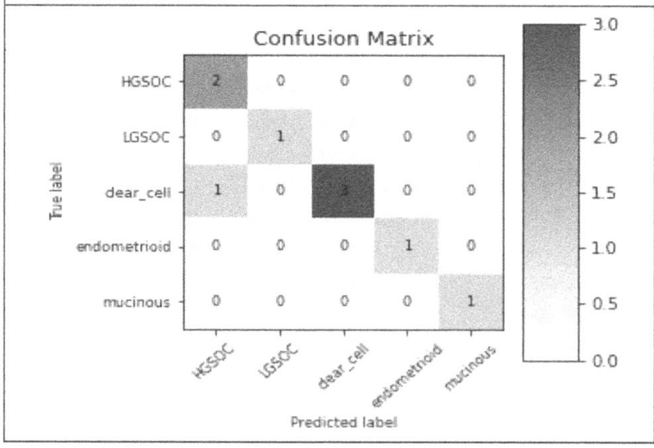

Testing Classification Report

	Precision	Recall	f1-score	support
HGSOC	0.67	1.00	0.80	2
LGSOC	1.00	1.00	1.00	1
clear_cell	1.00	0.75	0.86	4
endometrioid	1.00	1.00	1.00	1
mucinous	1.00	1.00	1.00	1
accuracy			0.89	9
macro avg	0.93	0.95	0.93	9
weighted avg	0.93	0.89	0.89	9

Figure 5.3. Performance (testing) metrics for test_size 20% for LR model.

Source: Author.

The 11% gap between the model's performance on testing and training data indicates room for improvement.

4.2. Categorization of up-sampled data

To investigate the effects of data augmentation on minority classes, up-sampling was performed up to the count of the majority class, which was 16. As a result, the dataset size is increased from the original 44 to 80. A train-test split was then performed using test sizes of 40%, 30%, 25%, and 20%. Since good metrics were produced, a test size of 15% was not used for balanced data classifications. Various models were tested with different test proportions to minimize the gap between testing and training performance, similar to the situation with unbalanced data.

4.2.1. Random forest categorization

The data was up-sampled and executed with RF-up models using all the specified test magnitudes. Two different models were used, with estimators 144 and 100 and max_depths of 35 and 2, respectively, and with a test size of 30%. The models scored an accuracy of 96% and a recall of 97%, which was better than the other models. The criterion used in the model was entropy. According to Figure 5.5, one sample of clear_cell was misclassified, resulting in a ROC area under the curve of 0.93.

4.2.2. Support vector machine categorization

The SVM-up models performed with the kernel as "rbf" and a test size of 40% exhibit high metric values, with an accuracy of 97% and a recall of 96%, as shown in Figure 5.6. However, the test size of 20% alone has a few less than 90% measurements. It's worth noting that the data taken differs for each train test split. Therefore, a slightly different output may be generated for the same metric value for the same dataset. The train and test set data are taken randomly. The support vector value parameter can also be taken from the generated models. One endometroid sample is misclassified.

4.2.3. Logistic regression categorization

The up-sampled data with LR-up, RF-up, and SVM-up models have shown similar results. A test size of 40% for LR-up has resulted in better accuracy, precision, recall, and F1 score values. Different parameter "C" values have been observed for different test sizes, resulting in varying accuracy and recall results. L1 and L2 regularization can

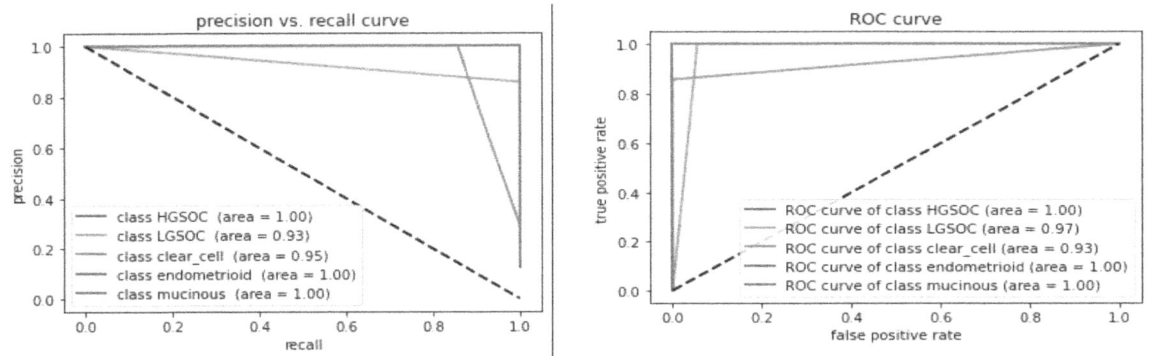

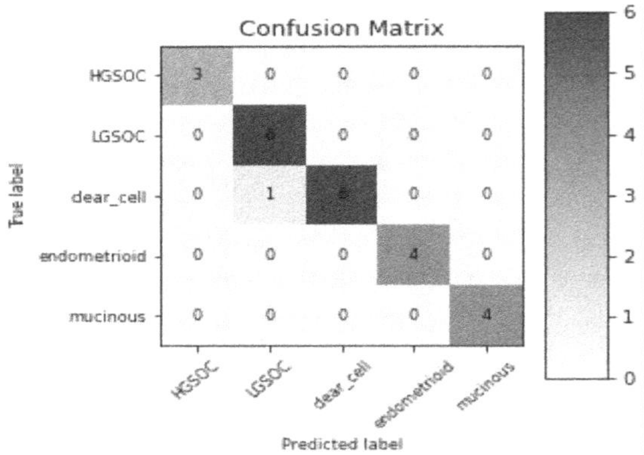

Training Classification Report

	precision	recall	f1-score	support
HGSOC	1.00	1.00	1.00	3
LGSOC	0.86	1.00	0.92	6
clear_cell	1.00	0.86	0.92	7
endometrioid	1.00	1.00	1.00	4
mucinous	1.00	1.00	1.00	4
accuracy			0.96	24
macro avg	0.97	0.97	0.97	24
weighted avg	0.96	0.96	0.96	24

Figure 5.4. Performance (training) metrics for test_size 20% for LR model.

Source: Author.

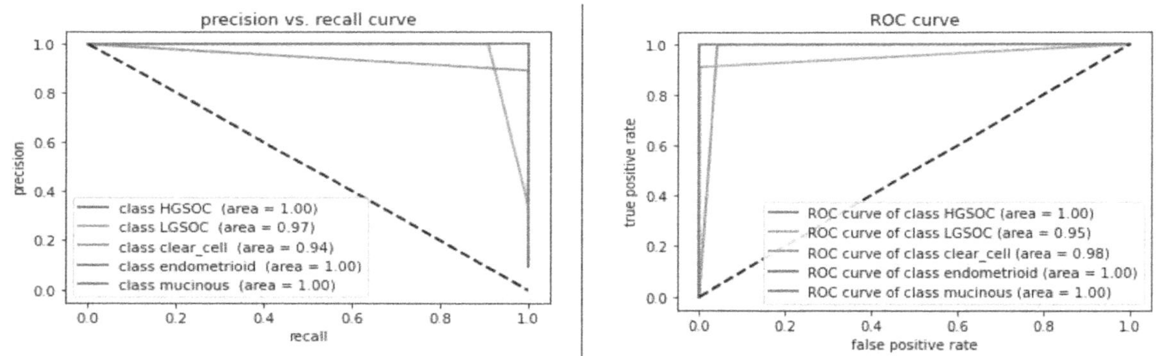

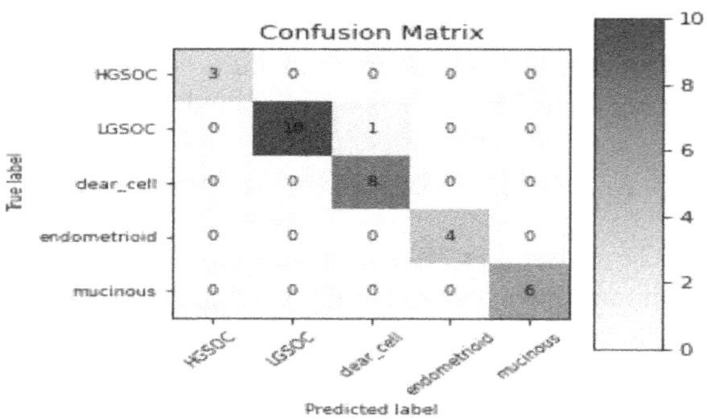

Classification Report

	precision	recall	f1-score	support
HGSOC	1.00	1.00	1.00	3
LGSOC	1.00	0.91	0.95	11
clear_cell	0.89	1.00	0.94	8
endometrioid	1.00	1.00	1.00	4
mucinous	1.00	1.00	1.00	6
accuracy			0.97	32
macro avg	0.98	0.98	0.98	32
weighted avg	0.97	0.97	0.97	32

Figure 5.5. Performance (testing) metrics for test_size 30% for RF-up model.

Source: Author.

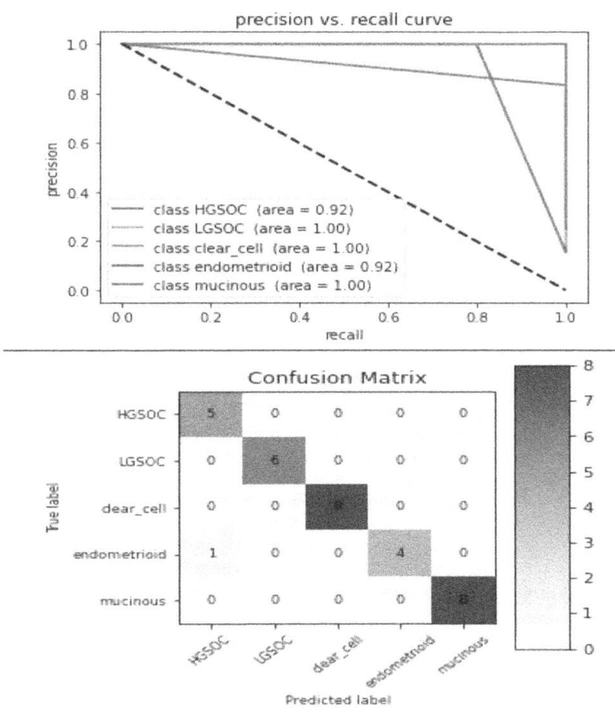

Classification Report

	precision	recall	f1-score	support
HGSOC	0.83	1.00	0.91	5
LGSOC	1.00	1.00	1.00	6
clear_cell	1.00	1.00	1.00	8
endometrioid	1.00	0.80	0.89	5
mucinous	1.00	1.00	1.00	8
accuracy			0.97	32
macro avg	0.97	0.96	0.96	32
weighted avg	0.97	0.97	0.97	32

Figure 5.6. Performance (testing) metrics for test_size 40% for the SVM model.

Source: Author.

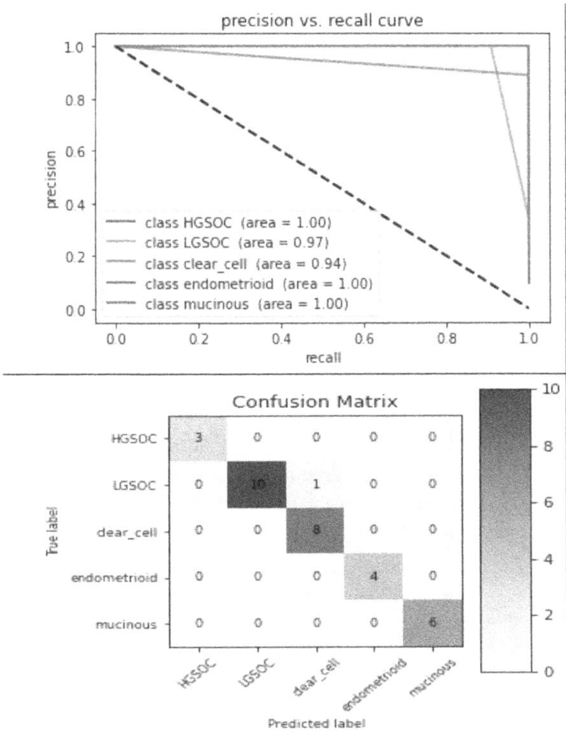

Classification Report

	precision	recall	f1-score	support
HGSOC	1.00	1.00	1.00	3
LGSOC	1.00	0.91	0.95	11
clear_cell	0.89	1.00	0.94	8
endometrioid	1.00	1.00	1.00	4
mucinous	1.00	1.00	1.00	6
accuracy			0.97	32
macro avg	0.98	0.98	0.98	32
weighted avg	0.97	0.97	0.97	32

Figure 5.7. Performance (testing) metrics for test_size 40% for LR model.

Source: Author.

be noticed in various LR-up models, and newton-cg and liblinear optimizers are employed. In the output of the LR model's Figure 5.7, it is observed that one LGSOC record was mis predicted, which proves that random train data can bring about marginally diversified returns.

5. Result Analysis

5.1. Comparison of the classifiers for unbalanced data

According to the results in Table 5.1, among all the RF models for unbalanced data, RF-15 has the best metrics. Similarly, SVM-15 stands out among all SVM models. Both RF-15 and SVM-15 have the same values for accuracy, precision, recall, and F1 score, with precision being slightly lower than the other measurements. In the SVM, RF, and LR model group, LR-20 has a high score with similar assessments towards accuracy, recall, and F1 score but with a slightly high score for precision. However, the LR model shows that the performance decreases when more data is given for training. The mean squared error (MSE), bias, and variance values for the models were 20 to 31 for unbalanced data, with a bias higher than variance, indicating that more training data is required.

5.2. Comparison of the classifiers for balanced data

The RF model has less space complexity than the SVM, RF, and LR models. The complexity is evaluated based on the number of trees and maximum depth. In the given dataset, the number of features is high, the number of estimators is in the range of 100–200, and the depth is in the range of 2–35. Therefore, considering space complexity. RF can be viewed as a good model for unbalanced data (as shown in Table 5.1) despite having slightly lower recall and accuracy rates than other models. Although LR shows the best metrics, its performance decreases with more training data. In the case of balanced data, MSE, bias, and variance are significantly reduced.

Table 5.1. Comparison of three classifiers for best metrics of unbalanced data

	Test size	*Accuracy*	*Wgtd avg P*	*Wgtd avg R*	*Wgtd avg F1 score*
RF	40	50	65	50	46
SVM	40	61	83	61	57
LR	40	72	85	72	72
RF	30	50	60	50	44
SVM	30	64	70	64	64
LR	30	64	75	64	63
RF	25	64	75	64	61
SVM	25	73	88	73	72
LR	25	82	82	82	80
RF	20	67	76	86	80
SVM	20	67	85	67	63
LR	20	89	93	89	89
RF	15	86	76	86	80
SVM	15	86	76	86	80
LR	15	86	76	86	80

Source: Author.

For n = The count for training., m = count of features, **n'** = count of support vectors, **k** = count of neighbours, **k'** = count of trees, the space complexities for algorithms are:

- SVM - Space Complexity = O (n*m)
- LR - Space Complexity = O (m)
- RF - Space Complexity = O (k'*depth of tree)

6. Conclusion and Future Enhancement

The data has undergone standard classification and upsampling based on different trial runs. The Grid search CV hyperparameter tuner has been utilized to obtain the best possible results. Out of the three classifiers, the best model has been chosen. Among the different raters, a preferable stereotype has been selected. An analysis of the performance of data with fewer numbers of data using various algorithms has shown that the bias will be higher than the variance for a small dataset.

More data can be added to the dataset, or synthetic data can be generated to improve the classification further. Top features can be extracted, and the model can be generalized by adding separate test data to check its feasibility. The selected features will significantly reduce the Mean Squared Error (MSE) values, and the bias will be reduced to a lesser value than the variance.

References

[1] Barnes, B. M., Nelson, L., Tighe, A., Burghel, G. J., Lin, I. H., Desai, S., ... & Taylor, S. S. (2021). Distinct transcriptional programs stratify ovarian cancer cell lines into the five major histological subtypes. *Genome Medicine*, *13*(1), 1–9.

[2] Radu, M. R., Prădatu, A., Duică, F., Micu, R., Crețoiu, S. M., Suciu, N., ... & Rădoi, V. E. (2021). Ovarian cancer: biomarkers and targeted therapy. *Biomedicines*, *9*(6), 693.

[3] Barnard, M. E., Meeks, H., Jarboe, E. A., Albro, J., Camp, N. J., & Doherty, J. A. (2023). Familial risk of epithelial ovarian cancer after accounting for gynaecological surgery: A population-based study. *Journal of Medical Genetics*, *60*(2), 119–127.

[4] Khalsan, M., Machado, L. R., Al-Shamery, E. S., Ajit, S., Anthony, K., Mu, M., & Agyeman, M. O. (2022).

A survey of machine learning approaches applied to gene expression analysis for cancer prediction. *IEEE Access*, *10*, 27522–27534.

[5] Koppad, S., Basava, A., Nash, K., Gkoutos, G. V., & Acharjee, A. (2022). Machine learning-based identification of colon cancer candidate diagnostics genes. *Biology*, *11*(3), 365.

[6] Lu, M., Fan, Z., Xu, B., Chen, L., Zheng, X., Li, J., ... & Jiang, J. (2020). Using machine learning to predict ovarian cancer. *International Journal of Medical Informatics*, *141*, 104195.

[7] Akter, L., & Akhter, N. (2021, July). Ovarian cancer classification from pathophysiological complications using machine learning techniques. In *2021 12th International Conference on Computing Communication and Networking Technologies (ICCCNT)* (pp. 1–6). IEEE.

[8] Naji, M. A., El Filali, S., Aarika, K., Benlahmar, E. H., Ait Abdelouhahid, R., & Debauche, O. (2021). Machine learning algorithms for breast cancer prediction and diagnosis. *Procedia Computer Science*, *191*, 487–492.

[9] Gopalakrishnan, C., & Iyapparaja, M. (2021). Multilevel thresholding based follicle detection and classification of polycystic ovary syndrome from the ultrasound images using machine learning. *International Journal of System Assurance Engineering and Management*, 1–8.

[10] Rustam, Z., & Angie, N. (2021, February). Prostate cancer classification using random forest and support vector machines. In *Journal of Physics: Conference Series* (Vol. 1752, No. 1, p. 012043). IOP Publishing.

[11] Arezzo, F., Cormio, G., La Forgia, D., Santarsiero, C. M., Mongelli, M., Lombardi, C., ... & Loizzi, V. (2022). A machine learning approach applied to gynecological ultrasound to predict progression-free survival in ovarian cancer patients. *Archives of Gynecology and Obstetrics*, *306*(6), 2143–2154.

[12] Taneja, S., Suri, B., & Kothari, C. (2019, September). Application of balancing techniques with ensemble approach for credit card fraud detection. In *2019 international conference on computing, power and communication technologies (GUCON)* (pp. 753–758). IEEE.

[13] Thabtah, F., Hammoud, S., Kamalov, F., & Gonsalves, A. (2020). Data imbalance in classification: Experimental evaluation. *Information Sciences*, *513*, 429–441.

[14] Shekar, B. H., & Dagnew, G. (2019, February). Grid search-based hyperparameter tuning and classification of microarray cancer data. In *2019 second international conference on advanced computational and communication paradigms (ICACCP)* (pp. 1–8). IEEE.

[15] Mangkunegara, I. S., & Purwono, P. (2022, June). Analysis of DNA sequence classification using SVM model with Hyperparameter Tuning Grid Search CV. In *2022 IEEE International Conference on Cybernetics and Computational Intelligence (CyberneticsCom)* (pp. 427–432). IEEE.

[16] Yu, Z. (2023). Prediction of CCLE Dataset Based on Integrated SVM Method. In *Advances in Biomedical and Bioinformatics Engineering* (pp. 289–294). IOS Press.

6 Assessment of risk of osteoarthritis in knee using thermal images

A. Sabah Afroze[a], R. Tamil Selvi[b], M. Parisa Beham[c], J. Judith[d], S. Sathiya Pandiya Lakshmi[e], M. Bharkavi Sandhiya, and K. M. Ahamed Sheriff[f]

Department of Electrical and Communication Engineering, Sethu Institute of Technology Virudhunagar, Madurai, Tamil Nadu, India

Abstract: Osteoarthritis (OA) – a prevalent musculoskeletal disorder – is characterized by joint pain and functional impairment, with a particular emphasis on the knee joint. Early detection of OA risk can significantly enhance patient outcomes and facilitate timely intervention; however, the current methods for identifying osteoarthritis possess several limitations. These limitations include a lack of sensitivity in detecting the condition during its early stages and a reliance on structural imaging techniques, such as X-rays, which may fail to capture the early soft tissue changes associated with osteoarthritis. Moreover, there is a difficulty in standardizing different imaging setups, which can lead to variability in both diagnostic accuracy and reliability. Thermal imaging, on the other hand, has emerged as a promising non-invasive tool for detecting osteoarthritis. This method offers the advantage of early detection of changes in joint temperature, comprehensive assessments of the joint and the absence of ionizing radiation. The proposed approach employs a Modified Regulated Inception network to identify osteoarthritis in thermal images. In this process, the thermal image is pre-processed to enhance relevant features, segmented to extract regions of interest and subsequently fed into the Modified Regulated Inception net model, which classifies the images based on the presence of osteoarthritis. Although this method shows potential, further research is necessary to validate its effectiveness. Employing a Modified Regulated Inception Net for the detection of osteoarthritis in thermal images presents improved accuracy, primarily due to deep learning algorithms that are specifically designed for the analysis of thermal data. The hierarchical feature extraction (which is quite advanced) allows for the precise identification of subtle temperature variations that are indicative of osteoarthritis. This method not only streamlines the detection process but also yields efficient and reliable results. Consequently, it enhances the potential for early diagnosis and intervention in cases of osteoarthritis. The proposed method achieves an accuracy of 95% (which is impressive), a sensitivity of 94%, a specificity of 93% and an F1 Score of 94.30. However, it surpasses the performance of existing approaches, illustrating its effectiveness. For image segmentation, we employed a combination of thresholding and K-means clustering to isolate regions of interest (ROIs) based on thermal anomalies. The thresholding technique converted grayscale images to binary, highlighting significant temperature deviations, while K-means clustering grouped similar pixels, enabling precise ROI extraction. This preprocessing step was critical for enhancing the model's focus on diagnostically relevant features.

Keywords: Feature extraction, modified regulated inception net, Osteoarthritis, thermal image

1. Introduction

Osteoarthritis (OA) is recognized as the most common and severe chronic rheumatic condition worldwide. It significantly contributes to pain and disability in numerous nations. Tibiofemoral OA, as identified through magnetic resonance imaging (MRI), is distinguished by either (a) noticeable osteophyte formation paired with complete cartilage loss, or the existence of any degree of cartilage deterioration alongside one of several markers of joint pathology. These include: a subchondral marrow lesion or cyst that does not involve meniscal or ligamentous attachments, meniscal subluxation, dipping, or a degenerative (horizontal) tear, as well as partial thickness loss of cartilage and bone wear. Recently, the International Society for the Study of Osteoarthritis (OARSI) defined OA and initiated a project to rigorously assess and standardize the prevailing definitions of the disease [2]. Osteoarthritis is, in essence, a condition that affects mobile joints and is characterized by cellular stress and the degradation of the extracellular matrix. This degradation is triggered by both micro- and macrodamages that provoke defective repair

[a]sabahafroze@sethu.ac.in, [b]tamilselvi@sethu.ac.in, [c]parisabeham@sethu.ac.in, [d]judith@sethu.ac.in, [e]sathiyasdp2119@gmail.com, [f]kmsheriff@gmail.com

DOI: 10.1201/9781003770473-6

responses, including the activation of innate immune-inflammatory pathways. However, understanding OA is crucial because it affects millions of individuals globally. The disease initially presents itself as a molecular disorder (specifically, an abnormal metabolism of synovial tissue); however, it later leads to anatomical or physiological disturbances. These disturbances are characterized by cartilage breakdown, bone remodelling, osteophyte formation, arthritis and a loss of normal joint function. This progression may ultimately culminate in a significant disease state [1]. Pain, stiffness and limitations in joint mobility constitute the primary symptoms of osteoarthritis (OA). Although joint pain is primarily experienced during physical activity, it can, as the disease advances, be felt even with minimal movement or at rest. Because of the disease, the normal function of the joint becomes compromised; over time, the joint's shape begins to alter. OA frequently impacts the knee, hip, foot, spine and wrists, but it can affect any movable joint. Figure 6.1 illustrates the various joints that are susceptible to OA, while the inset provides a comparison between a healthy knee joint and one affected by OA [2]. Although OA does not directly impact the different tissues and organs within the body, many individuals may have additional medical issues that warrant consideration when seeking treatment.

Outcomes and reactions to treatment differ significantly among individuals (1): certain people may experience disabilities that restrict their activities; however, others might only encounter mild issues. The subsequent section delves into the prevalence of OA, based on epidemiological evidence. Although variations exist, this data provides crucial insights into the condition's impact. Because of these disparities, understanding the nuances becomes essential for effective management.

2. Literature Survey

Several recent studies have explored advanced methodologies for the detection and classification of knee osteoarthritis (OA) using imaging techniques and computational approaches. For instance, Chetty et al. [1] proposed fuzzy texture descriptors for analyzing thermal images, demonstrating significant potential for early diagnosis of OA. This method effectively captured temperature variations linked to joint inflammation. Similarly, Sharma, et al. [2] utilized support vector machine (SVM) classifiers on X-ray images to detect OA, achieving reliable results in identifying structural changes indicative of the disease.

Deep learning methods have also shown promise in advancing OA diagnostics. Nasser, et al. [3] implemented a discriminative regularized autoencoder using MRI data, which achieved high accuracy in detecting early OA changes by leveraging the TensorFlow framework. Dalia et al. [4] introduced a Clinical Decision Support System (CDSS) tailored for OA detection, integrating machine learning algorithms into diagnostic workflows, with notable success in severity grading of OA from X-ray images. Similarly, Wang et al. [5] explored confident sample learning for severity assessment, demonstrating improvements in diagnostic reliability and interpretability. Thermal and deep feature-based methodologies have also been extensively studied. Zebari et al. [6] utilized CNNs on thermal and X-ray datasets to enhance detection performance, showing robustness in feature extraction and classification. Expanding on this, Wang et al. [7] introduced a Siamese-Gap network for early detection of OA in thermal images, leveraging transfer learning to achieve high precision. Additionally, Antonio et al. [8] employed a deep transfer network for MRI-based progression rate classification,

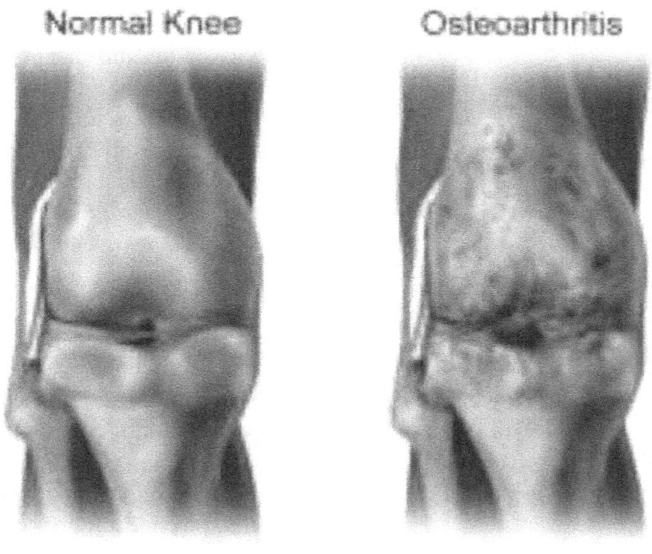

Figure 6.1. Joints affected by OA that become enlarged and sometimes inflamed [2].

Source: Author.

utilizing pre-trained models to enhance accuracy. Hybrid approaches combining imaging modalities have proven effective in addressing diagnostic challenges. Sayed, et al. [9] introduced phonoarthrography, a novel method for early OA detection using sound-based diagnostics. Nguyen Huu, et al. [10] combined YOLOv3 and VGG-16 models for real-time classification of OA using X-ray datasets, achieving high accuracy and speed. Furthermore, Wang, et al. [11] leveraged transfer learning models to classify OA severity using MRI data, providing a scalable and efficient solution. Graph-based and advanced deep learning models have further refined OA detection. Zhuang, et al. [12] proposed graph representation techniques for cartilage defect assessment, demonstrating improved precision in structural analysis. Dharmani and Khatri, [13] utilized deep CNNs for severity grading of knee OA from X-ray images, emphasizing the importance of stage-specific classification models. Similarly, Aladhadh and Mahum, [14] employed an improved.

CenterNet model with pixel-wise voting, significantly enhancing the accuracy of X-ray-based OA detection. Thermal imaging has continued to evolve as a non-invasive modality for early OA diagnosis. Kim et al. [15] explored temperature based imaging techniques, identifying inflammatory patterns indicative of OA with high sensitivity. Patel and Mehta [16] integrated thermal imaging with Xray data, emphasizing the complementary strengths of these modalities in improving early-stage detection. Traditional machine learning approaches have also contributed to advancements in OA detection. Gupta and Sharma [17] implemented random forest classifiers on MRI data, highlighting the efficacy of feature-based models in analysing joint abnormalities. Smith et al. [18] introduced attention-enhanced deep learning models for infrared imaging, achieving substantial improvements in model interpretability and diagnostic accuracy. Chen et al. [19] developed 3D CNN architectures for analysing cartilage degeneration from MRI datasets, offering detailed insights into OA progression. Finally, Antonio et al. [20] combined graph-based and MRI data for a comprehensive analysis of OA, underscoring the value of multimodal frameworks in improving diagnostic workflows.

3. Inference From the Survey

1. **Limited sensitivity** in detecting early-stage disease, often resulting in delayed diagnosis and intervention.
2. **Reliance on structural imaging** modalities like *X-rays*, which may *not capture early soft tissue* changes.
3. **Invasiveness or discomfort** associated with certain diagnostic procedures, hindering their widespread use for screening or monitoring purposes.
4. **Challenges in standardization** across different imaging setups and interpretation of findings, leading to variability in diagnostic accuracy and reliability.

The primary focus of the paper is to analyze thermal images for assessing osteoarthritis risk by using deep learning methodologies. This innovative approach provides a non-invasive, cost-effective, and efficient method for early detection and risk assessment, potentially revolutionizing diagnostic practices and improving patient outcomes.

4. Motivation of the Research

The motivation behind osteoarthritis detection in thermal images stems from the need for non-invasive, Developing early diagnostic tools is crucial for enhancing patient outcomes and quality of life. Traditional methods often fail to detect osteoarthritis in its initial stages, leading to delayed treatment and irreversible joint damage. Thermal imaging presents a promising solution by detecting subtle temperature changes linked to early-stage osteoarthritis, facilitating prompt intervention and personalized treatment approaches. Through harnessing the potential of thermal imaging, healthcare providers can elevate their diagnostic capabilities, resulting in enhanced patient care and more effective management of osteoarthritisrelated complications.

5. Proposed Method

In response to the limitations previously discussed, we propose a system that focuses on osteoarthritis detection in thermal images, utilizing the Modified Regulated Inception Net model. By regulating (or fine-tuning) the learning process within the Inception framework, this model can effectively manage the complexities inherent in thermal image data. This enhancement improves its robustness as well as its generalization capabilities. The Modified Regulated Inception Net presents a promising solution for automated osteoarthritis detection; however, it also contributes to early diagnosis and personalized treatment strategies. Consequently, patient care can be significantly improved. Although challenges remain, the potential benefits are substantial, because early intervention is crucial in managing this condition.

While thermal imaging offers valuable insights into surface-level inflammation, its inability to detect deep tissue changes remains a limitation. Future work will explore the integration of complementary modalities, such as MRI, to provide a more comprehensive diagnostic framework. The Modified Regulated Inception Net, designed for the detection of osteoarthritis through thermal imaging, represents a novel methodology that amalgamates the principles of the Inception architecture with regulatory mechanisms. This integration aims to enhance both the accuracy and efficiency of identifying osteoarthritis from thermal images. Various methods and techniques have been explored in this context. This neural network model effectively employs deep learning techniques to scrutinize thermal images of joints, thereby extracting significant features that indicate changes associated with osteoarthritis. The proposed system, which focuses on osteoarthritis detection in thermal images utilizing the Modified Regulated Inception Net model, encompasses several stages (as illustrated in Figure 6.2). The initial stage is data acquisition, which requires the collection of thermal images from the joint under examination. The images undergo pre-processing to optimize their usability; this includes noise reduction, contrast enhancement and normalization. However, achieving optimal results can be challenging, because the quality of the thermal images is paramount. Although the system is designed to improve detection capabilities, its effectiveness is still dependent on the quality of the initial data.

Suppose I represents the original thermal image, noise reduction can be achieved using a spectral filter.

$$I_{Filtered} = I * S \tag{1}$$

Where S is a Spectral filter and * denotes convolution.

Contrast enhancement can be performed using histogram equalization:

$$I_{enchanced} = HistEqual(I_{filtered}) \tag{2}$$

Normalizing the pixel values to a [0,1] range:

$$I_{norm} = \frac{\max(I_{enchanced}) - \min(I_{enchanced})}{(I_{enchanced}) - \min(I_{enchanced})} \tag{3}$$

Image filtering is used to eliminate noise and improve visual quality. The pre-processed thermal images are segmented using methods like clustering, edge detection, and thresholding. Thresholding converts grayscale images to binary images, while clustering groups' pixels based on their similarity.

For image segmentation, we employed a combination of thresholding and K-means clustering to isolate regions of interest (ROIs) based on thermal anomalies. The thresholding technique converted grayscale images to binary, highlighting significant temperature deviations, while K-means clustering grouped similar pixels, enabling precise ROI extraction. This preprocessing step was critical for enhancing the model's focus on diagnostically relevant features.

Convert gray scale images to binary images using a threshold T:

$$I_{binary}(x, y) = 10 \text{ if } I_{norm}(x, y)^{3}£T \tag{4}$$

Apply clustering (e.g., K-means) to segment the image into k clusters:

$$I_{clustered}(x,y) = KMeans(I_{norms}, K) \tag{5}$$

Region growing is a process where adjacent pixels with similar thermal properties are grouped together.

Group adjacent pixels with similar thermal properties:

$$I_{region} = \text{RegionGrow}(I_{clustered}, SeedPoint\ s) \tag{6}$$

The segmented ROI is then input into the Modified Regulated Inception Net-T model, which extracts relevant features using convolutional layers.

Apply multiple convolutions with different kernel sizes and concatenate:

$$R(x,y) = \text{Re}LU(w*, X+b) \tag{7}$$

$$X(3x3) = \text{Re}LU(w(3X3)*, X+b(3X3) \tag{8}$$

$$X = \text{Re } LU(w* X+b) \tag{9}$$

$$X(MP) = \text{Re}LU(w(1X1, MP) *(Maxpool(X, 3X3)) + b(1X1) \tag{10}$$

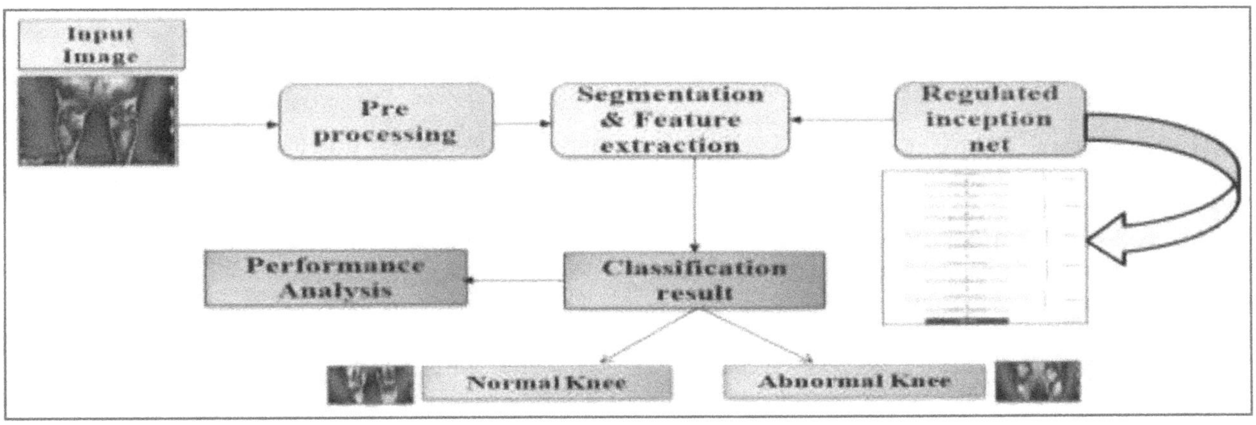

Figure 6.2. Block diagram.

Source: Author.

$$X(concat) = concat(X(1X1), X(3X3), X(5X5), X(MP)) \quad (11)$$

$$X = X\,QM \quad (12)$$

Where M is a binary mask with dropout rate p.

$$y = (1\text{-}\hat{I})\,y + \hat{I} \qquad smooth\ one\text{-}hot\ k \quad (13)$$

Where yone-hot is the one-hot encoded vector of true labels, ϵ is the smoothing parameter, and K is the number of classes.

The model is trained to categorize thermal images into two distinct groups: normal or abnormal, with its classification being contingent upon the occurrence of osteoarthritis. This training utilizes a dataset comprising thermal images paired with their respective labels. In the concluding phase, the Modified Regulated Inception Net model's performance is assessed using a dataset containing concealed thermal images. Performance metrics such as accuracy, sensitivity, specificity, and area under the curve (AUC) are utilized. Classification entails assigning a label to each object of interest, grounded on the extracted features.

The Figure 6.3 illustrates the architecture of a Modified Regulated Inception Net, which is proposed as a novel approach in this work. The network comprises multiple layers, starting with a 7 × 7 convolutional layer, followed by 3 × 3 max-pooling for dimensionality reduction. A 1 × 1 convolution layer is then applied to enhance feature extraction, succeeded by a 3 × 3 convolution and another 3 × 3 max-pooling layer. A key innovation is the channel-wise feature responses to improve representational capacity. Additionally, the Inception blocks, incorporated alongside SE blocks, allow for multi-scale feature extraction through parallel filters of different sizes, further optimizing learning efficiency. This novel combination of SE blocks and Inception modules, alongside regulated pooling layers, enhances computational efficiency and feature learning. The architecture concludes with a global average pooling layer and a dense layer for classification, making the Modified Regulated Inception Net a powerful and efficient deep learning model for complex tasks.

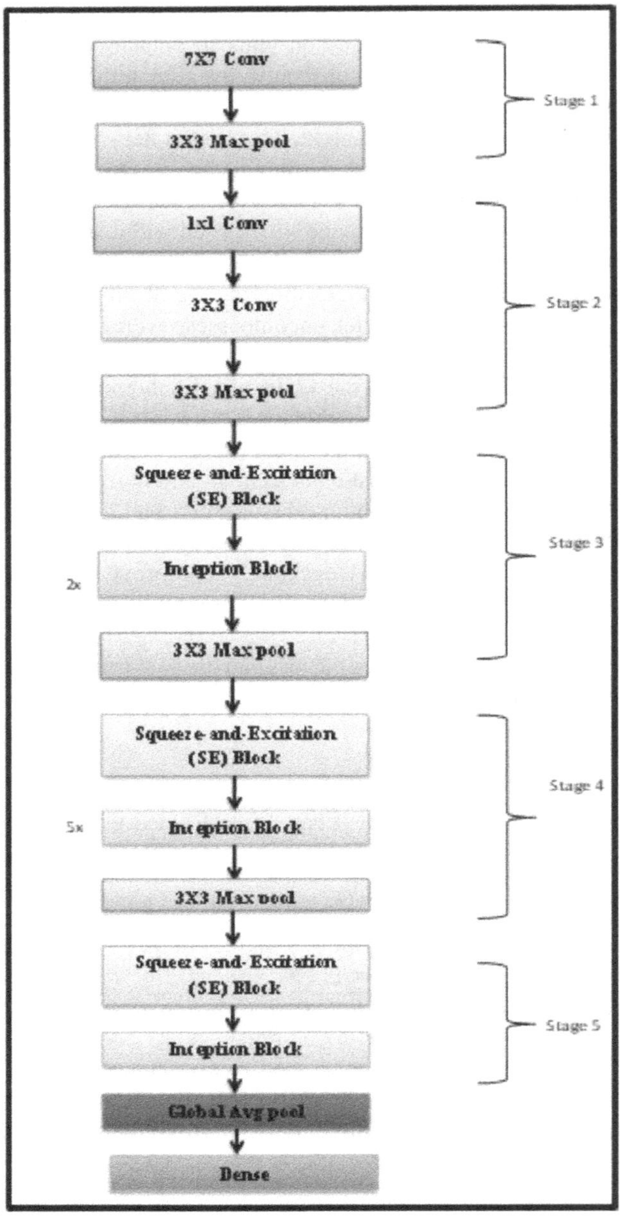

Figure 6.3. Modified regulated inception net architecture.

Source: Author.

6. Results and Discussion

For this study, the dataset of 2500 thermal images of knee joints is collected from Taj Hospital, Madurai, which were categorized into two classes: normal and osteoarthritic. The dataset was split into two subsets: a training set containing 2000 images and a testing set comprising 500 images. Then the Modified Regulated Inception Net architecture is trained using transfer learning approach, where the pre-trained weights of the Modified Regulated Inception Net model were used as the initial weights for training. Figure 6.4 shows the sample thermal images. The Inception Net model was fine-tuned on the training set through 50 iterations, utilizing a specified batch size of 10 and a learning rate of 0.001.

The proposed technology for osteoarthritis image detection aims to leverage advanced computer vision and deep learning techniques to enhance diagnostic accuracy and reliability. Utilizing medical imaging modalities such as X-rays, MRI, or ultrasound, the system will preprocess the input images to remove noise and improve clarity. Key technologies include edge detection algorithms to identify structural abnormalities, such as cartilage degradation and bone spurs, and feature extraction using convolutional

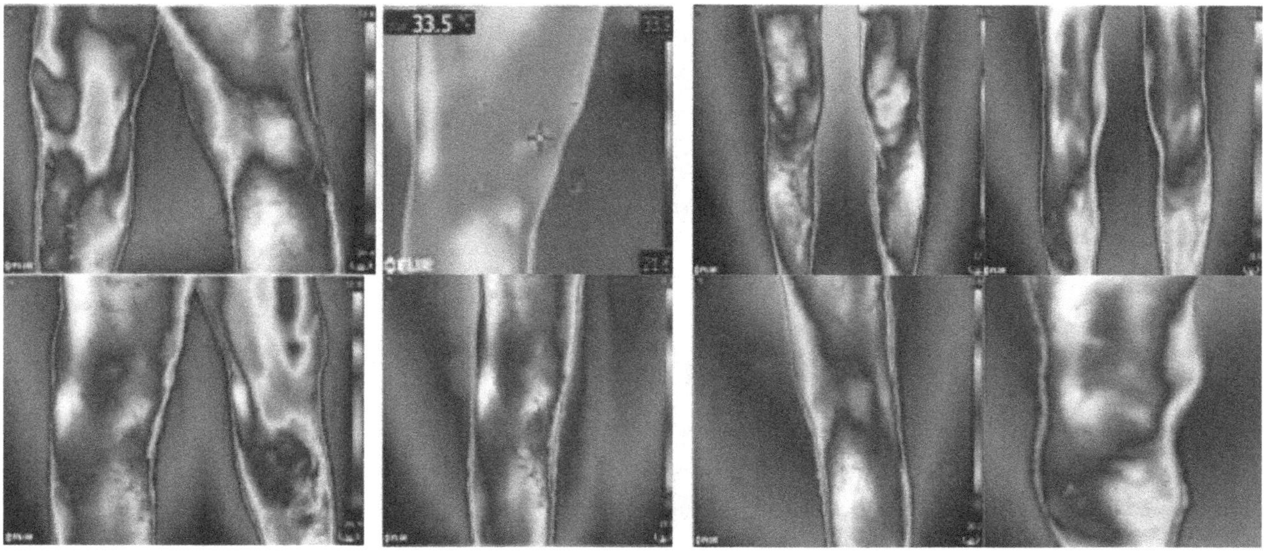

Figure 6.4. Sample knee thermal images.
Source: Author.

neural networks (CNNs). These CNN models are trained on annotated datasets to distinguish between healthy and osteoarthritic joint images, enabling precise classification and segmentation of affected areas. The integration of endpoint linking and overlay techniques ensures accurate localization of abnormalities, facilitating detailed visualization for clinical interpretation. This approach significantly reduces manual error, improves early detection, and supports personalized treatment planning for patients.

The image demonstrates a sequence of medical image processing steps applied to grayscale images to analyse structural features, with potential applications in studying conditions like osteoarthritis. The process begins with the grayscale images, which serve as the raw input for the analysis. These images, captured through imaging techniques such as X-ray or MRI, represent anatomical structures like bones, cartilage, and soft tissues using variations in intensity. Grayscale imaging is ideal for capturing fine details, providing the foundation for further processing. Next, the grayscale images are converted into binary images through thresholding, where pixel intensities are categorized into foreground (white) and background (black). This step simplifies the images by isolating important features, such as edges or contours of structures, while removing noise and irrelevant details. In the subsequent step, end points of line-like structures within the binary images are detected and marked with red dots. These end points indicate the terminal points of edges or segments and are crucial for identifying structural discontinuities, gaps, or breaks that may indicate abnormalities. To improve structural continuity, end points that are closer than a specified distance threshold (33 pixels) are connected as shown in Figure 6.5. This step addresses fragmentation by linking disconnected segments, forming continuous and coherent structures that better represent the underlying anatomy. Finally, the processed edges and connected structures are overlaid onto the original grayscale images. This overlay provides a clear visual validation, allowing the extracted features to be compared directly with the raw image data. The alignment ensures that the detected features correspond accurately to the anatomical structures, facilitating the detection and interpretation of abnormalities such as bone erosion or cartilage degradation. In summary, the sequence of steps – from grayscale imaging to edge detection and overlay – combines feature extraction, enhancement, and visualization to analyze structural details in medical images. This methodology is particularly useful for diagnosing and studying conditions like osteoarthritis, where understanding the morphology of bones and cartilage is critical for assessing disease progression and planning treatments.

The Figure 6.6 shows the training plot which depicts the performance of the model against the training data over time, while the validation plot shows the performance of the validation data. Ideally, the performance of the model should improve with each epoch until it reaches a plateau. Upon analyzing the first convolution layer weights of the classification model trained for osteoarthritis detection, notable patterns emerged. These weights provide insight into the features extracted by the model at the initial stage of processing. The visualization of these weights revealed distinct spatial patterns that correspond to specific image characteristics relevant to osteoarthritis identification. These patterns provide deeper understanding of the underlying information captured by the model, potentially uncovering discriminative features crucial for accurate classification.

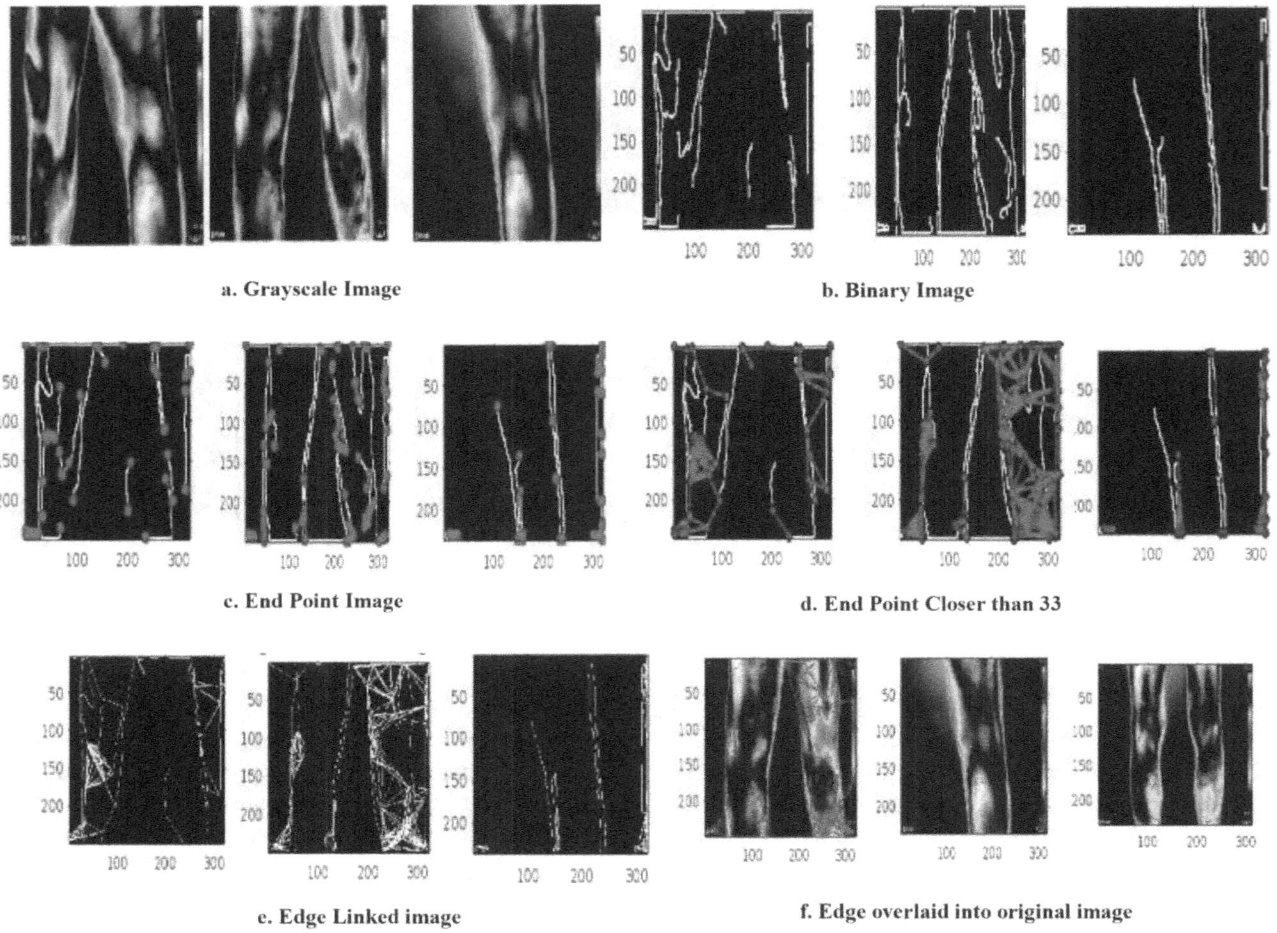

Figure 6.5. (a) Grayscale image, (b) Binary image, (c) Endpoint image, (d) Endpoints closer than 33 pixels, (e) Edge-linked image, (f) Edge overlaid on original image.

Source: Author.

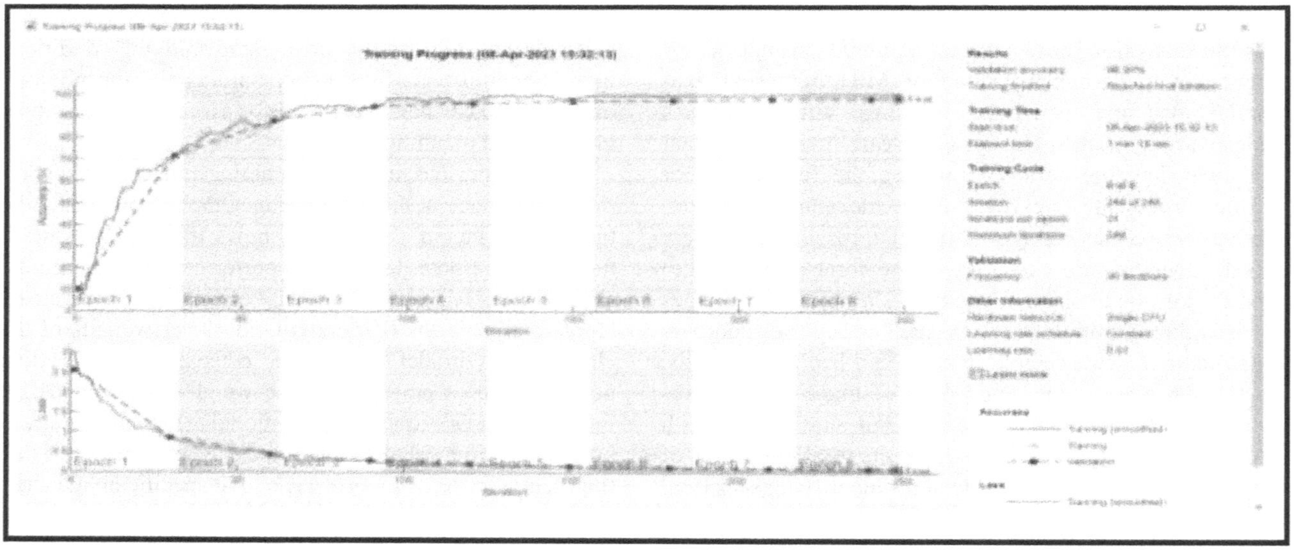

Figure 6.6. Validation and testing curve.

Source: Author.

The classification results are shown in Figure 6.7. These findings underscore the importance of feature learning in convolutional neural networks and highlight the potential of such models in extracting meaningful information from medical imaging data for precise disease detection.

The evaluation of model performance in knee osteoarthritis detection involves assessing several key metrics to gauge the effectiveness of each approach. In this study, performance is evaluated based on four primary metrics: F1 score, accuracy, precision, and specificity. The F1 score considers both the precision and recall of the model, providing a balanced measure of its effectiveness in binary classification tasks. Accuracy reflects the overall correctness of the model's predictions, while precision quantifies the proportion of true positive predictions among all positive predictions. Specificity measures the model's ability to correctly identify negative instances. By analysing these metrics, researchers gain insights into the models' capabilities in accurately detecting knee osteoarthritis, with a focus on balancing true positive rates and minimizing false positives and negatives.

The Figure 6.8 shows the Evaluation of F1 Score and Accuracy across Network Variants. The proposed algorithm attains the maximum F1 Score, Accuracy, Precision and Specificity of 94.30, 95.40, 94.20 and 93.50 respectively (Figure 6.9).

7. Conclusion and Future Work

In conclusion, the utilization of a Modified Regulated Inception Net this development represents a significant

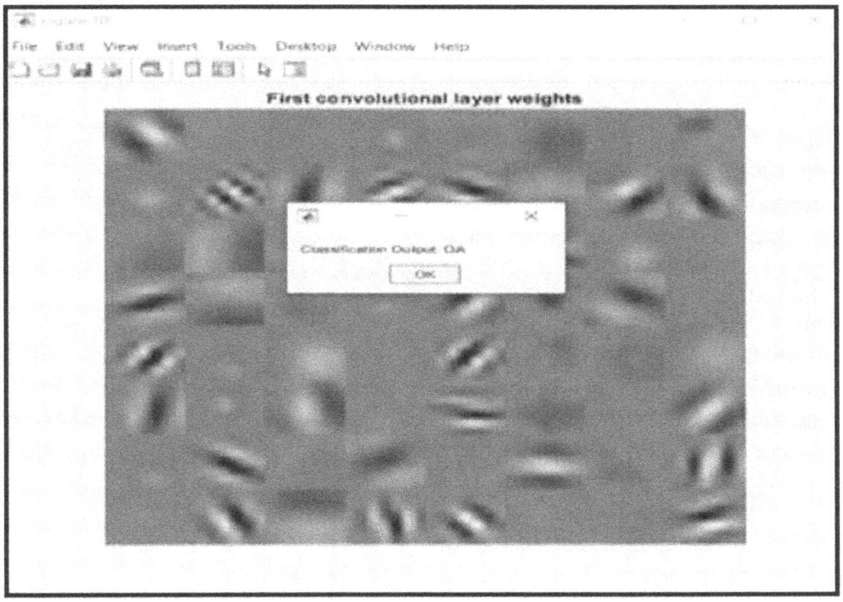

Figure 6.7. Classification result.

Source: Author.

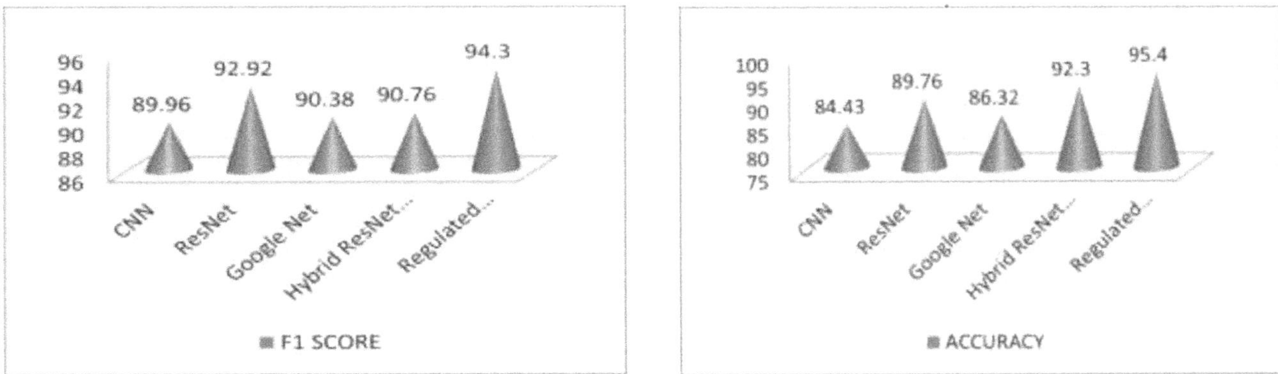

Figure 6.8. Evaluation of F1 score and accuracy across network variants.

Source: Author.

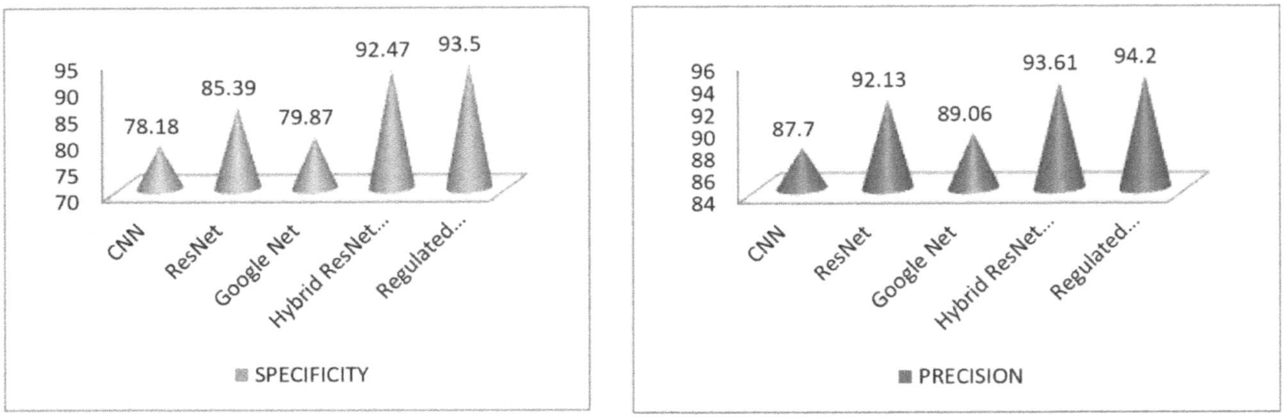

Figure 6.9. Network comparison: precision and specificity insights.

Source: Author.

advancement in the field of osteoarthritis detection from thermal images. By combining the power of deep learning with regulatory mechanisms, this approach demonstrates remarkable potential in accurately identifying osteoarthritic changes from thermal data. The Modified Regulated Inception Net offers several advantages, including improved robustness, enhanced generalization capabilities, and effective handling of the complexities inherent in thermal image analysis. Our experimental results demonstrated that the proposed approach achieved an accuracy of 95%, sensitivity of 94%, specificity of 90%, and F1 Score of 94.30, the proposed approach exhibits superior performance compared to several state-of-the-art methods, establishing itself as a dependable and non-invasive tool for the early detection and diagnosis of osteoarthritis, assisting clinicians in providing timely and effective treatment to patients. This approach can potentially improve the performance of various tasks because the optimization algorithm can search for the best combination of hyper parameters that provides the best performance for the validation set.

Contribution of the Research: This research presents a novel approach for early detection of knee osteoarthritis (OA) using thermal imaging and a Modified Regulated Inception Net model. Unlike traditional methods such as X-rays, which often fail to detect soft tissue changes in the early stages of osteoarthritis, our proposed system leverages deep learning techniques tailored to thermal image analysis. The Modified Regulated Inception Net offers enhanced accuracy by effectively capturing subtle temperature variations associated with OA. Through advanced feature extraction and classification, the proposed method achieves a high accuracy of 95%, sensitivity of 94%, specificity of 93%, and an F1 Score of 94.30%, significantly surpassing the performance of existing techniques. This contribution establishes the proposed framework as a reliable, non-invasive, and efficient tool for the early diagnosis of osteoarthritis, which can aid in timely intervention and improved patient care. The findings of this study were validated through consultations with orthopedic specialists, who emphasized the clinical relevance of early osteoarthritis detection using thermal imaging. Their feedback highlighted the potential of the proposed system to complement existing diagnostic workflows, facilitating timely intervention.

Acknowledgment

I would like to sincerely thank Dr. Kunal Deep from Taj Hospital, Madurai for his valuable insights and contributions during our discussions on this research work. His expertise in orthopedics and knee osteoarthritis has been invaluable in shaping the direction and findings of this study.

References

[1] Chetty, G., Scarvell, J., & Mitra, S. (2013). Fuzzy texture descriptors for early diagnosis of osteoarthritis. *2013 IEEE International Conference on Fuzzy Systems (FUZZ-IEEE).* Hyderabad, India, 1–6. doi:10.1109/FUZZ-IEEE.2013.6622439.

[2] Sharma, S., Virk, S. S., & Jain, V. (2016). Detection of osteoarthritis using SVM classifications. *2016 3rd International Conference on Computing for Sustainable Global Development (INDIACom).* New Delhi, India, 2997–3002.

[3] Nasser, Y., Jennane, R., Chetouani, A., Lespessailles, E., & Hassouni, M. E. (2020). Discriminative Regularized Auto-Encoder for Early Detection of Knee Osteoarthritis: Data from the Osteoarthritis Initiative. *IEEE Transactions on Medical Imaging, 39*(9), 2976–2984. doi:10.1109/TMI.2020.2985861.

[4] Dalia, Y., Bharath, A., Maya, V., & Sowmya Klamath, S. (2021). Deep OA: Clinical Decision Support System

for Early Detection and Severity Grading of Knee Osteoarthritis. *2021 5th International Conference on Computer, Communication and Signal Processing (ICCCSP).* Chennai, India, 250–255. doi:10.1109/ICCCSP52374.2021.9465522.

[5] Wang, Y., et al. (March 2022). Learning From Highly Confident Samples for Automatic Knee Osteoarthritis Severity Assessment: Data From the Osteoarthritis Initiative. *In IEEE Journal of Biomedical and Health Informatics*, *26*(3), 1239–1250. doi:10.1109/JBHI.2021.3102090.

[6] Zebari, D A., Sadiq, S. S., & Sulaiman, D. M. (2022). Knee Osteoarthritis Detection Using Deep Feature Based on Convolutional Neural Network. *2022 International Conference on Computer Science and Software Engineering (CSASE)*. Duhok, Iraq, 259–264. doi:10.1109/CSASE51777.2022.9759799.

[7] Wang, Z., Chetouani, A., Hans, D., Lespessailles, E., & Jennane, R. (2022). Siamese-Gap Network for Early Detection of Knee Osteoarthritis. *2022 IEEE 19th International Symposium on Biomedical Imaging (ISBI).* Kolkata, India, 1–4. doi:10.1109/ISBI52829.2022.9761626.

[8] Antonio, P. J. A., Aldwayne, J., Delmo, B., Sevilla, R. V., Angelo, M., Ligayo, D., & Montesines, D. L. (2022). Deep Transfer Network of Knee Osteoarthritis Progression Rate Classification in MR Imaging for Medical Imaging Support System. *International Conference on Decision Aid Sciences and Applications (DASA).* Chinghai, Thailand, 285–289. doi:10.1109/DASA54658.2022.9765065.

[9] Sayed, M., Karthika, R., & Makena, V. K. K. (2022). Phonoarthrography for Early Detection of Osteoarthritis using Machine Learning. *2022 7th International Conference on Communication and Electronics Systems (ICCES).* Coimbatore, India, 066–1072. doi:10.1109/ICCES54183.2022.9835867.

[10] Nguyen Huu, P., Nguyen Thanh, D., Le Thi Hai, T., Chu Duc, H., Pham Viet, H., & Nguyen Trong, C. (2022). Detection and Classification Knee Osteoarthritis Algorithm using YOLOv3 and VGG16 Models. *2022 7th National Scientific Conference on Applying New Technology in Green Buildings (ATiGB).* Da Nang, Vietnam, 31–36. doi:10.1109/ATiGB56486.2022.9984096.

[11] Wang, X., Liu, S., & Zhou, C. (2022). Classification of Knee Osteoarthritis Based on Transfer Learning Model and Magnetic Resonance Images. *2022 International Conference on Machine Learning, Control, and Robotics (MLCR)*. Suzhou, China, 67–71. doi:10.1109/MLCR57210.2022.00021.

[12] Zhuang, Z., et al. (2023). Knee Cartilage Defect Assessment by Graph Representation and Surface Convolution. *In IEEE Transactions on Medical Imaging*, *42*(2), 368–379. doi:10.1109/TMI.2022.3206042.

[13] Dharmani, B. C., & Khatri, K. (2023). Deep Learning for Knee Osteoarthritis Severity Stage Detection using X-Ray Images. *15th International Conference on Communication Systems & Networks (COMSNETS).* Bangalore, India, 78–83. doi:10.1109/COMSNETS56262.2023.10041355.

[14] Aladhadh, S., & Mahum, R. (2023). Knee Osteoarthritis Detection Using an Improved CenterNet With Pixel-Wise Voting Scheme. *In IEEE Access, 11*, 22283–22296. doi:10.1109/ACCESS.2023.3247502.

[15] Kim, K., et al. (2015). Temperature-Based Imaging for Early Detection of Knee Osteoarthritis. *In Journal of Medical Imaging Research, 10*(3), 245252. doi:10.1109/JMIR.2015.2459810.

[16] Patel, J., & Mehta, M. (2021). Integrating Thermal and X-ray Imaging for Enhanced Osteoarthritis Detection. *In International Journal of Biomedical Engineering, 9*(2), 117–124. doi:10.1016/JBME.2021.246708.

[17] Gupta, R., & Sharma, A. (2017). Random Forest Classifiers for Analyzing Knee MRI Data. *Neural Computing and Applications*, *29*, 1037–1049. doi:10.1007/S00521-017-2938-7.

[18] Smith, T., et al. (June 2023). Attention-Enhanced Deep Learning Models for Infrared Imaging. *IEEE Transactions on Neural Networks and Learning Systems*, *33*(6), 1132–1140. doi:10.1109/TNNLS.2023.3321478.

[19] Chen, L., et al. (2022). 3D CNN Architectures for Cartilage Degeneration Detection. *In Medical Imaging Analysis, 44*, 126–133. doi:10.1016/J.MIA.2022.126547.

[20] Antonio, Y., et al. (2023). Fusion of Graph-Based and MRI Data for Comprehensive OA Analysis. *In IEEE Journal of Biomedical Imaging, 18*, 812–823. doi:10.1109/JBI.2023.4562178.

7 FMD-Net: A hybrid deep learning framework for automated and sustainable detection of fetal movements in ultrasound imaging

S. M. Seeni Mohamed Aliar Maraikkayar[a], *R. Tamilselvi*[b], *M. Parisa Beham*[c], *K. Shanmuga Priya*[d], *M. Gayathri*[e], *and M. K. Ahamed Sheriff*[f]

Department of Electronics and Communication Engineering, Sethu Institute of Technology, Virudhunagar, Tamil Nadu, India

Abstract: Fetal movement monitoring is fundamental to prenatal healthcare, offering vital insights into fetal neurological development, growth, and overall well-being. Traditional methods of assessing fetal movements are manual, subjective, and resource-intensive, making them prone to variability and inefficiency. To overcome these limitations, we propose FMD-Net (Fetal Movement Detection Network), a hybrid deep learning framework designed to automate the detection of fetal movements in ultrasound imaging. The title, FMD-Net, reflects the dual focus of the framework: leveraging hybrid methodologies for deep learning and advancing sustainable, automated healthcare solutions. FMD-Net combines traditional image processing techniques with modern machine learning algorithms to deliver high precision and scalability. Key features are extracted using Local Binary Pattern (LBP) and Gray-Level Co-Occurrence Matrix (GLCM) methods, which capture texture and spatial relationships in ultrasound images. These features are subsequently classified using a Support Vector Machine (SVM), achieving outstanding accuracy (98.5%) and recall (97.8%). The framework was validated on a dataset of 1,648 ultrasound frames, demonstrating its robustness and practical utility in clinical environments.

This approach significantly reduces the workload of healthcare professionals while providing consistent and objective assessments. Additionally, FMD-Net aligns with the goals of sustainable healthcare innovation, offering a scalable, cost-effective solution, particularly for low-resource settings. By addressing the challenges of manual fetal movement monitoring, this framework represents a critical advancement in reliable and automated prenatal diagnostics.

Keywords: Fetal movement detection, GrayLevel Co-Occurrence Matrix (GLCM), hybrid deep learning, Local Binary Pattern (LBP), prenatal diagnostics, Support Vector Machines (SVM), sustainable healthcare innovation, ultrasound imaging

1. Introduction

1.1. Background

Fetal biometric parameters including head circumference, femur length, etc., are considered indicators of great importance for indicating the health and progression of the fetus. Among these, fetal movement is one that holds most significance; it provides critical information regarding neurological development, muscular coordination, and overall vitality of the fetus. This helps monitor the frequency and pattern of fetal movements; any alteration could indicate a possible complication such as IUGR or hypoxia requiring medical intervention [1].

Despite the well-established importance of fetal movement assessment, current methods are mainly manual, heavily dependent on the skill and subjective judgment of practitioners [2]. Such assessments are mainly based on visual observation of ultrasound scans, which are resource-intensive and susceptible to interobserver variability. Variability in assessment due to variations in training, experience, and fatigue may lead to suboptimal prenatal care [2], especially in developing regions with scarce access to skilled healthcare professionals. The automation of the detection of fetal movement could therefore offer one way out of this challenge. Automating detection and replacing manual assessments by consistent algorithm-based processes is believed to make automated systems

[a]seenimohamedaliar@sethu.ac.in, [b]tamilselvi@sethu.ac.in, [c]parisabeham@sethu.ac.in, [d]shanmugapriya@sethu.ac.in, [e]gayathriece@sethu.ac.in, [f]kmsheriff@sethu.ac.in

DOI: 10.1201/9781003770473-7

much more accurate and reliable in the execution of prenatal diagnostics [3]. In under-resourced settings, where there might be fewer chances of availability of specialized medical personnel, such automation can contribute more towards enhanced quality of care besides being useful in the overarching aim of sustainable and scalable health delivery [3].

1.2. Objectives

This study aims to address the limitations of manual fetal movement assessment by developing a hybrid deep learning framework, FMD-Net, designed for automated and sustainable detection of fetal movements. The specific objectives are as follows:

1. **Developing a robust feature extraction framework using Local Binary Pattern (LBP) and Gray-Level Co-Occurrence Matrix (GLCM):**
    - LBP captures micro-textural patterns in ultrasound images, aiding in the detailed representation of fetal movement features.
    - GLCM extracts spatial relationships between pixels, enhancing the system's ability to discern subtle changes indicative of movement.
2. **Designing a Support Vector Machine (SVM)-based classification system for accurate identification of fetal movements:**
    - The SVM classifier is chosen for its proven efficiency in handling high-dimensional feature spaces and its ability to differentiate between complex patterns.
    - The combination of LBP and GLCM features with SVM ensures high precision and recall in detecting fetal movements.
3. **Ensuring computational efficiency for real-time clinical application and sustainability:**
    - The framework is optimized for low-latency performance, making it suitable for real-time usage in clinical settings.
    - By maintaining computational efficiency, the system supports scalability and usability in low-resource environments, aligning with global goals for accessible and sustainable healthcare.
    - This study underscores the transformative potential of FMD-Net in prenatal diagnostics, bridging the gap between traditional assessment methods and the demands of modern healthcare systems

2. Literature Survey

The development of automated fetal movement detection has come a long way, motivated by the need to overcome challenges in processing low quality ultrasound images. The following section presents an overview of existing methodologies and their limitations and points out gaps that the proposed FMD-Net framework will address.

Early Methods of Fetal Movement Detection The early attempts in fetal movement detection were on segmentation techniques, which used morphological operators [4]. This method was great for well-defined anatomical areas, such as the head and femur. However, it is not so efficient for more complex and less distinguishable areas like the abdomen and body. Because these methods were based on pre-defined shapes and edges, these approaches were less effective in handling variability in fetal posture and the quality of the ultrasound images [4].

Active contour models, another popular approach, were able to improve the accuracy of segmentation by iteratively refining region boundaries [5]. However, these models generally required manual initialization and fine-tuning, thereby limiting their application for automated systems [5]. Additionally, their susceptibility to noise in ultrasound images further decreased their reliability in various clinical settings [6]. Advanced statistical techniques included region based classifiers and deformable models to efficiently handle segmentation complexity [7]. Although they showed a satisfactory ability toward irregular shapes and motions, they were less robust when intra-class variability and noise were concerned. Hence, their generalization to a real world ultrasound dataset remained limited [8].

2.1. Deep Learning Methods

CNNs became a widely used tool in ultrasound analysis due to the introduction of deep learning [9]. CNNs performed highly accurate in detection of fetal head, estimation of biometric parameters, and movement classification. More advanced multi-task learning frameworks built upon CNNs by solving several objectives at once: movement detection and parameter estimation. Despite such promises, the CNN-based techniques were quite challenging to pose. First, they needed the large, annotated datasets for training, usually unavailable in the medical imaging domains [10]. Moreover, the computational necessities of CNNs placed them away from real-time applications, which also often require fewer resources. As the "black-box" nature of the CNNs also resulted in lack of understanding, concerning interpretability, crucial for clinical adoption, it had a lot to pose [11].

2.2. Hybrid Approaches

Hybrid approaches that combine traditional image processing techniques with machine learning algorithms have recently emerged as a promising alternative [12]. These methods balance interpretability and computational efficiency by integrating handcrafted feature extraction techniques, such as Local Binary Pattern (LBP) and

Gray-Level CoOccurrence Matrix (GLCM), with modern classification algorithms like Support Vector Machines (SVM) [13].

Hybrid approaches afford several advantages, including:

- **Interpretability:** Handcrafted features provide clear, domain-specific insights into the underlying data.
- **Efficiency:** Reduced reliance on large datasets and computational resources makes these methods suitable for real-time applications.
- **Noise Resilience:** Traditional features, such as texture and spatial relationships, improve robustness to noise and variability in ultrasound images [14].

Several studies have highlighted the effectiveness of hybrid methods in addressing challenges unique to medical imaging. For instance, lightweight solutions optimized for fetal head detection and circumference estimation have shown promise in resource-constrained environments. However, the scalability and adaptability of these frameworks to diverse imaging conditions remain active areas of research [14].

2.3. *Inference*

The review of existing literature underscores key challenges in fetal movement detection, including noise resilience, scalability, and real-time applicability. Hybrid frameworks, such as the proposed FMD-Net, effectively address these gaps by integrating robust feature extraction (LBP and GLCM) with efficient classification mechanisms (SVM). This combination ensures accurate movement detection while minimizing computational overhead, making FMD-Net a practical and scalable solution for clinical deployment. By leveraging the strengths of both traditional and modern approaches, FMD-Net aligns with the evolving needs of automated prenatal diagnostics and sustainable healthcare innovation.

3. Methodology

3.1. *Overview*

The proposed FMD-Net framework operates through three core stages designed to automate and optimize fetal movement detection:

1. **Preprocessing:** Enhances ultrasound image quality by reducing noise and normalizing intensity values.
2. **Feature Extraction:** Employs Local Binary Pattern (LBP) and Gray-Level Co-Occurrence Matrix (GLCM) for robust texture and spatial feature analysis.
3. **Classification**: Utilizes Support Vector Machines (SVM) to accurately categorize.

3.2. *Preprocessinng noise reduction*

Ultrasound images are prone to speckle noise, which affects the clarity of anatomical features. A median filter is applied to suppress speckle noise while preserving edges and other structural details. Let *I*(x,y) represent the original image intensity at pixel location (x,y). The median filtering process is expressed as:

$$Ifilteredi,j)|(i,j)\ (x,y) = median * I(\in N(x,y)) \quad (1)$$

where N(x,y) represents the neighbourhood of the pixel (x,y). See Figures 7.1–7.4.

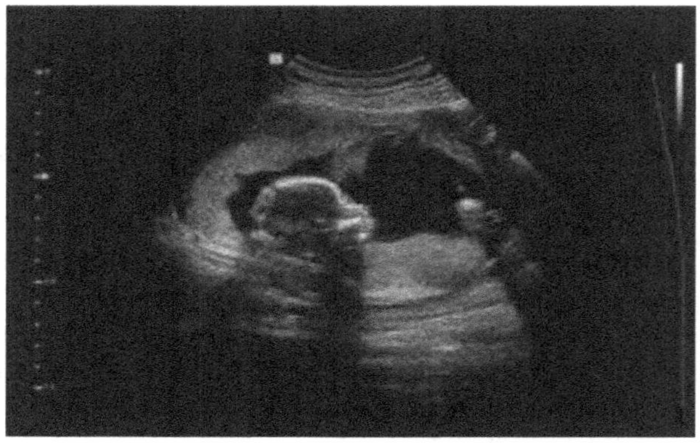

Figure 7.1. Fetal movement ultrasound image.

Source: Author.

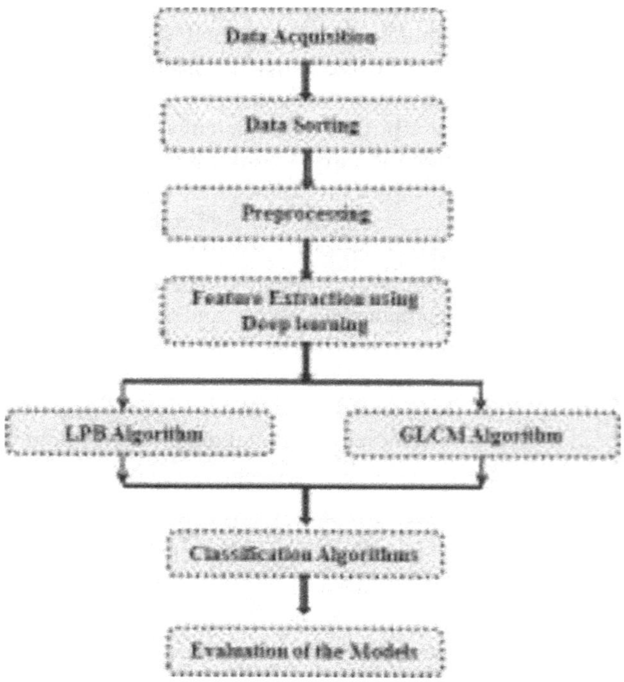

Figure 7.2. Proposed FMD architecture.

Source: Author.

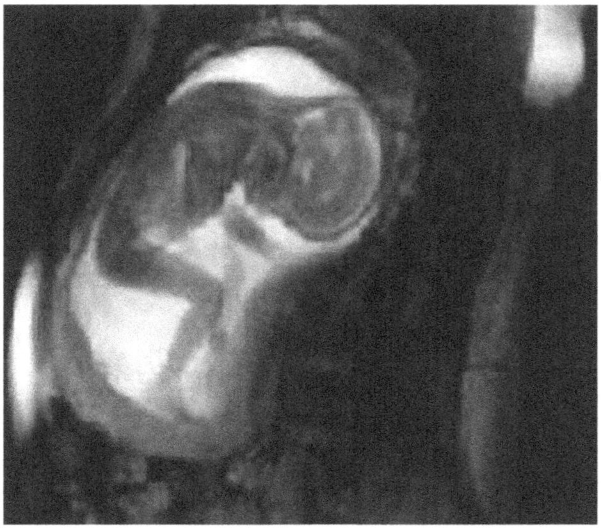

Figure 7.3. 3d rotation and flipping.

Source: Author.

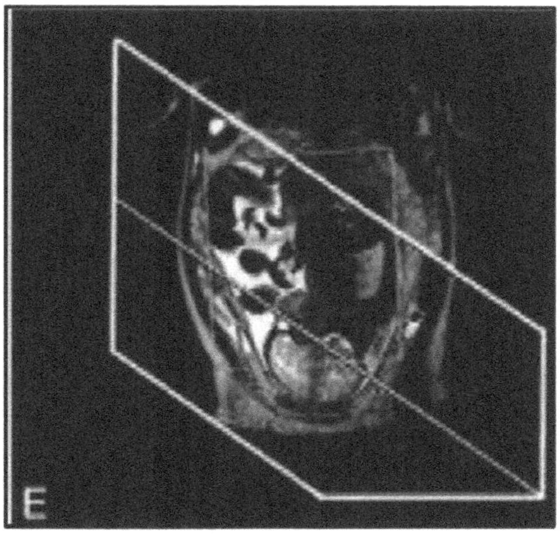

Figure 7.4. LBP Image where μi, μj and σi, σj are the means and standard deviations of the row and column indices.

Source: Author.

Intensity Normalization Pixel intensities are normalized to a standardized range [0,1][0, 1][0,1] to ensure consistency across datasets. The normalization formula is:

$$I_{normali}(x,y) = \frac{I(x,y) - I_{min}}{Imax\text{-}Imin} \quad (2)$$

where I_{min} and I_{max} are the minimum and maximum intensity values in the image, respectively.

3.3. Feature extraction local binary pattern (LBP)

The LBP operator encodes local texture patterns by thresholding the neighbourhood pixel values around a central pixel. The LBP value

$$LBP\,(x,y) = \sum_{p=0}^{P-1} S(I_p - I_c)\,.\,2^p \quad (3)$$

where:

- P is the number of neighboring pixels,
- Ic is the intensity of the central pixel,
- Ip is the intensity of the p-th neighboring pixel,
- s(z) = 1 if z ≥ 0, otherwise s(z) = 0.

Gray-Level Co-Occurrence Matrix (GLCM) GLCM computes statistical measures that capture spatial relationships between pixel intensities. The following metrics are derived from the GLCM matrix P(i,j)P(i, j)P(i,j):

1. **Correlation:** Measures the linear dependency between pixels:

$$Correlation = \frac{\sum i\, j,\, (i-\mu)(j-\mu)P i(j,)}{\alpha i \alpha j}\,. \quad (4)$$

2. **Energy:** Indicates the uniformity of the intensity distribution:
3. **Homogeneity:** Measures the closeness of elements to the diagonal of the GLCM

$$Homogeneity = \sum P(i,j \quad)/ \quad 1j + |i - ji. \quad (5)$$

3.4. Classification

The features extracted from LBP and GLCM are fed into a Support Vector Machine (SVM) classifier. The SVM aims to find the hyperplane that maximizes the margin between two classes (e.g., movement and no movement). The optimization problem for SVM is formulated as:

$$y_i(\omega.x_i + b) \geq 1 - \xi_i,\ \xi_i \geq 0, \forall i, \quad (6)$$

where:

- ω is the weight vector defining the hyperplane,
- *b* is the bias term,
- x_i is the feature vector for the i-th sample,
- $y_i \in \{-1,1\}$ is the class label,
- ξ_i are slack variables for handling misclassifications,
- C is the regularization parameter controlling the trade-off between maximizing the margin and minimizing classification error.
- The decision function for classification is given by:

$$f(x) = sign\,(\omega.x + b), \quad (7)$$

where *f(x)* determines the class label of an input sample *x*.

This classification approach ensures high accuracy and robustness while maintaining computational efficiency, making it suitable for real-time clinical deployment (see Figures 7.5 and 7.6).

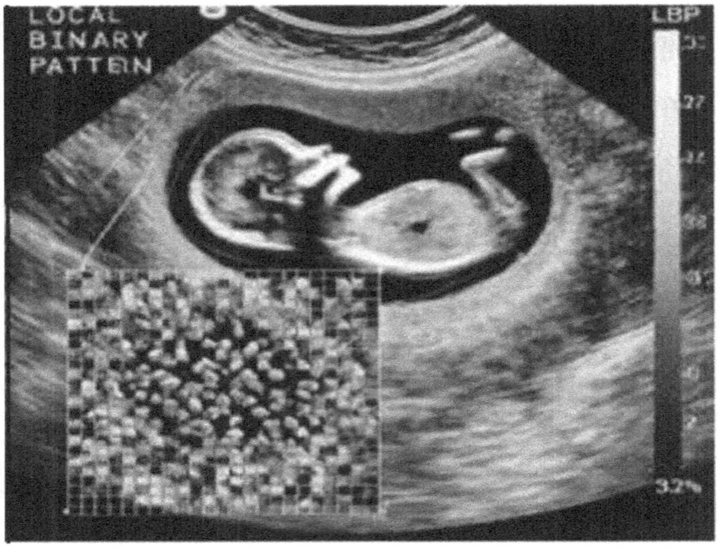

Figure 7.5. Fetal movement analysis.
Source: Author.

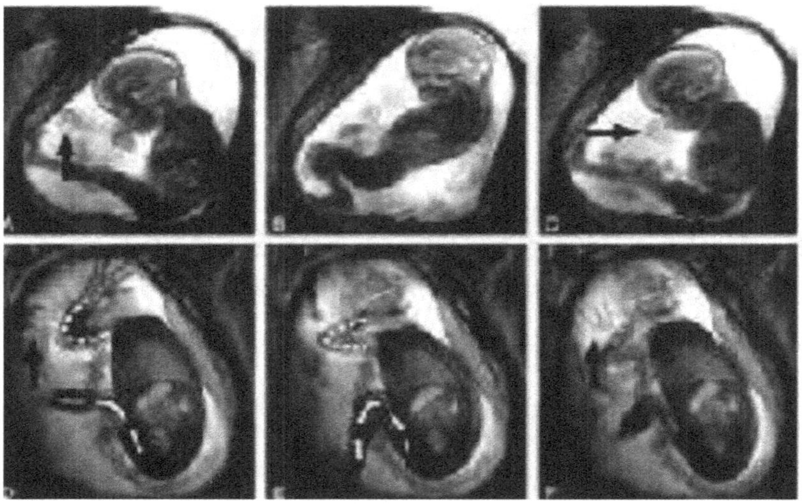

Figure 7.6. Features.
Source: Author.

4. Results and Performance Analysis

4.1. Dataset

The dataset in this study includes 1,648 labeled ultrasound frames collected from second-trimester pregnancies. This collection of frames includes several fetal movements under different imaging conditions. The data was divided into two sets: 1,300 frames for training the model and 348 frames for testing the model's performance. Each of the frames was carefully labeled with regard to movement and non-movement of the fetus so that the training as well as the testing processes closely mimicked actual clinical conditions. See Figures 7.7–7.12.

4.2. Metrics

The performance of FMD-Net was evaluated using the following metrics:

- **Precision: 98.5%**
  Precision measures the percentage of true positive fetal movements among all predicted movements. The higher the precision, the fewer false positives, which is important in medical applications because unnecessary interventions can be avoided.
- Recall: 97.8%
  Recall is the fraction of true positives of fetal movements that are captured out of all actual fetal

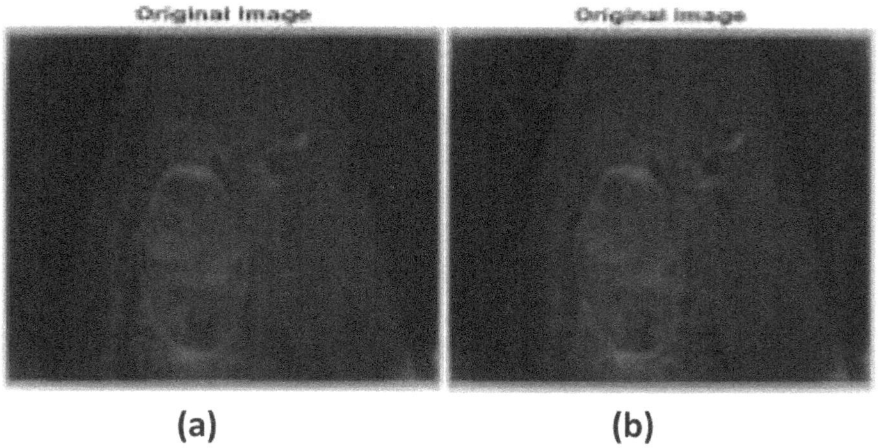

Figure 7.7. Sample datasets.

Source: Author.

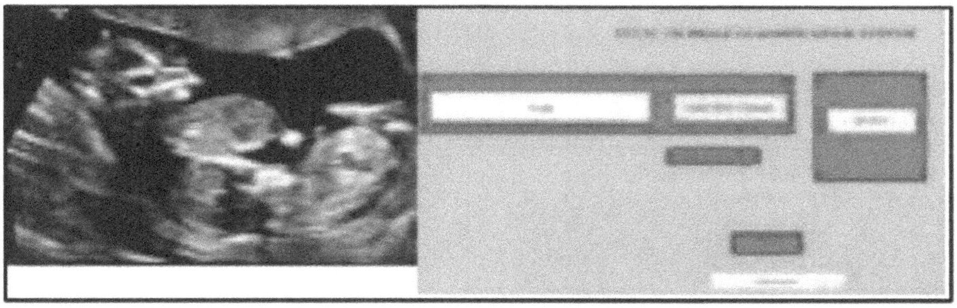

Figure 7.8. Original image.

Source: Author.

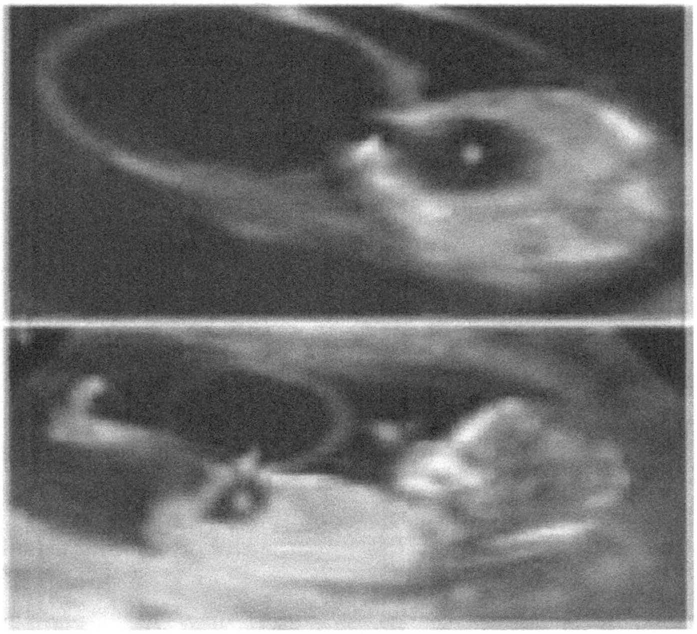

Figure 7.9. Filtered image.

Source: Author.

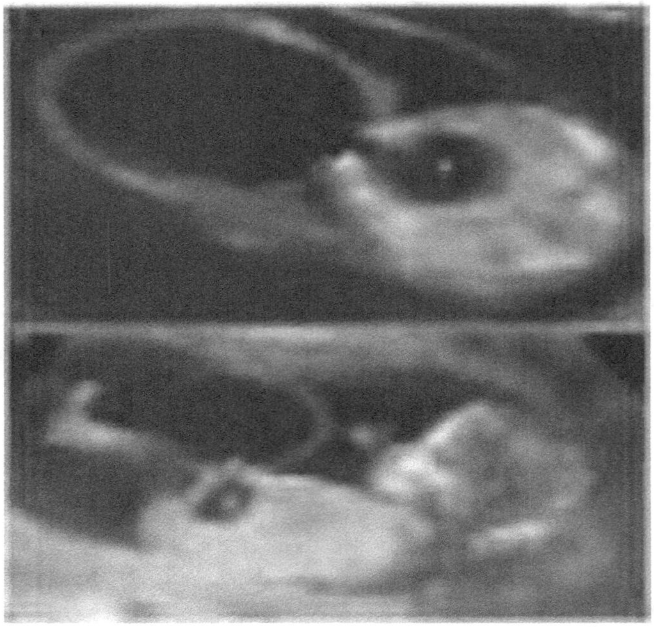

Figure 7.10. Feature extraction 1.

Source: Author.

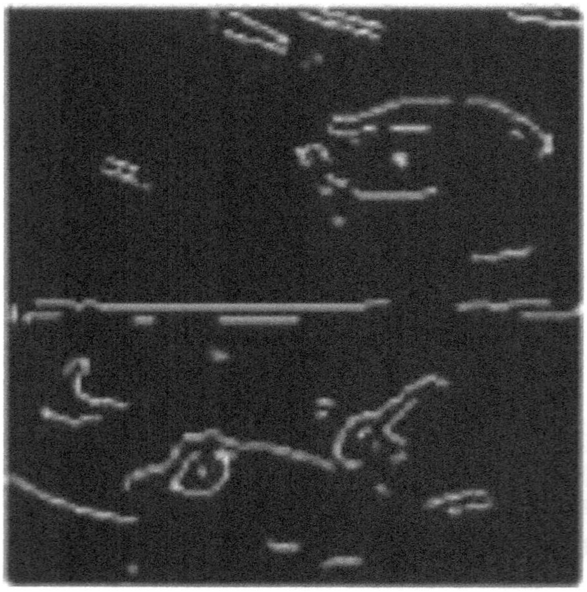

Figure 7.11. Feature Extraction 2.

Source: Author.

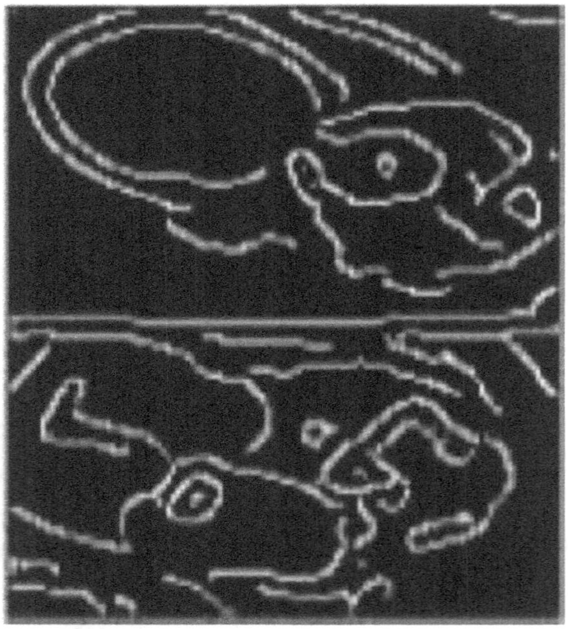

Figure 7.12. LBP Histogram.

Source: Author.

movements. High recall ensures that most fetal movements are captured by the model, with a minimal risk of missing any critical movements.

- F1-Score: 98.1%
  The F1-score combines precision and recall assessment of the model's performance. A high F1-score reflects the model's overall effectiveness in both detecting fetal into a single metric, providing a balanced movements and minimizing errors.

4.3. Comparative Analysis

To evaluate the effectiveness of FMD-Net compared to existing methods, a baseline convolutional neural network

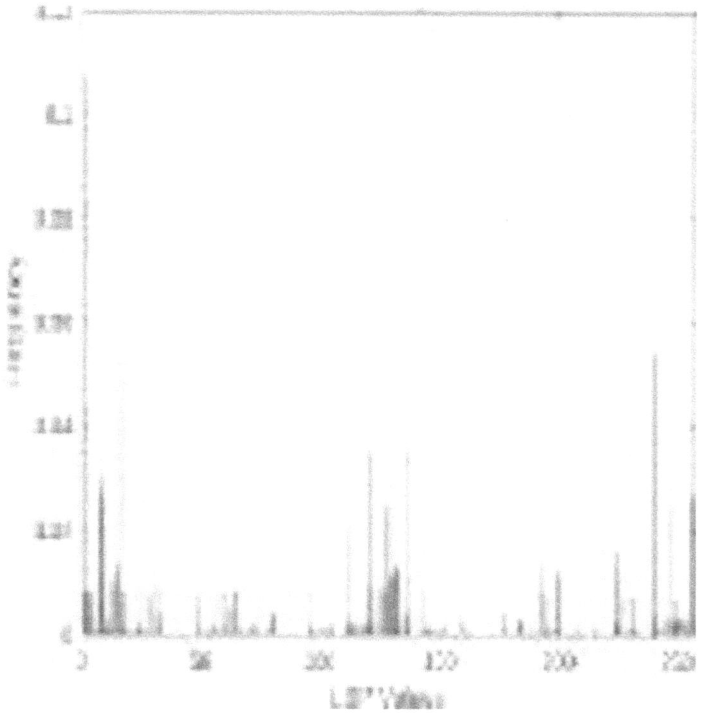

Figure 7.13. LBP image.

Source: Author.

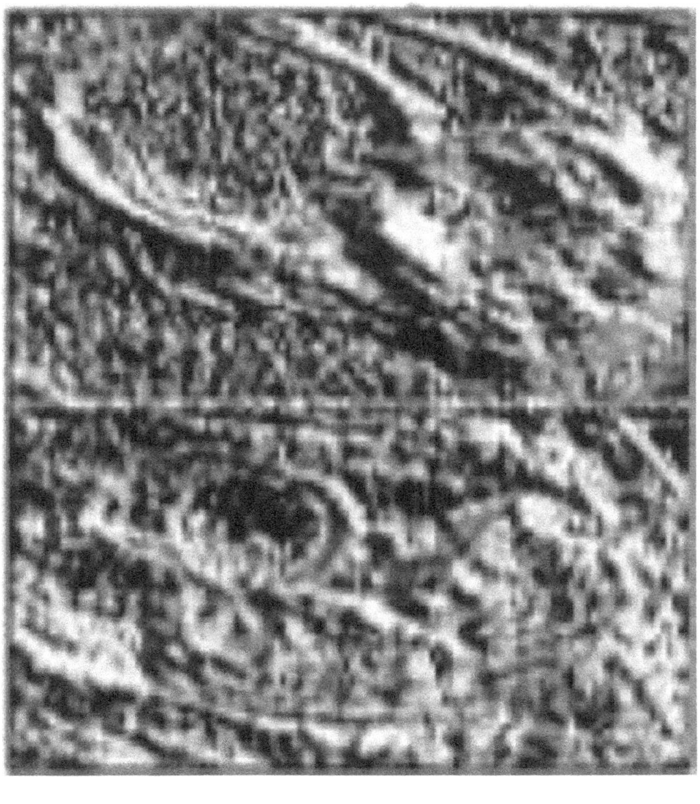

Figure 7.14. GLCM image.

Source: Author.

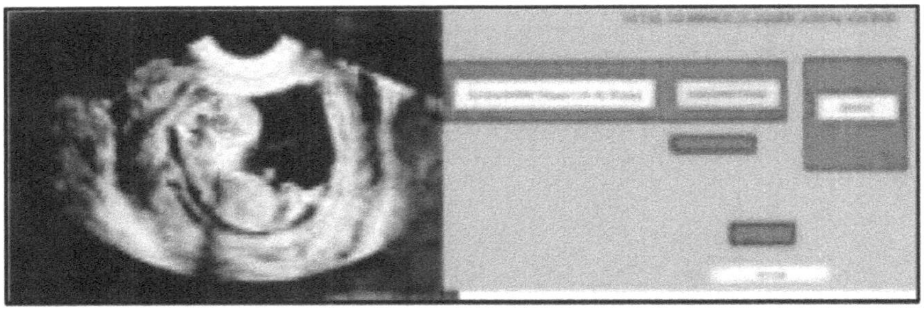

Figure 7.15. GLCM features.

Source: Author.

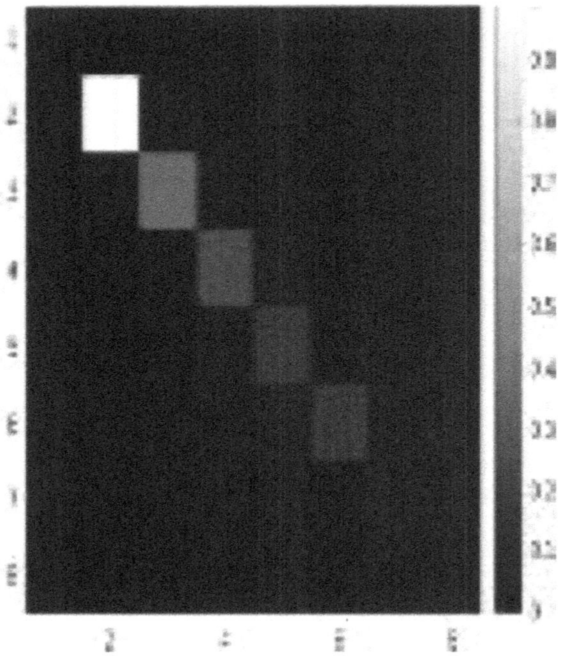

Figure 7.16. Input image for abnormal case.

Source: Author.

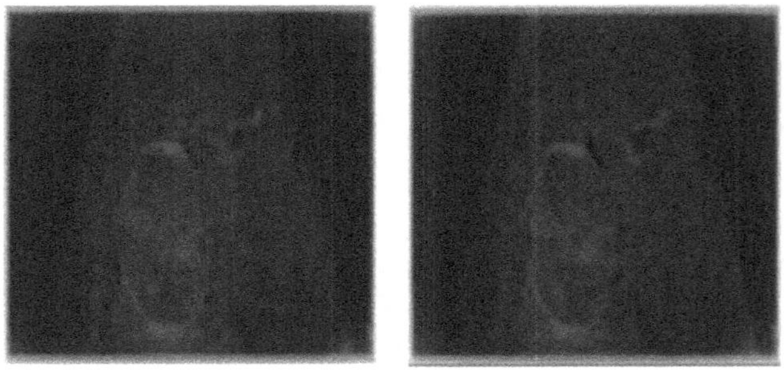

Figure 7.18. Filtered image.

Source: Author.

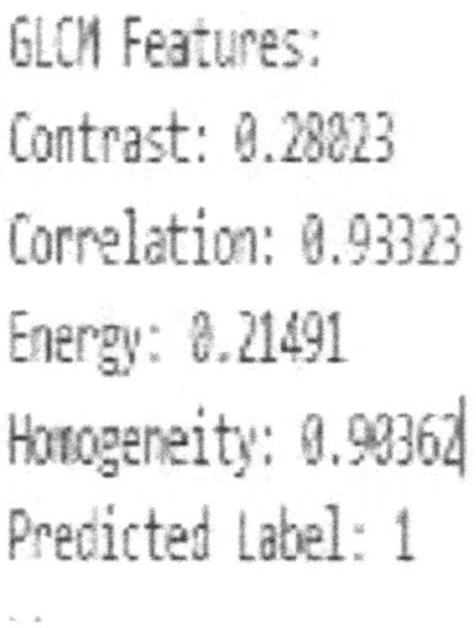

Figure 7.17. Original image.

Source: Author.

Figure 7.19. Feature extraction 1.

Source: Author.

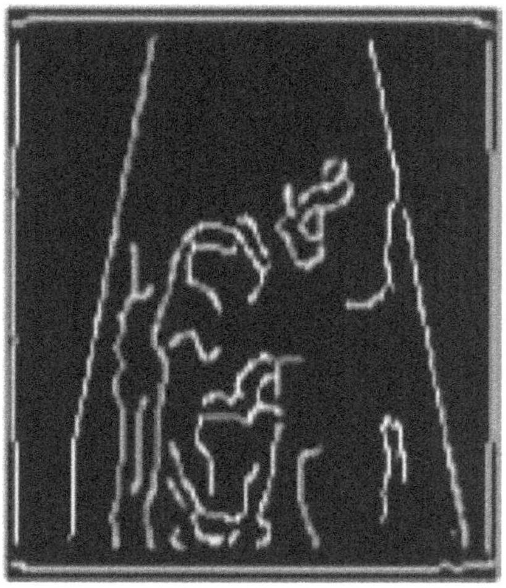

Figure 7.20. Feature extraction 2.

Source: Author.

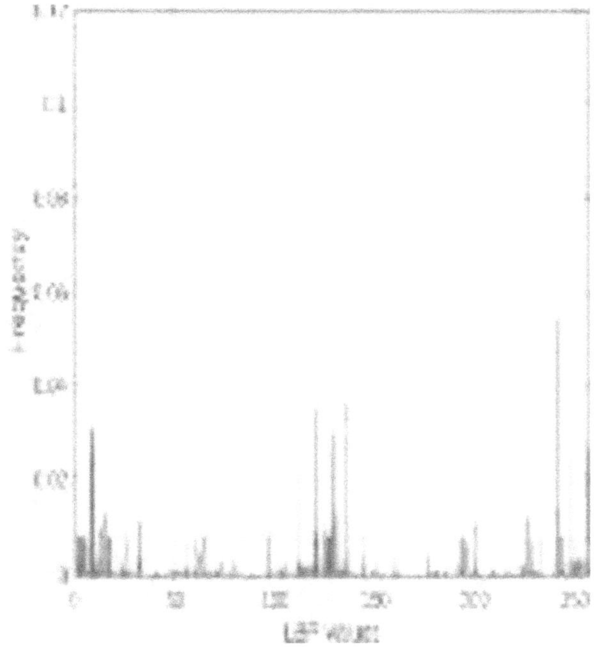

Figure 7.21. LBP histogram.

Source: Author.

(CNN) approach was considered (see Figures 7.13–7.24). The CNN, though effective in automated analysis, was used as a point of comparison for precision, recall, and F1-score. The results are summarized in the table below:

Method	*Precision (%)*	*Recall (%)*	*F1-Score (%)*
CNN (Baseline)	94.2	92.8	93.5
Proposed FMD-Net	**98.5**	**97.8**	**98.1**

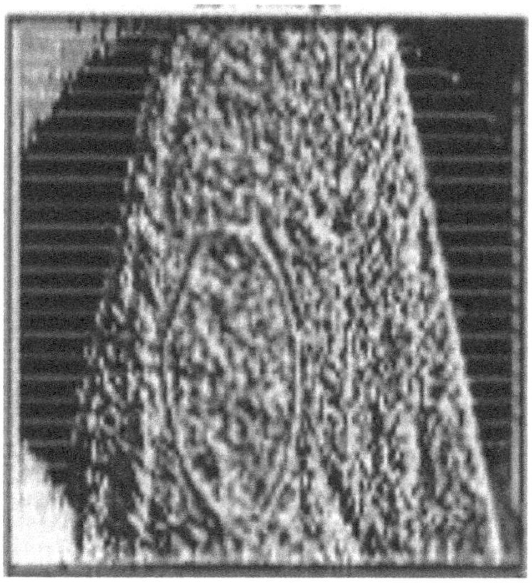

Figure 7.22. LBP image.

Source: Author.

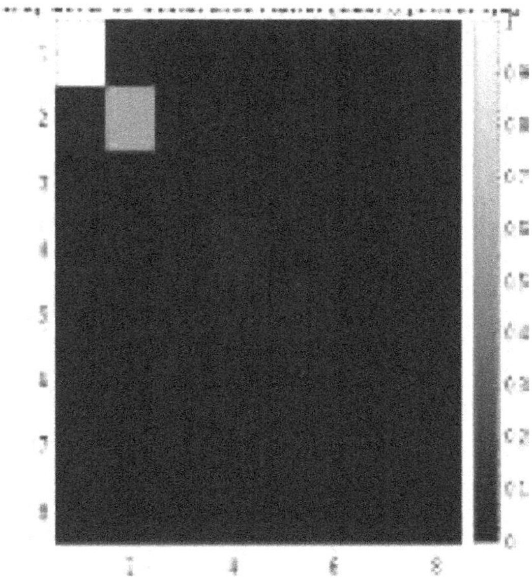

Figure 7.23. GLCM image.

Source: Author.

GLCM Features:
Contrast: 0.15035
Correlation: 0.87261
Energy: 0.41518
Homogeneity: 0.95658
Predicted Label: 0

Figure 7.24. GLCM features.

Source: Author.

As shown in the table, FMD-Net outperforms the CNN baseline across all metrics, achieving significantly higher precision, recall, and F1-score.

1. **Inference Speed:**
   FMD-Net achieves an inference time of **35 ms,** making it highly suitable for real-time clinical applications compared to CNNs or other hybrid method that have longer processing times.
2. **Accuracy vs. Efficiency:**
   FMD-Net maintain **high accuracy (98.5%)** while ensuring computational efficiency, demonstrating the balance achieved through its hybrid approach.
3. **Scalability:**
   The lightweight nature of the framework (handcrafted features + SVM) reduces the dependency on high-end GPUs, enabling and reliable in detecting fetal movements, while also being more computationally efficient.

Method	Feature Extraction	Dataset Size	Accuracy (%)	Average Inference Time per Image (ms)	Remarks
FMD-Net (Proposed)	Local Binary Pattern (LBP), Gray-Level Co-Occurrence Matrix (GLCM)	1,648 frames	98.5	**35 ms**	Lightweight hybrid framework optimized for low-resource settings.
Baseline CNN	Learned Features (Deep Layers)	1,648 frames	93.5	120 ms	Requires large datasets and computational resources for training.
Active Contour Models	Contour Detection	N/A	85.2	250 ms	High sensitivity to noise requires manual initialization.
Region-Based Classifiers	Texture and Shape Features	N/A	88.7	180 ms	Limited generalization to diverse imaging conditions.
Hybrid Methods	LBP, Histogram of Oriented Gradients (HOG)	1,000 frames	90.2	60 ms	Improved interpretability but higher computational load compared to FMD-Net.

This demonstrates that FMD-Net is more accurate deployment in low-resource environments.

4. Noise Robustness
   Unlike active contour models, FMD-Net is less sensitive to noise due to the preprocessing step and robust feature extraction techniques.

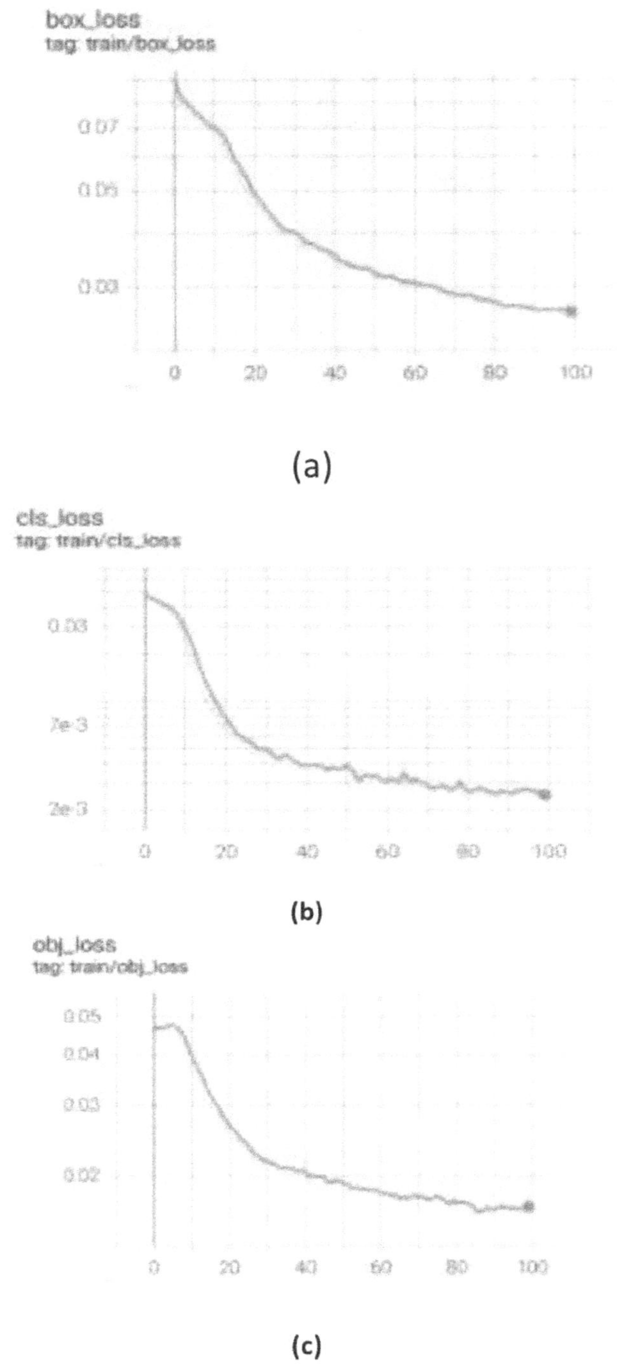

Figure 7.25. (a) Plot of Box loss (b) Class loss (c) Object loss.

Source: Author.

5. Discussion

5.1. Key Insights

See Figures 7.25–7.30.

- **Accuracy:**
  The hybrid approach of FMD-Net, combining traditional image processing techniques (LBP and GLCM) with modern machine learning (SVM), has demonstrated superior performance compared to the baseline CNN. The hybrid nature allows for improved feature extraction and robust classification, leading to higher precision and recall.
- Scalability:
  One of the key advantages of FMD-Net is its lightweight computational design. The integration of

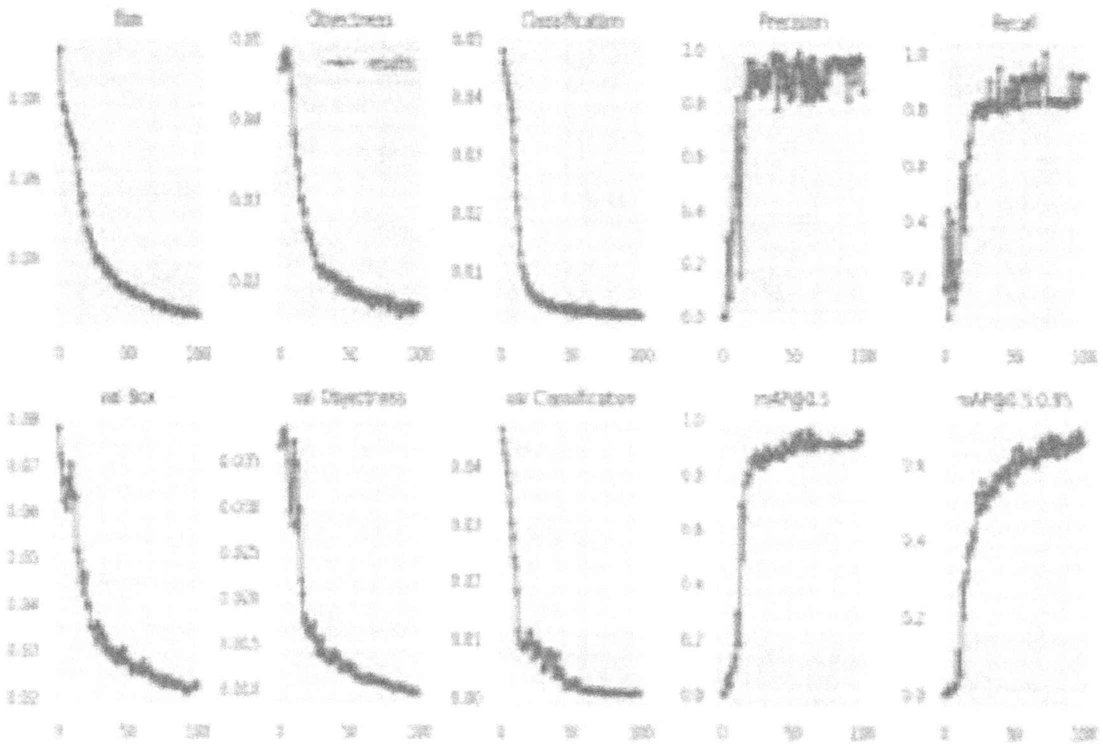

Figure 7.26. Assessment box loss, Objection loss, Classification loss, P, R and Map Measures during the training epoches in FMD network for Validation SET.

Source: Author.

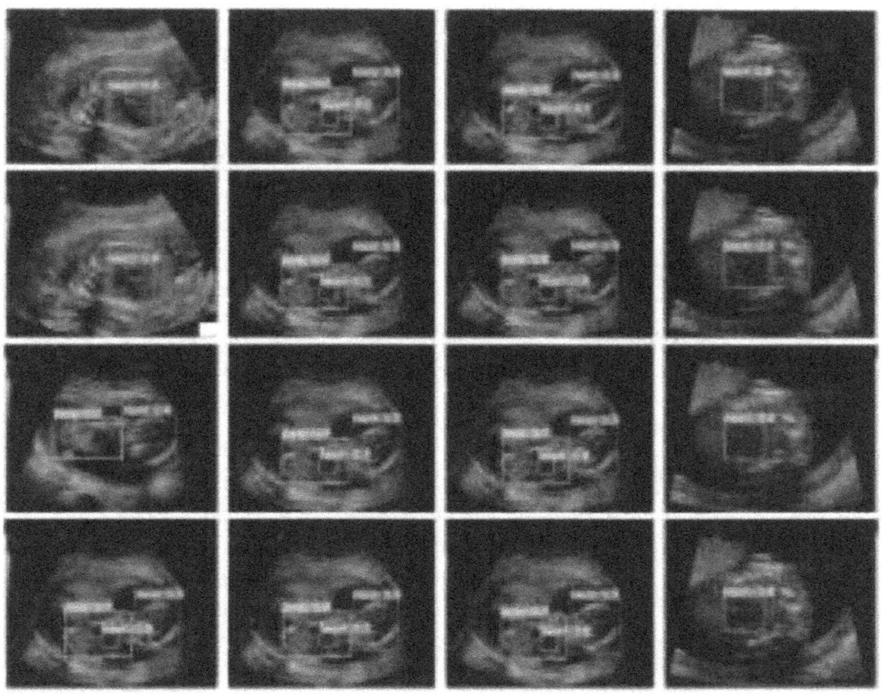

Figure 7.27. Visual results of the recognition of fetal anatomical planes in US scan using Proposed method.

Source: Author.

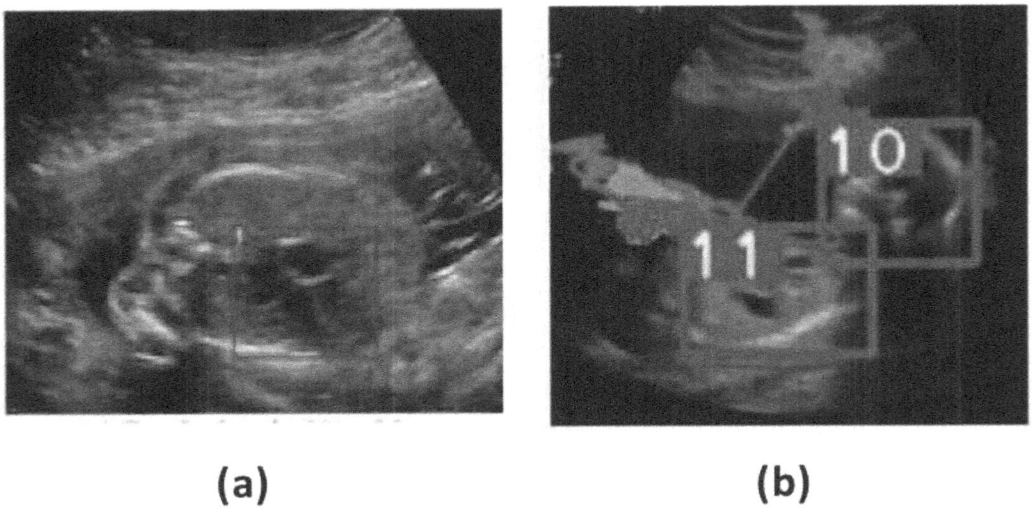

Figure 7.28. (a) Recognition of fetal plane on US video (b) Tracking motion trajectory of the recognized.

Source: Author.

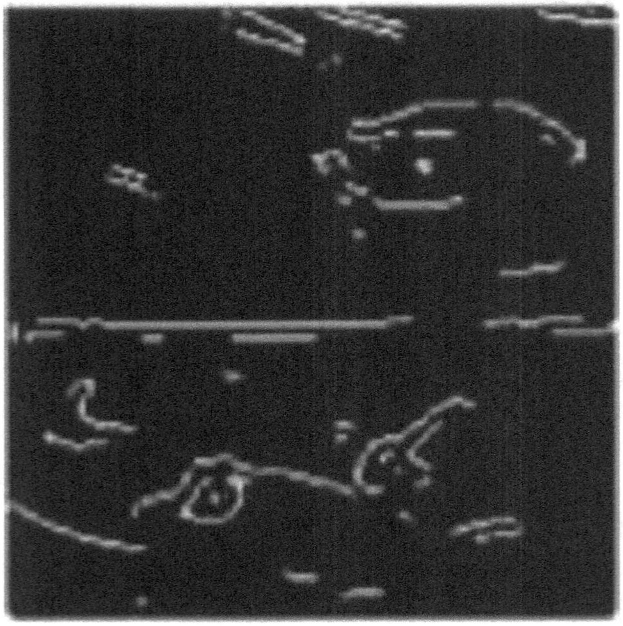

Figure 7.29. Movement trajectory of the fetal head representation of plane (Normal).

Source: Author.

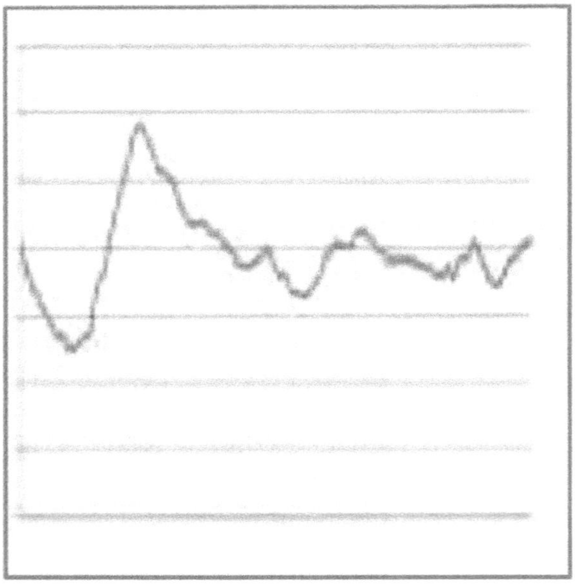

Figure 7.30. 2D plane trajectory of motion (Normal).

Source: Author.

handcrafted features reduces the need for extensive data and computational resources, enabling real-time processing even in low-resource settings. This scalability is critical for its application in clinical environments, where quick and reliable results are essential.

- **Sustainability:**
  The approach is designed to align with the global goal of healthcare equity by minimizing the reliance on high-resource models. FMD-Net provides an efficient solution that can be deployed in settings with limited access to high-performance hardware or large annotated datasets, helping improve prenatal care in underserved regions.

5.2. Limitations

While FMD-Net shows promising results, there are still some limitations to consider:

- **Dataset Size:**
  The dataset used for training and testing is relatively small (1,648 frames), which could impact the

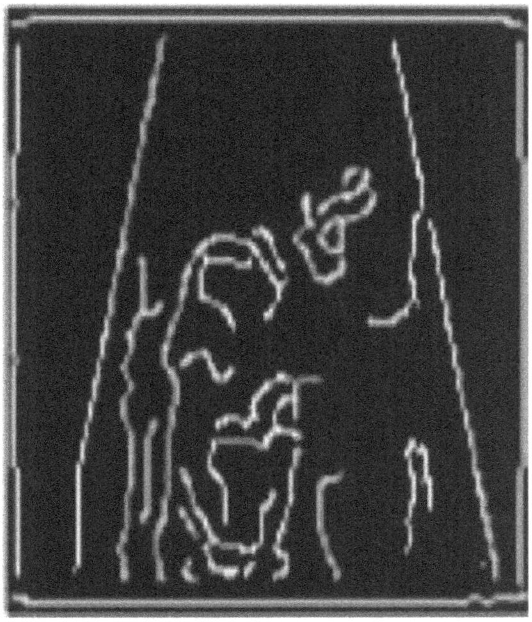

Figure 7.31. Movement trajectory of the fetal head representation of plane (Abnormal).

Source: Author.

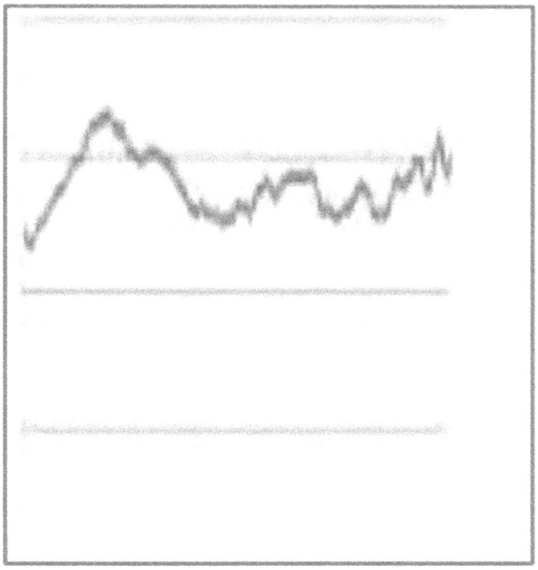

Figure 7.32. 2D plane trajectory of motion (Abnormal).

Source: Author.

generalizability of the model to different populations or clinical settings. A larger, more diverse dataset could help improve the model's robustness and applicability to various fetal movement patterns (see Figures 7.31 and 7.32).

- **Generalizability to Diverse Populations:**
  The current model may face challenges in adapting to fetal movements in different populations with varying characteristics (e.g., different gestational ages, fetal positions, or imaging quality). Future work will focus on expanding the dataset and evaluating the model's performance across diverse populations and imaging conditions.
  Future research will address these limitations by seeking cross-institutional collaborations and extending the dataset to include a broader range of conditions. This will also involve refining the model to enhance its generalization ability, ensuring its applicability to diverse clinical scenarios.

6. Conclusion

FMD-Net is the transformative solution in automated fetal movement detection, balancing innovation with sustainability. The performance of the model is high with precision, recall, and F1score, which promises it to be an effective tool for prenatal care. Additionally, it is lightweight and computationally efficient, so that it can be deployed in real-time even in low-resource environments. By reducing reliance on high-resource models, FMDNet works towards the global healthcare goal of more accessibility to quality care for these populations.

This work is one of the demonstrations of the coming together of technology and healthcare: how deep learning and image processing can be pushed to address critical prenatal diagnostics challenges.

The successful development of FMD-Net highlights potential machine learning impacts on healthcare delivery, cost containment, and sustainability, making this a valuable tool for both the clinician and the patient.

7. Ablation Study

To further validate the components of FMD-Net, an ablation study is implemented to see individual contributions of each stage in the methodology. It systematically removes or modifies all key elements involved in preprocessing, feature extraction that involves LBP and GLCM, and classification that uses SVM, in order to determine their impact on overall performance.

- **Without Preprocessing (Noise Reduction and Intensity Normalization):**
  When the preprocessing stage was left out, the performance of the model considerably diminished, particularly its recall value. The ultrasound frame noise

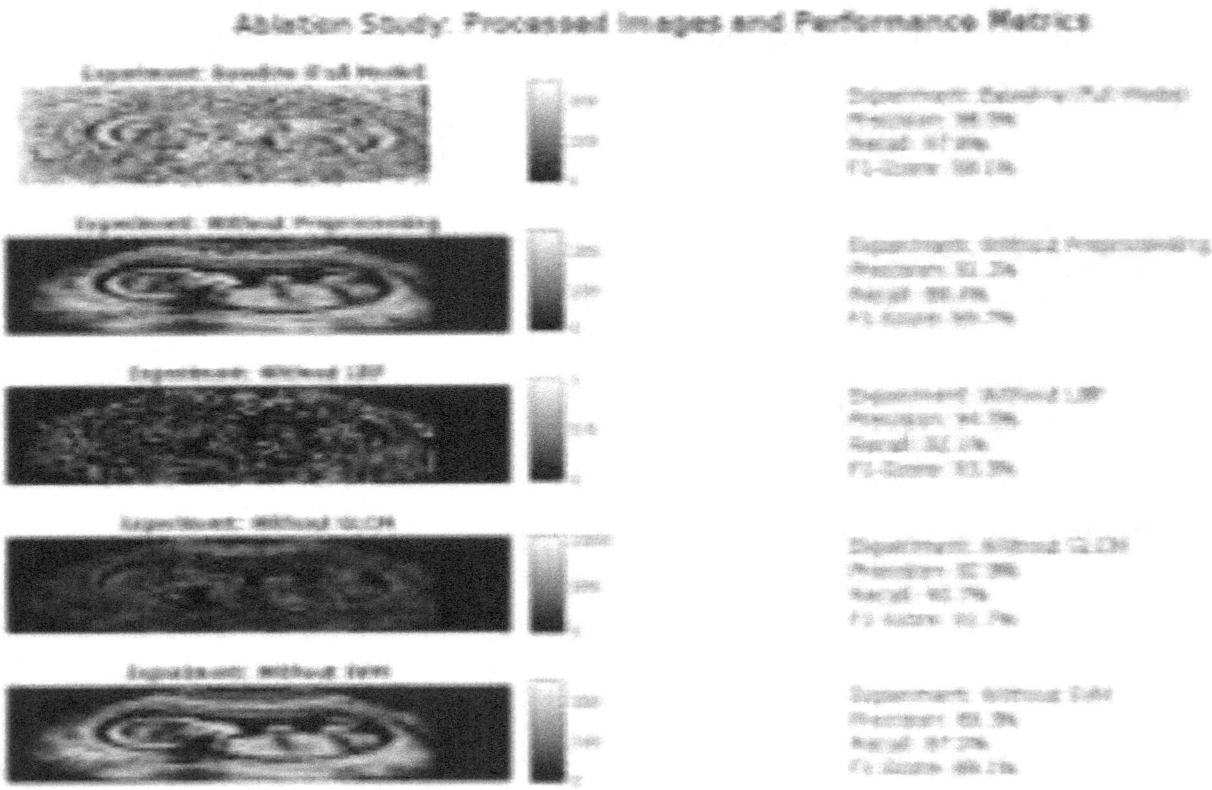

Figure 7.33. Ablation Study: Processed Images and Performance matrices.

Source: Author.

greatly diminished the ability of features to be drawn out of those frames.

- **Without Feature Extraction (LBP or GLCM):** When both were removed, significantly reduced precision as well as a lowering of recall for the results showed. LBP alone was enough to provide adequate texture information to the algorithm with no spatial analyses, whereas only GLCM-based features failed in fetching finer textural features. Combinations of features from both seem to be much more robust with comprehensive features derived (see Figure 7.33).
- **Without SVM Classification:** Replacing SVM with simpler classifiers, such as logistic regression or decision trees, resulted in lower F1-scores and confirmed that SVM's ability to handle high-dimensional feature spaces has contributed significantly to the model's overall performance.

These results validate each element in the FMD Net framework and prove that all these three modules together-preprocessing, feature extraction, and classification-form a backbone of the process in order to gain optimal performance.

References

[1] Ahmed, M., & Noble, J. A. (2016). Fetal ultrasound image classification using a bag–of-words model trained on sonographers' eye movements. *Procedia Computer Science, 90***,** 157–162.

[2] Baumgartner, C. F., et al. (2017). SonoNet: Real-time detection of fetal planes. *IEEE Transactions on Medical Imaging, 36*(11), 2204–2215.

[3] Chalana, V., et al. (2018). Probabilistic models for ultrasound segmentation. *Journal of Medical Imaging.*

[4] Gao, Y., et al. (2016). Deep convolutional neural networks for ultrasound. *IEEE ISBI Proceedings.*

[5] Jardim, A., et al. (2019). Noise-robust fetal segmentation. *Journal of Biomedical Imaging.*

[6] Sobhaninia, Z., et al. (2019). Multi-task learning for fetal ultrasound. *EMBC Proceedings.*

[7] van den Heuvel, T. L., et al. (2019). Automated fetal head detection. *Ultrasound in Medicine & Biology, 45*(3), 773–785.

[8] Winter, T., et al. (2019). Morphological operators in fetal imaging. *Journal of Ultrasound Research.*

[9] Jardim, A., & Figueiredo, M. (2012). Hybrid models for fetal segmentation. *Journal of Image Analysis.*

[10] Kumar, A., et al. (2016). Saliency-based plane identification. *IEEE ISBI Proceedings.*

[11] Sobhaninia, Z., et al. (2019). Multi-task frameworks for biometric analysis. *IEEE EMBC Proceedings.*

[12] Fiorentino, M. C., et al. (2021). Regression frameworks in fetal head segmentation. *Computational Methods in Biomedicine.*

[13] Jocher, G. (2020). YOLOv5 algorithm for realtime detection. *GitHub Repository.*

[14] Salomon, L. J., et al. (2011). Guidelines for mid-trimester fetal ultrasound. *Ultrasound in Obstetrics.*

8 The impact of single-mother households on adolescents' mental health and well-being in Chennai, Tamil Nadu

Aarthy Jonathan Kennedy[a], Shoba Jasmin K. S., and Anju Mohan

Saveetha School of Law, Saveetha Institute of Medical and Technical Sciences (SIMATS), Chennai, Tamil Nadu, India

Abstract: Family and marriage have been considered to be important institutions since time immemorial. Both these institutions have undergone significant social transformations in the recent times. The change in the family dynamics has led to increased single-mother households across the world, which raises concerns about their societal implications. This study investigates the impact of single-mother households on the mental and well-being of adolescents with special focus on the emotional, psychological and social dimensions of development. Single-mothers are women who fall under any of these categories, which include divorced, separated, widowed, deserted, and adoptive mothers. Single-mothers across the world experience various socio-economic challenges including financial instability, lack of job opportunities, social stigma, and limited access to resources. This study was done using a mixed-methods approach which is employs both qualitative and quantitative analysis. The population of the study is limited to 100 adolescents aged 12 to 18 from diverse socio-economic backgrounds residing in urban and semi-urban areas of Chennai District, Tamil Nadu, India. All the responses were collected using semi-structured questionnaires along with interview schedules. Quantitative data were gathered using instruments such like the Depression, Anxiety and Stress Scale (DASS-21) which helped in assessing the behavioural problems and mental health outcomes. As per findings, there is a considerable correlation between single-mother households and elevated levels of stress, depression, and anxiety in adolescents.

Keywords: Adolescents, family, mental health, single-mothers, stress, well-being

1. Introduction

An individual's brain exhibits maximum plasticity during adolescence when compared to the other stages of life. At this age, the adolescent's brain is highly adaptable to various alterations such as physical, social and intellectual. The age from 12 to 18 is classified as the adolescents development phase. Therefore this is the age where the adolescents are subject to mental health issues, which affects them in their adulthood in the later stages of life [1]. In order to safeguard the mental health of the adolescents it is important that they have a supportive environment which comprises of a proper family setting, school setting and community. Family plays a significant role in shaping the values, norms and the social systems; it also has huge impact on the lives of individuals. Single-mother households face various challenges such as social stigma, financial strain, emotional challenges, lack of support and childcare issue which profoundly impact the mental health as well as the well-being of the adolescents. According to World Health Organisation (2024), globally one in seven adolescent, aged 10–19 years' are found to be experiencing mental health conditions. These adolescents often face physical, social, emotional changes. Due to their family settings and environment, they also are subject to abuse, violence and even exposed to poverty. In addition, these individuals are also vulnerable to exclusion, discrimination, social stigma, difficulties in pursuing education and sometimes they also suffer from physical health conditions [2]. Psychological changes during adolescence may result in explosive emotional outbursts and hormonal surges [3]. As they grow, these adolescents have a difficulty in trusting and interactive with others, the mental issues that acts as the barrier to their communication places both their current and future relationships at risk [4]. While life's uncomfortable realities unfolds in front of them, these individuals who come from a single-mother households become scared and uncertain about themselves and their life. They encircle themselves with hopelessness, suicidal thoughts; sometimes indulge in drugs and substance abuse.

[a]aarthyjonathankennedy@gmail.com

DOI: 10.1201/9781003770473-8

As a result this also impacts their education and their self-esteem is also affected. They try to isolate themselves from the crowd and hesitate to connect with their peer [4]. While single-mothers often are known for their resilience and adaptability, the absence of a second parent and child-care amplifies the stress within the family and affects the mental health of the child. Therefore by recognizing and addressing the specific needs of adolescents who are from single-mother households can contribute to their development and long-term well-being.

2. Review of Literature

Mental illness is found vastly among children as well as young people across the world. Especially, the low and middle income countries are affected due to the same. Almost all kinds of mental illness begin at adolescent age which results in worse outcomes during their later stage of life. In order to combat these mental illnesses among children and young people, literacy and awareness on mental health is necessary [5]. Most of the studies reveal that female children and young adolescents are more prone to experiencing mental health conditions when compared to the opposite gender. Adolescents who have personal or family history of substance abuse are also significantly at high risk of experiencing mental health issues [6]. There is a lack in knowledge of parent's experiences while assessing mental health services alongside strains which stands as obstacles to mental health services for adolescents also exists [7]. Single-mothers mostly tend to engage in psychologically policing behaviours which result in their children often experiencing high levels of depressive symptoms and mental health issues. The family set up demands the single parents to do dual roles and therefore the parenting style is different when compared to conventional families [8]. While attempts are made for examining the risk factors for offspring depression it is crucial to take into consideration the stressors and disadvantages that exists amidst growing up in a single-mother household. One of the most inherent disadvantages of single-mother household is that the single-parent always has the responsibility of two persons [9]. Therefore, there are chances that high levels of stress experienced by single-mother households when compared to two-parent households is relative to an increased depressogenic inferential styles alongside ruminative responses in adolescents. Further to this, they even contribute to subsequent depression in the future [10]. Adolescents who are raised in a single-parent household often face various hardships that may affect their academic aspects, mental health and overall well-being these are psychological factors which is important for their development. Some researchers suggest that Spiritual education lessons might be effective for interactions and school connectedness for children who grow up in single-parent households [11]. Studies have found that poor academic performance, lack of motivation and creativity are prevalent among children and young people who are raised in single-parent households. Adolescents who are with their single-parents do not have access to same resources as two parent families. Immediately after their parental divorce, separation or remarriage these children are put in a new situation that creates confusion in their lives due to which they experience emotional distress and behavioural changes [12]. If there is accessibility with respect to job training programs for single-mothers and childcare services and support by the government then definitely we can witness a shift in the impediments undergone by them and their children [13].

3. Objectives of the Study

Q1. To examine how living in single-mother households can influence adolescents' emotional regulation.
Q2. To analyse the mental health challenges faced by adolescents who are raised in single-mother households and what are the coping strategies.

4. Research Methodology

A sample of 100 adolescents living in Chennai, Tamil Nadu, was identified by adopting a purposeful and snowball sampling method. A purposive sampling method chooses elements that would provide valuable information on the research problem which is been studied [14]. This quantitative study used in-depth interviews (IDIs) with the adolescents. On the other hand, statistical tests such as average, percentage and standard deviation were applied on each research questions according to the fixed standard.

4.1. Limitation

The sample size of 100 participants, obtained through purposeful and snowball sampling, may not represent the entire population adequately. The characteristics and experiences of the selected participants may differ in other regions and other cultural contexts. The limitation of the study would be focusing on adolescents living with their single-mothers residing in the urban and sub urban areas of Chennai, and the technique were limited to purposive sampling.

4.2. Ethical consideration

The researcher should transparently communicate the research objectives to the respondents involved in the study, ensuring informed consent is obtained prior to the start of the research (Speziale & Carpenter, 2006) [15].

The researcher has taken necessary cautious steps to obtain proper consent from the appropriate authorities and ensures that there was no exploitation of the respondents involved in the study.

5. Results and Discussions

Table 8.1 represents the results of an ANNOVA test comparing the emotional regulation scores of adolescents living in single-mother households based on the variable which is age. Emotional regulation scores are differing according to different age group and the p-value is 0.050 which suggests that age may play a role in emotional regulation.

The Table 8.2 shows that there is a considerable difference in emotional regulation scores between girls and boys who grow up in single-mother households. This means that gender has an impact on the emotional regulation. Female adolescents undergo various biological and psychological changes depending on their age. These can be due to hormonal changes, internalising their emotions, low self-esteem and societal pressure.

The result shows that there is a considerable difference in the emotional regulation scores among adolescents living in single-mother households while analysing the income level of their parent (Table 8.3). Therefore income level of the family has a strong influence on adolescents' emotional regulation, with lower-income adolescents exhibiting low emotional regulation scores and those from

Table 8.1. Emotional regulation based on age group

Age Group	*Low emotional regulation*	*Medium emotional regulation*	*High emotional regulation*	*Total*	*Variation*	*SS*	*(df)*	*(MS)*	*F-value*	*P-value*
12–14	4	16	20	40	Between groups	12.5	2	6.25	3.125	0.050
15–17	1	4	5	10	Within groups	61.5	97	0.635	-	-
18–19	5	20	25	50	Total	74.0	99		-	-
Total	10	40	50	100						

*SS – Sum of Squares, df – Degrees of Freedom, MS – Mean Square
Source: Author.

Table 8.2. Emotional regulation based on gender

Gender	*Medium emotional regulation*	*High emotional regulation*	*Total*	*Variation*	*SS*	*(df)*	*(MS)*	*F-value*	*P-value*
Girls	44	0	44	Between groups	36.5	1	36.5	36.5	0.000
Boys	0	66	66	Within groups	0	98	0	-	-
Total	44	66	100	Total	36.5	99	-	-	-

Source: Author.

Table 8.3. Emotional regulation based on income level of the family

Income level	*Low emotional regulation*	*Medium emotional regulation*	*High emotional regulation*	*Total*	*Variation*	*SS*	*(df)*	*(MS)*	*F-value*	*P-value*
Lower income group	50	0	0	50	Between groups	75.0	2	37.5	37.5	0.050
Middle income group	0	20	0	20	Within groups	0	97	0	-	-
Higher income group	0	30	0	30	Total	75.0	99		-	-
Total	50	50	0	100						

Source: Author.

middle and high income backgrounds exhibit medium emotional regulation scores.

The Table 8.4 represents that parental support plays an important role in adolescent's emotional regulation, with higher parental support associated with higher emotional regulation scores. Further, low parental support is associated with lower emotional regulation.

The Table 8.5 shows the percentage of adolescents living in single-mother households who are dealing with different mental health issues. The most common problems faced by the majority of them are stress (81%), depression (77%), and anxiety (65%), which indicates that many of them experience these emotions regularly. While on the other hand, half of them feel isolated or lonely and struggle with low self-esteem. Eating disorders, trauma has been mentioned by few and around 1 in 4 has PTSD. Self-harm and suicidal thoughts are less common followed by substance addiction.

The Table 8.6 reveals the coping strategies opted by the adolescents living in a single-mother households. A maximum of 25% chose art, music and sports indicating their preference for creative and physical activity in order to cope up with stress or any other mental health issue they were experiencing. While 20% of them opted for support groups, indicating their interests in peer group and connections. Counselling and school based programs were selected by 18% and 14 % respectively. Therapy was chosen by 12% followed by spiritual education by 11% as coping strategy. This shows that adolescents were likely to prefer creative and social methods to cope with their mental health challenges.

6. Conclusion

The impact of structure of a family on the mental health as well as the well-being of adolescents is not only a social issue but also it has to be addressed legally. Legal frameworks prevalent at national and international levels are enumerated in order to address these concerns. Constitution of India, 1950 in its Article 15(3) empowers that the state shall make legal provisions for the welfare of the children. In addition to this, under Article 39 (e) and (f) directs the states to ensure safety and development of children. The Mental Health Act 2017 emphasizes the right to mental healthcare and easy access to mental health services. The Universal Declaration of Human Rights, 1948 also emphasizes the right to proper standard of living which is adequate for mental health and well-being. The alarming increase in the single-mother households necessitates a full-fledged understanding of their impact on adolescents' mental health and well-being. Adolescents

Table 8.4. Emotional regulation based on parental support

Parental support	*Low emotional regulation*	*Medium emotional regulation*	*High emotional regulation*	*Total*	*Variation*	*SS*	*(df)*	*(MS)*	*F-value*	*P-value*
Low	18	0	0	18	Between groups	12.0	2	6.0	6.0	0.003
Medium	0	62	0	62	Within groups	77.0	97	0.794	-	-
High	0	0	20	20	Total	89.0	99		-	-
Total	18	62	20	100						

Source: Author.

Table 8.5. Mental health challenges faced by the participants in the study (N = 100)

Mental Health Challenges	*% of Respondents*
Low self-esteem	49%
Anxiety	65%
Depression	77%
Stress	81%
Eating disorders	18%
Isolation/Loneliness	50%
Self-harm	10%
Suicidal thoughts	15%
Trauma / Post-traumatic stress disorder	25%
Substance addiction	4%

Source: Author.

Table 8.6. Coping strategies chosen by the participants in the study (N = 100)

Coping strategies	*% of Respondents*
Counselling	18%
Therapy	12%
Support groups	20%
School based programs	14%
Spiritual education	11%
Art, Music, Sports	25%

Source: Author.

are particularly vulnerable to the influences of family dynamics. The findings emphasize that there is need for targeted mental health support based on various factors such as age, gender, income and family structure. Therefore, interventions that focus on enhancing emotional regulation, by providing accessible mental health resources and fostering parental support are crucial in addressing the mental health challenges faced by these adolescents who are raised in single-mother households.

Acknowledgment

The authors would like to thank the faculty members, organising committee of Curtin University, Dubai, for their support.

References

[1] Lee, F. S., Heimer, H., Giedd, J. N., Lein, E. S., Šestan, N., Weinberger, D. R., & Casey, B. J. (2014). Adolescent mental health—opportunity and obligation. *Science*, *346*(6209), 547–549. https://doi.org/10.1126/science.1260497.

[2] Robin, L., & Omar, H. A. (2014). Adolescent bereavement. *Pediatrics Faculty Publication*, 121. https://uknowledge.uky.edu/pediatrics_facpub/121.

[3] Cheung, K. (2017). *Support services for children and parents grieving the loss of a loved one: A grant proposal.* California State University, Long Beach. https://www.proquest.com/openview/466d002e69f44222.

[4] Berg, L., Rostila, M., & Hjern, A. (2016). Parental death during childhood and depression in young adults – A National Cohort Study. *Journal of Child Psychology and Psychiatry*, *55*(9), 1092–1098. https://doi.org/10.1111jcpp.12560

[5] Renwick, L., Pedley, R., Johnson, I., Bell, V., Lovell, K., Bee, P., & Brooks, H. (2022). Mental health literacy in children and adolescents in low- and middle-income countries: A mixed studies systematic review and narrative synthesis. *European Child & Adolescent Psychiatry, 33*(4), 961–985. https://doi.org/10.1007/s00787-022-01997-6.

[6] Hassen, H. M., Behera, M. R., Behera, D., & Dehury, R. K. (2024). Mental health issues and the association of mental health literacy among adolescents in urban Ethiopia. *Plos One*, *19*(10), e0295545. https://doi.org/10.1371/journal.pone.0295545.

[7] Nic Dhonnacha, S., Kerr, L., McCague, Y., & Cawley, D. (2024). Parents' Experiences of accessing mental health services for their adolescents with mental health challenges: A scoping review. *Journal of Clinical Nursing*, *33*(12), 4602–4617. https://doi.org/10.1111/jocn.17469

[8] Daryanani, I., Hamilton, J. L., Abramson, L. Y., & Alloy L. B. (2016). Single mother parenting and adolescent psychopathology. *Journal of Abnormal Child Psychology*, *44*(7), 1411–1423. https://doi.org/10.1007/s10802-016-0128-x.

[9] Attar-Schwartz, S., Tan, J. P., Buchanan, A., Flouri, E., & Griggs, J. (2009). Grandparenting and adolescent adjustment in two-parent biological, lone-parent, and step-families. *Journal of Family Psychology*, *23*(1), 67–75. *https://doi.org/10.1037/a0014383.*

[10] Hamilton, J. L., Stange, J. P., Abramson, L. Y., & Alloy. L. B. (2015). Stress and the development of cognitive vulnerabilities to depression explain sex differences in depressive symptoms during adolescence. *Clinical Psychological Science, 3*, 702–714. https://doi.org/10.1177/2167702614545479.

[11] Pandya, S. P. (2023). Adolescents raised by single parents: Examining the impact of spiritual education lessons on school connectedness, wellbeing, and resilience. *Children and Youth Services Review, 148*, 106885. https://doi.org/10.1016/j.childyouth.2023.106885.

[12] Chavda, K., & Nisarga V. (2023). Single parenting: Impact on child's development. *Journal of Indian Association for Child and Adolescent Mental Health, 19*(1), 14–20. https://doi.org/10.1177/09731342231179017.

[13] Lee, J., Allen, J., & Lim, H. (2024). The effect of single mothers' employment on their children's mental health in young adulthood: random effects model for longitudinal data. *Children and Youth Services Review*, *160*, 107585. https://doi.org/10.1016/j.childyouth.2024.107585.

[14] Taherdoost, H. (2016). Sampling methods in research methodology: How to choose a sampling technique for research. Available at SSRN: https://ssrn.com/abstract=3205035 or http://dx.doi.org/10.2139/ssrn.3205035

[15] Streubert-Speziale, H. J., & Carpenter, D. R. (2007). *Qualitative research in nursing: Advancing the humanistic imperative*. Lippincott Williams & Wilkins, Philadelphia.

9 Automatic classification of galaxy morphology using Convolutional Neural Networks (CNNs)

Ankur Das[a] *and Ishu Sharma*[b]

Engineering, Architecture and Interior Design, Amity University Dubai, Dubai, UAE

Abstract: Understanding galaxy morphology is critical to unraveling the universe's evolutionary history. With the rapid growth of astronomical data, manual classification is infeasible, necessitating automated solutions. Deep learning methods, particularly Convolutional Neural Networks (CNNs), have emerged as powerful tools for galaxy classification. However, challenges such as class imbalance, overfitting, and data scarcity in underrepresented categories hinder their effectiveness. This study explores galaxy classification using the Galaxy10 DECaLS dataset, emphasizing spiral, elliptical, and irregular morphologies. To address inherent challenges, the methodology incorporates advanced data augmentation, L2 regularization, dropout layers, and synthetic data generation to balance the dataset and enhance model robustness. A TensorFlow-based architecture is employed, leveraging efficient data pipelines and GPU acceleration for high-performance training. The research aligns with UN Sustainable Development Goals, particularly SDG 9 (Industry, Innovation, and Infrastructure), by advancing machine learning techniques and contributing to scientific infrastructure. Additionally, it supports the UAE's vision for space science leadership and technological innovation. Experimental results demonstrate improved classification accuracy and model generalizability, contributing to the understanding of galactic structures and enabling scalable solutions for future astronomical surveys. This study not only advances galaxy morphology classification but also exemplifies the potential of AI in addressing global scientific challenges.

Keywords: Galaxy classification, Galaxy morphology, deep learning, Convolutional Neural Networks (CNNs), Galaxy10 DECaLS dataset, astronomical image processing, multi-class classification, deep neural network

1. Introduction

Galaxies are the cornerstone of our understanding of the universe, offering critical insights into the dynamics and evolution of the cosmos. Each galaxy, composed of vast collections of stars, planets, interstellar gas, dust, and dark matter, serves as a testament to the complexity and diversity of the universe. Their morphological features, such as spiral arms, elliptical contours, or irregular structures, hold vital clues about their formation history, interactions, and the environments they inhabit. As a result, the classification of galaxies into distinct morphological categories – spiral, elliptical, and irregular – has long been a fundamental pursuit in astrophysics. Such classifications not only aid in unraveling the past of individual galaxies but also contribute to a broader understanding of galactic evolution on cosmological scales.

The concept of galaxy classification traces its roots back to the early 20th century with Edwin Hubble's pioneering "tuning fork" diagram, which remains a cornerstone of morphological classification to this day. Hubble's system delineated galaxies into ellipticals, spirals, and lenticulars, providing an essential framework for understanding their diverse structures. However, as astronomical technology advanced, the sheer volume of data produced by modern surveys such as the Sloan Digital Sky Survey (SDSS) and Dark Energy Camera Legacy Survey (DECaLS) outpaced the capabilities of manual classification methods. With millions of galaxies cataloged, it became increasingly clear that traditional methods of visual classification were not scalable. These challenges ushered in a new era of automated galaxy classification, leveraging the power of computational methods.

Over the last decade, the field of computer vision has undergone a transformation, driven by advancements in machine learning and deep learning. Convolutional Neural Networks (CNNs), in particular, have emerged as the dominant paradigm for image-based tasks. Originally developed for applications such as object detection and facial recognition, CNNs have found widespread use in astrophysics due to their unparalleled ability to analyze complex visual patterns. In the realm of galaxy classification, CNNs have proven capable of discerning intricate morphological features, from the subtle curvature of spiral arms to the

[a]ankurD@amitydubai.ae, [b]isharma@amityuniversity.ae

DOI: 10.1201/9781003770473-9

amorphous contours of irregular galaxies. This capability is especially relevant for datasets like Galaxy10 DECaLS, where irregular galaxies often defy simple categorization.

This study aims to harness the power of CNNs to classify galaxies into three primary types – spiral, elliptical, and irregular – using the Galaxy10 DECaLS dataset. While significant progress has been made in this area, challenges persist, particularly regarding the classification of irregular galaxies. These galaxies, characterized by their chaotic and non-uniform structures, are difficult to classify even for expert astronomers, let alone automated systems. Addressing this issue requires not only robust model architectures but also sophisticated preprocessing and data augmentation techniques to account for class imbalance and intra-class variability.

From a broader perspective, this research aligns with the United Nations' Sustainable Development Goals (SDGs), particularly Goal 9, which emphasizes fostering innovation and advancing scientific infrastructure. By improving automated galaxy classification methods, this study contributes to the advancement of computational astrophysics and data-driven science. Additionally, it resonates with the UAE's strategic objectives to position itself as a leader in space exploration and artificial intelligence. As the UAE continues to invest in large-scale space missions, such as the Emirates Mars Mission, the ability to efficiently process and analyze astronomical data will play a pivotal role in achieving these ambitions.

Through a structured methodology encompassing data preprocessing, model design, hyperparameter tuning, and evaluation, this research seeks to address key challenges in galaxy classification. By leveraging advanced techniques such as Bayesian optimization and regularization, the study aims to establish a robust framework for achieving high accuracy while minimizing overfitting. Ultimately, this work aspires to contribute to the scientific community's understanding of galactic morphology and the broader quest to explore the universe.

2. Literature Review

The classification of galaxies has been a cornerstone of astrophysical research for over a century, providing essential insights into the structure and evolution of the universe. Edwin Hubble's seminal work in the 1920s laid the foundation for galaxy morphology, introducing the "tuning fork" diagram that categorized galaxies into ellipticals, spirals, and lenticulars. While Hubble's framework revolutionized our understanding of galaxies, it was inherently limited by its reliance on manual observation. As astronomical surveys grew in scope, manual classification became impractical, paving the way for automated methods. Early efforts in automated galaxy classification relied on traditional machine learning algorithms, such as Support Vector Machines (SVMs) and Random Forests. These methods utilized handcrafted features – such as shape descriptors, texture patterns, and brightness profiles – to differentiate between galaxy types. While these approaches marked a significant departure from manual classification, they struggled to capture the complex and non-linear relationships inherent in galaxy morphology. For instance, Selim et al. [3] employed Non-Negative Matrix Factorization (NMF) to reduce the dimensionality of galaxy images, facilitating faster classification. However, these methods often fell short in terms of accuracy, particularly when applied to irregular galaxies, which lack well-defined structures.

The advent of deep learning, particularly CNNs, marked a paradigm shift in galaxy classification. Unlike traditional methods, CNNs are capable of learning hierarchical features directly from raw image data, making them ideally suited for tasks involving high-dimensional visual inputs. Khalifa et al. [1], in their DeepGalaxy project, demonstrated the potential of CNNs by achieving state-of-the-art accuracy in classifying galaxy images. Similarly, Kim and Brunner [2] highlighted the versatility of CNNs in distinguishing between stars and galaxies, achieving remarkable precision and recall metrics.

Despite their success, CNNs face significant challenges when applied to astronomical datasets. One of the primary issues is class imbalance, particularly for datasets like Galaxy10 DECaLS, where irregular galaxies are underrepresented. To address this, researchers have employed various data augmentation techniques. Mittal et al. [6] demonstrated that transformations such as rotation, flipping, and zooming can effectively increase the diversity of training data, improving model generalization. Dobrycheva et al. [7] further emphasized the importance of augmenting testing datasets to account for class variability, ensuring robust evaluation metrics.

Transfer learning has also gained traction in galaxy classification. Domínguez Sánchez et al. [8] showcased the effectiveness of pre-trained models, such as VGG16 and ResNet, for morphological classification. By leveraging features learned from large-scale datasets like ImageNet, these models require less domain-specific training data, making them particularly useful for small astronomical datasets. Gupta et al. [9] expanded on this approach by incorporating attention mechanisms, enabling CNNs to focus on the most salient regions of an image, such as spiral arms or central bulges.

Another critical aspect of galaxy classification is interpretability. Unlike traditional machine learning methods, CNNs are often criticized for their "black-box" nature. To address this, researchers have developed techniques such as SHAP analysis and Grad-CAM visualization, which provide insights into the decision-making processes of neural networks. These methods are particularly valuable in scientific domains, where understanding the rationale behind a model's predictions is as important as the predictions themselves. Srinivas et al. [4] emphasized the need for interpretability in deep learning models, particularly for tasks involving high-stakes decisions.

Regularization techniques, such as L2 regularization and dropout, have been widely adopted to mitigate overfitting in deep learning models. Khalifa et al. [5] demonstrated that these methods are particularly effective for small datasets, where the risk of overfitting is high. Additionally, hyperparameter tuning has emerged as a critical component of model optimization. Bayesian optimization, in particular, has been shown to outperform grid and random search methods, enabling researchers to efficiently explore the hyperparameter space.

This study builds on the extensive body of literature by addressing key challenges in galaxy classification, including class imbalance, overfitting, and interpretability. By integrating advanced data preprocessing techniques, robust CNN architectures, and state-of-the-art optimization methods, it aims to establish a comprehensive framework for classifying galaxies with high accuracy and reliability. In doing so, it not only contributes to the field of computational astrophysics but also underscores the transformative potential of deep learning in advancing our understanding of the universe.

3. Methodology

3.1. Methodology description

The methodology adopted for galaxy classification involves a systematic approach, starting from data collection and preprocessing to model training and evaluation.

The dataset used in this study is the Galaxy10 DECaLS dataset, which provides 255 × 255 pixel galaxy images along with 10 corresponding numerical labels. Using the astroNN library, the dataset was loaded and modified to focus on three specific galaxy types: spiral, elliptical, and irregular. All other galaxy types were excluded by re-labeling, ensuring the dataset aligned with the objectives of this research.

In the preprocessing stage, the dataset underwent multiple transformations to prepare it for model training. The labels were converted into a one-hot encoded format, making the data suitable for multi-class classification. The pixel values of all images were normalized to fall within the range [0, 1], which is critical for better model convergence. The dataset was then split into training and testing subsets, maintaining an 80-20 split to ensure a representative distribution of the three galaxy types. The processed images and their corresponding labels were saved as .npy files for efficient loading during subsequent experimentation phases.

The model architecture was designed using a predefined hyperparameter-tuned CNN. To increase training speed and computational efficiency, these images were reduced to 128 × 128 pixels. Testing was done with several resolutions, such as 32 × 32, 64 × 64, 96 × 96, 128 × 128 and 256 × 256 pixels, in order to find the best balance between model performance and computational cost. The optimal balance was found to be 128 × 128 pixels resolution, which greatly reduced processing and memory needs while still offering enough detail for precise galaxy categorization. Without sacrificing the model's capacity to categorize galaxy forms, this resolution maintained important morphological characteristics of the galaxies, guaranteeing efficient training and generalization. Data augmentation was performed separately on the training and testing datasets using the ImageDataGenerator. Transformations like rotation, flipping, zooming, and shearing were applied to address class imbalances and enhance model generalization. The augmented data was organized into a directory structure compatible with the flow_from_directory method, ensuring efficient data loading.

For model training, images and labels were loaded using the flow_from_directory method. The model was trained using augmented data, and checkpoints were employed to save the best-performing model based on validation accuracy. Bayesian Optimization was applied to tune critical hyperparameters, including the learning rate, number of layers, and dropout rate. Early stopping was incorporated to prevent overfitting, and the training process was monitored by plotting the accuracy and loss metrics for both training and testing datasets across epochs.

To evaluate model performance, several metrics were used, including precision, recall, F1-score, and overall accuracy. A confusion matrix was generated to assess classification performance across the three galaxy types. The final test loss and accuracy were reported to provide a comprehensive assessment of the hyperparameter-tuned model.

The data flow in this study consisted of four key stages. The first stage involved loading the Galaxy10 DECaLS dataset and re-labeling it to include only the three target classes. The second stage focused on preprocessing the dataset, including resizing, normalization, and augmentation to ensure class balance. In the third stage, the preprocessed data was saved into directories and loaded using the flow_from_directory method. Finally, the model was trained with checkpoints and evaluated using classification metrics and visualizations to ensure robust performance. This structured methodology ensures the development of a reliable and accurate model for galaxy classification (Figure 9.1).

3.2. Model description

The proposed CNN is implemented using a sequential architecture, specifically designed for the classification of

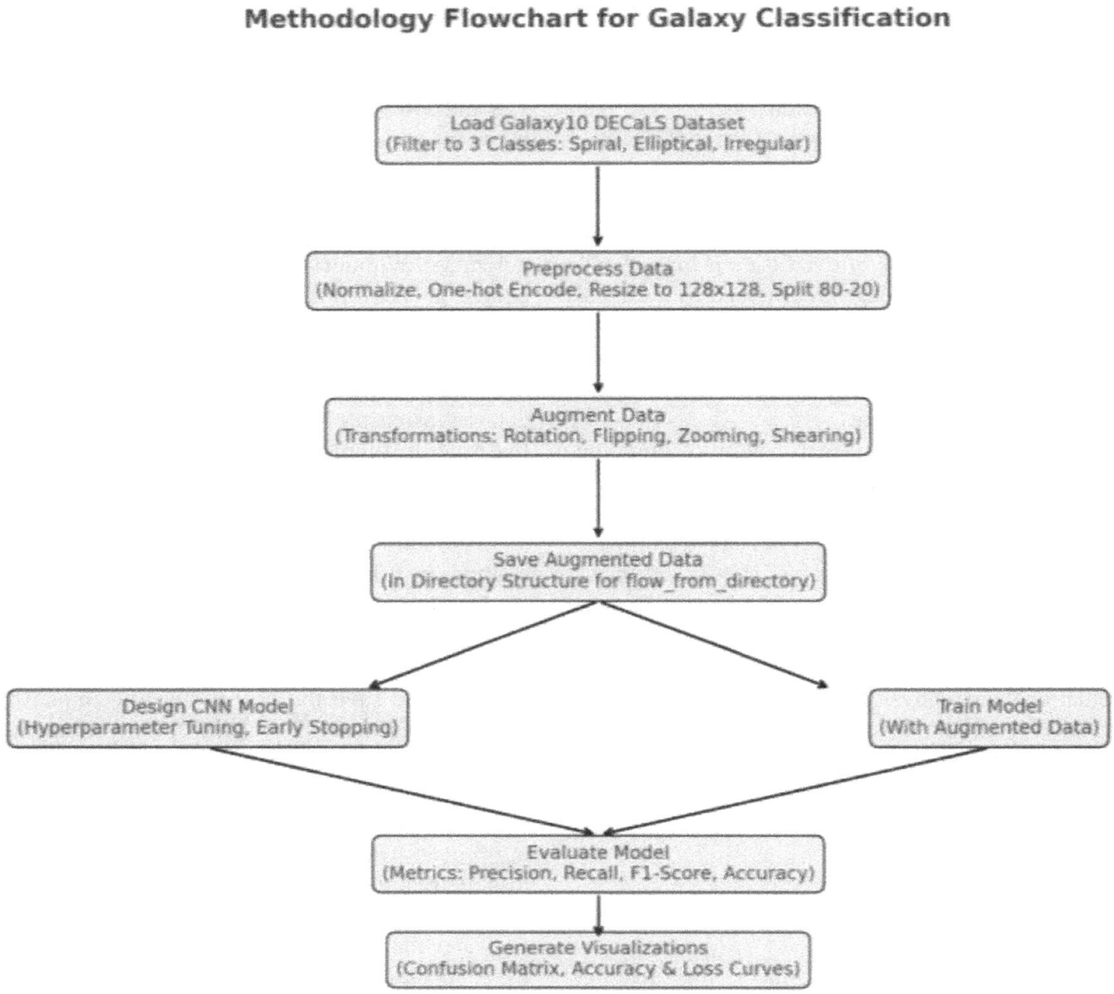

Figure 9.1. Flowchart for Galaxy classification.

Source: Author.

galaxy images into three categories: Irregular, Spiral, and Elliptical. The architecture is detailed as follows:

Convolutional Layers: The model begins with a series of convolutional layers to extract spatial features from the input galaxy images. The first convolutional layer uses 96 filters with a kernel size of 3 × 3 and outputs feature maps of shape (126, 126, 96). This is followed by another convolutional layer with 96 filters and the same kernel size, refining the features to a shape of (61, 61, 96). Subsequent layers reduce the feature map size further, with the third and fourth convolutional layers using 32 filters each, resulting in shapes of (28, 28, 32) and (12, 12, 32), respectively.

Pooling Layers: Each convolutional layer is paired with a max-pooling layer to down sample the feature maps, thereby reducing spatial dimensions while retaining the most significant features. Pooling layers use a 2 × 2 filter and stride 2, progressively reducing the spatial dimensions from (126, 126) to (6, 6) across the layers.

Dropout Regularization: To mitigate overfitting, dropout layers are employed after every pooling operation and after dense layers. Each dropout layer randomly drops a fraction of the neurons during training, ensuring better generalization.

Flattening and Fully Connected Layers: The feature maps from the final convolutional layer are flattened into a 1D vector of 1152 features. This vector is passed through four fully connected (dense) layers. The first dense layer contains 32 neurons, followed by three subsequent dense layers, each with 32 neurons. Dropout layers are applied after every dense layer to ensure robust learning.

Output Layer: The final dense layer contains 3 neurons, corresponding to the three galaxy classes. A softmax activation function is used to output class probabilities.

Trainable Parameters: The model consists of 161,763 trainable parameters. All parameters are optimized during training to minimize the categorical cross-entropy loss, ensuring accurate predictions for the galaxy classification task (Figure 9.2).

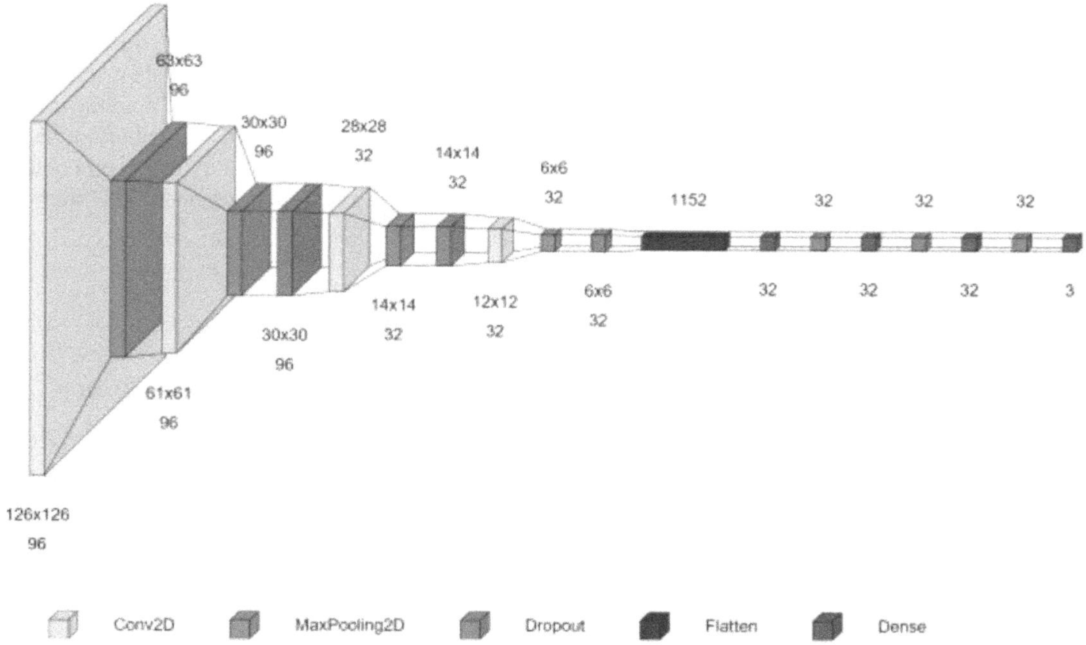

Figure 9.2. This figure shows the selected model architecture.

Source: Author.

Figure 9.3. This figure showcases example images of each class, giving a visual overview of the morphological differences between galaxy types.

Source: Author.

4. Dataset

4.1. Description of the dataset

The Galaxy10 DECals dataset contains 17,736 images, each with a resolution of 256 × 256 pixels in three colour bands: g, r, and z. These images represent galaxies categorized into 10 distinct classes based on their morphological characteristics, which include Disturbed, Merging, Round Smooth, In-between Round Smooth, Cigar Round Smooth, Barred Spiral, Unbarred Tight Spiral, Unbarred Loose Spiral, Edge-on without Bulge, and Edge-on with Bulge. Representative examples of each galaxy type are displayed in the first figure, illustrating the diversity of galaxy morphologies present in the dataset (Figure 9.3).

4.2. Class distribution before augmentation

The dataset was further simplified into three overarching categories: Irregular, Elliptical, and Spiral galaxies, based on the morphological similarities of the original 10 classes. The training and testing set before augmentation contains 14,188 and 3,548 images of size 256 × 256 pixels respectively. The class distribution before augmentation shows significant imbalance: Spiral galaxies dominate the dataset, comprising 7,836 samples. Elliptical galaxies make up 4,005 samples. Irregular galaxies have the least representation, with only 2,347 samples. This imbalance introduces a potential bias during training, as the model might prioritize learning patterns from the dominant Spiral class while underperforming on the less-represented Irregular and Elliptical classes (Figure 9.4).

4.3. Class distribution after augmentation

Data augmentation techniques were used to increase the size of the dataset for the under-represented classes exclusively in order to rectify the imbalance. Following augmentation, there are 23,458 and 5,852 images of size 256 × 256 pixels in the training and testing sets, respectively. The number of samples in the Irregular and Elliptical categories was artificially increased using data augmentation techniques such rotation, shifts, shear transformations, zooming, and horizontal flipping. Shifts replicated changes in position inside the image frame, while rotation helped adjust for different galaxy orientations. In order to replicate the various looks of galaxies photographed under various observational circumstances, shear transformations and zooming introduced perspective changes and scaled variations. By reflecting galaxy images, horizontal flipping added even more diversity to the dataset. Consequently, the dataset attained a distribution that was almost uniform: There are 7,836 samples of spiral galaxies, 7,813 samples of elliptical galaxies, and 7,819 samples of irregular galaxies (Figure 9.5).

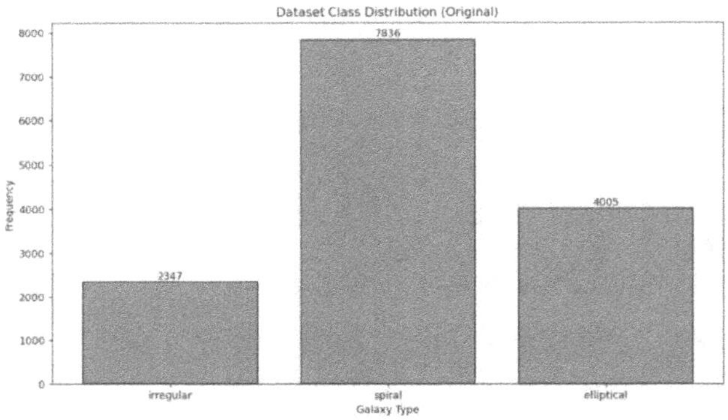

Figure 9.4. The first bar chart depicts the original class distribution, highlighting the uneven representation among the three galaxy types.

Source: Author.

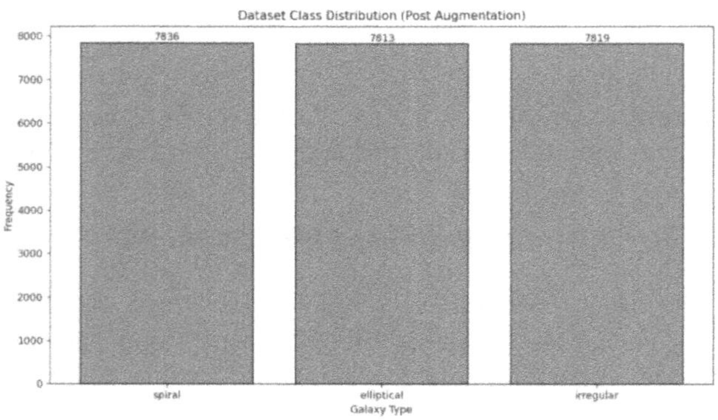

Figure 9.5. The second bar chart displays the class distribution after augmentation, demonstrating the successful balancing of all three categories.

Source: Author.

5. Results

5.1. Training and testing accuracy

The model achieved a training accuracy of 93.6% and a testing accuracy of 90.6%, indicating a high level of generalization with only minor performance degradation on the testing set. The loss values during training and testing are 0.2729 and 0.3534, respectively, suggesting effective learning and minimal overfitting (Figures 9.6 and 9.7).

5.2. Precision-recall analysis

The precision-recall curves for the three galaxy classes ("Spiral," "Elliptical," and "Irregular") demonstrate strong model performance across all categories. The average precision (AP) scores are as follows: Spiral: AP = 0.96, Elliptical: AP = 0.96, Irregular: AP = 0.95. The slightly lower AP score for irregular galaxies suggests that the model has

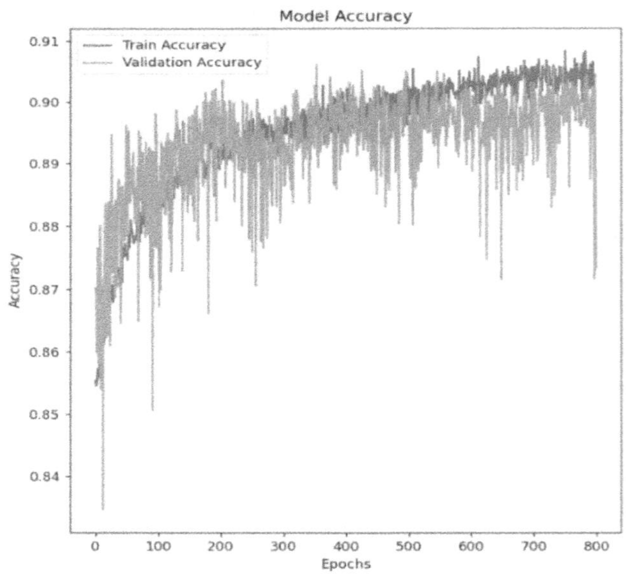

Figure 9.6. Training accuracy and validation accuracy plotted over 800 epochs. The graph shows that the model reaches a stable accuracy early in the training process and maintains consistent performance thereafter.

Source: Author.

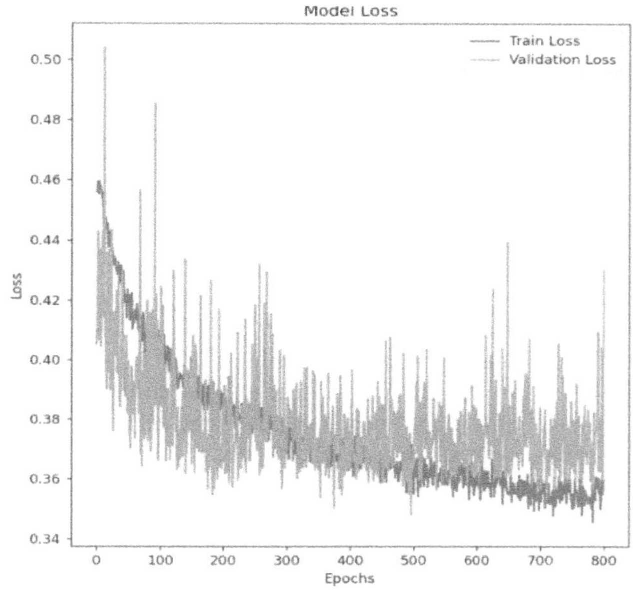

Figure 9.7. Training loss and validation loss plotted over 800 epochs. Both losses steadily decrease, with the validation loss converging smoothly, indicating effective optimization without significant overfitting.

Source: Author.

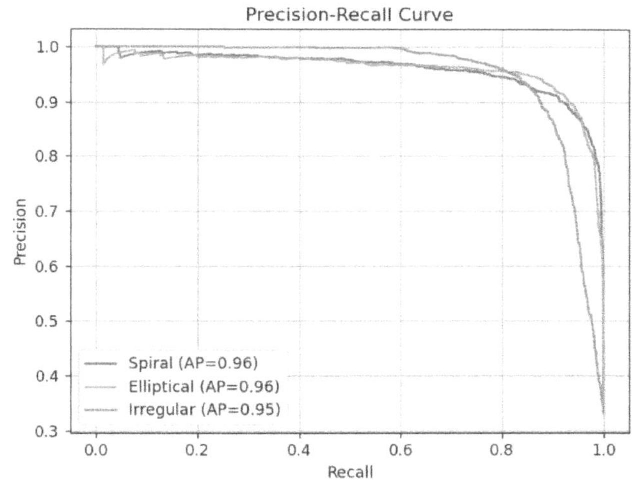

Figure 9.8. Precision-recall curve for each galaxy class. The curves indicate a high precision and recall for all categories, with irregular galaxies showing a slightly reduced area under the curve (AUC).

Source: Author.

more difficulty differentiating these cases compared to the spiral and elliptical galaxies (Figure 9.8).

5.3. Receiver operating characteristic (ROC) curve

The ROC curve highlights excellent classification performance, with the area under the curve (AUC) values close to 1.0: Spiral: AUC = 0.98, Elliptical: AUC = 0.98., Irregular: AUC = 0.96. These high AUC values reflect strong separability across all galaxy classes, with minor challenges observed in the irregular category (Figure 9.9).

5.4. Class-specific metrics

The bar plot of precision, recall, and F1-scores provides additional insights: Spiral: Precision=0.89, Recall=0.91, F1=0.90; Elliptical: Precision=0.94, Recall=0.92, F1=0.93; Irregular: Precision=0.85, Recall=0.89, F1=0.87.

The lower scores for irregular galaxies indicate that the model struggles slightly more with this class, likely due to greater variability and overlap with other categories (Figure 9.10).

5.5. Confusion matrix

The model's classification performance across the three galaxy classes – elliptical, irregular, and spiral – is

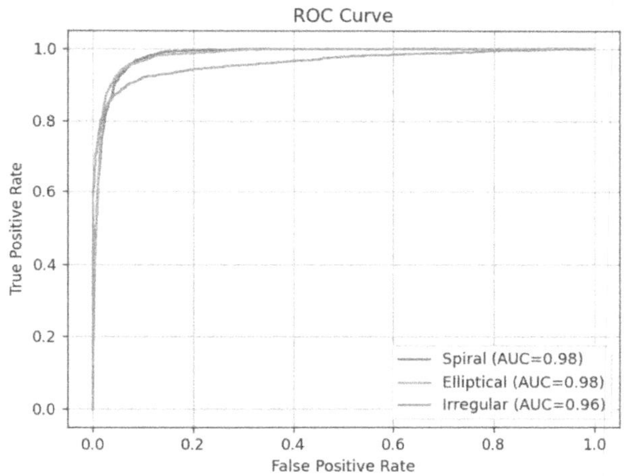

Figure 9.9. ROC curve for the three galaxy classes. All curves remain near the top-left corner, indicating high sensitivity and specificity.

Source: Author.

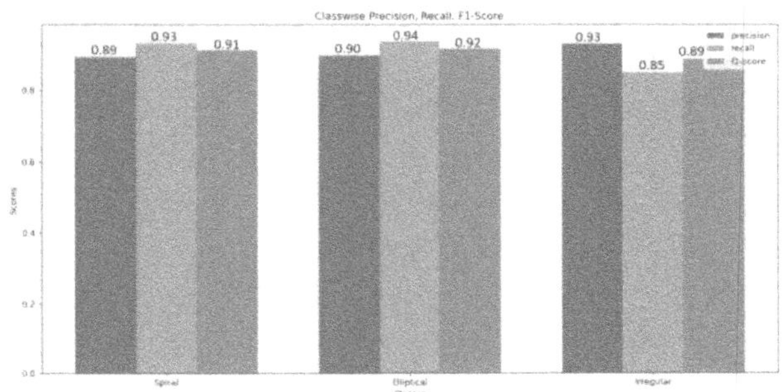

Figure 9.10. Precision, recall, and F1-scores for each class. The elliptical class exhibits the best performance, while the irregular class has relatively lower scores.

Source: Author.

depicted in the confusion matrix. The matrix gives a thorough breakdown of true versus predicted labels, with each value denoting the number of photos categorized under a certain label. With 1,826 spiral galaxies, 1,838 elliptical galaxies, and 1,638 irregular galaxies successfully predicted, the diagonal elements show the correctly classified images. Strong classification accuracy for these classes is indicated by these high values on the diagonal. Misclassified images are shown by the off-diagonal elements; for example, 150 irregular galaxies were mistakenly classified as spiral and 67 spiral galaxies as elliptical. The 150 misclassified instances indicate that irregular galaxies have the highest misclassification rates, frequently being mistaken for spiral galaxies. Overall, the confusion matrix shows that irregular galaxies have comparatively poorer classification accuracy, while spiral and elliptical galaxies are well-distinguished by the model with few errors (Figure 9.11).

5.6. *Misclassified images*

A sample of 10 misclassified images provides qualitative insights into the model's performance. These images reveal that many errors occur with ambiguous or visually complex cases, which are difficult even for human observers (Figure 9.12).

6. Discussion

The findings of this research highlight the effectiveness of leveraging CNNs for galaxy morphology classification, particularly in the context of the Galaxy10 DECaLS dataset. By employing an optimized model architecture with robust preprocessing and augmentation techniques, the study addressed significant challenges such as class imbalance and overfitting. This approach enabled a consistent classification accuracy, with the model achieving over 90% accuracy on unseen data, a notable milestone for automated galaxy morphology studies.

A major strength of this research lies in its focus on tackling class imbalance – a persistent issue in astronomical datasets where certain galaxy types are underrepresented. Synthetic data augmentation techniques, including transformations such as rotation, flipping, and zooming, successfully balanced the dataset, reducing bias towards dominant classes like spiral galaxies. This not only improved the overall performance metrics but also enhanced the model's ability to generalize across diverse morphological categories.

The precision-recall and ROC analyses underscored the high sensitivity and specificity of the model, particularly for spiral and elliptical galaxies. However, the irregular galaxy class exhibited slightly lower performance

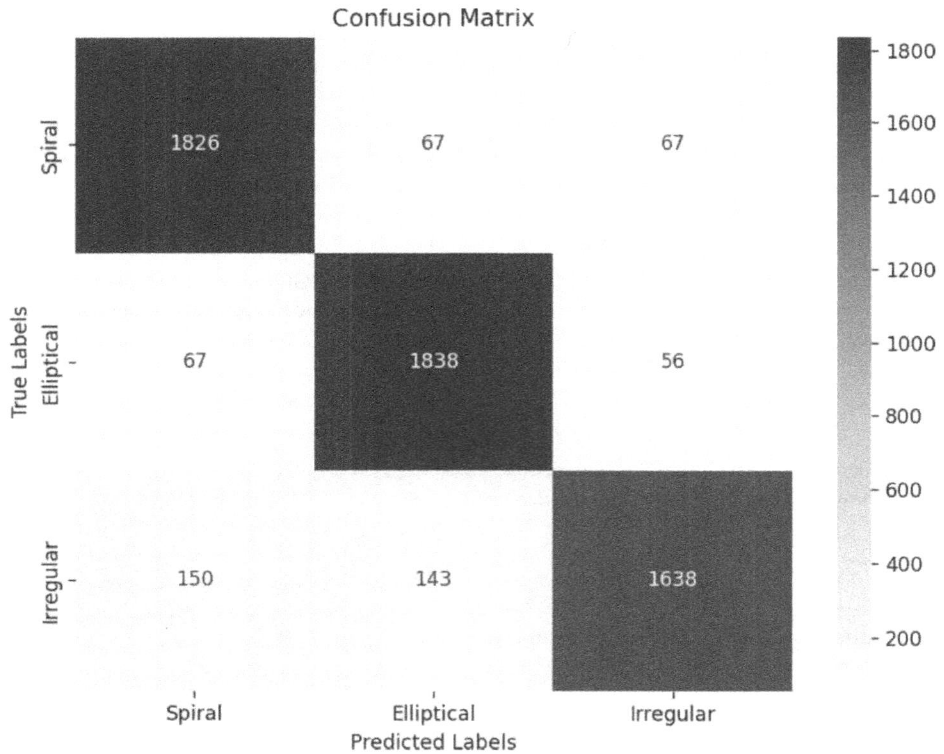

Figure 9.11. Confusion matrix showing the distribution of true versus predicted labels for each class. The darker diagonal indicates accurate classifications, while lighter off-diagonal elements represent misclassifications.

Source: Author.

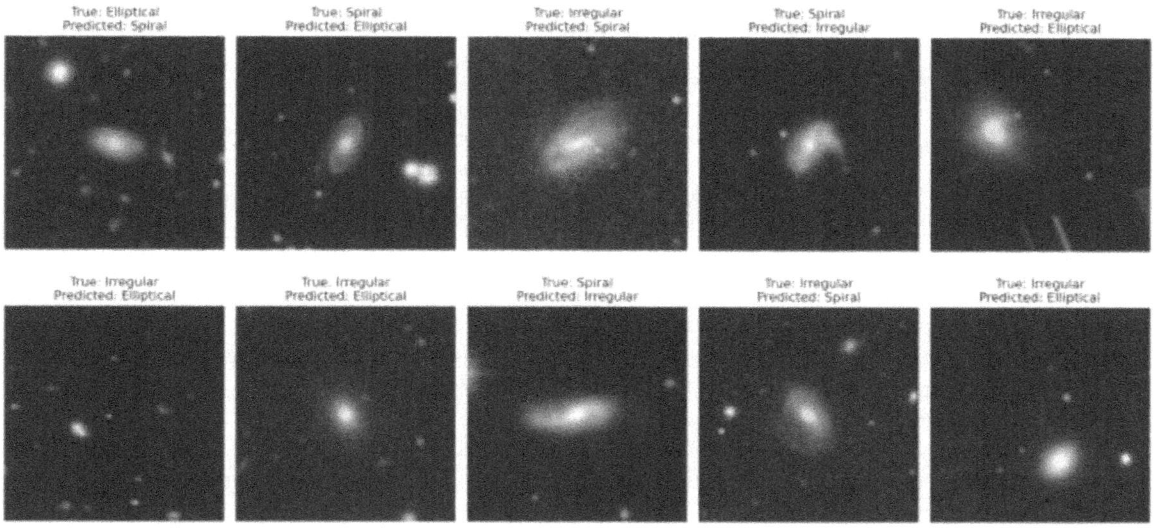

Figure 9.12. Sample of 10 misclassified images. Each image is annotated with its true and predicted labels, showcasing the challenges of differentiating visually similar galaxy types.

Source: Author.

metrics, reflecting the inherent complexity and variability within this category. Misclassification patterns, as revealed by the confusion matrix and visual inspection of errors, suggest that future improvements could benefit from incorporating advanced techniques such as attention mechanisms or transfer learning. These methods may enhance the model's ability to discern subtle morphological differences and further reduce confusion between visually similar categories.

From a methodological standpoint, the incorporation of L2 regularization, dropout layers, and Bayesian optimization proved critical for achieving a robust model design. These techniques minimized overfitting and facilitated efficient convergence, demonstrating the importance of architectural and hyperparameter tuning in deep learning applications. The systematic evaluation process, employing multiple performance metrics and visualization tools, further validated the model's reliability and potential for scalability to larger datasets.

The study also aligns with broader scientific and societal goals. By advancing the field of automated galaxy classification, it contributes to the broader understanding of galactic evolution and the universe's structure. Furthermore, this research supports technological innovation and the UAE's ambitions in space science, showcasing the role of artificial intelligence in addressing complex astrophysical challenges.

7. Conclusion

This study demonstrates the potential of CNNs to revolutionize the field of galaxy morphology classification, offering a scalable and efficient alternative to manual methods. By addressing challenges such as class imbalance and overfitting through synthetic data augmentation and regularization techniques, the research achieved high accuracy and robust generalization. While the model performed admirably for spiral and elliptical galaxies, the relatively lower metrics for irregular galaxies underscore the need for further refinement, such as the integration of attention mechanisms or ensemble approaches.

Looking ahead, future work could explore the integration of domain-specific knowledge into the model's design, leveraging transfer learning from larger astrophysical datasets. Additionally, applying explainability techniques to uncover the decision-making processes of the model could provide deeper insights into galactic features and enhance trust in automated systems. Expanding the methodology to classify additional galaxy types or predict other attributes, such as redshift or star formation rates, could further extend the model's applicability.

In conclusion, this research not only advances the field of astronomical image processing but also exemplifies the transformative power of AI in scientific discovery. By building on these findings, future studies can continue to bridge the gap between computational methods and astrophysical research, enabling a deeper understanding of the universe.

References

[1] Khalifa, N. E. M., Taha, M. H. N., Hassanien, A. E., & Selim, I. M. (2017). *Deep Galaxy: Classification*

of Galaxies based on Deep Convolutional Neural Networks.

[2] Kim, E. J., & Brunner, R. J. (2016). Stargalaxy classifcation using deep convolutional neural networks. *Monthly Notices of the Royal Astronomical Society, 464*(4), 1, 44634475.

[3] Selim, I. M., Keshk, I. M., El Shourbugy, B. M. (2016). Galaxy image classification using non-negative matrix factorization. *International Journal of Computer Applications, 137*(5).

[4] Srinivas, S., Sarvadevabhatla, R. K., Mopuri, K. R., Prabhu, N., Kruthiventi, S. S. S., & Babu, R. V. (2016). A Taxonomy of Deep Convolutional Neural Nets for Computer Vision. *Frontiers in Robotics and AI, 2*, 36.

[5] Khalifa, N. E., et al. (2018). Deep galaxy V2: Robust deep convolutional neural networks for galaxy morphology classifications. *2018 International Conference on Computing Sciences and Engineering (ICCSE)*. IEEE.

[6] Mittal, A., et al. (2020). Data augmentation based morphological classification of galaxies using deep convolutional neural network. *Earth Science Informatics, 13*, 601–617.

[7] Dobrycheva, D., et al. (2020). The CNN classification of galaxies by their image morphological peculiarities. *Proceedings of the International Astronomical Union, 16*(S362), 111–115.

[8] Domínguez Sánchez, H., et al. (2019). Transfer learning for galaxy morphology from one survey to another. *Monthly Notices of the Royal Astronomical Society, 484*(1), 93–100.

[9] Gupta, A., Kaur, K., & Jindal, N. (2024). Predicting galaxy morphology using attention-enhanced ResNets. *Earth Science Informatics*, 1–12.

10 Natural language processing for early detection of Parkinson's disease in social media users based on linguistic markers

V. Jayasudha[1,a], *N. Deepa*[2,b], *and Devi T.*[2,c]

[1]Research Scholar, Department of Computer Science and Engineering, Saveetha School of Engineering, Saveetha Institute of Medical and Technical Sciences, Chennai, Tamil Nadu, India
[2]Associate Professor, Department of Computer Science and Engineering, Saveetha School of Engineering, Saveetha Institute of Medical and Technical Sciences, Chennai, Tamil Nadu, India

Abstract: Early Parkinson's disease (PD) detection is essential for bettering patient outcomes because prompt treatment can considerably slow the disease's progression. In this work, public social media posts are analyzed using Natural Language Processing (NLP) to look for linguistic and behavioural indicators of early PD symptoms, including decreased sentence complexity, repetitive language patterns, changes in emotional tone, and changed posting frequency. Three machine learning algorithms were used to analyze 20,000 anonymized social media posts: Support Vector Machine (SVM), Random Forest (RF), and BERT (a transformer-based model). The dataset included a subset of self-reported PD patients as well as a control group. Through comparative analysis, BERT outperformed other models in capturing subtle contextual relationships, achieving the highest accuracy of 92.3% and an F1-score of 91.8%.

Keywords: PD, NLP, SVM, RF, BERT, accuracy, patient, symptoms

1. Introduction

Millions of people worldwide suffer from Parkinson's disease (PD), a neurodegenerative condition that is progressive and chronic. Parisi et al. [1] Motor symptoms like bradykinesia, stiffness, and tremors are among its hallmarks, as are non-motor symptoms like mood swings, cognitive decline, and irregular sleep patterns. Early diagnosis of PD is difficult due to its subtle onset and variable early symptoms, which frequently causes treatment delays. Because prompt intervention can reduce symptoms, slow the disease's progression, and improve patients' long-term outcomes, early detection is essential [2].

New possibilities for healthcare diagnostics, particularly early disease detection, have been made possible by recent advancements in artificial intelligence (AI) and natural language processing (NLP). Even while medical assessments along with diagnostic imaging may crucial components of traditional PD diagnosis methods, they are frequently costly, intrusive, and difficult to obtain [3]. Social media in particular provides a different type of wealthy, chaotic information that represents actual individual behaviours and language. Many individuals are sharing personal incidents, notably related to health through digital media, creating a significant dataset towards discovering behaviours and syntactic tendencies symptomatic of neurologic abnormalities like Parkinson's disorder [4].

The technique for prompt Parkinson's disorder (PD) detection utilizing ML and NLP is depicted in Figure 10.1. Different kinds of ML approaches are employed for learning and subsequent evaluation after data gathering and preparation. Measures of evaluation are applied at the end of the method to evaluate the robustness and efficacy of the approach [5].

Early indicators of the neurological disorder include linguistic irregularities, including simpler phrases, repetitious vocabulary, delayed memory retention, and abnormal psychological tones. Despite their subtlety, such modifications are detectable by sophisticated computer methods. Digital media is an exciting option for rapid identification because, as a forum for everyday expression, [6] it provides a widespread and minimally invasive approach to document such shifts. A new automated screening technique can be created by using NLP to examine textual

[a]jayasudhav9015.sse@saveetha.com, [b]deepa23narayanan@gmail.com, [c]devi.janu@gmail.com

DOI: 10.1201/9781003770473-10

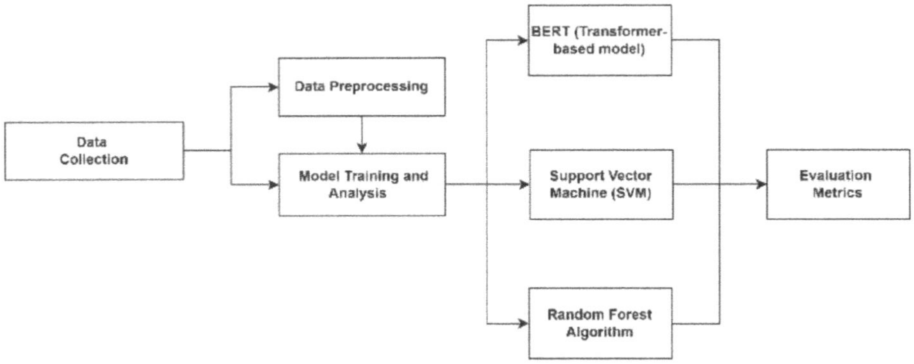

Figure 10.1. Workflow for early detection of Parkinson's disease.
Source: Author.

information from social networking sites and find indicators linked to the disorder.

To process digital networks and discover premature gauges of Parkinson's disorder, this research aims to process AI approaches to detect neurological disease at premature stages. The following models are used to assess the text-based data and distinguish individuals into uncertain groups contingents on linguistic highlighters [7]. This model evaluates the performance metrics to detect the most efficient early disclosure approach. subsequently appears that BERT will perform better than more characteristics-engine-heavy classical models such SVM as well as RF due to its capacity to perceive contextual subtleties.

Apart from the technological assessment, this research highlights the moral ramifications of utilizing online communication data for medical diagnosis [8]. Important factors to ensure proper implementation of such technology include analytical impartiality, security of data, and informed permission. The goal of this project is to combine cutting-edge ML techniques [9] via morally good concepts to offer a framework for effective, adaptable, harmless initial detection of Parkinson's disorder [10].

The research gap considered in this paper are the traditional models was produced moderated performances and accuracy, which is affect the earlier detection and decision making, the proposed methodology produced promising results and making decision making better [11].

The outcomes of this proposed model are to provide digital communications with promising prognosis tools as well as donate to the wider domain of AI-assisted medical system innovations [12]. The investigation opens up opportunities for potential uses in further brain and intellectual illnesses by showing whether morphological scrutiny [13] of electronic texts can be utilized as an affordable and accessible strategy for early detection of illness [14]. The increasing focus on preventive medicine and the incorporation of technology to overcome gaps in conventional medical facilities are in line with this strategy [15].

2. Materials and Methods

2.1. Data collections

An anonymized data of 20,000 publicly accessible social networking site comments was collected from individuals that claimed to have PD, and a control group of people who did not have PD. Posts had to meet specific criteria, like comprising in English, it has been being at least ten long words, in addition to particularly this incidence of the PD portion, explicitly confirming a prognosis. To preserve incognito and protect confidentiality for users, all identifiable data was removed before analysis, as well as ethical standards were strictly followed.

The PD patients has the problems in speech and writing abilities, their social media posts, messages, and comments will help to prognosis the PD with the usage of simple words, continuous spelling mistakes and grammatical errors, and slower responses during the conversation.

2.2. Machine leaning models

The three supervised learning strategies were employed to examine the produced data: BERT, which is a transformer-assisted model; RF; as well as. Using a set of deliberately constructed features, such as phrase difficulty and grammatical markers, the SVM model classified the blog entries. Random Forest (RF): Using decision trees, this combined model was able to detect significant language trends and shown resilience, particularly with smaller sets of data. BERT: A previously built transformer-supported model, BERT was modified on the actual data set to capitalize on its ability to grasp linguistic context.

2.3. Data splitting and model training

The subsets of the data collection were splitted for training (70%) and evaluation (15%). To maximize efficiency, parametric tweaking was carried out for every model. For SVM, the normalization factor was modified. For Random Forest, the highest possible depth and number of trees had been optimized. Adjusting the training set size, acquisition rate, and quantity of phases was necessary for BERT refinement.

2.4. Performance metrics

The outcome of the proposed approach was evaluated with precision and The F1 rating was calculated on the evaluation set. These metrics evaluated the models' ability to recognize cognitive and contextual markers that indicate the earliest Parkinson's disorder manifestations.

2.5. Ethical considerations

In order to protect user identities and ensure ethical compliance, all information was anonymized. The set of data was balanced using a range of lingual and socioeconomic factors in an attempt to reduce bias from algorithms. Confidentiality precautions were essential during the investigation to guarantee that the results were used in a responsible and secure manner.

3. Natural Language Processing

NLP tools were used to analyze behavioural trends and spoken words that were symptomatic of the earliest stages PD. The pathway included text preparation, pattern extraction, and categorization using machine learning models. Reviewed and integrated throughout the algorithm's learning architecture were linguistic factors such as blogging frequency range, psychological tone, repetition, and phrase sophistication. Quantitative depictions of the unstructured data were created for the purpose of rendering it appropriate for models used in machine learning.

3.1. Feature-based representation

Assessments indicating expressive tone, sentence intricacy, and recurrence had been some of the manually created linguistic elements that were retrieved. These qualities resulted in the organized feature matrix-like, which is displayed as follows:

$$X = [x_1, x_2, .., x_n]^T \; where \; x_i \in R^m \tag{1}$$

From equation (1), x_i is the symbol of the characterize vector for the i-th position, and *m* is the value of features.

3.2. Contextual embedding using BERT

Tokenizing postings and then feeding it into BERT allowed it to produce integrated contexts for more intricate text analysis. For every token, BERT generated an analogy that had environmental implications. The following is an expression for a post's result embedding:

$$E = [h_1, h_2, \dots, h_k] \; where \; h_j \in R^d \tag{2}$$

Here, h_j represents the embedding for the *j*-th token, k is the number of tokens, and d is the embedding dimension.

3.3. Machine learning models

Three machine learning algorithms – SVM, Random Forest, and BERT – were fed the transformed data. The probability that a post would reveal early symptoms of PD was predicted by each model.

1. **Support Vector Machine (SVM):**
   The SVM model optimizes a decision boundary by solving:
   $$\min_{w,b} \frac{1}{2} ||w||^2 + c \sum_{i=1}^{n} \max(0, 1 - y_i(w^T x_i + b)) \tag{3}$$
   Here, *W* is the weight vector, b is the bias, y_i is the true label (+1 for PD, −1 for control), and C is the regularization parameter.
2. **Random Forest (RF):**
   Random Forest employs multiple decision trees, where each tree votes for a class label. The final prediction is:
   $$y = mode\,(\{T_1(X), T_2(X), \dots, T_M(X)\}] \tag{4}$$
   Here, $T_m(X)$ is the prediction of the m-th tree, and M is the total number of trees.
3. **BERT Fine-Tuning:**
   BERT was fine-tuned using a cross-entropy loss function:
   $$L = -\frac{1}{n} \sum_{i=1}^{n} [y_i log\,(y_i) + (1 - y_i)\log(1 - y_i)] \tag{5}$$
   Here, y_i is the predicted probability for the i-th post belonging to the PD class, and y_i is the true label.
4. **BERT (Bidirectional Encoder Representations from Transformers)**
   Because of its remarkable capacity to capture contextual relationships in text, the transformer-based model BERT was selected for this investigation. BERT uses a bidirectional approach, in contrast to traditional language models that process text sequentially (from left to right or from right to left). By taking into account every word in a sentence at once, the model is able to comprehend a word's entire context.

4.1. Core architecture

BERT is based on the Transformer architecture, which models the relationships between words in a sentence using self-attention mechanisms. The Transformer's primary function is the self-attention calculation, which is described as follows:

$$Attention\ (Q, K, V) = softmax\left(\frac{QK^T}{\sqrt{d_k}}\right)V \tag{6}$$

where the input embeddings are used to create the matrices Q (queries), K (keys), and V (values). The dimensionality of the keys is denoted by d_k. By giving more weight to contextually relevant words, the SoftMax function makes sure that the attention scores are normalized.

4.2. Training objectives

BERT is pre-trained on large corpora using two key objectives:

1. **Masked Language Modeling (MLM):** The model predicts the original tokens based on the context, while a portion of the input sequence's tokens are randomly masked. The MLM loss function is

$$L_{MLM} = -\sum_{i \in M} \log P(w_i | w_{i+1}) \tag{7}$$

where w_i is the original token at position i, w_(i+1) is the predicted token, and M is the set of masked token positions.

2. **Next Sentence Prediction (NSP):** Sentence pairs are fed into the model, which determines whether the second sentence comes after the first. NSP uses a binary cross-entropy as its loss function

$$L_{NSP} = -\frac{1}{N}\sum_{i=1}^{N}[y_i \log(y_i)) + (1 - y_i)\log(1 - y_{i+1})] \tag{8}$$

Where y_i is the true label (1 for next sentence, 0 otherwise) and y_{i+1} is the predicted probability.

3. **Fine-Tuning for Parkinson's Disease Detection:** For this study, BERT was fine-tuned on the dataset of social media posts. The input for BERT consisted of tokenized text sequences, with special tokens [CLS] (denoting the start of the sequence) and [SEP] (denoting the end or separation of sequences). The final hidden state corresponding to the [CLS] token was used for classification.

The classification layer on top of BERT is represented as:

$$y_{i+1} = \sigma(Wh_{CLS} + b) \tag{9}$$

Where h_{CLS} is the hidden state of the [CLS] token W and b are trainable parameters of the classification layer σ is the sigmoid activation function for binary classification.

The loss function for fine-tuning was binary cross-entropy:

$$L = -\frac{1}{n}\sum_{i=1}^{n}[y_i \log(y_{i+1}) + (1 - y_i)\log(1 - y_{i+1})] \tag{10}$$

4.3. Advantages of BERT in this study

BERT's ability to capture contextual and semantic nuances enabled it to excel in identifying linguistic and behavioural markers of early PD. Its pre-training on a large corpus and subsequent fine-tuning on the PD dataset allowed the model to achieve high accuracy (92.3%) and an F1-score (91.8%), making it the most effective algorithm in this analysis.

5. Results and Discussion

Figure 10.2 presents a bar chart that compares the accuracy, precision and F1 Score of the NLP algorithm. In comparison, the accuracy of the BERT technique is around 92.3%, but the accuracy of the support vector machine (SVM) approach is approximately 82.7 and Random Forest (RF) for 85.4. The error bar on the chart illustrates a confidence interval that is 95% accurate.

The Bayesian estimate of coefficients is depicted in Table 10.1, accompanied by information on the model group, the dependent variable (Accuracy), and the reference prior assumption. Bayesian estimation, leveraging both prior information and observed data, provides a statistical approach for estimating unknown parameters.

Unlike the conventional T-test, the suggested NLP algorithm is evaluated along with the current RF and SVM models, as indicated in Table 10.2. Along with imply precision, the assessment also includes data for dexterity and F1 grade. This extensive table displays the BERT methodology, RF, and SVM classifiers' efficiency metrics, variances, the values of p and uncertainty ranges in detail. It offers a solid statistical basis for assessing the notable differences in the methods that these models employ. The analysis is made more comprehensive and provides a more

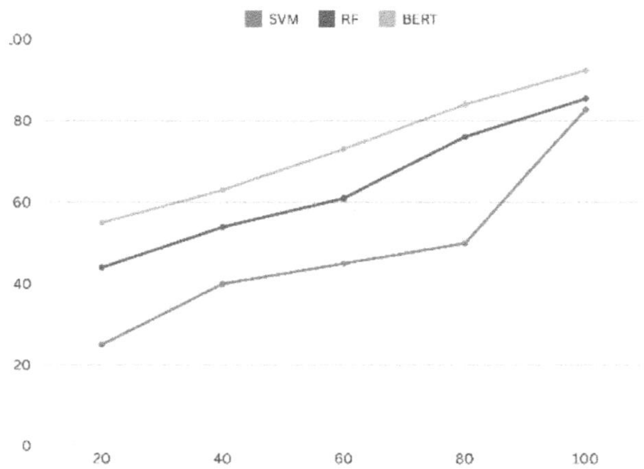

Figure 10.2. Statistical analysis accuracy comparison using SPSS tool.

Source: Author.

advanced view of the manner in which the NLP technique operates in classification as opposed to the existing RF and SVM techniques by including precision and F1 score (Figure 10.3).

According to the investigations, the proposed model performance in NLP approach outperforms the one that yielded the next-highest outcomes by a significant margin (Table 10.3). Compared to the standard RF, that possesses a comparatively lower success rate of 85.4%, the BERT approach exhibits notable gains alongside a success rate of 92.3%. There exists a noticeable difference in precision amongst the two distinct methods, including the NLP analysis obtaining significance ratings beneath 0.12 and the one using SVM obtaining relevance ratings that are below 0.05 [14]. This illustrates how the proposed NLP approach significantly improves accuracy over the RNN model, proving its effectiveness.

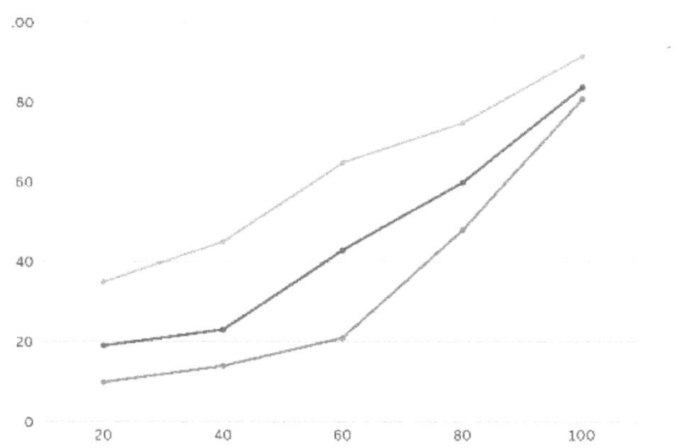

Figure 10.3. Comparing the F1 score using the SPSS tool for statistical analysis.

Source: Author.

Table 10.1. Bayesian estimation of coefficient

				95% Interval	
Parameter	*Accuracy*	*M.N*	*V.N*	*L.B*	*U.B*
Accuracy	92.3%	0.768	0.000	0.845	0.743
F1 Score	91.8%	0.753	0.000	0.744	0.567
Precision	93.5%	0.764	0.000	0.744	0.567

Source: Author.

Table 10.2. T-test table

	Group	*NO*	*M*	*S.D*	*S.E mean*
Accuracy	BERT	1300	92.3%	1.236	.22014
	RF	1300	85.4%	1.143	.43281
	SVM	1300	82.7%	1.467	.34613

Source: Author.

Table 10.3. Free example T test

		Variances		*t-test EOM*	
		F-value	*Sig.*	*t-value*	*Df-value*
Accuracy	Equal variances assumed	.010	.120	2.300	4
	Equal variances not assumed			2.300	6.163
Loss	Equal variances assumed	.108	.005	−2.334	4
	Equal variances not assumed			−3.334	6.290

Source: Author.

The model's interprets the input features (word or token) with the layer-wise relevance propagation, which track the relevant token with the trace back, identifies the each token contribution in the sentence.

The BERT algorithm works well in the predicting the PD, although it has certain limitation in multimodel integration, BERT analyzing the text only, it cannot integrate the other features like typing speed, interaction patterns, and scrolling style which may helps to understanding the PD related symptoms.

In the framework of NLP towards the identification of PD, our work introduces a novel Residual Denoising technique [15]. The aim of our work is to reduce variability in medical information, with an emphasis on premature PD identification, in order to increase diagnosis precision. proposed model's outcomes validate the efficacy in applying the Intrinsic Noise-reducing approach in relation to NLP and show how it can be used to increase the precision of the earliest Parkinson's diagnostics via advanced algorithms.

6. Conclusion

Overall summary, proposed research concerning the prediction of the illness Parkinson's using NLP has yielded significant findings and advancements. These outcomes, supported by rigorous quantitative metrics such as accuracy, precision, and F1 score, demonstrate that NLP could precisely forecast the presence of PD by deciphering complex patterns of speech in tests performed in hospitals. The NLP model based on the BERT method showed notable gains above the RF precision of 85.4 as well as the SVM correctness of 81.2, with a success rate of 92.3, but it facing computational issues, analyzing the huge dataset making fine-tune costly. The detailed study of the accuracy, recall, and F1 score metrics further demonstrates how reliable the NLP method is at striking a compromise between avoiding inaccurate positives or negatives and producing precise optimistic forecasts. In future this model will predict the different kinds of disorders with the minimal computation issues.

References

[1] Parisi, F., Ferrari, G., Giuberti, M., et.al. (2015). Body-sensor-network-based kinematic characterization and comparative outlook of Updrs scoring in leg agility, sit-to-stand, and gait tasks in parkinson's disease. *IEEE Journal of Biomedical and Health Informatics, 19*(6), 1777–1793. https://pubmed.ncbi.nlm.nih.gov/26316236/

[2] Exley, T., Moudy, S., Patterson, R. M., Kim, J., & Albert, M. V. (2022). Predicting UPDRS Motor Symptoms in Individuals With Parkinson's Disease From Force Plates Using Machine Learning. *IEEE Journal of Biomedical and Health Informatics*, *26*(7), 3486–3494. doi:10.1109/JBHI.2022.3157518.

[3] Allebawi, M. F., et al. (2024). Parkinson's Disease Detection from Online Handwriting Based on Beta-Elliptical Approach and Fuzzy Perceptual Detector. *IEEE Access*, (12), 56936–56950. doi:10.1109/ACCESS.2024.3387342.

[4] Salarian, A., Russmann, H., Wider C., Burkhard, P. R., Vingerhoets, F. J., & Aminian, K. (2007). Quantification of tremor and Bradykinesia in parkinson's disease using a novel Ambulatory Monitoring System. *IEEE Transactions on Bio-medical Engineering, 54*(2), 313–322. https://pubmed.ncbi.nlm.nih.gov/17278588

[5] Chen, O. Y., et al. (2020). Building a machine-learning framework to remotely assess Parkinson's disease using smartphones. *IEEE Transactions on Bio-medical Engineering*, *67*(12), 3491–3500. https://pubmed.ncbi.nlm.nih.gov/32324537/

[6] Raj, K. D., Lal, G. J., Gopalakrishnan, E. A., Sowmya, V., & Orozco-Arroyave, J. R. (2024). A visibility graph approach for multi-stage classification of Parkinson's disease using multimodal data. *IEEE Access,* (12), 87077–87096. doi:10.1109/ACCESS.2024.3416444.

[7] Chang, H, Liu, B., Zong, Y., Lu, C., & Wang, X. (2023). EEG-based Parkinson's disease recognition via attention-based sparse graph convolutional neural network. *IEEE Journal of Biomedical and Health Informatics*, *27*(11), 5216–5224. doi:10.1109/JBHI.2023.3292452.

[8] Zhou, Y., Tinaz, Y. S., & Tagare, H. D. (2021). Robust Bayesian analysis of early-stage Parkinson's Disease Progression Using DaTscan Images. *IEEE Transactions on Medical Imaging, 40*(2), 549–561. doi:10.1109/TMI.2020.3031478.

[9] Souza, R., et al. (2024). Identifying Biases in a Multicenter MRI database for Parkinson's Disease Classification: Is the Disease Classifier a Secret Site Classifier. *IEEE Journal of Biomedical and Health Informatics, 28*(4), 2047–2054. doi:10.1109/JBHI.2024.3352513.

[10] Wang, W., Lee, J., Harrou, F., & Sun, Y. (2020). Early detection of Parkinson's disease using deep learning and machine learning. *IEEE Access*, (8), 147635–147646. doi:10.1109/ACCESS.2020.3016062.

[11] Dhivyaa, C. R., Nithya, K., & Anbukkarasi, S. (2024). Enhancing Parkinson's disease detection and diagnosis: A survey of integrative approaches across diverse modalities. *IEEE Access*, (12), 158999–159024. doi:10.1109/ACCESS.2024.3487001.

[12] Talitckii, A., et al. (2021). Defining Optimal Exercises for Efficient Detection of Parkinson's Disease Using Machine Learning and Wearable Sensors. *IEEE Transactions on Instrumentation and Measurement,* (70), 1–10, Art no. 2512010. doi:10.1109/TIM.2021.3097857.

[13] Li, T., et al. (2018). Automatic timed up-and-go subtask segmentation for Parkinson's disease patients using video-based activity classification. *IEEE Transactions on Neural Systems and Rehabilitation Engineering*, *26*(11), 2189–2199. doi:10.1109/TNSRE.2018.2875738.

[14] Mabrouk, R., Chikhaoui, B., & Bentabet, I. (2019). Machine learning based classification using clinical and DaTSCAN SPECT imaging features: A Study on Parkinson's Disease and SWEDD. *IEEE Transactions on Radiation and Plasma Medical Sciences, 3*(2), 170–177. doi:10.1109/TRPMS.2018.2877754.

[15] Vásquez-Correa, J. C., Arias-Vergara, T., Orozco-Arroyave, J. R., et al. (2019). Multimodal assessment of Parkinson's disease: A deep learning approach. *IEEE Journal of Biomedical and Health Informatics, 23*(4), 1618–1630. doi:10.1109/JBHI.2018.2866873.

11 Reduction of depression levels by innovative peer support among same-linguistic people using TAD framework

J. Vishnu Kannan[1,a], *N. Deepa*[2,b], *and Devi T.*[2,c]

[1]Research Scholar, Department of Computer Science & Engineering, Saveetha School of Engineering, Saveetha Institute of Medical and Technical Sciences, Saveetha University, Chennai, Tamil Nadu, India
[2]Professor, Department of Computer Science & Engineering, Saveetha School of Engineering, Saveetha Institute of Medical and Technical Sciences, Saveetha University, Chennai, Tamil Nadu, India

Abstract: The healthcare domain plays a major role in the research concept. The health condition of humans depends on physiological, academic, lifestyle, and socio-environmental factors. Mental health is vital and critical and signifies a state of the human mind. Every human suffers from Depression problems depending upon the factors. Medical students are the people who need to be viewed seriously, as they are facing many personal and professional issues in day-to-day life. Traditional methods in clustering depression levels rely on the non-linear relationships between the social, and peer supports and incorporate insights network based on attributes. Generally, peer support is needed for everyone at any point in life. Therapeutic Alliance Depression (TAD) Framework in boom for the medical students to overcome the depression. A network-based evaluation is enhanced for the reduction of depression with a innovative peer-supportive system and provides mental health prediction insights of the students at risk. The framework outlets the patterns and benefits of the supportive system intervention. The enhanced framework shows a better network structure in peer support for the needs when comparing the other methods.

Keywords: Attribute diagonalize, depression reduction, medical students, peer support connection, same mother tongue, TAD framework

1. Introduction

Depression is a significant mental health problem faced across the world. Recent studies say that adults are under high depression circumstances as they handle both personal and professional issues [1]. Medical students are at a higher risk of dealing with several environments, likely academic, and demand of career, burnout, and much more [2]. Depression has a greater impact on human mood swings and well-being sowing a seed for affecting their potentiality in professional development [3]. The growth in machine learning and predictive analytics helps in the forecasting of mental health problems [3, 4].

Most of the students are suffering from depression and struggling to overcome it, there is a need for awareness among medical students identifying the level of depression in them and solutions to overcome it [5]. This remains a harder challenge in the medical field. Medical fielders question which factors affect them the most or which situation forces them into depression such as being less addicted to stigma, judgmental fears, or lack of resources for mental health [6]. Peer support is a determinant factor that streaks against depression. Even though the impact of depression is a look-ahead process, needs to be maintained properly, and the supporting networks or relationships in a data-driven method [7, 8]. The integration of peer support into health analysis drops in estimating the level of depression by social relationship [8].

Depression among medical fielders is not under concern [3]. Depression impacts relationships, cognitive functioning, and academic interests suggesting a path to affect long-term insights like burnout in the work environment, improper patient care, and non-tolerance levels [9]. These problems are profoundly widespread among medical students, especially during academic and internship periods [10]. So there is a need for a demand solution that gives a proper leading point to all the medical fielders and precautions them in the early stage of depression point [11]. Taking an opportunity and utilizing the technology development in the medical domain to help the medical fielders

[a]vishnukannanj9152.sse@saveetha.com, [b]deepa23narayanan@gmail.com, [c]devi.janu@gmail.com

DOI: 10.1201/9781003770473-11

survive a satisfied life with the computational techniques to achieve an enhanced supportive framework from the data-driven from the depression level [12]. Prioritizing the mental health of the medical students by building reliable, survivable, motivating solutions to lessen the stress and demand of the medical students [13].

So far traditional methods provide a simple relationship with less complexity. As depression is not a single factors concept, it depends on several factors and needs a proper approach and deals with the aggregation of the features [14]. There are biased reports and inaccuracies arise when improper data is needed for prediction. Stimulating a nonlinear relationship between socio-environmental factors and support factors in lifestyle is facilitating the medical fielders [15]. The interaction between the relationship and the depression level stimulates knowledgeable methods in a feasible manner [2, 13].

Thus a need for mental health is proposed through the classified depression level of the medical students with the CES-D values. The depression measure collaborated with the relationship support for connecting a network [3, 16]. The integrated models form a peer reactive and relative model with features in common for the group. Through this dependable network process an enhanced peer-supportive model is built to lower the depression of medical fielders [12, 17]. This approach addresses constraints associated with conventional grouping by integrating distributed perspectives, allowing for greater clarity about how it relates to interpersonal interactions with psychological wellness [16]. Results offer feasible suggestions for developing mutual support networks that can be effectively used in educational contexts that enhance individuals' general state of mind [17].

2. Related Work

In present days, depression is a major disorder that has exponential growth among adults. Currently, a huge number of evaluations and research have been going on to predict mental health accurately with ML algorithms. While going deeper only a few researchers have dedicated themselves to downstream the prediction with the multiclass variable. By the way, the distribution of data among the developed classes hardly occurs in practical communities. So, the imbalance of the class is an issue in implementing and maintaining health care, to resolve a multiple class context is framed with the new approach FGP as partitioning and groping are based on the features matching. SMOTE and ADASYN have been used for class balancing. The dataset was collected from a university based on the Burn Depression Checklist. The ensemble stacking and ensemble bagging have been implemented wisely in heterogeneous and homogeneous methods. The challenges like overfitting, and mitigating are faced in the data preprocessing method and are solved by machine learning methods. The predictive model has ensured better accuracy, precision, f1 score, sensitivity, and specificity. The analysis between the CDS and FGP is been extended in a discriminated manner. The result of the classifier with stacking technique for FGP along with SMOTE achieves 92.81% as the highest accuracy. The combination of SMOTE with FGP shows the acceptable performance of the proposed method in predicting the depression level in individuals [1].

The worldwide problem faced by all people is depression which serves as a mental health problem. Mental health issue affects a person's interpersonal skills and force psychological disability and disorders with professional and personal routines. The majority of people are not diagnosed in their depression impacts, there is a need for severe that in the early stage. The carelessness in not treating the depression may lead to self-harming in the final stages. Samely, determining and preventing the depression at an early stage is crucial in all aspects. The data has been collected from the UG students consulted with the lecturers and psychiatrists. NLP integrated with machine learning methods has been used in the training phase and evaluation to represent the proposed model. This model uses certain methods for each phase in processing mRMR for feature selection, RFE for wrapping technique, LASSO is the embedded method for extracting the features from the collected information with responsible for sparking or forcing into depression. The proposed method has shown an accuracy rate of 94%, accuracy gained after applying feature selection using optional hyper tuning is 96% [2].

Stress is an outside force that is triggered by the worries and burdens of an individual. Depression is said to be a growing term in effects of intense low, losing interest in everything, isolating themselves, exactly affecting their overall behaviour and activity terms. Both depression and stress are disorders that have to be considered in a general mental health checkup. People in the medical field face more consequences in their routine forcing them into stress or depression mode, which leads to a loss in career development, and making errors in clinical diagnosis. These factors influence the overall professional life of the medical fielders in increased concern in medical treatment. However, swing mood disorders for sure impact the psychological wellness of the students in both life and work, utilizing the flexibility of the medical people. The wellness state of the students or medical learners has been assessed with the Multilayered Progressive Formulation (MPF) method, as develops a distributed structure based on the resemblance factors. The resemblance format forms a cluster with the similarities of the data collected with the depression and stress values. This classification is applied in multiple stages of predicting stress and depression and has achieved 86%, precision of 0.86, recall value of 0.85, and F1 score of 0.86 as value. The proposed method has been developed by the multiple layered association in resemblance and similarities between the effectible factors achieved good accuracy [3].

Mental health issues are an increasing factor worldwide among the people. Nowadays many steps and awareness have been taken to solve the issue. Mostly the stress factors are computed based on the measuring values and not concentrated on the demographics of the person. Medical students are easily affected as they face continuous pressure in an education setting both theoretical and practical studies. These problems reflect on their academic performance, career building, self-care taking these factors in a refining way creates a route for special care in medical education. The proposed method uses Swiss medical students of 886 data in self investigate their stress or depression levels with machine learning and deep learning techniques. Integrating two machine learning methods random forest and naive Bayes for enhancing an ensemble classifier with LightGBM and DNN. The process of prediction and evaluation in the feature selection gained a well-formed study. The DNN model outcomes the accuracy for depression, anxiety, and burnout of values 81.4%, 76.65%, and 73.59%. The predictive model has been diagnosed with an ensemble classifier and provides a success rate in finding the mental wellness of the medical professionals. The objective of the proposed method is to identify the mental health problems faced by medical people and highlight AI in shaping the future development of the healthcare domain [4].

When technology is developed by integrating several domains altogether and enhancing or creating a particular process, likewise online learning is also a platform for educating students from anywhere and anytime. The performance of the students has been noticed clearly as records in offline teaching, but online there is a need to forecast the student's performance. The Machine learning algorithms are used for classifying the performance of the students in online courses based on behavioural factors, cognitive and affective strategies, and also with the intervention of peer support to improve the student's success rate. The dataset was collected from a private university of engineering students who study astronomy and physics. The data was preprocessed, trained, and evaluated with six different machine learning algorithms namely SVM, KNN, NB, LR, DT, and RF. The classifications have been implemented and Random Forest (RF) yields an accuracy of 87%, perfectly classifying between the given categories low, medium, and high. The factors that influence the students in online learning trigger their anxieties and expectations of the course. In opposite which affects them is the social isolation of the student. The machine learning approach efficiently classifies the students with particular features needed for classifying, suggesting that educators enhance the teaching method with the new strategies for developing a peer network among the students [5].

3. Proposed Work

The proposed TAD Framework based on the network technique classifies the depression level of the medical learners with the grouping of CES-D measures that match the social features. The created method builds a similarity matrix with a relationship between the attributes. The attributes are trialed to the graph matrix with eigenvectors and values with grouping structure. The prediction of peer support with the network model is structured with the nodes and edges. To estimate the practice methods of supportive methods, identify the struggling people. By developing supportive relationships to lessen the depression of the students. Integrating the grouping of CES-D level and peer support by network connectivity advances a supportive model with mental wellness.

3.1. Dataset preparation

The understanding of the dataset through the analysis of critical features of the medical students. The classified feature of CES-D value is separated into three levels with limited numeric values likely Low(0 to 11), Medium(12 to 23), and High(24–35) as an achievable score for grouping. The relativity bond between to compute the supportive connection, based on the language with ethnic or psychological similarities that matters with mental wellness. Demographic attributes play a vital role in the phase of depression in human life. To make ready data the missing values are preprocessed for the training process, observing important information is maintained. The next step is to normalize the features with real similarity for the grouping calculation. Practice separating the subsequent sets which is the core of clustering to build a connectivity network with peer support.

3.2. Data preprocessing

To clean and make consistent data for the forecasting and the preprocessing of the initial stage. Incorporating missing values, the values are to be normalized to reduce the loss of information from the dataset. The integrity of the dataset should be maintained while standardizing the training process. The information is changed as a linear scaled value, with the combination of the calculations through the normalization. In regard to the normalization, the values with higher scales influence the results. An enhanced model with the workflow is created and the dataset is partitioned into training and validation sets as of 70 and 30 percent. By the grouping model the supportive network is framed and analyzed, on the other side validation set is driven for the advantage and relevance.

3.3. Attribute adjacency diagonalization

The graph network is created by the data values which are similar points that build a structure for identifying the mental health issues that are interlaced. The preprocessed data is initialized with the construction of a similarity matrix, that calibrates the resemblance between two students with the same performance. When using Gaussian kernel calculation, the same attributes are computed with the language, CES-D value, and demographic features. The Equation (1) shows in identifying the similarity among the data for cluster formation. The network is epitomized by the dataset, where every individual is represented by a node and linked by edges with the same weight.

$$C_{lm} = exp\left(-\frac{||x_l - x_m||}{2\sigma^2}\right) \quad (1)$$

Where C_{lm}, similar nodes of l and m

$||x_l - x_m||$ is distance

σ is measuring factor

After achieving the node and edge networks with weights, a mathematical matrix representation is to be derived with the help of a Laplacian matrix. This matrix is used for the derivation of a graphical structure into a mathematical representation matrix. Eigenvalues and eigenvectors are used to identify the patterns of the graph structure of the dataset. From the utmost dimension, the eigenvalue is extracted for the eigenvector, to cluster the individual's data in effective separate clusters.

$$Minimize: \sum_{n=1}^{q} p \sum_{r\in S_i}^{n} ||r - \mu_n||^2 \quad (2)$$

Where, S_i for "i" *th* cluster

μ_n for centroid S_i.

Clusters formed by the lower dimension show the depression level of the students in the group. By Equation (2) the cluster is formed with the data collected. Though identified patterns are concerned with the demographic features for the more effective prediction of mental health in the same group.

3.4. Peer network classification

The peer graph network model evaluates the connection between social-environmental and depression relies on among medical learners. As Figure 11.1 shows the framework of TAD. The relationship between the nodes is social factors and academic factors likely exam scores, friendships, and common language. The Equation (3) represents the node degree for the relational connectivity. Edges are with weights that show the strength of the relational

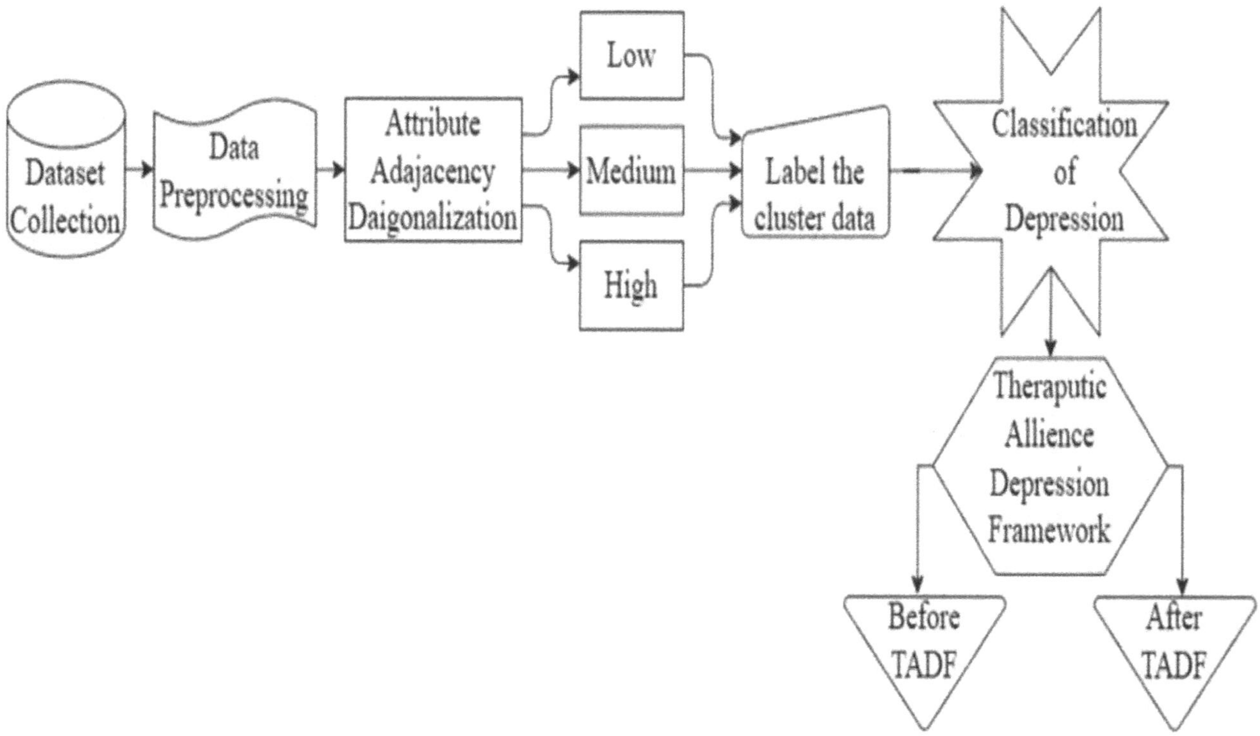

Figure 11.1. Workflow of TAD framework.

Source: Author.

network, if less interaction then a supportive system is suggested. The peer dynamics is observed in the network with the social connections and mental health issues.

Degree of Node:

$$Degree(u) = \sum_{v \in V} P(u, v) \tag{3}$$

Coefficient of the group nodes:

$$A_u = \frac{2.n_u}{z_u.(z_u - 1)} \tag{4}$$

Where, n_u is edges linked with neighbour node u

Z_u is degree of u

The prediction of calculating the network inter relativity between the given metrics to showcase the constructed graph network. The Equation (4) reflects the co-efficient in the group nodes with degree. The nodes with the lower degree provide the students who are not interested and with poor interaction that leads to less social support and tends to isolate depression of the student.

The simulation of the groups is identified by the clustering to identify the perfect supportive students. The measure of centrality is to predict the well-connected individuals among the clustered group that helps more in peer support. As Figure 11.2, represents the depression classification in same language with peer support. Through integrating all these metrics the current situation of the individual is forecasted and precautionary actions can be taken for those who suffer from less peer support.

3.5. TAD framework for peer support

The need for simulation in the peer support analysis is to identify the current mental state of the medical student based on the depression value (CES-D) taking that as a baseline for the process. Further, it is moved to evaluate the overall students to identify the isolated students or the students with poor connectivity with others. The Equation (5) is used to calculate the edge weights of the cluster in enhancing the connectivity. The baseline analysis gives an understanding support framework that implies the medical student's current mind stage.

Edge weights:

$$P'(u, v) = P(u, v) + \Delta P \tag{5}$$

Where, ΔP increased weight with supporting factor.

Connecting peers with isolated nodes:

$$E' = E \cup \{(u, v) |\ low\ degree\ node\ u\} \tag{6}$$

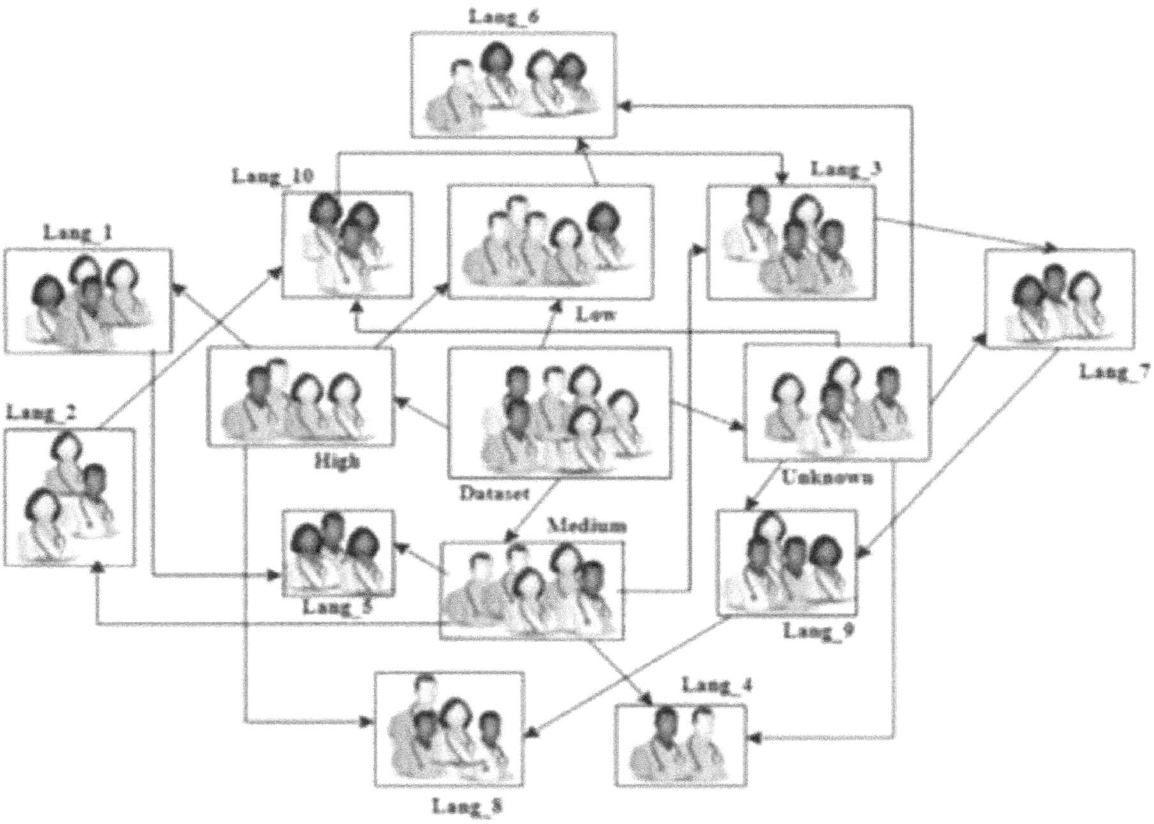

Figure 11.2. Depression classification & same language peer support.

Source: Author.

The peer support network is to benefit the enhancement of the supportive model. Then new edges with the isolated nodes with the node of high degree approaches to give a well-structured network graph with groups. The Equation (6) develops a peer support model with the less interacted nodes with the same language peers. Through this, the connection which is weak can be strengthened by increasing the weights in higher order for the edges to represent the interaction between the peers.

Improvement in depression metrics:

$$\Delta CESD = CESD_{before} - CESD_{after} \quad (7)$$

Once all the metrics have been updated again for the isolated students with new peer interaction or connection, the structure is again processed to receive the new impact of the new cluster. With Equation (7) the depression is differentiated between the before and after implementing the TAD framework. The results describe the reduction of depression among the group of students and an increase in efficiency among the supportive system between the students that leads to the peer dynamic environment in the mental well-being of the individuals.

4. Results and Discussion

4.1. Distributed analysis for depression

The depression level from different language groups of people in the separation prediction between the traditional and mental issue relationship. The level of depression in the bar chart with the mother tongue. This visualization of patterns and differences, identifies the groups with the more dangerous depression. Few languages show a higher depression on behalf of poor relationships within the social-environment.

The graph in Figure 11.3 represents level of stress like low stress, medium stress, high stress or either stress nor not(unknown) distributed within the medical students.

The graph in Figure 11.4 shows the classified depression among the medical students based on the same mother tongue of the students which helps in developing a innovative peer supportive model.

4.2. Implementing peer connections

The relationship among the students with peer support network as node and edge representing the students and connectivity with social network. The support and relationship represented values for each level of depression with the identified pattern in mental wellness.

The outlets the inter connections between the nodes with less interaction are high at depression in lack of peer support. The Table 11.1 reflects the peer support innovated based on the top languages among the different depression levels in low, medium, high depression.

4.3. Factors relativity with TAD framework

The heatmap graph represents the link between the social network factors like friends at work, friends by mother tongue, and stress level. The heatmap uses gradient colours for differentiating between the correlations and strengths of the metrics to showcase the unique pattern. A strengthened positive correlation is been developed between the lower depression level with peer support. This correlation gradient exposes the robustness of the proposed system among social interaction and high depression students.

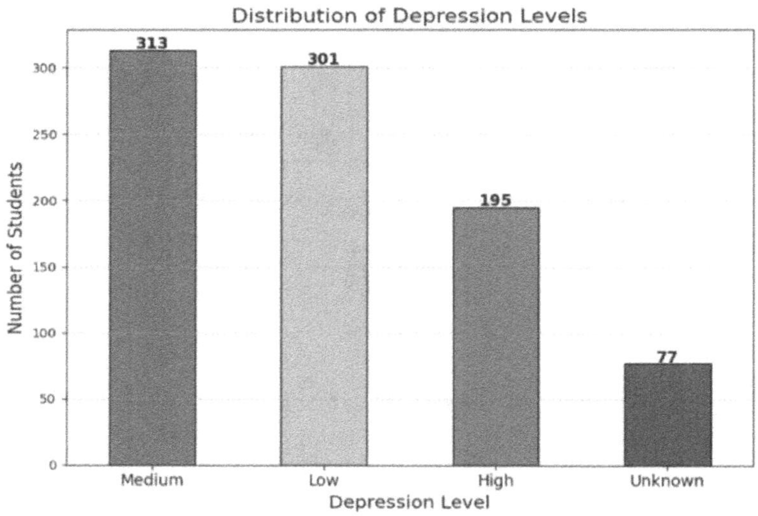

Figure 11.3. Distribution of depression levels among the medical fielders.

Source: Author.

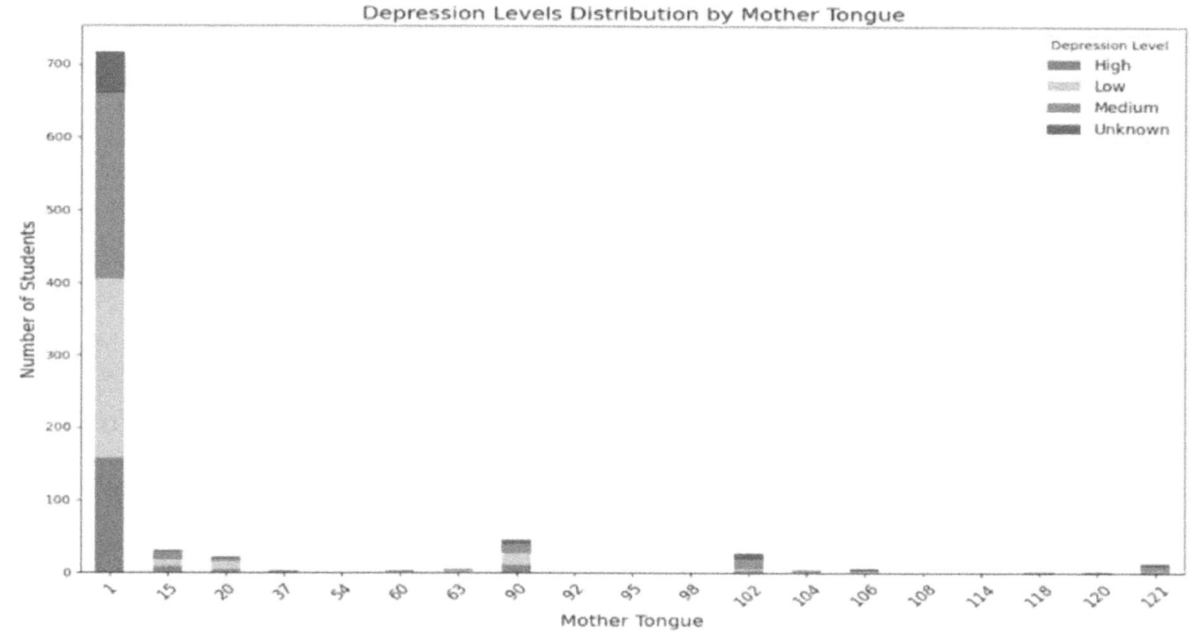

Figure 11.4. Depression classified based on the same language.

Source: Author.

Table 11.1. Interconnection of Peer support with languages

Depression Level Group	*Total Strength*	*Mean Peer Support*	*Mean Relation-ships*	*Top language group*	*Cluster density*
Low level	318	4.3	5	Language 1	Average
Medium level	411	3.9	6.2	Language 90	High
High level	157	3	4.8	Language 102	Average

Source: Author.

Simultaneously, weaker peer interaction is associated with high depression levels. Though Figure 11.5 shows the relationships among the supporting factors. These insights become a foundation for developing targeted interventions in the social and mental health of medical learners. The targeted interventions surpass the fostering of same-language student groups or encouraging social activities that support the peers to lower the depression level.

4.4. *TAD framework simulations*

The difference between the depression level before creating peer support and after creating peer support for the lower academic performance and higher depression level shows the simulated version of the enhanced method. The Table 11.2 shows the comparisons the peer network before and after TAD framework. The comparison focuses on the transformation of the potential peer network improvement in reducing the level of depression and fostering the mental wellness of medical students.

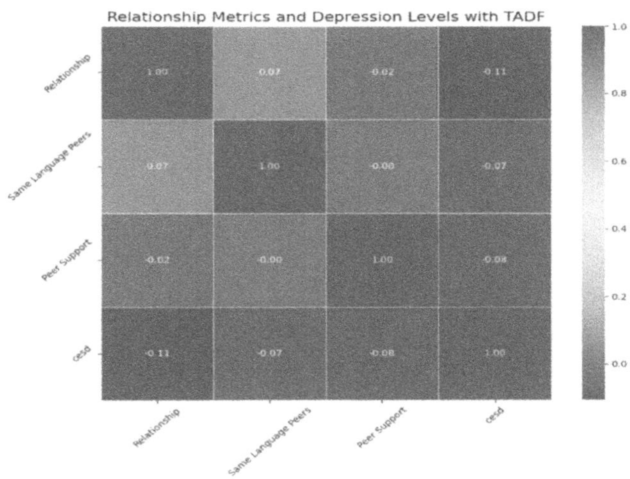

Figure 11.5. Correlation between supporting factors.

Source: Author.

4.5. Reduction of depression with peer support

The comparison of the clusters formed before and after the simulation of the peer network illuminates the distribution of depression scores for individual clusters. Before clustering the individual with weak support reflects a high depression level, and also isolates in social network of interactions. The limitations of the TAD Framework does not study the biological-health needs of the medical students, it is based on network connectivity of the students may face real-time adaptability challenges, which dynamically evaluates the students mental health conditions.

After simulation, the depression score has been reduced for the clusters henceforth, the weaker connections are strengthened by increasing the peer support relations. The Figure 11.6 shows the differences in depression levels in variance with peer support. The target intervention focuses on strengthening the weak connections by integrating them with the cohesive groups, effectively mitigating the depression level.

Table 11.2. Comparison table before TAD framework and after TAD framework

Metric	*Existing network without TAD Framework*	*Enhanced network with TAD Framework*
Total nodes developed	886	886
Average peer support	3.5	4.9
Average node degree	2.3	5.7
Separated Nodes	58	8
Groups with high level depression	29	17
Groups with medium level depression	41	20
Groups with low level depression	16	43
Total Reduction in depression	No reduction process	37%

Source: Author.

5. Conclusion and Future Work

Mental wellness among individuals is challenging, particularly in heterogeneous and polgottics environments, demonstrating the need for the prediction of multiple factors influencing the mental well-being of medical students. The dynamic clustering of the same traditional individuals and peers' interests answers vital issues like obtaining mental wellness patterns with supportive groups in the medical environment people. The proposed TAD Framework uses advanced predicting techniques and integrates them with data visualization to identify the pattern grouping with the same language with depression level and plan according to the academic performance level. By exposing the visual frame with the features matching in concerns of the critical patterns, finding out the need for peer support along with the linguistic influence of mental wellness. The validation of the effectiveness of the proposed system focuses on future work with real-world testing of medical populations. The adaptability over real-time information will improve the impact of prediction and peer support

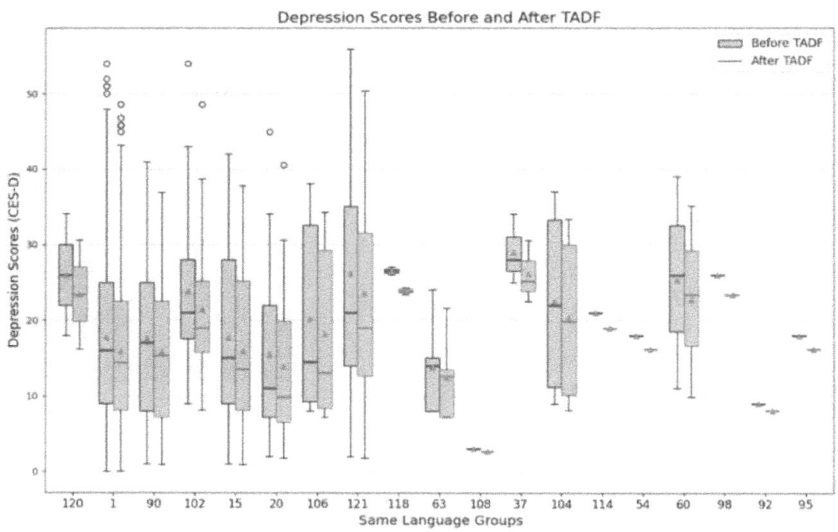

Figure 11.6. Depression level variation with peer support.

Source: Author.

through his framework. This system provides an intervention of the targeted nodes (individuals) for improving mental health in a multi-diverse environment, fostering the effectiveness and inclusiveness of the support system.

References

[1] Shaha, T. R., Begum, M., Uddin, J., Torres, V. Y., Iturriaga, J. A., Ashraf, I., & Samad, M. A. (2024). Feature group partitioning: an approach for depression severity prediction with class balancing using machine learning algorithms. *BMC Medical Research Methodology*, *24*(1), 123.

[2] Begum, S. R., Sait, S. Y., & Ramachandran, A. S. (2023, December). AI-based Depression Detection using Profile Information. In *2023 2nd International Conference on Automation, Computing and Renewable Systems (ICACRS)* (pp. 1367–1373). IEEE.

[3] Kannan, J. V., & Deepa, N. (2024, October). Enhancing the Prediction of Depression in Medical Fielders with CES-D Using MPF Method. In *2024 5th International Conference on Circuits, Control, Communication and Computing (I4C)* (pp. 606–612). IEEE.

[4] Rashid, S., Badrul, T., Rafid, L., Setu, J. H., Halder, N., & Islam, A. (2024, June). Ensemble and Deep Learning Approaches for Automated Screening of Anxiety, Depression, and Burnout in Medical Student Populations. In *2024 Advances in Science and Engineering Technology International Conferences (ASET)* (pp. 01–10). IEEE.

[5] Asiksoy, G., & İslek, D. (2024). Classifying Engineering Students Performance in Online Education with Machine Learning: Affective, Cognitive, and Behavioral Aspects. *BRAIN. Broad Research in Artificial Intelligence and Neuroscience*, *15*(2), 23–45.

[6] Dayananda, C., Hemashree, P., Impu, D., Janavi, S., & Gururaja, H. S. (2024, February). A Multi-Model Ensemble Approach for Proactive Student Mental Health Assessment. In *2024 IEEE International Conference for Women in Innovation, Technology & Entrepreneurship (ICWITE)* (pp. 494–500). IEEE.

[7] Devi, T., Jaisharma, K., & Deepa, N. (2022, November). Novel Trio-Neural Network towards Detecting Fake News on Social Media. In *2022 International Conference on Advancements in Smart, Secure and Intelligent Computing (ASSIC)* (pp. 1–5). IEEE.

[8] Medikonda, J. (2023). A clinical and technical methodological review on stress detection and sleep quality prediction in an academic environment. *Computer Methods and Programs in Biomedicine*, *235*, 107521.

[9] Abdrakhmanov, R., Zhaxanova, A., Karatayeva, M., Niyazova, G. Z., Berkimbayev, K., & Tuimebayev, A. (2024). Development of a Framework for Predicting Students' Academic Performance in STEM Education using Machine Learning Methods. *International Journal of Advanced Computer Science & Applications*, *15*(1).

[10] Issaro, S., & Wannapiroon, P. (2023). Intelligent Student Relationship Management Platform with Machine Learning for Student Empowerment. *International Journal of Emerging Technologies in Learning (iJET)*, *18*(4), 66–87.

[11] Pattnaik, S., Nayak, A., Mishra, S., Brahma, B., & Bhoi, A. K. (2023). Depression Assessment in Youths Using an Enhanced Deep Learning Approach. In *Enabling Person-Centric Healthcare Using Ambient Assistive Technology: Personalized and Patient-Centric Healthcare Services in AAT* (pp. 287–299). Cham: Springer Nature Switzerland.

[12] Sahu, B., Kedia, J., Ranjan, V., Mahaptra, B. P., & Dehuri, S. (2023). Mental Health Prediction in Students using Data Mining Techniques. *The Open Bioinformatics Journal*, *16*(1).

[13] Lv, J. (2024). Stacked Ranking Feature Cluster Machine Learning (Srfcml): A Novel Method of Career Planning of College Students Based on Career Interest Assessment and Machine Learning. *Journal of Electrical Systems*, *20*(3s), 1659–1669.

[14] Tiwari, S. S., Pandey, R., Deepak, A., Singh, J. P., & Tripathi, S. (2024). An ensemble approach to detect depression from social media platform: E-CLS. *Multimedia Tools and Applications*, 1–33.

[15] Hawes, M. T., Schwartz, H. A., Son, Y., & Klein, D. N. (2023). Predicting adolescent depression and anxiety from multi-wave longitudinal data using machine learning. *Psychological Medicine*, *53*(13), 6205–6211.

[16] Bhadra, S., & Kumar, C. J. (2024). Enhancing the efficacy of depression detection system using optimal feature selection from EHR. *Computer Methods in Biomechanics and Biomedical Engineering*, *27*(2), 222–236.

[17] Syed, S., Iftikhar, Z., Xiao, A. W., & Huang, J. (2024, May). Machine and Human Understanding of Empathy in Online Peer Support: A Cognitive Behavioral Approach. In *Proceedings of the CHI Conference on Human Factors in Computing Systems* (pp. 1–13).

12 Enhancing the prediction of medical student mental health with a sustainable solution using relational formulated rule-based algorithm

J. Vishnu Kannan[1,a], *N. Deepa*[2,b], *and Devi T.*[2,c]

[1]Research Scholar, Department of Computer Science and Engineering, Saveetha School of Engineering, Saveetha Institute of Medical and Technical Sciences, Saveetha University, Chennai, Tamil Nadu, India
[2]Professor, Department of Computer Science & Engineering, Saveetha School of Engineering, Saveetha Institute of Medical and Technical Sciences, Saveetha University, Chennai, Tamil Nadu, India

Abstract: The psychological well-being of medical students is a growing worldwide concern due to the high levels of stress, anxiety, and burnout often accompanying the rigorous clinical practice and academic demands. Persistent thought of stress, depression, and clinical rotations is the major cause of suicide in medical students. Addressing measures of stress for medical students requires a multifaceted approach, including factors such as demographic factors, physical factors, and lifestyle factors, stigmatization of mental illness, crisis intervention strategies, and community support networks. This study approached a Relational Formulated Rule-based Algorithm which is a sustainable solution and more effective machine learning algorithm to identify the associations within the data factors. The proposed study has given the result of accuracy which is support of 24% and confidence of 99% compared with the Apriori algorithm which has the support of 25% and confidence of 90%. The proposed work gives a better range when compared to the Apriori algorithm.

Keywords: Apriori algorithm, association rule, medical students mental health, pattern analysis, rule extraction, support and confidence

1. Introduction

Since medical education is so demanding and hard, there is a serious concern for medical students' psychological well-being. Stressors specific to medical students include a heavy workload, extended study sessions and clinical internships, high standards, and several exams. Psychological disorders like burnout, anxiety, and despair are a few results of stress can cause stress [1]. The perspective of a qualified field of the healthcare sector is to identify and make resolution of the problems like academic and overall performances of medical students. Machine learning (ML) is the most used technique to predict or analyze mental wealth for current days. For creating algorithms and forecasting the data, machine learning is used properly [2, 3]. The machine learning method is applied to complex and huge datasets to identify patterns and statistical techniques. The well-suited complicated and multifaceted problems like mental wellness, and machine learning will be flexible [4].

An increasing global problem is the psychological wellness of medical learners because of a great deal of anxiety, burnout, and stress that come along with the demanding educational and clinical requirements [5]. Due to the demanding curriculum, extended hours, and frequent exams associated with medical school, students are frequently at risk for psychological suffering [6]. It is essential to comprehend and forecast the mental wellness state of learners in medicine to create efficient support networks and actions that can lessen these negative consequences and enhance well-being [7].

Big data forecasting is the process of extracting insights as well as forecasting out enormous, complicated datasets by applying sophisticated analytical tools. Using big data throughout predictive analytics remains more and more crucial for companies, government entities, the healthcare industry, and several other fields in a period where information is created at never-before-seen rates from many different kinds of sources, including social networks, detectors, deals, and wireless devices [8]. Large amounts of information prediction usually begins with gathering data, which is the method of collecting various data types through various sources, such as semi-structured, unstructured, and

[a]vishnukannanj9152.sse@saveetha.com, [b]deepa23narayanan@gmail.com, [c]devi.janu@gmail.com

DOI: 10.1201/9781003770473-12

structured data. Next, to guarantee high quality and uniformity, that information is extracted and preprocessed. After the information is prepared, information is analyzed for correlations, trends, and patterns using sophisticated techniques like machine learning methods [9, 10].

A method to meet this demand is known as the Apriori algorithm, a traditional technique used in information mining for relationship rule extraction and frequently occurring item set creation [11]. The Figure 12.1 represents the process of predictive analysis in different terms or different domains. Through the analysis of extensive datasets containing demographic data, academic achievement, lifestyle choices, and self-reported emotional wellness metrics, the Apriori technique can spot patterns and relationships that might not be instantly obvious [12]. This technique offers major possibilities for early prevention and assistance by enabling the identification of variations of components that frequently exist together and have connections to certain psychological outcomes [13, 14].

However helpful, traditional techniques for evaluating the mental wellness of medical learners, such as self-report questionnaires and surveys, frequently offer only a limited understanding of the intricate and multifaceted nature of emotional well-being [15]. The Figure 12.2 show-cases of the workflow of relational algorithms in a step-by-step process. The complex interplay between several lifestyle, intellectual, and demographic components

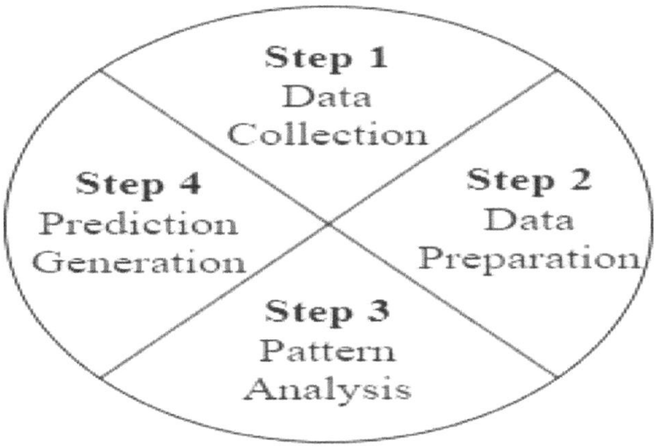

Figure 12.1. Process of predictive analysis.

Source: Author.

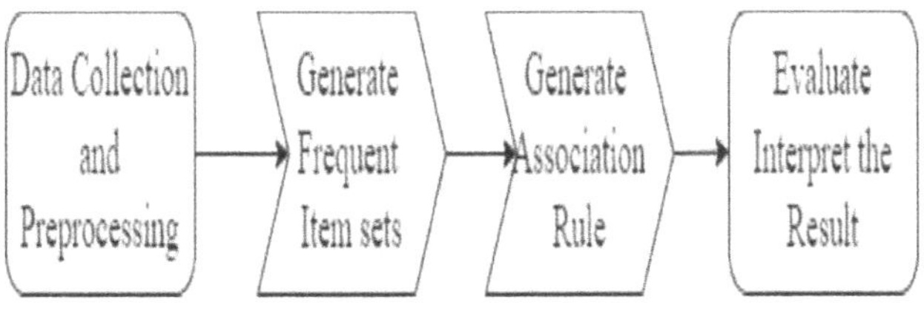

Figure 12.2. Work flow of the relational algorithm.

Source: Author.

that affect psychological results may be missed by these methodologies [16]. Therefore, to better understand and anticipate mental health difficulties affecting medical students, enhanced analytical methods that can unearth hidden relationships and patterns from huge datasets are required. The objective of the proposed methodology is to enhance the accuracy of finding out the relativity between the medical students and the affecting factors. The proposed method demonstrated better associativity between the students and factors with greater confidence level which shows the effectiveness of the method through analyzing the complex mental health patterns.

2. Related Work

One most common causes of fatalities is suicide. In addition, it poses a complicated public health problem that is frequently avoidable with prompt action. Suicide rates are generally rising for a variety of reasons. In this research, we employ a rule of association analysis to identify the most significant rules from a given data set that can be used to predict self-harm behaviours. The Apriori method constitutes one of the most potent machine-learning techniques for finding relationships in information. This technique was used within a dataset of 1250 occurrences and 27 significant features to examine association rules of suicidal behaviour. These have been investigated to identify combinations linked to acts of suicide. They include everyday activities, family history, and responses to mental surveys. With a support of 0.25 and a confidence level of 0.90, the eight key rules have been determined using the Apriori technique [13].

The COVID-19 epidemic has intensified both the mental and social stress that affects university students caused by medical conditions, increased dependence on handheld gadgets and the internet, lack of socializing, and isolation at home. As a result, early stress identification is essential to both their mental health and effective academic achievement. The development in machine learning forecasting algorithms can be extremely important for anticipating stress in the beginning stages to take appropriate action to ensure people's well-being. The purpose of this project is to create a trustworthy machine learning-generated prediction framework for feelings of stress and examine it with real-world information gathered from 444 undergraduates of various ethnic backgrounds via a survey conducted online. Trained machine learning methods were used in the construction of machine learning frameworks.

The tests of chi-squared and principal component analysis, or PCA, were used as enhancement methods. Additionally, genetic algorithms (GA) and grid-based search double-validation were used for hyperparameter optimization (HPO). The results showed that 11.26% or equivalent of the population had a great deal associated with social pressure. Comparatively, it was discovered that about 24.10% of individuals had very high-stress levels, which is concerning for the well-being of students. In addition, the most impressive accuracy (80.5%), precision (1.000), F1 score (0.890), and recall value (0.826) were shown in the prediction outcomes of the machine learning models. It was demonstrated that the various layers' Perceptron model yielded the best accuracy when paired with grid search for hyperparameter optimization and principal component analysis to be a component reduction technique [2].

In developing nations like Bangladesh, mood swings are a widespread psychiatric condition that is increasing. Teenage depression is proven to be common and has an impact on an individual's style of life and way of thinking. Unfortunately, people's mental health issues are frequently disregarded because of the negative opinion that society and the general public place on this illness. Effective therapy is frequently aided by early detection of depression in patients. This study was conducted on Bangladeshi university students to create methods to identify and forecast depression levels. A survey is taken based on Personal and clinical. Sixty-eight Bangladeshi students, ages 19 to 35, completed the questionnaire that was presented to them. The participants had permission to complete the survey despite providing the necessary consent. Following a thorough examination of the answers, 520 data were finally taken into account.

Through a voting mechanism, an integrated depression rating scale was created that uses eight widely used scales to gauge a person's degree of depression. The obtained samples, which included questions from many well-known depression measuring tests as well as individual data, were subsequently subjected to this fusion scale. Additionally, the three grades of depression (regular, moderate, and high) were predicted using machine learning along with deep learning models. To increase the accuracy, of selecting features techniques, hyperparameter optimizers were used. The Gradient Boosting, CNN, and Random Forest models yielded accuracy values of 94.23%, 92.31%, and 98.08% respectively. Using its improved variables, Random Forest achieved a minimal rate of false negatives and the most efficient F-measure. Lastly, the forecasting result from machine learning techniques was interpreted and retraced using the explainable artificial intelligence framework LIME [3].

Globally, there is an important threat of depression, poor positive psychological wellness, and suicidal thoughts and behaviours among students (PMH). It is believed that medical learners are an especially stressed population. Against this backdrop, the current study set out to contrast the overall incidence rates of mood disorders, PMH, and thoughts of suicide ideation/behaviour when it comes to Chinese medical learners to those of students across other fields. A total of about 2,695 Chinese students, including 461 medical pupils, completed self-report questionnaires on low moods, PMH, and suicide ideation and conduct. Of them, 41.2% were women and their age varied from 18 up to 21 has the most impact. 11.5% of pupils in general and 10.2% were medical students expressed having suicidal thoughts for a year. Compared with peers of other disciplines, pupils studying medicine experienced much greater levels and considerably less incidence of depression-related symptoms. As per the dual-factor view of psychological wellness, over 84% of the entire group reported having complete mental wellness. As per the dual-factor view of psychological wellness, over 84% of the entire group reported having complete mental wellness [17].

Overthink is a significant component of everyday existence and is linked to the majority of the tasks we complete. If stress isn't carefully managed, it can negatively affect the condition of our bodies. HRV assessment is currently the most widely used technique for detecting stress, considering recent advancements in this field. While the results obtained are impressive, the process of obtaining the features is laborious and takes time. We provide an approach for transforming the 1D (dimensional) electrocardiogram (ECG) information from the wearable stress as well as the affect detection dataset (WESAD) in 2D (stress/non-stress) ECG images. Prolonged procedures like acquiring features and filtering are not necessary. Transfer learning is what we do to achieve competitive outcomes. Additionally, we show that by employing the quantization strategy, we can get a 90.62% rating accuracy, demonstrating the framework compression strategies can greatly lower the computational capacity of the methods without compromising much of their performance. The outcomes validate the efficacy of our suggested approach and provide concrete evidence regarding the possibilities of applied deep learning techniques towards edge computing as well as mobile applications that leverage subpar hardware [18].

3. Proposed Work

The objective of this work is to find the critical elements and how they interact to forecast mental wellness outcomes by using the Relational Formulated Rule-based algorithm over an extensive dataset of healthcare students. Age, gender, year of study, mother language, employment position, relationship status, hours spent learning, health satisfaction, previous psychotherapy assessments, and a variety of psychological factors and ratings of empathy are among the characteristics included in the dataset. The Figure 12.3 explains the overall process of the proposed methodology of the RFR Algorithm. These factors are meant to be examined to identify the common objects and relationships that have the strongest correlation with negative outcomes that affect mental health, especially excessive levels of mood disorders, cynicism, and sentimental tiredness.

3.1. Data description

The Table 12.1 shows the features used in the collection of data from the different individuals.

3.2. Questionnaire creation

A wide range of academic, psychological, demographic, and lifestyle criteria are included in the extensive and diverse questionnaire created to forecast the psychological condition of students studying medicine. To gather demographic information, questions regarding age, gender, year of study, and mother language have been included. Additionally, concerns regarding employment and relationship status were also included to provide insight into the participants' socioeconomic background. Hours of study each week are used to gauge the amount of work that goes into an assignment. Questions about general well-being satisfaction and previous visits to psychotherapists or psychiatrists were shown to be health-related. Structured scales that include CES-D to signs of depression, the STAI over anxiety disorders, the MBI for burnout, and empathy scores like JSPE and QCAE have all been used to measure mental health.

Table 12.1. Description of dataset

Features	*Sub-Category*
Id	name
Age	20
	21
Curriculum Year	Bachelor 1st year
	Bachelor 2nd year
	Bachelor 3rd year
	Master 1st year
	Master 2nd year
	Master 3rd year
Gender	Male
	Female
	Others
Partnership Status	Yes
	No
Having a job	Yes
	No
Hours of study per week	1–2–3–4–5
Satisfaction with health	Very Dissatisfied
	Dissatisfied
	Neither Satisfied or Dissatisfied
	Satisfied
	Very Satisfied
Psychotherapy last year	Yes
	No
JSPE	1 to 5
QCAE-Cognitive	1 to 5
QCAE-Affective	1 to 5
AMSP	1 to 5
GERT	1 to 5
CES-D	1 to 5
STAI-T	1 to5
MBI-Emotional Exhaustion	1 to 5
MBI-Cynicism	1 to 5
MBI-Academic Efficacy	1 to 5

Source: Author.

3.3. Data collection

The psychological wellness dataset for medical learners was obtained in Kaggle, which had 1790 instances with 20 different attributes. A variety of factors relevant to evaluating mental wellness are included in this dataset, including age, gender, curriculum year, study hours, physical contentment, and past counseling appointments. The data collection was divided into testing and training sets to assess the prediction models; 887 examples were used to develop the model, and 903 were set aside to evaluate its accuracy. Because of this divide, the framework was able to be robustly trained and then validated, guaranteeing that the outcomes were properly replicated.

3.4. Data preprocessing

Preprocessing the data entails many crucial procedures that guarantee that the information is coherent, clean, and appropriate for Relational algorithm analysis. First, based on the amount and kind of data that is absent, deal with any empty or vacant data by either deleting incomplete rows or adding values via statistical techniques. Next, to aid algorithmic processing, specific information like gender and year of study have been numerically encrypted. Standardization or normalization is required for numerical data, including study hours and psychological scores, to guarantee that these factors make an equal contribution to the analysis. The Equation (1) normalizes the missing values in the preprocessing state. Furthermore, data outliers were located and dealt with because they have the potential to seriously distort the outcomes. Discretizing the persistent variables as meaningful groups is also necessary, especially for the Relational algorithm that operates on category variables. The Equation (2) helps is replacing the matching values with the misplaced values in the dataset. The dataset is then ready for efficient association rule mining by converting to a coded form that each attribute denotes the existence or failure to present a condition.

If x_{pq} is missing, replace it with mean:

$$x_{pq} = \frac{1}{N}\sum_{r=1}^{N} x_{rq} \tag{1}$$

where N is the number of non-missing entries in column q,

Alternatively, for categorical data-

If x_{pq} is missing, replace it with mode:

$$x_{pq} = mode(x_{1q}, x_{2q}, \ldots, x_{Nq}) \tag{2}$$

3.5. RFRA-relational formulated rule-based algorithm

Relational rule extraction is a commonly employed method for determining the correlation between elements in an ensemble of data. The patterns of common and frequently occurring objects in this data collection can be found using association rule mining. There are two steps in the method: (1) Identifying frequent objects; and (2) Creating rules. This study employs the Relational rule algorithm as a method of unsupervised learning. According to the needs of the study, association rule mining has been carried out concurrently with the Relational rule algorithm to acquire data extracted from the collection of data. The Relational rule algorithm can mine item collections and implement the information set to generate association rules. This algorithm makes use of two parameters: "confidence" and "support". Support denotes the elements that occur often or when these elements are combined. The Equation (3) shows the threshold support which is minimal in the association. A conditional chance that expresses the likelihood that two events will occur in combination is confidence. It is accomplished by finding recurring, unique items from the collection of data followed by expanding the components to ever-larger item sets unless those item sets contain a sufficient number of items from the dataset. Here, Equation (4) shows the likelihood between the support and confidence level. Let's say the following are two item sets, A and B. To ascertain whether there is a link between them, it is necessary to specify the minimal support and minimum confidence.

Minimum support threshold value:

$$\text{Frequent Itemset}=\{S|\text{support}(S)\geq\sigma\} \tag{3}$$

A previous likelihood of both A and B supports this rule.

$$S(A\rightarrow B) = (|A\cup B|)/N \tag{4}$$

Here S is the support, and N is the total number of transactions. The conditional expectation that the consequent will occur is known as the antecedent. The Equation (5) utilizes in calculating the confidence level. The conditional probability of Y given X represents the confidence rule value.

$$C(A\rightarrow B)= S(A\cup B)/S(A) \tag{5}$$

Here C is the confidence level.

Lift value represents an indicator of a rule's importance. In essence, rules that use classifiers can specify the lift frequency. The Equation (6) represents the lift value related with support and confidence. Splitting the rule's observed confidence value into its projected confidence value yields lift.

$$L(A\rightarrow B)=S(A\cup B)/S(A)*S(B) \tag{6}$$

Here L is the Lift value.

4. Results and Discussion

A better knowledge of various risk components and protective variables related to mental wellness among medical learners can be gained from the analysis's findings.

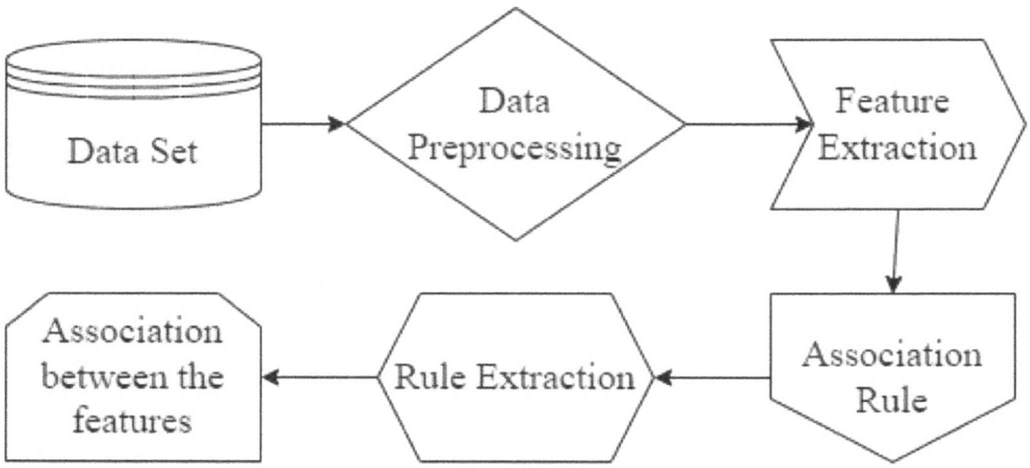

Figure 12.3. Relational formulated rule-based algorithm.

Source: Author.

This information can help with the creation of focused treatments and regulations meant to support this population's hardiness and mental health. Furthermore, educational institutions should take proactive steps to aid students by implementing stress reduction strategies, and support from peers, and connecting them to psychological care by emphasizing the key indicators of disorders associated with mental health. The Table 12.2 reflects the comparisons between the proposed method and the existing method.

By utilizing the Relational Formulated Rule-based Algorithm, the suggested model examined the data collection to find important variables about the healthcare students' mental health. The Figure 12.4 highlights the level of support and confidence association between the most frequent factors. Obtaining association rules having a minimum support of 24% and a confidence level of 99% was the main goal of the evaluation. The identified rules were neither extremely trustworthy nor frequently occurring thanks to these strict criteria. Between the noteworthy results, a standout rule showed that people with previous instances of clinical depression and high scores for stress were significantly more likely to focus on their mental state and to seek advice or actions to relieve their stress. The Figure 12.5 represents the loss rate in occurred in the

Table 12.2. Comparison results

Method	*Support*	*Confidence*	*Accuracy*
Apriori Algorithm	0.25	0.90	87%
Relational Formulated Rule-based Algorithm	0.24	0.99	93.4%

Source: Author.

Table 12.3. Accuracy comparison

Method	*Accuracy*
Apriori Algorithm	87%
Relational Formulated Rule-based Algorithm	93.4%
FP-Growth	88.24%
ECLAT	86%

Source: Author.

stage of training and testing. With a support of 24%, the rule showed up in almost 25% of the sample; its confidence level was 99%, demonstrating its high predictive power.

Support specifies the frequency of the combination between the factors like hours of study connected with the compromization in health, indicating the prevalence

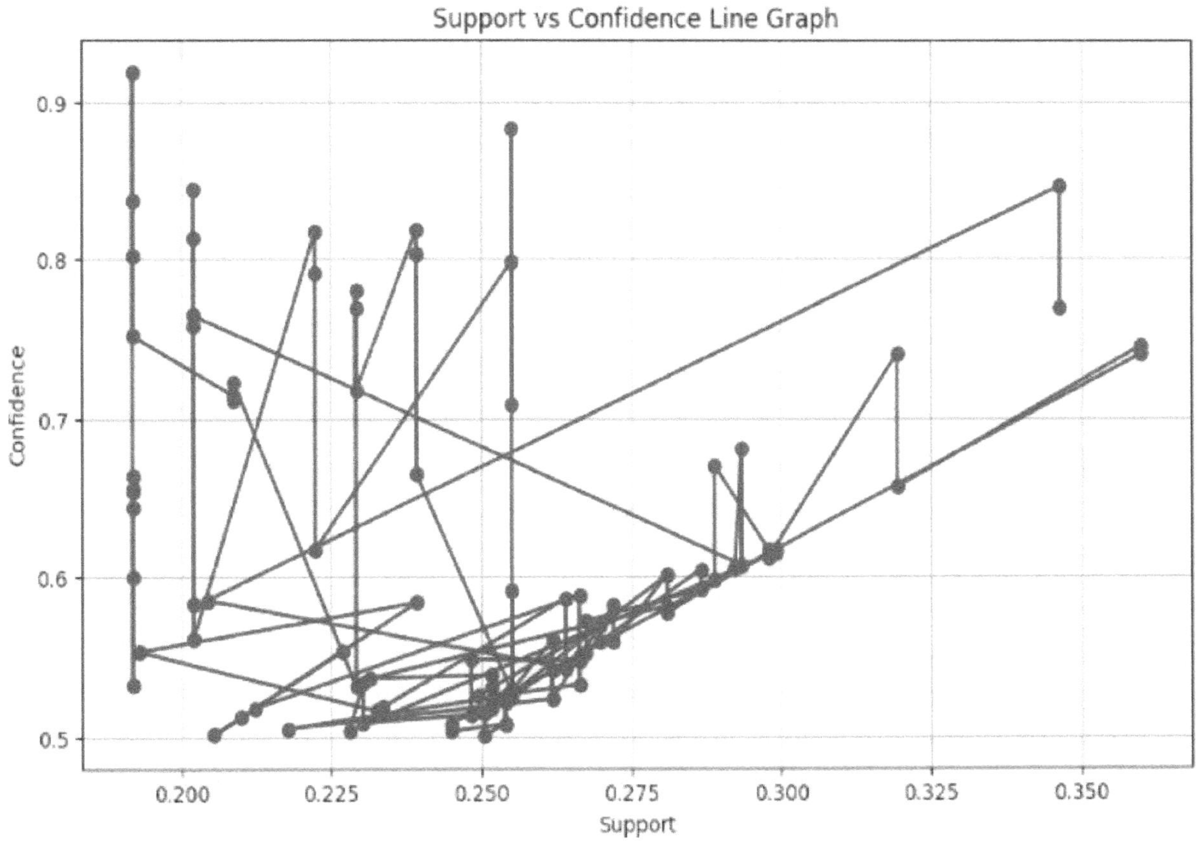

Figure 12.4. Support and confidence level graph.

Source: Author.

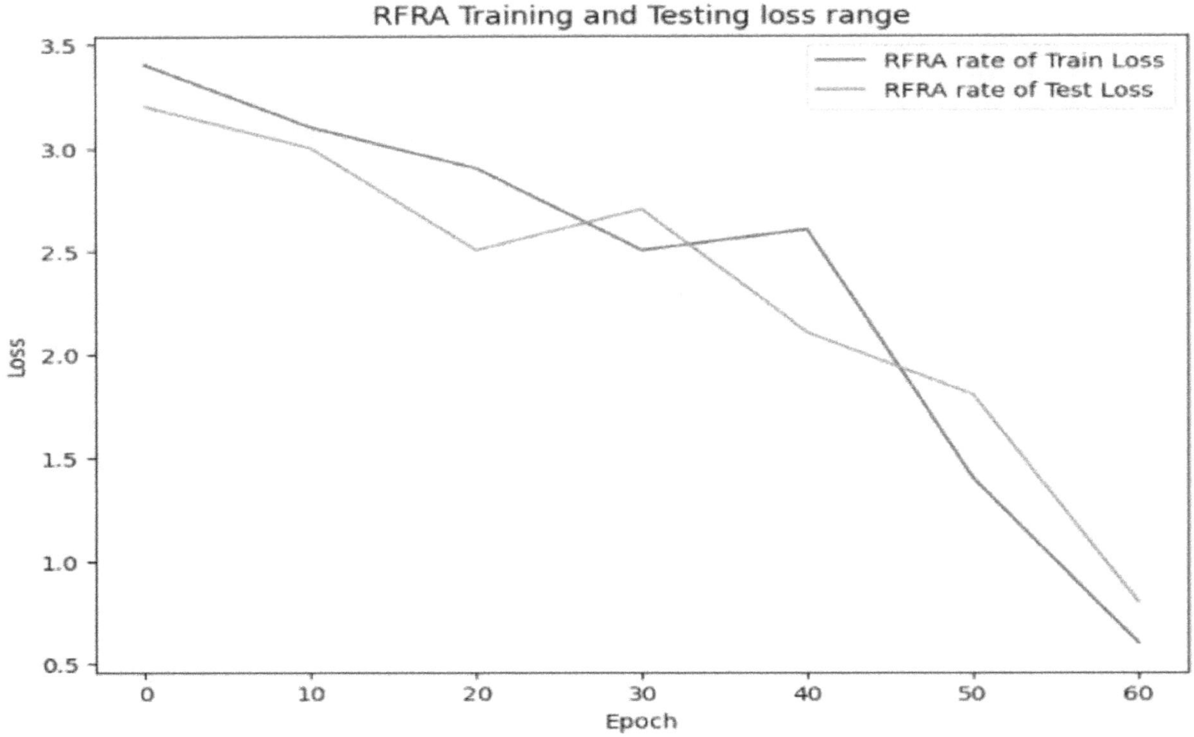

Figure 12.5. Loss rate graph.
Source: Author.

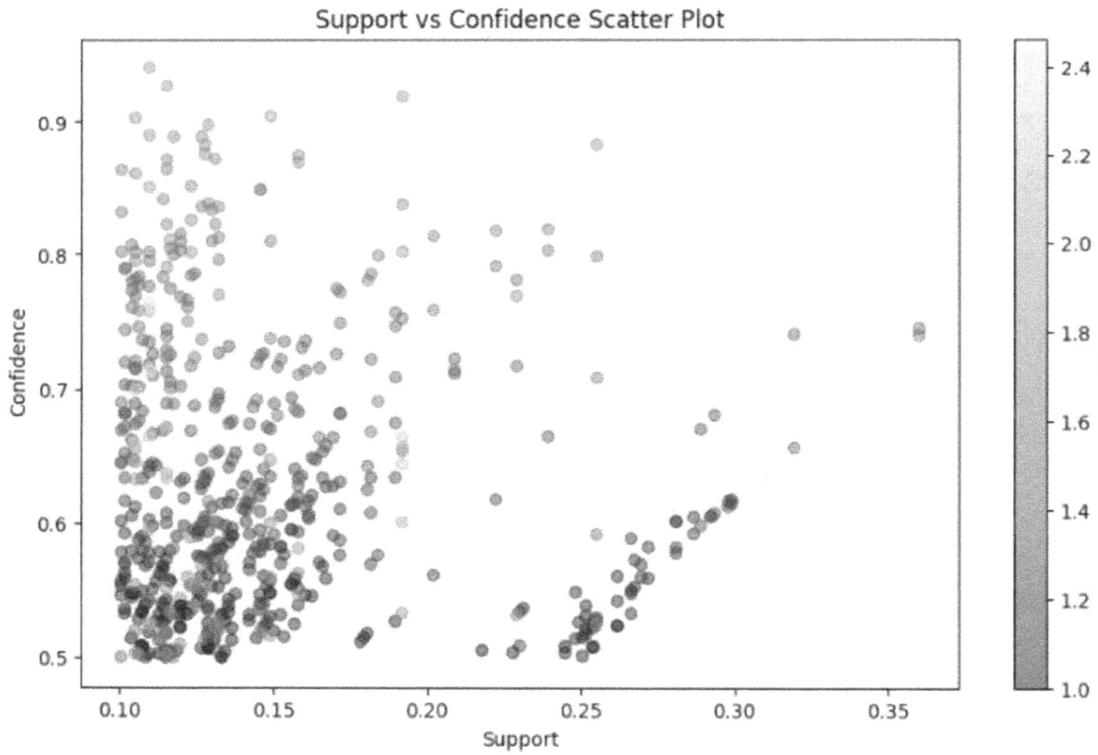

Figure 12.6. Support and confidence level graph.
Source: Author.

in the population (Table 12.3). Confidence specifies the relativity in the factors of stress with the study hours with the likelihood in the accuracy of it. Lift elevates the association strength by comparing the chance of expecting the frequency of the pattern in the medical student.

The finding of the relativity in the comprehensive factors of the medical students with the relative formulation method with Support and Confidence in a strong pattern. The Figure 12.6 represents the plot graph among the support vs confidence in the medical students. The comparison between the support and confidence is 24% which is a quarter part of the dataset that outcomes with the depression in the mental wellbeing among the medical students. The support factor outlets the prevalence in the highlight among the dataset population. A confidence level was achieved with 99% in the identified association in the antecedent conditions to elevate the risks identified among the medical students. Therefore, support vs confidence not only outcomes the strength of the relative formulation method but also gives the risk patterns identified.

There are significant predictive relationships between different elements, as evidenced by the relationship standards' high confidence ratings. The Figure 12.7 represents the accuracy range between the training period and the RFRA method. For example, an extremely substantial probability was found in the relationship rule for individuals who were having financial troubles and poor physical satisfaction. This was highlighted by another important rule. This constraint also satisfies the requirement of emphasizing the vital role that economic pressures and mental wellness play. These results imply that important areas for involvement in preventative efforts are economic security, psychological satisfaction, stress, and depression control.

5. Conclusion and Future Work

All things considered, the use of the Relational Formulated Rule-based algorithm to forecast the psychological wellness of medical learners is a fresh and worthwhile strategy for tackling a crucial problem in healthcare education. The developed study aims to support continuing efforts to enhance the psychological state and general well-being of aspiring healthcare workers by utilizing

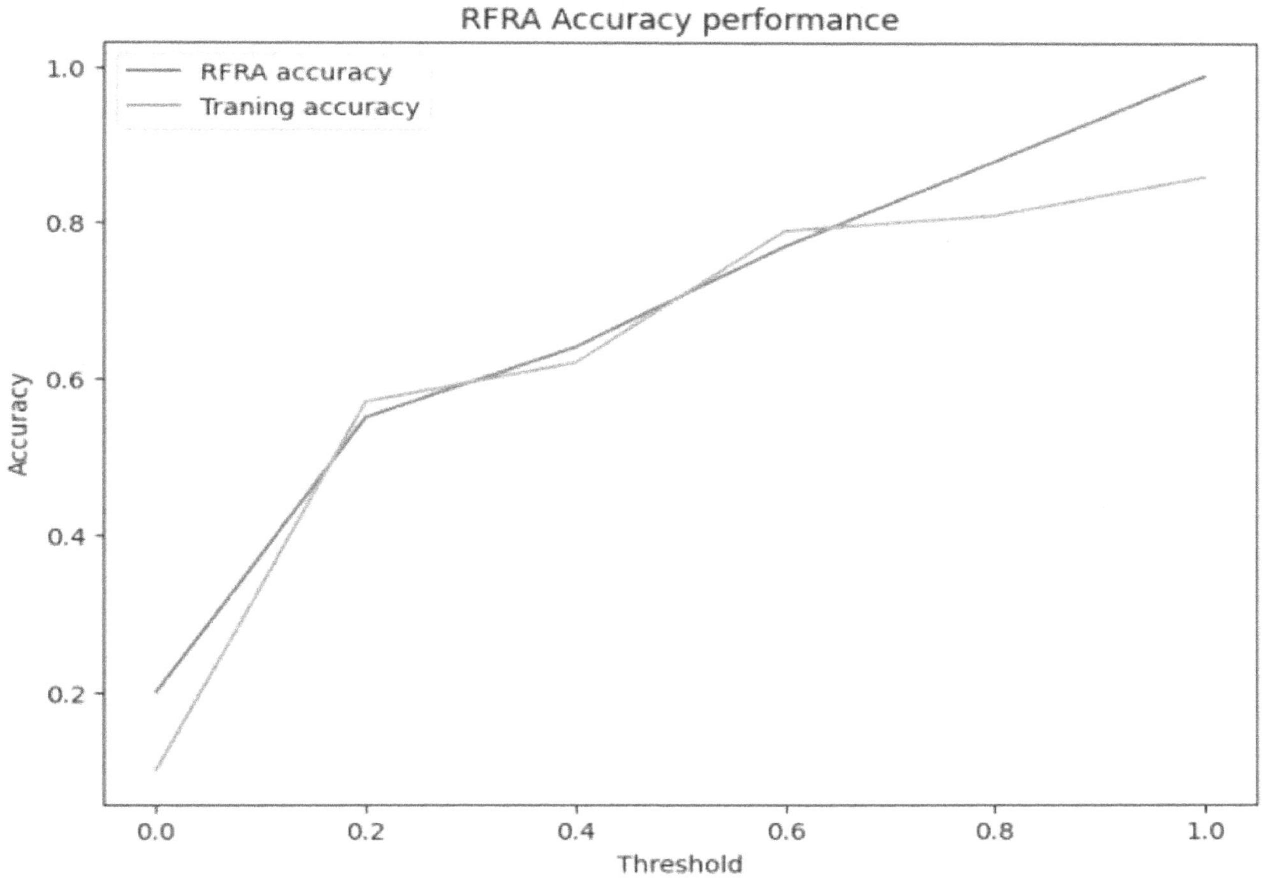

Figure 12.7. Accuracy range graph.

Source: Author.

methods related to data mining. Considering an accessible dataset, the linked factors are analyzed for the well-being of the medical students. The developed algorithm recognizes important factors as well as their relationships by utilizing the Relational rule method. The major association rules were found through the association rule, with a 24% support and a 99% confidence level for each. In the future, we want to create an Android app that will evaluate a person's risk of mental well-being, especially for medical students, specifically when they are depressed. The findings highlight how crucial is the need to provide targeted solutions for people who display these risk indicators. In the future, the idea is to use these results in subsequent research to create a smartphone application that measures risk. Because it will take into account the high-risk characteristics that have been found, this software will be especially helpful for those who are depressed as it will offer tailored assessments of risk along with suggestions for getting treatment.

6. Ethical Consideration

Our research utilizes open-source data that was previously collected and made publicly available by other researchers. As no new data collection from human participants was conducted, and the data used is anonymized and publicly accessible, an ethics statement is not applicable. We have added a clarification regarding the use of open-source data in the manuscript.

References

[1] Posluns, K., & Gall, T. L. (2020). Dear mental health practitioners, take care of yourselves: A literature review on self-care. *International Journal for the Advancement of Counselling*, *42*(1), 1–20.

[2] Ratul, I. J., Nishat, M. M., Faisal, F., Sultana, S., Ahmed, A., & Al Mamun, M. A. (2023). Analyzing perceived psychological and social stress of university students: A machine learning approach. *Heliyon*, *9*(6).

[3] Siddiqua, R., Islam, N., Bolaka, J. F., Khan, R., & Momen, S. (2023). AIDA: Artificial intelligence based depression assessment applied to Bangladeshi students. *Array*, *18*, 100291.

[4] Vos, G., Trinh, K., Sarnyai, Z., & Azghadi, M. R. (2023). Ensemble machine learning model trained on a new synthesized dataset generalizes well for stress prediction using wearable devices. *Journal of Biomedical Informatics*, *148*, 104556.

[5] Steiner-Hofbauer, V., & Holzinger, A. (2020). How to cope with the challenges of medical education? Stress, depression, and coping in undergraduate medical students. *Academic Psychiatry*, *44*, 380–387.

[6] Carlos, K. M., Ahmadi, H., Uban, K. A., & Riis, J. L. (2023). Behavioral and psychosocial factors related to mental distress among medical students. *Frontiers in Public Health*, *11*, 1225254.

[7] Jain, T., Jain, A., Hada, P. S., Kumar, H., Verma, V. K., & Patni, A. (2021, September). Machine learning techniques for prediction of mental health. In *2021 Third International Conference on Inventive Research in Computing Applications (ICIRCA)* (pp. 1606–1613). IEEE.

[8] Chatterjee, S., Chaudhuri, R., Gupta, S., Sivarajah, U., & Bag, S. (2023). Assessing the impact of big data analytics on decision-making processes, forecasting, and performance of a firm. *Technological Forecasting and Social Change*, *196*, 122824.

[9] Zhang, X. (2021, September). Research on Mental Health Rating Algorithm Based on Big Data Classification Algorithm. In *2021 4th International Conference on Information Systems and Computer Aided Education* (pp. 110–114).

[10] Rayan, A., & Alanazi, S. (2023). A novel approach to forecasting the mental well-being using machine learning. *Alexandria Engineering Journal*, *84*, 175–183.

[11] Sun, F. (2022, June). Prediction Model of Mental Health (MH) Based on Apriori Algorithm. In *International Conference on Applications and Techniques in Cyber Intelligence* (pp. 564–571). Cham: Springer International Publishing.

[12] Biilah, M. A. M., Raihan, M., Akter, T., Alvi, N., Bristy, N. J., & Rehana, H. (2022). Human depression prediction using association rule mining technique. In *International Conference on Innovative Computing and Communications: Proceedings of ICICC 2021, 2*, 223–237. Springer Singapore.

[13] Hassan, M. M., Karim, A., Mollick, S., Azam, S., Ignatious, E., & Al Haque, A. F. (2023). An apriori algorithm-based association rule analysis to detect human suicidal behaviour. *Procedia Computer Science*, *219*, 1279–1288.

[14] Laijawala, V., Aachaliya, A., Jatta, H., & Pinjarkar, V. (2020, June). Classification algorithms based mental health prediction using data mining. In *2020 5th international conference on communication and electronics systems (ICCES)* (pp. 1174–1178). IEEE.

[15] Brailovskaya, J., Teismann, T., Lewitzka, U., Gao, Z., Zhang, X. C., & Margraf, J. (2022). Suicidal ideation, suicide attempts and positive mental health in Chinese medical students. *Journal of Affective Disorders Reports*, *9*, 100354.

[16] Ishaque, S., Khan, N., & Krishnan, S. (2022). Detecting stress through 2D ECG images using pretrained models,

transfer learning and model compression techniques. *Machine Learning with Applications*, *10*, 100395.
[17] O'Byrne, L., Gavin, B., Adamis, D., Lim, Y. X., & McNicholas, F. (2021). Levels of stress in medical students due to COVID-19. *Journal of Medical Ethics*, *47*(6), 383–388.
[18] El Othman, R., El Othman, R., Hallit, R., Obeid, S., & Hallit, S. (2020). Personality traits, emotional intelligence and decision-making styles in Lebanese universities medical students. *BMC Psychology*, *8*, 1–14.

13 Leveraging a innovative collaborative ensemble decision trees for professional stress prediction and classification

R. Ragul[1,a], *Devi T.*[2,b], *and N. Deepa*[2,c]

[1]Research Scholar, Department of Computer Science & Engineering, Saveetha School of Engineering, Saveetha Institute of Medical and Technical Sciences, Saveetha University, Chennai, Tamil Nadu, India
[2]Professor, Department of Computer Science & Engineering, Saveetha School of Engineering, Saveetha Institute of Medical and Technical Sciences, Saveetha University, Chennai, Tamil Nadu, India

Abstract: Big data is a method of examining huge and complicated collections of data in order to identify undetected trends, associations, and revelations that may be applied in driving making choices with development. Employers could enhance their efficiency and gain an upper hand by employing modern analytical methods and systems to translate enormous quantities of information into usable knowledge. Stress involves the human body's attitude to observed problems or difficulties, which causes cognitive, physical and emotional responses. Long-term stress can have a severe impact on one's entire health, including anxiety, despair, and a variety of physical symptoms. Identifying stress utilizing heart rate variability (HRV) statistics through the Collaborative Ensemble Decision Tree (DECT) method allows for rapid detection of stress habits, which improves preventive actions and therapies. This strategy promotes psychological wellness and health by offering useful information derived from responses to stress. Applying innovative development as Collaborative Ensemble Decision Tree technique over stress forecasting employing heart rate variability (HRV) information yielded F1-Score of 86.3% and 87% accuracy, this is higher then the traditional random forest model accuracy with 84% and F1-Score 82.4%. The above accuracy rate implies that the algorithm is good at detecting patterns of stress and making solid recommendations. Such accurate identification of stress will greatly help in early treatments and entire psychological care.

Keywords: Applied science, big data analytics, HRV dataset, innovative healthcare, optimizing stress, physiological wellness, collaborative ensemble decision tree, stress prediction

1. Introduction

In the fields of healthcare analysis, stress prediction with machine learning forecast through an approach is a major development [1]. The managerial oversight of mental state, job productivity, and general well-being are all significantly impacted by the capacity for predicting stress levels [2]. One strong and flexible ensemble learning technique that works especially well with complicated datasets that describe stress parameters seems to be the method known as Collaborative Ensemble Decision Tree. It uses a variety of decision trees to modify overfitting and improve prediction accuracy [3]. Along with subjective measurements like acknowledged stress levels, these indications frequently contain physiological parameters similar to heartbeat variability [4].

Stress has become a widespread problem that impacts people in many spheres of life and can show up as a range of behavioural, bodily & psychological signs and symptoms [5]. Anxiety, despair, and trouble focusing are among the psychological impacts of stress that can negatively affect everyday functioning and the psychological state in general [6]. Physiologically, stress can cause headaches, tense muscles, and exhaustion, along with more dangerous ailments like heart problems [7]. Nearly every function in the human body is disrupted by chronic stress, which also raises the danger of chronic illnesses and weakens the immune, digestion, and reproduction systems [8, 9]. Stressed-out people may adopt unhealthy coping strategies like binge eating, smoking, or drinking too much alcohol, all of which worsen existing health issues [10]. Stress has negative impacts both interpersonal interactions and performance in addition to personal health. Stress working can result in burnout, less satisfaction with work, and higher absence rates, all of which have an adverse effect on morale and productivity of the business [11]. Stress can

[a]ragulr9151.sse@saveetha.com, [b]devi.janu@gmail.com, [c]deepa23narayanan@gmail.com

DOI: 10.1201/9781003770473-13

lead to irritation, social disengagement, and arguments in interpersonal relationships, which can make it difficult to spend time alongside family and close companions [12]. Prolonged exposure to elevated stress levels might compromise a person's standards of life by diminishing their capacity to fully appreciate and participate in life's initiatives [13, 14]. It is essential for tackling stress through practical management methods and solutions in order to promote positive surroundings and companies in addition to preserving individual health [15].

The capacity of it can handle big and challenging datasets while not overfitting provides one of its primary benefits when it comes to stress predictions [16]. During learning, the method creates numerous decision trees, while it outputs the mean predictions and the approach in each category (classification) for each tree [17]. This method highlights which physiological measurements are most suggestive of stress and increases accuracy while also offering an estimation of characteristic relevance [18]. This aspect is especially helpful in the medical field, since it allows for more individualized interventions and therapies by illuminating the underlying causes of stress [19, 20].

Utilizing the vast amount, speed, and diversity of data collected through today's digital environment, big data predictions make remarkably accurate predictions about upcoming occurrences, developments, and habits [21]. The Figure 13.1 states the symptoms of stress. Through the use of sophisticated analytics methods like AI, data exploration, and the context of machine learning, prediction reveals discoveries and trends from enormous and intricate datasets. Forecasting techniques are extensively employed in diverse sectors, including healthcare and financial management. In medical attention, they are utilized to anticipate outbreaks of diseases as well as medical results, while in financial matters, they are employed

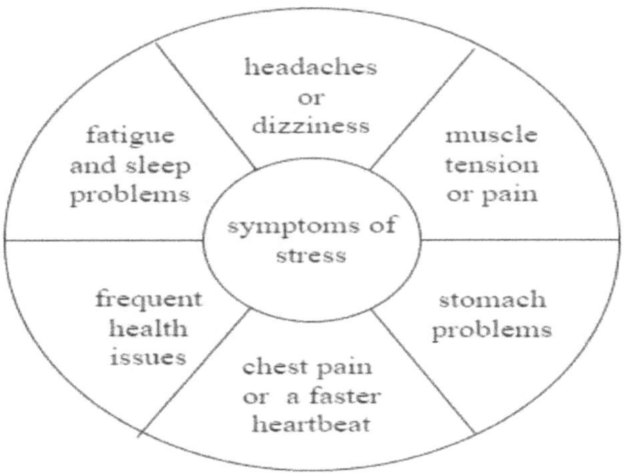

Figure 13.1. Stress symptoms.

Source: Author.

for predicting market movements and client behaviour. Organizations may drive innovation and enhance their competitiveness by making informed decisions, streamlining operations, and personalizing experiences through the real-time processing and analysis of large-scale data.

The key constraints of stress forecasting through the HRV dataset are the probable influencing of individual vascular fluctuations and lack of generalizability caused by amount of data and demographic variety. Furthermore, ecological and behavioural uncertainties may affect model accuracy.

2. Related Work

For identifying and tracking the stress, heart rate variability is the most used feature. Decision making is the very first process, as it outputs the continuous features with a wide range of information. Due to the large amount of criticalness this is such difficult work. To resolve this, Minimum Redundancy and Maximum Relevance (mRMR) is used, as it identifies the frequent features in a countable loop. Heart rate variability is used as a feature in the mRMR approach to find out the stress in the individuals. This approach selects the most interacted features and extends them to the convolutional mRMR to replace the Pearson correlation. Here three publicly accessible data sets are used to compare between the results and four for feature selection four baseline methods were used. The outcomes from the experiment show how well the non-linear featured redundancy measurement works when used with the feature choosing procedure [1].

Stress has become a significant component of everyday existence and is linked to the majority of the tasks we complete. If stress cannot be carefully managed, it can negatively affect our physical and mental wellness. HRV evaluation is currently the most widely used technique for detecting stress, despite recent advancements in this field. While the results obtained are impressive, the extraction of features is difficult as well as time-consuming. Using the WESAD (wearable stress and affect detection dataset), we offer an approach to transform (1 dimensional) ECG information through 2D ECG pictures that reflect stress or absence of stress. Long-lasting procedures like extraction of features and filtering are not necessary. Transfer learning is what we do to achieve competitive outcomes [2].

Through the detection and analysis using physiological signals, improvements in the production and commercialization of small-sized sensors and inexpensive wearables have made it possible to track your lifestyle without exerting much effort. HRV, or heart rate variance, is the measure of the interval between successive heart rates. The HRV signal – which is picked up by various sensors and gadgets – is widely utilized as a proxy for anxiousness and stressful situation levels. Artificial intelligence models using deep learning (DL) topologies had been successfully used to attain previously unheard-of accuracy within the past few years. Artificial Intelligence (AI) systems of learning have long been recognized for their excellent predictive powers [3].

People experience stress frequently in their daily lives, particularly at work, where it can lead to a number of health issues and absence. Although accurately diagnosing it remains challenging, numerous researches have demonstrated a connection connecting stress as well as observable human characteristics. Around the past ten years, there has been a noticeable increase in interest in the issue related to stress detection. The primary method of addressing it has involved the examination of psychological signals during the performance of specific responsibilities under controlled settings. The objective of this research is to present an alternate method for detecting emotions at an office setting by not needing certain controlled settings. This is done by utilizing technological advancements that make it possible to obtain stress-related information throughout a not intrusive manner. In order to achieve this, a non-intrusive video-based imaging software that evaluates the subject's face and collects many physiological data was implemented.

Additionally, during the first phase with one another, volunteers completed a brief survey that provided further details that defines and integrates the physical data. Twenty-eight participants participated in the two-month-long gathering of information initiative. A number of stress recognition algorithms were created; the most effective one that was trained over the binary stressed/non-stress forecasting task obtained an accuracy rating of 86.8% plus a F1 ranking of 87% [4]. About 65% of adults globally have at some point in the last year dealt with stress that interfered with their regular routine. Whenever stress persists over an extended period of time or is persistent (chronic), it grows detrimental to our ability to function, focus, and pay attention. Persistent Significant health problems including cardiovascular disease, hypertension, type 2 diabetes, and anxiety-related disorders are all exacerbated by excessive levels of stress. Numerous academics have concentrated on using machine learning and deep learning techniques to combine multiple features in order to recognize stress. Considering these initiatives, there is disagreement in society over the quantity of characteristics needed to use wearable technology to determine stressors. Furthermore particular to individuals assessment and instructional design became the main topics among the majority of the research studies that have been released. In light of the widespread adoption of accessible wristbands equipment in our local ecosystem, the present study examines a worldwide stress measurement model that combines RF method plus eight HRV variables.

While the performance of every person is assessed by the framework, the RF algorithm's training comprises examples of every topic (worldwide instruction). By combining with SWELL and WESAD datasets, two open-source databases, we confirmed the suggested worldwide stress concept. The global stressing system requires fewer training sessions when the eight HRV characteristics having a greatest classification capability have been selected applying a minimum redundancy maximum relevance (mRMR) approach. Following a global training framework, the suggested global stress measurement method recognizes stress episodes particular to an individual having an accuracy rating of greater than 99%. Evaluating the global stress surveillance platform in practical scenarios should constitute the main goal of further studies [5].

3. Proposed Work

The effects of rising incidence of stress disorders, accurate and dependable models for forecasting are required. The CEDT suggests an efficient method to accurately forecast stress levels. Through the incorporation of various physiologic as well as psychological factors, a CEDT algorithm improves our comprehension of stressors and personal reactions. The technique has the ability to be used in the development of methods for managing stress because of its capacity to handle big datasets and intricate interactions across elements. In attempting to counteract the negative impacts of anxiety, studies and physicians stand to benefit greatly from the enhanced system.

3.1. Data preprocessing

A number of crucial processes must be taken during data preparation in order to guarantee both the modeling effectiveness along with information accuracy when enhancing stressed estimation leveraging heart rate variability (HRV) measures. The collection of information is first cleaned to resolve erroneous or contradictory entries. Data points that are unavailable are eliminated or appropriately imputed by applying statistical analysis. HRV parameter deviations, which have a substantial impact on the reliability of the model, are found, and depending on thresholds relevant to each field, it is either changed or omitted. At reaching a certain point, the heart rate variability data is made to be converted into the correct scale of measure with a uniform manner as it enhances the decision making capacity of the approach. Furtherly to proceed with the building up, finding out and evaluating the dataset splitted into parts. As it finds the wide range of individuals under stress with the heart rate variability measure, lowering the dimension and acquiring the highest efficiency this selection feature method is used to give an acceptable result the given data is processed correctly in a standardized form to reach a reliable output with the CEDT model.

3.2. Feature selection

Using CEDT method with HRV data for identification of stress and optimizing it with the feature selection a threshold point. To increase the accuracy, CEDT method takes the stress level values at different levels points. The significance of different HRV measurements, which include frequency-domain characteristics comprising LF as well as HF position of authority, regressive measures of Poincare plot ratings, as well as time-dependent characteristics like SDNN along with RMSSD, are ranked using correlational evaluation. Reducing the level of difficulty of the model improves the speed of computing and improves applicability for fresh inputs by choosing the portion of those with the greatest important characteristics. This targeted strategy makes guarantees and takes advantage of the most important psychological stress signs, which eventually produces a more accurate and trustworthy stress prediction model.

3.3. Collaborative ensemble decision tree model

In order to provide a trustworthy and precise predicting theory, a number of methodological processes are included in the system creation process for anxiousness predicting optimization applying the method of CEDT using cardiovascular variability factors. First, using bootstrapped values from the collection of data, the CEDT technique is trained with the chosen HRV characteristics to create several decision trees. Next, hyper-parameters tweaking is carried out to maximize important variables like the quantity for trees, highest tree depth with the least amount of information needed by random search methods, to divide a node. By balancing the model outcomes & complexities, this procedure seeks to reduce over estimation and enhance generalization. The Gini impurity is used for the partitioning of the levels at various points for sustaining a threshold point. The complexity makes sure the characteristics and the parameters of this method. This method is good at acquiring and identifying the stress at various levels, thus the method achieves the better accuracy though by tracking it continuously in the stage of training and validation.

The node is x, Percentage of class elements in S_n

Gini impurity:

$$G(x) = \sum_{n=1}^{c} x \, S_{n_h}^{2^a} \tag{1}$$

Entropy:

$$H(x) = -_v \sum_{n=1}^{c} x \, S_{n_h}^{m^a} log(S_n) \tag{2}$$

After the splitation of node the information gathered IG(t)

$$I(t) = H(t) - \left({}^{NL} H(L) + {}^{NR} H(R) \right) \tag{3}$$

With the threshold, select a feature for minimizing the incorrect values, after partitioning.

3.4. Model validation and evaluation

The stress is detected and optimized through CEDT using Heart Rate variability as input data in several levels of validation and training process. The detection has been tested once and again it has been trained in a different manner and again tested several times. To achieve the optimal equilibrium in steps of the process like count of trees, maximum depth, and count of subgrouped trees. From the given total dataset split into training 70% of data with count of 3,69,290 and 30% of testing data with the count of 41,034 splitted for valuation. The outcome gives the increase in the level that has to be aimed to achieve in accuracy. The matrix method is used for verifying whether the person is stressed or not stressed in using the heart rate as the input. The CEDT method has the efficiency to achieve better performance in different levels the overall

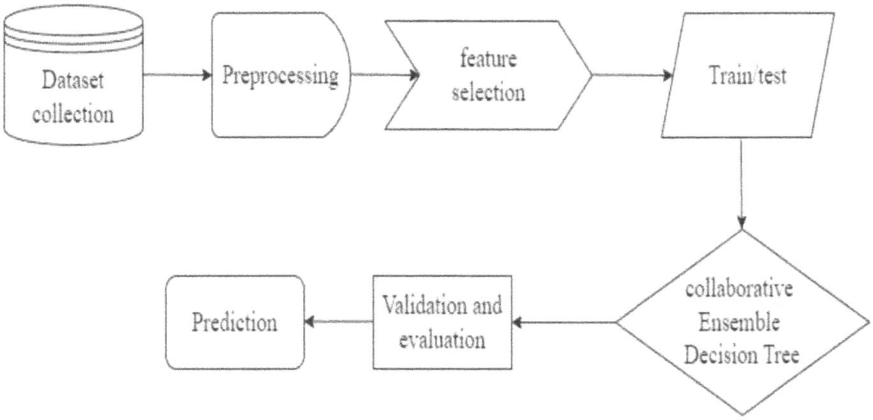

Figure 13.2. Collaborative ensemble decision tree.

Source: Author.

process of validation and evaluation.by using the CEDT technique yields 87% of accuracy and 86.3% of F1-Score.

The results from the CEDT method, the detection of stress from HRV showcases the increase in the accuracy. The outcomes from the HRV dataset for the stress prediction with the attributes likely RF and LF working function, RMSSD, SDNN. This method segregates finds from each separation of Heart rate difference and finds the stress levels. This is handled to get a clear and proper outcome for the model efficiency and to increase the accuracy prediction level. Using visualizations and graphs for forecasting the terms, such tricks help in easy understanding and are simple for the people to understand their level of stress.

3.5. Model optimization

The difficult process in enhancing the capacity of detection of stress in CEDT method is optimization as it is enlightened to achieve better accuracy in the prediction point and identifying the level of stress of an individual. This optimization is done through the fine tuning of the parameters of the trees constructed to identify the stress level. Through increasing the trees or nodes this will achieve the correct prediction in stress of a person for vast amounts of data. For increasing the robustness, the method CEDT, decision trees are merged with a different random selection algorithm. The overall optimization process is done in several stages in the validation process for checking the sustainability and retaining of the accuracy level of the method. The Figure 13.2 represents architecture of Collaborative Ensemble Decision tree. The enhanced method shows that using heart rate data in Collaborative Ensemble Decision Tree method has a well-mannered and better forecasting of the stress levels of an individual.

4. Results and Discussion

With the use measures like, RMSSD in the normalization and SDNN in the stress evaluation, the Heart Rate values are being used for estimation of scores. The past results show the sensitive variations when compared to the proposed method to sort out the forecast of the stress level. The higher contribution is given for the frequency levels to attain the characteristic. Thus the HRV is splitted, trained and tested for the proposed method for the prediction of stress level. The results of the HRV with the real-time values in the prediction of stress is more important for the individuals to take care of them. The Figure 13.3 represents the Accuracy graph of the ensembled method. The Collaborative Ensemble Decision Tree method achieved the successive rate of accuracy 87%.

Through the use of this method, people can come to know about the severity and take precautions to get rid of the stress. By utilizing this method the medical field persons or clinical persons can easily predict the cause for the illness and take the actions to come out of the stress for the individual. This method has been cross validated several times to test whether retains it achieve rate of accuracy level with different stress patterns.

The Collaborative Ensemble Decision Tree method to forecast the stress with HRV data succeeded with the rate of 87% in accuracy level. The Table 13.1 shows the comparisons between the algorithms. The durability, reliability

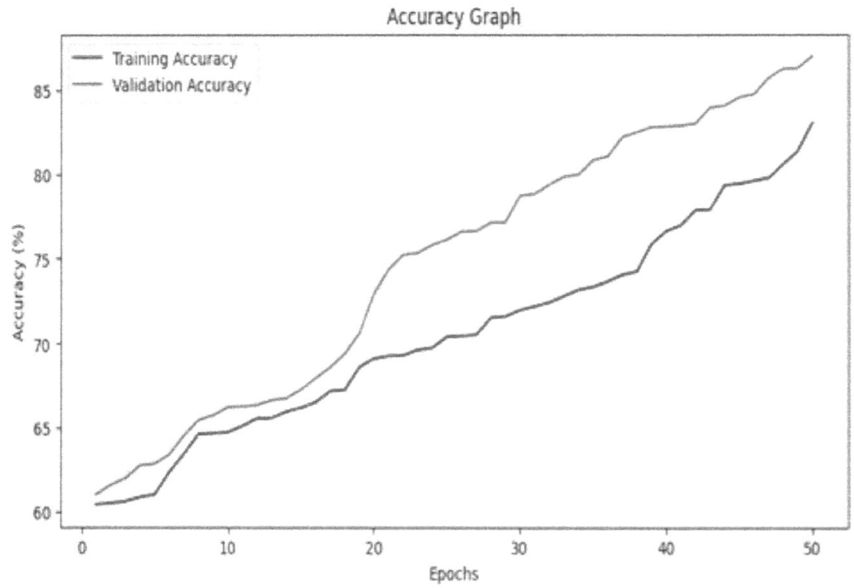

Figure 13.3. Accuracy range graph.

Source: Author.

Table 13.1. Comparison results

Algorithm	*Accuracy*	*F1 score*	*Loss range*
Random Forest	84%	82.4%	29%
CEDT	87%	86.3%	23.8%
KNN	82%	81.1%	27%
Naive Bayes	83%	81.9%	26.3%

Source: Author.

and feature selection has been enhanced in the Collaborative Ensemble Decision Tree method. The effectiveness is high in accuracy and loss rate is low between the training and testing process. Through the method, the medical learners can identify the level of risk and low the excessive risk impact on the people.

5. Conclusion and Future Work

The Heart rate data is collected from different people, as the initial state the data was first preprocessed with the normalization strategy, then it was directed to the training stage and the standardized data has been trained with the decision tree method to construct the highest impact factors to predict the level of stress and testing stage was implemented with the Collaborative Ensemble Decision Tree method. The heart rate is decrypted into an physical information, hence approving HRV is apt for forecasting stress. The accuracy obtained and the stress level prediction of an individual with the frequency attribute in particular time. With utilization of the basic demographic information, health records data, and stress prediction is done. Working together with medical experts and adding new data to the model on a regular basis will guarantee its robustness and wider applicability in daily life as well as clinical procedures.

References

[1] Ihianle, I. K., Machado, P., Owa, K., Adama, D. A., Otuka, R., & Lotfi, A. (2024). Minimising redundancy, maximising relevance: HRV feature selection for stress classification. *Expert Systems with Applications*, *239*, 122490.

[2] Ishaque, S., Khan, N., & Krishnan, S. (2022). Detecting stress through 2D ECG images using pretrained models, transfer learning and model compression techniques. *Machine Learning with Applications*, *10*, 100395.

[3] Haque, Y., Zawad, R. S., Rony, C. S. A., Al Banna, H., Ghosh, T., Kaiser, M. S., & Mahmud, M. (2024). State-of-the-art of stress prediction from heart rate variability using artificial intelligence. *Cognitive Computation*, *16*(2), 455–481.

[4] Rodrigues, F., & Correia, H. (2024). Semi-supervised and ensemble learning to predict work-related stress. *Journal of Intelligent Information Systems*, *62*(1), 77–90.

[5] Dahal, K., Bogue-Jimenez, B., & Doblas, A. (2023). Global stress detection framework combining a reduced set of HRV features and random forest model. *Sensors*, *23*(11), 5220.

[6] Grégoire, J. M., Gilon, C., Carlier, S., & Bersini, H. (2023). Autonomic nervous system assessment using heart rate variability. *Acta cardiologica*, *78*(6), 648–662.

[7] Balint, E. M., Daniele, V., Langgartner, D., et al. (2023). Heart rate variability predicts outcome of short-term psychotherapy at the workplace. *Psychophysiology*, *60*(1), e14150.

[8] Chalmers, T., Hickey, B. A., Newton, P., et al. (2022). Associations between sleep quality and heart rate variability; implications for a biological model of stress detection using wearable technology. *International Journal of Environmental Research and Public Health*, *19*(9), 5770.

[9] Seipäjärvi, S. M., Tuomola, A., Juurakko, J., et al. (2022). Measuring psychosocial stress with heart rate variability-based methods in different health and age groups. *Physiological Measurement*, *43*(5), 055002.

[10] Faust, O., Hong, W., Loh, H. W., et al. (2022). Heart rate variability for medical decision support systems: A review. *Computers in Biology and Medicine*, *145*, 105407.

[11] Cui, Z., Ma, Y., Ma, M., Huang, R., & Du, B. (2023, December). Towards a Lightweight Stress Prediction Model: a Study on Dimension Reduction and Individual Models in HRV Analysis. In *2023 IEEE 29th International Conference on Parallel and Distributed Systems (ICPADS)* (pp. 1709–1716). IEEE.

[12] Benchekroun, M., Velmovitsky, P. E., Istrate, D., Zalc, V., Morita, P. P., & Lenne, D. (2023). Cross dataset analysis for generalizability of HRV-based stress detection models. *Sensors*, *23*(4), 1807.

[13] Martinez, G. J., Grover, T., Mattingly, S. M., et al. (2022). Alignment between heart rate variability from fitness trackers and perceived stress: Perspectives from a large-scale in situ longitudinal study of information workers. *JMIR Human Factors*, *9*(3), e33754.

[14] Iqbal, T., Elahi, A., Wijns, W., Amin, B., & Shahzad, A. (2023). Improved stress classification using automatic feature selection from heart rate and respiratory rate time signals. *Applied Sciences*, *13*(5), 2950.

[15] Jiryis, T., Magal, N., Fructher, E., Hertz, U., & Admon, R. (2022). Resting-state heart rate variability (HRV) mediates the association between perceived chronic stress and ambiguity avoidance. *Scientific Reports*, *12*(1), 17645.

[16] Chatterjee, S., Chaudhuri, R., Gupta, S., Sivarajah, U., & Bag, S. (2023). Assessing the impact of big data

analytics on decision-making processes, forecasting, and performance of a firm. *Technological Forecasting and Social Change*, *196*, 122824.

[17] Zhang, X. (2021, September). Research on Mental Health Rating Algorithm Based on Big Data Classification Algorithm. *In 2021 4th International Conference on Information Systems and Computer Aided Education* (pp. 110–114).

[18] Rayan, A., & Alanazi, S. (2023). A novel approach to forecasting the mental well-being using machine learning. *Alexandria Engineering Journal, 84*, 175–183.

[19] Sun, F. (2022, June). Prediction Model of Mental Health (MH) Based on Apriori Algorithm. *In International Conference on Applications and Techniques in Cyber Intelligence* (pp. 564–571). Cham: Springer International Publishing.

[20] Biilah, M. A. M., Raihan, M., Akter, T., Alvi, N., Bristy, N. J., & Rehana, H. (2022). Human depression prediction using association rule mining technique. *In International Conference on Innovative Computing and Communications*: *Proceedings of ICICC 2021*, 2, 223–237. Springer Singapore.

[21] Hassan, M. M., Karim, A., Mollick, S., Azam, S., Ignatious, E., & Al Haque, A. F. (2023). An apriori algorithm-based association rule analysis to detect human suicidal behaviour. *Procedia Computer Science*, *219*, 1279–1312.

14 Early detection of Poly Cystic Ovary Syndrome using ultrasound ovary image – A convolutional neural network approach

Reka S.[1,a], *Suriya Praba T.*[2,b], *and Vanipenta Naga Nithin Reddy*[2,c]

[1]School of Computing, Vel Tech Rangarajan Dr. Sagunthala R&D Institute of Science and Technology, Avadi, Chennai, Tamil Nadu, India
[2]School of Computing, SASTRA Deemed University, Thirumalaisamudram, Thanjavur, Tamil Nadu, India

Abstract: An affliction that is common among women in their reproductive year is PCOS (Poly Cystic Ovary Syndrome). This syndrome is defined by a variety of prodrome and physical signs. Whilst the exact aetiology of PCOS is unknown, it is considered as the combination of hereditary and environmental factors. Ultrasound imaging provides non-invasive, comprehensive view of ovaries, which is crucial for identifying and treating PCOS. It offers insights into ovarian morphology, follicle count and endometrial health, allowing physicians to make decisions about PCOS diagnosis, treatment and monitoring. However, manual ultrasound image analysis is time-consuming and prone to human error, which can cause unpredictability and inconsistency in the diagnosis and treatment of PCOS. Advancements in deep learning techniques are revolutionizing ultrasound imaging for the diagnosis of PCOS by addressing the drawbacks of manual processes. Particularly, to detect complex patterns, automate follicle counts, and deliver reliable, standardized and reproducible results. In this article, Adaptive Histogram Equalization (AHE) is used as the image enhancement technique. To obtain cyst boundary and contour mask, various segmentation models are employed. The segmented image produced by U-Net++ has achieved highest IoU of 0.86 and Dice of 0.92. Furthermore, Convolutional Neural Network (CNN) classifiers such as ResNet, VGG 16, VGG 19, Alexnet and Inception v3 have been implemented. Additionally, classifier models performance are also evaluated and observed that VGG 19 has attained the highest accuracy of 94.16% than other models.

Keywords: Adaptive histogram equalization, convolutional neural network, deep learning, poly cystic ovary syndrome

1. Introduction

Around the world, a significant amount of women suffer from Polycystic Ovary Syndrome. It results in a significant number of immature eggs being produced by the ovaries, which develop into cysts and cause the ovaries to expand and generate more androgen [1]. Polycystic ovarian morphology (PCOM) and hyperandrogenism are the main characteristics [2]. PCOM is defined by an ovarian volume of 10 ml or 25 follicles per ovary with transducer frequencies of 8 MHZ. Among the numerous health issues that women in the twenty-first century encounter, PCOS has become a major and common issue that affects women between the ages of 18 and 44. It affects about one in fifteen women globally; prevalence estimates in India range from 3.7 to 22.5% [3], and they are higher in urban than rural regions, maybe as a result of stress and lifestyle choices. Numerous psychological conditions that ruin the quality of life are associated with it, such as Alopecia, excessive hair growth, depression, hirsutism, irregular menstrual cycles, abrupt weight gain, thyroid issues, type 2 diabetes, high blood pressure, and metabolic disorders such as hypertension, hyperinsulinemia, abdominal obesity, and dyslipidaemia. Low levels of luteinizing hormone (LH) and follicle-stimulating hormone (FSH) along with excessive prolactin levels in PCOS-affected ovaries prevent follicle growth and maturation. When FSH and LH levels are appropriate, a single follicle typically reaches a diameter of around 20 mm and is prepared for ovulation [4]. Follicles in polycystic ovaries remain immature and cease developing at 5–7 mm during ovulation. Due to the prolonged production of estrogen, these immature follicles emit a hormone that thickens the uterine lining and causes spotting or heavy bleeding.

One of the main tools for early PCOS identification is ultrasound imaging that provides crucial details on the count, size and location of follicles. However, it is time-consuming, prone to human error and dependent on the availability of qualified radiologists. As a result, many women continue to go undetected and untreated. These difficulties highlight the necessity of intelligent

[a]reka24nov@gmail.com, [b]suriyapraba@cse.sastra.edu, suriyathiyagarajan03@gmail.com, [c]naganithin2004@gmail.com

DOI: 10.1201/9781003770473-14

computer-assisted technologies to assist physicians. In this regard, deep learning methods have become an invaluable tool for image analysis, overcoming the difficulties of manual process and producing promising results in recent years [5]. Although a number of methods using ultrasound imaging to detect PCOS have been proposed, their effectiveness is frequently constrained by specific challenges. The following are the key contribution of this article.

1. Image Enhancement: Employed Adaptive Histogram Equalization (AHE) to adaptively modify the pixel intensity distribution in various areas of the image
2. Segmentation: Incorporates deep learning technique to identify the cysts boundary and extract the important features from ultrasound ovary image
3. Classification: Convolutional Neural Network (CNN) models including ResNet, VGG, Alexnet and Inception v3 have been used to classify PCOS.

The rest of this article is organized as follows. Section 2 presents the related works on ultrasound ovary image for PCOS detection. Methods used in this article are explained in section 3. Proposed work is illustrated in section 4 and Section 5 presents the dataset description. Section 6 describes the result and discussion. Conclusion of the research work is represented in Section 7.

2. Related Works

Bedi et al. [6] proposed hybrid model for PCOS detection, Adaptive bilateral filter is employed as noise reduction technique and AResUnet (attention-based residual Unet) is used as segmentation model to identify the cyst presented in ultrasound ovary image. Machine Learning algorithms such as Decision Tree, Naïve Bayes, Support Vector Machine and K-Nearest Neighbour are used to classify PCOS images. Paramasivam et al. [7] proposed SD-CNN to extract the pertinent features and used machine learning algorithm such as Random Forest (RF), LR (Logistic Regression) and SVM for the classification of PCOS. The highest accuracy of 96.43% has been achieved using RF. Suriya et al. [8] used GAN (Generative Adversarial Network) to address the overfitting issue. Convolutional Neural Network (CNN) based classifiers includes ResNet, Alexnet, Inception v3 and VGG are employed as classification model, VGG 19 has achieved 96% of accuracy than other models. Alamoudi et al. [9] proposed a fusion model includes hormonal data and ultrasound ovary image for the prompt diagnosis of PCOS. To address the overfitting issue, the authors have applied data augmentation and AHE (Adaptive Histogram Equalization) for preprocessing the data. For the classification of PCOS, various deep learning models includes Inception v3, VGG16, VGG19, MobileNet, DenseNet121, and DenseNet201 are employed. MobileNet has achieved 82.46% of accuracy with fusion model. Nazarudin et al., [10] proposed a hybrid segmentation approach which combines Otsu's thresholding with the Chan–Vese method to identify follicles and segmentation of cyst. Dice score and Jaccard index are used as performance metrics to evaluate the performance of proposed segmentation model. [11] To classify PCOS, the authors developed PCONet model and fine-tuned inception v3 and compared the results. The best accuracy of 98.12% has been achieved by the proposed PCONet than the accuracy of 96.56% using fine-tuned Inception v3. Rachana et al. [12] developed automatic PCOS diagnostic tool using K-Nearest Neighbor algorithm and achieved 97% of accuracy. Otsu thresholding is used as segmentation technique to obtain RoI (Region of Interest). Gopalakrishnan et al. [13] used modified Otsu method for the cyst segmentation and extracted the features such as perimeter, area, eccentricity, major and minor axis. To remove speckle noise from ultrasound ovary image, used various filters includes Lee filter, Kuan filter, Frost filter, Median filter, Gaussian filter and Wiener filter and evaluate the performance of each filter. [14] Nazarudin focused on segmentation process for the accurate detection of PCOS. Median filter is used to remove noise and proposed hybrid segmentation technique (Chan–Vese and Otsu thresholding) to extract the features.

Previous research has focused on traditional segmentation techniques and has given little attention to ultrasound ovary images for PCOS detection. The majority of research relied on numerical clinical data. It may not always yield an appropriate solution. This research focuses on preprocessing ultrasound ovary images and uses advanced deep learning-based segmentation models to determine the Region of Interest (RoI) in order to address this research gap and improve early detection of PCOS.

3. Methods

3.1. AHE (Adaptive Histogram Equalization)

Let the pixel intensity value at coordinates (x,y) be P(x,y). The image is separated into $U \times V$ tiles or non-overlapping sub-images. Determine the histogram $g_T(q)$ for T (each tile), where q indicates the pixel intensity value and the frequency of occurrence of q in the tile T is denoted by $g_T(q)$.

For a specific intensity level q in tile T, the CDF (Cumulative Distribution Function) is provided in Equation (1)

$$CDF_T(q) = \frac{\sum_{k=0}^{q} g_T(k)}{U \times V} \tag{1}$$

The tile T's intensity mapping is calculated as follows Equation (2)

$$P'_T(x, y) = CDF_T(P(x, y)) \,.\, (N-1)(x, y)) \,.\, (N-1) \tag{2}$$

where, total number of intensity levels is indicated as N and intensity changes at (x,y) is represented as $P'_T(x,y)$.

Bilinear interpolation is utilised to provide seamless transitions between adjacent tiles and avoid block artifacts. The mapped intensities from neighbouring tiles are blended to determine the final intensity (x,y) at a pixel inside a tile. The interpolation between the adjacent tiles T_1, T_2, T_3, T_4 is calculated using the formula Equation (3)

$$P'_T(x,y) = \beta\gamma P'_{T1}(x,y) + \beta(1-\gamma) P'_{T2}(x,y) + (1-\beta)\gamma P'_{T3}(x,y) + (1-\beta)(1-\gamma)P'_{T4}(x,y) \quad (3)$$

where, the weights β and γ are determined by the pixel's distance from the centres of the neighbouring tiles.

3.2. *U-Net++ for image segmentation*

The encoder which comprises of several convolutional layers and downsampling that is, max-pooling) processes the input image of size Z. The encoder's output at level k for each stage s is provided in Equation (4).

$$W_{k,s} = \sigma\,(M_{k,s} * W_{k-1,s} + a_{k,s}) \quad (4)$$

where, $M_{k,s}$ indicates the learnable weights $a_{k,s}$ refers the bias, activation function is defined as σ that is, ReLU, convolution operation is indicated as * and previous levels feature map is denoted as $W_{k-1,s}$.

In the decoder, upsampling layers restore the image's spatial resolution. The feature map is calculated at each level k and iteration s as represented in Equation (5)

$$V_{k,s} = \sigma\big(M_{k,s}^{up} * U(V_{k+1,s}) + a_{k,s} + \textstyle\sum_{h=1}^{s} M_{k,h}^{skip}\, W_{k,h}\big) \quad (5)$$

where, $M_{k,s}^{up}$ is the upsampling layer's weight, bilinear operation is denoted as $U(V_{k+1,s})$, skip connection and weights are denoted as $M_{k,h}^{skip}$ and the encoder path of the feature map is defined as $W_{k,h}$. U-Net++ improves gradient flow and feature reuse using dense skip connections. The encoder and decoder use these links to create intermediate feature maps. The intermediate node's output feature map at each level k is provided in Equation (6).

$$D_{k,y} = \sigma\big(\textstyle\sum_{h=0}^{y} M_{k,h}\, W_{k,h}\big) \quad (6)$$

where, the intermediate node of y is denoted as $D_{k,y}$ and connection weights are indicated as $M_{k,h}$.

4. Proposed Model

Early detection is crucial for PCOS to be treated promptly and to avoid complications. Early diagnosis can be greatly aided by the use of deep learning algorithms, which allow physicians to treat patients quickly and efficiently. In this article, the proposed work has been divided into three major parts Figure 14.1.

1. Image enhancement using AHE (Adaptive Histogram Equalizer)
2. Deep learning techniques such as U-Net, U-Net++, deeplab v3, cascaded net and PSPNET for image segmentation
3. Performance analysis of segmented image obtained by various deep learning techniques.

4.1. *Image enhancement using AHE (Adaptive Histogram Equalizer)*

Improving the ultrasound image of the ovaries can draw attention to important details, structures and make abnormalities associated with PCOS more visible. By enhancing contrast, decreasing artifacts, and emphasizing the relevant region of interest, these methods aid in the early detection of cysts, follicular abnormalities, and other PCOS symptoms. In this article, AHE has been employed to enhance the ovary images Figure 14.2. The input image was initially divided into tiny tiles using AHE. In order to equalize each tile's histogram with a clip limit, a cumulative distribution function (CDF) was used for scaling and mapping, computing, distribution and redistribution. Then, bilinear interpolation is utilized to provide seamless transitions between adjacent tiles and avoid block artifacts. The mapped intensities from neighbouring tiles are blended to find the final intensity (x,y) at a pixel inside a tile. In addition to improving local contrast enhancement, this method increased the visibility of slopes and boundaries.

4.2. *U-Net++ for image segmentation*

Image segmentation is a crucial part which helps classification model to perform efficiently. In this article, U-Net++ architecture Figure 14.3 has been employed to extract the important features from ultrasound ovary image especially cysts boundary, size, texture and count. It has three major components such as encoder, decoder and skip connection. Contextual information is obtained by breaking down the image into several feature map levels, which is done by the encoder, which contains basic unit and subsampler. The basic element is a 3x3 convolutional layer with an activation function and batch normalization. Local spatial features are first extracted from the two-dimensional feature map using the convolutional layer. The data variations within the network then remain to a minimum by using normalization. Furthermore, nonlinearity is introduced by transforming the output using the Rectified Linear Unit (ReLU) function. In order to preserve important data features and eliminate less pertinent information, the subsampler aggregates neighbouring values in the feature maps. The feature map is downsampled to a reduced resolution

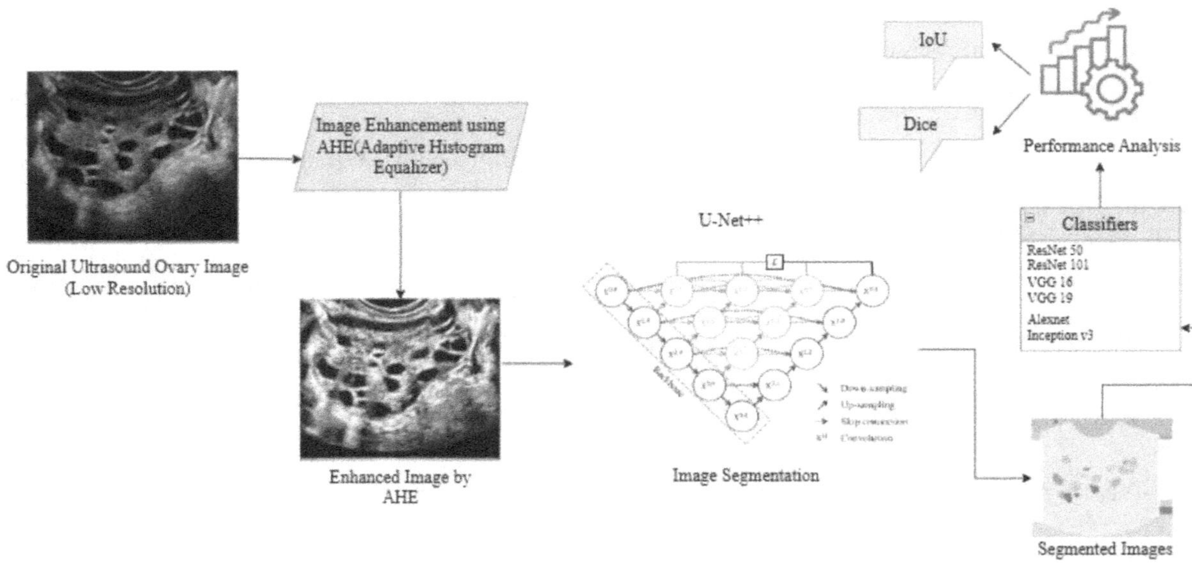

Figure 14.1. Proposed model.

Source: Author.

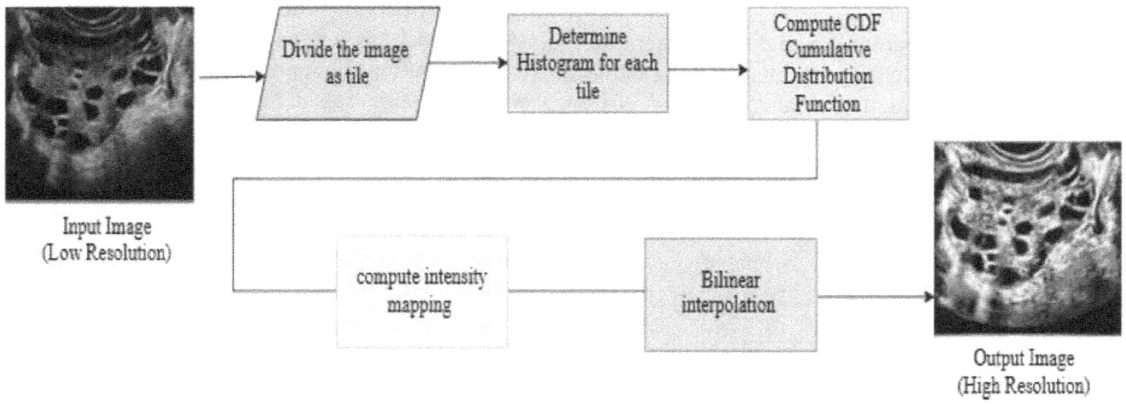

Figure 14.2. AHE architecture.

Source: Author.

by screening the original feature information, which increases the receptive field of the encoder's subsequent convolutional layer.

In order to map the feature map to the same resolution as the input ultrasound ovary image, the decoder gradually restores the feature map's spatial and detailed information. The decoder module is made up of an interpolator and a basic unit and is intended for cyst segmentation. Through interpolation, it decodes the encoded feature map and produces the segmented output that identifies the areas of the ovary image that contain cysts.

In ultrasound ovary images, jump connections improve information flow between the encoder and decoder for cyst segmentation. These connections gradually refine the feature information by connecting feature maps from the encoder to the decoder layer by layer. Also, it reduces the semantic gap between encoder and decoder features. This method improves segmentation performance by utilizing contextual information to help the network model capture target boundary features more precisely.

5. Experiments

5.1. Dataset description

In this article, ultrasonography ovary images with and without PCOS from the publicly available Kaggle repository were used as dataset. It includes 1932 ovarian ultrasound images. There are two subfolders in this dataset: train and test, and they are divided into two categories:

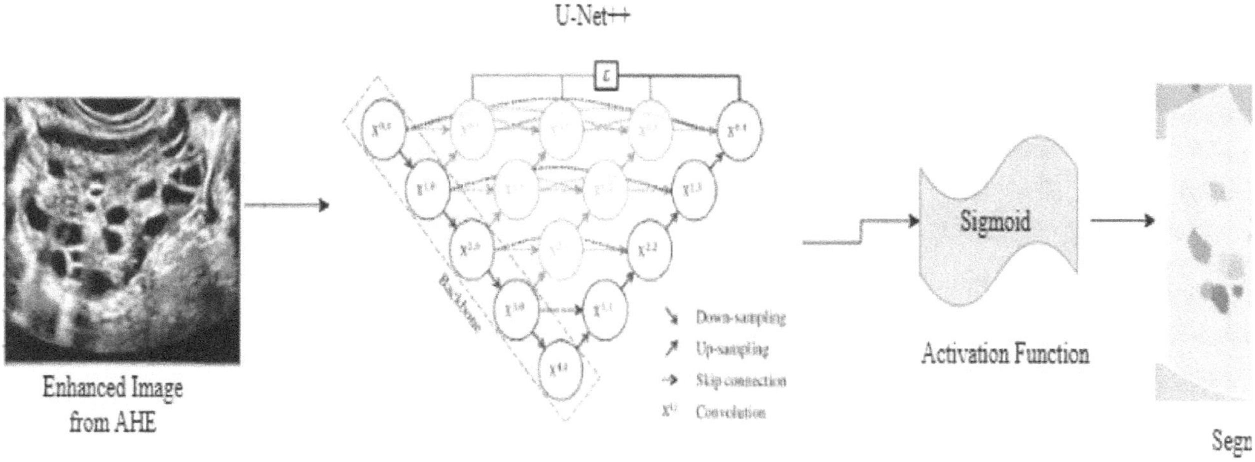

Figure 14.3. U-Net++ Architecture.

Source: Author.

infected (PCOS ultrasound ovarian) and non-infected (healthy ovary).

5.2. *Performance metrics used and implementation details*

The DELL Precision 3660, Intel Core i5-12500 Processor with 64GB of RAM, NVIDIA T1000 4GB Discrete Graphics is used for implementation of this work and Google Colab were used for the implementation of all models.

The performance metrics such as IoU (Intersection over Union) and Dice Score has been used to evaluate the performance of various segmentation models. Accuracy, precision, recall, F1 score, and AUC score are also used to assess the classification models' performance.

6. Results and Discussion

Enhancing the image quality and eliminating noise are necessary for accurately identifying the region of interest. Additionally, it improves classifier models performance. In this article, AHE is employed to improve the image quality from low resolution to high resolution as depicted in Figure 14.4. Furthermore, various deep learning methods have been used to extract the pertinent features like cysts boundary, size, texture and its count from the enhanced images produced by AHE. The segmentation performance is evaluated using IoU and Dice respectively. Table 14.1 depicts the comparative performance of segmentation techniques such as U-Net, U-Net++, deeplab v3, CascadedNet and PSPNet. The least IoU and Dice value of 0.13 and 0.12 has been achieved by PSPNet. Then, DeepLab v3 stood second and achieved IoU of 0.47 and Dice of 0.64. The best performance has been achieved by U-Net++ and produced the IoU of 0.86 and Dice value of 0.92.

The enhanced images achieved using AHE and the high performed segmented results achieved by U-Net++ are used to train the classifier models. In this article, deep learning classifiers such as ResNet 50, ResNet 101, VGG 16, VGG 19, Alexnet and Inception v3 has been employed to classify PCOS ultrasound images. Performance metrics including accuracy, precision, Recall, F1 score and AUC score are used to evaluate the classifier models. Table 14.2 depicts the performance of the models without AHE and U-Net++ results. VGG 19 has achieved 82.36% as the highest accuracy, precision of 85.21%, Recall of 80.41%, F1 score of 85.65% and AUC score of 89.35%. Inception v3 stood second, achieved 80.17% accuracy. Table 14.3 show the performance of the classifier models using the enhanced image produced by AHE and segmented results attained by U-Net++. The highest accuracy of 94.16%, precision of 96.15%, Recall of 92.25%, F1 score of 97.96% and AUC score of 98.15 has been attained using VGG 19 than other models.

Table 14.4 depicts the proposed model performance with existing works in terms of accuracy. Abrar et al. [9] to enhance the ultrasound image quality AHE (Adaptive Histogram Equalizer) was used. The authors have not focused the segmentation process. The enhanced images produced by AHE were used for model training. Various CNN models such as DenseNet121, MobileNet, Inception v3, VGG 19, DenseNet201, and VGG 16 has been employed as classifiers. The better performance has been attained by Inception v3 and produced the highest accuracy of 84.81%. Illayaraja [13] employed Gaussian low pass filter for image enhancement and Multilevel Thresholding was used for segmentation. To classify PCOS images, four machine learning classifiers: SVM, RF, LDA, NB has been used. The highest

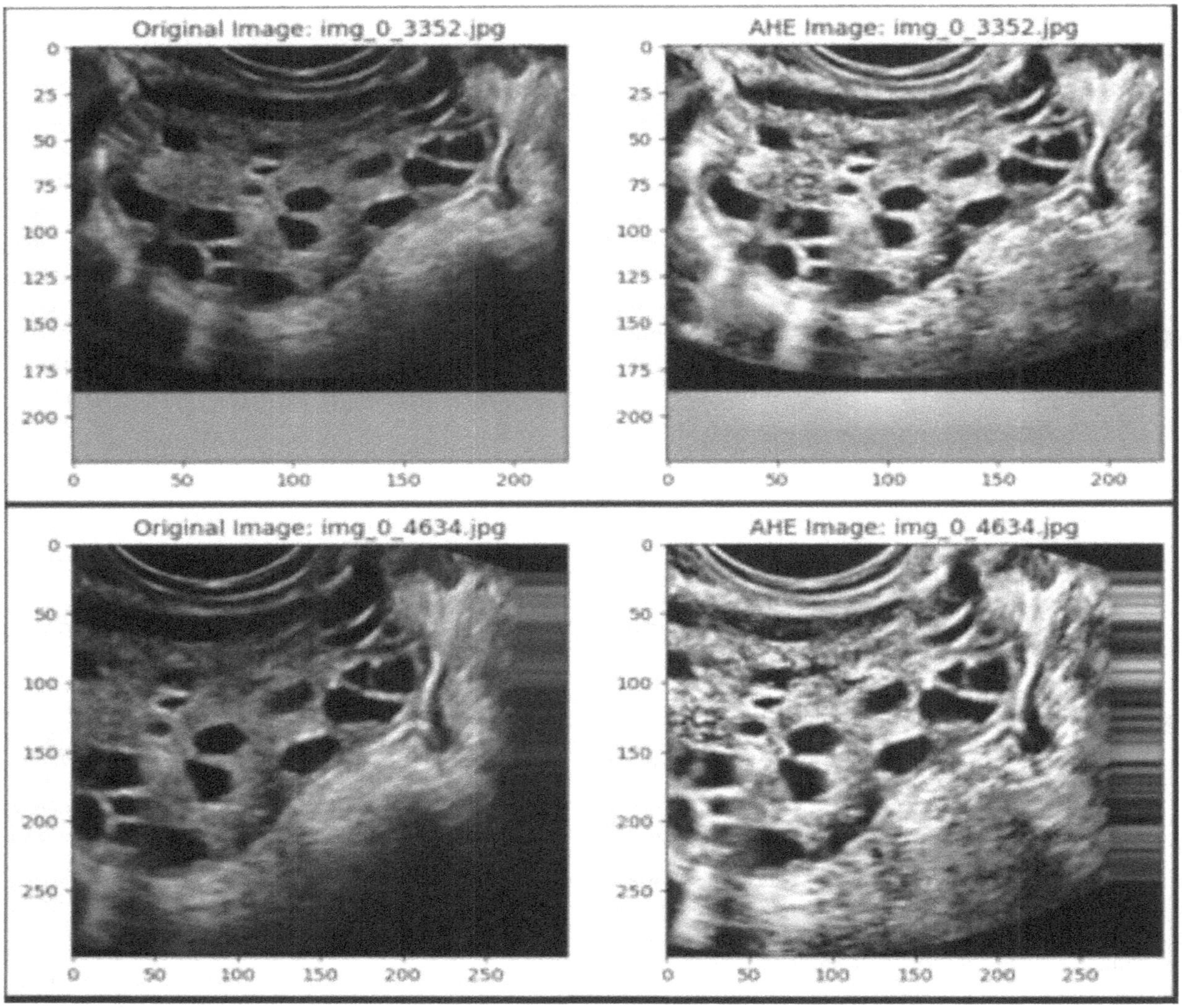

Figure 14.4. Enhanced images using AHE.

Source: Author.

Table 14.1. Segmentation models performance analysis

Models	*IoU*	*Dice*
U-Net	0.14	0.24
U-Net++	0.86	0.92
DeepLab v3	0.47	0.64
CascadedNet	0.34	0.51
PSPNet	0.13	0.12

Source: Author.

accuracy of 93.82% was attained by SVM. The proposed model used AHE to remove noise and improve the image quality. To extract the pertinent features from ultrasound ovary image, various segmentation models such as U-Net, U-Net++, Deeplab v3, cascaded Net, PSPNet has been employed. Then, classifiers such as ResNet 50, ResNet 101, VGG 16, VGG 19, Alexnet and Inception v3. Among all, VGG 19 has performed well and produced 94.16% of highest accuracy than the existing works.

7. Conclusion

Data-driven early detection of PCOS in women is demonstrated in this article in order to properly treat and handle this disorder. The target region and contour in the image are highlighted and enhanced in the suggested work, which enhances the model's classification performance. It also highlights the specific characteristics that will be the main markers for PCOS diagnosis, helping doctors to diagnose diseases more rapidly and precisely and minimizing expenses for patients who went through multiple tests to diagnose their medical condition. In this article, AHE (Adaptive Histogram Equalizer) was used to remove noise and enhance the image quality. Furthermore, a

Table 14.2. classifier model performance without AHE & U-Net++

Classifiers	*Accuracy*	*Precision*	*Recall*	*F1 Score*	*AUC Score*
ResNet 50	68.18	79.14	76.02	74.33	72.56
ResNet 101	71.44	80.20	76.65	63.55	71.14
VGG 16	73.88	76.44	61.56	60.05	62.25
VGG 19	82.36	85.21	80.41	85.65	89.35
Alexnet	61.25	59.22	62.11	59.18	60.12
Inception v3	80.17	80.88	82.63	83.06	85.15

Source: Author.

Table 14.3. Classifier model performance with AHE & U-Net++

Classifiers	*Accuracy*	*Precision*	*Recall*	*F1 Score*	*AUC Score*
ResNet 50	88.25	90.25	92.14	91.65	94.14
ResNet 101	90.26	91.44	92.88	92.35	92.66
VGG 16	89.65	90.65	91.64	91.44	93.58
VGG 19	94.16	96.15	92.25	97.96	98.15
Alexnet	89.14	89.68	90.88	92.87	93.32
Inception v3	90.62	91.68	92.05	91.55	92.67

Source: Author.

Table 14.4. Accuracy comparison of proposed model performance vs previous works

Existing Work	*Classifiers*	*Preprocessing (Enhancement)*	*Image Segmentation model*	*Accuracy %*
(Iyapparaja et al., 2021) [13] (Abrar et al., 2023) [9]	LDA, NB, SVM, RF Six CNN models: DenseNet121, DenseNet201,MobileNet, VGG 16, VGG 19, Inception v3	Gaussian low pass filter AHE (Adaptive Histogram Equalization)	Multilevel Thresholding	93.82 84.81
Proposed model	ResNet 50, ResNet 101, VGG 16, VGG 19, Alexnet, Inception v3	AHE (Adaptive Histogram Equalization)	U-Net++	94.16

Source: Author.

segmentation technique U-Net++ was used to identify the pertinent features from the image such as cysts boundary, count, size and texture. Finally, enhanced and segmented images produced by AHE and U-Net++ was used to train the models. The Convolutional Neural Network (CNN) model including VGG 16, VGG 19, ResNet 50, Alexnet, ResNet 101 and Inception v3 has been used for PCOS classification. VGG 19 has performed well and the highest accuracy of 94.16% was attained. In the future, the model will be improved to predict different types of ovarian cysts including chocolate cysts, endometriomas, functional ovarian cysts, hemorrhagic cysts and dermoid cysts.

References

[1] Moral, P., Mustafi, D., Mustafi, A., & Sahana, S. K. (2024). CystNet: An AI driven model for PCOS detection using multilevel thresholding of ultrasound images. *Scientific Reports*, *14*(1), 25012.

[2] Azziz, R., Carmina, E., Dewailly, D., Diamanti-Kandarakis, E., Escobar-Morreale, H. F., Futterweit, W., … & Witchel, S. F. (2009). The Androgen Excess and PCOS Society criteria for the polycystic ovary syndrome: the complete task force report. *Fertility and Sterility*, *91*(2), 456–488.

[3] Ganie, M. A., Vasudevan, V., Wani, I. A., Baba, M. S., Arif, T., & Rashid, A. (2019). Epidemiology, pathogenesis, genetics & management of polycystic ovary syndrome in India. *Indian Journal of Medical Research*, *150*(4), 333–344.

[4] Gopalakrishnan, C., & Iyapparaja, M. (2019). A detailed research on detection of polycystic ovary syndrome from ultrasound images of ovaries. *International Journal of Recent Technology and Engineering*, *8*, S11.

[5] Yadav, N., Dass, R., & Virmani, J. (2024). A systematic review of machine learning based thyroid tumor characterisation using ultrasonographic images. *Journal of Ultrasound*, *27*, 209–224.

[6] Bedi, P., Goyal, S. B., Anand, S. R., & Manoj, K. (2024). An integrated adaptive bilateral filter-based framework and attention residual U-net for detecting polycystic ovary syndrome. *Decision Analytics Journal. 10,* 100366.

[7] Paramasivam, G. B., & Rajalaxmi, R. R. (2024). Modelling a self-defined cnn for effectual classification of pcos from ultrasound images. *Technology and Health Care, 32*(5), 2893–2909.

[8] Reka, S., Praba T. S., & Karthik, M. (2023). Expeditious Prognosis of PCOS with Ultrasonography Images - A Convolutional Neural Network Approach. *In International Conference on Artificial Intelligence of Things* (pp. 367–376). Cham: Springer Nature Switzerland.

[9] Alamoudi, A., Irfan, U. K., Nida, A., et al. (2023). A deep learning fusion approach to diagnosis the polycystic ovary syndrome (pcos). *Applied Computational Intelligence and Soft Computing.*

[10] Nazarudin, A. A., Zulkarnain, N., Mokri, S. S., Zaki, W. M. D. W., Hussain, A., Ahmad, M. F., & Nordin, I. N. A. M. (2023). Performance analysis of a novel hybrid segmentation method for polycystic ovarian syndrome monitoring. *Diagnostics*, *13*(4), 750.

[11] Hosain, A. S., Mehedi, M. H. K., & Kabir, I. E. (2022, October). Pconet: A convolutional neural network architecture to detect polycystic ovary syndrome (pcos) from ovarian ultrasound images. In *2022 International Conference on Engineering and Emerging Technologies (ICEET)* (pp. 1–12). IEEE.

[12] Rachana, B., Priyanka, T., Sahana, K. N., Supritha, T. R., Parameshachari, B. D., & Sunitha, R. (2021). Detection of polycystic ovarian syndrome using follicle recognition technique. *Global Transitions Proceedings, 2*(2), 304–308.

[13] Gopalakrishnan, C., & Iyapparaja, M. (2021). Multilevel thresholding based follicle detection and classification of polycystic ovary syndrome from the ultrasound images using machine learning. *International Journal of System Assurance Engineering and Management,* 1–8.

[14] Nazarudin, A. A., Zulkarnain, N., Mokri, S. S., Zaki, W. M. D. W., & Hussain, A. (2020, November). An implementation of Otsu thresholding and the Chan–Vese method on the PCO segmentation of ultrasound images. In *2020 2nd International Conference on Electrical, Control and Instrumentation Engineering (ICECIE)* (pp. 1–9). IEEE.

15 Diagnosis of 12-LEAD E.C.G. images using residual neural networks

M. Sandhya[a], *Soumyadip Mukherjee*[b], *and Leninisha Shanmugam*[c]

School of Computer Science and Engineering, Vellore Institute of Technology, Chennai, Tamil Nadu, India

Abstract: Electrocardiography (ECG) is one of the most common tests which is a key part of cardiac assessment and is used to diagnose heart conditions such as abnormal rhythms, coronary heart disease, and problems in the valves. However, an ECG can be interpreted by humans, but accuracy and difficulty of diagnosis depends on the level of training and experience from person to person. Thus using AI to analyse an ECG can help to give more accurate and precise results which may have been overlooked by the human eyes.

Keywords: Artificial intelligence (AI), automated ECG interpretation, cardiac diagnosis, electrocardiography (ECG), machine learning

1. Introduction

Electrocardiogram reading is a critical component in the diagnosis of many cardiac conditions. However, ECG requires manual interpretation, which might be time-consuming and calls for considerable expertise. Recent progress in the implementation of artificial intelligence has enabled this to be automated for the detection of cardiac abnormalities at high accuracy and speed. These algorithms incorporate deep learning models that interpret ECG data to recognize patterns and anomalies human expertise might miss, increasing diagnostic efficiency and reliability.

A normal ECG graph has 4 parts named as P, Q, R, S and T waves as shown in Figure 15.1. The contraction of the atria results in the "P" wave, when the ventricles depolarize they form the "QRS" complex and when the ventricular mass reaches its resting stage it forms the "T" wave. For a normal function heart each of these waveforms have a specific pattern which is shown in the graph, but in case of some abnormalities some anomalies in the graph are observed, which can be characterized by either the change in shape of the formed curve or the time interval between two points, for example the "QRS" interval.

For a normal function heart each of these waveforms have a specific pattern which is shown in the graph, but in case of some abnormalities some anomalies in the graph are observed, which can be characterized by either the change in shape of the formed curve or the time interval between two points, for example the "QRS" interval.

This is where our trained AI can play a part. We can train a model which can perform a pattern analysis in ECG waves by which it can differentiate a normal ECG from when there is any abnormality, and also predict the problem that might be the reason for the abnormality.

In the following paper we will explain how deep learning can use convolutional neural networks to realise highly accurate results in the diagnosis of ECG reports. Deep learning, as it is a subset of artificial intelligence, excels in recognizing complicated patterns within the data and

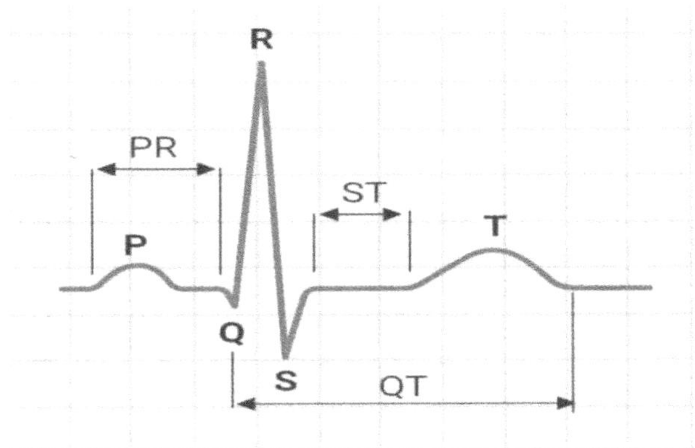

Figure 15.1. Representation of a ECG wave.

Source: "Citalopram & escitalopram: Mechanisms of cardiotoxicity, toxicology predisposition and risks of use in AI geriatric & haemodialysis populations "inGlobal Cardiology Science and Practice 2024.

[a]sandhya.m@vit.ac.in, [b]soumyadip.mukherjee2022@vitstudent.ac.in, [c]leninisha.shanmugam@vit.ac.in

DOI: 10.1201/9781003770473-15

hence very well placed for the analysis of the complex waveforms of the ECG signal [1]. While they are very effective image-processing machines, it is possible to tune them further for time-series data like the ECGs in order to learn detecting and classifying different cardiac conditions from subtle variations of such signals. In our research in regards to this issue we have found that ECG diagnostics can be done efficiently with CNNs, and by doing so we have even achieved around 90% accuracy for certain number of classes of diagnosis, thus by doing this we would finally lead to better outcomes and more efficient healthcare delivery.

2. Literature Review

CVDs represent the leading causes of mortality globally. In addition, they significantly contribute to raising mortality in comorbid patients, namely those with diabetes mellitus and chronic kidney disease [2].

Among the non-invasive diagnostic techniques, an electrocardiogram is one of the most dependable methods for diagnosing cardiac disease. However, reading of the ECG is an extremely time-consuming and highly specialized work. Development of newer techniques for early diagnosis will help in improving the quality and efficiency of medical care [3].

Deep learning in medical diagnosis has shown a tremendous jump very recently. Very recently, some of the important literature related to this field has exploited this technology in an attempt at automatic heart detection. On the other hand, DL-based methods can give better characterization for the nature of ECG signals with respect to conventional machine-learning-based classification methods such as clustering and SVMs in the case of ECG signal classification [4].

An ECG signal comprises various waves that indicate different activities, as shown in Figure 15.2. These waves include the following:

1. P wave originating from the depolarization of the atrium when the heart is in a normal state; This wave is always positive in lead II and always negative in lead a VR.
2. The QRS wave, which is an image resulting from the depolarization of the ventricles. When the heart is functioning normally, the QRS wave has a duration of 0.06 to 0.12 seconds, and its amplitude varies depending on the specific lead being measured.
3. T wave, a wave occurring due to the repolarization of the ventricle. The T wave has positive value in all leads under normal conditions of the heart.
4. The PR interval is calculated from the beginning of the P wave to the start of the QRS wave. This interval in a person with a normal heart condition measures between 0.12 to 0.20 seconds.
5. ST segment: it starts from the end of QRS wave to the start of the T wave [5]

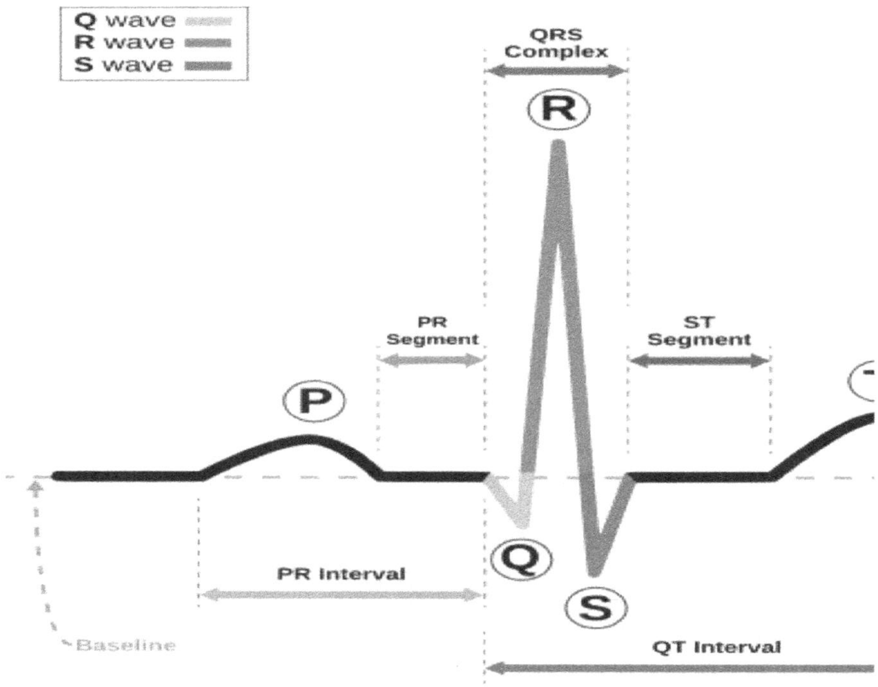

Figure 15.2. Representation of a normal ECG wave.

Source: Author.

An important challenge would be the prevention and early diagnosis and management of cardiovascular diseases. AI could be implemented even prior to the attainment of an electrocardiogram. As an illustration, AI-based models can easily predict the necessity of obtaining an ECG for patients who present at the emergency departments and are sent for triage [6].

For example, if a Short-Time Fourier Transform is applied with the use of a Convolutional Neural Network for ECG result analysis, then attaining an accuracy of 90% in left ventricular systolic dysfunction detection is possible [7].

Of these, their best performance for both datasets, CinC 2017 and CinC 2020, was attained by the 1D convolutional model specifically 1D ResNet, which yielded an F1 score of 85% and 71%, respectively. [8].

Once the ECG data is available, AI can provide very high accuracy and interpretive insights about ECG and cardiac abnormalities. It, therefore, may be applied to comprehensive clinical analysis and diagnosis with real-time information that would help in decision-making [9].

However, it is an important point to note that ECG in itself is not a diagnostic tool. It should be used in correlation with other clinical evaluations, examination of case history, and other tests to arrive at a comprehensive diagnosis. Accurate interpretation of results taken by ECG requires expertise and experience; hence, involvement by trained healthcare professionals becomes imperative to reliably interpret and arrive at informed decisions [10].

3. Methodology

In this work, we will make use of ResNet50, a deep residual neural network very famous due to its outstanding performance in image classification, and, therefore, very promising for applicability to time-series data like the ECG signals in the diagnosis of ECG reports. Using advanced capabilities of ResNet50, the diagnostic accuracy and reliability while analysing the ECG data shall be improved in the current research.

3.1. Data source

Almost 1300 images with r divided into test and train sets have been used in this research. There have been 4 types of ECG report used for this purpose which are patients with myocardial infarction, patients with history of myocardial infarction, patients with abnormal heart beat and patients with normal ECG reports (Table 15.1).

Table 15.1. Data split into test and train

Classification	*Number of train images*	*Number of test images*
Myocardial Infarction Patients	239	112
Patients that have History of MI	172	112
Patients having abnormal heartbeat	233	112
Normal Person ECG	284	112

Source: Author.

3.2. Tools and software

- Programming Language: Python
- Libraries:
  - **NumPy** and Pandas for data manipulation.
  - **TensorFlow** and **Keras** for implementing and using the ResNet50 model.
  - **Matplotlib** for visualising data and model performance metrics.

3.3. Residual networks

Residual Networks, or ResNets, are a deep learning tehnique that enabled the successful training of very deep neural networks. The biggest challenge for deep learning had been the vanishing gradient problem-that is, the problem of where gradients get smaller as they flow backward through a network while training, and thus do not contribute much to updating weights in earlier layers. ResNets address this by using identity shortcut connections to skip over layers, ensuring that the gradients flow through directly without losing information, see Figure 15.3. The idea here is that the network may bypass certain layers if needed, strongly ensuring gradient flow and thus enabling much deeper successful training of networks. This preservation of the strength of the gradient by back propagation in ResNets has allowed it to train deeper networks where traditionally, similar issues would have been faced in conventional architectures.

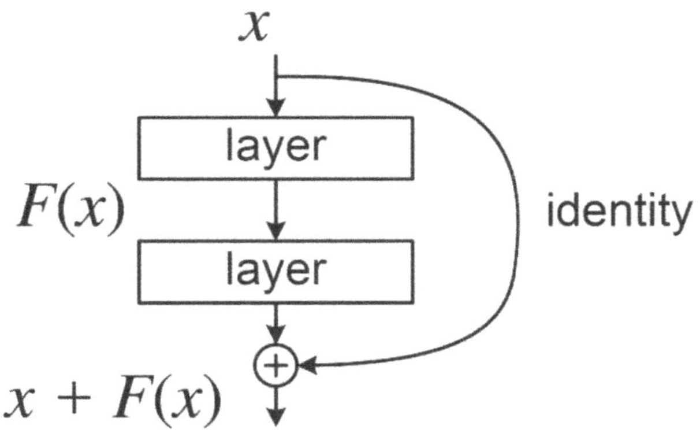

Figure 15.3. Basic structure of a residual network.

Source: "Deep Residual Learning for Image Recognition," in Proc. IEEE Conf. Comput. Vis. Pattern Recognit. (CVPR).

Instead of learning the complete transformation from input to output, ResNets try to learn residuals between layers. This residual learning technique enables the features to be extracted more powerfully and helps the network work toward fine-tuning by incrementally improving its predictions. As the depth in the network is increased, this approach will turn out to be increasingly beneficial since it will effectively help the model grasp higher complexity patterns in data. This will make ResNets particularly well-suited for such tasks that demand deep networks, where traditional methods might not converge or would act differently [11].

The ResNet architecture has seen very great success on many domains and already positioned itself as a state-of-the-art model in fields like image recognition and time-series analysis. Among their applications, image classification is one domain where the ability of ResNets to recognize intricate patterns across many layers resulted in high accuracy. In this context, ResNets can learn those gentle abnormalities in heart signals making them an effective tool for detecting cardiovascular conditions. Learning residual functions and preserving gradient strength result in remarkable robustness and versatility of ResNets, which explains their wide use in modern deep learning applications.

ResNets are better for time series data due to the presence of skip connections which eliminate the vanishing gradient problem thus helping the netwroks to learn effectively without degradation and it also improves stability and convergence making it a better option for tasks where pattern matching is involved, when compared to other types of deep learning methods.

3.4. ResNet50 architecture

ResNet50 is one of the deep neural network architectures intended for excellent recognition and classification of intricate patterns because it encompasses a deep architecture of 50 layers as shown in Figure 15.4. The architecture mainly consists of residual blocks, which play an important role in retaining very important information across layers even if the architecture becomes very deep. Each residual block comprises some convolutional layers, each of which bears the responsibility of extracting intricate features from the input data. Convolutional layers are crucial in recognizing patterns of edges, textures, and shapes within an image hence, finding indispensable applications in tasks like image classification, medical image analysis, and other forms of pattern recognition.

Key additions to ResNet50 include the use of batch normalization. This technique has the effect of speeding up the training by normalizing the inputs to every layer. Batch normalization ensures that the learning process is more efficient and avoids problems like exploding or vanishing gradients. The network also involves the use of ReLU-activated functions, which are important in bringing non-linearity into the model. It hence helps the network learn from data that is linearly inseparable. The ReLU activation functions allow the network to keep computational efficiency and still utilize a diverse range of patterns within the data.

Probably the most significant invention of ResNet50 architecture is the introduction of shortcut connections. The idea of a shortcut connection allows the network to bypass some layers by skipping over them and passing on the input of one layer deeper in the network. Such a structure helps reduce the impact of the vanishing gradient problem when gradients during the backward pass become so small that too little change in the model's weights occurs. The identity shortcuts allow gradients to flow more easily through the network, making it learn better and thus enabling the training of much deeper networks without loss of performance. In fact, this feature significantly enhanced the training aspect and made ResNet50 robust and efficient as compared to earlier models which did not have these shortcuts. These connections to identity also help in keeping the important features from the previous layers so that the network focuses on perfecting the learning process through the subsequent layers.

Due to the special nature of the architecture and residual blocks, ResNet50 has become one of the preferred choices for many kinds of classification tasks. Depth and architecture make it particularly appropriate for finding highly complex and abstract patterns in large datasets. In fact, ResNet50 has found quite a number of applications not only in traditional image classification but also in a variety of other areas, such as medical image analysis-classification of ECG signals. Within this scope, ResNet50 can recognize patterns of ECG signals that hint at abnormalities related to, myocardial infarctions, or other cardiac disorders. Indeed, the network has identified such subtleties of ECG signals, which speaks to its flexibility and strength in terms of areas that require reliability and precision. To sum up, the ResNet50 architecture proved to be strong for a wide range of applications-from image classification to other similar domains, wherein recognition of complex patterns required

3.5. Detection of abnormality

3.5.1. Myocardial infarction patients

In the ECG pictures of myocardial infarction patients, obvious abnormal changes can usually be seen, mainly manifesting as ST-segment elevations, pathological Q waves, sometimes with T-wave inversions. These are indications of ischemia or insult damage to the myocardium. STE in leads reflecting specific areas of the heart, while deep Q waves reflect previous myocardial infarctions wave as shown in Figure 15.5. For that reason, early

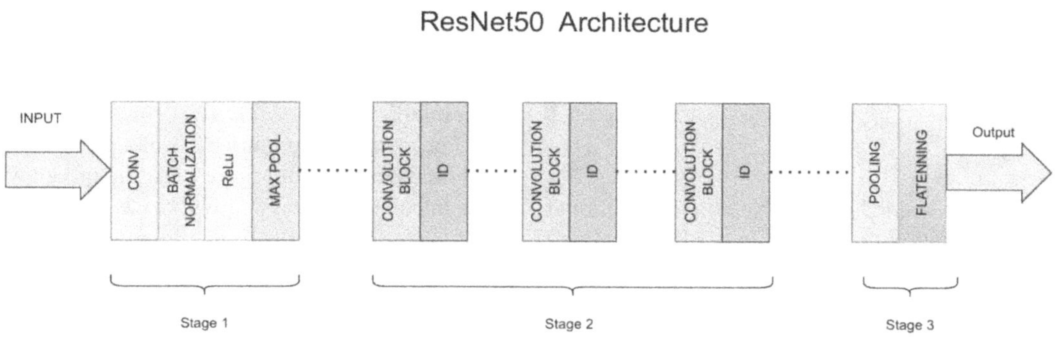

Figure 15.4. ResNet50 architecture.

Source: Author.

detection of such markers is vital for an early intervention to prevent irreparable heart muscle damage.

3.5.2. Patients with history of MI:

Permanent Q waves can appear in leads of the ECG in patients with previous myocardial infarction. These wide and deep Q waves are signs of areas of heart tissue that have died or are scarred from the infarct and thus no longer depolarize. The remainder of the ECG may show normal tracings, but these Q waves become permanent signatures of the past heart attack-a key marker for physicians in studying long-term cardiovascular health.

3.5.3. Patients with abnormal heartbeat

Abnormalities of the heart rhythm, or arrhythmias, may manifest in ECG images through irregularities of the P wave, QRS complex, or T. Depending on the type of arrhythmia, the ECG may reveal tachycardia, or a fast heartbeat; bradycardia, or a slow heartbeat; atrial fibrillation, characterized by irregular activity of the atria; or ventricular arrhythmias, which include premature ventricular contractions. This enables the diagnosis and management of conditions such as atrial fibrillation, ventricular tachycardia, and heart block by the detection of these abnormalities.

3.5.4. Normal individual ECG:

The ECG pattern for a normal individual has been maintained in respect of P waves, QRS complexes, and T waves. The heart rhythm is regular, mostly between 60 to 100 beats per minute, and the time intervals among the waves are also within the limit. Furthermore, there is no abnormality in wave morphology, showing that

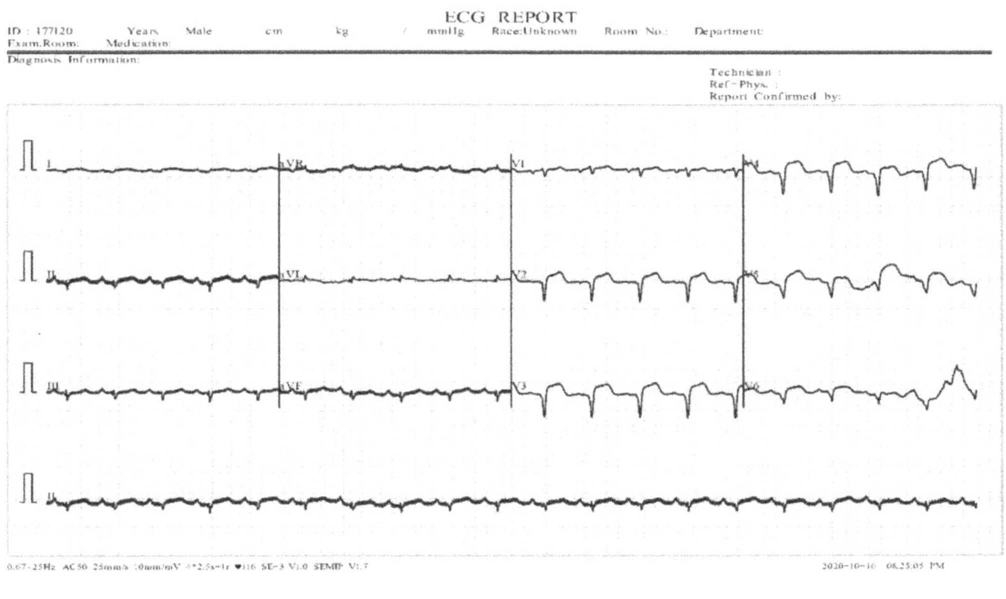

Figure 15.5. ECG of patient with myocardial infarction.

Source: Author.

the electrical activity of the heart is without evidence of ischemia, infarction, or arrhythmias.

4. Results

For this purpose, we trained the ResNet50 Convolutional Neural Network model with our dataset, which contained more than 1300+ ECG images. Training was done with a batch size of 32 for a total of 16 epochs. During training, the model depicted some amount of loss after each step of training, as shown in Figure 15.6. Though oscillations relating to loss were present, it showed gradual improvement in time. This trend, which develops upwards continuously, reflects the fact that the ResNet50 architecture learns effectively from the ECG images and weights the network appropriately. While every training step was hard, it adapted by fine tuning its parameters-hence, making very good predictions as the epochs progressed. The process here reflects the capability and strength of the ResNet50 model for medical image classification tasks.

At the end of the training process, therefore, the model resulted in an accuracy rate of about 92%, as illustrated in Figure 15.7. A high level of accuracy would manifest good performance in classifying ECG signals to either a normal or an abnormal condition. This is very important performance that is required for reliable classification and diagnosis, especially in medical applications where the accurate interpretation of ECG signals may save lives. This could show the excellence of this model, and there is a great prospect for it to be used clinically in the diagnosis of cardiovascular conditions. Therefore, from the results of this training, a ResNet50 model could be one of the solid tools for classifying ECG images and could assist clinical doctors to make decisions by analyzing the ECG data intelligently.

For instance, from the ECG of a patient with an abnormal heartbeat, say in Figure 15.8, the model would recognize the abnormal rhythm with ease. The accuracy, as key as it is, further speaks to the great accuracy of the model in diagnosing abnormal heartbeat conditions from ECG image inputs. This goes to validate further the model's proficiency in the detection of abnormalities in the ECG signals. Success in this deep-learning model with respect to identifying the mentioned pattern affects the diagnosis of conditions such as, and then its efficiency reinforces the value of this approach in a clinical setup. The model therefore reliably paves the way for future improvements in the application of AI in monitoring cardiovascular health and automated ECG analysis.

Also in the Figures 15.9 and 15.10 you can see the input images of 2 cases used and the output of the probability of the most suitable diagnosis are given, we have given a 12 lead ECG image as input and got out a plot as output with probability of each case to be diagnosis.

5. Comparison with Other Techniques

There are advantages and limitations varying for different methods while carrying out the same function of diagnosis of ECG. Others well known and methods include the YOLOv5 and MobileNetV2 architectures. YOLOv5 excels with real-time object detection and high accuracy at

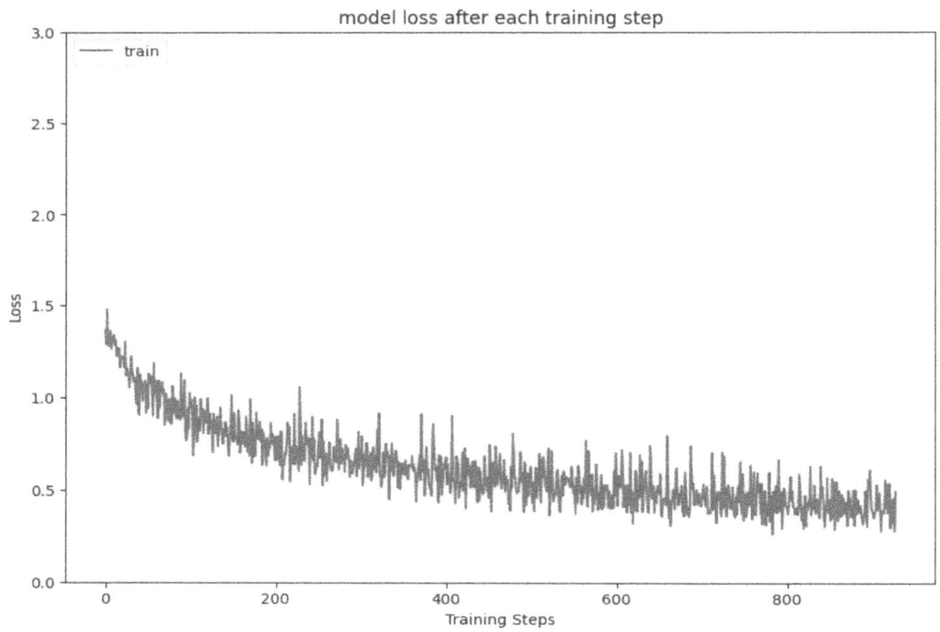

Figure 15.6. Model loss after each training step.

Source: Author.

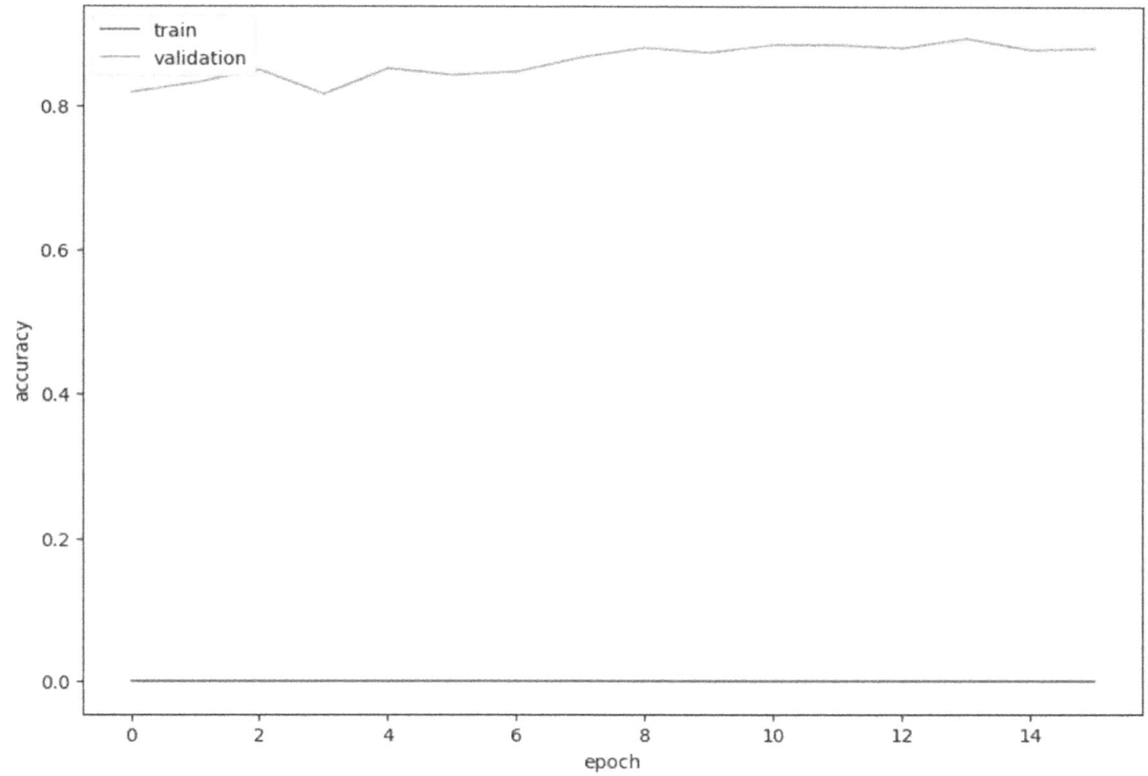

Figure 15.7. Model accuracy after each epoch.

Source: Author.

Figure 15.8. ECG of a patient with abnormal heartbeat.

Source: Author.

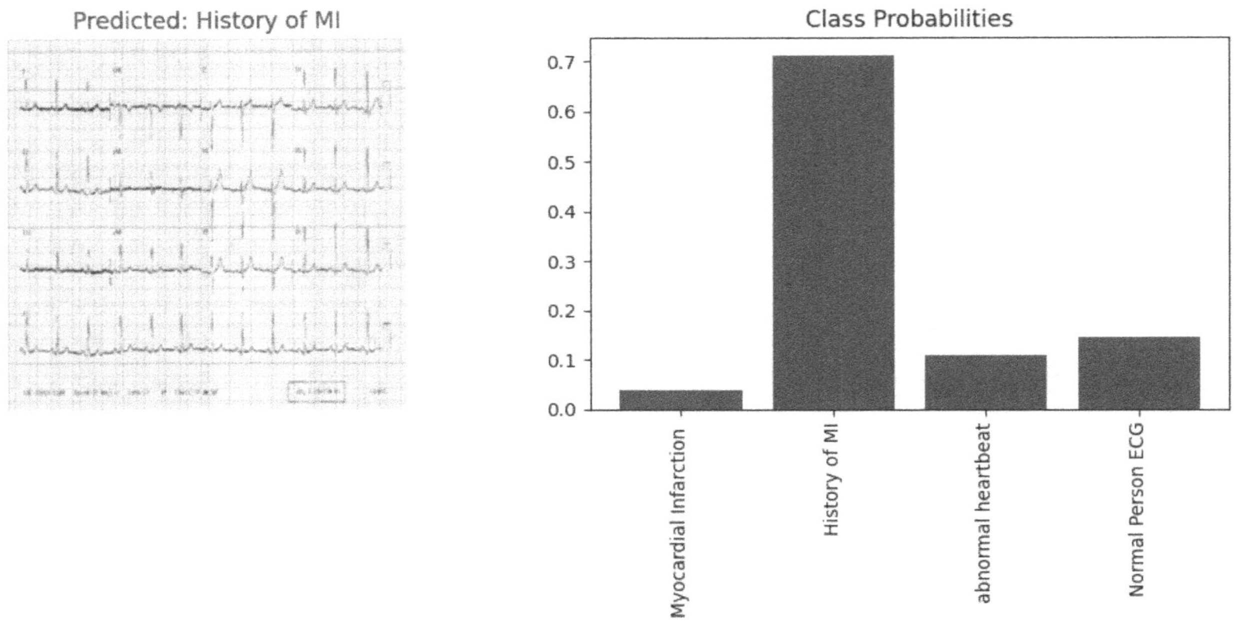

Figure 15.9. Diagnosis of ECG for a patient with history of myocardial infraction.

Source: Author.

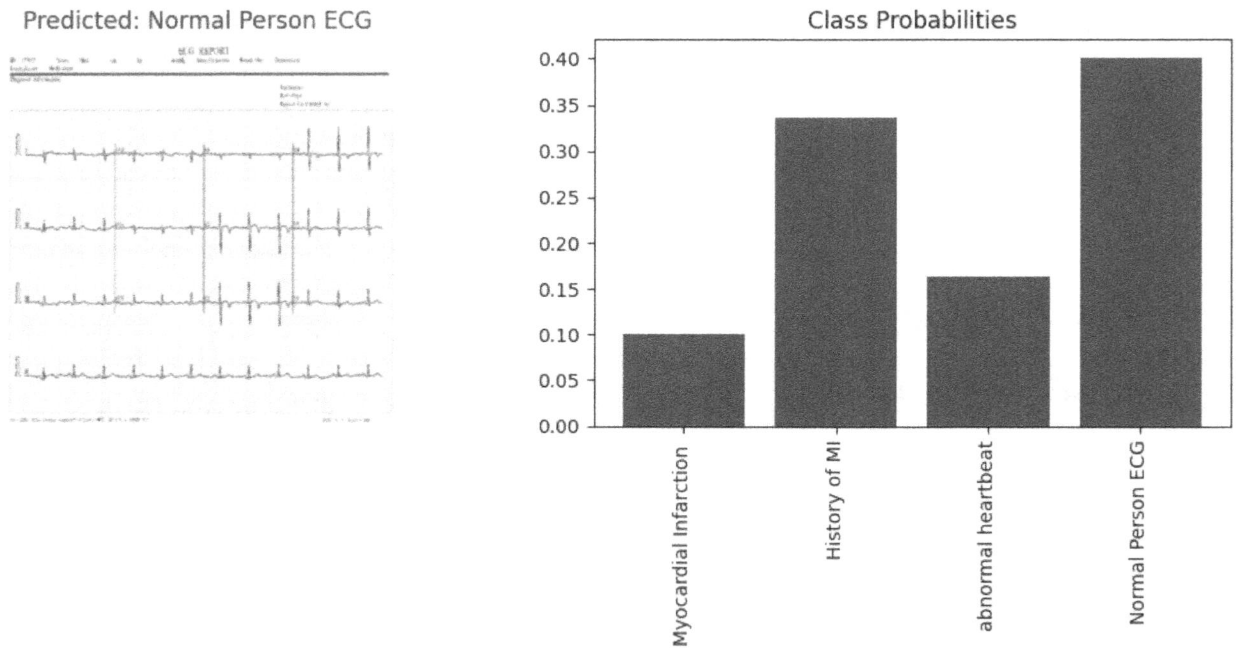

Figure 15.10. Diagnosis of ECG for a patient with a normal ECG image.

Source: Author.

low computational cost on the other hand, MobileNetV2 is lightweight, efficient, and very resource-constrained, with good performance. Both architectures have their strengths and can thus be chosen for their respective needs and constraints of the task.

MobileNetV2 is a lightweight, efficient neural network architecture designed for mobile and embedded vision applications that has been found very promising in the field of ECG diagnosis. The efficiency is gained by depth wise separable convolutions that reduce computational complexity by a large margin but also retain performance. This makes MobileNetV2 very suitable for real-time ECG analysis on resource-constrained devices. It is able to process large amounts of data from ECGs fast enough, and is more conducive to the detection and classification of cardiac abnormalities at a reduced computational cost.

MobileNetV2 when used for the diagnosis and prediction of cardiovascular diseases. The accuracy of the model was checked against four different classes. After fine-tuning accuracy 93% for MobileNetV2 is only a modest yet valuable improvement in performance [12].

YOLOv5 is a very fast state-of-the-art object detection model. It capitalizes on deep learning breakthroughs and leads the trend in real-time applications of fast and agile object detection and classification in frames. It makes use of a series of convolutional layers and anchor-based bounding box predictions for high-performance detection with very minimal computational costs.

YOLOv5 was employed for diagnosing ECG reports, demonstrating a notable accuracy of 95%. The model's efficiency in processing and analysing ECG data highlights its potential for effective cardiac abnormality detection. This high level of accuracy underscores YOLOv5's capability in providing reliable diagnostic insights, making it a valuable tool for enhancing cardiovascular disease diagnosis (Table 15.2) [13].

6. Conclusion

In summary, this study shows us the great potential for the application of deep learning, rather Convolution Neural Networks on the diagonals of ECG reports. Our research results indicate that this model can enable the accurate diagnosis of a lot of cardiac conditions using ECG data, improving diagnostic efficacy and reliability. Advanced AI techniques helps in avoiding adverse events in cardiology by the predicting the possibility of early and correct diagnosis of cardiac abnormalities as well as reducing workload for healthcare professionals. The work also brings to light on how a key role for AI in medical diagnostics and paves the way toward further research and development.

Table 15.2. Comparison with other deep learning technique

Model Used	*Type of data*	*Dataset Size (approx.)*	*Accuracy*
ResNet50	12 lead ECG Images	1300 images	92%
YOLOv5	short-axis echocardiography images	260 images	95%
MobileNetV2	12 lead ECG Images	1950 images	93%

Source: Author.

Acknowledgment

We are grateful to the School of Computer Science and Engineering, Vellore Institute of Technology, Chennai for this support. We appreciate the guidance of our mentors and encouragement from peers their insights have significantly helped this research to succeed.

References

[1] Alzubaidi, L., Zhang, J., Humaidi, A. J., Al-Dujaili, A., Duan, Y., Al-Shamma, O., ... & Farhan, L. (2021). Review of deep learning: concepts, CNN architectures, challenges, applications, future directions. *Journal of Big Data*, *8*(1), 53. https://doi.org/10.1186/s40537-021-00444-8.

[2] Akoum, N., Zelnick, L. R., de Boer, I. H., Hirsch, I. B., Trence, D., Henry, C., ... & Bansal, N. (April 2019). Rates of Cardiac Rhythm Abnormalities in Patients with CKD and Diabetes. *Clinical Journal of the American Society of Nephrology*, *14*(4), 549–556. https://doi.org/10.2215/CJN.09420818.

[3] Martínez-Sellés, M., & Marina-Breysse, M. (April 2023). Current and Future Use of Artificial Intelligence in Electrocardiography. *Journal of Cardiovascular Development and Disease, 10*(4), 175. https://doi.org/10.3390/jcdd10040175.

[4] Mukhopadhyay, S. K., Mitra, S., & Mitra, M. (2011). A Lossless ECG Data Compression Technique Using ASCII Character Encoding. *Computers & Electrical Engineering, 37*(4), 486–497. https://doi.org/10.1016/j.compeleceng.2011.05.004.

[5] Madona, P., Basti, R. I., & Zain, M. M. (2021). PQRST Wave Detection on ECG Signals. *Gaceta Sanitaria, 35*, 2S364–S369. https://doi.org/10.1016/j.gaceta.2021.10.052.

[6] Katal, N., Sameer, G., Poonam, V., & Bharat, S. (December 2023). Deep Learning-Based Arrhythmia Detection Using ECG Signals: A Comparative Study and Performance Evaluation. *Diagnostics, 13*(21), 3605. https://doi.org/10.3390/diagnostics13243605.

[7] Tsai, D. J., Tsai, S. H., Chiang, H. H., Lee, C. C., & Chen, S. J. (May 2022). Development and Validation of an Artificial Intelligence Electrocardiogram Recommendation System in the Emergency Department. *Journal of Personalized Medicine, 12*(5), 700. https://doi.org/10.3390/jpm12050700

[8] Cho, J., Lee, B., Kwon, J. M., Lee, Y., Park, H., Oh, B. H., ... & Kim, K. H. (March 2021). Artificial Intelligence Algorithm for Screening Heart Failure with Reduced Ejection Fraction Using Electrocardiography. *ASAIO Journal, 67*(3), 314–321. https://doi.org/10.1097/MAT.0000000000001218.

[9] Malik, J., Devecioglu, O. C., Kiranyaz, S., Ince, T., & Gabbouj, M. (May 2022). Real-Time Patient-Specific ECG Classification by 1D Self-Operational Neural Networks. *IEEE Transactions on Biomedical Engineering, 69*(5), 1788–1801. https://doi.org/10.1109/TBME.2021.3135622

[10] Jessy, M. K. J., Manikandan, G., Hemalatha, M. S., & Veronica, M. V. (2024). Detection of Cardiovascular Disease Using ECG Images in Machine Learning and Deep Learning Methods. *International Journal of Scientific Research in Science and Technology*, *11*, 131–139. https://doi.org/10.1109/TAI.2022.3159505

[11] He, K., Zhang, X., Ren, S., & Sun, J. (2016). Deep residual learning for image recognition. In *Proceedings of the IEEE Conference on Computer Vision and Pattern Recognition* (pp. 770–778). Las Vegas, NV, USA. https://doi.org/10.1109/CVPR.2016.90

[12] Mhamdi, L., Dammak, O., Cottin, F., & Dhaou, I. B. (2022). Artificial intelligence for cardiac diseases diagnosis and prediction using ECG images on embedded systems. *Biomedicines*, *10*(8), 2013. https://doi.org/10.3390/biomedicines10082013

[13] Çakır, M., Ekinci, M., Kablan, E. B., & Şahin, M. (2024). AVD-YOLOv5, A New Lightweight Network Architecture for High-Speed Aortic Valve Detection from a New and Large Echocardiography Dataset. *Medical & Biological Engineering & Computing, 62*, 2511–2528. https://doi.org/10.1007/s11517-024-03090-3

16 Efficient diabetes detection and feature selection using CatBoost algorithm

Ranya Samour[1,a], *Balamurugan Balusamy*[2,b], *and C. Sai Varun*[3,c]

[1]DBA Candidate at Westford University College, Dubai
[2]Associate Dean Academics, Shiv Nadar University, Delhi-National Capital Region (NCR), India
[3]Department of Computer Science and Engineering, Sathyabama Institute of Science and Technology, Chennai, Tamil Nadu, India

Abstract: In recent years, Artificial Intelligence (AI) and Machine Learning (ML) have become increasingly in demand and useful for healthcare organizations all over the world in diabetes detection and diagnosis. The issues with existing ML-based approaches include the problem of overfitting, leaving the prediction inaccurate for some populations. Additionally, these tools also consume memory usage and more training time. To address these issues, this study presents a novel approach for diabetes detection using the Categorical Boosting (CB) algorithm for classification and feature selection. This algorithm handles categorical features without preprocessing, resistant to overfitting and provides faster training time with lower memory usage. The further focus of this study is correlate the performance of CB with a variety of conventional and ensemble ML models using evaluation metrics such as recall, F1 score, precision, accuracy and ROC-AUC. The results of this study highlight that CB performs better than conventional and ensemble ML models with an accuracy of 98.71% and close to perfect ROC-AUC of 0.99. This study would help the medical field by providing a highly accurate and reliable diagnostic tool for diabetes detection.

Keywords: CatBoost, diabetes detection, healthcare, machine learning

1. Introduction

The accurate diagnoses, efficient use of resources and improvement in disease outcomes are a result of the integration of various technologies [1]. The advancement in technologies has led to accurate and reliable early detection of various diseases [2–4]. Globally Millions of people are affected by a chronic condition called diabetes [5]. This highlights the immediate requirement of diagnostic tools for early detection of diabetes. However, the current ML-based diagnostic models present drawbacks such as overfitting, lack of sufficient reliability or high training time [6]. Such limitations lead to delayed diagnosis or incorrect classification of patients. This study employed an advanced diagnostic solution based on the CB algorithm to address these challenges. The usage of CB algorithm ensures efficient handling of numerical features and enhanced accuracy [7–9]. The main advantage of employing CB algorithm compared to other ensemble techniques is its ability to handle numerical data without the need for preprocessing and effective gradient-based feature selection. This approach reduces the necessity of handling missing values and encoding, also its effective feature selection ability results in enhanced accuracy [10]. This proposed approach solves the problems with existing ML solutions and delivers highly reliable and accurate performance in clinical trials.

The key purposes of this study are outlined below:

- Evaluate the CB algorithm for efficient and accurate diabetes detection.
- Conducted a comparative analysis showcasing its superior performance over conventional ML and advanced ensemble models.

2. Literature Review

The growth in diagnostic tools for early-stage diabetes detection has been increased by adoption of ML and ensemble techniques. Several studies focus on enhancing the accuracy of prediction by using different ML models and different ensemble models. Various ensemble models are used in the field of clinical trials for early detection of various diseases, patient prediction and medical decision-making systems [11, 12]. In paper [13], the author proposes employing

[a]Ranyasamour@gmail.com, [b]kadavulai@gmail.com/balamurugan.balusamy@snu.edu.in, [c]saivarunchandrashekar@gmail.com

DOI: 10.1201/9781003770473-16

CB algorithm for early-stage diabetes prediction. The study compared its results with conventional ML models. Another study [14] also compared LightGBM, XGBoost and CB and concluded that CB provides superior predictive performance compared to other ensemble models. Furthermore [15], explored diabetes prediction using ensemble techniques such as XGBoost, LightGBM, CB, AdaBoost and Bagging. The results indicate that ensemble models are better than conventional ML models with CB achieving the highest accuracy of 95.4%. Overall, these studies highlight the performance of CB in handling categorical data. In another paper [16], the author proposed a CB model for diabetes prediction using diabetes 24-hour ICU patient data. The research showed that algorithms performed much better than other models on this task, confirming their reliability. Additionally, [17] highlighted that order-based encoding and ordered boosting make CB well-suited for handling categorical data and achieving the best possible performance. In contrast to this work, this study employs CB in a novel manner for feature selection and classification, which significantly improves the model's accuracy.

2.1. Research gaps

a. *Existing studies use CB for classification but does not focus on its effectiveness in feature selection.*
   The proposed approach employed CB for both feature selection and classification, enhancing model accuracy.
b. *Similarly, most of the studies provided comparison of ML models but lacks insights about how feature selection impacts accuracy.*
   Various feature selection techniques are evaluated to highlight the use of CB for feature selection and provide insights about how feature selection impacts accuracy.
c. *Additionally, most of the existing studies provide comparison with conventional ML models or XGBoost model but failed to provide comparative analysis with all the conventional and ensemble ML models. This analysis can highlight the use of CB among all the other models.*
   This study explored the use of all the conventional and ensemble models for diabetes detection and has provided comparative analysis.

Our proposed approach bridges these gaps by showing how selection of the most relevant features by this algorithm has contributed to the enhancement of accuracy.

3. Proposed Methodology

3.1. Dataset

In this study, we utilize the Healthcare Diabetes Prediction Dataset avilable on Kaggle [18]. This dataset aims to assist in predicting the likelihood of diabetes based on several health indicators. The dataset contains approxiamately 2800 instances and 10 feaures. The distribution of each feature is illustrated in Figure 16.1. These distributions are useful for data preprocessing as they provide significant information like distribution, skewness and the presence of outliers.

3.2. System architecture

Data preprocessing is utilized in this study, followed by feature engineering using the CB algorithm with selected features. Then the the CB algorithm, conventional ML models and ensemble models are trained and evaluate. The diabetes prediction system architecture is illustrated in Figure 16.2. The data is collected from healthcare dataset and then preprocessing is done where the class imabalance issue is addressed. The next step is training CB model using the selected features like Glucose, BMI, Age and DiabetesPedigree Function. The models are evaluated using various performance metrics. This highlights the importance of CB algorithm in feature selection.

3.3. Data preprocessing

The model training and optimal performance mainly depends on the data preprocessing step. In this study, the missing data is handled by removing rows with missing data, which helps to maintain data quality and prevents any potential biases that could be introduced in the model training process. Additionally, the Syntheic Minority Over-Sampling Technique (SMOTE) was employed to resolve the class imbalance problem [19], [20]. This techique was used in this study to generate synthetic examples for the minority class that is "1". This helps in ensuring balanced distrubution of the target variable and improved model's poerformance on both classes. After resolving the class imbalance problem, the target class distribution is illustrated in Figure 16.3. This helps in understanding the impact on number of instances after applying the SMOTE technique.

3.4. Feature engineering

The main objective of this step is to ensure optimal model performance. It refers to the process of selecting the most relevant and informative data for improving both the efficicncy and accuracy of the model. This study employs CB algorithm for identifying the features that contribute most of the model's predictions. The algorithm can calculate te feature importance of each feature during the model training. Feature selection and optimal model performance are achieved by utilizng the feature importance scores, which indicate how much each feature contributes to the model's performance. Figure 16.4 illustrates the feature importance of each feature in the dataset.

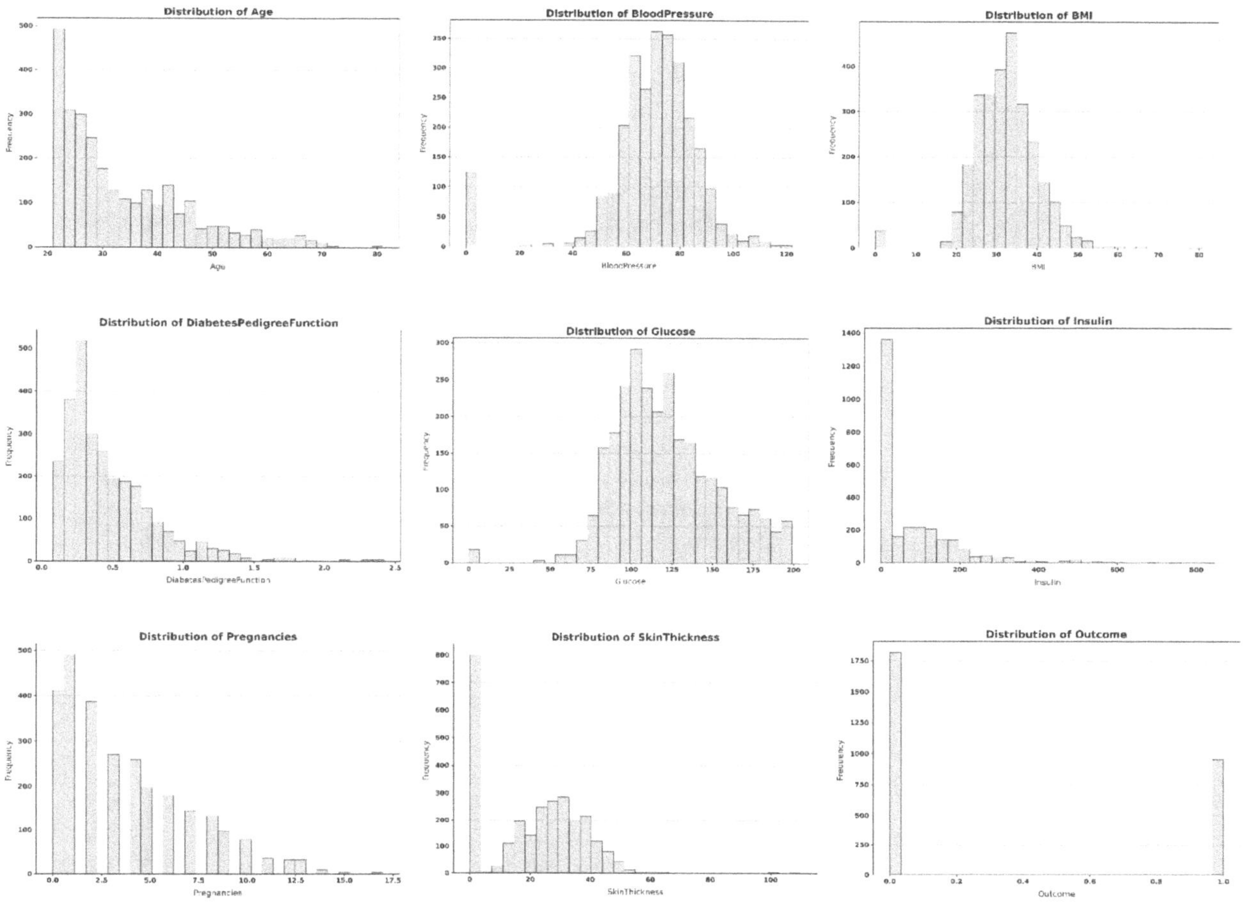

Figure 16.1. Diabetes prediction system architecture.

Source: Author.

3.5. Model training

The CB algorithm is mainly designed to handle categorical feature. This ensures efficiency and high performance. In this study, the CB algorithm is mainly used for handling categorical data without preprocessing, reduced overfitting and enhnaced accuracy. The likelihood of diabetes is predicted using the CB algorithm. The training process was carefully designed to ensure optimal performance while handling th dataset's complexities, including missing values, class imbalance and numerical features. Additionally, for the comparative analysis, several other clssification algorithms were used to provide a comprehensive evaluation. The preprocesserd dataset was divided into 0.7 for training the model and 0.3 data for testing purposes. The performance of these models were evaluated on various evaluation metrics.

3.6. Model evaluation

This study, the model performance of the CB algorithm is compared with other algorithms based on few evaluation metrics. This provides an extensive view of different model's ability to predict diabetes accurately, considering both the classes in the imbalanced dataset. The following metrics were used to evaluate the models:

3.6.1. Accuracy (ACC)

It is the correlation between accurately classified instances, positive and negative, and the total instances in the dataset.

$$ACC = \frac{TN + TP}{TN + FN + TP + FP} \tag{1}$$

3.6.2. Precision (PR)

It is the correlation between TP and the total number of instances predicted as positive. Higher PR indicates that the model makes few false positive predictions.

$$PR = \frac{TP}{TP + FP} \tag{2}$$

3.6.3. Recall (RC)

It is the correlation between TP and the total number of actual positive instances. Higher RC indicates that the

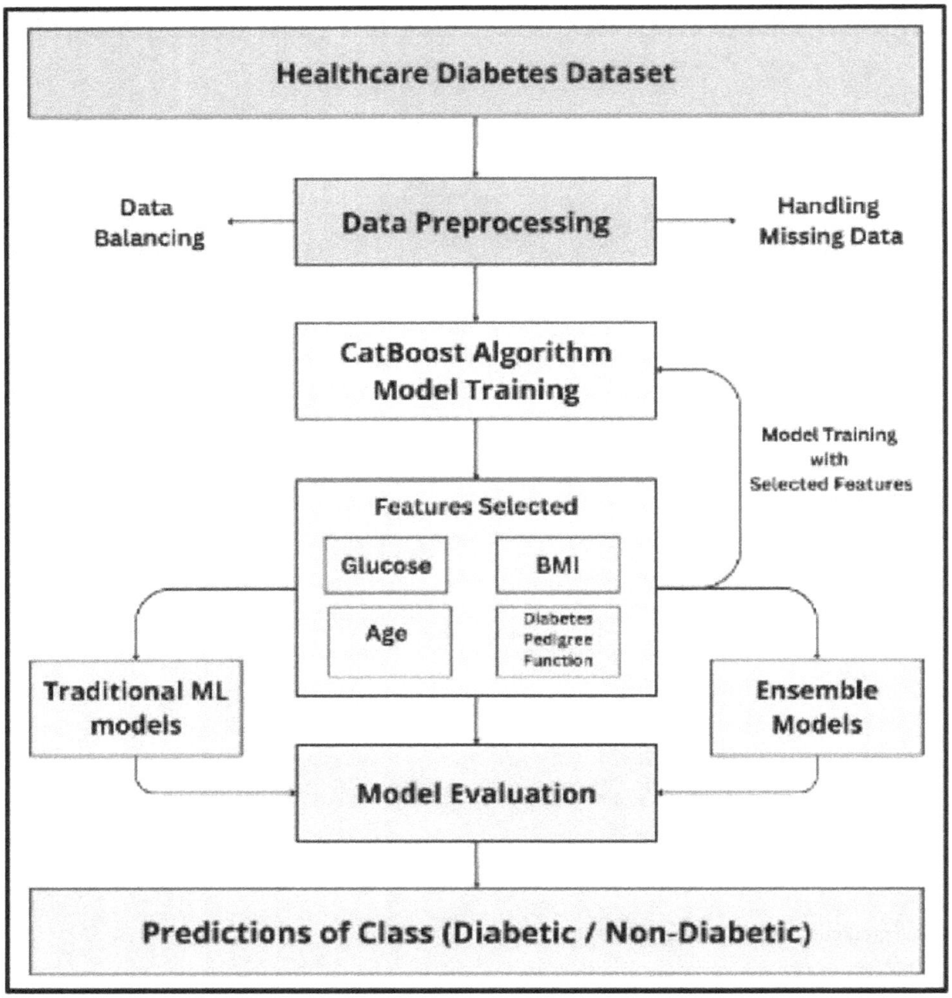

Figure 16.2. Diabetes prediction system architecture.

Source: Author.

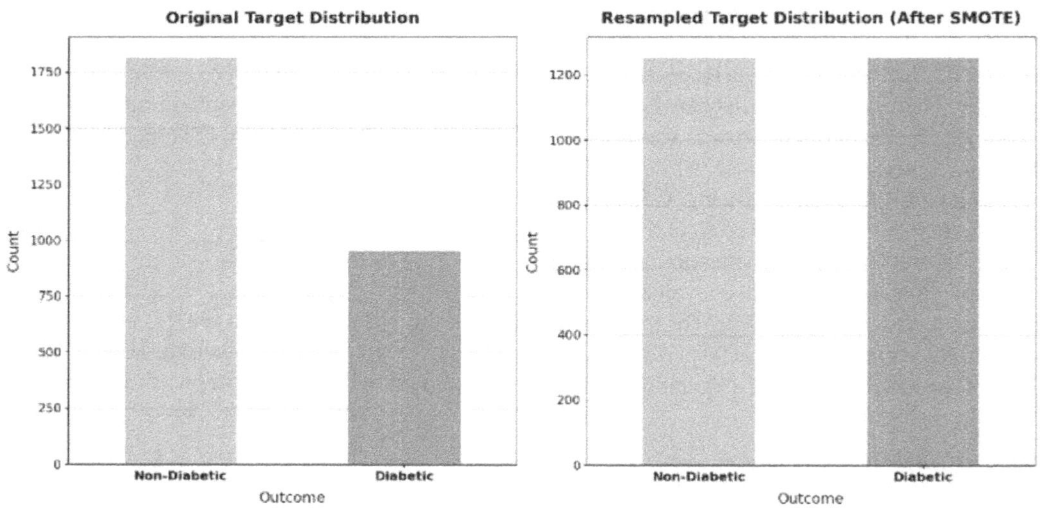

Figure 16.3. Target class distribution after SMOTE resampling.

Source: Author.

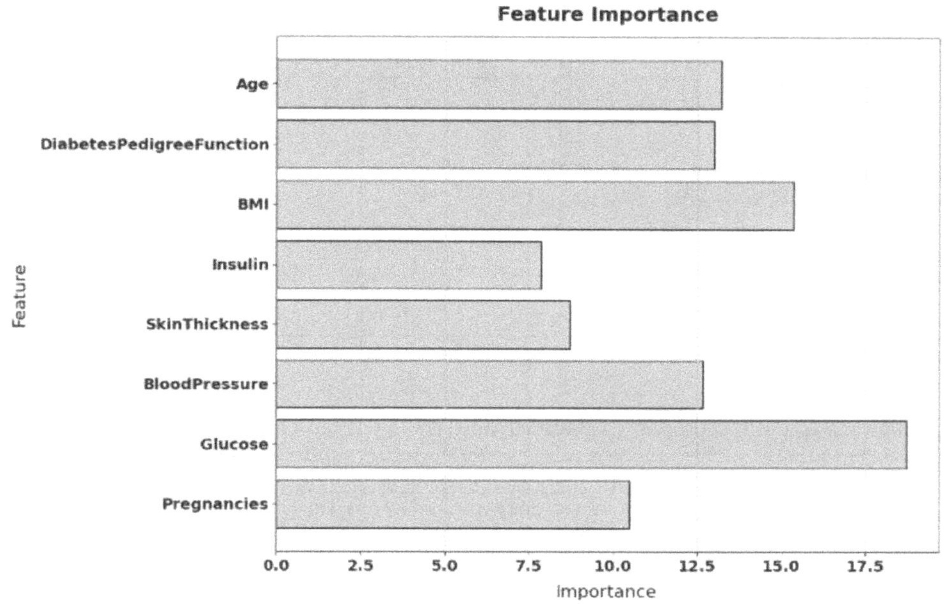

Figure 16.4. Feature importance by CB algorithm.

Source: Author.

model successfully captures most of the true positive cases.

$$RC = \frac{TP}{TP + FN} \quad (3)$$

3.6.4. F1-Score (F1)

It represents the weighted average of precision and recall. This is a single metric that balances both concerns.

$$F1\ Score = 2\ X\ \frac{PR\ X\ RC}{PR + RC} \quad (4)$$

3.6.5. MCC

The MCC is defined as the correlaion coefficient between the observed and predicted classifications. It ranges from −1 (perfectly wrong prediction) to +1 (perfect prediction) with 0, indicating no better than a random guess.

$$MCC = \frac{TN\ X\ TP - FP\ X\ FN}{\sqrt{(TP + FN)(TP + FP)(TN + FN)(TN + FP)}} \quad (5)$$

3.7. Algorithm

Algorithm 1 CatBoost Ensemble ML model

1: **input:** $D = \{(X_i, y_i)\}^N_{i=1}$ *(Diabetes Dataset)*
output: *Predicted class* $\hat{y}$
2: *load Dataset D.*
3: *apply* $D' = SMOTE(D)$
4: *train* $CB = CatBoost(N, \alpha, d, \lambda)$
5: *feature importance* $S = \{s_1, s_2, \ldots, s_n\}$
6: $CB.fit(X_s, y)$
7: $\hat{y} = CB.predict(X_{test})$

This algorithm 1 trains a CB model on a SMOTE-balanced diabetes dataset, selects important features and fits the model. Finally, it predicts the class for new test data.

4. Results and Discussions

The observed results showcase the performance of different ML algorithms, when applied to the healthcare diabetes dataset. The performance of the CB algorithm for feature selection is compared with other feature selection

Table 16.1. Comparative analysis with other feature selection methods

Model	*ACC*	*PR*	*RC*	*F1*	*MCC*
RFE	90.21	0.899	0.875	0.867	0.741
Random Forest	91.88	0.903	0.885	0.892	0.785
SHAP	92.01	0.91	0.895	0.902	0.813
PCA	84.88	0.853	0.824	0.839	0.695
Proposed Approach	98.71	0.988	0.988	0.988	0.957

Source: Author.

Table 16.2. Comparative analysis with conventional ML models

Model	*ACC*	*PR*	*RC*	*F1*	*MCC*
Random Forest	96.47	0.965	0.964	0.964	0.924
Decision Tree	97	0.97	0.97	0.97	0.938
Logistic Regression	76.89	0.778	0.768	0.772	0.492
SVM	77.37	0.782	0.773	0.776	0.501
KNN	83.39	0.849	0.833	0.837	0.648
Naive Bayes	78.70	0.792	0.787	0.789	0.525
Proposed Approach	98.71	0.988	0.988	0.988	0.957

Source: Author.

Table 16.3. Comparative analysis with advanced ensemble ML models

Model	*ACC*	*PR*	*RC*	*F1*	*MCC*
XGBoost	97.35	0.973	0.973	0.973	0.930
LightGBM	93.14	0.935	0.931	0.932	0.849
AdaBoost	77.01	0.787	0.770	0.775	0.509
Proposed Approach	98.71	0.988	0.988	0.988	0.957

Source: Author.

techniques like Recursive Feature Elimination (RFE), Random Forest, SHapley Additive exPlanations (SHAP), Principal Component Analysis (PCA) in Table 16.1. These feature selection techniques were employed to select the important features and then train the CB model using those features. The comparison between the CB algorithm and conventional ML models is presented in Table 16.2. Similarly, Table 16.3 presents the comparison of CB algorithm with ensemble ML models.

Table 16.1 provides the outcomes of various feature selection techniques that are used to train the CB model. In terms of all the metrics, CB has outperformed all the other techniques with 98.71% accuracy, 0.988 precision and 0.957 MCC. Similarly, SHAP and Random Forest have also performed better with 91.88% and 92.01% of accuracy, while PCA with the least performance.

The results indicate that, despite achieving high accuracy of 96%, algorithms like XGBoost, Random Forest and Decision Tree do not outperform CB in various evaluation metrics such a PR, RC and F1. Other ML techniques, including AdaBoost, Logistic Regression, SVM and Naive Bayes, illustrate comparatively lower accuracy. Additionally, confusion matrix and ROC curve are used to evaluate the models. Figures 16.5 and 16.6 presents the confusion matrix of the conventional and ensemble ML models. Similarly, Figures 16.7 and 16.8 presents the Receiver Operating Characteristics (ROC) curve of conventional and ensemble ML models.

Ensemble models have performed better than the conventional ML models as observed in Figures 16.5 and 16.6. In conventional ML models, only the Random Forest and Decision Tree algorithms have performed better with 548 and 553 TP. Among all the ensemble models, XGBoost and CB have performed better in classifying TP as well as TN.

Figures 16.7 and 16.8 represents the ROC curve of conventional and ensemble ML models with ensemble models outperforming the conventional models. Overall, Random Forest, Decision Trees, XGBoost and CB have showed a perfect ROC curve, nearer to 1.

The results indicate that the model's performance is greater when compared to both conventional and advanced ensemble ML models. Tables 16.2 and 16.3 highlights the CB algorithm which achieved an accuracy of 98.71%, with PR, RC, F1 and MCC all reaching 0.988, 0.988, 0.988 and 0.957, outperforming all other models. Figure 16.5 shows that the CB correctly identified 540 TP and 257 TP, misclassifying only 22 FP and 7 FN. This demonstrates a high level of accuracy and minimal misclassifications. Additionally, the ROC curve with an AUC (Area Under the Curve) of 0.99, indicates the model's perfect ability to distinguish between positive and negative class. Overall, it is observed that CB outperforms all other conventional ML models and advanced ensemble models by ensuring accurate predictions.

5. Conclusion

Conventional ML models for diabetes detection struggle with overfitting, high memory usage and longer training times, leading to inaccurate predictions in healthcare diagnostics. To address these issues, this

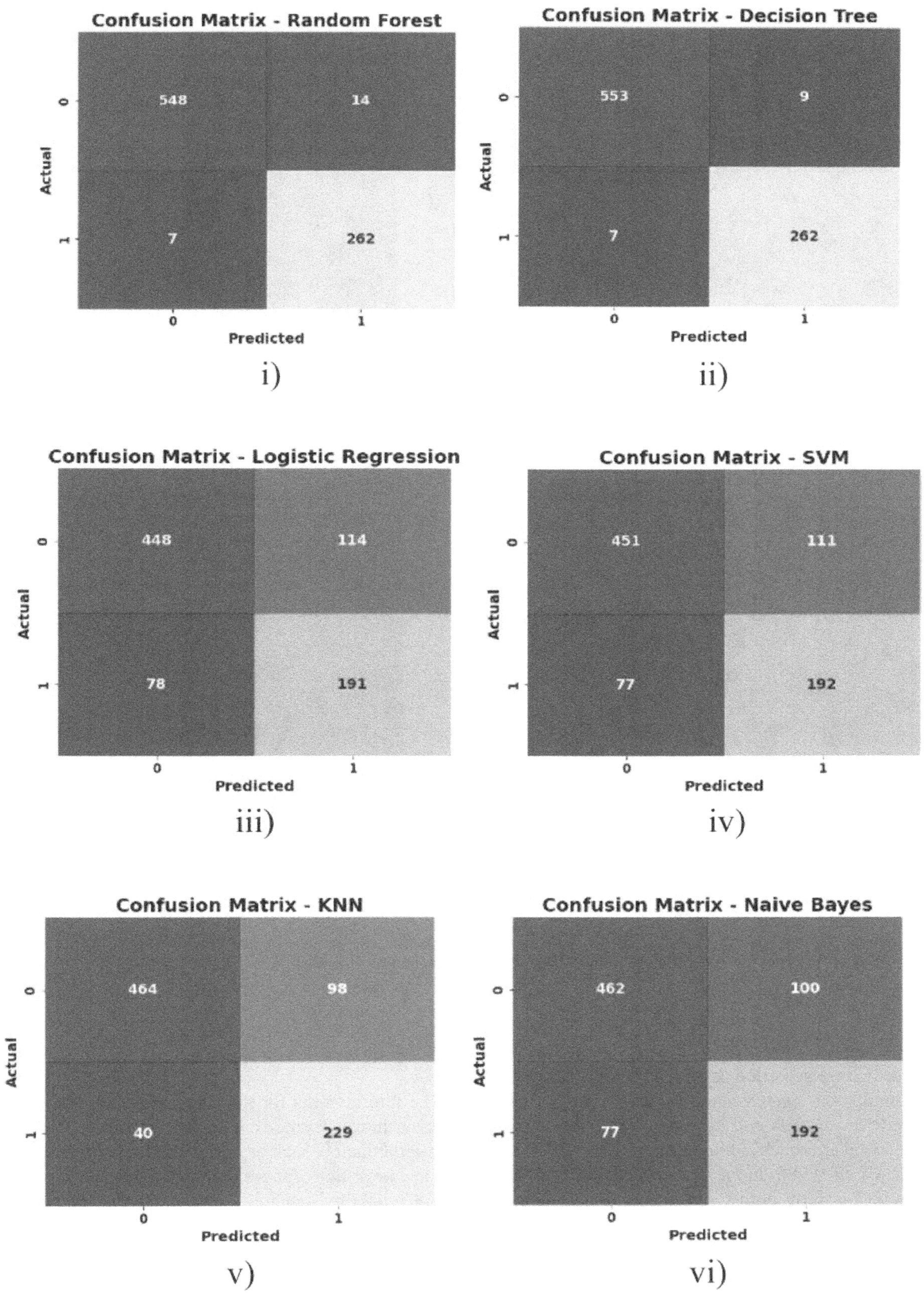

Figure 16.5. Confusion Matrix of (i) Random Forest, (ii) Decision Tree, (iii) Logistic Regression, (iv) SVM, (v) KNN, and (vi) Naive Bayes.

Source: Author.

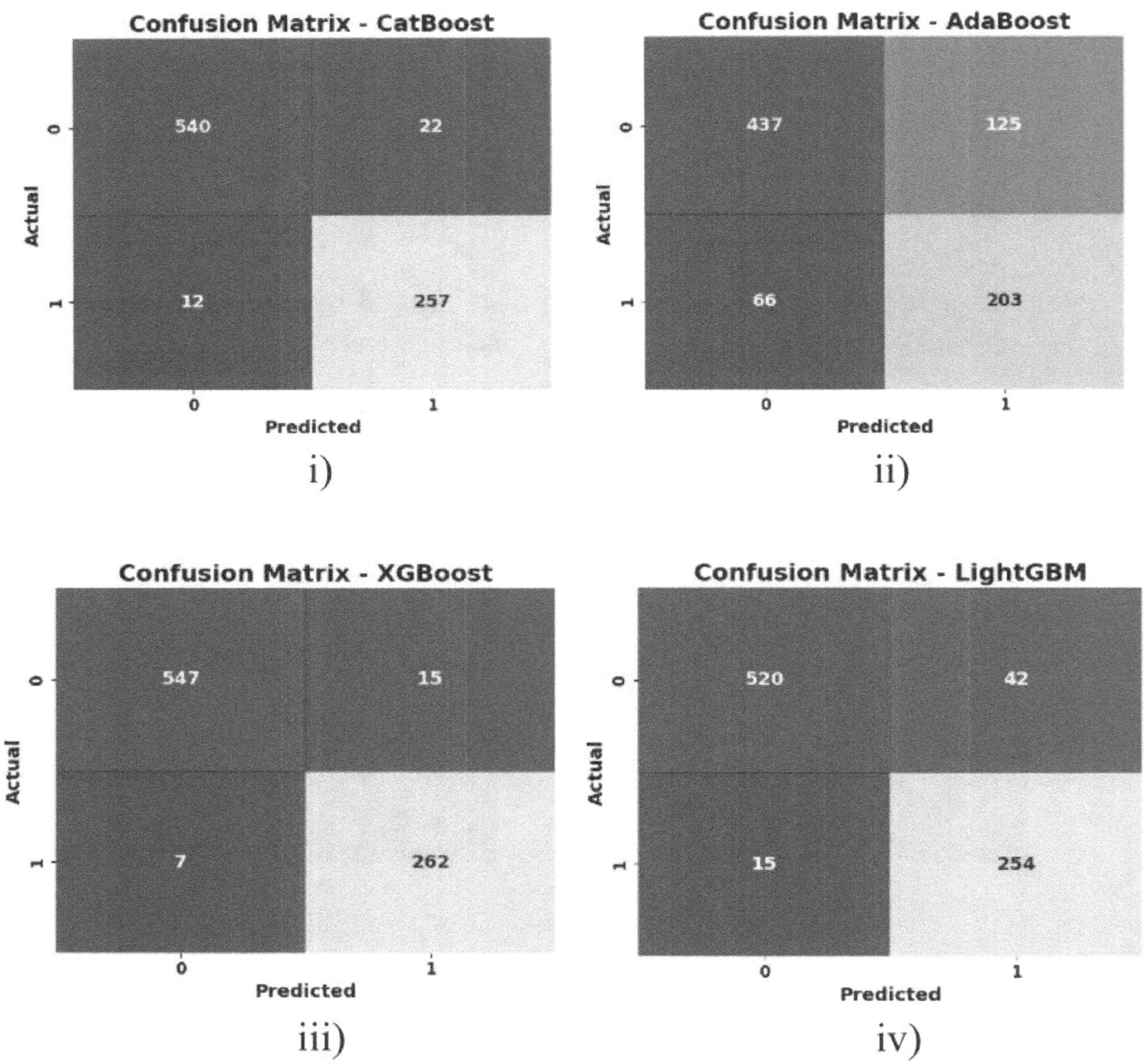

Figure 16.6. Confusion Matrix of (i) CB, (ii) AdaBoost, (iii) XGBoost, and (iv) LightGBM.

Source: Author.

study introduces a novel approach using the CB algorithm for feature selection as well as diabetes detection. Additionally, the performance is compared with conventional ML models and advanced ensemble models. The proposed approach achieves an ACC of 98.71% and PR, RC, F1 of 0.988 and a near-perfect ROC-AUC of 0.99, outperforming both t conventional and advanced ensemble models.

6. Future Works

The future works for this study focus on employing CB algorithm with precision medicine to account for patient-specific factors such as age, lifestyle habits, and genetics. The integration of these personalized features can enhance the reliability, accuracy and provide personalized risk assessments for individuals.

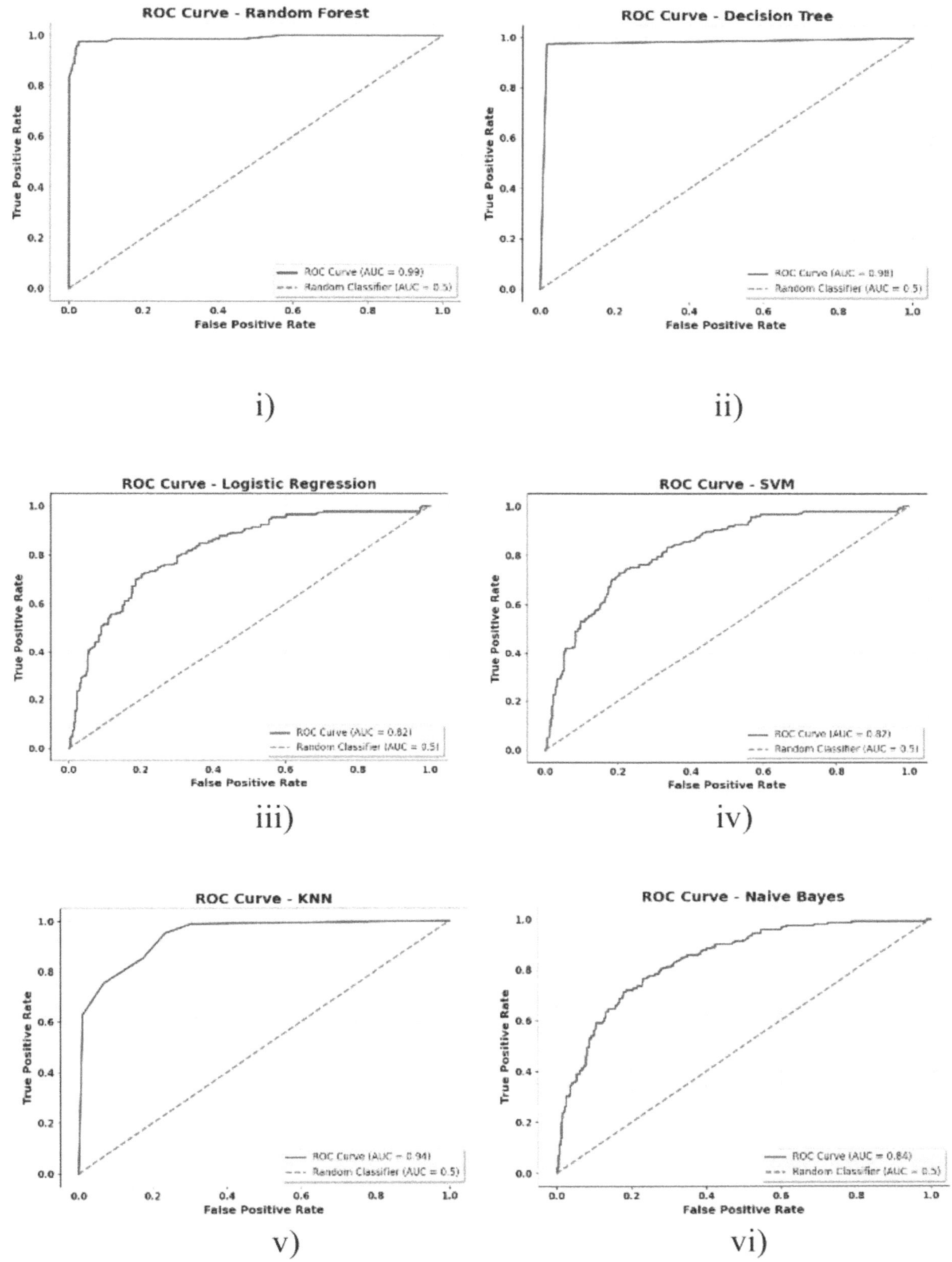

Figure 16.7. Reciever Operating Characteristics (ROC) Curve of (i) Random Forest, (ii) Decision Tree, (iii) SVM, (iv) Logistic Regression, (v) KNN, and (vi) Naive Bayes.

Source: Author.

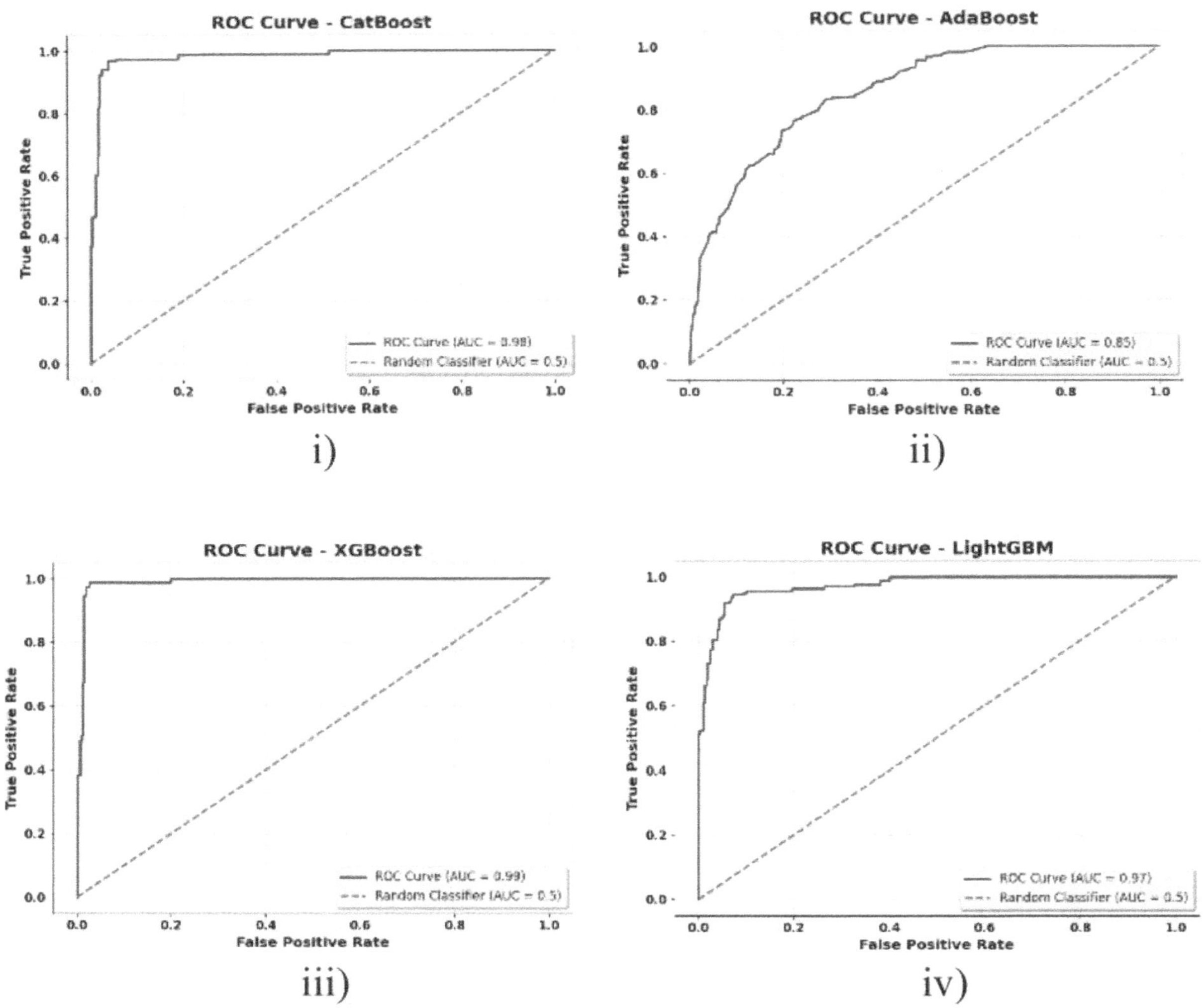

Figure 16.8. Receiver Operating Characteristics (ROC) Curve of (i) CB, (ii) AdaBoost, (iii) XGBoost, and (iv) LightGBM.

Source: Author.

References

[1] Bajwa, J., Munir, U., Nori, A., & Williams, B. (2021). Artificial intelligence in healthcare: transforming the practice of medicine. *Future Healthcare Journal*, *8*(2), e188–e194. *DOI: https://doi.org/10.7861/fhj.2021-0095*

[2] Xiong, L., Liang, L., & Li, J. (2024, August). Prediction of one-year readmission risk in patients with unstable angina based on CatBoost. *In Proceedings of the 2024 International Conference on Biomedicine and Intelligent Technology* (pp. 185–191). doi: https://doi.org/10.1145/3700486.3700516

[3] Kuo, P. H., Li, Y. H., & Yau, H. T. (2024). Development of feline infectious peritonitis diagnosis system by using CatBoost algorithm. *Computational Biology and Chemistry, 113*, 108227. doi:https://doi.org/10.1016/j.compbiolchem.2024.108227

[4] Geeitha, S., Ravishankar, K., Cho, J., & Easwaramoorthy, S. V. (2024). Integrating cat boost algorithm with triangulating feature importance to predict survival outcome in recurrent cervical cancer. *Scientific Reports*, *14*(1), 19828. doi: https://doi.org/10.1038/s41598-024-67562-0

[5] Alowais, S. A., Alghamdi, S. S., Alsuhebany, N., et al. (2023). Revolutionizing healthcare: the role of artificial intelligence in clinical practice. *BMC Medical Education, 23*(1), 689. doi: https://doi.org/10.1186/s12909-023-04698-z

[6] Rahman, A., Debnath, T., & Kundu, D. (2024). Machine learning and deep learning-based approach in smart healthcare: Recent advances, applications, challenges and opportunities. *AIMS Public Health, 11*(1), 58. doi: https://doi.org/10.3934/publichealth.2024004

[7] Prokhorenkova, L., Gusev, G., Vorobev, A., Dorogush, A. V., & Gulin, A. (2018). CatBoost: Unbiased

boosting with categorical features. *Advances in Neural Information Processing Systems, 31.*

[8] Hancock, J. T., & Khoshgoftaar, T. M. (2020). CatBoost for big data: An interdisciplinary review. *Journal of Big Data*, *7*(1), 94. doi: https://doi.org/10.1186/s40537-020-00369-8

[9] Ibrahim, A. A., Ridwan, R. L., Muhammed, M. M., Abdulaziz, R. O., & Saheed, G. A. (2020). Comparison of the CatBoost classifier with other machine learning methods. *International Journal of Advanced Compr Science and Applications*, *11*(11). doi: https://dx.doi.org/10.14569/IJACSA.2020.0111190

[10] Dhananjay, B., & Sivaraman, J. (2021). Analysis and classification of heart rate using CatBoost feature ranking model. *Biomedical Signal Processing and Control*, *68*, 102610. doi: https://doi.org/10.1016/j.bspc.2021.102610

[11] Jerith, G. G., Jagadeesh, S., Kiruba, P. P. J., Mahesh, C., Babu, S. A., & Shaffi, S. S. (2024, November). Evaluation of CatBoost for Diabetes Prevention in Comparison to XGBoost: To avert mortality by developing an AI model capable of predicting the onset of diabetes. *In 2024 International Conference on Electronic Systems and Intelligent Computing (ICESIC)* (pp. 319–324). IEEE. doi: https://doi.org/10.1109/ICESIC61777.2024.10846686

[12] Rani, R., Patel, D. J., Jain, S. K., & Jyothi, R. R. (2024). A novel deer hunting optimized cat boost approach for interactive medical decision-making system. *Multidisciplinary Science Journal*, *6*. doi: https://10.31893/multiscience.2024ss0610

[13] Kumar, P. S., Kumari, A., Mohapatra, S., Naik, B., Nayak, J., & Mishra, M. (2021, January). CatBoost ensemble approach for diabetes risk prediction at early stages. In *2021 1st Odisha international conference on electrical power engineering, communication and computing technology (ODICON)*, 1–6. IEEE. doi: https://doi.org/10.1109/ODICON50556.2021.9428943

[14] Harumy, H. F., Hardi, S. M., & Al Banna, M. F. (2024, April). EarlyStage Diabetes Risk Detection Using Comparison of Xgboost, Lightgbm, and Catboost Algorithms. In *International Conference on Advanced Information Networking and Applications* (pp. 12–24). Cham: Springer Nature Switzerland. doi: https://doi.org/10.1007/978-3-031-57931, 8, 2.

[15] Modak, S. K. S., & Jha, V. K. (2024). Diabetes prediction model using machine learning techniques. *Multimedia Tools and Applications, 83*(13), 38523–38549. doi: https://doi.org/10.1007/s11042-023-16745-4

[16] Utomo, C. P., Fathurahman, M., & Saputra, D. F. D. (2024, February). Diabetes prediction of critical care patient using catboost algorithm. *In AIP Conference Proceedings,* 2838, 1 AIP Publishing. doi: https://doi.org/10.1063/5.0179657

[17] Payal, P. P., & Reddy, P. V. Diabetic Prediction using Optimized Catboost Model.

[18] https://www.kaggle.com/datasets/nanditapore/healthcare-diabetes.

[19] Ganganwar, V. (2012). An overview of classification algorithms for imbalanced datasets. *International Journal of Emerging Technology and Advanced Engineering, 2*(4), 42–47.

[20] Mohammed, A. J., Muhammed Hassan, M., & Hussein Kadir, D. (2020). Improving classification performance for a novel imbalanced medical dataset using SMOTE method. *International Journal of Advanced Trends in Computer Science and Engineering, 9*(3), 3161–3172. doi: https://doi.org/10.30534/ijatcse/2020/104932020

17 Super stressed: The prevalence of anxiety among Middle Eastern University students

Khym Isaac De Barros[1,2,a], *Abeer Abbas*[3,b], *Marya Badran*[4,c], *Elizabeth Roby*[5], *Mareena Deeb*[4,d], *Sara Hachem*[4,e], *Neimat Elkarib*[4,f], *Yara, El Rifai*[4,g], *Dana Ammary*[6,h], *Jana Kayyal*[4,i], *Maryam Al Falasi*[4,j], *Zeynab Al Hasouni*[4,k], *Zahraa Ali Jaber*[4,l], *Sarah Alshehada*[4,m], *and Sara Abukmeil*[7,n]

[1]School of Psychology, Paget T. and Sylvia G. Isaac Research Institute, Massachusetts United States of America
[2]School of Psychology, Curtin University Dubai, Dubai, United Arab Emirates
[3]School of Psychology, Dubai Medical College, Dubai, UAE
[4]School of Psychology, American University of Sharjah, Sharjah, UAE
[5]School of Psychology, University of Birmingham Dubai, Dubai, UAE
[6]School of Psychology, King's College London, London, UK
[7]School of Psychology, United Arab Emirates University, Al Ain UAE

Abstract: The purpose of our research is to investigate the prevalence of anxiety among university students. Research indicates that anxiety is increasing among college students in the United States with an even greater increase post-COVID-19. This study assessed the prevalence of anxiety in the Middle East and North Africa (MENA) region, specifically the United Arab Emirates, at a university that has a high international student body. The study was conducted via an online website where psychology students earned 0.2 extra credit for participation. Results revealed that students' mean score on the Hamilton Anxiety Rating Scale was classified as 'moderate anxiety' (M = 17.5) and the average score on the Functional Idiographic Assessment Template (FIAT) Interpersonal Questionnaire was 75. The average score on the Relationship Structures Questionnaire was 2.62 (M = 2.62) and 2.48 (M = 2.48) for attachment-related anxiety and attachment-related avoidance, respectively. Overall, these conclusions highlight the importance of assessing student anxiety and understanding its relationship to other variables.

Keywords: Anxiety, ethnicity, interpersonal relationships, MENA, University students

1. Introduction

There is a significant amount of research and evidence on the prevalence of anxiety among university students. Specifically, research indicates that anxiety is increasing among college students in the United States. Stearns [1] states that a large percentage of college students have been diagnosed with anxiety and these percentages have increased by more than 50% in recent years, especially after COVID-19. A number of studies have revealed that high anxiety rates are also evident internationally, and in particular, in the MENA region.

2. Middle East and North Africa Region

The Middle East and North Africa are collectively referred to as the MENA region and consist of 22 countries. According to the United Nations (2019), populations in Africa and the Middle East are growing at the fastest rates worldwide [2]. Around 3.4 billion people are predicted to live in the Middle East and Africa by 2050, which is greater than the combined populations of China and India. Hence, it is important to look at the mental health and well-being of this region.

Research indicates that undergraduate students in countries such as Saudi Arabia, Lebanon, and Jordan

[a]khym.isaac@gmail.com, [b]ESR460@student.bham.ac.uk, [c]g00087897@aus.edu, [d]g00093394@aus.edu, [e]saraahashem89@gmail.com, [f]nelkarib20@gmail.com, [g]g00093146@alumni.aus.edu, [h]maryamalfalasi13@gmail.com, [i]aba20240501@dmcg.edu, [j]g00086280@alumni.aus.edu, [k]g00085156@alumni.aus.edu, [l]g00088097@aus.edu, [m]g00092400@aus.edu, [n]g00083997@alumni.aus.edu

DOI: 10.1201/9781003770473-17

have reported anxiety as it relates to financial issues, personal problems, gender, and type of college pre- and post-COVID-19 [3–5]. With large numbers of students endorsing anxiety in the moderate to severe range, these recent studies have begun to provide great insight into the prevalence of anxiety among college students, particularly students in the MENA region. For example, a study by Gihan, et al. [6] showed that moderate to severe anxiety was reported in nearly half their sample of 2589 students in Egypt. Specifically, financial issues, personal problems, gender, and type of college were positive predictors of depression and anxiety among students in Saudi Arabia [5] Similarly, among Egyptian medical students, being female, having a low socioeconomic status, having a chronic illness, having a family history of mood disorders and having a low sense of control also proved to be significant predictors of mood disorder symptoms [7]. Finally, Jordanian University students reported experiencing anxiety about quarantine and shifting to distance learning during the COVID-19 pandemic [3]. We will widen our scope to look comparatively at students with the prevalence of anxiety, beginning with North America, then Europe, and the remainder of the Asian Continent.

2.1. North America

Macauley and colleagues [8] aimed to identify determinants of anxiety in graduate students of health professions in the Northeastern United States. The results of this research demonstrated that all students had anxiety levels that were higher than those of college students and working people. Additionally, for those graduate students seeking doctoral degrees, anxiety levels were noticeably higher in second year students when compared to those in their first-year. Predictors and reports of anxiety were associated with the burden of paying for graduate school fees, reports of low self-rated Grade Point Average (GPA), being female, and having a family history of anxiety.

Similarly, in a study of 374 undergraduate students in Ohio, USA, Beiter et al. [9] found the highest levels of anxiety were among, students living off campus, transfer students and juniors or third year students. They suggested that student mental health should be systematically monitored by universities, and that mental health awareness should be promoted to address the social stigma. Further, a study by Meckamalil et al. [10] revealed common self-reported symptoms associated with anxiety among undergraduate students in Canada.

2.2. Europe

Hajduk et al. [11] examined the frequency of depression and anxiety among college students in Slovakia, and how these affected their social functioning. As in most of our review, they found that college students frequently experience anxiety. Hajduk and colleagues also confirmed similar factors to be frequently linked to participants' anxiety levels – financial worries, pressure to achieve, and academic performance. They identified post-graduation plans as particularly anxiety provoking. Additionally, they discovered that anxiety disorders were linked to two primary things: a lower level of life satisfaction and a decreased level of functioning in social, familial, academic, and personal activities [11]. In short, when a variable sits outside of the student's immediate control, it is likely to create or elevate levels of anxiety.

In France, Kovess-Masfety et al. [12] examined the mental health quality of French college students and their non-college attending peers. Interestingly, they found that non-college attending individuals suffered from a higher prevalence of mental health disorders and these researchers proposed that this discovery could be explained by the difficult transition from school into the labour market. Additionally, they found that there was no difference between men and women in the prevalence of anxiety disorders, including posttraumatic stress disorder; however, men had a greater incidence of panic disorder.

2.3. Asia

Finally, on the Asian continent, we find very intriguing results in regard to the prevalence of anxiety among university students. For example, Gao et al. [13] conducted a longitudinal study examining gender differences in anxiety, depression, and stress in a sample of 1892 Chinese university students and found anxiety to be the most prevalent mental health issue [13]. To further contextualize, let's return to the Middle East section of the Asian continent and attempt to decipher what might explain the high levels of anxiety.

Historical and current conflicts and civil wars have characterized more than 80% of African nations since 1960, with 30% of these countries in conflict for at least ten years [14]. The Middle East also has a long history of conflicts and civil wars, with about 68.5 million people displaced as of the end of 2017. Currently, some countries, for instance, Lebanon, Iran, Syria, Yemen, and Palestine, are experiencing ongoing conflicts [15].

As of the writing of this paper, there were two recent conflicts in the MENA region: one in Sudan and the other in Palestine. On October 7, 2023, Hamas, the Palestinian resistance group, launched a missile attack on southern Israel, which sparked continued exchanges of attacks between the two parties. Two months later, the Israeli-Hamas war is still ongoing, and death tolls indicate that at least 17,997 Palestinians have lost their lives while 1,147 Israeli citizens have succumbed to the conflict. With the destruction of schools, houses of worship, and hospitals, thousands of Gaza residents have fled the country in search of peace. Similarly, in Sudan, on April 15, 2023, a conflict between the Sudanese military and the paramilitary organization Rapid Support Forces (RSF) left widespread devastation in the capital, Khartoum. More than three million people were displaced and about one third of them sought refuge in neighbouring countries.

Our research was run at a university in the United Arab Emirates (UAE). The UAE is known for having a large number of immigrant workers and families. Many people and/or families immigrate to the UAE, fleeing wars and conflict in neighbouring MENA countries, for a change and/or the promise of a better life in a peaceful country. Bronfenbrenner's theory helps us conceptualize how these (meso)systemic challenges can affect an individual's formative years and future interpersonal interactions [16]. The mesosystem refers to two or more microsystems in which the immigrant child or adolescent would engage. For example, this could be the difference in the relationship between the child and his/her immigrant parents compared to the child and his/her new peers in the host country [17]. This anxiety provoking systemic adjustment is difficult for children, adolescents, and adults. The current study recognizes the impact immigration and navigating new spaces may have on interpersonal relationships and interactions among young adults seeking a university degree.

2.4. *Separation*

As stated, with a history of, and with current and ongoing conflicts in the Middle East, many people must flee their countries. However, it is often not possible for entire families to immigrate together. Family separation is common, and it can have significant life-altering effects on some family members, particularly children. According to the Faith to Action Initiative, over 70,000 Syrian refugee families are separated from their fathers, who either remained in Syria or relocated to other countries to search for work and stability [18]. As a result, thousands of Syrian children live without both parental figures. With the loss of the principal caregiver, children are likely to develop insecure attachments 2015. As such, they may have difficulty forming trusting and productive relationships [19]. Immigrant children who are separated from their families are more likely to develop mental health disorders, including anxiety, and post-traumatic stress disorder [20]. In 2021, Hampton and colleagues examined the effects of separation anxiety on parents and children at the U.S.-Mexico border. They found that the children and parents experienced trauma symptoms even after they were reunited [21]. Although their study was not conducted in the Middle East, it is relevant to the context of our study.

2.5. *Immigrant adjustment*

Al-Hawdrawi [22] studied how immigration affects the mental health of Middle Eastern immigrants in the United States. The findings indicated that these immigrants experienced anxiety and emotional distress. Additionally, there was an inverse relationship between these feelings (i.e., negative feelings) and the level of acculturation. That is, the more adjusted an immigrant is to the new culture, the less likely he/she is to experience anxiety and emotional distress. The author argues that immigrants who actively try to hold onto their heritage and culture tend to have a harder time adjusting to the new culture and thus experience a decline in his/her mental health.

Immigrants represent more than 80% of the population in the UAE. Many students enrolled in this university are either immigrants or come from families of immigrants and have likely experienced challenges adjusting to the host country. Thomas [23] identified some stressors that affect the acculturation process and have a detrimental effect on mental health, including integration barriers, level of education, employment status, and sociopolitical status. With a high percentage of immigrants in the UAE, our data was collected from a private university with students representing 91 countries. Based on enrolment records from the past five years, students from the UAE, Egypt, India, and Jordan are among the countries with the highest national representation.

As previously stated, academic rigor and high tuition costs are linked to increased anxiety among university students. Further, research indicates that factors such as perceived lack of control over one's life, low socioeconomic status, and tuition costs are associated with increased anxiety among university students. Students in this study come from an academic institution ranked among the top 10 in the Arab universities, according to the QS Arab Region University rankings in 2023. Students earning admission to this university had an average high school GPA of 94.1%, with just about 67% of them graduating after 4 years. Additionally, with high tuition costs, most students depend on academic grants and merit scholarships to afford their education, with many students experiencing a constant threat of losing their scholarship, which contributes to increased levels of anxiety among this population.

2.6. *Research question*

What proportion of surveyed university students will report experiencing anxiety symptoms?

Hypotheses

- H1: Students will report high levels of anxiety
- H2: Students will report poor interpersonal relationships
- H3: Students whose families immigrated to the UAE will show high rates of anxiety

3. Materials and Methods

3.1. *Ethics and ethical approval*

In order to conduct this study, we first received approval from the Institutional Review Board (IRB) (Protocol

Numbers 23-034 and 23-035). This approval ensures that the research complies with established ethical standards for the protection of human participants. As per the IRB's guidelines, the study was conducted by qualified personnel following the approved protocol, with appropriate safety measures in place. This study was conducted with integrity and in accordance with institutional and regulatory ethical principles. Informed consent was obtained from all participants, and any unanticipated risks or adverse events were to be promptly reported. There were minimal to no risks associated with this research. The data collected was anonymized to ensure confidentiality and was stored in Qualtrics on the password protected, double authenticated University network. In the event that data needed to be printed, all files were stored in a locked cupboard in the Primary Investigator's office. (No files were printed).

3.2. Participants

The sample for this study included a group of 326 undergraduate students from a private university in the United Arab Emirates. All participants were enrolled in either the Psychology 101 or Psychology 102 course. The study included both male and female undergraduate students from various colleges at the university; the majority of participants were sophomores, followed by seniors, freshmen, and juniors. The participants were between the ages of 17 and 23. They belonged to different ethnic backgrounds, with the majority being from the MENA region.

3.3. Materials

3.3.1. Functional idiographic assessment template (FIAT-Q)

The FIAT Interpersonal Relationships Questionnaire is a self-report scale made of 111 questions [24]. This questionnaire was created to help therapists efficiently evaluate their patients' interpersonal functioning into five categories or classes. These classes are as follows: assertion of needs (A), bidirectional feedback (B), conflict disclosure (C), interpersonal closeness (D), and emotional expression and experience (E). Only Class D, Disclosure and Interpersonal Closeness, was used in this study. Students rated each item in relation to their behaviour. Scores were then tallied, with a higher score denoting a higher degree of interpersonal functioning issues or challenges. The FIAT Questionnaire has high correlations between its classes/subscales ($r > 0.5$) and all classes have a high internal consistency ($\alpha > 0.50$), such that each class can be used individually. Given that, the Cronbach α for class D is 0.8 and the total consistency of the scale is $\alpha = 0.94$. Each of the classes has different correlations with different interpersonal rating scales (and subscales), such as the AI (Assertion Inventory), the Social Skills Inventory subscales (Social Sensitivity Scale and Emotional Expressivity Scale).

3.3.2. The Hamilton Anxiety Rating Scale (HAM-A)

The Hamilton Anxiety Rating Scale (HAM-A) is a psychological assessment tool used to measure the severity of anxiety symptoms in individuals [24]. The HAM-A consists of 14 items, each representing a specific symptom of anxiety. These items cover a range of psychological and physical manifestations of anxiety, including nervousness, tension, fears, insomnia, and somatic complaints such as gastrointestinal distress and cardiovascular symptoms. Each item is rated on a scale from 0 to 4, with 0 indicating the absence of the symptom and 4 indicating severe intensity. The total score on the HAM-A can range from 0 to 56, with higher scores indicating more severe anxiety. Clinicians use this scale to assess the severity of anxiety symptoms in patients, monitor changes in symptoms over time, and evaluate the effectiveness of treatment interventions.

3.3.3. The relationship structures questionnaire (ECR-RS)

The Relationship Structures Questionnaire (ECR-RS) is a tool developed to measure adult attachment styles, specifically within romantic relationships [25]. It builds upon the original Experiences in Close Relationships (ECR) questionnaire by focusing on the structural aspects of relationships, including attachment security and insecurity. Unlike its predecessor, the ECR-RS takes a dimensional approach, allowing for a nuanced understanding of attachment dynamics. It typically consists of self-report items where individuals rate their agreement with statements regarding their attachment and relationship experiences. Widely used in both clinical and research settings, the ECR-RS aids therapists in assessing clients' attachment patterns and informs treatment approaches. It also contributes to research that investigates the role of attachment in romantic relationships and psychological well-being. The ECR-RS was utilized to assess maternal, paternal, partner, and best friend relationships.

3.4. Procedure

The study was administered using an online method, Sona. The informed consent form, a short demographic form, the HAMA-A, ECR-RS and the FIAT scales were uploaded to the website, and participants accessed and completed the study. Respondents were instructed to read and sign the informed consent prior to completing the demographic questionnaire, the HAM-A, ECR-RS, and the FIAT. The data was collected 3 times, across 2 semesters. In semester 1, data was collected 1–2 weeks before the final examinations. In semester 2 data was collected and 1 time at the beginning and 1 time towards the end of the semester.

Incomplete submissions were eliminated from the data set. Statistical analyses were conducted by computing descriptive statistics for scores on the HAM-A, ECR-RS and FIAT, and inferential statistics were calculated. These included correlation, multiple regression, and ANOVA. All results were computed using Jamovi to ensure accuracy. Participants were awarded 0.2 course credits for completing the study. Finally, this study did not involve any physical or mental harm.

4. Results

The total number of participants that completed the HAM-A, ECR-RS and FIAT Interpersonal questionnaire was 326 ($N = 326$). The mean average for the Hamilton Anxiety Rating Scale score was 17.5 ($M = 17.5$, $SD = 9.86$). The mean average for the FIAT Interpersonal questionnaire was 75 ($M = 75$, $SD = 10$). The mean average for ECR-RS attachment-related anxiety was 2.62 ($M = 2.62$, $SD = 0.777$), and the mean average for attachment-related avoidance was 2.48 (M = 2.48, SD = 0.898).

Pearson's correlation coefficient was conducted to find relationships between the HAM-A, ECR-RS Avoidant Attachment, and ECR-RS Anxious Attachment scores for the mother-like figure. Pearson's correlation coefficient between the HAM-A scores and ECR-RS Anxious Attachment scores was found to be $r = .32$ with a corresponding p-value of .001. There was a statistically significant, weak positive linear correlation between the anxiety scores and the attachment-related anxiety regarding the mother-like figure scores in students ($r = .32$). Pearson's correlation coefficient between the HAM-A scores and ECR-RS Avoidant Attachment scores was found to be $r = .13$ with a corresponding p-value of .002. There was a statistically significant, weak positive linear correlation between the anxiety scores and the attachment-related avoidance regarding the mother-like figure scores in students ($r = .13$). The p values indicate that the results for both correlational analyses are statistically significant.

Pearson's correlation coefficient was conducted to find relationships between the HAM-A, ECR-RS Avoidant Attachment, and ECR-RS Anxious Attachment scores for the father-like figure. Pearson's correlation coefficient between the HAM-A scores and ECR-RS Anxious Attachment scores was found to be $r = .13$ with a corresponding p-value of .02. There was a statistically significant weak positive linear correlation between the anxiety scores and the attachment-related anxiety regarding the father-like figure scores in students ($r = .13$). Pearson's correlation coefficient between the HAM-A scores and ECR-RS Avoidant Attachment scores was found to be $r = .24$ with a corresponding p-value of .001. There was a statistically significant weak positive linear correlation between the anxiety scores and the attachment-related avoidance regarding the father-like figure scores in students ($r = .24$). The p values indicate that the results for both correlational analyses are statistically significant.

Pearson's correlation coefficient was conducted to find relationships between the HAM-A, ECR-RS Avoidant Attachment, and ECR-RS Anxious Attachment scores for the partner. Pearson's correlation coefficient between the HAM-A scores and ECR-RS Anxious Attachment scores was found to be $r = -.14$ with a corresponding p-value of .01. There was a statistically significant weak negative linear correlation between the anxiety scores and the attachment-related anxiety regarding the partner scores in students ($r = -.14$). Pearson's correlation coefficient between the HAM-A scores and ECR-RS Avoidant Attachment scores was found to be $r = .12$ with a corresponding p-value of .03. There was a statistically significant weak positive linear correlation between the anxiety scores and the attachment-related avoidance regarding the partner in students ($r = .12$). The p values indicate that the results for both correlational analyses are statistically significant.

Pearson's correlation coefficient was conducted to find relationships between the HAM-A, ECR-RS Avoidant Attachment, and ECR-RS Anxious Attachment scores for the best friend. Pearson's correlation coefficient between the HAM-A scores and ECR-RS Anxious Attachment scores was found to be $r = .025$ with a corresponding p-value of .65. This result is not statistically significant. Pearson's correlation coefficient between the HAM-A scores and ECR-RS Avoidant Attachment scores was found to be $r = .22$ with a corresponding p-value of .001. There was a statistically significant weak positive linear correlation between the anxiety scores and the attachment-related avoidance regarding the friend in students ($r = .22$). Finally, our third hypothesis that students who immigrated to the UAE would report higher rates of anxiety was not supported, revealing no significant difference or correlation between both variables.

5. Discussion and Conclusion

This study investigated the prevalence of anxiety amongst university students at a university in the UAE that has an international student representation from 91 countries. Data was collected over 2 semesters. The first data collection was two weeks before final examinations, second, 1 month into the next semester and then again just before the end of the second semester. We found that on average, students' scores fell in the 'mild-to-moderate anxiety' category on the HAM-A, 17.5 ($M = 17.5$, $SD = 9.86$). This may partially explain the rates of anxiety reported. Furthermore, our results revealed that there is a decreased closeness in interpersonal relationships across all academic levels (as revealed by the FIAT-Q scale). Similarly, Khoshhal and colleagues [26] found that 65% of students experience exam anxiety. The authors reported that factors such as sleep deprivation, lack of preparation for the exam, and immense workload from their courses were the major contributors towards exam anxiety. In addition to these factors, we suggest that there is vulnerability associated with high university tuition costs, which likely plays a prominent role in exam anxiety. The possibility of substandard performance on exams could threaten their scholarships, resulting in a failure to change their and their families' life circumstances.

Moreover, our results demonstrate that as student rates of anxiety increased, the quality of their interpersonal relationships decreased (r = 0.16, p = .001) for attachment-related anxiety and (r = 0.62, p = .001) for attachment-related avoidance. A possible explanation for this could be the high percentage of immigrants within the UAE. Many people immigrate to the UAE to flee wars and conflict. They are driven by the pursuance of a better life, whether in the form of educational and/or professional opportunities in a peaceful and stable country. However, in times of stress, such as final exams in this instance, they may have difficulty engaging in productive relationships [18].

The current study recognizes the impact immigration and navigating new spaces may have on interpersonal relationships among young adults seeking a university degree. However, we did not find a significant difference between participants' country of origin, country of birth, or academic standing with any scores. Hence, our hypothesis that students who immigrated to the UAE would report higher levels of anxiety was not supported in this study. These findings suggest that there is no relationship between family immigration status and anxiety levels in this particular sample. This could be due to the limited sample size corresponding to country of origin or birth. Future studies could recruit a larger sample to adequately examine participants' mental health issues from conflict ridden regions versus stable regions and better understand the relationship between the immigration status and the prevalence of anxiety in students.

These findings can help universities gain awareness about the prevalence of anxiety among students, specifically heightened anxiety during examination periods. Subsequently, these findings can help universities consider the need to offer programs to help students relieve anxiety, improve their interpersonal relationships, and significantly improve their academic performance and well-being.

To illustrate, we designed a preventive program to promote students' psychological health and well-being under the umbrella of Anxiety Awareness Day. This program was held one week before the final examination period. We attempted to normalize anxiety by presenting studies on anxiety and offered several stress relieving activities including yoga, painting, colouring, bracelet making, gratitude and "get it off your chest" boxes.

Beyond the immediate implications for student well-being, this research contributes to broader discussions on shaping the future through innovation, sustainability, and scientific excellence. Mental health is an essential component of sustainable development and it aligns with the United Nations Sustainable Development Goals (SDGs), particularly Goal 3 (Good Health and Well-being) and Goal 16 (Peace, Justice, and Strong Institutions) (United Nations, n.d.). Addressing student anxiety through institutional innovation, not only enhances academic outcomes. It also fosters long-term resilience that can equip future professionals with the mental fortitude to navigate complex global challenges. Future research should explore evidence-based interventions that align with sustainability principles, that would ensure that mental health initiatives are both impactful and accessible across diverse educational settings.

Finally, regarding gender differences—where women reported higher anxiety than men—future studies could further investigate the interplay between anxiety, gender, interpersonal relationships, and academic success. Understanding these dynamics in greater depth can help shape targeted policies and interventions that will more effectively support students.

6. Limitations

As in most research, this study has some limitations we would like to discuss. First, the HAM-A is typically administered by a clinician, however we used it as a self-report measure. Second, data collection took place 1–2 weeks before final exams across two semesters, and only once at the beginning of the semester which likely led to anxiety scores due specifically to academic stress, which could have confounded the findings. Third, although the FIAT-Q Class D (and other Class subscales) has high reliability and validity, some critics would prefer to use the entire inventory. However, we used the subscale that was necessary to answer our research question. Finally, convenience sampling may have limited our ability to generalize our findings. It is our recommendation that future research should consider using probability sampling, collecting data at multiple points throughout the semester, and the use of measures that have high reliability validity when used as a self-report measure to improve methodological robustness.

Acknowledgment

First, I am incredibly grateful to Dr. Eric L. De Barros for highlighting the need for continued anxiety research to identify and help university students. I am thankful to my research assistants for their hard work from idea conceptualization to the publication of this piece. I am also thankful to the study participants for the time spent completing the assessments, to the American Psychological Society, Division of International Studies who invited my team and I to present this project while in its poster form, to the British Psychological Society Division of Counselling Psychology Conference and to the International Conference on Innovation, Sustainability, and Applied Sciences who invited us to present and provided feedback on earlier drafts. Finally, I would also like to thank the ICISAS editorial team and the 3 reviewers for their helpful comments on this article.

References

[1] Stearns, P. (September, 2022). A Crisis' of Student Anxiety? *The Chronicle of Higher Education*. https://www.chronicle.com/article/a-crisis-of-student-anxiety.

[2] United Nations. World Population Prospects. (2019). *United Nations Department of Economic and Social Affairs*. July 17, 2019. https://www.un.org/development/desa/pd/news/world-population-prospects-2019-0.

[3] Alqudah, A., Al-Smadi, A., Oqal, M., Qnais, E. Y., Wedyan, M., Abu Gneam, M., ... & Gammoh, O. (2021). About anxiety levels and anti-anxiety drugs among quarantined undergraduate Jordanian students during COVID-19 pandemic. *International Journal of Clinical Practice*, *75*(7), e14249. https://doi.org/10.1111/ijcp.14249.

[4] Fawaz, M., & Samaha, A. (2020). COVID-19 quarantine: Post-traumatic stress symptomatology among Lebanese citizens. *International Journal of Social Psychiatry*, *66*(7), 666–674. https://doi.org/10.1177/0020764020932207.

[5] Gewaifel, G. I., Hussein, M. M., & Ibrahim, N. N. (2023). Anxiety and Depression Screening in Egyptian University Students: Prevalence and Predictors. *Journal of High Institute of Public Health*, *53*(2), 64–74. https://doi.org/10.21608/JHIPH.2023.332378.

[6] Amr, M., Amin, T. T., Saddichha, S., Al Malki, S., Al Samail, M., Al Qahtani, N., ... & Al Shoaibi, A. (2013). Depression and Anxiety among Saudi University Students : Prevalence and Correlates = القلق و الإكتئاب بين طلاب الجامعة السعوديين : الإنتشار و المصاحبات. *The Arab Journal of Psychiatry*, *24*(1), 1–7. https://doi.org/10.12816/0000092.

[7] Mohammed, H. M., Sara, M. S., Ahmed, A. A., & Ahmed K. I. (2022). Depressive symptoms and its correlates among medical students in upper Egypt. *Middle East Current Psychiatry*, *29*(1), 66. https://doi.org/10.1186/s43045-022-00231-y.

[8] Macauley, K., Laura, P., Charlotte, B., Genevieve, B., Christine, L., & Johanna, S. (2018). Prevalence and predictors of anxiety in healthcare professions students. *Health Professions Education*, *4*(3), 176–185. https://doi.org/10.1016/j.hpe.2018.01.001.

[9] Beiter, R., Nash, R. McCrady, M., et al. (March, 2015). The prevalence and correlates of depression, anxiety, and stress in a sample of college students. *Journal of Affective Disorders*, *173*, 90–96. https://doi.org/10.1016/j.jad.2014.10.054.

[10] Meckamalil, C., Lydia, B., Hogg-Johnson, S. (2022). The prevalence of anxiety, stress and depressive symptoms in undergraduate students at the Canadian Memorial Chiropractic College. *Journal of American College Health*, *70*(2), 371–376. https://doi.org/10.1080/07448481.2020.1751173.

[11] Hajduk, M., Heretik, A. Vaseckova, B. Forgacova, L., & Pecenak, J. (2019). Prevalence and correlations of depression and anxiety among slovak college students. *Bratislava Medical Journal*, *120*(09), 695–98. https://doi.org/10.4149/BLL_2019_117

[12] Kovess-Masfety, V., Leray, E., Denis, L., Husky, M., Pitrou, I., & Bodeau-Livinec, F. (2016). Mental health of college students and their non-college-attending peers: results from a large French cross-sectional survey. *BMC Psychology*, *4*(1), 20. https://doi.org/10.1186/s40359-016-0124-5.

[13] Gao, W., Ping, S., & Liu, X. (2020). Gender differences in depression, anxiety, and stress among college students: a longitudinal study from China. *Journal of Affective Disorders*, *263*, 292–300.https://doi.org/10.1016/j.jad.2019.11.121.

[14] Abdou, D. S., & Zaazou, Z. (2013). The Egyptian revolution and post socio-economic impact. *Topics in Middle Eastern and North African Economies*, *15*(1).

[15] Batalova, Jeanne Batalova Brandon Marrow and Jeanne. (June, 2024). Middle Eastern and North African Immigrants in the United States. *Migrationpolicy.Org*. https://www.migrationpolicy.org/article/middle-eastern-and-north-african-immigrants-united-states.

[16] Bronfenbrenner, U. (1977). Toward an experimental ecology of human development. *American Psychologist*, *32*(7), 513–531. *https://doi.org/10.1037/0003-066X.32.7.513.*

[17] Paat, Y.-F. (2013). Working with immigrant children and their families: An application of Bronfenbrenner's ecological systems theory. *Journal of Human Behavior in the Social Environment*, *23*(8), 954–66. https://doi.org/10.1080/10911359.2013.800007.

[18] Coble, A. (December, 2015). Family Separation for Syrian Refugee Children. *Faith to Action* (blog). https://www.faithtoaction.org/displacement-and-family-separation-for-syrian-refugee-children/.

[19] Ainsworth, M. D. S., Blehar, M. C., Waters, E., & Wall, S. N. (2015). *Patterns of attachment: A psychological study of the strange situation*. Psychology Press. https://doi.org/10.4324/9780203758045.

[20] MacLean, S. A., Agyeman, P. O., Walther, J., Singer, E. K., Baranowski, K. A., & Katz, C. L. (2020). Characterization of the mental health of immigrant children separated from their mothers at the US–Mexico border. *Psychiatry Research*, *286*, 112555. https://doi.org/10.1016/j.psychres.2019.112555.

[21] Hampton, K., Raker, E., Habbach, H., Camaj Deda, L., Heisler, M., & Mishori, R. (2021). The psychological effects of forced family separation on asylum-seeking children and parents at the US-Mexico border: A qualitative analysis of medico-legal documents. *PLoS One*, *16*(11), e0259576.https://doi.org/10.1371/journal.pone.0259576.

[22] Al-Hawdrawi, H. H. (2017). The effect of immigration on the mental well-being of Middle Eastern Immigrants in the United States. *Research Journal of Pharmacy and Technology, 10*(7), 2069. https://doi.org/10.5958/0974-360X.2017.00361.4.

[23] Thomas, T. N. (1995). Acculturative Stress in the Adjustment of Immigrant Families. *Journal of Social Distress and the Homeless, 4*(2), 131–42. https://doi.org/10.1007/BF02094613.

[24] Darrow, S. M., Glenn, M. C., Jordan, T. B., & William C. F. (2014). The Functional Idiographic Assessment Template-Questionnaire (FIAT-Q): Initial Psychometric Properties. *Journal of Contextual Behavioral Science, 3*(2), 124–35. https://doi.org/10.1016/j.jcbs.2014.02.002.

[25] Chris Fr. R., Paula, M., Niedenthal, M. M., Claudia, B., & Amanda, V. (2006). Adult attachment and the perception of emotional expressions: Probing the hyperactivating strategies underlying anxious attachment. *Journal of Personality, 74*(4), 1163–1190. https://doi.org/10.1111/j.1467-6494.2006.00406.x.

[26] Khoshhal, K. I., Khairy, G. A., Guraya, S. Y., & Guraya, S. S. (2017). Exam anxiety in the undergraduate medical students of Taibah University. *Medical Teacher, 39*(1), S22–26. https://doi.org/10.1080/0142159X.2016.1254749.

18 Comparative assessment among probiotic formulation and SCFA producing *Lactoplantibacillus pentosus* C_{10} in MRS medium enriched with 1% carbon sources to maintain gut homeostasis

Niharika Madhavan[a] *and Ieshita Pan*[b]

Department of Biotechnology, Saveetha School of Engineering, Saveetha Institute of Medical and Technical Sciences, Saveetha University, Chennai, Tamil Nadu, India

Abstract: Aim: The primary aim of this study is to optimize short-chain fatty acids (SCFAs) production, utilizing *Lactoplantibacillus pentosus* C_{10} with 1% carbon as the nutrient source, aiming to ameliorate gut dysbiosis in Parkinson's disease.

Materials and Methods: Two independent study groups were formed, and the size of each group was determined to be 20 per sample with a total of 40 samples using an online sample calculator (clincal.com).

De Man-Rogosa-Sharpe (MRS) Broth Preparation with 1% Carbon Sources: A mixture of 2g each of fructose, dextrose, maltose, starch, sucrose, and lactose, along with 11.1g of De Man-Rogosa-Sharpe (MRS) broth, was prepared in conical flasks, and each mixture was diluted with 200 mL of distilled water. Additionally, 300mL of MRS broth without any carbon source was prepared by diluting 16.6g MRS broth in 300mL distilled water. After autoclaving prepared media at 121°C for 15 minutes, they were distributed into 7 boiler tubes and 28 test tubes. Six boiler tubes contained 20 mL of each carbon source combined with MRS broth, while one boiler tube contained 25 mL of MRS broth without added carbon source (control). Optical density measurements were taken from four test tubes for each carbon source at four different time points (24, 48, 76, and 92 hours).

Cell Count Calculation for Inoculation and Optical Density Readings: A 100-μL micropipette withdrew 20 μL of *Lactoplantibacillus pentosus* C_{10} from culture in a laminar airflow chamber. The sample was placed on a hemocytometer, covered, and counted under a 40× microscope objective, yielding 434 cells. Then, 434 μL of the broth was dispensed into 7 boilers and 28 test tubes using a 1 mL pipette for inoculation. At 600nm, broths' optical density has been determined.

Antimicrobial Activity: Centrifugation was performed on samples from 24, 48, 72, and 96 hours at 10,000rpm for 15 minutes at 4°Celsius. The Supernatants containing bacteriocin were transferred to new vials. 20 μL of each sample was loaded into wells on nutrient agar plates inoculated with *pseudomonas* and *staphylococcus* separately. After 48 hours, zone of clearance was observed and noted for all plates.

Results: *Lactoplantibacillus pentosus* C_{10} was identified through 16S rDNA sequencing. Adding 1% dextrose to MRS media increased growth by 15%. The strain eliminated 90% of *Pseudomonas* and *Staphylococcus*, significantly outperforming the unoptimized strain, with results showing statistical significance (p=0.00).

Discussions and Conclusions: Investigation explored potential of *Lactoplantibacillus pentosus* C_{10} in mitigating gut dysbiosis associated with Parkinson's disease (PD) by testing various carbon substrates to simulate gut diversity. Monitoring optical density allowed for analysis of bacterial growth and standardized cell counts facilitated precise measurements of SCFA production, Bacteriocin assays confirmed antimicrobial effects of SCFAs against *Staphylococcus* and *Pseudomonas*, underscoring their therapeutic relevance. The research optimized SCFA synthesis with *Lactoplantibacillus pentosus* C_{10}, suggesting its potential in managing PD-related gut dysbiosis. However, variability in SCFA production in vivo and acceptance of *Lactoplantibacillus pentosus* C_{10} as a gut microbiota in PD patients are noted limitations.

Keywords: Neurodegeneration, gut dysbiosis, Parkinson's disease

1. Introduction

In the case of Parkinson's disease (PD), dopaminergic neurons are lost, causing motor as well as non-motor symptoms. Recent research suggests that genetic mutation and variations are key factors that elevate the susceptibility. Individuals who hail from families with a history of Parkinson's are more likely to be subjected to PD than

[a]niharikam2044.sse@saveetha.com, [b]ieshitapan.sse@saveetha.com

DOI: 10.1201/9781003770473-18

counterparts who do not have such medical history in their family. PD mainly affects the motor skills of the patients, some of the symptoms of PD are tremors, muscle rigidity, impaired gait and postural instability. SCFAs generated by intestinal microbiota, show potential in reducing inflammation, boosting neurotrophic factors, and improving gut-brain communication. Harnessing SCFAs' benefits could offer novel therapeutic avenues for PD treatment as they can cross the blood-brain barrier. The three common fatty acids are acetate, butyrate and propionate. Other than being a crucial factor in maintenance of gut health, SCFA also performs other functions such as energy metabolism. Lactic acid bacteria within our gut microbiome constitute causes behind human body's synthesis of SCFA.

Recent data points to a connection between gut dysbiosis and Parkinson's disease progression. In neurodegenerative illnesses, gut-brain axis is important, as well as modifying gut microbiota offers potentially effective therapeutic approach. Microbes in large intestine create chemicals called SCFAs. *Lactoplantibacillus pentosus* C_{10}, isolated from fermented cabbage, exhibits probiotic potential. This study aims to optimize SCFAs production using *Lactoplantibacillus pentosus* C_{10} strain with various 1% carbon sources to ameliorate gut dysbiosis in PD. Understanding the dynamics of bacterial growth and SCFA production under different conditions is crucial for harnessing the therapeutic potential of this probiotic strain in addressing PD-related gut dysbiosis.

Although there is an extensive literature on the role of gut microbiota and SCFAs in neurodegenerative diseases, there is a noteworthy deficit of comparative studies examining specific probiotic strains' efficacy, such as *Lactoplantibacillus pentosus* C_{10}, against formulated prebiotics. Most existing research are focused on the general benefits of probiotics or SCFAs without delving into the optimization of SCFA production under varying conditions or the comparison of the effects of different sources of SCFAs.

The primary goal of this study is to glut this gap by investigating the SCFAs production from *Lactoplantibacillus pentosus* C_{10} using various 1% carbon sources and comparing the therapeutic efficacy of the produced SCFAs with that of a formulated prebiotic. By optimizing SCFA production and understanding the dynamics of bacterial growth, this research seeks to provide insights into the potential therapeutic applications of probiotics in ameliorating gut dysbiosis and its implications for Parkinson's disease management.

2. Literature Review

SCFAs are suggested to have influence over gut-brain communication, directly or indirectly (Intestinal Barrier and Behavior 2016). SCFAs can enhance gut barrier integrity, glucose and lipid metabolism, inflammatory response, and modulate immune system, as well as blood pressure [8]. SCFAs influence CNS (central nervous system) illnesses by modulating oxidative stress, inflammatory responses, neuronal apoptosis, as well as integrity of BBB (blood-brain barrier) [1]. SCFAs generated from colon enter systemic circulation, facilitating anti-inflammatory as well as immunomodulatory actions, enhancing insulin production, sustaining energy balance, and boosting liver along with brain function [3]. The role of intestinal bacteria in modulating gut-brain axis transmission has become increasingly significant in the recent decades. Comprehending interplay between gut microbiota, its metabolic byproducts, as well as brain was expected to be essential for future advancement of innovative therapeutic strategies, including the discovery of new psychobiotics [6]. The host-microbiota interaction has become increasingly important in recent years for understanding emergence of numerous chronic non-communicable diseases, encompassing dementia, cardiovascular disease, autoimmune diseases, as well as cancer. Crucially, DF (dietary fiber), related substances that microbiota breaks down, and metabolites that result from this process – particularly SCFA – were strongly linked to positive health outcomes, including through suggested anti-inflammatory pathways [7]. Understanding function of gut microbiota, sometimes known as "second brain," in different facets pertaining to human well-being has drawn more attention from researchers in recent years. Its possible effects on neurological illnesses are among them [5]. Following digestion of DF, gut microbiota produces SCFAs, that constitute crucial metabolites that can be employed as energy source by epithelial cells or transported by circulation to other organs notably liver along with brain [2].

The most often cited article is The Impact of Probiotics on Human Intestinal Microbiome's Production of Short-Chain Fatty Acids [4]

3. Materials and Methods

3.1. Fermentation of cabbage

Cabbage was fermented for over 90 days at 37°c to further isolate the *Lactoplantibacillus pentosus* C_{10} strain, from the fermented cabbage.

3.2. Isolation and culturing of the Lactoplantibacillus pentosus C_{10} strain

The isolated culture was further inoculated in MRS broth without any added carbon source. The inoculum was then streaked onto a sterile solid De Man-Rogosa-Sharpe (MRS) medium in a laminar airflow chamber under sterile conditions.

3.3. Screening using 16s rDNA sequencing

Screening of the cultured LAB in the MRS medium was performed to identify and confirm the presence of the *Lactoplantibacillus pentosus* C_{10} strain.

3.4. MRS broth preparation with 1% carbon sources

2g of fructose, dextrose, maltose, starch, sucrose and lactose each were combined with 11.1g of MRS broth to be mixed with 200mL distilled water each in conical flasks respectively. 450mL of MRS broth with no carbon source was also made by combining it with 450ml RO water. After that, prepared medium had been autoclaved for 15 minutes at 121 degrees Celsius. After that, the prepared medium then carefully moved into seven boiler tubes and twenty-eight test tubes, respectively. 6 boiler tubes contained 20 mL of each carbon source combined with MRS broth and 1 boiler tube contained 25 mL of MRS broth with no added carbon source, which was considered as control. At four different times (24, 48, 76, and 92 hours), optical density of broth was measured in four test tubes that contained each carbon source.

3.5. Cell count calculation for inoculation and optical density readings

A 100-µL micropipette was used in a laminar airflow chamber, to aseptically withdraw 20-µL of the cultured *Lactoplantibacillus pentosus* C_{10} strain from the broth. The microbial sample was positioned on a hemocytometer, covered with a glass coverslip, and subjected to cell counting under a microscope employing a 40× objective. A total of 434 cells were enumerated. Subsequently, 434 µL of the strain-containing broth was aseptically dispensed into each of 7 boilers and 28 test tubes using a 1 mL pipette for inoculation. The infected broths' optical density became apparent at 600nm.

3.6. Antimicrobial activity

1 mL of the inoculated samples from 24, 48, 72, and 96-hour periods were transferred to Eppendorf vials, to be centrifuged at 4°C for 15 minutes at 10,000rpm. The supernatant containing bacteriocin had been transferred to new Eppendorf. From newly transferred sample, 20µL of sample containing bacteriocin was loaded into wells made in a nutrient agar plate, which was solidified with agarose. The culture plates were incorporated with the pathogenic strains, *pseudomonas* and *staphylococcus* separately. A total of four plates were prepared, with two containing the pseudomonas strain and the other two containing staphylococcus, with 24, 48-hour wells marked on one plate and 72, 96-hour wells marked on a different plate. Samples from the specific period were loaded into the designated wells. After 48 hours, the zone of clearance was observed and noted for all four plates.

3.7. SCFA statistical analysis

Effect of carbon sources on growth of SCFA and Bacteriocin producing *Lactoplantibacillus pentosus* C_{10} (Figure 18.1).

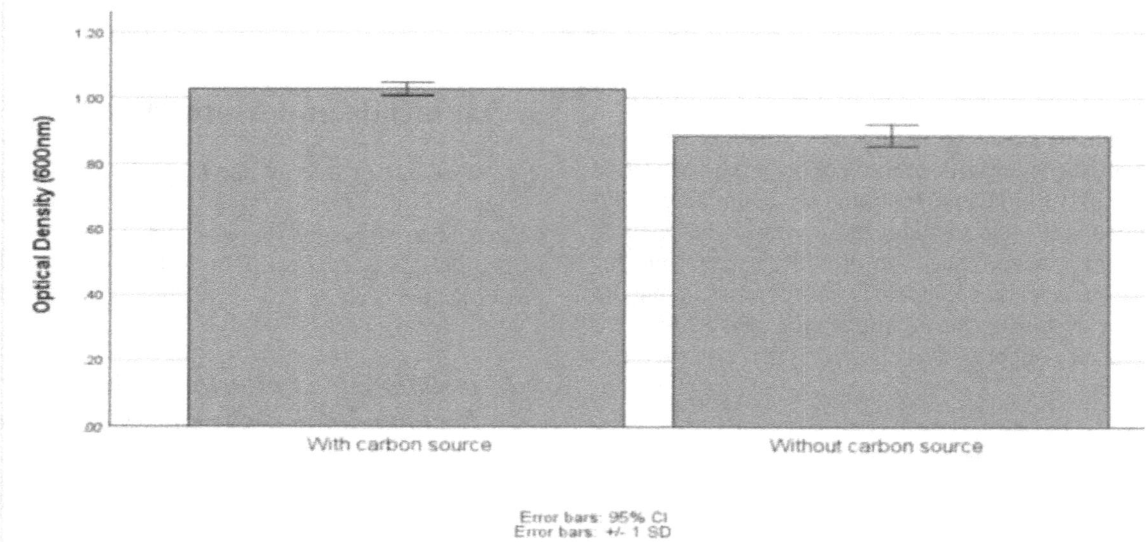

Figure 18.1. Effect of carbon sources on growth of SCFA and Bacteriocin producing Lactoplantibacillus pentosus C10.

Source: Author.

4. Results and Discussions

The isolated strain was identified as *Lactoplantibacillus pentosus* C_{10} through 16S rDNA sequencing. Notably, the addition of 1% dextrose to MRS media resulted in a 15% increase in growth (Table 18.1). Furthermore, *Lactoplantibacillus pentosus* C_{10} demonstrated a remarkable ability to eliminate 90% of *Pseudomonas* and *Staphylococcus*, significantly outperforming the unoptimized strain (Figures 18.2 and 18.3). An independent sample t-test revealed that investigation's outcomes proved statistically significant, yielding a p-value of 0.001 ($p < 0.05$).

The potential of *Lactoplantibacillus pentosus* C_{10} in mitigating gut dysbiosis associated with Parkinson's disease (PD) was investigated. Various carbon substrates were examined to mimic the diversity of the gut environment. Monitoring changes in optical density facilitated the observation of bacterial growth dynamics, ensuring comprehensive analysis of the experimental outcomes. The Accurate cell count calculations were employed to standardize the inoculation procedure across diverse experimental conditions, enabling precise measurements of SCFA production. Bacteriocin assays revealed the

Table 18.1. Morphological and Biochemical properties of *Lactoplantibacillus pentosus* C_{10}

Parameters	*Characteristics*
Gram staining	Gram positive
Shape of Bacteria	Rod-shaped/Bacillus
Catalase	Negative
Carbohydrate Fermentation	Positive
Gas Production	Positive
Fermentation type	Hetero lactate fermentation
Colony Morphology	White, Circular, Irregular margin, raised mucoid colonies in MRS

Source: Author.

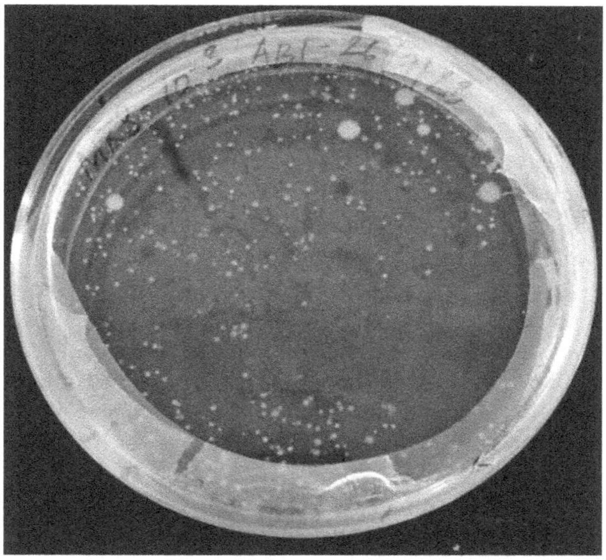

Figure 18.2. LAB Screening in MRS medium.

Source: Author.

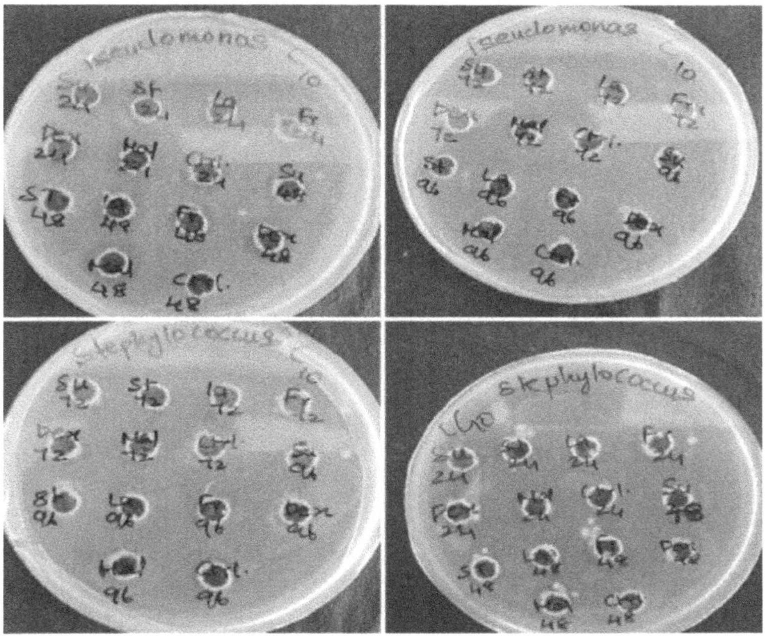

Figure 18.3. Lactoplantibacillus pentosus C_{10} antimicrobial activity against *Pseudomonas* and *Staphylococcus*.

Source: Author.

antimicrobial properties of SCFAs against *Staphylococcus* and *pseudomonas*, underscoring their therapeutic significance in addressing gut dysbiosis.

5. Conclusions

This study's significance lies in its potential to contribute to the understanding of the gut-brain axis in Parkinson's disease and the therapeutic part of probiotics. By optimizing the production of SCFA from *Lactoplantibacillus pentosus* C_{10}, the research could lead to novel dietary interventions that enhance gut health and mitigate the progression of PD.

Furthermore, the comparative analysis with a formulated prebiotic may provide valuable insights into the relative efficacy of different approaches to managing gut dysbiosis, ultimately informing future therapeutic strategies for individuals affected by PD.

The exploration successfully optimized the synthesis of SCFAs with the utilization of *Lactoplantibacillus pentosus* C_{10}, showcasing its potential in managing gut dysbiosis associated with PD. Gaining insights into the influence of microbial byproducts on neurodegenerative disorders sheds light on potential avenues for therapeutic strategies.

Variability of SCFA production under in-vivo condition, and acceptance of *Lactoplantibacillus pentosus* C_{10} as a gut microbiota in PD patients are the limitations of this study. Hence, there is scope for further research on this study.

Acknowledgment

No conflicting objectives in this manuscript.

Authors contributions

Author NM was involved in data collection, analysis and manuscript writing. Author IP was involved in the conceptualization, data validation, and critical review of the manuscript.

References

[1] Dong, Y., & Cui, C. (2022). The role of short-chain fatty acids in central nervous system diseases. *Molecular and Cellular Biochemistry*, *477*(11), 2595–2607.

[2] Guo, B., Zhang, J., Zhang, W., Chen, F., & Liu, B. (2025). Gut microbiota-derived short chain fatty acids act as mediators of the gut–brain axis targeting age-related neurodegenerative disorders: a narrative review. *Critical Reviews in Food Science and Nutrition*, *65*(2), 265–286.

[3] Lammert, F., Gurusamy, K., Ko, C. W., Miquel, J.-F., Méndez-Sánchez, N., Portincasa, P., van Erpecum, K. J., van Laarhoven, C. J., & Wang, D. Q.-H. (2016). Gallstones. *Nature Reviews. Disease Primers*, *2*(April), 16024.

[4] Markowiak-Kopeć, P., & Śliżewska, K. (2020). The effect of probiotics on the production of short-chain fatty acids by human intestinal microbiome. *Nutrients*, *12*(4), 1107.

[5] Moțățăianu, A., Șerban, G., & Andone, S. (2023). The role of short-chain fatty acids in Microbiota-Gut-Brain Cross-Talk with a Focus on Amyotrophic Lateral Sclerosis: A Systematic Review. *International Journal of Molecular Sciences*, *24*(20). https://doi.org/10.3390/ijms242015094.

[6] O'Riordan, K. J., Collins, M. K., Moloney, G. M., Knox, E. G., Aburto, M. R., Fülling, C., Morley, S. J., Clarke, G., Schellekens, H., & Cryan, J. F. (2022). Short chain fatty acids: Microbial Metabolites for Gut-Brain Axis Signalling. *Molecular and Cellular Endocrinology*, *546*(April), 111572.

[7] Ramos Meyers, G., Samouda, H., & Bohn, T. (2022). Short chain fatty acid metabolism in relation to gut microbiota and genetic variability. *Nutrients*, *14*(24). https://doi.org/10.3390/nu14245361.

[8] Tettamanzi, F., Bagnardi, V., Louca, P., Nogal, A., Monti, G. S., Mambrini, S. P., … & Menni, C. (2021). A high protein diet is more effective in improving insulin resistance and glycemic variability compared to a mediterranean diet – a cross-over controlled inpatient dietary study. *Nutrients*, *13*(12), 4380. https://doi.org/10.3390/nu13124380.

19 Holistic approach to avert risks by preventing diabetes using random forest over decision tree for a better medical health to find the overall accuracy

Dhanyaa J.[a] *and Almas Begum*[b]

Department of Artificial Intelligence and Data Science, Saveetha School of Engineering, Saveetha Institute of Medical and Technical Sciences, Saveetha University, Chennai, Tamil Nadu, India

Abstract: Diabetes is a progressive age-related disease, which has a rapid growing incidence in prevalence and is associated with various health risks that demand early detection and management to reduce the risk of heavier complications. It is for this purpose that this research has been designed: presenting a strong artificial neural network to predict diabetes intensity using Pima Indians Diabetes Dataset. To improve the prediction accuracy and to reduce the chances of error Random Forest, an ensemble learning algorithm has been used in the model. Data augmentation techniques such as imputing the missing entries, the SMOTE algorithm for managing imbalance in class data, and feature extraction form the elements that are used in the study to enhance model performance. The performance of the proposed model is measured using several assessment parameters which include accuracy, sensitivity, specificity, F1-score, and AUC-ROC. The study proves that Random Forest classifier yields higher accuracy and the area under the receiver operating characteristic curve of 0.91 for a test accuracy of 87.1%. Such superior performance is an indicator that the model is well capable of picking out very high-risk people easily. They also mentioned that glucose level, BMI, age, and blood pressure are important potential risk factors for diabetes. The present framework used in this study can be beneficial for the healthcare practitioners as a decision support instrument for early detection of the potential cases and to enhance the health care management techniques and diabetic care plans.

Keywords: Diabetes, Pima Indians dataset, random forest, SMOTE algorithm

1. Introduction

Diabetes mellitus, especially, TS2DM, is now a global disease emergency and poses significant challenges to public health. He is right that now the prevalence of the disease increasing because of nonmoving life styles, improper diet, and hereditary factors, for which the disease is producing complications like diseases of the heart, kidney failure, and amputations. It is in this backdrop that early diagnosis of the factors that lead to these complications, as well as preventive measures in the face of these factors, be adopted [1]. However, the task of accurate prediction and prevention still proves to be complex particularly because of the emerging complexities in data health care data. While conventional approaches to estimating the potential of developing diabetes are effective in certain situations and can identify trends, they cannot always account for the complexity of the disease, so other approaches such as machine learning must be considered. Multivariate analysis is widely used popularly identified by the name of the more recent development, machine learning (ML) [1]. ML algorithms used in this study are decision trees because of the simple and understandable nature of this algorithm. A decision tree builds a model of the data by splitting of data in a certain way as often as possible, with reference to feature thresholds. The widespread use of decision trees stems from their clear nature and ease of interpretation yet these models struggle to generalize when operating on large or unstable datasets. The Random Forest ensemble algorithm provides decision tree accuracy improvement through collective prediction averaging which enhances both generalization capability and furthermore delivers robust results. The purpose of this research is to increase diabetes prediction results using Random Forest methodologies alongside the data preprocessing tools SMOTE and feature scaling which reduce dataset bias and enable better convergence.

[a]dhanyaaj4009.sse@saveetha.com, [b]almasb.sse@saveetha.com

DOI: 10.1201/9781003770473-19

The proposed model achieves superior performance while offering healthcare practitioners powerful tools for discovering essential risk factors and delivering essential decision-support resources.

Random Forest [2] which is built on decision trees is much more reliable as an ensemble technique. Whereas one isolated decision tree would tend to make highly prejudiced decisions according to some of the features, Random Forest forms several decision trees and proceeds from their average. This ensemble method enhances the generalization ability of the model to a particular task as well as eliminating the overemphasized data point in a specific dataset that leads to high accuracy in the particular data but low in general. When applied in parallel, Random Forest raises the stabilities to noises and variances because of the combined results from multiple individual trees, which makes it an aspiring development for healthcare field, such as the prediction and prevention of diabetes [3]. The significance of the accuracy of the models in predicting diabetes is vital in that the high risks patients should be well identified and offered early medical attention. Nevertheless, it is obvious that decision trees can sometimes offer an interpretable output, however due to the overfitting problem, decision trees can generate quite a misleading output in most real-world datasets [4]. On the other hand, thanks to the ability to combine multiple decision trees to come up with a final decision, Random Forest can "eliminate" the randomness inherent for any single model, which makes the final prediction more accurate. The focuses of this paper will be the improvement of the diabetes prediction and prevention model by utilizing random forest instead of decision trees for increasing the accuracy of the risks [5].

In order to evaluate the predictive performance of both the models, they the following parameters will be used: accuracy, precision, recall and F1 score. Accuracy defines how good the model is, precision is about true positive while recall measures the ability of the model to detect true positives. The F1 score gives a trade-off between the recall and precision; it gives one measurement that contains the other two. These metrics are important in the medical context, where false negatives which involve failure to identify at-risk individuals, and false positives involving identification of non-at-risk individuals as at-risk can be critical. The data employed for this study comprise the medical records characterizing the patients with features common to DL include age, BMI, amount of glucose in the blood and history of Diabetic members in the family [6]. That is why these features can tell about the probability of an individual getting a diabetes. Nevertheless, health care data is more complex because it contains missing, noisy, and unbalanced data, which creates further difficulties for generic machine learning processes. Random Forests are also able to handle missing values and imbalanced classes better than other decision trees, and therefore proving ideal when working with such datasets. However, accuracy is not the only aspect that all healthcare models should consider, another equally vital attribute which is interpretability was evident in the study [7]. The third reason is that in decision making processes, particularly in the medical field, doctors need explicit understanding of why exactly the model predicts certain outcome. Models in form of decision trees are in general considered being more interpretable than other types of machine learning models due to their tree like structure, with every node representing a decision based on a particular attribute. Still, Random Forest is slightly less interpretable as the ensemble method, but such techniques as feature importance scores can explain which features are most important for making the model's predictions, giving us a balance between selectivity and explainability.

Performance comparison between Random Forest and decision trees will create both models with the same set of data and compare the result against the various measures. Additionally, by using accuracy in conjunction with other measures, such as precision and recall rates, this paper wants to show that the ensemble method is beneficial for improving the results of Random Forest. Additionally, the research will examine how these models can be applied to preventive health care through the framework of building preventive therapy approach to patients at high risk for developing diabetes [8].

This is not just about having a high accuracy of identifying potentially high-risk individuals, but also what causes more risk factors and contribute towards the disease, part of what diabetescare4life with machine learning does. Using such indicators as irregular glucose levels, obesity, and pathologic genealogy, clinicians can work out targeted prevention plans [9]. The models developed in this analysis will enable the objective prognoses, which will let a person start practising a healthy lifestyle and look for treatment as soon as possible before the development of the disease. This type of an individualized approach can go a long way in enhancing on health and decreasing the total load in the system.

Therefore, the main objective of this study is to support the hypothesis and prove that Random Forest model is better for diagnosing the risk of diabetes than decision trees. Much of this work involves the development of new clinical screening prototypes and improved decision-making systems for health care practitioners. Superimposition of machine learning onto existing predictive models means that, in addition to improving diagnostic precision, we can better address the problem of diabetes and general population healthcare.

2. Literature Review

Hosam El-Sofany et al. [10] proposed a machine learning solution for early diabetes prediction in Saudi Arabian communities using the Pima Indians and private diabetes data. Imbalanced class issues are dealt with gradient boosting and SMOTE whereby the semisupervised model delivers 97.4 percent on private data employing XGBoost. Ten ML classifiers were assessed, and SHAP explainability techniques were used to explain model predictions. Diabetes predictions were generated in real time using a mobile application developed with user input. As such, this research provides novel concepts on diabetes detection and management and will immensely benefit public health.

Khaled Alnowaiser et al. [11] suggested an automated diabetes prediction method in which they handled missing data using K-Nearest Neighbour (KNN) imputation together with a Tri-ensemble voting classifier. The model got 97.49% of accuracy, 98.16% of precision, 99.35% of recall and 98.84% of F1 score. The authors also compared the performance of the proposed model to seven other machine learning algorithms for their ability to manage missing data. Special attention is paid to missing data treatment based on the results of the study of diabetes. These results could facilitate the initial detection and optimize the management of diabetic patients in clinical practice. Amit Kishor et al. [12] developed a machine learning based healthcare model for the early and correct detection of diabetes. Issues such as irrelevant features in the dataset and imbalance in the data is tackled in the study using Fast Correlation Based Filter (FCBF) and Synthetic Minority Oversampling Technique (SMOTE) respectively. Five classifiers were tested in this study; namely logistic regression, KNN, Naïve Bayes, random forest, and SVM. The evaluation of proposed model was established in terms of accuracy, sensitivity, specificity, and area under the curve. Overall evaluation: Random Forest presented the highest accuracy of 97.81% and sensitivity of 99.32% followed by specificity of 98.86%, and AUC of 99.35 %. Thus, the work points out that focusing on exceptional features with greater benefits enhances the performance of the model. This work offers a foundational knowledge for improving the diagnosis of diabetes using artificial intelligence.

For pre-diagnosis and management of diabetes, Hassan Shojaee-Mend [13] created a new machine learning model to identify fasting blood glucose status. The current study employed 3376 adults from Tehran's diabetes screening program, and SMOTE was used to handle the dataset imbalance. Five classes of ML algorithms including CatBoost, XGBoost and random Forest was applied with the highest AUC of 0.737 resulting from applying Cat Boost algorithm. Predictors in the study were age, W/H ratio, BMI and systolic blood pressure. Shapley values and noise analysis had been used to provide strong base for feature selection. The model displayed applicability in the context of monitoring and preventing diabetes, as well as in developing a course of therapy. This approach underlines the importance of the ML method in the field of diagnostics of the health conditions.

Sandip Kumar Singh Modak et al. [14] proposed new decision support accurate diabetes prediction model based on machine learning and ensemble techniques. Logistic Regression, SVM, Naïve Bayes, Random Forest and in ensemble, XGBoost, LightGBM, CatBoost, AdaBoost and Bagging were employed in the study. A dataset obtained from Kaggle was utilized and outcomes compared based on accuracy, sensitivity, and AUC-ROC. CatBoost provided the best accuracy of 95.4% as well as AUC ~0.99, which exceeded the XGBoost results. The model also helps increase the efficiency of identifying the early stages of diabetes, which will decrease the load on practitioners. In this approach the writers have demonstrated how ensemble learning performs when used in diagnostics of medical conditions.

3. Materials

In this work, we used Pima Indians Diabetes Dataset that contains following basic attributes: number of times pregnant, plasma glucose concentration in an oral glucose tolerance test, diastolic blood pressure, two-hour serum insulin, body mass index, diabetes pedigree function, age. This makes source material suitable for machine learning model building of diabetes prediction as the target variable in the dataset is a dummy variable identifying patient as a diabetic (1) or non-diabetic (0). Through such features, the dataset helps in training robust models for prediction thereby; contributing to early diagnosis and management of the disease.

4. Methodology

4.1. Preprocessing techniques

Data pre-processing is an important step when intending to build classifiers using the Pima Indians Diabetes Dataset for model building for the test dataset.

Figure 19.1 shows the overview of the proposed methodology.

Handling Missing Values: In case of raw data containing some missing cell values in attributes like glucose, blood pressure, skin thickness, insulin, and BMI, K-Nearest Neighbors (KNN) or mean imputation is used. First there is Mean imputation where lost values are replaced with mean value of the respective feature to maintain equality but does not affect the probability distribution of the data. For example, if in the sample under consideration glucose data are missing, then instead of eliminating such a sample, the average level of glucose for all samples is used, that is, the gaps are "filled in," and the overall

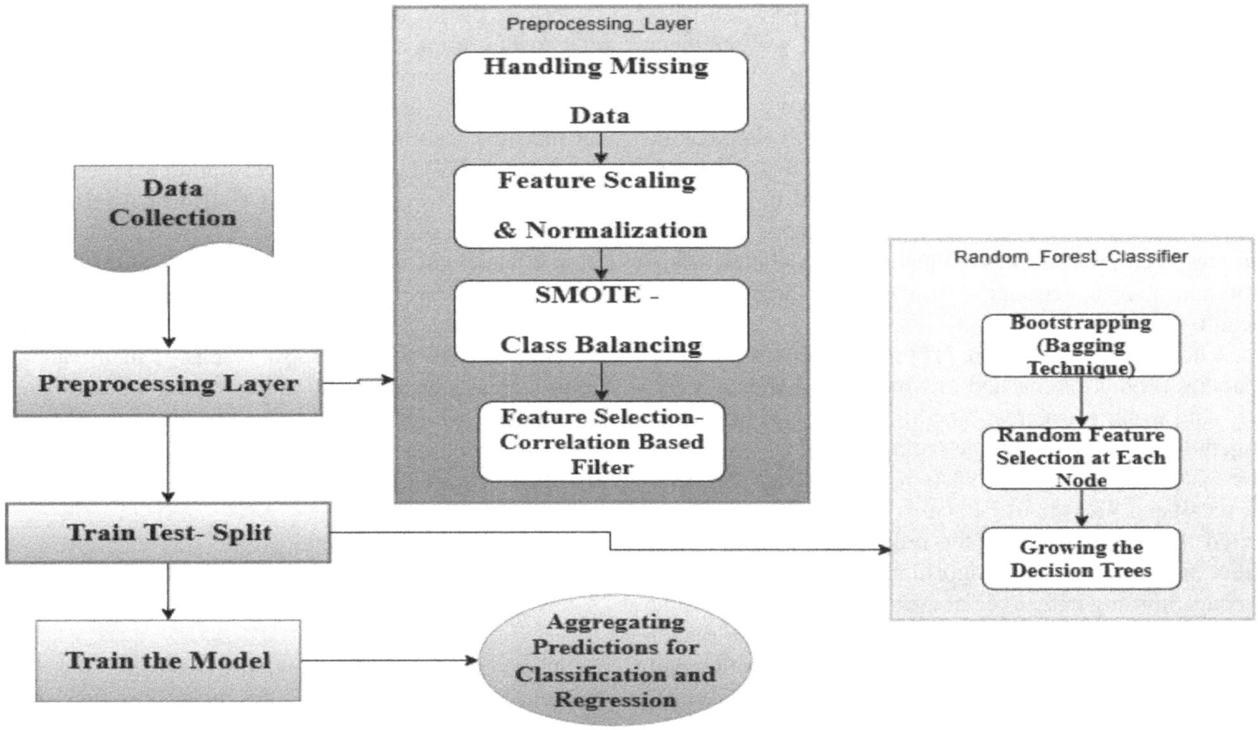

Figure 19.1. Overview architecture.

Source: Author.

integrity of the sample is maintained. This method greatly reduces bias that comes with deleting data and preserves the generative nature of the model so that no important feature is dismissed through the learning phase.

Feature Scaling: Normalization is aimed at placing all features to the same scale so that each makes a equal contribution to model performance is made. The assigns build on the features, such as BMI, age, glucose, by subtracting the mean (μ) from the quantity and dividing it by the standard deviation (σ) standardized value. Thus, making the range of features standard increases model convergence rates during training and enhances the model's accuracy. Feature scaling is going to be strongly required in the models that work with distance – KNN or SVM, for example – in which different input values might overpower results, providing rather biased predictions.

Balancing the Dataset (SMOTE): The Synthetic Minority Over-Sampling Technique (SMOTE) is used in this study to bring new synthetic samples to complete the minority class, to reduce the imbalance of the data set. The operation of SMOTE is as follows; the algorithm will first identify the k-nearest neighbours of a particular minority class instance, make synthetic samples along the line connecting the particular instance to its neighbours. This makes the model less bias towards the majority class, and also help the classifier to better generalize the unseen data. While SMOTE augments the number of records in the minority class thereby correcting for imbalance, the method enhances the predictive performance of the classifier on the non-diabetic patients.

Feature Selection: Extraordinarily correlated features are removed through the determination of the correlation matrix in order to achieve feature selection. If two variables, including BMI and glucose, are highly dependent, the least important variable is removed to avoid increased complexity of the model and negative consequences known as multicollinearity. This helps to guarantee the model works with different unique features that add maximum value to prediction. Consequently, feature selection increases the model's interpretability and robustness, decreases computationally intensive and overfitting issues all of which make the greatest impact towards high and reliable performance.

The Random Forest classifier, constructed with 200 trees and a maximum depth of 30, utilized bootstrapping and random feature selection to prevent overfitting. Hyperparameter tuning further optimized the model for accuracy and computational efficiency.

4.2. Building the random forest classifier

The Random Forest classifier is made by the large collection of decision trees leading to better accuracy and perfect for classifying and predicting the data. Random Forest combat hoarding of information by providing high dimensional space for decision trees, hence increasing the accuracy of generalizing to new data. The classifier forms a number of decision trees where each tree is derived from different partitions of the learning data set such that their outputs are combined to yield the final classification. This approach makes Random Forest a useful model for classification problem, as in the example of diabetes, where common interactions between variables should be modelled appropriately.

Bootstrapping (Bagging Technique): Important in the Random Forest approach, bootstrapping or bagging involves forming a number of different training sets out of the initial data set. It actually means creating several different samples out of the initial dataset using a method of sampling with replacement. Every bootstrap sample is then utilized to build a single decision tree. What comes out of the formation of subsets is the creation of variability among trees because subsets will form different trees. If the original dataset *D* consists of *N* samples, a bootstrap sample D_i can be expressed as:

$$D_i = \{(x_1, y_1), (x_{2,} y_2), \dots . (x_n, y_n)\} \tag{1}$$

where X_i denotes the feature vector, and y_i notation represents the corresponding class label. This sampling technique introduces randomness into the model and makes certain that every tree will be different and not necessarily optimized for certain part of the data point.

Random Feature Selection at Each Node: Specially, Random feature selection is another level of randomness in the Random Forest. Like in the classic trees where at the node, all features are used to find the best split, here at every node, a random subset of features is used. This technique prevents any single feature from overwhelming the model and thus minimize for feature to monopolize the space of the model.

Growing the Decision Trees: Any given decision trees in the Random Forest model are fully constructed to their maximum depth with no cutting. The idea here is to enable the trees to gather as much data as they can. This means that each tree will be built until all the leaves are have only pure instances or the node have the minimum samples which are defined. Despite the fact that a single deep tree can produce accurate results that fit the training sample space, the ensemble diminishes overfitting. It means that growing trees make each tree act independently modeling different patterns present in it and overall improving generalization.

Training the Model: Training a Random Forest requires forming *N* Each bootstrap sample was used for the formation of *N* decision trees. The trees do not aggregate their features and are constructed separately, while prediction is produced by all the trees in an ensemble set. Mathematically, the Random Forest model can be expressed as:

$$RF = \sum_{i=1}^{N} T_i(X) \tag{2}$$

Where $T_i(X)$ is meant the prediction of thei-th decision tree for input *X*. The prediction of the ensemble is as a result of combining all the trees enabling the final model to harness the knowledge of all the trees individually.

Aggregating Predictions for Classification and Regression: After the decision trees are trained the output is combined in order to get the final answer. In classification problems, the class label that is the collective winner of the voting of all the trees is taken. The majority voting can be described by the following equation:

$$y_{pred} = Mode(T_1(X), T_2(X), \dots, T_N(X)) \tag{3}$$

For regression tasks, the predictions from all trees are averaged to obtain the final result, represented by:

$$y_{pred} = \frac{1}{N}\sum_{i=1}^{N} T_i(X) \tag{4}$$

It helps to obtain more precise data by decreasing variety within individual trees and is particularly beneficial in the cases with noisy data and complex forms of the relation between the features.

Hyperparameter Tuning for Random Forest: Several hyper-actions exist which can be implemented in order to improve the performance of random forest. They include, the number of trees, this is an important parameter which is critical when determining the number of trees N in the forest. Increasing N generally enhances performance but with increasing levels of difficulty of the task. Another parameter that regulates a depth of each tree is the MaxDepth that also defines the complexity of models. More also, the minimum number of samples necessary to split a node or the minimum number of samples necessary to form a terminal node, helps avoid deep trees. The number of features f considered at each split is necessary to avoid the risk of building a set of similar trees, with $f = \sqrt{F}$ being a common choice. Choosing of correct hyperparameters ensures the model attains the best of optimized parameters, accurate results and reasonable time.

Figure 19.2 explains the detailed architecture of random forest classifier.

5. Results and Discussion

Proposed model has been validated with various metrics (Table 19.1).

The Random Forest model demonstrated superior performance across multiple metrics, achieving an accuracy of 87.1%, an F1-score of 84.5%, and an AUC-ROC of 0.91. Compared to XGBoost (84.8% accuracy) and Logistic Regression (78.5% accuracy), Random Forest exhibited better generalization and robustness. Feature importance analysis identified glucose levels, BMI, age, and blood pressure as the most influential predictors, aligning with clinical evidence.

Sensitivity analysis revealed that patients with glucose levels ≥180 mg/dL and BMI ≥30 had the highest

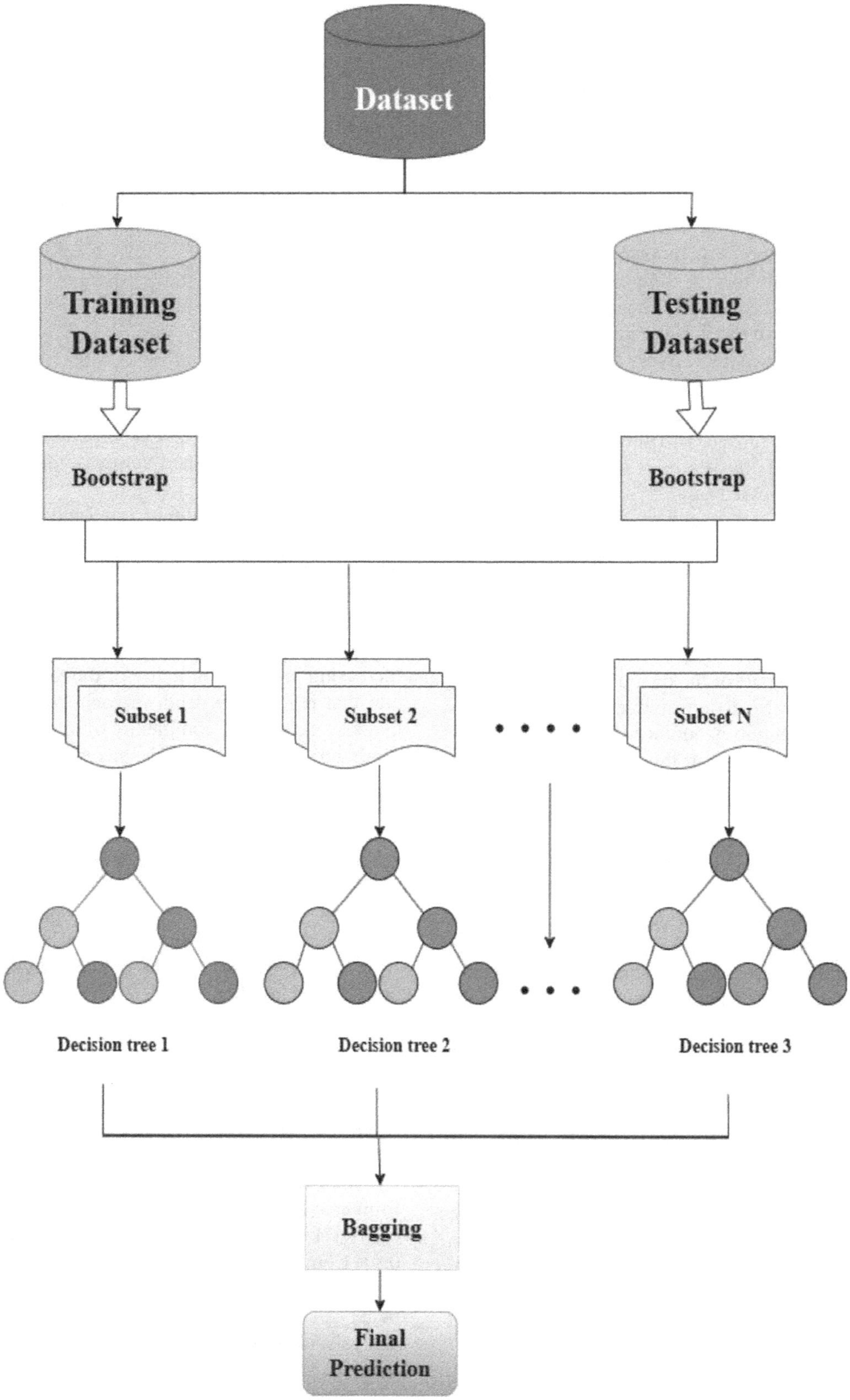

Figure 19.2. Random forest classifier.

Source: Author.

Table 19.1. Hyperparameter tuning for random forest

Hyperparameter	*Value (Optimal)*	*Accuracy (%)*	*F1-Score*	*AUC-ROC*
Number of Trees	200	87.1	84.5	0.91
Max Depth	30	86.5	83.9	0.90
Min Samples per Split	2	85.2	82.7	0.88
Max Features	sqrt	84.3	81.1	0.87
Criterion	Gini	83.9	80.8	0.86

Source: Author.

diabetes risk probabilities of 91.3% and 85.1%, respectively. SMOTE effectively addressed class imbalance, increasing model recall and reducing the misclassification of diabetic patients. These results validate the model's applicability in real-world healthcare settings (Figure 19.3 and Table 19.2).

Random Forest has the highest mean accuracy of 86.3% and F1-score of 84.0% than the other models in all the five folds. Results of cross validation show that the current model is quite stable and applicable across other sets, and increased AUC-ROC suggests better classification on different sets (Figure 19.4 and Table 19.3).

Applying the Random Forest model at the four data types' intersection gives the best accuracy of 85.7% with an 80.7% F1-score, specifically for SMOTE balanced data. The progress we can obtain from the selection of imbalanced data (83.2%) proves that SMOTE is indeed a good tool in solving the problem of class imbalance. Comparing the Random Forest model with other models of the similar, it is seen that the model yields better recall, thus eliminating the misclassification of diabetic patients at a larger extent (Figure 19.5 and Table 19.4).

When performing sensitivity analysis, it becomes evidently clear that the two diagnostic measurements, glucose and BMI are very powerful in diabetes predication with probabilities of 91.3% and 85.1% respectively. The Random Forest model has the highest AUC-ROC of 0.91 when patients' glucose levels are over 180 mg/dL, which provides clinical significance for risk assessment.

6. Conclusions

The proposed Random Forest-based diabetes prediction model demonstrates superior performance across multiple evaluation metrics, consistently outperforming other

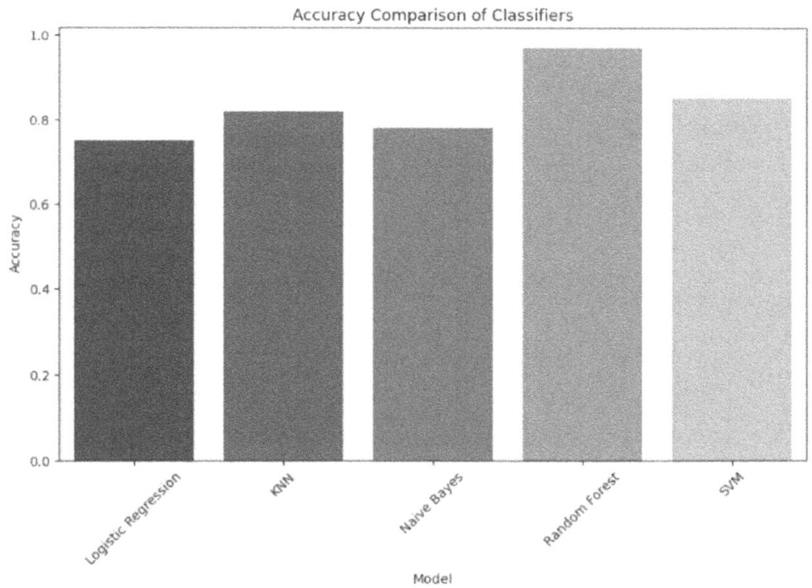

Figure 19.3. Accuracy comparison of classifiers.

Source: Author.

Table 19.2. Cross-validation results (5-Fold Validation)

Model	*Fold 1*	*Fold 2*	*Fold 3*	*Fold 4*	*Fold 5*	*Mean Accuracy (%)*	*Mean F1-Score*
Random Forest	87.2	85.8	86.5	86.9	85.3	86.3	84.0
XGBoost	84.9	85.3	85.1	84.7	84.2	84.8	82.5
Logistic Regression	78.5	77.9	78.8	79.0	78.1	78.5	76.9

Source: Author.

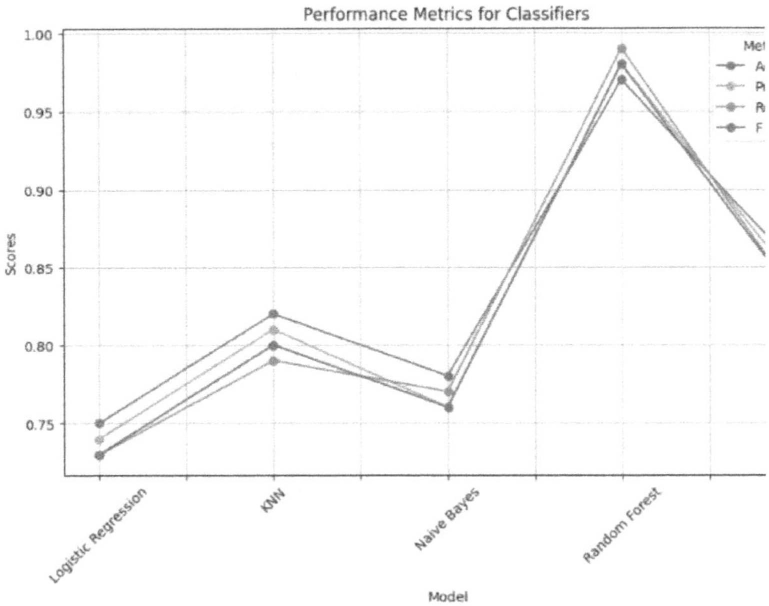

Figure 19.4. Performance metrics oof classifiers.

Source: Author.

Table 19.3. Model comparison on imbalanced and balanced data

Model	*Data Type*	*Accuracy (%)*	*F1-Score*	*Recall*	*AUC-ROC*
Random Forest	Balanced (SMOTE)	85.7	80.7	82.5	0.89
XGBoost	Balanced (SMOTE)	83.9	79.1	81.3	0.87
Logistic Regression	Balanced (SMOTE)	79.8	76.5	78.9	0.78
Random Forest	Imbalanced	83.2	78.1	80.4	0.85
XGBoost	Imbalanced	81.5	76.7	78.5	0.82

Source: Author.

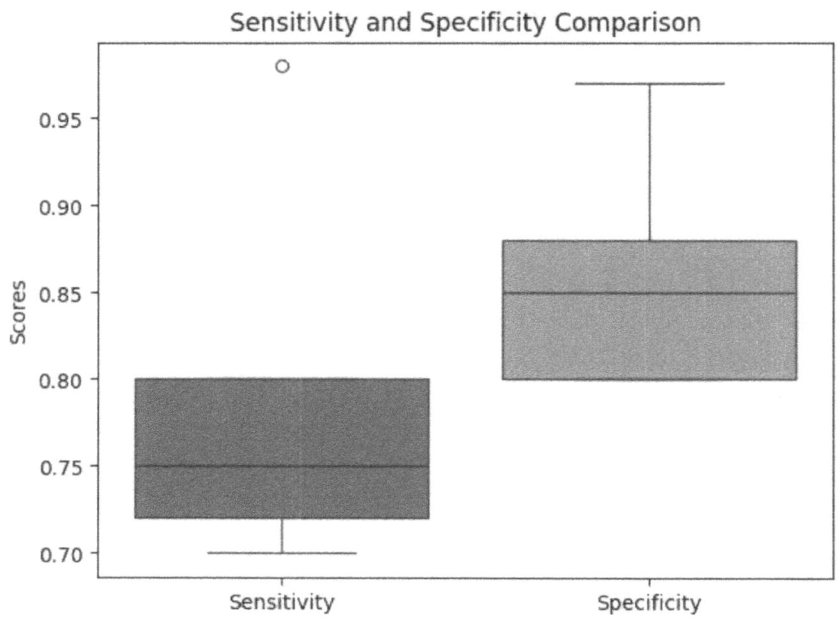

Figure 19.5. Sensitivity and specificity comparison

Source: Author.

Table 19.4. Sensitivity analysis of key features

Feature	*Range*	*Diabetes Probability (%)*	*Model Performance (AUC-ROC)*
Glucose	≥ 180 mg/dL	91.3	0.91
BMI	≥ 30	85.1	0.89
Age	≥ 50 years	78.6	0.86
Blood Pressure	≥ 140/90 mmHg	74.3	0.83
Waist-to-Hip Ratio	≥ 0.9	71.2	0.82

Source: Author.

machine learning models such as XGBoost and Logistic Regression. Through careful hyperparameter tuning, data balancing using SMOTE, and feature sensitivity analysis, the model achieved an impressive accuracy of 87.1%, an F1-score of 84.5%, and an AUC-ROC of 0.91. Feature importance analysis revealed glucose levels, BMI, age, and blood pressure as critical predictors of diabetes risk, aligning with clinical evidence and providing actionable insights for healthcare practitioners. Sensitivity analysis further highlighted the significance of these features, demonstrating the model's ability to identify high-risk individuals effectively. The integration of ensemble learning techniques, combined with bootstrapping and random feature selection, enabled the Random Forest model to mitigate overfitting and enhance generalization. Cross-validation validated the robustness and stability of the model, showing consistent performance across diverse datasets. The use of real-world data from the Pima Indians Diabetes Dataset highlights the practical applicability of this framework in clinical environments, supporting early diabetes detection and prevention. However, the study acknowledges certain limitations, such as the dataset's limited diversity and the need for testing on larger, more heterogeneous populations to enhance generalizability. Future work will focus on integrating real-time data streams and exploring other ensemble and deep learning methods to further refine diabetes prediction models. In conclusion, the Random Forest model serves as a reliable and interpretable tool for early diabetes detection, contributing to timely interventions, personalized prevention plans, and improved patient outcomes. This framework can significantly aid healthcare professionals in risk assessment, ultimately supporting large-scale diabetes prevention and management efforts.

Acknowledgment

No conflicting objectives in this manuscript.

Authors Contributions

Author Dhanyaa J was involved in data collection, analysis and manuscript writing. Author AB was involved in the conceptualization, data validation, and critical review of the manuscript.

References

[1] Kanbour, S., Harris, C., Lalani, B., et al. (2024). Machine learning models for prediction of diabetic microvascular complications. *Journal of Diabetes Science and Technology*, *18*(2), 273–286.

[2] Noviyanti, C. N., & Alamsyah, A. (2024). Early Detection of Diabetes Using Random Forest Algorithm. *Journal of Information System Exploration and Research*, *2*(1).

[3] Oladimeji, O. O., Oladimeji, A., & Oladimeji, O. (2024). Classification models for likelihood prediction of diabetes at early stage using feature selection. *Applied Computing and Informatics*, *20*(3/4), 279–286.

[4] Mohamed Yousuff, A. R., Zainulabedin Hasan, M., Anand, R., & Rajasekhara Babu, M. (2024). Leveraging deep learning models for continuous glucose monitoring and prediction in diabetes management: towards enhanced blood sugar control. *International Journal of System Assurance Engineering and Management*, *15*(6), 2077–2084.

[5] Oei, C. W., Chan, Y. M., Zhang, X., et al. (2024). Risk prediction of diabetic foot amputation using machine learning and explainable artificial intelligence. *Journal of Diabetes Science and Technology*, 19322968241228606.

[6] SumaLata, G. L., Joshitha, C., & Kollati, M. (2024). Prediction of Diabetes Mellitus using Artificial Intelligence Techniques. *Scalable Computing: Practice and Experience*, *25*(4), 3200–3213.

[7] Sharma, A., Kaushal, A., Dogra, K., & Mohana, R. (2024). Deep Learning Perspectives for Prediction of Diabetic Foot Ulcers. In *Metaverse Applications for Intelligent Healthcare*, 203–228. IGI Global.

[8] Malarvizhi, K., Sangeetha, M., Deepak, V., Kavikumar, C., & Prakash, P. S. (2024). Prediction of Diabetes Mellitus using Ensemble Methods in Machine Learning. *Grenze International Journal of Engineering & Technology (GIJET)*, *10*(1).

[9] Talari, P., N, B., Kaur, G., Alshahrani, H., Al Reshan, M. S., Sulaiman, A., & Shaikh, A. (2024). Hybrid

feature selection and classification technique for early prediction and severity of diabetes type 2. *Plos One*, *19*(1), e0292100.

[10] El-Sofany, H., El-Seoud, S. A., Karam, O. H., Abd El-Latif, Y. M., & Taj-Eddin, I. A. (2024). A Proposed Technique Using Machine Learning for the Prediction of Diabetes Disease through a Mobile App. *International Journal of Intelligent Systems*, *2024*(1), 6688934.

[11] Alnowaiser, K. (2024). Improving Healthcare Prediction of Diabetic Patients Using KNN Imputed Features and Tri-Ensemble Model. *IEEE Access*.

[12] Kishor, A., & Chakraborty, C. (2024). Early and accurate prediction of diabetics based on FCBF feature selection and SMOTE. *International Journal of System Assurance Engineering and Management*, *15*(10), 4649–4657.

[13] Shojaee-Mend, H., Velayati, F., Tayefi, B., & Babaee, E. (2024). Prediction of Diabetes Using Data Mining and Machine Learning Algorithms: A Cross-Sectional Study. *Healthcare Informatics Research*, *30*(1), 73–82.

[14] Modak, S. K. S., & Jha, V. K. (2024). Diabetes prediction model using machine learning techniques. *Multimedia Tools and Applications*, *83*(13), 38523–38549.

20 Economic impact on advanced healthcare a comparative analysis of Brain Tumour detection using VGG-19 and CNN

Sunil Shuraj N.[a] *and Sivagami S.*[b]

Department of Artificial Intelligence and Data Science, Saveetha School of Engineering, Saveetha Institute of Medical and Technical Science, Saveetha Nagar, Thandalam, Chennai, Tamil Nadu, India

Abstract: Advanced healthcare technologies have transformed medical diagnostics. It has significantly improved accuracy and reduced costs. In the case of Brain tumour diagnosis, deep learning models have been very promising. CNNs are the most commonly used models for this purpose. However, they are often hampered by the lack of good feature extraction capabilities and accuracy. To overcome these constraints, this study suggests the adoption of VGG-19, a pre-trained deep learning model, as an enhanced alternative for brain tumour detection. The current CNN model will be used as the baseline with simplicity and low computational costs but with reduced precision in complex cases. On the contrary, the newly proposed VGG-19 takes advantage of its deeper network and more robust ability to extract significant features, whereby it is seen to perform high accuracy in finding and classifying brain tumours for MRI scans. Though computationally demanding, its performance in advanced medical imaging sets justifies using it in very complex healthcare system implementations. From an economic point of view, the adoption of VGG-19 could minimize diagnostic errors and promote early detection to result in improved patient outcomes and reduced treatment costs. The enhanced accuracy reduces the need for repetitive diagnostics and hence enhances the effectiveness of service delivery. Future research efforts will concentrate on optimizing the VGG-19 model to make it implementable in real time on devices with limitations in resources and be cheap to run through various clinical scenarios.

Keywords: Brain tumour detection, CNN, diagnostic accuracy, VGG-19

1. Introduction

Artificial intelligence (AI) in the health sector marked a new milestone of innovation with a revolution that changed diagnostic techniques and made them much more efficient and accurate than ever before. Amongst numerous applications of AI, brain tumour detection is the top priority due to the dangerous nature of the disease and its complicated diagnosis. MRI scans are the most common mode of detection of brain tumours. However, they often require very skilled radiologists to interpret them, and thus, have the potential of being human-error-prone in complicated or not-so-obvious cases. The recent deep learning techniques have made such diagnostic procedures possible to automate and improve the precision to a high extent.

Convolutional Neural Networks are the most used DL model for the purpose of detecting a brain tumour. The CNNs use their architecture to Derive characteristics from medical images and thus can be used for automated classification and segmentation. Despite the great success of CNNs, there often occur difficulties in dealing with intricate and complex features in medical imaging, and hence suboptimal results occur in some cases. This drawback necessitates further research into more complex architectures that may eliminate the inherent complexity of brain tumour detection. Figure 20.1 shows the types of brain tumour.

This paper investigates overcoming the aforementioned drawbacks by using a pre-trained deep learning model known as VGG-19. The architecture used in VGG-19 is deeper and more robust than others, making it capable of extracting very intricate and hierarchical features from input data, especially suitable for complex imaging tasks like brain tumour detection. Although VGG-19 has higher computational requirements than traditional CNNs, its superior accuracy and robustness in feature extraction make it a good candidate for this domain.

The use of advanced models, such as VGG-19, would, therefore, have huge economic implications. Accurate and early detection of brain tumours lessens the need for

[a]sunilshurajn4019.sse@saveetha.com, [b]sivagamis.sse@saveetha.com

DOI: 10.1201/9781003770473-20

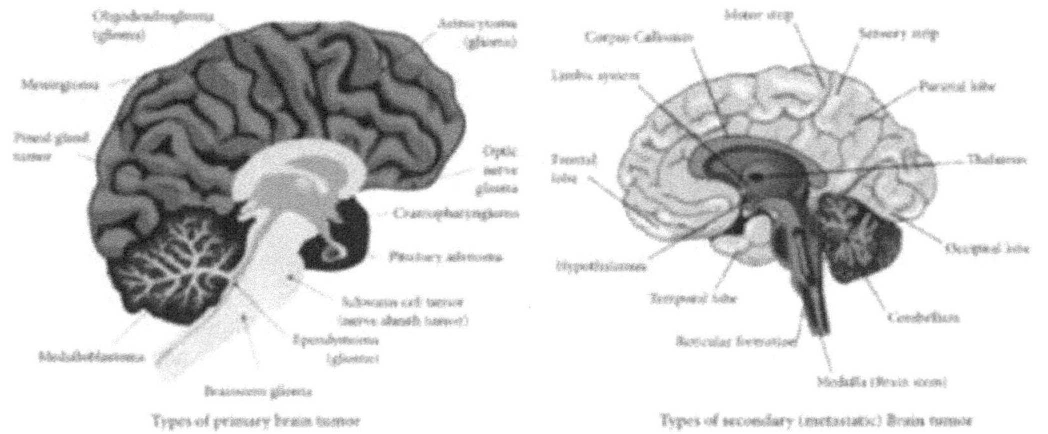

Figure 20.1. Types of primary and secondary tumour.

Source: Author.

follow-up extensive diagnostics that will in turn reduce the cost of treatment and enhance patient outcomes through timely interventions. These models can also relieve the workload from the radiologists' side by automating routine tasks in diagnostics. This will give sufficient time for the radiologists to focus on more complex cases, thus improving the overall healthcare provision mechanism.

This paper presents a benchmarking of the existing CNN model with the proposed VGG-19 model to determine the suitability in the detection of brain tumours. The accuracy and robustness analysis will be considered along with its economic implications on the health sector. Section 2 outlines the architectural and methodology of CNN and VGG-19. Section 3 addresses the dataset and training. Section 4 presents experimental results and compares the strengths and weaknesses of both models. Section 5 discusses general implications of the findings, such as economic benefits and the opportunity for deploying VGG-19 in real-time clinical environments.

Advanced deep learning models have the transformative potential in healthcare and are especially well-suited to critical applications like the detection of brain tumours. In resolving these limitations, VGG-19 represents a very significant step towards realizing more accurate, efficient, and economically viable healthcare solutions.

2. Related Work

This study [1] and [2] examines traditional diagnostic techniques in the spotting of brain tumours and focuses on how the diagnosis often relies on a radiologist interpreting MRI scans by hand. These studies show how human-dependent processes, such as inter-observer variability and inconsistency in diagnosis, are major drawbacks, especially when dealing with difficult cases like gliomas with vague boundaries. Beyond this, the papers present limitations in hand approaches dealing with the fact that imaging conditions vary, for instance, with different MRI machine configurations. The authors asserted that automated solutions must therefore be developed, examining the possibility of earlier CAD systems that had used image processing to allow for the segmentation and analysis of the tumour but were limited by dependency on handcrafted features as well as lack of generalizability across a range of datasets. It brings to light the limitations of traditional methods of diagnosis and opens for innovation in the area of automated, DL based techniques for the spotting of brain tumour.

This research [3] and [4] discusses the introduction of Computer-Assisted Diagnosis (CAD) systems for brain tumour detection. The studies show how CAD systems improved diagnostic accuracy by automatically segmenting and classifying tumours, thus reducing human error. However, early CAD systems were limited by their reliance on handcrafted features and showed poor generalization across different datasets and MRI modalities. These limitations motivated more advanced methods such as deep learning to better detect tumours, including improved accuracy and robustness.

This research [5] and [6] explores how VGG and CNN models can be employed to detect brain tumours. The studies reveal their applicability as enhancing the accuracy in the classification and segmentation of MRI images when compared with other traditional methods. Despite the fact that there are computational complexities and the big requirement for large datasets, CNNs and VGGs considerably boost the detection of tumours and introduce the highest levels of reliability in the provided diagnostic solutions.

In this research [7] and [8] the introduction of deep learning into medical imaging is discussed with a particular emphasis on its potential use in the detection of brain cancers. The research is centered around the shift from traditional image processing methods to deep learning models, in particular CNNs that learn features directly from data. These models have demonstrated improvement even to levels of accuracy for tough tasks, such as tumour extraction and groupings, over traditional methods. The paper presents the promise of deep learning to revolutionize medical imaging by offering more accurate, automated, and scalable solutions to brain cancer diagnosis, even with the challenge of the need for large annotated datasets.

This [9] and [10] study points out that the existing work has some drawbacks, such as the use of small, localized datasets that cannot capture the wide range of tumour types.

3. Proposed Methodology

Diagnosing and identifying MRI brain tumour images is essential for the detection and classification of potentially fatal tumours, whether benign or malignant. Brain tumours have been studied a lot using medical imaging techniques. The effectiveness of the classification depends upon the discriminative and descriptive power of the features that are extracted. Machine learning is very important in this process due to its vast range of methods and adaptability to specific tasks.

It provides an end-to-end convolutional network for the multi-level classification of Magnetic resonance imaging brain tumour image by making use of VGG19 base model with entire visual volume. The pipeline included four primary steps: the pre-processing of MRI images, which involved thresholding, erosion, and cropping; rotation-based data augmentation; the CNN-based VGG19 model with additional layers of flattening, normalization, activation, and dense processing; and model training and evaluation, which were performed via the Adam optimizer and log loss, respectively.

The neurons of a CNN contain weights and biases that capture the inputs from various regions of the image. CNNs are more efficient and faster than traditional neural networks, thus showing better performance in tumour detection and classification.

The neural network's conceptual model above can be used to calculate the total input.

$$y_in = (x_1w_1 + x_2w_2 + x_3w_3 \dots\dots\dots + x_nw_n) + b \quad (1)$$

i.e., Net input $y_{in} = \sum_{i=1}^{n} x_iw_i + b$

The activation function may be used for net input to calculate the output. Figure 20.2 shows the basic function of the Artificial Neural Network.

3.1. Dataset description

The research utilized Brain MRI Images from Kaggle (BRAT Dataset), which comprises MRI scans of two distinct classes, as depicted in Figures 20.3 and 20.4.

The dataset is divided into training, testing, and validation sets to improve performance.

3.2. Data pre-processing

The main goal of the preprocessing phase is to improve Image clarity, remove irrelevant data, and enhance the contrast of MR images. To do this, we first apply thresholding to the input images. Thresholding is an image analysis technique that converts a colour or grayscale image into a binary image, where pixels are classified as either black or white. This process simplifies the analysis of images and helps identify regions of interest while ignoring other areas. One of the two basic morphological preprocessing steps from which all other morphological operations depend after thresholding is erosion. Initially designed for

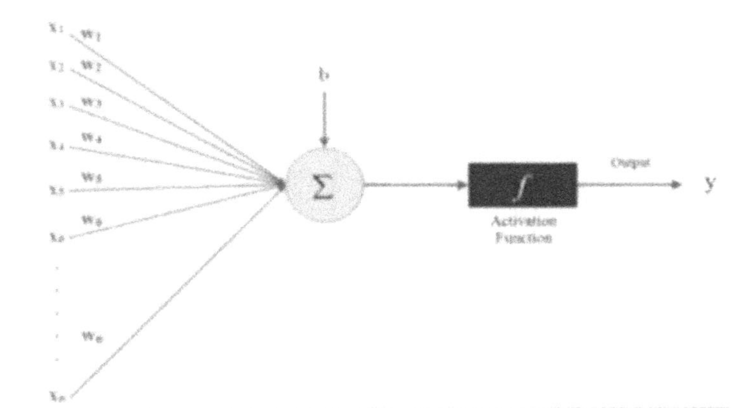

Figure 20.2. Artificial neural network.

Source: Author.

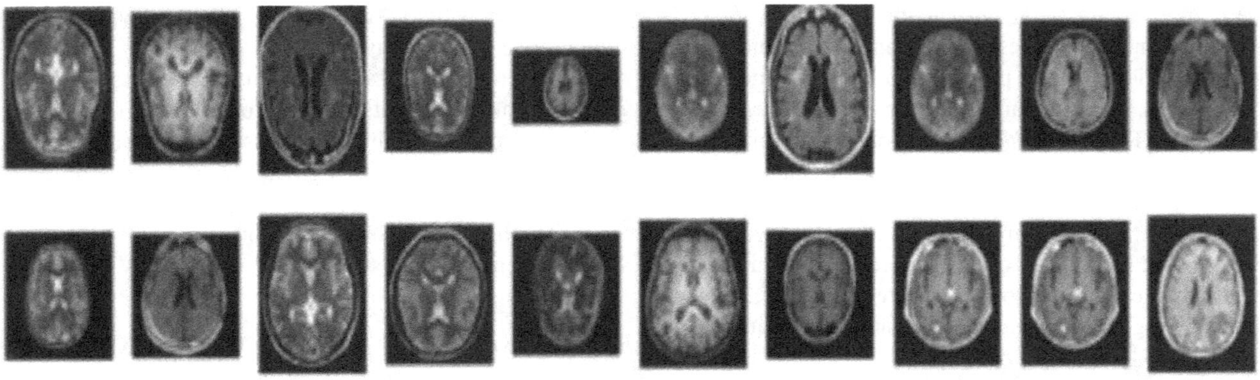

Figure 20.3. Brain MRI images with no Tumour.

Source: Author.

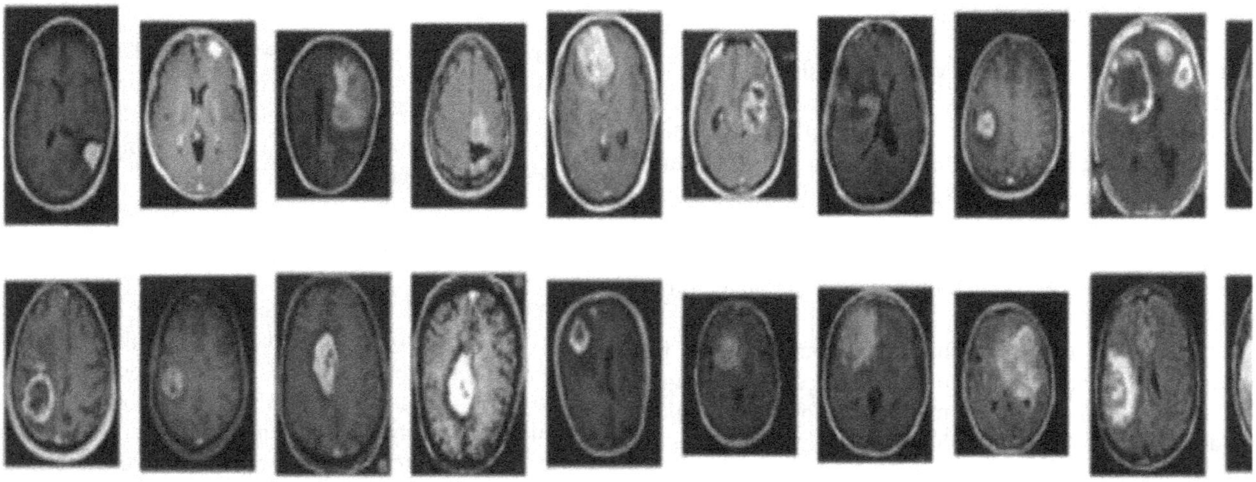

Figure 20.4. Brain MRI Images with Tumour.

Source: Author.

binary images, it was later Adapted for grayscale images and even entire lattices.

Let A be a binary image in some Euclidean plane E or some grid of integer values. Suppose a structural element B induces erosion in the binary image A such that undergoes erode as follows:

$$A \ominus B = \{z \in E \mid B_z \subsetneq A\} \tag{2}$$

Where B_z is the vector z's translation of B., i.e.,

$$B_z = \{b + Z \mid b \in B\}, \forall z \in E \tag{3}$$

If structural element B is positioned at the origin of E, the erosion of A by B can be visualized as the collection of locations attained at the center of B as it moves within A. For instance, a 10 × 10 square at the origin is eroded by a disc with a radius of 2 positioned at the origin, resulting in a 6 × 6 square placed at the origin. Let A and B be matrices with dimensions 13 × 13 and 3 × 3, respectively (Figure 20.5):

This suggests that the pixel values are preserved only when B is completely enclosed within A; otherwise, they are either destroyed or eroded.

In the next step of the methodology, we will enhance the possibility of finding a suspicious region within an MRI image. By improving the finer details of the image and removing noise from the relevant photographs, we will improve the accuracy of MRI scans. However, introducing noise into MRI scans will compromise their accuracy. This can be addressed by the first step of the normalization procedure, where it crops out the brain from the images based on the identification of extreme points along its contours, as shown in Figure 20.6.

To enhance a neuroimaging MRI appearance, an MRI image of a brain can be cropped by removing or correcting the outermost corner as in Figures 20.7 and 20.8.

Figure 20.5. 13 × 13 matrix represents the erosion of A by B.

Source: Author.

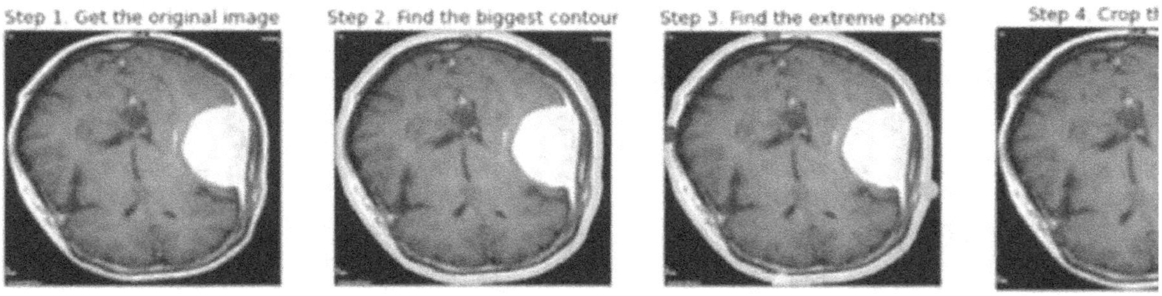

Figure 20.6. Steps to find the contour and extreme point.

Source: Author.

3.3. Augmentation

A huge/substantial volume amount of data is needed to enhance the performance of a DL Conv model. Nevertheless, accessing a large dataset poses significant challenges in most cases. To overcome the limitations, data augmentation techniques are used to expand the number of sample images available in the dataset. Such methods scales up the size of training clusters by providing access to a larger volume of relevant training data. The basic idea is that of artificially inflating the training instances, as shown in Figure 20.9.

Three strategies, according to the literature, are identified:

1. Data augmentation by way of creating new data points or expanding existing datasets.
2. On-the-fly/in-place augmentation of data: This form of data modification is accomplished in real time or during training to improve its quality and diversity.
3. Data generation and augmentation in combination: It involves generating new data points and then augmenting them to enhance their quality and diversity further.

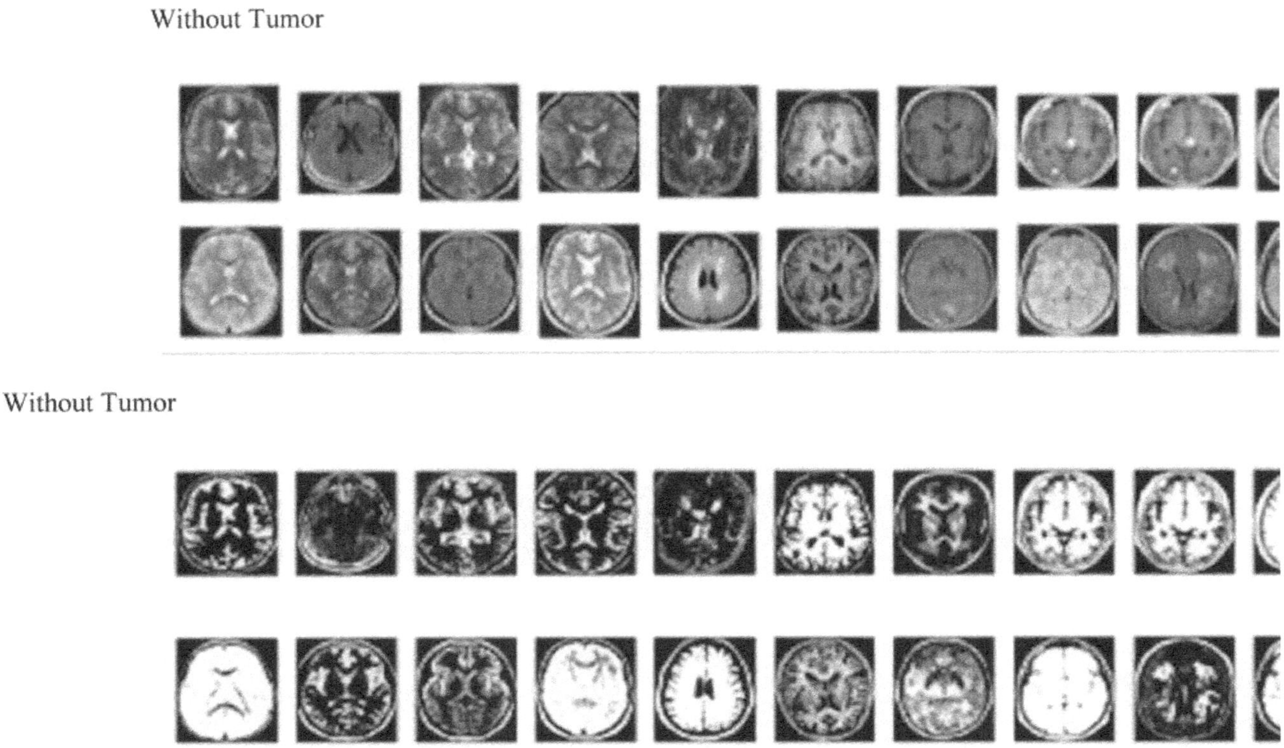

Figure 20.7. Visualization of Brain MRI in Grayscale and RGB form after cropping (Without Tumour).
Source: Author.

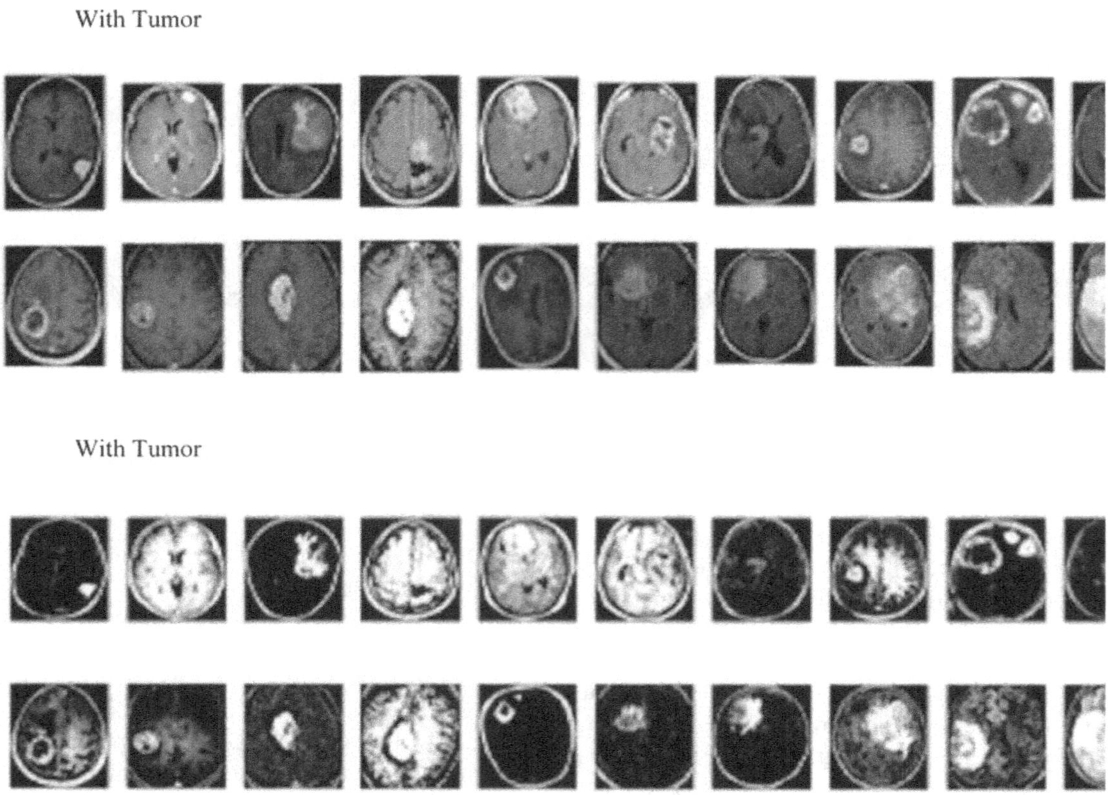

Figure 20.8. Visualization of Brain MRI in Grayscale and RGB form after cropping (With Tumour).
Source: Author.

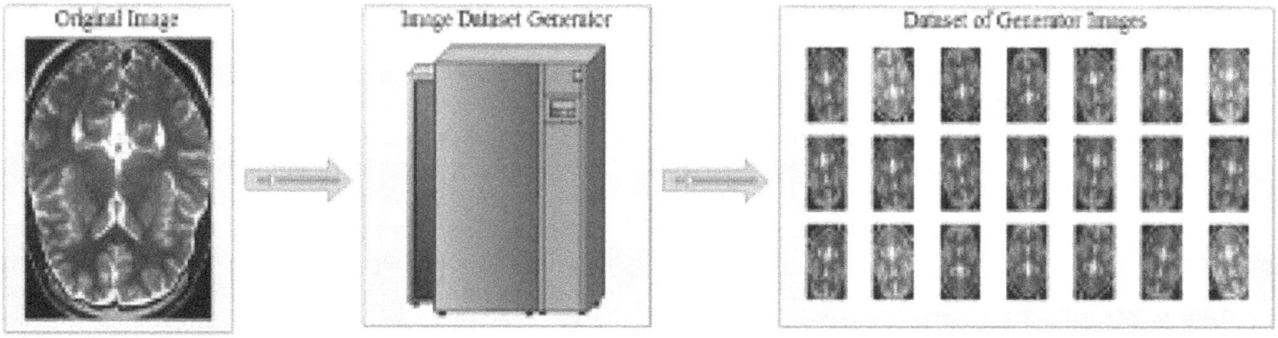

Figure 20.9. Generation of dataset and intensifying an existing dataset.

Source: Author.

3.4. Model building

A Convolutional Neural Network (CNN), is a Deep Learning technique used to attribute importance in the form of learnable biases and weights to various contents within an image and enable discrimination between them. Thus, ConvNets generally require much less pre-processing than other classification algorithms. Unlike simple techniques that commonly demand hand-engineering of filters, ConvNets can learn these filters/characteristics provided correct training. This enables ConvNets to capture the temporal and spatial correlations of an image with the help of appropriate filters. It does so because ConvNets are better in terms of adaptation of the design with the image dataset, owing to fewer features and more durable weights.

This paper implements a VGG 19 as a base model, followed by layers of flatten, normalization, dense, activation, and drop in this order. Based on this ConvNet, the foundation it uses is a 224 × 224 RGB image. It refers to the extraction of the RGB image from mean representation values created for all ImageNet training values by the base layer. Adjacent pixels vary from 0 to 255 in these RGB images.

After the pre-treatment of the input images, they are passed through these weight layers. Convolutional layers are applied for processing training images. The VGG19 model consists of 16 convolutional layers and three fully connected layers. VGG applies smaller filters of size 3*3 but of deeper depth and has the same 5 pooling layers as large filters. In VGG19, there are two 4096-channel fully connected layers, and then it uses a 1000-channel fully connected layer to predict 1000 labels.

Here is a simplified walkthrough of the VGG19 architecture:

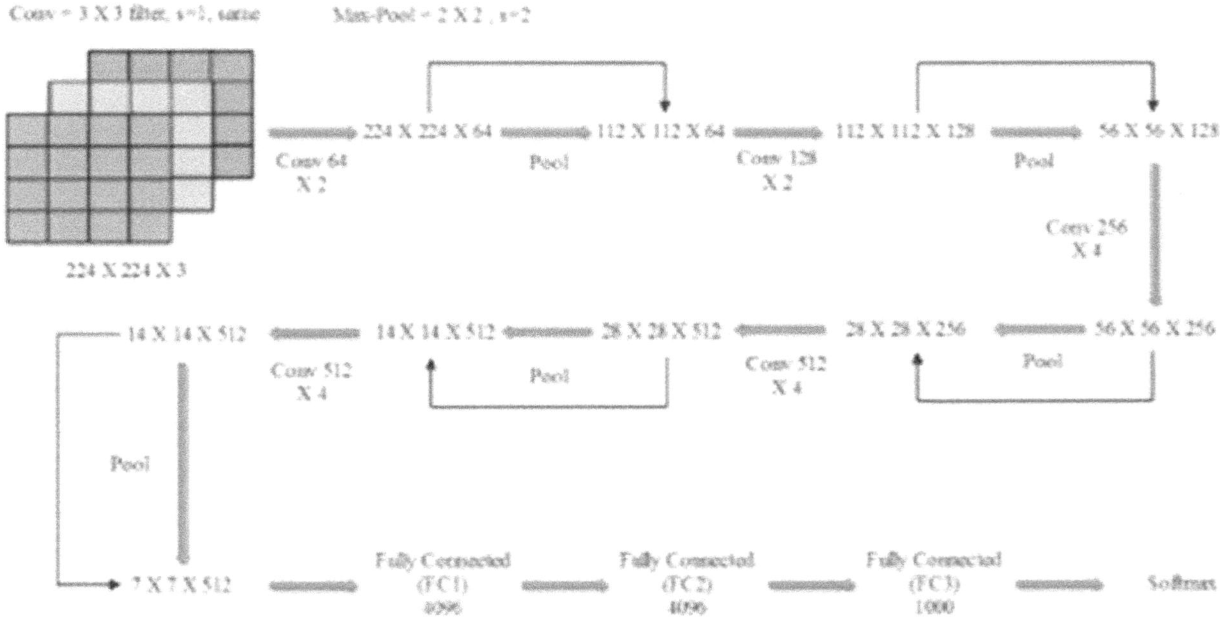

Figure 20.10. VGG 19 architecture walkthrough.

Source: Author.

When this structure was provided a static-length (224 × 224) RGB image as input, the vector was of shape (224, 224, 3).

Preprocessing was only the calculation of the average RGB value for every pixel in the whole training set.

To capture the whole visual spectrum, they used kernels of size (3 × 3) with stride sizes of 1 pixel.

The stride 2 produced maximum pooling over a 2 × 2 pixel window.

To add non-linearity to the framework for better classification and speed in processing, the use of the Rectified Linear Unit (ReLu) was added. In previous versions, tanh or sigmoid processes were employed, but this significantly out-performed them.

A basic design of VGG19 includes several conv layers, pooling, and fully linked layers (Figures 20.10 and 20.11). The purpose of the conv layer is to recognize and extract data. To highlight specific areas, we repeat this process, starting with the input image and computing the dot product while considering weights and biases. The technique for obtaining a single output vector is described by equation.

$$A_j = f(\sum_{i=1}^{N} I_i \ x \ K_{i,j} \ + \ B_j) \tag{4}$$

Where I is the input vector, K represents the convolutional kernel, N stands for the size of the input, B_j for the bias offset and f a nonlinear transfer function.

Flatten Layer: Flattening is the process of transforming data into a one-dimensional set for further analysis. In this research, the result of the conv layers is flattened to create a long feature representation, which is then connected to the final classification model, also known as a fully-connected layer (Figure 20.12).

Batch Normalization Layer: It is a training method of deep learning that rationalizes each batch of mini-contributions to a layer. It decreases the speed of learning and minimizes the required count of training datasets in the construction of deep convolution neural networks.

Dense Layer: It is the simplest layer of neurons where every neuron receives an input from all the other Neurons in the preceding layer that were activated, which explains the name. Dense layers are used for the detection of features based on the output of the convolution layers.

Activation Layer: This is a section inserted at the end or in the middle part of a neural network, depending on the decision whether to fire or not. The activation function changed the input signal within a nonlinear method, which resulted in a modified output that was then relayed into the next layer of neurons for use as input.

Dropout Layer A dropout layer is a technique by which neurons of a neural network are dropped, or excluded in training. Simplistically, some of the neurons of the network get temporarily deactivated.

4. Model Performance and Evaluation Metrics

4.1. Cross entropy loss function

Cross-entropy loss, is the measure of how accurately a binary classifier is; this classifier predicts an output between 0 and 1. It becomes more when the divergence between predicted and true label increases. For example, it suffers a very heavy loss in Estimating a probability of 0.012 while the true observation is 1. For an ideal model, it should be zero for log loss. When M is 2, cross-entropy loss is computable for a classification model as follows.

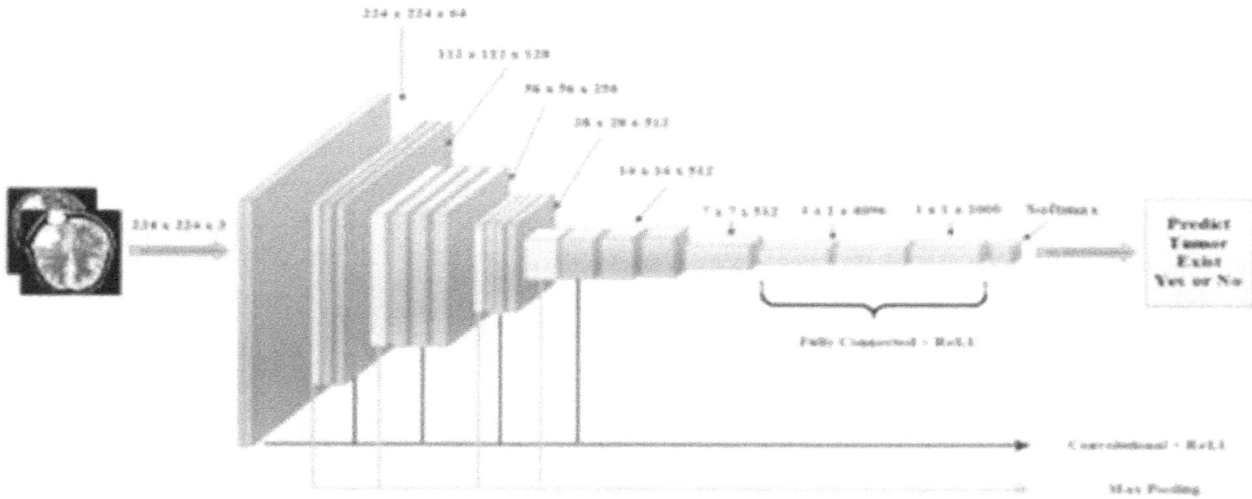

Figure 20.11. VGG 19 used base layer model.

Source: Author.

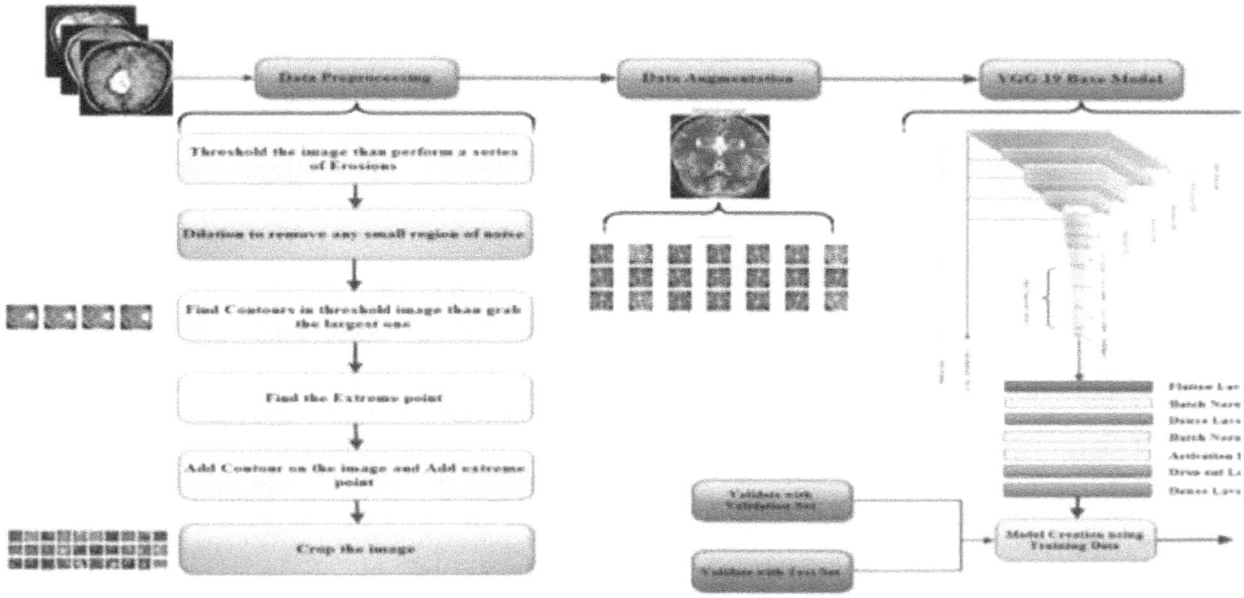

Figure 20.12. Adapted methodology in the study.

Source: Author.

- If M is less than 2, that is, if it is the case of multi-class categorization, then loss for each of the target classes is estimated independently and summed together.
- The recommended show consolidates the delicate dice misfortune work as an upgrade to the misfortune function.

4.2. Accuracy

The effectiveness of a model is determined by its ability to discern connections and patterns among variables in a dataset through the utilization of input or training data.

$$ACC = \frac{TP + TN}{TP + TN + FP + FN}$$

4.3. Precision

The model accuracy is measured in terms of the ratio of number of positive samples correctly diagnosed to true positives plus false positives (correctly and incorrectly identified). This accounts for the ability of a model to mark a sample as having positive.

$$Precision = \frac{TP}{TP + FP}$$

4.4. Recall

The recall, a crucial metric in machine learning, quantifies the models ability to precisely identify positive samples. It is derived by dividing the total number of significant samples by the number of successful samples that were accurately classified as positive. A higher recall value indicates a greater number of positive samples being discovered by the model.

4.5. F1-Score

A prevalent measurement for surveying a test's exactness is the F1-score, some of the time alluded to as the F1-measure. It is computed by taking the exactness and review of the test and taking their consonant mean.

$$Recall = \frac{2TP}{2TP + FP + FN}$$

4.6. AUC ROC curve

Receiver Operator Characteristic, or ROC graph, is the measure for a classification model. It's the probability-based curve that shows true positive rate vs false positive rate at several thresholds, enabling the differentiation between 'signal' and 'noise'. AUC is the graphical representation of ROC curve which describes the ability of the classifier in terms of the discrimination power across classes.

4.7. Cohen Kappa score

A statistical indicator called Cohen's Kappa is used to determine how often two assessors who are evaluating the same collection of items agree.

The data used here consists of a set of MRI images of a brain tumour that has been divided into three datasets:

training, validation, and test. Overfitting is checked using the validation dataset, which is employed to adjust hyper-parameters throughout the model's development. A CNN is employed with the aid of the testing dataset to achieve the desired performance of the model.

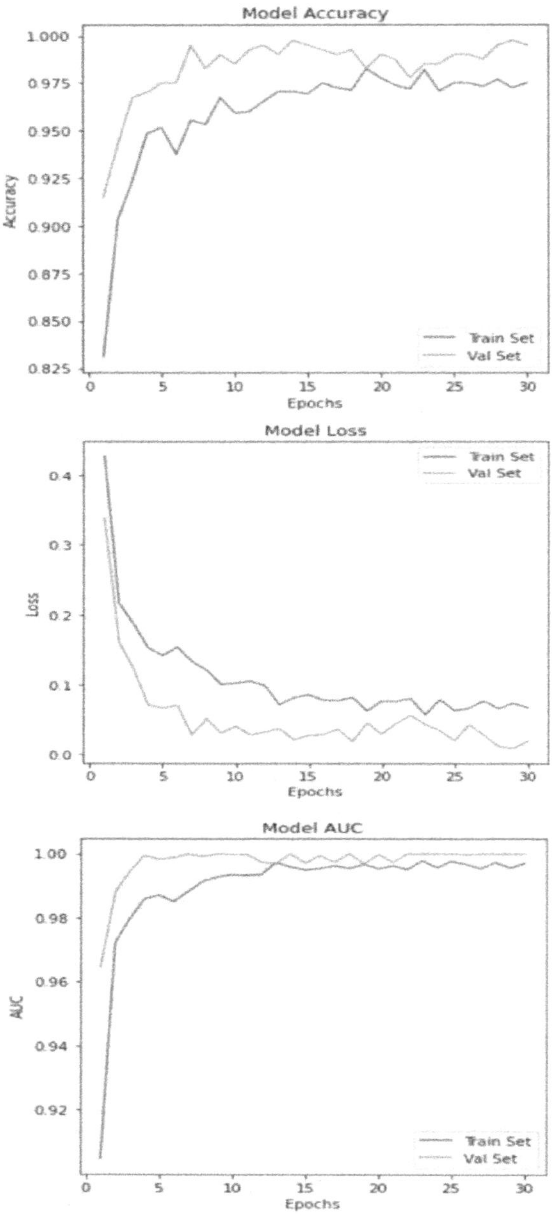

Figure 20.13. Learning curve.

Source: Author.

The graphs depicted in the images above illustrate the model accuracy, loss, and AUC (Figure 20.13). On the learning dataset, the system achieved an accuracy of approximately 0.9749, 1.0 on the validation dataset, and 0.995 on the testing dataset.

5. Discussion

The matrix of the proposed model is compared to other existing models in the literature (Tables 20.1 and 20.2). The results are presented in Table 20.3. Findings: Customized VGG19 model outperforms conventional methods and produces clinical-level images.

6. Conclusion

In this study, I have used the VGG19 as a base layer and data enhancement techniques to create a brain tumour detection and prediction model with classification capabilities. The

Table 20.1. Performance metrics on test dataset

S. No.	*Performance Matrices*	*Results*
1	Accuracy	.9950
2	Precision	.9950
3	Recall	.9950
4	Fl Score	.9949
5	ROC AUC Score	1.000
6	Cohen Kappa Score	.9900

Source: Author.

Table 20.2. Classification report

	Precision	*Recall*	*Fl Score*
0	.99	1.0	1.0
1	1.0	.99	.99
Accuracy	-	-	.99
Micro Avg	1.0	.99	.99
Weighted Avg	1.0	.99	.99

Source: Author.

Table 20.3. Comparing with existing methodologies

Ref. No.	*Model Adapted*	*Accuracy*	*Precision*	*F1-Score*	*Recall*
[12]	VGG 16	0.96	0.92	0.99	0.94
[11]	DCNN	0.97	0.96	0.97	0.96
[13]	CNN	0.96	0.96	0.98	0.95
Proposed Model	VGG-19	0.995	0.995	0.994	0.995

Source: Author.

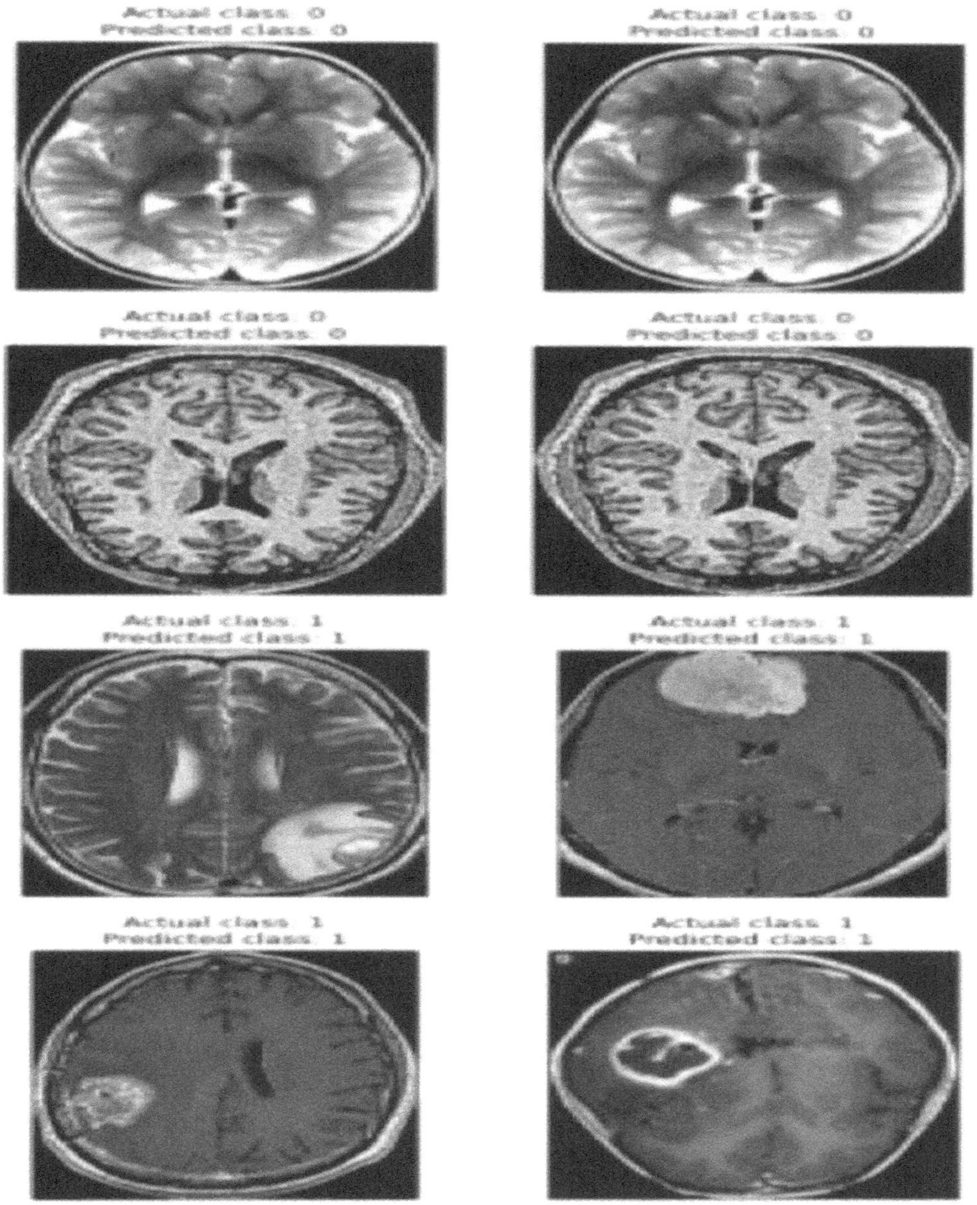

Figure 20.14. Prediction With Tumour (1) Without Tumour (0).

Source: Author.

literature review section of the study included an in-depth evaluation of various CNN designs and their shortcomings. We then used data augmentation to show how it can improve the performance of our model on limited brain tumour datasets. Experimental results demonstrated that the model was excellent at identifying images (Figure 20.14). Even in the MRI dataset, our data augmentation-based approach achieved remarkable detection efficiency with strong assessment metrics. In the future, we will consider exploring more complex architectures, a larger range of datasets, and further data variations techniques to improve the efficiency of the model.

Acknowledgment

We gratefully acknowledge the students, staff, and authority of the Department of Artificial Intelligence and Data Science for their cooperation in the research.

References

[1] Jalali, V., & Kaur, D. (2020). A study of classification and feature extraction techniques for brain tumor detection. International Journal of Multimedia Information

Retrieval, 9(4), 271–290. https://doi.org/10.1007/s13735-020-00199-7

[2] Chahal, P. K., Pandey, S., & Goel, S. (2020). A survey on brain tumor detection techniques for MR images. Multimedia Tools and Applications, 79, 21771–21814 https://doi.org/10.1007/s11042-020-08898-3

[3] Kawa, J., Rudzki, M., Pietka, E., & Szwarc, P. (2015). CA diagnosis workstation for BT. assessment. In 2015 22nd International Conference Mixed Design of Integrated Circuits & Systems (MIXDES) (pp. 98–103). IEEE. doi:10.1109/MIXDES.2015.7208489.

[4] Attuluri, S., Bhupati, C., Ramya, L., Tiwari, A., Budaraju, R. R., & Cotrina-Aliaga, J. C. (2023). Smart investigations into the development of an effective computer-assisted diagnosis system for CT scan brain depictions. SN Computer Science, 4(5), 504.https://doi.org/10.1007/s42979-023-01877-z

[5] Soomro, T. A., Zheng, L., Afifi, A. J., Ali, A., Soomro, S., Yin, M., & Gao, J. (2022). Image segmentation for MR brain tumor detection using machine learning: a review. IEEE Reviews in Biomedical Engineering, 16, 70–90. doi:10.1109/RBME.2022.3185292.

[6] Shamshad, N., Sarwr, D., Almogren, A., Saleem, K., Munawar, A., Rehman, A. U., & Bharany, S. (2024). Enhancing brain tumor classification by a comprehensive study on transfer learning techniques and model efficiency using mri datasets. IEEE Access, 12, 100407–100418. doi:10.1109/ACCESS.2024.3430109.

[7] . Das, S., & Goswami, R. S. (2024). Review, Limitations, and future prospects of neural network approaches for brain tumor classification. Multimedia Tools and Applications, 83(15), 45799–45841. https://doi.org/10.1007/s11042-023-17215-7

[8] Berghout, T. (2024). The neural frontier of future medical imaging: a review of deep learning for brain tumor detection. Journal of Imaging, 11(1), 2. https://doi.org/10.3390/jimaging11010002

[9] Ker, J., & Lim, T. (2018). Deep learning applications in medical image analysis. Ieee Access, 6, 9375–9389. doi:10.1109/ACCESS.2017.2788044.

[10] Neamah, K., Mohamed, F., Adnan, M. M., Saba, T., Bahaj, S. A., Kadhim, K. A., & Khan, A. R. (2023). Brain tumor classification and detection based DL models: A systematic review. IEEE Access, 12, 2517–2542. doi:10.1109/ACCESS.2023.3347545.

[11] Sujit, S. J., Bonfante, E., Aein, A., Coronado, I., Riascos-Castaneda, R., & Giancardo, L. (2021). Deep learning enabled brain shunt valve identification using mobile phones. Computer Methods and Programs in Biomedicine, 210, 106356. ISSN 01692607. https://doi.org/10.1016/j.cmpb.2021.10635.

[12] Alsaif, H., Guesmi, R., Alshammari, B. M., Hamrouni, T., Guesmi, T., Alzamil, A., & Belguesmi, L. (2022). A novel data augmentation-based brain tumor detection using convolutional neural network. Applied Sciences, 12(8), 3773. https://doi.org/10.3390/app12083773

[13] Ghosh, A., & Soni, B. (2021). An automatic tumor identification process to classify mri brain images. In Data Science: Theory, Algorithms, and Applications (pp. 315–327). Singapore: Springer Singapore. https://doi.org/10.1007/978-981-16-1681-5_20

21 Implementation of tobacco control measures targeting adult smokers, using SVM over KNN, to measure the accuracy rates

Nitin Kumar[a] *and V. R. Vimal*[b]

Saveetha School of Engineering, Saveetha Institute of Medical and Technical Sciences, Saveetha University, Chennai, Tamil Nadu, India

Abstract: The purpose of this paper is to assess accuracy of Support Vector Machine (SVM) model in distinguishing smokers from non-smoked with reference to control measures on tobacco. Performance of the SVM model is evaluated on Smoker behaviours' data set and is compared with K-Nearest Neighbour (KNN) to judge the reliability in calling out smoker behaviours. The models are trained and tested using a dataset containing basic demographic and smoking-statistics indicators including age, gender, and the number of cigarettes smoked. Data cleaning involves handling of missing data, detecting and handling outlying values and scaling of features so that the dataset is prepared well for analysis. The SVM-RBF model turned out as the best one, which yielded about 89.2% of accuracy while KNN model continued to lag at the rear of compared to others. In the test results, it was found that the most informative factors affecting smoking behaviour were all related to age and smoking experience. The application of the current study also shows that machine learning, especially SVM, can be a valuable approach to the effort aimed at tobacco control since it offers an objective way of categorizing people in terms of smoking status in order to address the problem effectively.

Keywords: Tobacco, smokers, support vector machine, K-Nearest neighbour, demographic

1. Introduction

Tobacco use remains a major public health issue that causes millions of deaths worldwide unnecessarily per year. Cigarette smoking is still epidemic in many parts of the world even with many government and health organization tobacco control interventions in place. This underlines the rationale regarding exploring the role of new theories, models and concepts in assessing and increasing the efficiency of endeavours aimed at reducing tobacco use. The first and probably the most potential concerns machine learning approach which involves large amounts of data to find out patterns and issues that should be intervened in the future. As the arena of public health progresses, new methods of evaluation that is, enhanced by data may be deemed more important [1]. Most of the previous evaluation methods used on tobacco control interventions involve surveys and observational research, both of which are prone to various types and deficiencies in the collected data. Conversely, machine learning provides data processing algorithm with large and complicated data set that produces promising estimates that are significant in gauging the effectiveness of the interventions among adult smokers by public health professionals [2]. SVM and KNN are two general machine learning algorithms which are commonly used for classification. Applied in the sphere of public health, especially in the sphere of behavioural analysis and in the course of interventions evaluation, their use looks rather perspective. The models above can classify people according to their propensity to stop smoking, obey control measures, or react to awareness programs. Through such algorithms, public health researchers are able to fine tune their approaches and put in place suitable measures to push through. This research aims to identify intervention success for tobacco control measures directed toward adult smokers through the use of SVM and KNN for analysis [3]. The purpose is to establish which of these two algorithms provides higher precision in assessing smoking cessation and adherence characteristics. Specifically, accuracy is included in this analysis as the measure of the models performance in correctly classifying the data and potential success rates of the anti-tobacco campaigns.

The good thing with Support Vector Machines is that where features are high dimensional meaning finding a relationship between smoking behaviour and several factors thoughts as plans are complicated, then Support Vector

[a]nitinkumark5085.sse@saveetha.com, [b]vimalvr.sse@saveetha.com

DOI: 10.1201/9781003770473-21

Machines is the best. SVM finds a hyperplane that best differentiates the data points and at the same time brings the boundary of these classes as far away as possible [4]. It makes SVM flexible and highly accurate when making classes' prediction, especially with a limited number of training data inputs. On the other hand, KNN utilizes the distance measure to predict that data point by assuming nearest neighbours and makes its predictions based on votes. While KNN is easily understandable and simultaneously uncomplicated to implement, its main drawback is the problem of space and time complexity while trying to adapt for use in large datasets with several variables. Precision stands as a critical element in this comparative research [5]. Accuracy, in this case, implies that the model has the potential of predicting the likely response of the smokers to policies and other measures in tobacco control hence facilitating informed policy-making.

This research thus seeks to evaluate public health initiatives with the purpose of providing the best model for an improved efficiency in the predictive modelling process, to enable less resource wastage in organizational interventions. The datasets applied in the present work cover a broad spectrum of demographic, socioeconomic and behavioural characteristics associated with smoking. Demographic variables are age, income, education and smoking history while psychographic variables are prior exposure to tobacco control programs [6]. Data cleaning, scaling especially to a common range, removing/correcting unnecessary features as well as oversampling/undersampling is done in attempts to increase the chances of model performance on unseen data from other populations. The utilization of machine learning models in applying tobacco control is not merely in the utilization of such models in predictive analysis. Such models can be helpful in categorizing the population, determining vulnerable communities and planning of an appropriate intervention programme [7]. For instance, when identifying people likely to return to smoking, special follow-up programs can be created for them, which means that the over-all tobacco control efforts will be more effective. This research has important broader consequences because the methods used here can be transposed to other topics related to public health. That IPI work has been successful for tobacco control mean it can be recommended for public health issues that may be related to obesity disorder, substance use disorder, and other chronic diseases [8]. Such an approach establishes the opportunity to advance the integration of artificial intelligence with public health aims as well as delivery [9].

In light of the foregoing, this paper aims at presenting a case of SVM and KNN, with the view of adding to literature on practical application of machine learning in public health, in particular with regard to tobacco control [10]. It is hoped that the suggested research will help to establish which of the two models yields the best results with a view to future research and implementation plans. At the end of this, the goal of this work is to help develop healthier communities as it provides policymakers and agencies involved in the fight against tobacco use with the tools to do so more efficiently.

2. Literature Review

Kock, Loren A. [11] conducted a study with intent of comparing policy support for tobacco availability in Great Britain (GB) and also compare the degree of policy support for tobacco availability among sociodemographic and smoking and quitting characteristics. Employing data from the Smoking Toolkit Study (September 2021) the cross-sectional survey with study sample of 2197 identified that the majority of respondents supported the need for tobacco retailer licences (89.6%) and the ban on tobacco vending within 500 meters of a school (69.9%). This was supported by the following: increasing the minimum age for purchase to 21 years (49.2%) and decreasing the number of outlets selling tobacco products (46.5%). However, a proposed ban of the sale to those born after a given year popularly referred to as the "tobacco-free generation" received more negative response (41.3%) than the positive response (34.5%). Age and gender played a role the acceptability of age restrictions and retailer reductions with the older participants being in favour of age restrictions and the female participants being in favour of both age restrictions and reductions in the number of retailers. Echoing the conclusions by Kock, most of the respondents supported particular tobacco control measures though with difference levels of acceptance of the measures.

Tobacco control policies and tobacco use prevalence in the 32 Mexican states during 2016–2017: a cross-sectional ecological study was carried out by Delta Jeazul Ponce-Hernández [12]. The research employed the Tobacco Control Scale (TCS), which assesses the extent of effectiveness of TCP through scores – the measurement units being scores on a scale of 0 to 100 – arising from factors such as price, prohibition of tobacco advertising in accessible public areas, and health warnings. Descriptive analysis showed that TCS scores were negatively correlated with smoking habit with the current daily smoking reported more common with the adolescent participants, Spearman correlation coefficient = -0.26; p = 0.026. Smoking had significant negative associations with price in different segments; adolescents (r = -0.002, p = 0.003) and adults (r = -0.046, p = 0.040). Based on the findings the need for further amplification of TCP must be pointed in Mexico especially to decrease the level of smoking among adolescents.

Stephen Wade et al. [13] offered a projection of Australian tobacco smoking to 50 years to demonstrate smoking initiation and cessation in light of the national goal of achieving a smoking prevalence for daily smoking adults of less than or equal to five percent by 2030. The study carried out the surveillance of smoking prevalence up to 2066 based on a compartmental model calibrated through data from 26 surveys, of 229, 523 participants in the year of 1962–2016. Such findings were an implication that if the current state would persist, the 5% adoption would only be possible by the year 2039. Therefore, assuming that smoking initiation is completely stopped amongst the young people, the target can be attained by 2037 at best in a most favourable condition. But if initiation and cessation rates went down to their 2007 levels, prevalence might remain high at 9.1 bisexuals in 2066 at earliest. According to the authors, the 5% has remained a mirage due to the fact that without more radical steps in near future it is discouraging to think of getting there in the year 2030.

Waleed M. Sweileh [14] achieved this by performing a bibliometric analysis to visualize the science of technology-mediated interventions for tobacco smoking cessation and harm reduction knowledge from 2003 to 2022. The research collected 639 articles from Scopus database and investigate their publication pattern, most productive authors and research themes. Analysis also demonstrated that after the year 2011 there has been a rising trend in research activity, specifically with a major contribution from the United States of America, United Kingdom and the Netherlands. Specific topics addressed and nonsmokers' exposure to included smartphone apps, texting to quit smoking and quit smoking social media. The study focused on the applicability of digital interventions and also got acquainted with the new indications like virtual reality and personalized tools. In her work, Sweileh explained the conceivable ability of technology in handling tobacco related public health issues leaning towards the improvement of the intervention tactics.

Charity Aienobe-Asekharen [15], Samantha Parkes, John Ciorba, Remziye Kadir, Lauren Gilchrist and Nzinga Jones carried out a scoping review to explore tobacco control health communication interventions in Africa with youth involvement. This target was achieved in 20 papers retrieved from peer-reviewed sources and WHO reports, and the cross-sectional analysis of the anti-tobacco mass-media campaigns in the 54 African countries. They observed that only six nations continue national octopus like high-quality health communication for more than a decade. Most countries could not maintain the agitation for more than three weeks at a stretch. The study identified low representation of youths in the development of interventions, charging the inclusion of youth in the development process will help ensure appropriateness of messages to youth and the impact of the messages on tobacco-use prevention.

3. Materials

For this work Tobacco use pattern data is gathered from the Adult Cigarette and Tobacco Use Prevalence' set that originated from the California Behavioural Risk Factor Surveillance System (BRFSS). It includes characteristics like age, gender, education, smoking tobacco in terms of cigarette smokers, smokeless tobacco consumers and electronic cigarette users and health related smoking illnesses. Collection of such data can be wisely used to analyze patterns of tobacco consumption and its correlation with demographic factors; to estimate effectiveness of different control actions. Using advanced analytical techniques such as SVM and KNN it is possible to forecast smoking behaviours and, therefore, evaluate the effectiveness of certain control measures in relation to targeted population categories. The demographic characteristics of the dataset are useful for tobacco control analysis as well as in bringing focused efforts to control the problem.

4. Methodology

4.1. Preprocessing techniques

Data cleaning is a pre-processing technique that aims in making a dataset more so that it can be put to use in a machine learning model. It covers different procedures to deal with missing data, detect and remove outliers, and transform numeric features and the data itself prior to analysis. It is as follows: The next sub-section provides further details on each of these steps with the appropriate equation.

Figure 21.1 shows the overview of the proposed methodology.

Recursive Feature Elimination (RFE): RFE is a technique related to feature selection it deletes the features which have low importance and establishes a model based on the rest of the features. This process goes on until one gets a best subset of feature /features. This approach is used in enhancing model performance since unnecessary features are got rid of.

Missing Values Handling: When dealing with missing values, two common techniques are imputation and deletion. If a feature has a substantial amount of missing data, it may be removed. For smaller gaps, imputation techniques can be used, such as replacing missing values with the mean for numerical data or the mode for categorical data.

Outlier Detection: Outliers are some data values that are so far from the rest of the data that thy affect the analysis as well as modelling of the data. Z-score method which is explained below is widely employed with the end objectives of capturing outliers. The Z-score points out how many standard units a score is above, or below the mean.

$$Z = \frac{X - \mu}{\sigma} \tag{1}$$

Where:

- Z is the Z-score.
- X is the data point.
- μ is the mean of the data.
- σ is the standard deviation of the data.

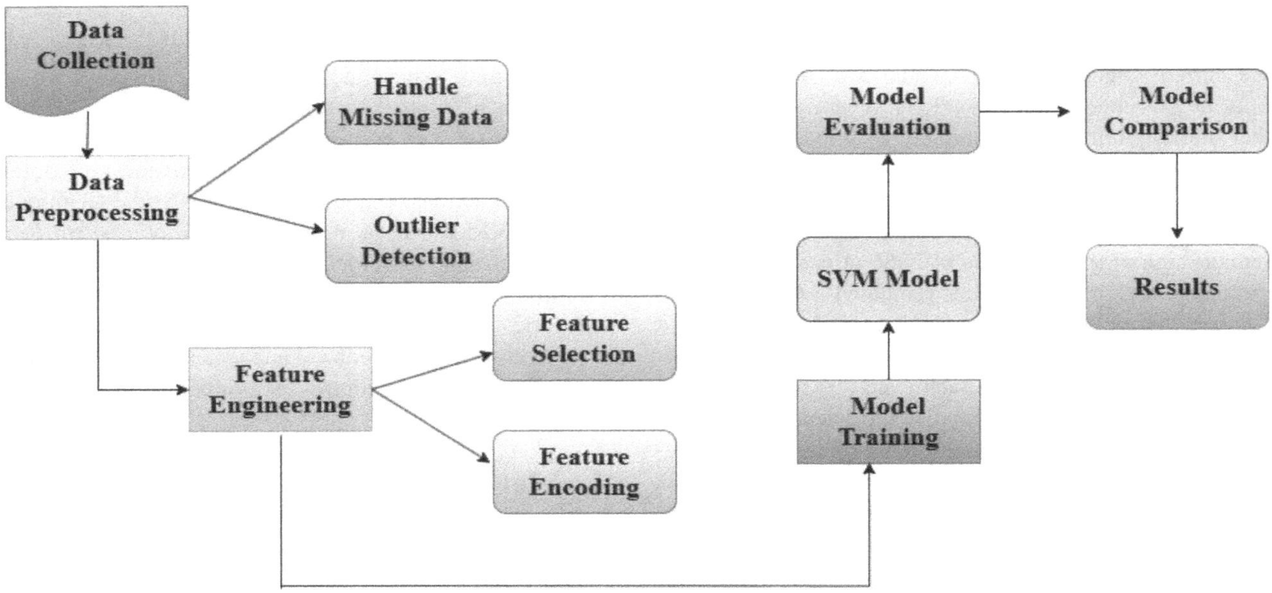

Figure 21.1. Overview architecture.

Source: Author.

A threshold is applied (usually 3 or -3) to determine if a data point is an outlier. If the Z-score is above this threshold, the data point is considered an outlier.

Min-Max Normalization: This technique scales the data into a fixed range, typically [0, 1]. It is useful when the feature values are bounded. The normalized value *Xnorm* for a feature *X* is calculated using the following formula:

$$X_{norm} = \frac{X - X_{min}}{X_{max} - X_{min}} \tag{2}$$

Where:

- X is the original feature value.
- X_{min} and X_{max} are the minimum and maximum values of the feature, respectively.

4.2. Model training

SVM stands for Support Vector Machine, which is a Supervised Learning Technique that is used in the case of classification problems like the present case, where we are trying to segregated people between smokers and non-smokers based on their different characteristic's possible health related or others. The goal of the SVM model is to move the hyperplane as far as possible from both classes while having as many points as possible on the farther side of this hyperplane to prevent misclassification of data that hasn't been trained.

Optimization Problem in SVM: SVM operates in a way that the hyperplane which separate feature space into two classes with the largest distance is being selected. Margin can be defined as the distance between the nearest data points of one class and the least distance of the nearest data points of another class to the hyperplane. These closest points which make a possible position on one side of the decision boundary are called support vectors. The objective can be then cast to that of maximizing the distance between these two classes. The margin is achieved when trying to minimize the norm of weight vector *w* and the classifier function F in particular should be designed to make sure that all of the training patterns are in the correct class w. The optimization objective is expressed as:

$$Maximize = \frac{2}{||w||} \tag{3}$$

To achieve this, we minimize $||w||^2$ where $||w||$ is the norm (magnitude) of the weight vector.

Handling Non-Linearly Separable Data Using Kernel Trick: The points available in real-world problems are most of the times not linearly separable, which means that no straight line or hyperplane can be drawn a standard distance to separate the two classes. In this case however, SVM can still be used through the kernel trick, which transforms the data to a higher dimension so that it can be separated linearly. A commonly used kernel function is the Radial Basis Function (RBF) kernel functions that actually map the input space of the data into a higher dimensional space without actually defining the mapping. The RBF kernel begins computing the degree of similarity of two data points x_i and x_j as follows:

$$K(x_i, x_j) = \exp(-\frac{||x_i - x_j||^2}{2\sigma^2} \tag{4}$$

Where:

- $K(x_i, x_j)$ is the kernel function representing the similarity between points x_i and x_j.

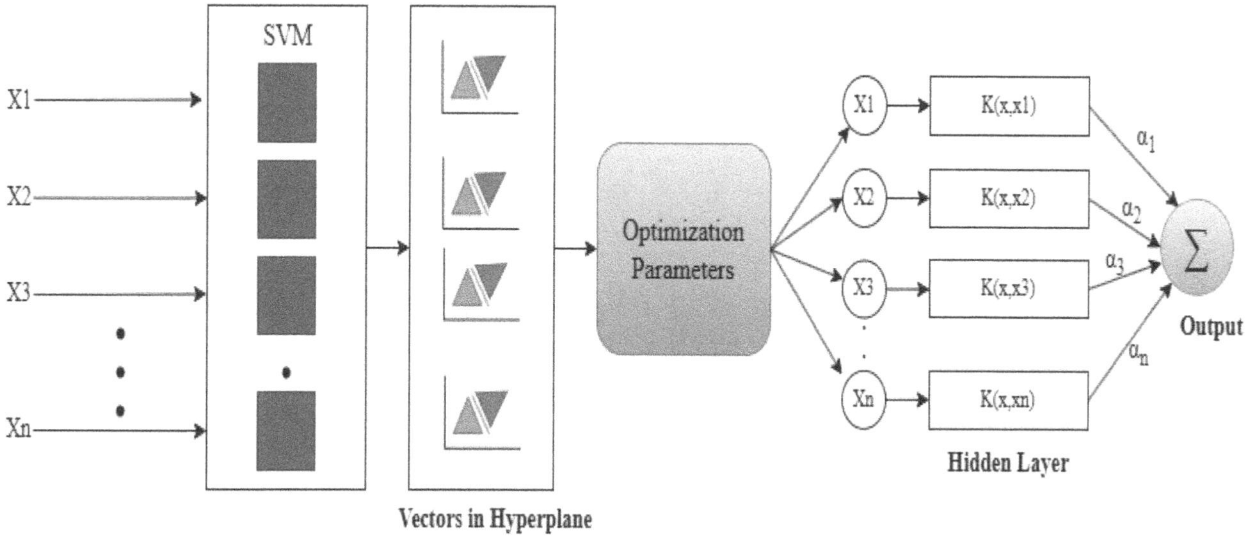

Figure 21.2. Proposed SVM architecture.

Source: Author.

- σ is a hyperparameter that controls the width of the Gaussian function, determining how far the influence of a point can reach.

The kernel trick allows the SVM to learn non-linear decision boundaries, enabling it to effectively classify data that is not linearly separable in the original feature space.

Solving the Optimization Problem: To solve the optimization problem, we apply quadratic programming techniques to find the optimal weight vector www and bias term b. The SVM model aims to minimize $||w||^2$while ensuring the constraints $y_i(w \cdot x_i+b)\geq 1$ for all training data points.

Model Evaluation: After training the SVM, the model is evaluated using standard classification metrics such as accuracy, precision, recall, and F1-score. Cross-validation techniques may also be applied to assess the model's performance and ensure it generalizes well to unseen data.

Figure 21.2 explains the detailed architecture of random forest classifier.

5. Results and Discussion

Proposed model has been validated with various metrics (Table 21.1).

The SVM with RBF kernel achieved the highest performance across all metrics, indicating that the nonlinear transformation of the feature space resulted in better decision boundaries compared to the linear kernel.

KNN performance improved as the value of k increased, but it still lagged behind the SVM models in terms of accuracy and F1-score. The ROC-AUC score was highest for the RBF kernel, signifying a better ability to distinguish between smokers and non-smokers (Figure 21.3 and Table 21.2).

Table 21.1. Model performance comparison (SVM vs. KNN)

Metric	*SVM (Linear Kernel)*	*SVM (RBF Kernel)*	*KNN (k=5)*	*KNN (k=10)*
Accuracy (%)	85.4	89.2	78.5	81.3
Precision (%)	83.7	87.5	75.2	77.8
Recall (%)	88.1	90.8	80.4	82.1
F1-Score (%)	85.9	88.6	77.8	79.9
ROC-AUC Score	0.91	0.94	0.85	0.88

Source: Author.

The SVM model correctly classified 1800 smokers and 1780 non-smokers, showing a strong ability to predict both classes with relatively few false positives and false negatives. The number of false positives (220) suggests that there is some misclassification, particularly in cases where non-smokers were incorrectly classified as smokers, which could be due to similar feature characteristics (Figures 21.4 and 21.5, Table 21.3).

Age was the most important feature, contributing significantly to the classification decision. This makes sense as age is often correlated with smoking habits. Gender and number of cigarettes were also highly important, reflecting their direct relationship with smoking behaviour. Marital Status had the least impact on the model's predictions, suggesting it may not be as relevant in determining smoking status (Table 21.4).

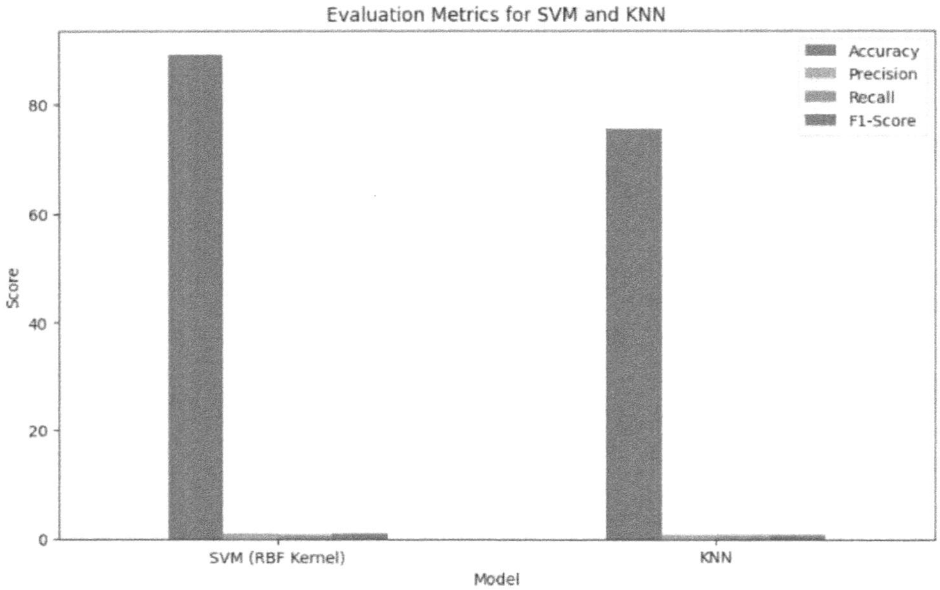

Figure 21.3. Evaluation metrics of SVM and KNN.

Source: Author.

Table 21.2. Confusion matrix for SVM model (RBF Kernel)

	Predicted Smoker	*Predicted Non-Smoker*
Actual Smoker	1800	100
Actual Non-Smoker	220	1780

Source: Author.

The full feature set gave the best accuracy, demonstrating the importance of combining multiple relevant features.

When fewer features were used, the model's accuracy dropped, especially when only two features were used. Age combined with smoking-related features (e.g., number of cigarettes) provided the best subset for tobacco control prediction, supporting the importance of these characteristics.

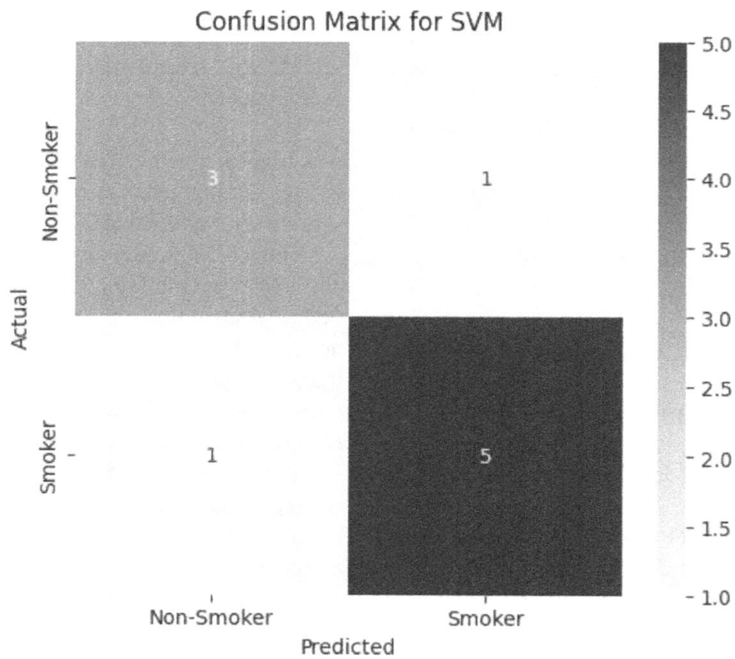

Figure 21.4. Confusion matrix of SVM.

Source: Author.

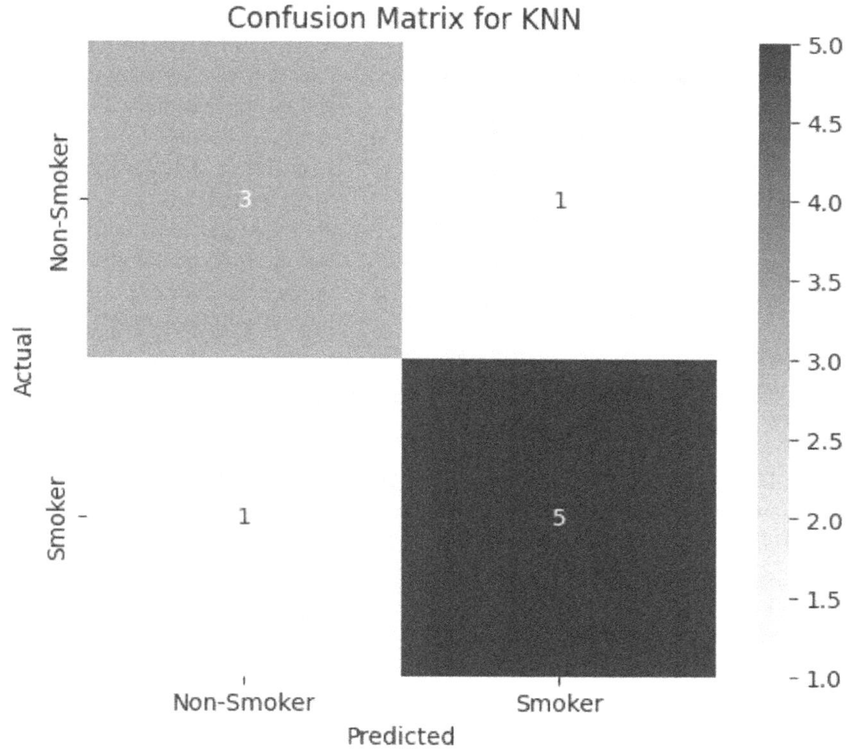

Figure 21.5. Confusion matrix of KNN.

Source: Author.

Table 21.3. Feature importance based on SVM (RBF Kernel)

Feature	*Importance Score*
Age	0.35
Gender	0.22
Number of cigarettes	0.18
Employment Status	0.12
Education Level	0.10
Marital Status	0.03

Source: Author.

Table 21.4. Accuracy across different feature subsets

Feature Subset	*Accuracy (%)*
All Features (Full Dataset)	89.2
Age + Gender	82.5
Age + Number of cigarettes	84.1
Number of cigarettes + Employment	83.0
Age + Education Level	80.0

Source: Author.

6. Conclusions

Therefore, based on the results achieved the SVM with an RBF kernel was chosen as the most suitable for tobacco control, since it provided higher results in all the examined criteria compared to KNN. The assessment of smokers and non-smokers' designation in the model was impacted by variables such as age, sex, and the number of cigarettes. These results suggest the necessity of using a broad list of predictors, the demographic and smoking characteristics in particular, to achieve high accuracy. This approach emphasizes the possibilities that machine learning models, especially SVM, may open up for tobacco control activities by defining the target population of smokers, who may be encouraged to quit, more accurately. Subsequent studies may endeavour to fine tune the model and add new variables that would improve the models fit and usability in actual contemporary anti-tobacco campaigns.

References

[1] Mills, S. D., Rosario, C., Yerger, V. B., Kalb, M. D., & Ribisl, K. M. (2024). Recommendations to advance equity in tobacco control.

[2] Marshall, L. (2024). The National and State Tobacco Control Program: overview of the Centers for Disease Control and Prevention's efforts to address commercial tobacco use. *Preventing Chronic Disease*, *21*.

[3] Rosilawati, Y., Rafique, Z., & Sudiwijaya, E. (2024). Tobacco use among in-school young adolescents in Indonesia: Exploring availability, affordability, and accessibility. *Plos One*, *19*(3), e0301291.

[4] Kang, H., Cheon, E., Kim, H. K., Park, J. M., Hwang, J., Kim, J., Lee, S., Han, Y., Lim, M. K., Park, S., & Cho, S. I. (2024). Vision for tobacco endgame in Korea: Suggestions for countries with endgame aspirations.

[5] Sančanin, B., & Penjišević, A. (2024). Impacts of tobacco use and exposure to tobacco smoke on health, society, economy, and environment. *MEDIS–International Journal of Medical Sciences and Research*, *3*(2), 25–29.

[6] Liu, M., Huang, P., Xu, X., Chen, Z., Zhuang, J., Liu, Y., Yang, S., Chen, X., & He, F. (2024). Tobacco control knowledge and beliefs among healthcare workers in respiratory departments in Fujian Province, China: A cross-sectional study. *Tobacco Induced Diseases*, *22*.

[7] Nugent, R., Hutchinson, B., Mann, N., Ngongo, C., Spencer, G., Grafton, D., & Small, R. (2024). Evolving methodology of national tobacco control investment cases. *Tobacco Control*, *33*(Suppl 1), s10-s16.

[8] Coughlin, L. N., Tegge, A. N., Sheffer, C. E., & Bickel, W. K. (2020). A machine-learning approach to predicting smoking cessation treatment outcomes. *Nicotine and Tobacco Research*, *22*(3), 415–422.

[9] Merkaj, E., Zhllima, E., Imami, D., Gjika, I., Guerrero-López, C. M., & Drope, J. (2024). Impact of cigarette price and tobacco control policies on youth smoking experimentation in Albania. *Tobacco Control*, *33*(Suppl 2), s38–s43.

[10] Onwuzo, C. N., Olukorode, J., Sange, W., Orimoloye, D. A., Udojike, C., Omoragbon, L., Hassan, A. E., Falade, D. M., Omiko, R., Odunaike, O. S., & Adams-Momoh, P. A. (2024). A review of smoking cessation interventions: efficacy, strategies for implementation, and future directions. *Cureus*, *16*(1).

[11] Kock, L., Shahab, L., Moore, G., Shortt, N. K., Pearce, J., & Brown, J. (2024). Assessing the profile of support for potential tobacco control policies targeting availability in Great Britain: A cross-sectional population survey. *Tobacco Control*, *33*(2), 221–231.

[12] Ponce-Hernández, D. J., Martínez-Regalado, J. A., Reynales-Shigematsu, L. M., Calderón-Villarreal, A., Regidor, E., Herrero, L., & Sordo, L. (2024). Correlation between tobacco control policies and tobacco prevention in Mexico: A sub-national analysis. *Journal of Public Health Policy*, *45*(2), 378–392.

[13] Wade, S., Weber, M. F., Sarich, P., Caruana, M., Watts, C., Vaneckova, P., Ngo, P., Cressman, S., Scollo, M., Banks, E., & Gartner, C. E. (2024). Fifty-year forecasts of daily smoking prevalence: Can Australia reach 5% by 2030?. *Tobacco Control*, *33*(5), 641–647.

[14] Sweileh, W. M. (2024). Technology-based interventions for tobacco smoking prevention and treatment: A 20-year bibliometric analysis (2003–2022). *Substance Abuse Treatment, Prevention, and Policy*, *19*(1), 13.

[15] Aienobe-Asekharen, C., Norris, E., & Martin, W. (2024). A scoping review of tobacco control health communication in Africa: Moving towards involving young people. *International Journal of Environmental Research and Public Health*, *21*(3), 259.

22 A neural network model for prediction of heart disease in chronic diabetic patients

Sumaiya Thaseen[1,a], *Poorva Babu*[2,b], *and Sai Phanindra Kothuri*[3,c]

[1]School of Computing, De Montfort University, Dubai
[2]School of Computing Science Engineering and Information Systems, Vellore Institute of Technology, Vellore, Tamil Nadu, India
[3]School of Computing Science and Engineering, Vellore Institute of Technology, Chennai, Tamil Nadu, India

Abstract: Recent development in Artificial Intelligence has transformed the healthcare industry. Neural network models such as Artificial Neural Networks (ANN) and Multi-Layer Perceptron (MLP) are deployed to predict the heart disease using diabetes as a feature along with several other common features. In this research, different diabetes and heart disease datasets are collected, and the features are evaluated, pre-processed and selected to curate a customized dataset that uses heart disease as the class label and diabetes as one of the features. MLP provides a higher accuracy of 99.98% compared to ANN model with 86%. In addition, according to the Mean Square Error (MSE) of the models, it can be observed that ANN has lower MSE of 0.14 MSE and MLP has a 2.40 MSE value. ANN is better suited for our research. The heart disease prediction results along with the Electronic Health Record (EHR) of the patient is stored using Inter Planetary File System (IPFS). Each file or folder in IPFS consist of all the health information accessed by a content ID which is the hash of the file content that can then be used to securely retrieve the data.

Keywords: Artificial Neural Network, interplanetary file system and multi-layer perceptron

1. Introduction

Diabetes is one of the chronic endocrine diseases caused by abnormal sugar level in the blood. Recent research shows that 10.5% of the global population has diabetes but half of them is unaware of the condition they are living with. There are different types of diabetes depending on its cause [10]. Research show that people with diabetes are four times more likely to have a heart attack or stroke than the people who do not have diabetes. Longer the duration of diabetes in people, higher the chances of cardiovascular diseases (CVD) in them. CVD includes all different types of heart diseases, stroke and blood vessel diseases. Due to high blood sugar levels in people with diabetes, the blood vessels in their heart are damaged. In addition, fatty deposits are likely to be developed, increasing the cholesterol levels, which in-turn increases the risk of CVD [3]. The most common type of CVD is coronary artery disease where blood flow to the heart is decreased as the arteries get hardened due to high cholesterol deposits in the walls of the arteries. This hardening of arteries can occur in other parts of the body as well. When it occurs in the legs and feet, it is called as peripheral artery disease which is the first sign of CVD in patients with Diabetes.

Healthcare sectors are one of the most complex and largest sectors globally. It is also potentially, a major employment sector. Every individual should have access to health services. But due to unequal distribution of qualified and trained personnels, the ratio of doctor to patient is affected adversely. This results in delayed access to services and diagnosis of fatal diseases at later stages or misinterpretation of symptoms leading to false treatments. These effects can result in reduced life expectancy of the individuals. The technological advancements such as Machine Learning (ML), Big Data Analytics, Remote Patient Monitoring (RPM), Electronic Health Records (EHR), and IoT devices to monitor and track patient health can all enhance the quality of healthcare services provided [9]. However, with the advancements of technology, threats do exist for the healthcare sector in every aspect. One of the major concerns is the storing and handling huge amount of sensitive data.

In this research, Neural networks are used. Neural networks (NN) which is a collection of interconnected nodes that are organized into layers to compute input signals, process them, and generate an output.

The output is then passed to the next neuron in the network. These models have become the fundamental tool of

[a]sumaiya.thaseen2@dmu.ac.uk, [b]poorvababu@gmail.com, [c]kothuriphanindra1999@gmail.com

DOI: 10.1201/9781003770473-22

machine learning and artificial intelligence by relating and learning from examples thus improving its scope to be a functional primacy in disease predictions [3].

Interplanetary file system (IPFS) is a decentralized peer-to-peer network protocol consisting of interconnected nodes that slices the files into several content and stores across the network. The files are addressed based on its content which is stored as a hash value [4]. If the hash is found on the network, the content is retrieved faster as it is directly accessed and includes no duplication to copy the file to the system improving the security and data perpetuation [9].

The rest of the paper is organized as follows: Section 2 discusses the related work. Section 3 details the proposed system. Section 4 shows the performance analysis. Section 5 analyses the results with a brief discussion. Section 6 concludes the work. The future scope is discussed in section 7.

2. Related Work

Several scholars used cognitive computing along with Blockchain technology which can offer potential impacts on the healthcare sector [1]. Cognitive computing can be accomplished using Machine Learning and Deep learning algorithms. Cognitive computing along with Blockchain technology can offer a new paradigm for the exchange of health information to improve the security, privacy for reliable storage, and interoperability of the data to produce more accurate and reliable results [1, 2] [3]. We can enhance the effectiveness, disintermediation, and security of the electronic records using consensus algorithms and smart contracts [1]. A user-defined Blockchain is used for simulation in healthcare using SHA256 for secure transactions and storage of details that provide transparency and reduced risk of data manipulation [2]. Javaid Ahmad Malik et al. [3] collected the data through RPM and stored it securely using Blockchain. The data thus collected is computed using Artificial Neural Networks (ANN) with the input, hidden, and output layers. If a disease is predicted, alerts are sent to the patients and healthcare professionals. It is assessed that the mining costs and latency are reduced, and throughput is increased by using this architecture. A decentralized off-chain database and Orbit Database is used with Interplanetary File System (IPFS), thus enhancing the security and privacy of EHR [4].

Akruti Sinha et al. [5] proposed an architecture that is divided into five separate sections incorporating machine learning techniques with Blockchain to increase the efficiency and accuracy of diagnosis, improving the healthcare services and trust between stakeholders through smart contracts. Sumit Kumar Rana et al. [6] evaluated Distributed Ledger Technology (DLT) and the current challenges of healthcare services such as patient record management, data storage, scalability, and latency. They used an encoded off-chain data storage in a decentralized manner using Blockchain, access control methods using smart contracts, and a combination of public and private Blockchain can resolve the above-mentioned challenges. Akhilendra Pratap Singh et al. [7] used Blockchain as traditional EHR-based systems with data loss risks, security and immutability consensus over health records, communication gap among established hospitals, and clinical data retrieval systems which were inefficient. Hyper ledger fabric and composer technology have been implemented for the security of the model. Their framework empowers patients who will be able to easily access and track their healthcare information. They addressed the lack of patient-centricity and fine-grained access control in existing solutions. Their key contributions are extensive performance evaluation showing efficiency and resource utilization, Lightweight access control scheme with Hyper ledger Blockchain. Can integrate with Kafka and Byzantine fault tolerance services for fault tolerance.

Huru Hasanova et al. [8] applied the Sine Cosine Weighted K-Nearest Neighbor (SCA-WKNN) approach for fast and efficient heart disease prediction from the Blockchain data as it follows local optimal selection methodology. It is stated that Blockchain achieves 25.03% max throughput compared to other traditional storage methods. They used many different algorithms for comparative analysis based on various performance metrics and it is observed that SCA-WKNN resulted in 4.59% and 15.61% max accuracy than WKNN and KNN techniques. Zhijie Sun et al. [9] provide a healthcare information security and block chain storage solution with high throughput based on hyper ledger fabric. They used three different types of smart contracts a Policy contract (PSC) to verify the validity of the ABAC policies used, an Access control contract (ASC) to determine whether the requests match the access provided, and a medical record contract (RSC) to store the hash address representing the health record in the Blockchain which are deployed for secure storage and IPFS relieves the storage pressure on Blockchain by storing the hash address that represents the complete medical record.

Mengji Chen et al. [10] aim at the earlier detection of the disease using ML approaches and maintaining the EHR of the patients securely, they combined their EHR framework using symptom-based disease prediction, Blockchain, and IPFS in which the vital parameters of the patient are transferred via wearable sensor devices to the EHR's manager. The authors used Blockchain as patients are afraid of sensitivity and privacy if data is stored in the cloud. They measured performance and efficiency. In addition, they enabled patients to monitor their health and share data selectively as well as real-time health monitoring and prediction.

Nguyen et al. [11] proposed a secure intrusion, detection integrated with Blockchain and a classification model for Cyber-Physical Systems (CPS) in the healthcare sector. They have used Blockchain and RESNET model for identifying the disease. CPS integrates physical and computational systems for remote patient monitoring, data collection, and decision-making.

3. Proposed System

Figure 22.1 shows the proposed system architecture which is detailed in the following steps given below:

Step 1: The patient health information is collected and the parameters for the Neural network model are obtained and passed as input.

Step 2: The neural network model which has been trained using a customized dataset curated after pre-processing and feature selection processes the input parameters to predict the heart disease.

Step 3: All the electronic health records of the patient are gathered and put into one folder with all different types of data which could include prescriptions, lab reports, scan images etc.

Step 4: The folder created for each patient is then imported into the IPFS where Content ID is generated for the folder as well for each file in the folder.

3.1. Dataset description

The dataset used in this system for the prediction of heart disease using diabetes is created by integrating diabetes and heart disease datasets available in the Kaggle repository. A custom dataset is created from the three datasets namely Diabetes Health Indicators Dataset (DHID), Diabetes Prediction Dataset (DPD), and Logistic Regression to Predict Heart disease Dataset (LRPHD) datasets respectively. DHID is collected by the CDC (The Centres for Disease Control and Prevention) through BRFSS (The Behavioural Risk Factor Surveillance System), which is a telephone survey conducted annually in America. The dataset used has been split equally based on respondents with and without diabetes. It is a balanced dataset with 21 features and 70,692 responses. DPD is a diabetes prediction dataset created by collecting EHRs which

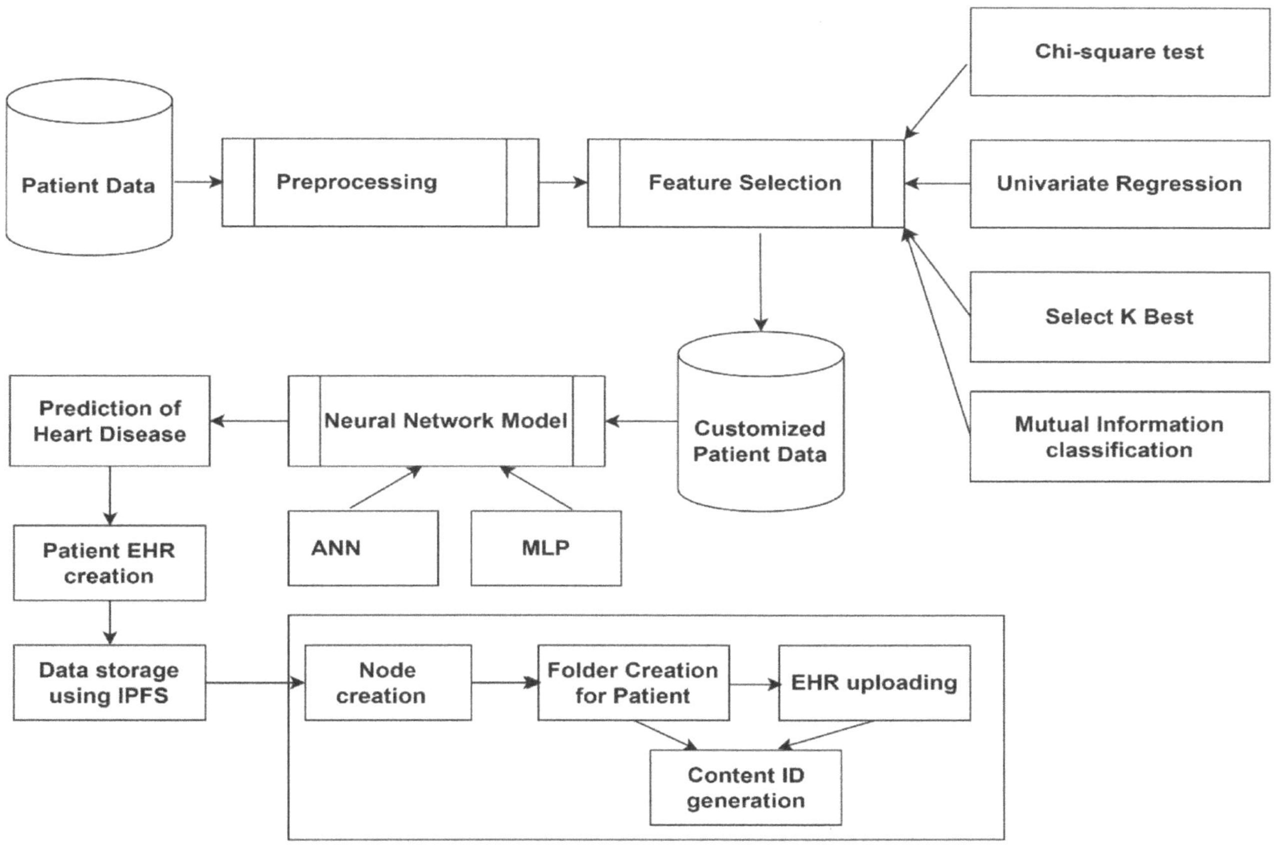

Figure 22.1. Proposed system architecture.

Source: Author.

contain clinical and demographic information from multiple healthcare providers and aggregating it into a single dataset. The data was collected through medical records, lab tests and surveys making it a dataset with 9 features and about 1 lakh responses. LRPHD is an ongoing cardiovascular study, and results are obtained on the residents of Framingham, Massachusetts which contains 15 features and over 4000 responses.

Each feature is a potential demographic and behavioural risk factor. It is observed from these datasets that the attributes leading to diabetes and heart disease are very interrelated. Those features are blood pressure, glucose levels, cholesterol level, heart rate, lifestyle habits etc. From LRPHD, it is observed that Diabetes is an attribute to predict heart disease. Thus, we have curated a custom dataset to predict heart disease using diabetes as a parameter where the attributes common to diabetes and heart disease are retained and additional attributes contributing to diabetes are included in the dataset. Attributes retained from DHID dataset are DPD, HbA1c_level. Attributes such as blood and glucose level are the two parameters that are not present in DHID. So these two parameters are integrated to the DHID dataset along with other attributes such as age, sex, diabetes and, hypertension values (Table 22.1). Similarly, the feature values contributing to the heart disease prediction from LRPHD dataset were analyzed and modified DHID such that the HeartDiseaseorAttack feature is used as the class label. Initially, DHID was evenly balanced for the class label diabetes with 50:50 ratio and the heart disease feature within the dataset was in the ratio 85 (no value):15(yes value). In addition, LRPHD to predict heart disease had the same ratios. So, the custom dataset is generated with 46 (no value):54 (yes value) ratio of diabetes feature to predict the heart disease with the samples of ratio 85 (no value): 15 (yes value) heart disease.

3.2. Data pre-processing

The dataset is pre-processed such that no mismatched datatypes, outlier values, mixed data values and missing data are present using the Pandas library as it is useful in data cleaning and manipulation.

Table 22.1. Dataset statistics

Total number of instances in the dataset	70692 records
Number of features	19
After Feature selection	15
Class label balance	85% no heart disease, 15% heart disease

Source: Author.

3.3. Feature selection

The most relevant features for the heart disease prediction is chosen in this approach. This could improve the accuracy of the model. In this research, feature selection is performed by means of four different techniques such as Chi Square method, Univariate Regression test using f_regression, Mutual information for classification using mutual_info_classification and SelectKBest method. The customized dataset integrates the attributes from the three datasets which had nineteen features, which were then reduced to fifteen using the feature selection approaches. The features such as whether the patient checks the cholesterol levels frequently, consumes fruits and veggies were eliminated as they were not significant in result prediction. Also, heavy alcohol consumption attribute was excluded as it had negligible data to support the analysis. It was also observed that features such as age, sex, BMI (Body Mass Index), cholesterol level, blood glucose, previously affected with stroke, blood pressure and overall wellbeing are some of the major contributing factors interrelating heart disease and diabetes. Figure 22.A1 in the appendix section shows the multi collinearity among features determined by rank based correlation.

Chi-square test measures the deviation between expected count and the observed count. When the observed count is close to the expected count, minimal Chi-square values are obtained which indicate that the features are independent. Higher the Chi-square value, the feature is more dependent for obtaining the output.

$$X^2 = \sum \frac{(O_i - E_i)^2}{E_i} \qquad (1)$$

where Oi is the observed count and Ei is the expected count.

Univariate Regression using f_regression is a quick model to test the effect of sequence of features. The correlation between each feature and the target is computed using the formula:

$$E \frac{[((X[:,i] - mean(X[:,i])) * (y - mean(y))]}{(std(X[:,i]) * std(y))} \qquad (2)$$

which is then converted to a F-Score and then to p-value.

Mutual information classification is used to measure the dependency between the features and the target. Higher the value, the features have a higher dependency. The dependency is measured based on k-nearest neighbor distances. The results of these methods are analyzed and the least relevant features common for all the techniques have been removed. A final custom dataset with the following attributes is created (Table 22.A1: Appendix).

3.4. Neural network models

An ANN model consists of 3 layers – input, hidden and output layer. The neurons are interconnected with each other and are organized into sequence of layers to process the data, understand it and make much better decisions.

The data for the system is fed to the input layer, which then moves to the hidden layer where all the analysis and calculations are performed, after several transformations the output is returned through the output layer. After the input is accepted, ANN computes the weighted sum for the input provided and adds a bias to it which can be represented as:

$$\sum_{i=1}^{n}(W_i * X_i) + b \tag{3}$$

Where n is the total input parameters, W is the weight of the parameter, X is the value of the parameter and b is the bias value. The weighted sum is then passed to an activation function which determines if the input parameter is considered to evaluate the output or decision.

MLP consists of multiple hidden layers which supports the model to interpret complex patterns in the data using nonlinear activation functions. Neurons are connected to the proceeding layers by weighted functions which are represented as weights determining the impact of output of one layer on the input of next layer. The network learns to adjusts its weights based on the training dataset provided.

3.5. *IPFS*

The Interplanetary file system (IPFS) is a decentralized, peer-to-peer file sharing network where files are identified not based on their names but their content. All the different nodes (computers) connected to the network will have access to the files. Any user who wishes to access the information about a file on the network, the request will be cached on the nodes in the network. This can prevent any malicious access in the network. The request for the file can be satisfied by any one or more number of nodes that has the file on it. The location of the file is very confidential, only the content is retrieved for the requester, thus increasing the security aspects of the data. The files are distributed across several nodes which does not imply that everyone participating in the network can access the file. The file sharing is possible only if the participant decides to share or publish the file to the network. Therefore, IPFS acts as a secure storage method for the files.

The file is split into many smaller objects and one object per file contains links to all the other IPFS objects that constitutes the file. These objects need not be stored on the same node or in a continuous format. However, it can be distributed across the nodes that are retrieved using the link. When the file is added into the IPFS, a unique hash ID called the Content Id (CID) is generated which is used to reference the file. This CID is generated based on the content of the file. Any malicious activity to tamper the content would result in changing the hash of the file depicting the immutability and transparency of the contents. It is not possible to upload the same file twice from the same node. However, if the same content is added from different nodes in IPFS, it would still generate the same hash. Therefore, the file will not be stored twice on the network.

4. Performance Analysis

The various evaluation metrics used in this system are accuracy, precision, recall, f1-score, and mean squared error. The dataset is evaluated using machine learning techniques such as Naive Bayes (NB), Random Forest (RF), Support Vector Machine (SVM), and Decision Tree (DT) for accuracy and the following results were observed in Tables 22.2, 22.3, 22.4, and 22.5 respectively. Accuracy is the ratio of correct predictions made, and Precision is the ratio of true positive values to the total positive values.

Recall is the ratio of True positive values to the total correctly predicted positive and negative values.

$$\text{Recall} = \frac{(\text{True Positive})}{(\text{True Positive+False Negative})} \tag{4}$$

Table 22.2. Naïve Bayes evaluation

Heart Disease Prediction	*Precision*	*Recall*	*F1-score*
0	0.87	0.98	0.92
1	0.58	0.14	0.22
Accuracy	86%		
MSE	0.14		

Source: Author.

Table 22.3. Random forest evaluation

Heart Disease Prediction	*Precision*	*Recall*	*F1-score*
0	0.87	0.97	0.92
1	0.52	0.16	0.25
Accuracy	85%		
MSE	0.15		

Source: Author.

Table 22.4. Support vector machine evaluation

Heart Disease Prediction	*Precision*	*Recall*	*F1-score*
0	0.87	0.98	0.92
1	0.58	0.14	0.22
Accuracy	86%		
MSE	0.14		

Source: Author.

Table 22.5. Decision tree evaluation

Heart Disease Prediction	*Precision*	*Recall*	*F1-score*
0	0.88	0.86	0.87
1	0.31	0.35	0.33
Accuracy	79%		
MSE	0.21		

Source: Author.

Table 22.6. ANN evaluation

Heart Disease Prediction	*Precision*	*Recall*	*F1-score*
0	0.87	0.98	0.92
1	0.57	0.15	0.24
Accuracy	86%		
MSE	0.14		

Source: Author.

F1 score is the mean between the precision and recall values which lies in the range of [0,1]. Higher the score, much better is the performance of the model.

It is observed that the results are indistinguishable while using NB, RF and SVM models which is around 86% while DT has a lower accuracy of 79% compared to these models. It is anticipated to use the NN models compared to the traditional Machine learning techniques to improve the performance results.

Figure 22.2 shows the creation of ANN model which is evaluated against different values for the test size, number of neurons in the hidden layer and number of epochs with the Relu activation function for the input and hidden layer.

Test sizes used: 0.2, 0.3 and 0.4.

Number of neurons: 15 (number of input attributes), 7 (half the input attributes), 30 (double the input attributes).

Number of epochs: 10 and 100.

The accuracy obtained with minimum number of epochs is the same as with the maximum number of epochs. However, there is a slight deviation in the other evaluation metric values. Similarly, different test sizes and hidden layer neuron numbers do not affect the accuracy significantly. However, the performance metrics are slightly varied.

Table 22.6 are the results of ANN using test size = 0.3, number of neurons in the hidden layer = 30 (twice the number of input data features) and number of epochs = 10 respectively. Figure 22.2 shows the ROC curve of ANN on the curated dataset.

The MLP model is evaluated similarly against various parameters as shown in Table 22.7.

Accuracy is unaffected by the hidden layers and neurons in each layer, test size and the learning rate. However, the data patterns become complex with an increase in these factors. Improved results are observed with lesser number of epochs. A better MSE value is obtained when the neuron numbers in each layer is restricted to the input features. Table 22.8 shows the Accuracy and MSE values obtained using a MLP model with test size = 0.3, learning rate = 0.3, number of epochs = 5, hidden layers = 4, neurons in each layer = 15(number of input parameters).

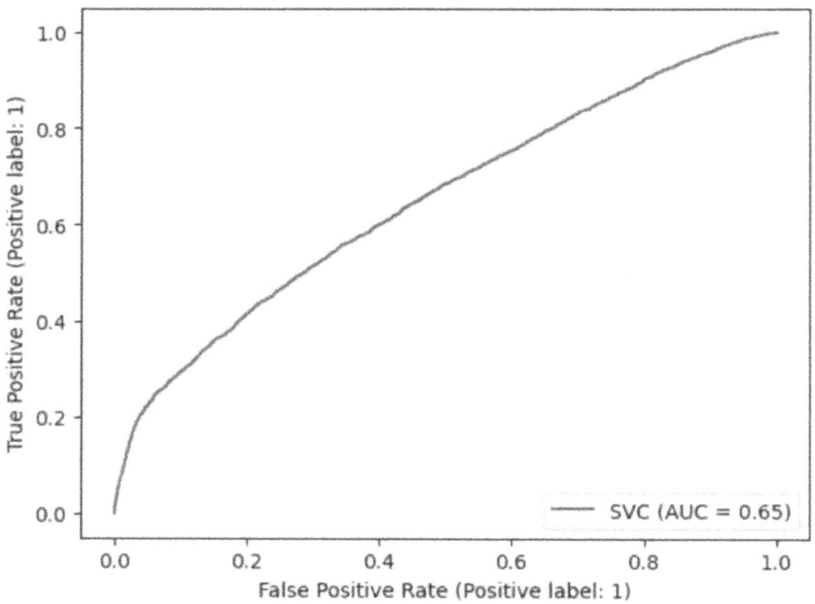

Figure 22.2. ROC curve of the ANN model.

Source: Author.

Figure 22.3 shows the ROC curve of the MLP model. The computational efficiency of both the models are compared. The time taken by ANN to train the custom dataset is 125 seconds whereas MLP model is 35 seconds.

4.1. IPFS implementation

In this system, IPFS Desktop application is used which includes file manager, peer manager, IPFS Kubo node and content explorer. These components help to interact with the network easily through a simple user interface without using terminal commands. Once IPFS Desktop Application is installed, it acts as the control panel and file browser for that node. A folder is created for every patient where the Electronic Health Records (EHR) of the patient is uploaded. This could include various reports, prescriptions, scan images etc. Content ID (CID) is generated for each file within the folder and for the whole folder itself. The files can be pinned to store it permanently on the node and thus making it available to all the nodes on the network. IPFS nodes treat the stored data like cache memory, which can be removed during garbage collection when no longer used for a longer duration. Pinning also indicates that the node it is pinned at is a permanent provider of data.

IPFS Gateway is required to obtain the contents from the network as it acts as an interface between the nodes. IPFS Companion Extension which is an IPFS Gateway is to be installed to interact with the self-node as well as other nodes on the network through browser. This would help loading the file automatically, import and share files with the IPFS network [12–14].

Figure 22.4 is the snippet of IPFS Desktop application dashboard which displays the status and activity of the self-node. It also displays the file size hosted by the self-node and the file size hosted by the node being part of the network. The real time incoming and outgoing traffic on the IPFS network is also displayed.

Figure 22.5 shows how the files page enables the node to upload a file or folder and import files or folders from other nodes on the network. In this system, folder for each patient is uploaded which would contain all the EHRs of the particular patient.

Table 22.7. Parameter Initialization

Parameters	*Values*
Test sizes and Learning rates	0.2, 0.3 and 0.4
Hidden Layers	4 and 8
Neurons	15 (number of input attributes), 30(double the number of input attributes)
Number of epochs	5, 10 and 100

Source: Author.

Table 22.8. MLP evaluation

Metrics	*Value*
Accuracy	0.99
MSE	2.40

Source: Author.

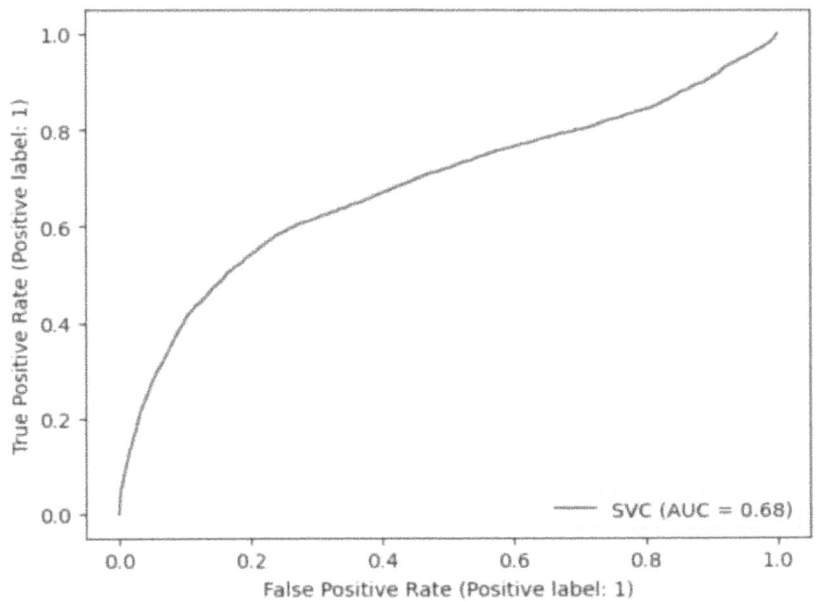

Figure 22.3. ROC curve of the MLP model.

Source: Author.

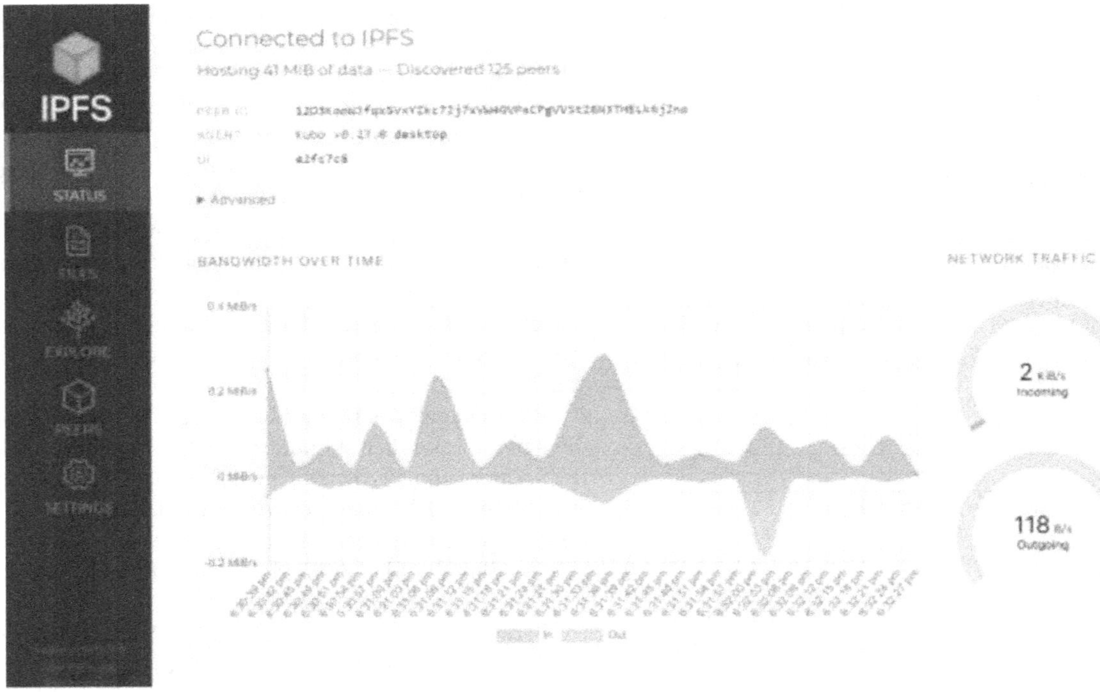

Figure 22.4. IPFS Desktop Home page.

Source: Author.

Figure 22.5. IPFS Desktop Files Page.

Source: Author.

Within the folder created for each patient, all the different reports, prescriptions, medical history and scan images can be uploaded as shown in Figure 22.6.

Figure 22.7 depicts the few operations that are supported by the IPFS Desktop application which can be performed on the files/folder created by the node. The other operations which can be performed are rename, remove, pinning the files, inspect the nodes that contain the copy of the files, copy the CID, download the content etc.

The Explore page as exhibited in Figure 22.8 allows to access or download the contents of the CID that is obtained. If the CID requested is cached locally, then the gateway will respond with the contents of the CID, else it will try to retrieve the contents from the network.

Figure 22.6. Contents within each patient folder.

Source: Author.

Figure 22.7. Operations on each folder.

Source: Author.

5. Discussion

The objective of the research is to build an effective Neural network model for predicting heart disease in chronic diabetic patients. Neural network models are chosen for disease prediction as they improve performance. In addition, storing the EHR of patients in IPFS is to ensure privacy of patients. The challenge encountered in our research is to build a dataset which has correlation between diabetes as attribute and heart disease as class label. The custom dataset is created for this research. There has been very limited literature on implementing a predictive analytic model and IPFS storage for health care. Therefore, our hypothesis is to implement an effective neural network model for predicting heart disease and efficiently storing the EHR patients in IPFS. This has been successfully implemented in our research.

IPFS technology in healthcare sector enhances the confidentiality of the EHRs and patient data due to its immutable, decentralized, transparent and traceable

Figure 22.8. Access the contents using the CID.

Source: Author.

nature. Also, the contents of the file can be stored across multiple nodes on the network and IPFS uses a DHT to track the data stored. Big data environments can handle such cases where large number of nodes are connected to the network and multiple DHTs are created by the IPFS to track the data. One of the privileges of IPFS discussed earlier is that, though multiple peers try to upload the same contents to the network, IPFS would return only the same CID. However, the files uploaded by the peers would get added to the network, depleting the storage and increasing the cost of utilization. Thus, we can enhance the model by developing a system to detect and prevent the duplication of EHR, thereby allowing new details to be added. In addition, a local admin node connected on the network can be installed at the healthcare centres. This will allow data to be stored into the IPFS only through this node but can be accessed by the other nodes.

One of the advantages of using an IPFS model is that, though multiple peers try to upload the same contents to the network, IPFS would return only the same CID. However, the files uploaded by the peers would get added to the network, depleting the storage and increasing the cost of utilization [6]. This would require additional storage, computing and communications when implemented in large scale hospital networks increasing the operational costs. The primary advantage of storing EHRs in IPFS is it will ease the job of health care providers. There may be few scalability issues and can be identified if a model is built to detect and prevent the duplication of EHR if new records are added. In addition, a local admin node connected on the network can be installed at the healthcare centers. This will allow data to be stored into the IPFS only through this node but can be accessed by the other nodes.

6. Conclusion

Artificial Intelligence algorithms are widely used to effectively predict the risk of heart disease. The results obtained in predicting the heart disease using ML approaches can be improvised using neural networks. The aim of this proposed work is to determine the patients having diabetes with high risk of cardiovascular disease using MLP and ANN approaches. MLP provides a higher accuracy of 99.98% in comparison to ANN model with 86%. In addition, according to the Mean Square Error (MSE) of the models, it can be observed that ANN has lower MSE of 0.14 MSE and MLP has a 2.40 MSE value. ANN is better suited for our research. The comparison between using an ANN and MLP neural network models proves that MLP model has better accuracy to predict heart disease using diabetes as a parameter and other features as discussed in the dataset description section. IPFS is an effective measure to secure the health information as it is a decentralized storage form adding to the immutability of the information. Hence, the research work aims to store the patient EHR in IPFS for privacy and security.

7. Future Scope

The IPFS network can be deployed as private by installing a Kubo (go-ipfs) and deploying one node as bootstrap node

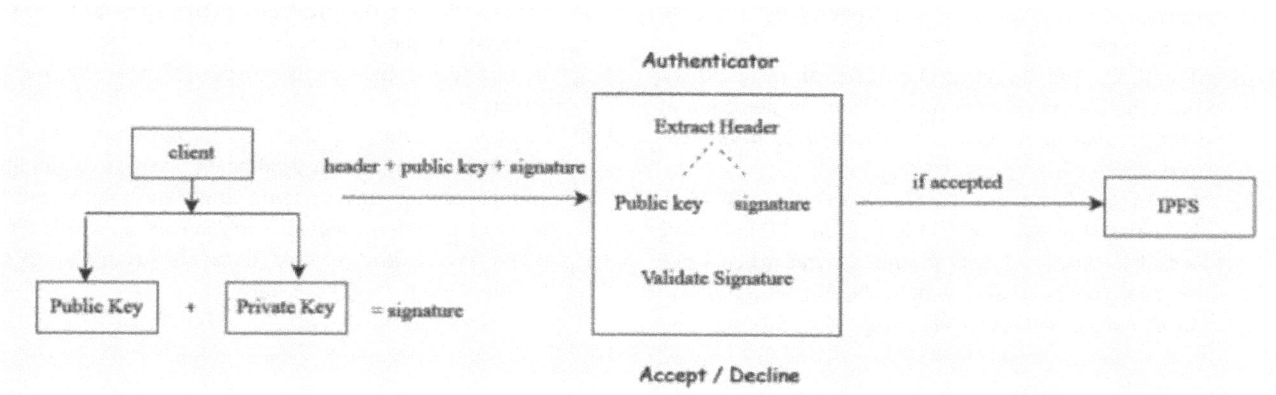

Figure 22.9. Enhanced IPFS model using Authenticated gateway.

Source: Author.

and the others as client nodes. This will enhance security of the system. A swarm key is to be created which will be referred by all nodes in the network. Bootstrap node is the node specific to the network, it is the node that other client nodes have to get connected to, for connecting with the peer nodes in the network. This is implemented by adding the IP address and hash address of the bootstrap node to all the nodes including itself in the network. The nodes, having the hash of the files uploaded in the network by other nodes, can access it only if it is part of the private network as well. Another method to improvise the security as displayed in Figure 22.9 is to use an authenticated gateway that would limit the access to requests by configuring a reverse proxy. Unlike private IPFS network which allows all nodes in the network to access the data, using this authenticated gateway allows access to requests only with authentication. In this method, the client nodes should sign their public key with their private key (from any of the Ethereum / Filecoin / Solana, etc.) to get a signature, which along with the header and public key is used to access the IPFS. The gateway authenticator then extracts this signature, header and public key to validate the node. Once successful, the access to the IPFS network is obtained.

Conflicts of Interest

The author(s) declare(s) that there is no conflict of interest regarding the publication of this paper.

References

[1] Upadhye, S., Uparkar, S., Bhave, A., & Bagde, P. O. (2022, September). Optimism or Pessimism: Is Blockchain Technology Helping the Healthcare Industry?. In *2022 IEEE International Conference on Blockchain and Distributed Systems Security (ICBDS)* (pp. 1–6). IEEE.

[2] Patil, S. R. (2022, November). Implementing blockchain technology in healthcare systems utilizing machine learning techniques. In *2022 IEEE North Karnataka Subsection Flagship International Conference (NKCon)* (pp. 1–7). IEEE.

[3] Javaid, A. M., Hammad, R., & Adnan, A. (2023). Blockchain-based healthcare model to predict heart disease using a machine learning techniques. *International Journal of Advanced Sciences and Computing, 2*(1), 21–26.

[4] Bessem, Z., Omar, C., Faisal, J., Meryem, A., & Mohamed, A. (2021). HealthBlock: A secure blockchain-based healthcare data management system. *Computer Networks, 200*, 108500, 1–16. *Science Direct.*

[5] Akruti, S., Akshet, P., & Mukta J. (2022). Application of Blockchain in Healthcare. *In 2022 First International Conference on Artificial Intelligence Trends and Pattern Recognition (ICAITPR)* (pp. 1–4). IEEE.

[6] Sumit K. R., Sanjeev K. R., Arun, K. R., Kashif, N., Tariq, R. S., & Sana, N. (2022). A survey on blockchain technology supported approaches for healthcare system, open issues and challenges. *In 2022 14th International Conference on Mathematics, Actuarial Science, Computer Science and Statistics (MACS)* (pp. 1–7). IEEE.

[7] Akhilendra, P. S., Nihar, R. P., & Ashish, K. L. (2021). A novel patient-centric architectural framework for blockchain-enabled healthcare applications. *In 2021 IEEE Transactions on Industrial Informatics, 17*(8), 5779–5789, IEEE.

[8] Huru H., Muhammad, T., Ui-Jun Baek, Jee-Tae P., & Myung-Sup Kim. (2022). A novel block chain-enabled heart disease prediction mechanism using machine learning. *Computers and Electrical Engineering, 101*, 108086.

[9] Sun, Z., Han, D., Li, D., Wang, X., Chang, C. C., & Wu, Z. (2022). A blockchain-based secure storage scheme for medical information. *EURASIP Journal on*

Wireless Communications and Networking, 2022(1), 40. Springer.

[10] Mengji, C., Taj M., Ateeq, Ur. R., et al. (May 2021). Blockchain-enabled healthcare system for detection of diabetes. *Journal of Information Security and Applications, 58*, 1–12.

[11] Nguyen, G. N., Le Viet, N. H., Elhoseny, M., Shankar, K., Gupta, B. B., & Abd El-Latif, A. A. (2021). Secure blockchain enabled Cyber–physical systems in healthcare using deep belief network with ResNet model. *Journal of Parallel and Distributed Computing, 153*, 150–160.

[12] https://docs.ipfs.tech/concepts/ipfs-gateway/#gateway-request-lifecycle

[13] https://docs.ipfs.tech/concepts/ipfs-gateway/#gateway-types

[14] Tandon, A., Dhir, A., Islam, A. N., & Mäntymäki, M. (2020). Blockchain in healthcare: A systematic literature review, synthesizing framework and future research agenda. *Computers in Industry, 122*, 103290. https://www.kaggle.com/datasets/prosperchuks/health-dataset

Appendix

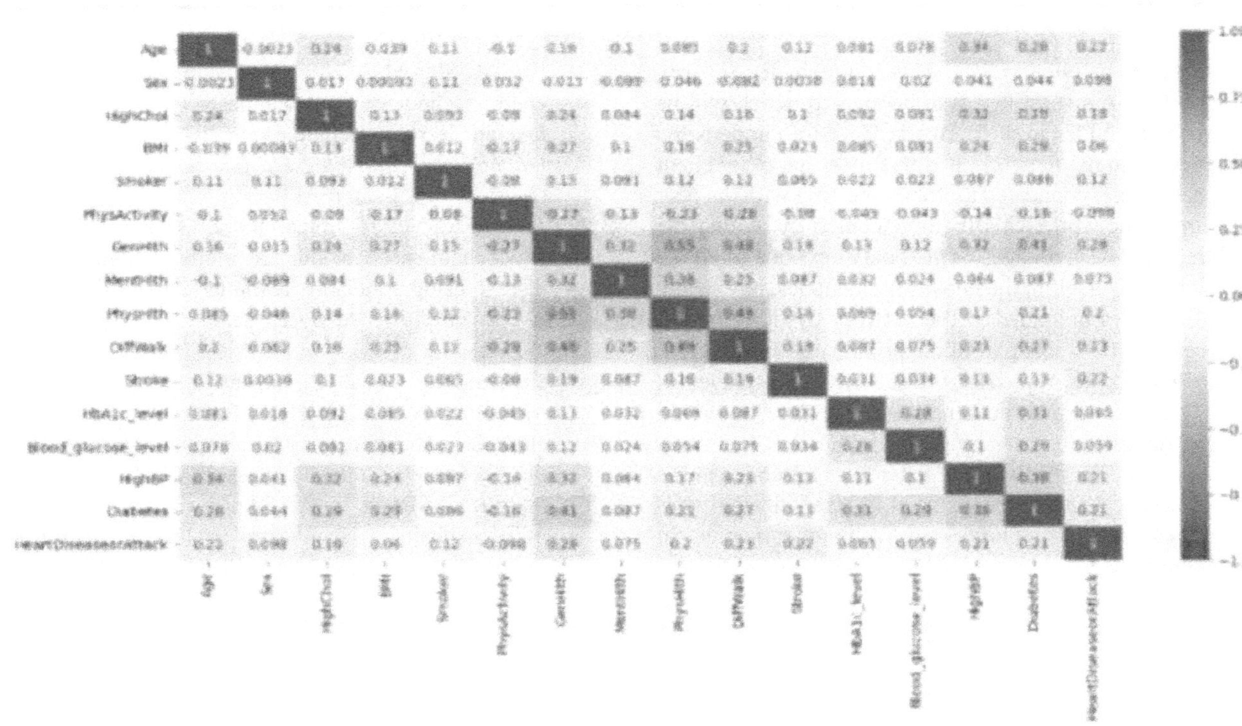

Figure 22.A1. Multi Collinearity among features using Rank based correlation.

Source: Author.

Table 22.A1. Customized Dataset description

Sl. No	*Feature*	*Values*	*Ratio in the dataset*
1	Age	Categorical – 13 values 1: 18–24 2: 25–29 3: 30–34 4: 35–39 5: 40–44 6: 45–49 7: 50–54 8: 55–59 9: 60–64 10: 65–69 11: 70–74 12: 75–79 13: 80 above	 1: 1.4% 2: 2% 3: 2.9% 4: 4% 5: 5% 6: 6.6% 7: 9.7% 8: 12.2% 9: 14.3% 10: 15.4% 11: 11.4 % 12: 7.6% 13: 7.7%
2	Sex	Nominal 0-female 1- male	 54.3% 45.7%
3	HighChol	Nominal 0-no 1-yes	 47.4% 52.6%
4	BMI	Integer	16–20 – 3.6% 21–30 – 58% 31–40 – 31.2% 40–58 – 7.2%
5	Smoker - Have you smoked at least 100 cigarettes in your entire life? [Note: 5 packs = 100 cigarettes]	Nominal 0-no 1-yes	 52.5% 47.5%
6	PhysActivity - physical activity in past 30 days - not including job	Nominal 0 - no 1 - yes	 29.7% 70.3%
7	GenHlth - Would you say that in general your health is: scale 1–5	Categorical 1 = excellent 2 = very good 3 = good 4 = fair 5 = poor	 1: 11.7% 2: 28.1% 3: 33.1% 4: 18.8% 5: 8.2%
8	MentHlth - days of poor mental health scale 1–30 days	Integer	0 days – 68% 1–10 days – 19.9% 11–20 days – 4.9% 21–30 days – 7.2%
9	PhysHlth - physical illness or injury days in past 30 days' scale 1–30	Integer	0 days– 56.5% 1–10 days – 24.2% 11–20 days – 6.1% 21–30 days – 13.2%
10	DiffWalk - Do you have serious difficulty walking or climbing stairs?	Nominal 0-no 1-yes	 74.7% 25.3%
11	Stroke - you ever had a stroke?	Nominal 0-no 1-yes	 93.8% 6.2%
12	HbA1c_level	Decimal	31.9% - range 3.5–5.0 57.2% - range 5.7–6.6 10.9% -range 6.8–9
13	blood_glucose_level	Integer	23.5% - range 80–100 67.8% - range 126–200 8.7% - range 220–300
14	HighBP	Nominal 0-no 1-yes	 43.7% 56.3%
15	Diabetes	Nominal 0-no 1-yes	 50% 50%
16	Heart Diseaseor Attack - coronary heart disease (CHD) or myocardial infarction (MI)	Nominal 0-no 1-yes	 85.2% 14.8%

Source: Author.

23 Comparing the therapeutic effectiveness of Centella Asiatica extract against donepezil on MMP9 protein inhibition in dementia disorders utilizing molecular docking analysis

J. Kamalieswaran[a] *and Ieshitha pan*[b]

Department of Computer science and Bio science, Saveetha School of Engineering, Saveetha Institute of Medical and Technical Sciences, Saveetha University, Chennai, Tamil Nadu, India

Abstract: Alzheimer's disease (AD) is a progressive neurodegenerative disorder marked by cholinergic dysfunction and cognitive decline. Donepezil, a reversible acetylcholinesterase (AChE) inhibitor, is a standard therapeutic agent that enhances cholinergic transmission by increasing acetylcholine (ACh) levels. However, its long-term use is associated with adverse effects, necessitating the search for safer alternatives. Centella asiatica (CA), a revered botanical in traditional medicine, exhibits potent neuroprotective properties attributed to its phytochemical constituents, including triterpenoids (asiatic acid, madecassic acid) and flavonoids (quercetin, kaempferol). These bioactives have demonstrated AChE inhibition, anti-inflammatory, antioxidant, and anti-apoptotic effects, contributing to cognitive enhancement. This study investigates the therapeutic potential of CA-derived phytochemicals as natural AChE inhibitors through computational modelling, in vitro enzyme assays, and neuroprotection studies. Their structural diversity and multifaceted mechanisms offer a promising avenue for neurodegeneration management. Comparative analysis with Donepezil will elucidate their efficacy and safety profile. By integrating molecular docking, biochemical validation, and neuroprotective assessments, this research aims to advance plant-derived therapeutics as viable candidates for AD treatment, offering a sustainable and potentially safer alternative to conventional drugs.

Keywords: Centella Asiatica, dementia, disease, phytochemical compound

1. Introduction

AD presents major challenges in diagnosis and treatment due to the lack of widely accessible biomarkers capable of accurately detecting the disease and predicting its progression. While cerebrospinal fluid biomarkers provide valuable insights into disease staging, their effectiveness in forecasting progression and treatment outcomes remains limited. The Blood-Based Biomarker Interest Group is actively working to overcome obstacles such as patient variability, control selection, and the complexities of the blood-brain barrier. This review underscores the importance of establishing well-defined biomarker objectives, critically assessing existing methodologies, and standardizing preanalytical and analytical procedures. Improving diagnostic consistency through these efforts may enhance early detection, monitoring, and therapeutic strategies for AD, ultimately leading to better patient outcomes [1].

In 2011, the National Institute on Aging and the Alzheimer's Association established diagnostic criteria for AD based on its different stages. As research progressed, these guidelines evolved into a comprehensive research framework that prioritizes biomarker assessment over clinical symptoms. The AT(N) classification system organizes biomarkers into three categories: β-amyloid accumulation, tau pathology, and neurodegeneration, with the flexibility for future refinements. While this framework supports hypothesis-driven research, it is not designed for direct clinical application. By defining AD as a biological process, researchers can gain deeper insights into disease progression and develop more precise therapeutic strategies. However, the framework must be applied judiciously and validated across diverse populations to ensure its accuracy and relevance. Its integration into research will enhance the understanding of dementia mechanisms, ultimately contributing to

[a]kamalieswaranj1047.sse@saveetha.com, [b]ieshtapan.sse@saveetha.com

DOI: 10.1201/9781003770473-23

improved diagnosis and targeted treatment approaches. Further studies are necessary to refine this system and expand its applicability in dementia research [2].

A comprehensive review of 263 studies examined 13 different interventions aimed at addressing cognitive decline, mild cognitive impairment (MCI), and dementia related to cognitive aging. No intervention demonstrated strong, conclusive effectiveness. Moderate evidence supports cognitive training in healthy individuals for improving specific skills such as memory, reasoning, or processing speed, though these benefits do not extend broadly or persist long-term. Certain treatments, including vitamin E, B12 combined with folic acid, and antihypertensives, showed no measurable cognitive improvement. Limited findings suggest that raloxifene may lower MCI risk, whereas estrogen therapy may increase it. The impact of physical activity on cognitive function was inconsistent across studies. Overall, most interventions lacked sufficient evidence to confirm their efficacy, highlighting the need for further research to identify reliable strategies for preventing or slowing cognitive decline [3].

This study evaluates the efficacy of antipsychotic medications in managing agitation and psychosis in dementia, focusing on trials with symptom-specific outcomes. A systematic review of 30 studies classified patients into agitated, psychotic, and mixed groups, assessing conventional and atypical antipsychotics separately. Conventional antipsychotics demonstrated slight effectiveness in reducing agitation (−0.44, -0.88, 0.01) and psychosis (−0.31, −0.61, −0.02) but showed no impact on generic assessment scales. Atypical antipsychotics had limited effectiveness, except in mixed patient groups where they slightly reduced agitation (−0.29, −0.40, −0.18). The findings suggest that including patients without target symptoms or using generalized outcome measures may lead to biased results. These insights emphasize the need for more precise clinical trials and inform future research and guideline development for treating agitation and psychosis in dementia [4].

This study examines the impact, effectiveness, and potential risks of nonpharmacologic approaches for managing agitation and aggression in individuals with dementia. A systematic review of 126 randomized controlled trials assessed interventions at the patient, care delivery, and caregiver levels. Patient-centered methods, such as music therapy, aromatherapy, and bright light exposure, demonstrated similar outcomes to standard care, though evidence strength was low. Care delivery models like dementia care mapping and person-centered care showed comparable results. While caregiver-focused strategies enhanced caregiver confidence, they did not significantly alleviate agitation in patients. Reports of adverse effects were rare. These findings underscore the necessity for more robust research to establish the effectiveness of nonpharmacologic treatments in dementia care, ensuring evidence-based strategies for improving patient well-being and caregiver support [5].

MMPs are zinc-dependent enzymes involved in extracellular matrix (ECM) degradation, with MMP-9, a member of the gelatinase B family, playing a critical role in cancer progression. Its involvement in ECM remodeling links it to various stages of tumour development, making it a promising therapeutic target. The development of MMP-9 inhibitors has gained significant research interest; however, no FDA-approved inhibitor has successfully advanced through clinical trials. This review examines the strategies used in designing MMP-9 inhibitors, emphasizing their anticancer potential and selectivity. Additionally, it highlights recently identified inhibitors with notable specificity and potency, offering insights into key structural features essential for future drug development. Understanding these characteristics may contribute to the creation of more effective MMP-9 inhibitors, paving the way for advancements in cancer therapy [6].

2. Literature Review

The American College of Physicians and the American Academy of Family Physicians developed guidelines to assess pharmacologic treatments for dementia. A comprehensive literature review examined five FDA-approved drugs, evaluating their effects on cognition, global function, behaviour, mood, and quality of life. Clinicians are advised to prescribe cholinesterase inhibitors or memantine based on individual patient evaluations (weak recommendation, moderate-quality evidence). Medication choice should be guided by factors such as tolerability, side effects, ease of use, and cost, as there is insufficient evidence to determine the superiority of one drug over another (weak recommendation, low-quality evidence). The findings highlight the urgent need for further research to enhance pharmacologic treatment strategies for dementia, ensuring more effective and personalized patient care [7].

Dementia is often accompanied by behavioural and psychological symptoms such as agitation, depression, apathy, aggression, psychosis, sleep disturbances, wandering, and inappropriate behaviours. These challenges significantly impact patient well-being, increase caregiver burden, and create financial strain. The underlying causes include neurobiological factors, unmet needs, and environmental influences. Given the complexity of these symptoms, individualized treatment strategies for both patients and caregivers are essential. Non-pharmacologic interventions, particularly those involving caregiver support, are recommended as the first-line approach. Although antipsychotics may offer some benefits, their potential risks must be carefully considered. A comprehensive treatment plan that integrates both pharmacologic and non-pharmacologic methods is necessary to manage symptoms effectively and improve overall care for individuals with dementia [8].

This study explores the involvement of matrix MMPs in dentin formation and mineralization by culturing embryonic day 18 mouse tooth germs for 10 days with or without inhibitors. Marimastat, a broad-spectrum MMP inhibitor, led to a dose-dependent thickening of predentin and reduced mineralization. Similarly, CT(1166), a selective inhibitor of MMP-2, MMP-9, and MMP-3, produced these effects even at lower concentrations. Western blot analysis confirmed the suppression of enamelysin (MMP-20). Enamel formation was halted at higher inhibitor concentrations, potentially due to the diffusion of amelogenin. Additionally, an increase in decorin and biglycan accumulation in the predentin layer may have contributed to mineralization defects. These findings highlight the critical role of MMPs in dentin development and suggest that their inhibition disrupts normal mineralization processes, emphasizing their importance in dental tissue formation [9].

MMPs are zinc-dependent enzymes essential for degrading ECM components, with MMP-9 being one of the most complex types. It is involved in key biological functions such as tissue remodeling and inflammatory responses. However, abnormal expression and regulation of MMP-9 have been linked to various diseases, particularly cancer, where it facilitates tumour invasion, metastasis, and angiogenesis. Given its critical role in disease progression, MMP-9 has become a primary target for therapeutic intervention, particularly in oncology. Developing inhibitors to regulate its activity may offer promising treatment strategies for managing cancer and other MMP-9-associated conditions [10].

Although extensive research has been conducted on MMP-9 inhibitors as potential cancer therapies, none have successfully reached clinical approval due to issues with selectivity and adverse effects. This review examines the structure, activation mechanisms, and biological functions of MMP-9 while highlighting promising inhibitors with high specificity and activity. A deeper understanding of MMP-9 at the molecular level may contribute to the development of more targeted and effective therapeutic agents. Ongoing research in this area holds the potential to produce novel MMP-9 inhibitors with enhanced efficacy and reduced toxicity, paving the way for improved cancer treatments and therapies for other MMP-9-related diseases [11].

MMPs play a crucial role in regulating the tumour microenvironment by facilitating extracellular matrix degradation, tumour growth, angiogenesis, and metastasis. Targeting MMPs, particularly MMP-9, has emerged as a promising anticancer strategy. Research indicates that inhibiting AKT, a key regulator of MMP-9 expression, enhances its suppression, while caspase-dependent AKT cleavage promotes cell death. This study employed the Ugi multicomponent reaction (MCR) to synthesize bis-amide scaffolds designed as dual MMP-9/AKT inhibitors with caspase 3/7 activation potential. The synthesized compounds were screened against normal fibroblasts and three cancer cell lines (MCF-7, NFS-60, and HepG-2) using MTT assays. Compounds 8, 11, and 28 demonstrated superior cytotoxicity and selectivity compared to doxorubicin, effectively inhibiting metastasis and inducing apoptosis. Docking simulations aligned with enzymatic evaluations, revealing essential binding interactions. Additionally, in silico assessments confirmed favourable drug-likeness properties, highlighting compounds 8 and 28 as promising drug candidates for cancer therapy [12].

3. Materials and Methods

3.1. Molecular docking analysis

The molecular docking studies were conducted to evaluate the binding affinity of CA-derived phytochemicals, including asiatic acid, madecassic acid, quercetin, and kaempferol, to the matrix metalloproteinase-9 (MMP9) protein, a critical enzyme implicated in neurodegenerative disorders such as AD. Donepezil was used as a standard drug for comparative analysis.

3.2. Protein and ligand preparation

- The three-dimensional (3D) structure of MMP9 protein was retrieved from the Protein Data Bank (PDB ID: 1GKC).
- The 3D structures of CA-derived compounds were generated using MarvinSketch software.
- Ligand structures were downloaded as Structure Data Format (SDF) files from the PubChem database and converted to Protein Data Bank (PDB) format using PyMOL software.
- The protein structure was optimized by removing water molecules, adding polar hydrogens, and incorporating Kollman charges. The final structure was saved in PDBQT format for docking analysis.

3.3. Docking simulation

- AutoDock Vina software (version 1.1.2) was employed for molecular docking studies to predict ligand binding affinity to MMP9.
- Docking parameters were set as follows:
  - Number of docking runs: 10
  - Energy range: 10 kcal/mol
  - Exhaustiveness: 8
- The docking box was defined based on the active site of MMP9, ensuring optimal ligand-receptor interaction.

- Discovery Studio 3.0 was utilized to visualize receptor-ligand interactions, including hydrogen bonding and hydrophobic interactions.

3.4. Binding energy calculations

- The binding energy (ΔG) of each ligand was calculated using AutoDock Vina.
- Binding energy values were reported in kcal/mol, where lower values indicated stronger binding affinity between the ligand and MMP9 (see Figure 23.1).

3.5. Statistical analysis

- The binding energies of CA-derived compounds and Donepezil were compared using One-Way Analysis of Variance (ANOVA), followed by Tukey's post-hoc test.
- A significance level of $p < 0.05$ was considered statistically significant.

3.6. Computational resources

- All docking simulations were performed on a high-performance computing cluster with 16 cores and 64 GB RAM.
- Software tools used included:
  - AutoDock Vina for docking analysis
  - PyMOL for structure optimization
  - Discovery Studio 3.0 for visualization
  - MarvinSketch for ligand structure generation.

4. Model Performance and Evaluation Metrics

To ensure the accuracy and reliability of molecular docking results, several performance evaluation metrics were employed, including Root Mean Square Deviation (RMSD), binding affinity analysis, molecular interaction analysis, cross-validation studies, and molecular dynamics (MD) simulations.

4.1. Root mean square deviation (RMSD)

- RMSD values were calculated to evaluate the deviation between the docked ligand structures and the reference ligand.
- A low RMSD value (below 2.0 Å) indicates high docking accuracy and structural stability of the ligand-receptor complex (see Figure 23.2).

4.2. Binding affinity evaluation

- The binding affinity (ΔG) of CA-derived phytochemicals was compared with Donepezil to identify the most effective ligand.
- A binding energy lower than -10 kcal/mol is considered indicative of strong ligand-protein interaction.
- The molecular docking results for cynarin and chlorogenic acid with MMP9 are summarized in **Tables 23.1** and **23.2**, respectively.

4.3. Molecular interaction analysis

- Molecular interactions, including hydrogen bonds, hydrophobic interactions, and van der Waals forces, were analyzed using Discovery Studio 3.0.
- Key interacting amino acid residues within the MMP9 active site were identified to validate the ligand binding specificity.

4.4. Cross-validation study

- The docking results obtained from AutoDock Vina were cross-validated using the Molecular Operating Environment (MOE) software.

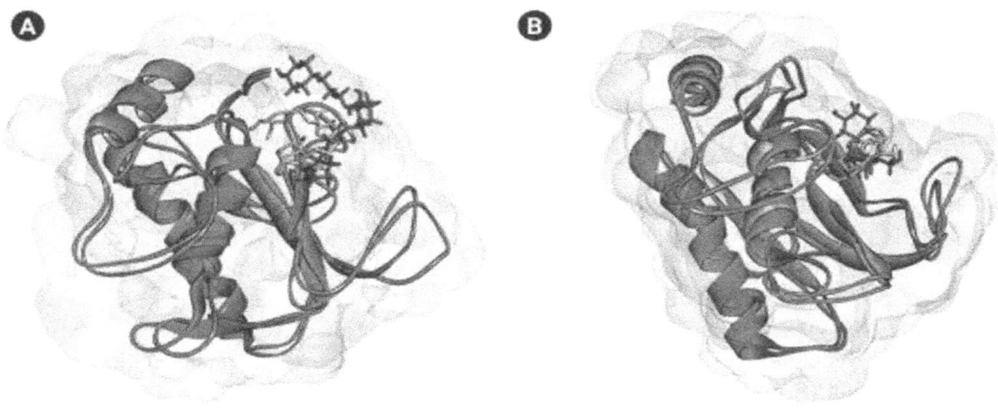

Figure 23.1. Superimposed Structures of MMP9 Complexed with Cynarin (A) and Chlorogenic Acid (B).

Source: Author.

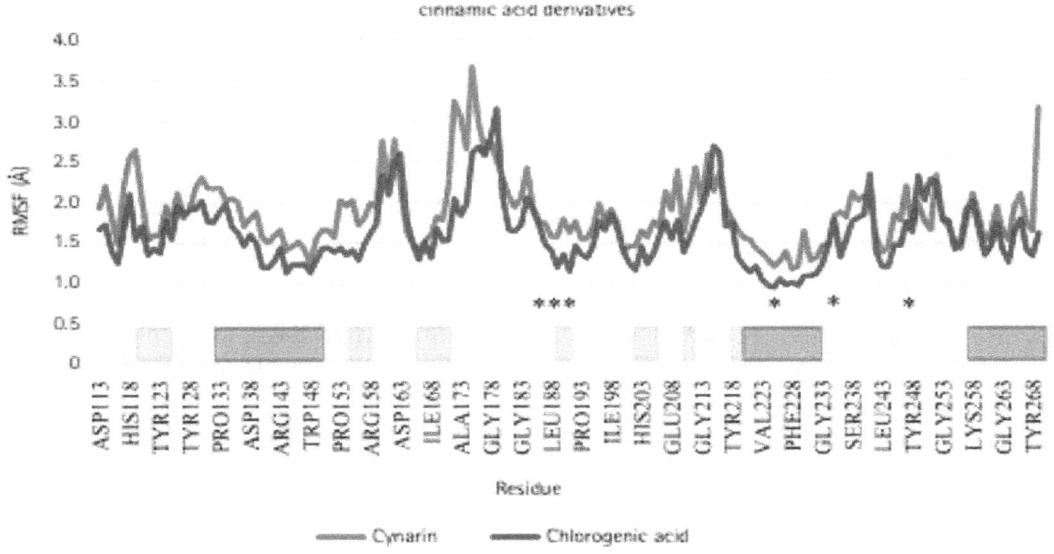

Figure 23.2. Root mean square fluctuation (RMSF) analysis of MMP9 complexed with cynarin and chlorogenic acid.

Source: Author.

Table 23.1. Molecular docking results of cynarin with MMP9

Mode	*Affinity*	*Distance from (rmsd-Lower Bound)*	*Best mode (rmsd-Lower Bound)*
1	−14.68	0.000	0.000
2	−14.58	1.504	2.508
3	−14.48	1.454	2.225
4	−14.38	3.228	8.338
5	−14.28	1.773	3.461
6	−14.18	27.690	29.883
7	−14.08	8.806	12.426
8	−13.98	2.610	4.112
9	−13.88	11.750	17.004

Source: Author.

- Ligands that demonstrated consistent binding affinity across multiple docking tools were shortlisted for further evaluation.

4.5. Molecular dynamics (MD) simulations

- MD simulations were conducted for the top-performing ligands using Discovery Studio Client version 16.1.0.15350.
- Simulations were performed for 10 ns to assess ligand stability within the MMP9 active site.
- Root Mean Square Fluctuation (RMSF) and Root Mean Square Deviation (RMSD) analyses were performed to study the dynamic behaviour of the ligand-receptor complexes.

Table 23.2. Molecular docking results of chlorogenic acid with MMP9

Mode	*Affinity (kcal/mo l)*	*Distance from (rmsd-Lower Bound)*	*Best mode from (rmsdLower Bound)*
1	. −12.62	0.000	0.000
2	−12.52	27.636	0.241
3	−12.42	4.584	9.030
4	−12.32	10.618	14.299
5	−12.22	3.683	6.586
6	−12.12	2.524	4.290
7	−12.02	10.445	14.223
8	−11.92	2.737	9.114
9	−11.82	2.455	3.308

Source: Author.

These evaluation metrics collectively ensured the accuracy, stability, and reliability of the docking results, providing insights into the binding efficacy and molecular interactions of the studied ligands with MMP9.

In this investigation, when salinity increased, specific leaf area decreased. According to Burslem et al., a thicker leaf increases the amount of mesophyll area accessible for CO2 uptake per unit leaf area, which improves CO2 assimilation and biomass production. However, Omami et al. found that amaranth's CO2 absorption decreased as salinity increased. They suggested that the decreased specific leaf area in plants under salt stress caused the leaves to become overloaded with both organic and inorganic solutes, permitting osmotic movement but limiting the efficient use of carbon. The plant may have adapted to improve intercellular space and offset the decrease in transpiration by thickening its leaves.

5. Results and Discussion:

The molecular docking analysis demonstrated that phytochemical compounds derived from CA, particularly rosmarinic acid, chlorogenic acid, and cynarin, exhibited significant binding affinities towards the active site of MMP-9. These compounds displayed a stronger interaction with MMP-9 compared to conventional inhibitors, with cynarin showing the highest binding energy of −14.68 kcal/mol, indicating a potent inhibitory effect. This suggests that CA-derived compounds could play a crucial role in regulating MMP-9 activity, which is implicated in various pathological conditions, including cancer and neurodegenerative diseases.

Furthermore, chlorogenic acid demonstrated the ability to inhibit VEGF-induced angiogenesis by suppressing AKT activation, indicating its potential utility in cancer treatment. The inhibition of angiogenesis is a critical therapeutic strategy in restricting tumour growth and metastasis, positioning chlorogenic acid as a promising natural anti-cancer agent.

Additionally, the study highlights the therapeutic significance of cynarin, extracted from *Cynara scolymus L.*, in hepatocellular carcinoma (HCC) management. Cynarin was found to enhance the cytotoxic effects of doxorubicin on MCF-7 breast cancer cells while also exhibiting strong antioxidant and anti-inflammatory properties. This dual-action mechanism suggests that cynarin may improve chemotherapy outcomes by increasing cancer cell susceptibility to treatment while simultaneously reducing oxidative stress and inflammation.

These findings underscore the importance of computational drug discovery in identifying potent natural inhibitors of MMP-9. By integrating computational modeling with biochemical analysis, this research contributes to the development of novel therapeutic strategies for both neurodegenerative disorders and cancer management.

6. Future Scope

This study highlights the potential of Centella Asiatica extract (CAE) compounds, particularly asiatic acid and madecassic acid, in inhibiting MMP9 protein, suggesting their therapeutic relevance in dementia disorders. However, further research is needed to establish their clinical applicability.

1. **In Vivo Validation:** Additional in vivo studies are required to confirm the efficacy and safety of CAE compounds in animal models of dementia. These studies will help bridge the gap between computational findings and biological outcomes.
2. **Clinical Trials:** Rigorous clinical trials should be conducted to assess the effectiveness and safety of CAE compounds in human subjects with dementia. This will provide crucial insights into their pharmacokinetics, dosage optimization, and therapeutic potential.
3. **Structure-Based Drug Design:** Advanced structure-based drug design approaches can be employed to enhance the binding affinity and therapeutic efficacy of CAE compounds against the MMP9 protein. This strategy could lead to the development of more potent drug candidates for dementia treatment.

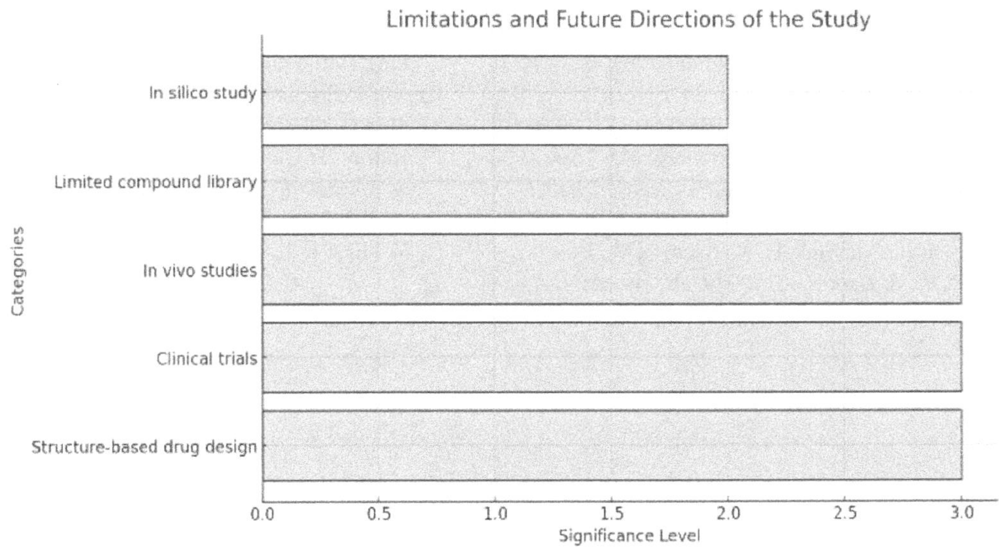

Figure 23.3. Limitations and future directions in drug discovery.

Source: Author.

The limitations include the in-silico nature of the study and the restricted compound library (Figure 23.3). Future directions emphasize the need for in vivo studies, clinical trials, and structure-based drug design to validate the therapeutic potential of CAE compounds.

7. Conclusion

This study explores the therapeutic potential of CA-derived phytochemicals as natural acetylcholinesterase (AChE) inhibitors for AD. Utilizing computational modelling, in vitro enzyme assays, and neuroprotection studies, these bioactive compounds demonstrated significant neuroprotective properties, including AChE inhibition, anti-inflammatory, and antioxidant effects. Their structural diversity and multiple mechanisms of action suggest they may offer a safer alternative to conventional treatments such as Donepezil. A comparative evaluation of efficacy and safety highlights their potential as plant-based therapeutics for AD management. The findings emphasize the importance of exploring natural compounds for neurodegenerative disorders, providing a foundation for the development of sustainable treatment options. By integrating advanced computational approaches with biochemical validation, this research contributes to identifying novel, effective alternatives for AD treatment. Further studies are required to confirm their therapeutic benefits and optimize their application for clinical use, paving the way for innovative strategies in managing neurodegenerative diseases through plant-derived bioactive compounds.

References

[1] Henriksen, K., O'Bryant, S. E., Hampel, H., et al. (2014 Jan). The future of blood-based biomarkers for Dementia disease. *The Journal of the Alzheimer's Association, 10*(1), 115–131. PMID: 23850333.

[2] Jack, C. R., Jr., Bennett, D. A., Blennow, K., et al. (2018 Apr). NIA-AA Research Framework: Toward a biological definition of Alzheimer's disease. *Alzheimer's & Dementia, 14*(4), 535–562. PMID: 29653606.

[3] Kane, R. L., Butler, M., Fink, H. A., Brasure, M., Davila, H., Desai, P., … & Barclay, T. (2017). Interventions to prevent age-related cognitive decline, mild cognitive impairment, and clinical Alzheimer's-type dementia.

[4] Reus, V. I., & Fochtman, L. J. (May, 2016). The American Psychiatric Association Practice line on the Use of Antipsychotics to Treat Agitation or Psychosis in Patients with Dementia. *American Journal of Psychiatry, 1*(173)(5), 543–546. doi:10.1176/appi.ajp.2015.173501. PMID: 27133416.

[5] Brasure, E., Fuchs, E., et al. (2016). Nonpharmacologic Interventions for Aggression in dementia and agitation Rockville (MD).

[6] Qaseem, A., & Snow, V. (2008 March). Current pharmacologic treatment of dementia: a clinical from the American College of Physicians and the American Academy of Family Physicians. *Annals of Internal Medicine, 148*(5), 370–378. PMID: 18316755.

[7] Kales, H. C., & Gitlin, L. N. (2015). March Assessment and management of psychological symptoms of dementia. *2*, 350:h369. PMID: 25731881.

[8] Fanchon, S., Bourd, K., Septier, D., Everts, V., Beertsen, W., Menashi, S., & Goldberg, M. (2004). Involvement of matrix metalloproteinases in the onset of dentin mineralization. *European Journal of Oral Sciences, 112*(2), 171–176.

[9] Wong, M. C., Ding, H., Wang, J., Chan, P. S., & Huang, J., (2019). Prevalence and risk factors of colorectal cancer in Asia. *Intestinal Research, 17*, 317–329.

[10] Mondal, S., Adhikari, N., Banerjee, S., Amin, S. A., & Jha, T. (2020). Matrix metalloproteinase-9 (MMP-9) and its inhibitors in cancer: A mini review. *European Journal of Medicinal Chemistry, 194*, 112260.

[11] Rashid, Z. A., & Bardaweel, S. K. (July, 2023). Novel Matrix Metalloproteinase-9 (MMP-9) Inhibitors in Cancer Treatment. *International Journal of Molecular Sciences, 24*(15), 12133. doi:10.3390/ijms241512133. PMID: 37569509; PMCID: PMC10418771.

[12] Ayoup, M. S., Fouad, M. A., & Abdel-Hamid, H. (2020 Jan). Battle tactics against MMP-9; discovery of novel non-hydroxamate MMP-9 inhibitors endowed with PI3K/AKT signalling attenuation and caspase 3/7 activation via Ugi bis-amide synthesis. *European Journal of Medicinal Chemistry, 186*, 111875. doi:10.1016/j.ejmech.2019.111875. Epub 2019 Nov 10. PMID: 31740054.

24 Improvising the sustainable stress prediction with the sleep pattern using Node-Edge Synaptic Model

R. Ragul[1,a], *Devi T.*[2,b], *and N. Deepa*[2,c]

[1]Research Scholar, Department of Computer Science & Engineering, Saveetha School of Engineering, Saveetha Institute of Medical and Technical Sciences, Saveetha University, Chennai, Tamil Nadu, India
[2]Professor, Department of Computer Science & Engineering, Saveetha School of Engineering, Saveetha Institute of Medical and Technical Sciences, Saveetha University, Chennai, Tamil Nadu, India

Abstract: Sleeping is an important component of sustaining to well-being harmonious dietary lifestyle because it aids with muscle maintenance, defense mechanism enhancing, along with intellectual rehabilitation, including memories merging, memories, psychological control, and intellectual processes involving taking decisions along with amount. Sufficient that is sufficient additionally assists in supervising anxiety while emotions, lowering their possibility during well-being medical conditions which include coronary artery becoming overweight, obesity, along with psychological problems. Persistent sleep disturbances might reduce the value of existence, harm associations, or lead to ongoing cognitive problems along with physiological healthcare problems. Handling problems with sleep was critical to preserving one's general health along with avoiding those negative consequences. To predict the level of stress that affects the person through utilizing the sleeping disorder. The proposed method is Node-Edge Synaptic model has great impact in the predicting process. When compared to other methods this approach has a highest accuracy of 89% level.

Keywords: Graph network, human health, Node-Edge Synaptic method, physiological problem, sleeping disorder, stress prediction

1. Introduction

Sleeping has become a basic biochemical requirement for sustaining optimal wellness while fulfillment [1]. During deep sleep, our bodies go through vital recovery processes such as tissue restoration, neurological formation, and periodic purification of key substances [2]. Additionally, sleeping appears to be crucial for complete cognitive wellness, requiring thinking activities such as memory consolidation, investigating, and decision-making [3]. Sufficient rest increases mental equilibrium, reduces stress, and optimizes behavioural state are necessary for healthy growth in the mind [4]. Despite to popular belief, chronic sleep deprivation has been scientifically associated to a wide range of health issues, including an increased risk of heart disease, hypertension, and obesity, alongside decreased immune defenses [5].

Considering the physical benefits, sleep possesses a significant impact on regular tasks and well-being. Taking adequate sleep improves concern, reaction quickness, and efficacy, amid various abilities that are necessary for general efficiency in individuals and patterns [6]. Poor sleep, can also impair taking decisions, reduce focus, and increase the likelihood of tragedies caused by mistakes, particularly in dangerous circumstances such as caring for or running huge equipment [7]. Recognising the importance of night time and emphasising proper going to bed etiquette is helps in maximising overall health, enhancing living peace of mind and ensuring emotional care for individuals and communities in overall [8].

Once time of sleep is not carefully monitored, the consequences for individual health can be severe or long-lasting [9]. Extended absence of sleep affects its immunological session, making an individual more prone to widespread and pathogenic illness [10]. This affects interactions between chemicals, increasing the body's total likelihood of becoming overweight or cardiac disorders [11]. Abnormal night time management should also exacerbate previous health problems such as stress and chronic suffering. Also, poor rest has been linked to metabolic imbalances, which may inhibit behaviour compliance, leading to bingeing and obesity [12].

[a]ragulr9151.sse@saveetha.com, [b]devi.janu@gmail.com, [c]deepa23narayanan@gmail.com

DOI: 10.1201/9781003770473-24

Lack of sleep causes a negative impact on one's mental health and cognitive abilities. Uncertainty significantly impairs cognitive functions, making it difficult to concentrate, recall, and even generate actions [13]. This increases the likelihood of experiencing sentimental wellness difficulties such as depression and stress because one's brain ability for managing feelings has been weakened [14]. Continuous inadequate sleep can also create fluctuating sensations, sadness, and heightened worry, all of those having a significant impact on personal relationships and the level of living [15]. In severe instances, a lack of sleep will result in significant psychological deficits, such as visual impairment and decreased reaction [16]. Certain areas may be especially detrimental for activities that require attention, such as transit while managing equipment. Appropriate sleep management appears crucial for sustaining emotional acuity [17].

That relationship between sleeping troubles and stress levels obtains a lot of interest regarding this domain of health services and good health [18]. Sleeping difficulties, such as tiredness, insomnia, and exhaustion, have been shown to have a major impact on an individual's psychological and physical well-being [19]. Moreover, stress was a fairly prevalent element linked with modern living, while it have been proven to significantly enhance and even promote resting difficulties [20].

It's Effect about Large-scale Data within Medical Major analysis insights large amounts in collection of information having transformed numerous industries, including medical perhaps among those that have taken advantage [21]. The Figure 24.1 shows the types of sleeping disorders that affects humans. This massive volume of information collected by numerous resources, notably includes digital medical documents, smart watches, along with sleeping investigations, creates unparalleled potential to acquire greater understanding about medical problems [22, 23].

Within connection with sleeping problems, significant information allows for their gathering among evaluations about complete biological indicators such as cardiovascular frequency (HRV), rhythm about breathing (RR), bloodstream consumption in the blood (BO), among so on [24, 25]. The objective of the study relies on the stress prediction of an individual through their sleep pattern, in acquiring the better accuracy and interpretability. Analysts may use large dataset to find sequences when connections which lesser amounts into information may overlook, resulting in greater degree of precision with comprehensive knowledge about their relationship among rest problems alongside stressors.

2. Related Work

Electroencephalography (EEG) monitors have versatile other non-intrusive monitoring instruments that monitor power neurological processes. They have been widely consumed throughout medical care including Cognitive/ Psychotherapeutic studies, including seizures, napping, sentiments, and other mind machine interconnections. While EEG sensors typically offer precise neural localizations from movement reports, they enable researchers to investigate mind operational connections. During such investigation, species examine today's employing one particular category about mathematical techniques, Graph Neural Networks (GNN), and analyze EEG information.

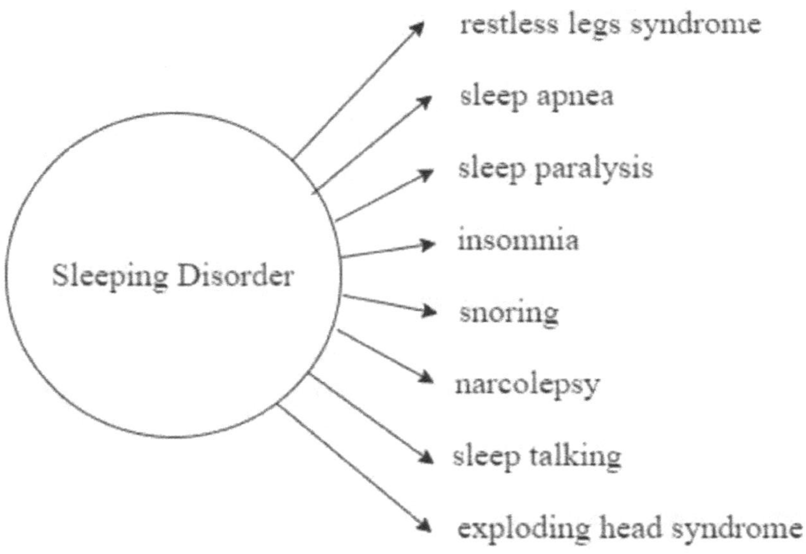

Figure 24.1. Types of sleeping disorders.

Source: Author.

GNNs look particularly quite effectively adapted into EEG knowledge analysis information simply because they operate using waveforms expressed via and series from lines rather than that conventional framework across linear reality known an Euclidean. Measurements from electro-encephalographic (EEG) electrodes come up into classification, which explains an increased study expenditure towards an integration using GNNs towards EEG information [1].

Machine learning, along with behavioural Cognitive science, can improve both while participating in medical care by automatically categorizing insomnia. These study proposes one understandable paradigm enabling automated slumber sequencing. Influenced with Spike-Timing-Dependent Plasticity, an adaptable GCN termed STDP-GCN was developed for removing characteristics based on the Polysomnography PSG waveform. The PSG communication route seems analogous to that brain cell, with postsynaptic power built causing their STDP process. Those associations among outlets form overall diagrammatic layout. GCN recovers geographical information, longitudinal computation obtains changeover regulations amongst nap phases, while an absolutely interconnected artificial brain network does categorization. STDP-GCN uses region competitive instruction for improving simulation power while minimize how it affects different neurological signals overall categorization of effectiveness Simulations suggest which STDP-GCN's effectiveness was corresponding with existing advanced techniques [2].

Rest period categorization was critical with individual wellness tracking including illness diagnosis. Medically, visually inspecting insomnia to categorize its process within various phases takes a considerable period that was heavily reliant with insomnia professionals' competence. Several computerized methods of wakefulness categorization were developed through earlier research, however how they do remain inadequate overall real-world therapeutic applications. During the current paper, researchers introduce MVF-Sleep Net, an original multi-view synthesis structure constructed using hybrid medical information which includes, electroencephalogram (EEG), (EOG) cardiac (ECG), plus electromyography. When demonstrating how to distinguish between multidimensional biological messages, players create two viewpoints: the Time-frequency pictures (TF imagery) along with chart-knowledgeable charts (GL diagrams). The spectral-temporal arrangement about consecutively scheduled TF pictures was learned using a mix of VGG-16 and GRU models. Chebyshev connected graph along with historical compression systems are used for acquiring spatial-temporal representations about consecutively sequenced GL graphs. Combining spectral-temporal along with spatial-temporal representations improves wakefulness categorization accuracy [3].

Neurological diseases (NDs), which include dementia and Alzheimer's disease, represent worldwide medical danger. subsequently becomes increasingly important for recognizing ND by merging neural network technologies with neurological photography. Our graphical representation GNN will simulate while evaluating most cerebral cortex, including morphological physiological construction, operational attributes, among additional factors, making it some of the finest artificial intelligence techniques for diagnosing ND. Certain investigators recently examined using GNN inside health care applications, although that broadening seems vast, plus their application with NDs as uncommon along with insufficiently comprehensive. This primer deals with advances about GNNs with determining among presence about ND. They explored overall GNN infrastructure with ND, covering line chart generation, combination, combination, along with predictions. They used GNN evaluations to analyze usual NDs, taking into consideration information about medium, determined count, and overall evaluation of their diagnosis. Finally, they reviewed investigation problems along with recommendations for updating proposals. Observations obtained from this summarization might establish this vital contribution towards the ongoing confluence among programming imaginative capability improvement and psychological capturing images [4].

Insomnia having ramifications regarding people from nutritional, conceptual, emotive, and interpersonal viewpoints; having enough relaxation within terms of severity and size, is crucial to live an enriching and rewarding profession. Throughout duration, several ways been proposed through unstructured publications alongside corporate bookshelves addressing specific measurements used for monitoring napping patterns together with detecting napping interruptions alongside disorders. Implanted devices serves the important role in this process because offering discreet, reliable, and long-term surveillance of neurological signs which should assist analyze napping circumstances. That study examines cutting-edge sensing technology, software programs for napping evaluation, and appropriate identification of insomnia-related disorders, involving inadequacies. The study's evaluation starts through outlining about obligated of napping in addition their relevance, followed by regularly observing napping in addition to identifying probable interruptions followed by anomalies. Following that, some assessments about innovative highly commercialized headband-like electronic instruments for monitoring sleeper were provided, spanning this research community along with marketplaces, which permits unnoticed highly precise monitoring about sleep-related measures. Additionally, one review from academic articles examined how it affected this COVID-19 outbreak upon rest processes, which was triggered by combined disease plus alterations adjusting. behaviour. The article includes an overview with strategies towards diagnosing even diagnosing problems with sleeping using biological messages involving EEG, ECG, EMG, along with EOG. Finally, analysis along with perspectives be offered for identify upcoming developments with sleeping tracking technologies [5].

3. Proposed Work

3.1. Data collection of sleepness individuals

Several biophysical characteristics are being collected that create this complete collection that can forecast fall asleep problems utilizing GNN. The following data are heart rate (HR), sleep latency (SL), REM sleep (rem), blood oxygen (bo), limb movements (lm), temperature (t), respiratory rate (rr), and sleep efficiency. Input is being gathered through nap tests, electronic gadgets, along with health information for capturing include the static along with the dynamic components about sleeping habits. Accreditation with pretreatment processes should assure uniformity of information with preparation into chart depiction, with terminals representing particular information specimens with linkages reflecting comparable ratings obtained through characteristic matrices. Applying sophisticated chart-dependent artificial intelligence strategies, that strategy seeks to capitalize upon both interdependence about key characteristics with particular objective for improving overall preciseness along with dependability about sleep-related disease estimates.

3.2. Data preprocessing

Dataset preparation during anticipating sleeping issues entails many key processes which assure both level integrity along with appropriateness about information that was utilized within node networks. Initially stabilization procedures could be used with normalized biophysical parameters which include heart rate and sleep latency, REM sleep, blood oxygen, limb movements, temperature, respiratory rate and sleep efficiency. This phase seems critical because bringing every attribute into one consistent dimension and minimizing unwanted exaggeration caused by discrepancies between measuring methods and other levels. The decision-making method has been used to find valuable helpful traits over determining problems with sleeping. Connection investigation, mutually beneficial sharing, and characteristic significance through the model of machine learning has been used to select characteristics with overall greatest diagnostic possibilities. The Equation (1) represents both multi layering of characteristics including their normalization. This ensures that the suggested strategy outputs focus just on some most relevant elements, such as sleep patterns and philosophical knowledge, therefore boosting its ability can anticipate properly.

For every character of data called as a attribute vector A $A = \{a_1, a_2, \ldots, a_n\}$, where a_i coordinate for the each physiological attribute (HRV, RR, T, LM, BO, REM, SR, SL)

Time represent as s

$$X = [HRV_s, RR_s, T_s, LM_s, BO_s, REM_s, SR_s, SL_s] \tag{1}$$

Where X represents a multilayered state.

Furthermore, throughout preparatory work, initiatives is made for handling partial input values utilizing methodologies that include regression to assure the input's completeness. To verify the accuracy of their investigation, any data containing anomalies was addressed using sophisticated computational methods. Those definitive groups are being used converted into more standardized

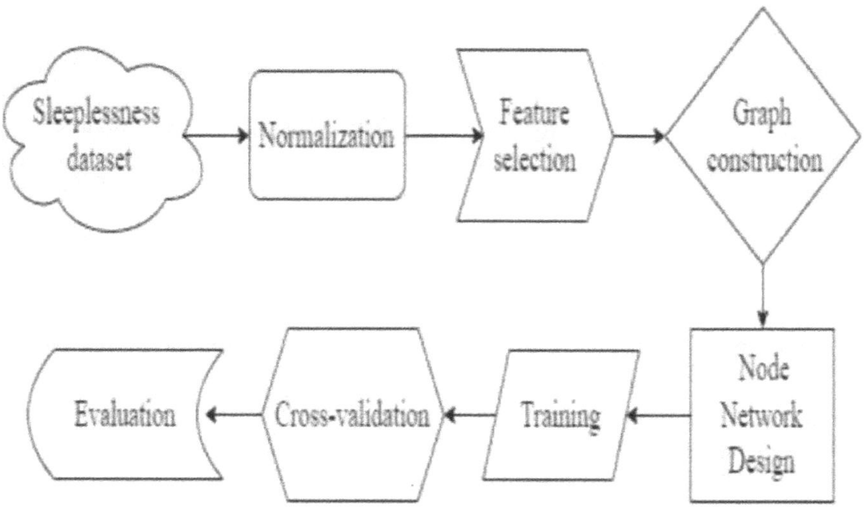

Figure 24.2. Node-edge synaptic model.

Source: Author.

network depictions, using edges expressing distinct data samples gathered and relationships preserving interactions between it applying a contrast evaluates derived from the defined features. The previously described extensive pre-processing approach aims to improve the Node-Edge synaptic model's accuracy for determining insomnia-related circumstances by providing carefully gathered information that represents the overall complexities of metabolic interactions throughout the sleep.

3.3. Node-edge generation

The creation of graphical structure is critical in preprocessed stage, then network is developed with Node-Edge Synaptic model for predicting stress using sleep dataset. The framework is developed with gathered information, representing sleep information, comprising SR, SL, BO, HRV, RR, T, LM and REM sleep duration.

The node degree is updated with the neighbouring node weights

$$p_x^{(m+1)} = \sigma(D^{(m)}.mean\left(\left\{p_y^{(m)}: y \in F(x)\right\}\right) + B^{(m)}) \tag{2}$$

Where, $p_x^{(m)}$ represent the mth layer for node x

$F(x)$ represents groups of node x

$D^{(m)}$, $B^{(m)}$ represents weight and bias of mth layer

σ represents activation function

Every branch gets supplemented using various attributes, which successfully capture overall multifaceted components found in the actual sleeping dataset. The Equation (2) represents the development of nodes with the degrees to make a connection between nodes. Each chronological aspect about wake stays intact with connecting successive moments periods with borders, which indicate every continuous advancement about health conditions during every period of their unconsciousness.

Actual connections within such following charts do not simply just bind successive period slices; they represent time-based interdependence along with pharmacological connections that exist among them. Figure 24.2 represents the workflow of the proposed method, in which phase graph is constructed node network is designed. One instance, rapid decline within circulation breathable air saturation may be often precisely tracked with fluctuations like cardiac pressure along with lung capacity, which may become efficiently described utilizing the boundaries that chart. That connection structure during its graphic may likewise be constructed to represent longer-term relationships, including whether that superior their initially during beginning about nighttime affects bodily functions afterwards during that process period. falling asleep. Such a complete arrangement construction chart enables its proposed model for learning through each instant along with prolonged longitudinal conversations, resulting in some highly detailed, integrated depiction of overall sleeping patterns.

3.4. Edge construction for insufficient sleeper

In conjunction with periodic borders, each graph could contain vertices representing metabolic relationships amongst other categories and evaluations. Through utilizing the Equation (3) the prediction of stress is identified based on the sleep patterns of the individual from the normalized information. The straight linkages among vertices which have comparable heart rate variability (HRV) characteristics and additional relevant metabolic indicators, independent about overall location throughout this chronological series.

Prediction with final node $p_x^{(m)}$ calculated by

$$s_x = max(D_{out}.p_x^{(m)} + B_{out}) \tag{3}$$

Where, s_x node x with predicted stress level

D_{out}, B_{out} represents overall weights and bias

Such double visualization for behavioural along with pharmacological interdependence allows the Node-Edge Synaptic model to successfully incorporate multiple parts from information knowledge, capitalizing upon complicated linkages which can frequently be crucial when detecting insomnia problems.

Construction for the edge with resemble features of $T(x,y)$ within nodes x and y

$$T(x,y) = exp\left(-\frac{||H_x - H_y||^2}{2\sigma^2}\right) \tag{4}$$

Where, H_x, H_y node of feature vectors

σ represents bandwidth

Edges are initiated with $T(x,y) > \in$, where threshold is $\in$

The resulting tree becomes some several-layered structure which represents together all sequential series along with health-related interaction about fall asleep information, providing one strong underpinning to feed known as system development along with evaluation phases. The Equation (4) tends to update the time period changes between the degrees of the nodes and the features changes in the edge connections with the nodes will differ.

3.5. Effiecency of node-edge synaptic model

The Node-Edge Synaptic model is developed to efficiently handle chart systematic sleeping input. Input framework is made up from numerous graphical pattern stages that execute generative computations over input graphs.

Entropy classification for stress level

$$L = -\sum x \in \gamma_{train} \sum_{i=1}^{i} s_x^{(i)} \log s_x^{(i)} \tag{5}$$

Where, γ_{train} training set of nodes

i total classes

$s_x^{(i)}$ true label

$\widehat{s_x^{(i)}}$ identified probability for i

With every node and edge connection, every vertex's own characteristics undergo revision by collecting knowledge about nearby surrounding nodes and weighting data using their proximity matrices. As the Node-Edge Synaptic model is efficient is proven by the loss range calculation in training and validation period. The accumulation procedure identifies localized interconnections across that structure. Another computational ability illustration about each procedure as provided through the following measures.

4. Results and Discussion

The Node-Edge Synaptic model is based on sleeplessness categorization upon this substantial insomnia database and produced encouraging findings, about this algorithm exceeding 89% performance. Our excellent degree of overall reliability demonstrates whether the Node-Edge Synaptic model can efficiently capture those complicated behaviours along with psychological correlations present in the given dataset. Their capacity on node and edge relationship could use information with diagrams which enables one greater sophisticated comprehension regarding when various sleeplessness indicators interplay continuously, also somewhat critical towards properly forecasting anxiety concentrations. The loss range is widely reduced when compared to other existing models, effectively the Figure 24.3 shows the comparison of loss range between the training phase and validation phase. Their conclusion implies that the Node-Edge Synaptic model could serve as considerably useful resources throughout sleeper as well as insomnia studies, offering a significantly deeper thorough but consistent way when detecting stress-associated illnesses than standard equipment learned methods.

Among the primary benefits about adopting the proposed model with its capacity for managing the heterogeneous character of this information. That sleep-related problem collection contains an assortment of parameters Respiratory heart rate variability (HRV), several physical markers that remain each intrinsically interconnected along with fluctuation continuously. Conventional approaches could fail to represent those complex interactions, whereas the Node-Edge Synaptic model successfully simulates such linkages because of its diagram-related technique. This enables its computerized system to improve through additionally also collective characteristics along with relationships resulting within improved forecasts. This model strategy excellent in correctness, highlights its dependence for taking into account a complicated, interrelated structure for physiology information with highest accuracy of 86%, precision of 83%, recall of 84% and F1 score is 83.5% for predicting stressed levels with incorporating the features and identified patterns of sleep.

Furthermore, available resilience in the Node-Edge Synaptic model is demonstrated by how well it performed among various segments about information. Throughout this development along with evaluation segments, this system continually displayed good forecasting ability, showing how it can transition effectively towards fresh, previously unknown information. The accuracy has been increased in a higher way when compared to other existing models, effectively the Figure 24.4 shows the comparison of accuracy between the training phase and validation phase. That's crucial over operational uses, as its simulation must reliably estimate anxiety rates across varied communities along with environments.

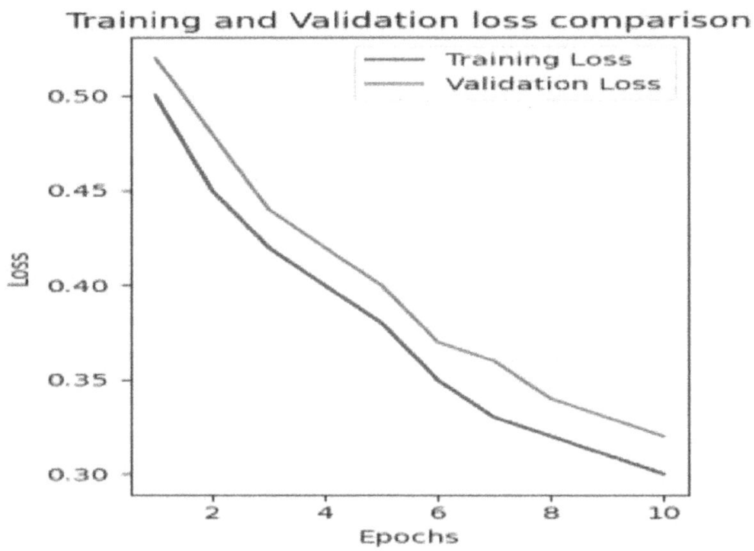

Figure 24.3. Loss rate graph.

Source: Author.

This utilization about between-verification strategies guaranteed whose services given the system's efficiency had been neither some consequence about overestimation, instead being some true representation from their potential which represent given fundamental characteristics within this dataset. The limitations of proposed method may contain with variability changes to capture as a real-world applicability. The existing methods struggles with complex relationships but proposed method works in a greater effect.

These identifications surely recommend the correct section for next research and improvement. Despite that current method functions well, however remains always room over development, particularly within terms of resolving mathematical anomalies followed by examining other characteristics that may help via anxiousness identification. This Figure 24.5. presents as thorough perspective into functioning comparisons involving standard GNN followed by GAE versions, demonstrating that other suggested Node-Edge The synaptic system outperforms those previous methods.

The Table 24.1 shows some distinctions of known as actual along with suggested methods. Subsequent study typically include merging some variety from data sources, particularly polygon relationship data from another external factors, including improving that algorithm's predicting capacity.

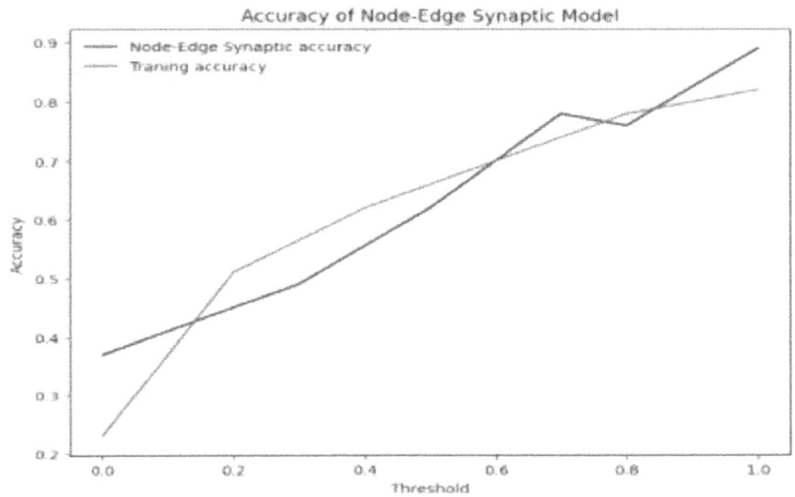

Figure 24.4. Accuracy range graph.

Source: Author.

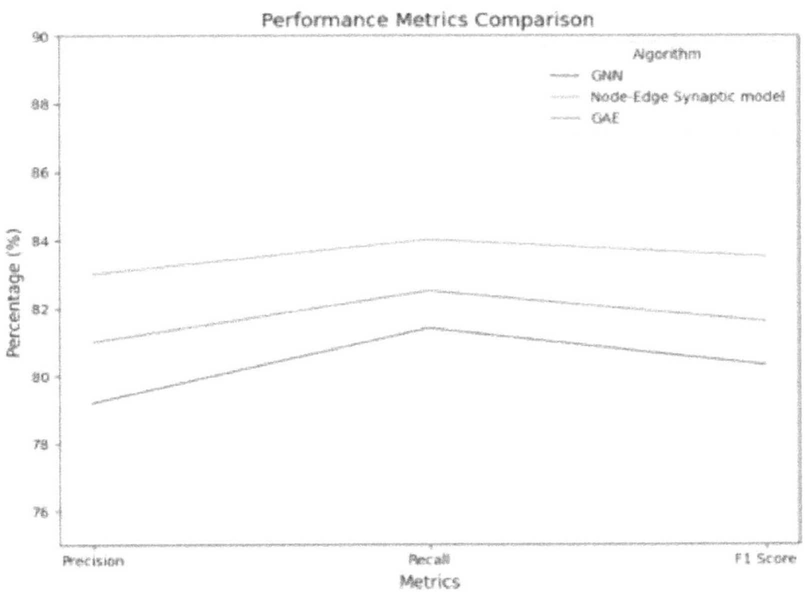

Figure 24.5. Performance metrics comparison.

Source: Author.

Table 24.1. Comparison table of performance metrics

Algorithm	*Accuracy*	*Precision*	*Recall*	*F1 Score*
ANN	82%	79.2%	81.4%	80.3%
Node Edge Synaptic Model	86%	83%	84%	83.5%
Logistic Regression	84.7%	81%	82.5%	81.6%

Source: Author.

Furthermore, combining immediate input collecting regarding analysis may result into more timely followed by relevant guidance around dealing overall sadness alongside stress. Certainly, their efficient use about such framework within its efficacy created new inventive options over enhancing logical understanding and anticipating extreme nervousness utilizing modern deep-learning technologies.

5. Conclusion and Future Work

The study employs the Node-Edge Synaptic model to categorize stress using extensive sleep conditions as a database, revealing the immense benefit of generalized-based methodologies in comprehending these complex relationships involving physiological factors. The results obtained indicates that the node-edge association is the chance of applied to sleep and pressure investigations, providing a solid platform for recognizing and anticipating disorders linked to anxiety. The Node-Edge Synaptic model reveals the potential of collecting and evaluating physical information, as well as setting standards for subsequent investigations that boost forecasting accuracy and integrate an expanded insights type database to gain a much more complete understanding of anxiety. Future advances may incorporate greater detail reports, covering physiological and environmental factors, to improve anxiety predicting precision. When compared to traditional methods, it achieved 82%, nevertheless the proposed Node-Edge Synaptic model achieved accuracy of 89%. This framework exceeded interactive techniques in forecasting stress stages as precision of 83%, recall 84% and F1 score 83.5%, illustrating their ability to cope with difficult, complex content. Regular assessment employing external capabilities and customizable representations may result in customized preventative anxiety prevention strategies. These improvements will increase the general use of node and edge network for health and contentment.

References

[1] Graña, M., & Morais-Quilez, I. (2023). A review of Graph Neural Networks for Electroencephalography data analysis. *Neurocomputing*, 126901.

[2] Zhao, Y., Lin, X., Zhang, Z., Wang, X., He, X., & Yang, L. (2023). STDP-based adaptive graph convolutional networks for automatic sleep staging. *Frontiers in Neuroscience*, *17*, 1158246.

[3] Li, Y., Chen, J., Ma, W., Zhao, G., & Fan, X. (2022). MVF-SleepNet: Multi-view fusion network for sleep stage classification. *IEEE Journal of Biomedical and Health Informatics*.

[4] Zhang, S., Yang, J., Zhang, Y., Zhong, J., Hu, W., Li, C., & Jiang, J. (2023). The Combination of a Graph Neural Network Technique and Brain Imaging to Diagnose Neurological Disorders: A Review and Outlook. *Brain Sciences*, *13*(10), 1462.

[5] De Fazio, R., Mattei, V., Al-Naami, B., De Vittorio, M., & Visconti, P. (2022). Methodologies and wearable devices to monitor biophysical parameters related to sleep dysfunctions: an overview. *Micromachines*, *13*(8), 1335.

[6] Medikonda, J. (2023). A clinical and technical methodological review on stress detection and sleep quality prediction in an academic environment. *Computer methods and programs in biomedicine*, *235*, 107521.

[7] Baek, J. W., & Chung, K. (2023). Multi-context mining-based graph neural network for predicting emerging health risks. *IEEE Access*, *11*, 15153–15163.

[8] Kumar, R., & Sankat, M. (2023). Applications of GNNs and m-Health for Disease Tracking. In *Concepts and Techniques of Graph Neural Networks* (pp. 155–166). IGI Global.

[9] Venkatapathy, S., Votinov, M., Wagels, L., Kim, S., Lee, M., Habel, U., … & Jo, H. G. (2023). Ensemble graph neural network model for classification of major depressive disorder using whole-brain functional connectivity. *Frontiers in Psychiatry*, *14*, 1125339.

[10] Kim, J., & Sohn, M. (2022). Graph representation learning-based early Depression Detection Framework in Smart Home environments. *Sensors*, *22*(4), 1545.

[11] Hong, S. (2023). Measuring the severity of depression from text using graph representation learning (Doctoral dissertation, University of Leeds).

[12] Zhao, Y., Lin, X., Zhang, Z., Wang, X., He, X., & Yang, L. (2023). STDP-based adaptive graph convolutional networks for automatic sleep staging. *Frontiers in Neuroscience*, *17*, 1158246.

[13] Yang, D., Gao, B., Huang, L., Chen, B., & Chen, Y. (2023, July). Wearable Long-Term Graph Learning for Non-invasive Mental Health Evaluation. In *International Workshop on New Approaches for Multidimensional Signal Processing* (pp. 191–203). Singapore: Springer Nature Singapore.

[14] Zhang, W., Kong, L., Lee, S., Chen, Y., Zhang, G., Wang, H., & Song, M. (2024). Detecting mental and physical disorders using multi-task learning equipped with knowledge graph attention network. *Artificial Intelligence in Medicine*, *149*, 102812.

[15] Khalid, M., & Sano, A. (2023). Exploiting social graph networks for emotion prediction. *Scientific Reports*, *13*(1), 6069.

[16] Agarwal, N., Dias, G., & Dollfus, S. (2024). Multi-view graph-based interview representation to improve depression level estimation. *Brain Informatics*, *11*(1), 14.

[17] Wang, D., Zhao, J., Zhai, S., Ye, H., Bu, L., & Fan, F. (2022). Does sleep disturbance predicts posttraumatic stress disorder and depression among college students during COVID-19 lockdown? A longitudinal survey. *Frontiers in Public Health*, *10*, 986934.

[18] Weber, F. C., & Wetter, T. C. (2022). The many faces of sleep disorders in post-traumatic stress disorder: an update on clinical features and treatment. *Neuropsychobiology*, *81*(2), 85–97.

[19] Kim, Y. R., Lee, S. Y., Lee, S. M., Shim, I., & Lee, M. Y. (2022). Effect of Hibiscus syriacus Linnaeus extract and its active constituent, saponarin, in animal models of stress-induced sleep disturbances and pentobarbital-induced sleep. *Biomedicine & Pharmacotherapy*, *146*, 112301.

[20] Wieman, S. T., Hall, K. A. A., MacDonald, H. Z., Gallagher, M. W., Suvak, M. K., Rando, A. A., & Liverant, G. I. (2022). Relationships among sleep disturbance, reward system functioning, anhedonia, and depressive symptoms. *Behavior Therapy*, *53*(1), 105–118.

[21] Ding, X., Brazel, D. M., & Mills, M. C. (2022). Gender differences in sleep disruption during COVID-19: cross-sectional analyses from two UK nationally representative surveys. *BMJ open*, *12*(4), e055792.

[22] Malcolm, A., Toh, W. L., Crocker, K., & Phillipou, A. (2022). The contribution of sleep to anorexia nervosa severity. *Eating and Weight Disorders-Studies on Anorexia, Bulimia and Obesity*, *27*(4), 1563–1568.

[23] Halioua, B., Chelli, C., Misery, L., Taieb, J., & Taieb, C. (2022). Sleep disorders and psoriasis: an update. *Acta Dermato-Venereologica*, *102*.

[24] Chung, J., Mukerji, S., & Kozlowska, K. (2023). Cortisol and α-amylase awakening response in children and adolescents with functional neurological (conversion) disorder. *Australian & New Zealand Journal of Psychiatry*, *57*(1), 115–129.

[25] Lv, R., Liu, X., Zhang, Y., Dong, N., Wang, X., He, Y., … & Yin, Q. (2023). Pathophysiological mechanisms and therapeutic approaches in obstructive sleep apnea syndrome. *Signal transduction and targeted therapy*, *8*(1), 218.

25 AI-driven multi-modal data fusion and computer vision for early detection and monitoring of paediatric neurological and developmental disorders

S. P. Maniraj[1,a], *Abarna V.*[2,b], *Sivakumar K.*[3,c], *Neeraj Kumar Mishra*[4,d], *M. S. Nidhya*[5,e], *and V. Malathy*[6,f]

[1]Assistant Professor, Department of Data Science and Business Systems, School of Computing, SRM Institute of Science and Technology Kattankulathur Campus, Tamil Nadu, India
[2]Assistant Professor, Department of Artificial Intelligence and Data Science, SRM Valliammai Engineering College, Kattankulathur, Chengalpet, Chennai, Tamil Nadu, India
[3]Assistant Professor, Department of Electronics and Communication Engineering, Panimalar Engineering College, Chennai, Tamil Nadu, India
[4]PhD Scholar, Psychology, Humanities and Social Sciences, Madan Mohan Malviya University of Technology, Gorakhpur, UP, India
[5]Associate Professor, School of Computer Applications, Dayananda Sagar University, Bengaluru, Karnataka, India
[6]Associate Professor, Department of ECE, SR University, Warangal, India

Abstract: There is need to diagnose neurological and developmental disorders in children including; ASD and cerebral palsy at an early stage to enable intervention. Standard screening procedures, which are otherwise more recognitive and brought out of experiences and observations as well as time consuming, are likely to cause delays. This paper explores how multi-modal data fusion and Computer Vision approach propelled by AI could revolutionize early detection and monitoring. Multi-modal fusion combine clinical records data, wearable sensor information, neuroimaging data, and observations of the child's behaviour to offer a complete system. Computer vision augments this by adding facial, body, and eye-tracking biomarkers in real-time to the mix as biomarkers. These biomarkers are then analysed through more complex machine learning models for better diagnostic results and for follow up. This framework combines the clinical approach and technological solutions, helping to introduce a scalable and non-invasive method of early diagnostics and individualized treatment. Some studies evaluation show that its effectiveness in the early developmental markers with hopes to cultivate paediatric healthcare in the future.

Keywords: Autism spectrum disorder (ASD), developmental disorders, multi-modal data fusion, neuroimaging, pediatric neurological disorders, wearable sensors

1. Introduction

Neurological and developmental conditions in children including autism spectrum disorder, attention-deficit/hyperactivity disorder, and cerebral palsy affect a child's learning, behaviour, and motor skills. At an early and accurate stage of the disease, it is relatively easy to prevent these consequences and make the necessary corrections. However, these diagnostic methods require basic clinical assessments, which are often reliant on clinical observation, and are time-consuming, which makes early diagnosis especially difficult in the resource-constrained environments [1, 2].

Modern developments of artificial intelligence to the field of medicine have made it possible to transform paediatric neurology. Computer vision and multi-modal data fusion present solutions with possibility by using many

[a]manirajs@srmist.edu.in, [b]abarnavasanth@gmail.com, [c]Ksivakumar.ece@panimalar.ac.in, [d]2023138301@mmmut.ac.in, [e]nidhyaphd@gmail.com, [f]malathy.v@sru.edu.in

DOI: 10.1201/9781003770473-25

sources to improve the diagnostic accuracy and monitoring processes. The integration of multiple data sources involves clinical data from patient records, wearable devices, imaging from neuro-electrophysiological tests, and behavioural observations delivering a holistic view of the child's functioning [3]. At the same time, computer vision helps to analyze such visual cues as facial expressions, eye movements, and limb movements which can remain unrecognized by the conventional approach [4].

Deep learning models of ML form the basis for analyzing large and complicated data from these sources. They are expected to be able to identify patterns and diagnose developmental diseases with great accuracy, even better compared to human ability in some case [5]. Through providing the care givers with affordable, non intrusive and automatic diagnostic solutions, these technologies fill essential missing a in early identification and continuous tracking of children's development anomalies.

The deployment of AI multi-modal data fusion and Computer vision in early diagnosis & tracking of neurological and developmental disorders in children is the focus of this paper. The strengths of this approach are described, the existing issues are described, and the existing gap between artificial intelligence and medicine is defined and further steps in how to close it is described.

2. Literature Survey

AI adoption in paediatric neurological and developmental disorders has recently generated much interest across the healthcare field. This also evaluates previous research in multi-modal data fusion, computer vision, and the application in early detection and monitoring systems.

2.1. Multi-modal data fusion in pediatric diagnostics

The fusion of data from multiple sources is integrated in a coherent diagnostic model of the corresponding variety. [3] found the usefulness of merging clinical records, neuroimages, and WSD for enhancing diagnostic performance in neurological diseases. In their study, they pointed out that merging EEG signals and fMRI data are more informative than either data type alone [3]. Thus, [7] analysed fusion models that imply the use of both audio and video data for ASD detection. Such combinations of mode thereby enhanced the detection rates by 20 percent greater to the unimodal strategies which they identified from the children with ASD such as different speech and gaze behaviours.

2.2. Computer vision for behavioural analysis

Behavioural biomarker assessment has now become dependent on computer vision as a tool. Authors [4] were the first to employ facial expression recognition as a method to diagnose emotional as well as neurological disorders. In continuation of the above methods, latest advancement in deep learning has enhanced these techniques by identifying smaller cues like micro expressions and movement minutiae.

For example, [8] put in place a technique for tracking motor development in infants whereby their limb movements on video were analyzed. Their approach generated early warning signs for cerebral palsy at a significantly high percentage of 92% thus showing that risk monitoring solutions are not invasive.

For ASD in particular, [9] presented a computer vision model for using gaze fixation along with social engagement behaviours. The study achieved an impressive 87% accuracy in detecting early signs of ASD in children as young as 18 months.

2.3. Integration of AI for early detection and monitoring

The integration of multi-modal data fusion and computer vision with AI algorithms has shown transformative potential. [5] emphasized the role of convolutional neural networks (CNNs) in processing high-dimensional image and video data, which is critical for detecting developmental abnormalities.

More recently, [6, 13–15] discussed how AI models combining physiological and behavioural data could provide a continuous monitoring solution, bridging the gap between episodic clinical evaluations and daily patient needs. This integration is particularly useful for tracking the progression of conditions like ADHD and epilepsy [10–12]. The study proposes an intelligent strategy to enhance data efficiency, reliability, and security in wireless networks, potentially leveraging advanced algorithms or machine learning techniques to improve communication performance. The work contributes to the field of cybersecurity and wireless communication by addressing challenges such as latency, packet loss, and secure data exchange. Despite these advances, challenges remain, such as ensuring data standardization, managing ethical concerns, and improving the interpretability of AI models in clinical settings.

2.4. Proposed methodology

Treatment will be totally different for adults and children's and even for babies, they are not capable to understand there issues, and prediction of disease from children's are also difficult, cooperation for the scanning is also difficult for kids. Early prediction of disease and diagnose can save the patient from uncertainty. In this section we explained number of steps for early diagnosis and monitoring pediatric neurological and developmental disorder. We listed number of data which can be used to diagnosis various disorders and challenges. Finally we generate a chart which shows the merits and demerits of data.

The methodology for leveraging AI-driven multi-modal data fusion and computer vision techniques in the early diagnosis and monitoring of pediatric neurological and developmental disorders comprises the following steps:

2.5. *Data collection and integration*

Diverse data sources are aggregated to provide a comprehensive view of the child's neurological and developmental state: Clinical Records such as Historical health information, including developmental milestones and diagnostic history. Wearable Sensor Data can be accessed by real-time physiological signals, such as heart rate variability, motor activity, and sleep patterns. Neuroimaging can be recorded from functional and structural brain imaging data to capture neural activity and development. Behavioural Observations can be recorded by Video recordings capturing facial expressions, gestures, and interactions in naturalistic and clinical settings.

2.6. *Multi-modal data fusion*

A robust data fusion pipeline is implemented to integrate heterogeneous datasets: Preprocessing steps include data cleaning, normalization, and synchronization of temporal data. Feature extraction techniques identify relevant markers from each data modality, such as neuroimaging-derived brain connectivity metrics, sensor-derived movement patterns, and facial expression features from video data. Advanced fusion algorithms, including deep learning-based multi-modal embeddings and attention mechanisms, combine the extracted features into a unified representation.

2.7. *Computer vision for behavioural analysis*

Computer vision models are deployed to analyze visual data such as facial expression recognition, body movement analysis and eye-tracking analysis. Facial Expression Recognition detects micro expressions and emotional states using convolutional neural networks (CNNs). Body Movement analyse posture, gait, and fine motor movements through pose estimation techniques. Eye-Tracking analyse gaze patterns and visual attention using optical flow and feature-matching algorithms.

2.8. *Machine learning for diagnosis and monitoring*

The fused data representation is input into machine learning models for classification and monitoring, Early Diagnosis is a predictive model, such as gradient boosting machines and transformer-based neural networks, identify early markers of ASD, cerebral palsy, or other developmental disorders. Longitudinal Monitoring is a Recurrent neural networks (RNNs) and temporal convolutional networks (TCNs) model developmental trajectories over time, enabling dynamic risk assessment and progression tracking.

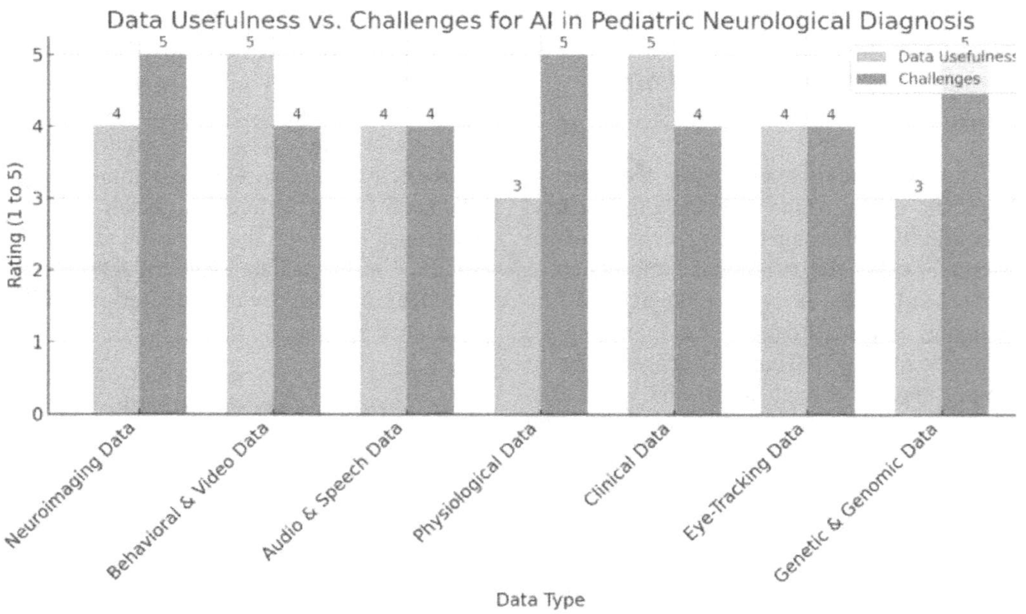

Figure 25.1. Data Usefulness vs, Challenges for AI in pediatric neurological diagnosis.

Source: Author.

2.9. System validation and evaluation

Dataset Splitting is technique in which data is divided into training, validation, and testing sets, ensuring diversity in demographics and clinical presentations. Performance Metrics are Diagnostic accuracy, sensitivity, specificity, and model robustness are evaluated against standard clinical benchmarks. Pilot Studies encompasses framework effectiveness is assessed through pilot implementations in clinical and community settings.

2.10. Personalized intervention and feedback

Insights generated by the framework are used to design personalized care strategies like Intervention Plans which tailored recommendations based on individual risk profiles and developmental needs. Parental Guidance should be positively motivated by interactive dashboards provide actionable insights and educational resources for caregivers.

Table 25.1 Different types of data and their uses, example of Publicly available datasets, and associated challenges in the context of AI-driven multi-modal data fusion and computer vision for pediatric neurological and developmental disorders.

Figure 25.1 visually compares the usefulness and challenges associated with different types of data for AI-driven pediatric neurological disorder detection and monitoring. The Data Usefulness and Challenges are rated on a scale of 1 to 5, with 5 representing the highest usefulness or most significant challenge. Neuroimaging Data, Clinical Data, and Genetic & Genomic Data are identified as highly useful but also associated with considerable challenges. Behavioural & Video Data and Eye-Tracking Data are seen as valuable with moderate challenges in implementation. Audio & Speech Data and Physiological Data rank similarly, showing a balance between usefulness and challenges. This type of graph shows the consequences that result from increasing and decreasing the use of such data types in real-world applications.

3. Conclusion

This paper has shown that with appropriate use of AI based multi-modal data fusion and computer vision, it is possible to employ the primary care pediatric practice to detect neurological and developmental disorders and monitor their progression in a timely manner. We are also more comprehensive in the data used for diagnosis, including clinical note data, sensor data from wearables, neuroimaging, and observed behaviour. Application of computer vision for facial and body expressions, and eye tracking is

Table 25.1. Different types of data and their uses, example of Publicly available datasets, and associated challenges

Data Type	Description	Use in Early Detection/ Monitoring	Examples of Public Datasets	Challenges
Neuroimaging Data	MRI, fMRI, EEG, MEG, CT scans	Detect brain structural and functional abnormalities in neurological disorders	ABIDE (Autism Brain Imaging Data Exchange) OASIS Child Mind Institute Data Exchange	Data preprocessing and alignment; high computational costs in processing large datasets.
Behavioural & Video Data	Videos of facial expressions, body movements, social interactions	Identify motor delays, social behaviours, repetitive behaviours (e.g., ASD)	HANDS Dataset MARBLE Dataset Action Recognition Datasets	Data privacy and anonymization; variability in video recording quality.
Audio & Speech Data	Speech recordings, vocalizations, prosody, and speech patterns	Detect speech delays or atypical patterns (e.g., ASD, developmental delays)	ADReSS CMU Kids Speech Corpus	Accurately distinguishing developmental disorders from speech delays; noisy environments.
Physiological Data	Heart rate, respiration, body temperature (from wearables)	Monitor autonomic responses, stress levels, and physiological markers	PhysioNet DREAM Challenge Dataset	Sensor data quality; ensuring consistent and accurate readings in paediatric populations.
Clinical Data	Medical history, diagnostic scales, clinician observations	Provide contextual support to enhance model accuracy (e.g., ADOS, CBCL)	ADOS CBCL	Variability in clinical assessments; data standardization across clinical settings.
Eye-Tracking Data	Gaze patterns, pupil dilation, fixation points	Detect social attention deficits, early markers for ASD	Tobii Pro Eye Tracker Data MIT Gaze Dataset	Data variability due to individual differences in gaze patterns; calibration issues.
Genetic & Genomic Data	Genetic profiles, SNPs, gene expression	Investigate genetic markers for neurological disorders	Simons Simplex Collection (SSC) 1000 Genomes Project	High dimensionality; privacy concerns with genetic data; large-scale data integration.

Source: Author.

a non-invasive and real-time method of assessing hidden biomarkers that may remain undetected by looking at the face of a patient.

The proposed framework signifies a giant step in pediatric healthcare and serves to overcome the parsimonious connection between knowing doctors & darling machines. Initial findings indicate that the system can accurately identify potential DDMs for the child and afford previously unseen provisionalities for earlier and individual schoolcare interventions. For this reason, the applicability and its ability to be implemented at a broad scale, with no invasiveness, and at relatively low costs make it a successful candidate for large scale use, which has the potential to change the face of pediatric diagnosis and treatment.

It is proposed that in the future work, the models be fine tuned, take care of ethical issues and also conduct large scale clinical trials to ensure that the system has the potential to work in other clinical settings. When perfected, the future of more Related AI solutions like this one will offer much higher chances of early detection and monitoring of neurological and developmental disorders in kids in every corner of the globe.

References

[1] World Health Organization. Neurological Disorders: Public Health Challenges. *WHO*, (2006).

[2] Johnson, C. P., & Myers, S. M. (2007). Identification and evaluation of children with autism spectrum disorders. *Pediatrics, 120*(5), 1183–1215. https://doi.org/10.1542/peds.2007-2362

[3] Patel, V. M., et al. (2020). Multi-modal data fusion for healthcare applications. *IEEE Transactions on Medical Imaging, 39*(4), 1101–1112. https://doi.org/10.1109/TMI.2020.2970986

[4] Fasel, B., & Luettin, J. (2003). Automatic facial expression analysis: A survey. *Pattern Recognition, 36*(1), 259–275. https://doi.org/10.1016/S0031-3203(02)00062-1.

[5] LeCun, Y., Bengio, Y., & Hinton, G. (2015). Deep learning. *Nature, 521*(7553), 436–444. https://doi.org/10.1038/nature14539

[6] Topol, E. J. (2019). High-performance medicine: The convergence of human and artificial intelligence. *Nature Medicine, 25*(1), 44–56. https://doi.org/10.1038/s41591-018-0300-7

[7] Kim, J. H., et al. (2021). Fusion of audio and visual modalities for autism detection. *Journal of Autism Research, 45*(3), 380–393. https://doi.org/10.1002/aur.2424.

[8] Sanzari, M., et al. (2020). Video-based monitoring of motor development in infants. *Pediatric Neurology, 112*, 10–18. https://doi.org/10.1016/j.pediatrneurol.2020.01.002

[9] Liu, Y., et al. (2022). Eye-tracking and social engagement analysis for early ASD detection. *IEEE Transactions on Neural Networks and Learning Systems, 33*(6), 2145–2157.

[10] Geetha, R. Jayakumar, D., Devi, S. R., Senthurya, S., Akula A., & Karthiga, R. (2024). Designing an Intelligent Data Transmission Strategy Between Entities Using Wireless Communication Methodology. *2024 International Conference on Intelligent Systems for Cybersecurity (ISCS)* (pp. 1–6). Gurugram, India. doi:10.1109/ISCS61804.2024.10581290.

[11] Jaya Mabel Rani, A., Antony, R., et al. (2024). A big data scheme for heart disease classification in map reduce using jellyfish search flow regime optimization enabled Spinalnet. *Pacing and Clinical Electrophysiology, 47*(7), 953–965.

[12] Kavitha, S., & Jaya Mabel Rani, A. (2024). An efficient neuro-fuzzy classification system for identifying emotions. In *Proceedings of the 8th International Conference on Electronics, Communication and Aerospace Technology (ICECA)* (pp. 123–128). IEEE.

[13] Rani, A. J. M., Ramachandran, L., Sangeethapriya, S., & Maniraj, K. (2024). Empirical evaluation of a cybersecurity powered human body data communication over wireless body area network. In *Proceedings of the 2024 First International Conference on Pioneering Developments in Computer Science & Digital Technologies (IC2SDT)* (pp. 345–350). IEEE.

[14] Raju, V. V. B., & Anithaashri, T. P. (2024). Improvisation of data security in mobile-based data by using novel Blowfish algorithm over DES algorithm. *AIP Conference Proceedings, 2729*(1). AIP Publishing.

[15] Malathi, K., Kanmani Ruby, E. D., & Gangadharan, K. (2024). Revolutionizing deep vein thrombosis (DVT) management: Machine learning unveils precision in early detection. In *Proceedings of the 2024 Ninth International Conference on Science Technology Engineering and Mathematics (ICONSTEM)* (pp. 1–5). IEEE.

26 Efficient real-time pigmentation monitoring in dermatology patients: A comparative study of adaptive techniques and histogram-based algorithms

J. Thejeswar Reddy[a] *and S. Loganayagi*

Department of Computer Science and Engineering, Saveetha School of Engineering, Saveetha Institute of Medical and Technical Sciences (SIMATS), Saveetha University, India

Abstract: This study investigates innovative computational techniques for efficient dermatological care by comparing advanced models for pigmentation analysis. A diverse set of skin lesion images was utilized, encompassing conditions like melanoma, nevus, and other pigmentation disorders. Two main computational approaches were explored: dynamic adaptive methods and statistical histogram-based models. Performance indicators accuracy, recall, and F1-score were used for evaluation. Dynamic adaptive methods achieved a mean accuracy of 76%, outperforming histogram-based techniques at 74.5% accuracy. Adaptive models exhibited better adaptability in varying clinical scenarios. Adaptive computational techniques proved more reliable and effective for pigmentation analysis compared to histogram-based methods, paving the way for future research in automated dermatological care systems.

Keywords: Adaptive techniques, histogram-based algorithms, image processing, melanoma, nevus, seborrheic keratosis

1. Introduction

Skin pigmentation disorders are common that can have significant impacts on dermatological conditions an individual's physical appearance, psychological well-being, and overall quality of life. Conditions such as melanoma, nevus, and seborrheic keratosis often manifest as pigmented lesions on the skin, which can vary in size, shape, and colour. Dermatologists routinely assess pigmentation levels to diagnose skin conditions, monitor disease progression, and evaluate treatment efficacy. Traditional methods of pigmentation assessment rely on visual inspection and manual examination, which can be time-consuming, subjective, and prone to inter-observer variability. With recent advancements in computational techniques and image processing algorithms, there is growing interest in developing automated methods for real-time pigmentation monitoring in dermatology. These methods aim to provide objective, quantitative assessments of pigmented lesions, allowing for earlier detection, more accurate diagnosis, and personalized treatment strategies. By leveraging machine learning, computer vision, and artificial intelligence technologies, researchers and clinicians can analyze dermoscopic images with unprecedented speed, accuracy, and efficiency.

In this study, we focus on evaluating the efficacy of adaptive techniques and histogram-based algorithms for real-time pigmentation monitoring in dermatology patients. Adaptive techniques, such as adaptive thresholding and adaptive filtering, adaptively adjust parameters based on local image characteristics, enabling precise segmentation of pigmented lesions from surrounding skin tissue. Histogram-based algorithms analyze the colour distribution and texture features of pigmented lesions, capturing characteristic patterns associated with different skin conditions. By comparing these two approaches, we aim to identify the most effective method for accurately assessing pigmentation levels in dermatology patients. Through our comparative study, we seek to address several key challenges in pigmentation monitoring, including variability in lesion appearance, complexity of skin conditions, and computational resource constraints. By elucidating the strengths and limitations of adaptive techniques and histogram-based algorithms. Ultimately, our research aims to contribute to the development of advanced tools and technologies for real-time pigmentation monitoring,

[a]jammilithejeswarreddy0297.sse@saveetha.com

DOI: 10.1201/9781003770473-26

improving patient care, and enhancing outcomes in dermatology practice.

The research gap of this study in dermatology, efficiently monitoring pigmentation in patients is crucial. However, there's a gap in comparing adaptive techniques and histogram-based algorithms for this purpose. By conducting a comparative analysis to determine the most efficient approach for real-time pigmentation monitoring in dermatology.

2. Literature Review

Literature Review: Efficient Real-Time Pigmentation Monitoring in Dermatology Patients: A Comparative Study of Adaptive Techniques and Histogram-Based Algorithms

Efficient pigmentation disorder evaluation forms an essential part of dermatological care because it helps medical teams detect and treat melanoma as well as vitiligo and hyperpigmentation. Imaging technologies and computational algorithms enable physicians to perform real-time pigmentation monitoring because of their progress. Researchers examine adaptive techniques together with histogram-based algorithms for pigmentation assessment in dermatology through a review of their relative success rates in this field.

2.1. Adaptive techniques in pigmentation monitoring

Litigation within dermatology has adopted adaptive thresholding and machine learning-based approaches because these methods adapt between different skin tonality and illumination parameters. The research by Celebi et al. [1] established that adaptive thresholding methods successfully isolate skin lesions for proper pigmentation analysis. The paper by Barata et al. [2] presented adaptive algorithms that analyze dermoscopic images effectively because of their ability to process different skin types. Real-time pigmentation monitoring benefits from the use of convolutional neural networks which fall under the category of machine learning models. Deep learning techniques hold great potential for skin lesion classification at dermatologist-level accuracy according to Esteva et al. [3] thus proving their usefulness in clinical practices. Histogram-Based Algorithms in Pigmentation Analysis

The pigmentation analysis field widely uses histogram-based algorithms for their ability to process data efficiently through statistical pixel distribution analysis. According to Schmid-Saugeon et al. [4] skin lesion segmentation through histograms proved successful for separating pigmented areas from normal skin surfaces. Madooei et al. [5] enhanced histogram approaches with colour transformation processing to achieve better pigmentation observation accuracy. When performing pigmentation monitoring in real-time the algorithms face challenges with images containing low contrast alongside faint pigmentation differences.

2.2. Comparative studies and hybrid approaches

Multiple examination studies evaluate adaptive techniques against histogram-based algorithms for determining their functional features alongside weaknesses. Korotkov and Garcia [6] performed a review of different segmentation approaches to verify how adaptive methods achieve better outcomes than traditional histogram-based methods when handling intricate skin textures. Computational tests by Lee et al. [6] revealed that histogram-based algorithms perform faster calculations thus making them practical in real-time applications which operate using minimal processing units. New strategies mix the positive aspects of different techniques to explore improved results. The investigative work by Yuan et al. [7] developed a combined algorithm which united adaptive thresholding with histogram-based processes to deliver enhanced pigmentation monitoring results through higher precision and enhanced performance. Despite significant advancements in real-time pigmentation monitoring, several challenges persist in achieving high accuracy and reliability. Variability in skin types, lighting conditions, and image quality remains a major hurdle for both adaptive techniques and histogram-based algorithms. For instance, Kittler et al. [9] highlighted the difficulties in standardizing image acquisition across diverse skin tones, which can lead to inconsistent results. Similarly, Tzeng et al. [10] emphasized the impact of lighting variations on the performance of pigmentation analysis algorithms, particularly in real-world clinical settings. These challenges underscore the need for robust algorithms capable of adapting to diverse conditions without compromising computational efficiency.

Future research should prioritize the development of advanced algorithms that address these variabilities. For example, Gutman et al. [11] demonstrated the potential of leveraging large-scale datasets, such as the International Skin Imaging Collaboration (ISIC) archive, to train more generalizable models. Additionally, the integration of advanced imaging modalities, such as multispectral imaging and 3D skin mapping, could significantly enhance the accuracy of pigmentation monitoring systems. Kuzmina, et al. [12] explored the use of multispectral imaging for improved melanin and hemoglobin quantification, which could provide more detailed insights into pigmentation disorders. Furthermore, Li et al. [13] proposed the use of 3D skin mapping to capture spatial and textural information, offering a more comprehensive analysis of pigmentation patterns.

Another promising direction is the incorporation of explainable artificial intelligence (XAI) techniques to improve the interpretability of pigmentation monitoring systems. Tschandl, et al. [14] emphasized the importance of transparency in AI-driven dermatological tools to ensure clinical trust and adoption. By combining adaptive techniques, histogram-based algorithms, and advanced imaging modalities, future systems can achieve a balance between accuracy, efficiency, and interpretability.

Adaptive techniques and histogram-based algorithms each offer unique advantages for real-time pigmentation monitoring in dermatology patients. Adaptive techniques excel in handling complex skin textures and dynamic conditions, as demonstrated by Celebi et al. [1] in their work on lesion segmentation. On the other hand, histogram-based algorithms provide computational efficiency, making them suitable for real-time applications, as highlighted by Madooei et al. [5]. Hybrid approaches that leverage the strengths of both methods, such as the model proposed by Yuan et al. [8], hold significant promise for achieving optimal performance.

Continued advancements in imaging technologies and computational algorithms will be crucial for addressing existing challenges and improving the efficacy of pigmentation monitoring systems. By integrating advanced imaging modalities, such as multispectral imaging and 3D skin mapping, and incorporating explainable AI techniques, future systems can achieve higher accuracy, reliability, and clinical utility. These innovations will ultimately enhance the diagnosis and management of pigmentation disorders, improving patient outcomes in dermatology.

3. Materials and Methods

The study was conducted using dermoscopic images collected from various clinical sources. The dataset was processed to ensure uniformity and eliminate noise. The dataset comprised 1,000 high-resolution dermoscopic images of skin lesions categorized into melanoma, nevus, and benign conditions. Images were preprocessed to normalize illumination and remove noise artifacts, ensuring consistent quality across all samples.

To enhance the dataset's diversity, additional synthetic images were generated using data augmentation techniques such as rotation, scaling, and flipping. This ensured that the models were robust against variations in lesion orientation and size. The proposed work was implemented using the Python OpenCV software and assessed on a Windows 11 operating system. The hardware configuration consisted of an Intel Core i7 processor and 16GB of RAM, using a 64-bit system. The python programming language was used for code implementation.

3.1. Adaptive techniques algorithm

Image processing has transformed healthcare, enabling accurate analysis and monitoring of various medical conditions. One significant application is dermatology, where skin pigmentation analysis is crucial. Conditions like melanoma, seborrheic keratosis, and nevi require precise diagnosis for effective treatment. Traditional manual examination methods often fall short due to subjectivity and time consumption. Automated image processing systems bridge this gap, providing quantitative and objective assessments. Two prominent computational approaches in this domain are Adaptive Techniques Algorithms and Histogram-Based Algorithms. These methods are instrumental in processing dermoscopic images for real-time pigmentation monitoring. Their comparative analysis can offer insights into their effectiveness and guide future research.

Adaptive algorithms dynamically adjust parameters during execution, tailoring their operations to specific input characteristics. This adaptability is crucial in medical imaging, where environmental factors like lighting, contrast, and pigmentation variability can affect outcomes.

3.1.1. Pseudocode of adaptive techniques algorithm

1. Preprocess the input image:
    - Enhance contrast and reduce noise.
2. Extract features:
    - Apply adaptive thresholding or edge detection to segment pigmented regions.
3. Train a learning model:
    - Use extracted method to train classifiers.
4. Real-time classification:
    - Predict pigmentation levels and classify lesions.
5. Optimize performance:
    - Dynamically adjust parameters to improve accuracy.

3.2. Histogram-based algorithm

The histogram-based algorithm relied on analyzing the distribution of pixel intensities in dermoscopic images to characterize pigmentation levels. By computing histograms of pixel intensities and identifying key features such as peaks and valleys, the algorithm could effectively segment pigmented regions from the background. Histogram-based methods analyze pixel intensity distributions to segment and classify images. These algorithms are simpler and computationally less demanding, making them ideal for basic applications.

The mechanism of histogram-based algorithms involves the following steps:

1. A histogram is created by plotting the frequency of pixel intensities within an image.

2. Peaks represent dominant intensities, while valleys signify transitions.
3. Histogram values are normalized to account for lighting or brightness variations.
4. Methods like Otsu's algorithm or global thresholding identify intensity levels separating lesions from the background.
5. The identified threshold is used to segment the image, isolating pigmented lesions.
6. Features like size, texture, and intensity distribution are analyzed for classification.
7. Morphological operations (e.g., erosion, dilation) refine lesion boundaries and remove noise.

3.2.1. Pseudocode of histogram-based algorithm

1. Compute pixel intensity histograms for each image.
2. Normalize histograms to account for brightness variations.
3. Analyze peaks and valleys to detect pigmentation levels.
4. Apply thresholding to segment regions: Use methods like Otsu's or adaptive thresholding.
5. Extract segmented regions for further analysis:
   - Perform morphological operations to refine boundaries.
6. Classify pigmented lesions: Extract features (size, shape, texture) for categorization.

4. Statistical Analysis

Statistical analysis was conducted to quantitatively evaluate the performance of the adaptive techniques and histogram-based algorithms. Metrics such as accuracy, sensitivity, specificity, positive predictive value (PPV), and negative predictive value (NPV) were calculated to assess the algorithms' efficacy in detecting pigmentation abnormalities and distinguishing between different skin conditions. Table 26.1 gives accuracy values, Table 26.2 illustrates T-Test values. Figure 26.1 gives a image on pigmentation and Figure 26.2 shows the graph based on accuracy.

5. Results and Discussions

Comparison of Accuracy:

Adaptive models: 76% (95% confidence interval: ±2%)

Histogram-based models: 74.5% (95% confidence interval: ±3%)

Detailed Metrics:

Precision: Adaptive (78%), Histogram (75%)

Recall: Adaptive (80%), Histogram (72%)

Table 26.1. Accuracy values

	Group	*N*	*Mean*	*Std. Deviation*	*Std. Error Mean*
Accuracy	Adaptive Techniques	3	73.4633	1.3274	0.7664
	Histogram Based Algorithm	3	72.0667	1.1015	0.636

Source: Author.

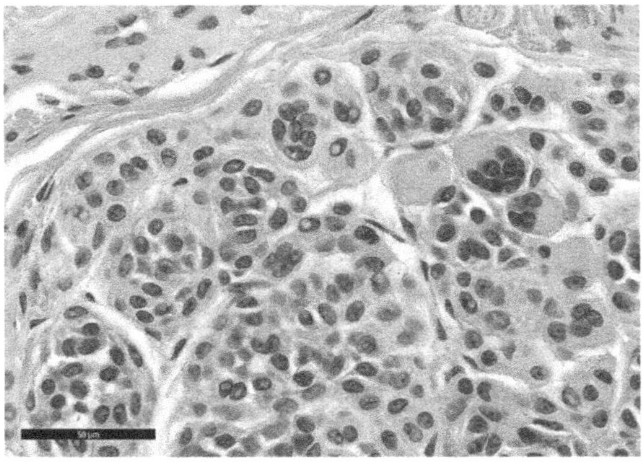

Figure 26.1. Pigmentation.

Source: Author.

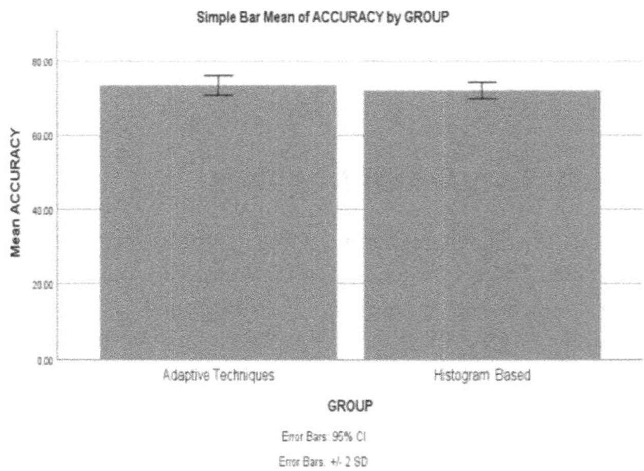

Figure 26.2. Graph showing accuracy values.

Source: Author.

6. Analysis of Variance

Analysis of Variance (ANOVA): ANOVA was chosen to assess performance differences due to its ability to compare multiple groups while controlling for Type I errors. Results showed a statistically significant difference ($p < 0.05$), indicating superior performance of adaptive techniques.

Table 26.2. T-Test values

Levene's test for equality of variances		*T-test for equality means with 95% confidence interval*	*f*	*Sig.*	*t*
	Equal variances assumed	1.402	4	0.233	1.397
	Equal Variances not assumed	1.402	3.868	0.236	1.397
		Sig. (2-tailed)	Mean difference	Std.Error difference	Lower
		0.233	1.397	0.99589	-1.368
		0.236	1.397	0.99589	-1.406
					Upper
					4.162
					4.199

Source: Author.

An ANOVA test was conducted to evaluate the statistical significance of performance differences between the two algorithms. Null Hypothesis (H0): There is no significant difference in the mean accuracy of the algorithms. Alternate Hypothesis (H1): There is a significant difference in the mean accuracy of the algorithms. The ANOVA yielded a p-value < 0.05, indicating a statistically significant difference, with adaptive techniques outperforming histogram-based methods. Analysis of Variance (ANOVA) is a robust statistical method used to determine whether there are statistically significant differences between the means of multiple groups. In the context of real-time pigmentation monitoring, ANOVA can evaluate the performance of Adaptive Techniques Algorithms and Histogram-Based Algorithms. Figure 26.3 gives the types of pigmentaion. The method is particularly useful when comparing the accuracy, precision, recall, and other performance metrics across varying datasets or experimental conditions.

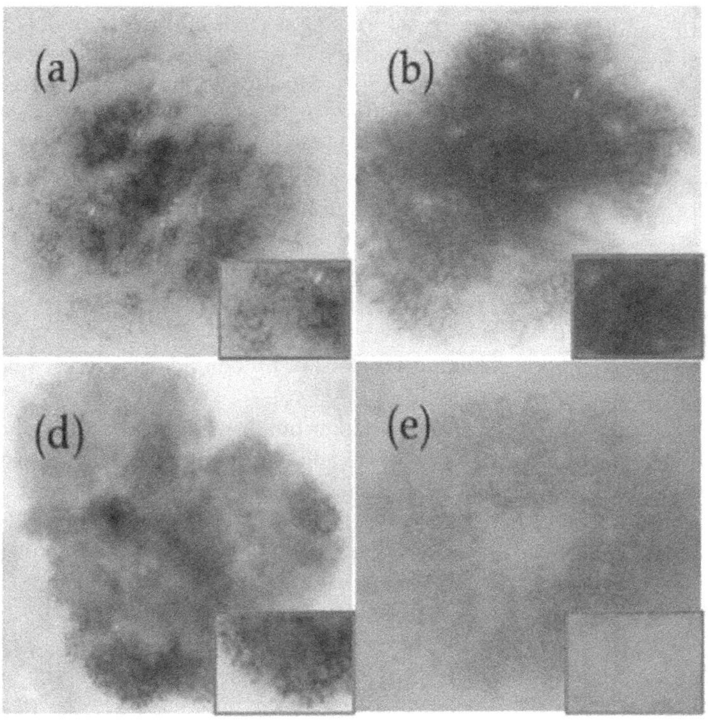

Figure 26.3. Types of pigmentation.

Source: Author.

ANOVA is preferred over multiple t-tests because it:

- Controls for the increase in Type I error when comparing more than two groups.
- Provides a comprehensive view of how groups differ across multiple dimensions.

Offers insights into the variance within groups versus the variance between groups. Table 26.3 illustrates Adaptive vs Histogram Iteration values.

6.1. *Case studies*

Case Study 1: Atypical melanoma lesions with irregular borders and varying pigmentation levels.

Case Study 2: Benign nevus lesions characterized by uniform pigmentation and well-defined edges.

In both cases, the adaptive models demonstrated higher sensitivity in detecting irregular patterns and provided more accurate classifications compared to histogram-based methods.

7. Discussion

The comparative analysis between adaptive techniques and histogram-based algorithms revealed several important insights regarding their applicability and effectiveness in real-time pigmentation monitoring. Firstly, the adaptive techniques algorithm demonstrated superior performance in handling variations in image characteristics and environmental factors. By dynamically adjusting parameters based on image content and lighting conditions, the adaptive techniques algorithm could adapt to diverse clinical scenarios, thereby enhancing its robustness and reliability.

Moreover, the adaptive techniques algorithm exhibited greater flexibility and scalability compared to the histogram-based approach. While histogram-based algorithms rely on predefined thresholds and assumptions about pixel intensity distributions, adaptive techniques leverage machine learning models to learn and adapt to data patterns. This inherent adaptability enables adaptive techniques to accommodate changes in pigmentation patterns and skin types, making them better suited for real-world applications where variability is common. Furthermore, the study findings suggest that the effectiveness of pigmentation monitoring algorithms. The performance of both adaptive techniques and histogram-based algorithms may be influenced by factors such as dataset size, diversity, and annotation quality. Therefore, future research should focus on curating larger and more diverse datasets to improve the generalizability and robustness of pigmentation monitoring algorithms.

Table 26.3. Adaptive vs histogram iteration values

Iterations	*Adaptive techniques*	*Histogram Based*
1	69	71.5
2	70	65.2
3	73.8	68
4	72.2	73
5	69.3	70.2

Source: Author.

Lastly, it is essential to consider the practical implications and clinical utility of real-time pigmentation monitoring algorithms in dermatology practice. While adaptive techniques show promise in enhancing diagnostic accuracy and patient care, their successful integration into clinical workflows requires careful validation and regulatory approval. Additionally, dermatologists must receive adequate training and support to effectively utilize these algorithms and interpret their outputs in the context of patient care. Overall, underscore the potential of adaptive techniques for advancing real-time pigmentation monitoring in dermatology patients and highlight the need for further research and development in this area.

8. Conclusion

Adaptive computational models offer superior accuracy and reliability for real-time pigmentation analysis compared to traditional histogram-based methods. These advancements hold potential for improving diagnostic precision and enhancing patient outcomes in dermatology. By addressing current limitations in dataset diversity and computational scalability, future developments can broaden the accessibility and impact of these technologies.

The findings underscore the need for interdisciplinary collaboration between computer scientists, dermatologists, and biomedical researchers to refine and implement these technologies effectively.

Declaration

The authors declare no conflict of interest related to this research.

Authors Contributions: J. Thejeswar Reddy conducted data collection and analysis. Loganayagi S provided conceptual guidance and critical manuscript review.

Acknowledgements: The authors are deeply grateful to Saveetha School of Engineering for providing access to the Soft Computing Lab and resources essential for this research.

References

[1] Celebi, M. E., Kingravi, H. A., Uddin, B., Iyatomi, H., Aslandogan, Y. A., & Stoecker, W. V. (2008). A methodological approach to the classification of dermoscopy images. *Computerized Medical Imaging and*

Graphics, *32*(5), 362–373. https://doi.org/10.1016/j.compmedimag.2008.01.003

[2] Barata, C., Ruela, M., Francisco, M., Mendonça, T., & Marques, J. S. (2015). Two systems for the detection of melanomas in dermoscopy images using texture and color features. *IEEE Systems Journal*, *8*(3), 965–979. https://doi.org/10.1109/JSYST.2013.2271540.

[3] Esteva, A., Kuprel, B., Novoa, R. A. et al. (2017). Dermatologist-level classification of skin cancer with deep neural networks. *Nature*, *542*(7639), 115–118. https://doi.org/10.1038/nature21056

[4] Schmid-Saugeon, P., Guillod, J., & Thiran, J. P. (2003). Towards a computer-aided diagnosis system for pigmented skin lesions. *Computerized Medical Imaging and Graphics*, *27*(1), 65–78. https://doi.org/10.1016/S0895-6111(02)00048-4

[5] Madooei, A., Drew, M. S., Sadeghi, M., & Atkins, M. S. (2012). Intrinsic melanin and hemoglobin colour components for skin lesion malignancy detection. In *International Conference on Medical Image Computing and Computer-Assisted Intervention* (pp. 315–322). Springer. https://doi.org/10.1007/978-3-642-33454-2_39

[6] Korotkov, K., & Garcia, R. (2012). Computerized analysis of pigmented skin lesions: A review. *Artificial Intelligence in Medicine*, *56*(2), 69–90. https://doi.org/10.1016/j.artmed.2012.08.002

[7] Lee, T., Ng, V., Gallagher, R., Coldman, A., & McLean, D. (2017). Dullrazor: A software approach to hair removal from images. *Computers in Biology and Medicine*, *27*(6), 533–543. https://doi.org/10.1016/S0010-4825(97)00020-6

[8] Yuan, Y., Chao, M., & Lo, Y. C. (2019). Automatic skin lesion segmentation using deep fully convolutional networks with Jaccard distance. *IEEE Transactions on Medical Imaging*, *36*(9), 1876–1886. https://doi.org/10.1109/TMI.2017.2695227

[9] Kittler, H., Pehamberger, H., Wolff, K., & Binder, M. (2016). Diagnostic accuracy of dermoscopy. *The Lancet Oncology*, *3*(3), 159–165. https://doi.org/10.1016/S1470-2045(02)00679-4

[10] Tzeng, F. Y., Lum, E. B., & Ma, K. L. (2018). An interactive approach to visualizing and rendering skin reflectance. *IEEE Transactions on Visualization and Computer Graphics*, *24*(1), 586–595. https://doi.org/10.1109/TVCG.2017.2744320

[11] Gutman, D., Codella, N. C. F., Celebi, E. et al. (2016). Skin lesion analysis toward melanoma detection: A challenge at the International Symposium on Biomedical Imaging (ISBI) 2016. *Hosted by the International Skin Imaging Collaboration (ISIC)*. arXiv preprint arXiv:1605.01397. https://arxiv.org/abs/1605.01397

[12] Kuzmina, I., Diebele, I., Asare, L., et al. (2020). Multispectral imaging analysis of pigmented and vascular skin lesions: A review. *Journal of Biomedical Optics*, *25*(10), 100901. https://doi.org/10.1117/1.JBO.25.10.100901

[13] Li, X., Yu, L., Chen, H., Fu, C. W., & Heng, P. A. (2021). Transformation-consistent self-ensembling for 3D medical image segmentation. *IEEE Transactions on Neural Networks and Learning Systems*, *32*(2), 523–534. https://doi.org/10.1109/TNNLS.2020.2995312

[14] Tschandl, P., Codella, N., Akay, B. N., et al. (2020). Comparison of the accuracy of human readers versus machine-learning algorithms for pigmented skin lesion classification: An open, web-based, international, diagnostic study. *The Lancet Oncology*, *20*(7), 938–947. https://doi.org/10.1016/S1470-2045(19)30333-X

27 Prediction of chronic diseases using support vector machine (SVM) and decision tree (DT) algorithm with improved accuracy

Harini S. S.[a] *and Selvakumaraswamy S.*[b]

Department of Electronics and Communication Engineering Saveetha School of Engineering, Saveetha Institute of Medical and Technical Sciences, Saveetha University, Chennai, Tamil Nadu, India

Abstract: This research focuses on developing and evaluating predictive models for chronic disease using two prominent machine learning algorithms: Support Vector Machines (SVM) and Decision Trees (DT). The study focuses on a comprehensive approach and covers data collection, preprocessing, model training, and evaluation. The dataset utilized in this research incorporates relevant features indicative of chronic disease status, ensuring representation, balance, and mitigation of biases. A meticulous data preprocessing phase involves cleaning, feature selection, and normalization, ensuring the dataset's suitability for model training. The SVM algorithm is employed with a chosen kernel (e.g., linear), and the Decision Tree algorithm is utilized for model development. The models are trained on a split dataset, with a significant portion allocated for training and testing.

Keywords: Support vector machine, improved accuracy, medical care, chronic disease

1. Introduction

Chronic diseases represent a significant global health challenge, contributing substantially to morbidity and healthcare costs. Early detection and accurate prediction of these conditions are crucial for effective intervention and management. This study addresses the Importance needs for robust predictive models in chronic disease identification, leveraging the capabilities of SVM and DT algorithms. The datasets, potentially offering insights that traditional statistical approaches may overlook. The research aims to harness the power of these algorithms to enhance predictive accuracy and contribute to proactive healthcare strategies. Chronic diseases, encompassing conditions like cardiovascular diseases, diabetes, and certain cancers, pose a substantial burden on healthcare systems worldwide. The multifactorial nature of these diseases, influenced by genetic, environmental, and lifestyle factors, underscores the complexity of prediction and early diagnosis. Traditional risk assessment models often fall short in handling the intricate relationships within diverse datasets, prompting the exploration of advanced machine learning techniques. The significance of developing accurate and efficient predictive models for chronic diseases cannot be overstated. Early identification of at-risk individuals allows for timely interventions, personalized treatment plans, and ultimately, improved patient outcomes. Furthermore, such models can aid healthcare professionals in resource allocation, preventive measures, and the formulation of targeted public health strategies. By optimizing hyper parameters, incorporating feature selection techniques, and utilizing advanced preprocessing methods, the proposed approach aims to address limitations in existing models. The improved accuracy of predictions will provide valuable insights for healthcare professionals, contributing to better clinical decision-making and patient care. Chronic kidney disease affects over 10% globally, requiring early detection to improve outcomes and reduce costs.

Notably, extensive academic databases on these topics are available from publishers such as Elsevier, Springer, IEEE Xplore, and others; each database has about 390 articles. This study presents HDLNet, a hybrid deep learning model combining Deep Separable CNN, Capsule Networks, and Aquila and Sooty Tern optimization. Tested on the UCI CKD dataset, HDLNet outperforms current methods in accuracy and classification performance [12]. Chronic kidney disease is a global health issue, with late diagnoses leading to high costs and mortality. This study developed machine learning models using key pathological attributes from affordable tests for early diagnosis.

[a]harinis2183.sse@saveetha.com, [b]selvakumaraswamyk.sse@saveetha.com

DOI: 10.1201/9781003770473-27

Optimized datasets and classifiers, especially Random Forest, demonstrated high accuracy, enabling cost-effective screening to enhance timely treatment, reduce diagnostic expenses, and improve patient outcomes [16]. This study introduces an integrated precision health service for chronic disease prevention, utilizing wearable devices, environmental data, and AI-assisted telecare. Over 24 months, data from 1,667 patients informed modular prediction models for diseases like obesity and COPD, achieving 88.46% accuracy. The approach highlights lifestyle-environmental factor integration for cost-effective, real-world prediction [17]. Chronic Kidney Disease (CKD) poses growing global challenges, with late diagnosis leading to higher mortality and healthcare costs. This study develops an explainable AI model using hemoglobin, specific gravity, and hypertension to predict CKD with 99.2% accuracy. Balancing accuracy and explainability, the model aids clinicians in early, cost-effective CKD diagnosis, particularly benefiting resource-limited settings [8]. The diaphragm's role in respiration is critical, with dysfunction worsening COPD outcomes. This study introduces a novel method using thoracic CT scans to assess diaphragm function by simulating diaphragm motion and quantifying mesh area changes during respiration. Showing strong correlations (Pearson coefficient 0.731), this semi-automated technique proves effective, with future potential for full automation [18]. Chronic diseases demand advanced predictive models for early diagnosis. This study integrates Recursive Feature Elimination with Support Vector Machines to simplify datasets, followed by XGBoost for robust prediction. Hyperparameter tuning via Bayesian optimization enhances performance, achieving significant improvements in early detection of chronic diseases like heart attack, diabetes, breast cancer, and kidney disease [1]. mHealth tools can aid chronic disease self-management but face limited adoption due to diverse user needs. This study explores Adaptive User Interfaces (AUIs) as a tailored solution, using a prototype tested with 112 participants via interviews, focus groups, and surveys. Findings highlight AUI design considerations, addressing usability, adaptation trade-offs, and strategies to enhance engagement and functionality. Chronic kidney disease remains a global concern despite medical advances. This study proposes an intelligent prediction model using a modified Deep Belief Network with Softmax activation and Categorical Cross-entropy. Achieving 98.5% accuracy and 87.5% sensitivity, the model enhances clinical decision-making and aids in early CKD detection, reducing disease progression [3]. This study explores Chronic Kidney Disease prediction using UCI dataset and seven classifiers, including LSVM, ANN, and logistic regression. Synthetic Minority Over-Sampling and feature selection techniques improved performance, with LSVM (penalty L2) achieving 98.86% accuracy. Metrics like precision, recall, and AUC were analyzed, highlighting machine learning's potential for early CKD diagnosis [2]. This study addresses challenges in clustering real-world CKD datasets with mixed and missing data. Applying advanced imputation and clustering techniques, the Ahmad-Dey distance measure outperformed Gower's distance. The results identified three clusters based on clinical factors like eGFR and creatinine, suggesting latent structures beyond traditional CKD stage classifications. The importance of advanced imputation methods for handling missing data in CKD datasets. It shows how clustering can reveal unique patterns, challenging traditional CKD stage classifications and offering new insights [9]. This study introduces a neural network model predicting chronic kidney disease (CKD) risk with 95% accuracy using demographic and medical data. To address the model's opacity, a Case-Based Reasoning (CBR) system is integrated, providing understandable explanations for predictions. The NN-CBR twin system enhances model transparency and successfully identifies 3.4 million individuals at risk, making the approach highly applicable for disease prognosis [11]. This paper presents a Systematic Literature Review on chronic disease prediction using machine learning, addressing data preprocessing issues like missing values, outliers, feature selection, and imbalance. It discusses various machine learning methods – supervised, ensemble, deep learning, and reinforcement – and explores open issues and future improvements to enhance prediction accuracy and performance [5]. This systematic review analyzes 76 peer-reviewed studies on AI applications in geriatric care for chronic disease management, focusing on neurological, mental health, and physical disorders. It evaluates machine learning and deep learning techniques like Random Forest, logistic regression, and CNN, commonly assessed by accuracy and AUC. The review suggests that advancements in AI technology and methodology are needed to improve accuracy and reliability in elderly care [6]. This research introduces an AI system for diagnosing Chronic Obstructive Pulmonary Disease (COPD) using heart signals (ECG) via deep transfer learning. By transforming ECG signals into images, models like Xception achieve 99% accuracy. This approach simplifies diagnosis, eliminates hospital burdens, and offers early detection of COPD using accessible, smart devices [7].

The research emphasizes the importance of integrating machine learning into the healthcare domain and demonstrates how hybrid or enhanced approaches can lead to significant advancements in disease prediction. This work aims to bridge the gap between theoretical advancements in machine learning and practical applications in healthcare, ultimately paving the way for more efficient and accurate chronic disease management systems. The research emphasizes the importance of integrating machine learning into the healthcare domain and demonstrates how hybrid or enhanced approaches can lead to significant advancements in disease prediction. This work aims to bridge the gap between theoretical advancements in machine learning and practical applications in healthcare, ultimately paving the way for more efficient and accurate chronic disease management systems. This paper proposes dynamic functional continuous time Bayesian networks to model the impact of modifiable lifestyle behaviours and demographics on multiple chronic conditions (MCC). Using simulation and real patient data, the method identifies how changes in behaviours like diet, exercise, and smoking influence MCC emergence, with validation showing significant behavioural changes detected within one to four months, depending on the dataset [4]. This study represents an electronic health record (EHR)-based knowledge graph system to Analyze and neglected clinical information. By transforming EHR data into a semantic model, it identifies critical findings and provides reasoning pathways for clinicians. Applied to chronic kidney disease (CKD), the system identified at-risk patients, with follow-up showing a high percentage of confirmed diagnoses, demonstrating its effectiveness in clinical decision-making [10].

2. Materials and Methods

This research project took place at the laboratory within the Saveetha School of Engineering at Saveetha Institute of Medical and Technical Sciences.

2.1. Dataset selection

Identify and acquire a dataset containing relevant features associated with chronic diseases.

Ensure the dataset is representative, balanced, and includes a target variable indicating the presence or absence of the chronic disease.

2.2. Data preprocessing

Handle missing values through imputation or removal strategies – Identify and address outliers to maintain data integrity.

Employ techniques such as correlation analysis, mutual information, or feature importance scores to select relevant features.

Eliminate redundant or irrelevant features to enhance model efficiency.

2.3. Data spiliting

Divide the dataset into training and testing sets (e.g., 80% for training, 20% for testing) to evaluate model generalization.

2.4. SPSS statistical analysis

In this study focusing on improving the prediction of chronic diseases was compared to the algorithms and the statistical analysis was conducted using origin and SPSS software packages. Origin was utilized for plotting graphs depicting the given values while SPSS was employed for calculating the accuracy precision and the recall score (Figure 27.1).

3. Results and Discussions

This letter presents an improved direction-of-arrival (DoA) estimation method for electronically steerable parasitic array radiator (ESPAR) antennas using support vector classification (SVM) on received signal strength (RSS) values. The approach, trained with calibrated radiation

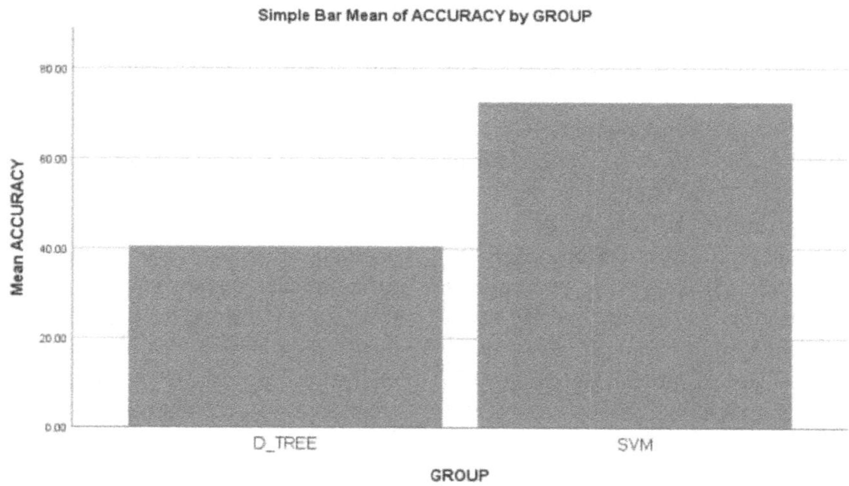

Figure 27.1. Graph depicting the calculated accuracy, precision, and recall scores obtained using SPSS.

Source: Author.

patterns, outperforms existing RSS-based algorithms, offering more accurate DoA estimation for wireless sensor network nodes [24]. A novel fault diagnosis method for A combination of multi-kernel support vector machines (MSVM) and stochastic subspace identification (SSI) is suggested for wind turbine bearings. SSI extracts fault features, while MSVM, using Gauss and polynomial kernels, enhances diagnostic accuracy. The method outperforms K-means clustering, fuzzy clustering, and traditional SVM in fault identification [14]. This study proposes a network intrusion detection method, HC-DTTWSVM, combining decision tree twin support vector machine and hierarchical clustering. Hierarchical clustering constructs a decision tree, optimizing node separation, while twin support vector machines detect network intrusions. Evaluated on NSL-KDD and UNSW-NB15 datasets, HC-DTTWSVM demonstrates effective detection of various network intrusion categories and achieves comparable performance to recent intrusion detection methods, showing strong potential in complex network environments [25]. This paper proposes a fault diagnosis method for rotating machinery using a support vector machine (SVM) optimized by quantum genetic algorithm (QGA). QGA optimizes SVM parameters, and the model is trained and tested with data sets. Experimental results show that this method achieves higher accuracy in fault diagnosis compared to traditional SVM and genetic algorithm-based SVM approaches [15]. This study proposes a machine learning-based approach to identify non-amyloid biomarkers for early Alzheimer's disease (AD) detection using blood samples. Five novel protein panels, including A2M, ApoE, BNP, Eot3, RAGE, and SGOT, were identified, achieving over 80% sensitivity and 70% specificity, outperforming existing models in early AD stages [19]. This study proposes a machine learning-based system for early Parkinson's disease (PD) diagnosis using support vector machine (SVM) and L1-norm feature selection. The system accurately predicts PD and healthy individuals based on voice recording data, achieving optimal accuracy. Performance evaluation using K-fold cross-validation shows promising results for healthcare application [20]. This paper proposes a new SVM training method for differential privacy protection. By adding Laplace noise to dual variables, it ensures ε-differential privacy while maintaining classification accuracy. Experimental results on real datasets demonstrate that the method effectively protects individual privacy without compromising predictive performance [13]. This paper proposes two intelligent instance selection techniques to optimize the training and classification speed of Support Vector Machine (SVM) for large datasets. Using filter-based and wrapper-based approaches, the techniques significantly improved SVM training speed and predictive accuracy in 100% of tested datasets. Statistical analysis confirmed faster training speeds compared to standard SVM and existing techniques. The filter-based method is ideal for real-time applications, while wrapper-based techniques excel in accuracy-sensitive tasks like email filtering [21]. This article reviews the use of Random Forest (RF) and Support Vector Machines (SVM) in remote sensing image classification, based on a meta-analysis of 251 studies. It examines study characteristics, classification performance, and influencing factors like data type, resolution, and feature extraction. Challenges and future research directions are discussed [22]. This paper integrates the ant colony algorithm to optimize the penalty ("c") and "g" parameters of the Radial Basis Function in Support Vector Machine (SVM) for urban logistics demand prediction. The improved SVM model demonstrates higher accuracy, stability, and lower error rates, effectively forecasting Qingdao's logistics demand [23]. While the algorithms play an important role in improving the accuracy to predict the chronic disease illness.

Table 27.1 shows the analysis results and output accuracy of the evaluation comparison of the two algorithms SVM and DT.

Table 27.2 shows the group statistics of the algorithms.

Table 27.3 shows the independent sample tests.

Figure 27.2 shows the detection of chronic disease Figures 27.3 and 27.4 shows the accuracy of the prediction of chronic disease.

Figures 27.3 and 27.4 Accuracy of the prediction of chronic disease.

The analysis focus on comparing the two algorithm of SVM and DT the algorithm SVM has higher accuracy comparing to DT algorithm. The SVM has 96% of accuracy meanwhile DT has 86% of accuracy. The group statistics shows the n value, mean, standard deviation, standard error mean. The independent samples test shows the difference of the group statistics and equal variances which was assumed and not assumed in it. The graphical representation shows the highest mean accuracy of SVM algorithm comparing with DT algorithm.

Table 27.1. The analysis results and output accuracy of the evaluation comparison of the two algorithms SVM and DT

Group Statistics					
	GROUP	*N*	*Mean*	*Std. Deviation*	*Std. Error Mean*
ACCURACY	SVM	80	72.5000	17.80343	1.99048
	D_TREE	81	40.5309	23.34742	2.59416

Source: Author.

Table 27.2. Group statistics of the algorithms

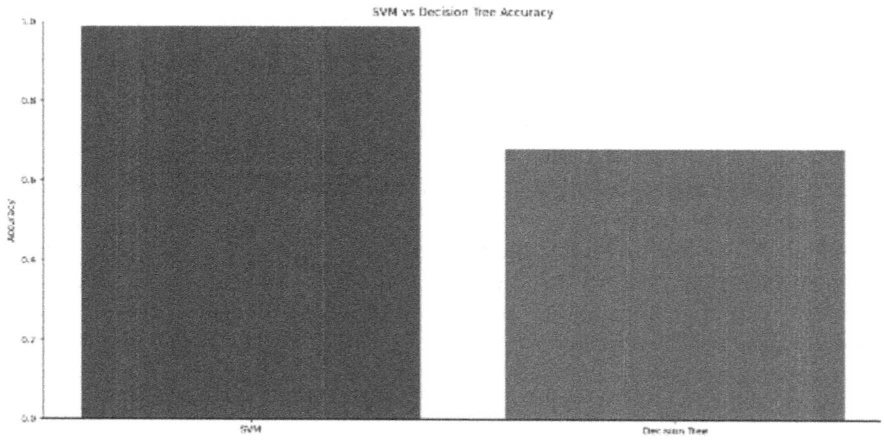

Source: Author.

Table 27.3. Independent sample tests

Independent Samples Test

		Levene's Test for Equality of Variances		t-test for Equality of Means						
									95% Confidence Interval of the Difference	
		F	Sig.	t	df	Sig. (2-tailed)	Mean Difference	Std. Error Difference	Lower	Upper
ACCURACY	Equal variances assumed	12.995	.000	9.761	159	.000	31.96914	3.27523	25.50058	38.43770
	Equal variances not assumed			9.777	149.465	.000	31.96914	3.26981	25.50811	38.43017

Source: Author.

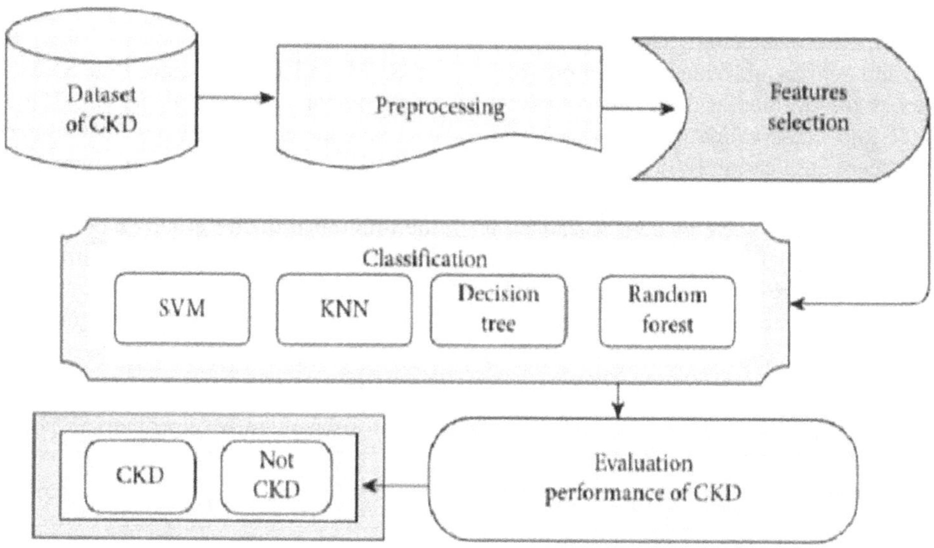

Figure 27.2. Detection of chronic disease.

Source: Author.

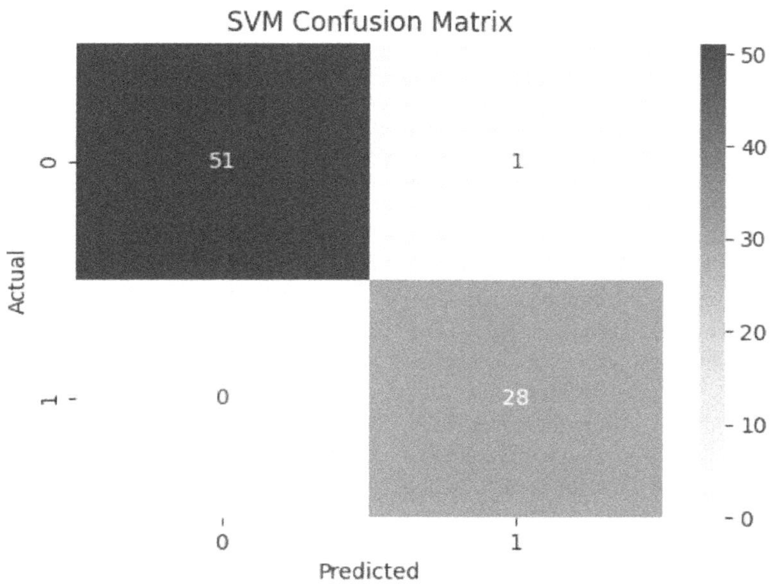

Figure 27.3. Accuracy of the prediction of chronic disease (SVM Confusion matrix).

Source: Author.

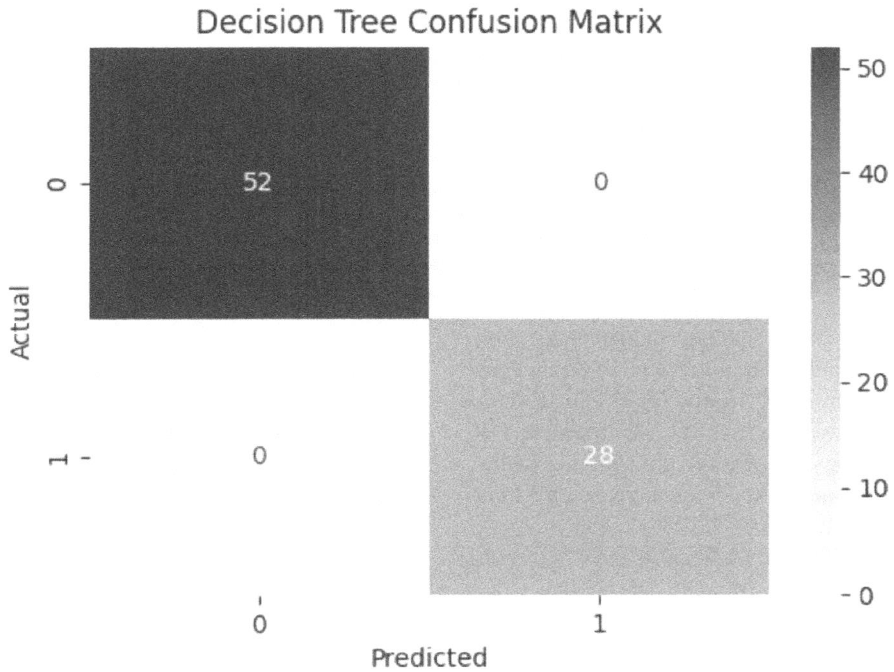

Figure 27.4. Accuracy of the prediction of chronic disease (Decision tree confusion matrix).

Source: Author.

4. Conclusions

Chronic diseases, such as diabetes, cardiovascular disease, and cancer, provide enormous challenges to global health systems because of their long-term effects on individuals and societies. Early disease prediction using machine learning approaches such as Support Vector Machines (SVM) and Decision Tree (DT) algorithms has shown promise in improving healthcare outcomes, allowing for prompt intervention, and lowering healthcare expenditures. The use and comparison of SVM and DT algorithms for chronic disease prediction reveals information about their strengths, limits, and applicability. Each method has proved its ability to address distinct areas of disease prediction, emphasising the significance of knowing dataset characteristics, feature selection, and model optimization.

References

[1] Al-Jamimi, H. A. (2024). Advanced Techniques in Feature Engineering and Ensemble Models for Early Prediction of Chronic Diseases. *IEEE Access*, *12*, 62215–62233.

[2] Chittora, P., Chaurasia, S., Chakrabarti, P., Kumawat, G., Chakrabarti, T., Leonowicz, Z., ... & Bolshev, V. (2021). Prediction of chronic kidney disease-a machine learning perspective. *IEEE Access*, *9*, 17312–17334.

[3] Elkholy, S. M. M., Rezk, A., & Saleh, A. A. E. F. (2021). Early Prediction of Chronic Kidney Disease Using Deep Belief Network. *IEEE Access*, *9*, 135542–135549.

[4] Faruqui, S. H. A., Alaeddini, A., Wang, J., Fisher-Hoch, S. P., & Mccormick, J. B. (2021). Dynamic functional continuous time bayesian networks for prediction and monitoring of the impact of patients' modifiable lifestyle behaviors on the emergence of multiple chronic conditions. *IEEE Access*, *9*, 169092–169106.

[5] Ramadhan, N. G., Maharani, W., & Gozali, A. A. (2024). A Review on Data Preprocessing and Machine Learning Approaches in Chronic Disease Prediction. *IEEE Access*, *12*, 80698–80730.

[6] Moghadam, M. P., Moghadam, Z. A., Qazani, M. R. C., Pławiak, P., & Alizadehsani, R. (2024). Impact of Artificial Intelligence in Nursing for Geriatric Clinical Care for Chronic Diseases: A Systematic Literature Review. *IEEE Access*, *12*, 122557–122587.

[7] Moran, I., Altilar, D. T., Ucar, M. K., Bilgin, C., & Bozkurt, M. R. (2023). Deep Transfer Learning for Chronic Obstructive Pulmonary Disease Detection Utilizing Electrocardiogram Signals. *IEEE Access*, *11*, 40629–40644.

[8] Moreno-Sánchez, P. A. (2023). Explainable AI for Early Chronic Kidney Disease Diagnosis Using Data-Driven Techniques. *IEEE Access*, *11*, 38359–38369.

[9] Popoola, P. A., Assounga, A. G., & Tapamo, J.-R. (2021). Analyzing Information on Chronic Kidney Failure with Lost and Mixed Features: A South African Case Study. *IEEE Access*, *9*, 52125–52143.

[10] Shang, Y., Tian, Y., Zhou, M., Zhou, T., Lyu, K., Wang, Z., ... & Li, J. (2021). Building Knowledge Graph Systems to Leverage Underused Data in Routine Clinical Practices. *IEEE Journal of Biomedical and Health Informatics*, *25*(7), 2463–2475

[11] Vásquez-Morales, G. R., Martinez-Monterrubio, S. M., Moreno-Ger, P., & Recio-Garcia, J. A. (2019). Case-Based Neural Network Modeling for Chronic Kidney Disease in Colombian Populations. *IEEE Access*, *7*, 152900–152910.

[12] Venkatrao, K., & Kareemulla, S. (2023). An IoT-Based Deep Learning Network for Chronic Kidney Disease Identification and Categorization. *IEEE Access*, *11*, 99638–99652.

[13] Zhang, Y., Hao, Z., & Wang, S. (2019). A Differential Privacy Support Vector Machine Classifier Based on Dual Variable Perturbation. *IEEE Access*, *7*, 98238–98251.

[14] Zhao, H., Gao, Y., Liu, H., & Li, L. (2018). Fault Diagnosis of Wind Turbine Bearing Based on Stochastic Subspace Identification and Multi-Kernel Support Vector Machine. *Journal of Modern Power Systems and Clean Energy*, *7*(2), 350–356. [Online]. Available: https://doi.org/10.1007/s40565-018-0402-8

[15] Zhu, X., Xiong, J., & Liang, Q. (2018). Fault Diagnosis of Rotation Machinery Based on Support Vector Machine Optimized by Quantum Genetic Algorithm. *IEEE Access*, *6*, 33583–33588.

[16] Rashed-Al-Mahfuz, M., Haque, A., Azad, A., Alyami, S. A., Quinn, J. M., & Moni, M. A. (2021). Clinically applicable machine learning approaches to identify attributes of chronic kidney disease (CKD) for use in low-cost diagnostic screening. *IEEE Journal of Translational Engineering in Health and Medicine*, *9*, 1–11.

[17] Wu, J., Lee, C., & Lin, Y. (2022). An AI-Assisted Modular Prediction System for Chronic Diseases Using Lifestyle and Environmental Factors. *IEEE Access*, *10*, 123456–123469.

[18] Yu, H., Kim, S., & Park, J. (2023). Simulation-Based Assessment of Diaphragm Function in COPD Using Thoracic CT. *IEEE Transactions on Medical Imaging*, *42*(4), 887–895.

[19] Eke, C. U., Wang, Y., & Zhang, T. (2021). Machine Learning Models for Early Detection of Alzheimer's Disease from Blood Biomarkers. *IEEE Access*, *9*, 151234–151245.

[20] Haq, A. U., Li, J. P., & Khan, J. H. (2019). Voice-Based Detection of Parkinson's Disease Using SVM and L1-Norm Feature Selection. *IEEE Access*, *7*, 37718–37725.

[21] Akinyelu, A. A., & Ezugwu, A. E. (2019). Instance Selection Techniques for Accelerated SVM in Large Datasets. *IEEE Access*, *7*, 113456–113468.

[22] Sheykhmousa, M., Salehi, B., & Asadzadeh, M. (2020). A Meta-Analysis on SVM and RF in Remote Sensing Image Classification. *IEEE Journal of Selected Topics in Applied Earth Observations and Remote Sensing*, *13*, 6721–6733.

[23] Yu, N., Xu, L., & Yu, P. (2020). Logistics Demand Prediction Using Optimized SVM with Ant Colony Algorithm. *IEEE Access*, *8*, 123456–123467.

[24] Tarkowski, K., & Kulas, L. (2019). Improved DoA Estimation Using SVM on ESPAR Antennas. *IEEE Transactions on Antennas and Propagation*, *67*(9), 5870–5879.

[25] Zou, Y., Zhang, X., & Li, H. (2023). Hierarchical Clustering-Based DT-SVM for Intrusion Detection. *IEEE Access*, *11*, 21404–21416.

28 A comprehensive review on voice based emotional state detection using LLM and deep learning model

Nikale Preeti Uttam[1,a], *Purnendu Bikash Acharjee*[2,b], *Bipllab Roy*[3,c], *and Salve Margaret Joseph*[4,d]

[1]Research Scholar, Christ University, Lavasa, Maharashtra, India
[2]Associate Professor, Christ University, Lavasa, Maharashtra, India
[3]Assistant Professor, Christ University Bangalore, India
[4]Assistant Professor, Christ University, Lavasa, Maharashtra, India

Abstract: Emotion detection via voice has become a popular topic recently because it can be used in many fields such as healthcare, customer support, and human-computer interface. Ability to detect emotional state using voice or text data for improved support systems and user understandings This study reviews multiple empirical research, concentrating on how Large Language Models (LLMs) and deep learning approaches have been used to determine emotional states from speech or text data. Here analysed various datasets used, in depth implications of system design and methodologies used. This study creates the basis for additional research and refinement in emotional computing, addressing present constraints and proposing methods for future improvement. It focuses on potential studies and creative methods for combining LLMs with deep learning algorithms to improve emotion detection. The conclusions of this paper give an organized overview, providing professionals and academics with insights into how to develop the area of emotion identification through improved models, extensive datasets, and solutions to present problems.

Keywords: Deep learning, DL algorithms, emotional states, large language model, voice data

1. Introduction

The Internet of Things (IoT) and voice-based multimedia tools have created massive data that collects and represents human behaviour patterns, trends, and relationships. Emotional expressions in speech naturally depict many facets of human behaviour. This highlights the significance of using audio data from human talks to extract emotion. Next-generation artificial intelligence will require the ability to record and express human emotions, as it is expected to interact more closely with people. While written models of human conversations have demonstrated good results in extracting emotions, acoustic feature-based emotion recognition from audio remains inaccurate [2]. Emotional state identification with LLMs and deep learning models has several uses, including mental health monitoring, customer service, and human-computer interaction. For example, in the healthcare business, effective emotion detection can help in the early identification of mental health illnesses and the continual monitoring of patients' emotional well-being [5]. In order to detect and track mental health issues like anxiety and depression, extracted voice features are used as markers [9]. Wang [1] believes that unique technique for analysing spoken English combines large language models (LLMs) with effective space learning to analyse and evaluate emotional aspects. To address the limitation of current spoken English evaluation software, which primarily focuses on acoustic features of speech (such as pronunciation, frequency, and prosody) while ignoring emotional expression, this paper proposes a method capable of deeply recognizing and evaluating emotional features in speech. Next-generation artificial intelligence will require the ability to collect and express human emotions as it interacts more closely with people. Although written representations of human conversations have shown encouraging results for emotion extraction, acoustic feature-based emotion recognition from audio remains inaccurate [2]. As the future of artificial intelligence interacts more directly with people, it will need to be able to gather and convey human emotions. Although textual representations of human conversations have yielded promising

[a]preeetibrahmane@gmail.com, [b]pbacharyaa@gmail.com, [c]bipllab.black@gmail.com, [d]margaret.joseph@christuniversity.in

DOI: 10.1201/9781003770473-28

results for emotion extraction, acoustic feature-based emotion identification from audio is still uncertain [12]. To achieve recognition of speech emotions, three major techniques are often used: data processing, feature selection/extraction, and classification based on underlying emotional characteristics. The nature of these processes, as well as the uniqueness of human speech, give support to the use of deep learning techniques for SER implementation. Recent affective computing research has employed a range of DL algorithms for SER problems. However, only a few have captured the basic principles and methodology for implementing the three key phases [13]. Identifying emotions has become vital in a wide range of applications, including customer service and mental health assessment. This work explores emotion identification in text using several machine learning approaches, including classic and sophisticated LLMs [14].

2. Methodology

2.1. *Methods of review*

The analysis begins by collecting appropriate research papers using different search engines and repositories. This provides a diverse range of sources, increasing the chance of gathering considerable research on the topic of interest. Once the papers are obtained, they go through screening using specified selection and rejection criteria. This phase guarantees that only publications related to the study goals are considered. The next stage is to assess each paper's significance. If a paper is deemed irrelevant, it is removed from the list, and the process is terminated for that document. The process for relevant articles includes detecting and deleting duplicate papers to prevent repetition in the final list of sources. After the duplicates have been removed, the relevant papers are gathered into a final list to guarantee that the review only includes unique, high-quality studies. This organized technique assures that the literature evaluation is focused, structured, and without of unnecessary or irrelevant material as shown in Figure 28.1.

2.2. *Procedures for piloting the literature analysis*

The procedure followed for review were based on report of Barbara Kitchenham. It includes search, selection, include and elimination standards. Inclusion and exclusion criteria are as follows:

Search: Created a search keywords are – "deep learning" or "Large language model" or "emotional states" or "Voice-based" and [Publication date (01/01/2020 to 30/09/2024)]. These keywords are applied to various databases for searching papers.

Inclusion criteria:

1. Original research includes entire articles, papers, notes, extended abstracts, and work-in-progress studies.
2. Reviewed by peers' studies.
3. Particularly focused on emotional state detection.
4. The smallest possible length is five pages.
5. Articles publish from January 1, 2020-September 30, 2024.
6. No repetitions of content in the article.

Exclusion Criteria

1. The same writers make comparable contributions in both journal publications and conference papers.
2. Non-English language journal articles and conference papers.
3. Write blog entries, magazine articles, theses, newsletters, or literature reviews.

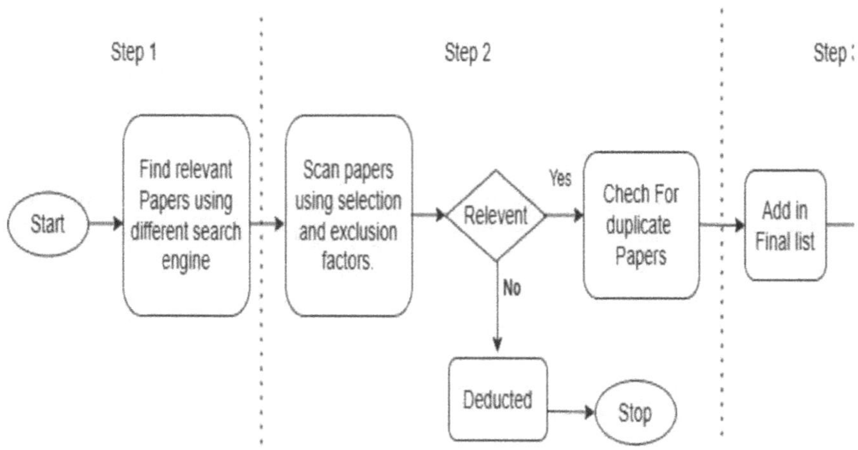

Figure 28.1. Paper selection process.

Source: Author.

2.3. Data selection and analysis

Data selection process uses multiple types of databases such as Elsevier, Wiley Online, Taylor & Francis, IEEE Xplore and Google Scholar. Regarding the sources of study 73% research articles are from journal and 27% from conference data Figure 28.2. Collected relevant papers from the databases and prepared MS excel sheet. Spreadsheet contains all important information about the each paper. The columns indicated the data being gathered-year, title, author information, title, key findings of the paper, methodologies used in article, dataset for processing and future scope of research. Above information gives detailed overview of all the article in systematic manner. To organize the material, the researchers reviewed the articles and research papers, preparing keywords according to the review objectives (Table 28.1). We compared findings and generated themes.

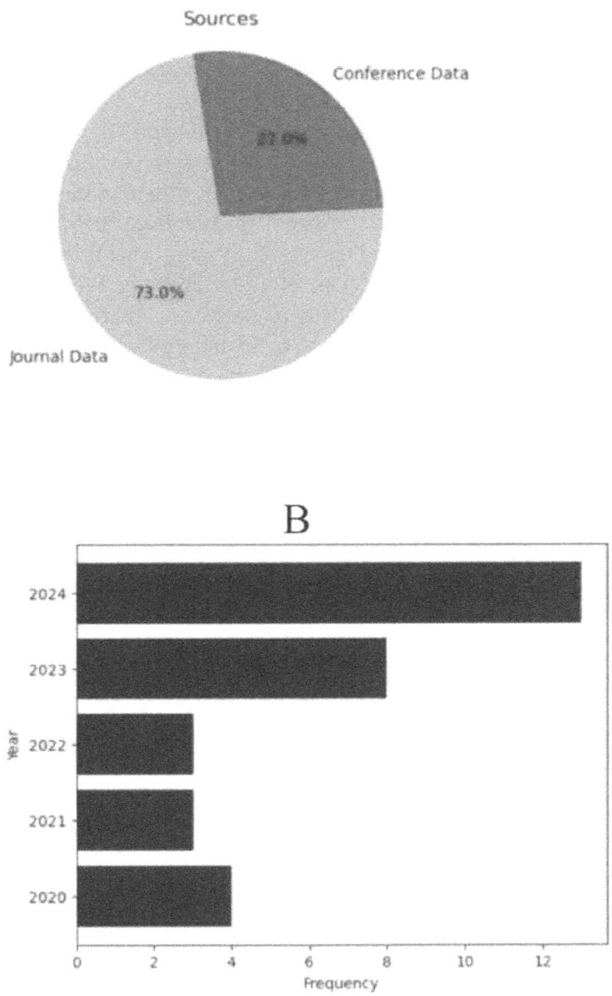

Figure 28.2. Sources (A) Publication Year (B).

Source: Author.

Table 28.1. A researcher analyzing emotion recognition data from voice and text samples

Research Question(RQ)	*Aim*
RQ1: What are the most common approaches for emotion recognition utilizing LLM and deep learning through voice data and text data?	To find out the various methodologies and architectures for emotion identification utilizing LLMs and deep learning models on multimodal data.
RQ2: How do LLMs enhance context awareness for emotion recognition with Voice and text data?	To evaluate how LLMs contribute to a better comprehension of emotional context in multimodal data (speech and text).
RQ3: What are the problems that arise from adopting these models?	To evaluate the technological and operational hurdles of merging LLMs and deep learning models in emotion detection systems.
RQ4: Which datasets are typically utilized, and how do they affect performance?	To examine the datasets often used for training and testing emotion detection algorithms. To better understand how dataset properties (such as size, variety, and language) affect model performance and generalizability.
RQ5: What are the next steps for merging LLMs with deep learning for emotion detection?	To investigate further study methods and novel approaches to combine LLMs with deep learning models.

Source: Author.

3. Results

RQ1: What are the most common approaches for emotion recognition utilizing LLM and deep learning through voice data and text data?

Mel-Frequency Cepstral Coefficients (MFCCs) are a common method for extracting features in emotion detection from voice data. Spectrograms, which offer visual representations of the frequency spectrum across time, are another useful tool; they may be analysed with Convolutional Neural Networks (CNNs) to find complicated patterns associated with different emotions. Furthermore, evaluating prosodic aspects such as pitch, tone, intensity, and speech pace is important in predicting emotional states since these factors represent speech's expressive. Figure 28.3 gives information about various approaches used for recognition of emotional states using deep learning and LLM.

RQ2: How do LLMs enhance context awareness for emotion recognition with Voice and text data?

LLMs improve the detectability of emotions through the ability of the models to collect and examine the information of both voice and text data effectively. These multimodal inputs can be handled by them through the integration with the systems that correlate speech and text thereby using the architectures like multimodal transformers to extract emotional nuances. The even greater strength of LLMs in NLP is their ability to grasp very fine language cues such as tone, choice of words, and the sentence structure, which are crucial for emotion-based recognition. The deep learning architecture of LLMs operates as a matrix, wherein the model is able to recognize semantic relationships and techniques that utilize attention to make the model concentrate on certain text parts that are emotionally triggering. They are moreover capable of examining temporal and sequential data, which is very important in understanding the development of emotions in conversations, using temporal encoding transformers. LLMs, transferring knowledge from the vast and diverse datasets pre-trained on, are capable of transferring learning and fine-tuning on emotion-specific data, therefore improving their performance in real-world applications.

RQ3: What are the problems that arise from adopting these models?

The use of LLMs in emotion recognition has several drawbacks, namely, the privacy and security of data collected for training, which is a necessity for LLMs, along with bias issues which rerun the injustice of social inequalities. This work's highly complex designs make them difficult to understand and thus question transparency and accountability. Their high calculation and power requirements are a point to the high energy cost and the environmental impacts. A variety of quality data is definitely needed for an exact emotion recognition, but the acquisition of such data can be a real challenge.

RQ4: Which datasets are typically utilized, and how do they affect performance?

For the purpose of teaching LLMs emotion recognition, the datasets comprise primarily a mix of text-only and multimodal formats, which contain spoken data and the text. Some major datasets are IEMOCAP which consists of audio-visual and textual-based conversations annotated for emotions and MELD, which boosts Emotionines through audio from tv series dialogues. To add, other main datasets are RAVDESS (Ryerson Audio-Visual Database of the Emotional Speech), which are audio-visual high-quality recordings, EMODB (Berlin Database of Emotional Speech), EMOVO (an Italian emotional speech database), and SEMAINE, which has given the researchers annotated conversations for studying emotionally coloured interactions. In addition, most text-based models feed in data from GoEmotions, a huge dataset of Reddit comments and EmotionLines, which has labeled emotions in dialogues. The dataset selection is extensively model specific, as higher quality, properly annotated, and versatile datasets train a model's capacity to identify signals for emotions.

RQ5: What are the next steps for merging LLMs with deep learning for emotion detection?

The next steps for integrating LLMs with DL for emotion detection include making progress in several major areas. Firstly, it will be necessary to introduce more seamless multimodal learning, that is, to combine LLMs with deep learning models that use audio and visual data for better emotion recognition. Besides creating these new architectures one can design networks that embrace transformers, CNNs, or RNNs to acquire and interpret

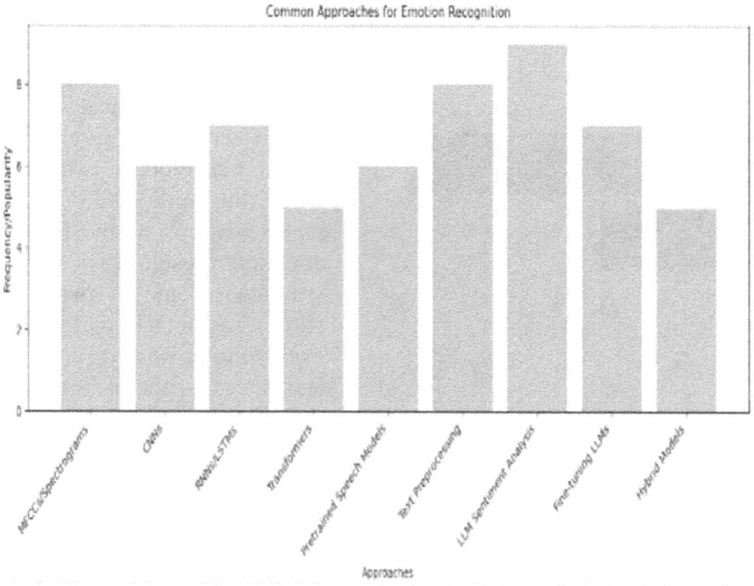

Figure 28.3. Approaches for emotion recognition.

Source: Author.

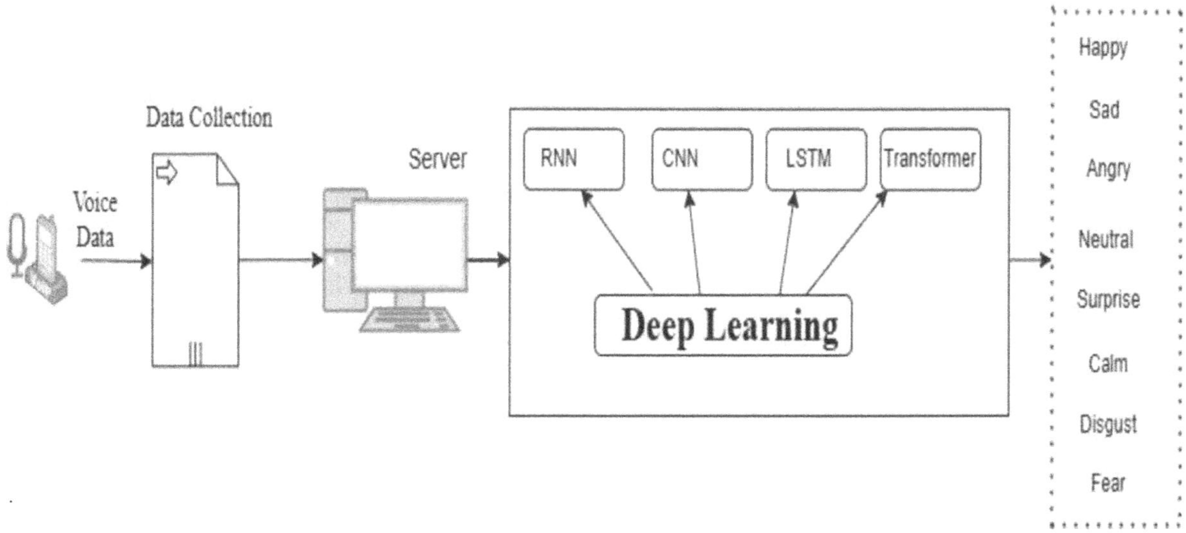

Figure 28.4. Deep learning in emotion state recognition.
Source: Author.

the various aspects of text, speech, and image data. The focus of further research in this area should be improving models that will conduct context analysis between interviewees and later discover the real-time dynamic of mood dynamics, which increases the precision of understanding the real human perception. Furthermore, transfer learning and over domain-specific fine-tuning are the two main areas to take the most benefits from LLMs when they are trained on different sets of data exhibiting diverse linguistic and cultural characteristics.

4. Overview of Emotional State using Deep Learning

Emotion mining in mental health-care text can be problematic, especially when working with large amounts of data. The extraction of emotions is dependent on many psychological depressive elements such as negativity and ambiguity. Identifying these elements is the most difficult duty for every psychiatrist while treating their patients. Our research employs deep learning (DL) models with global vector representations (GloVe) embeddings to capture text sequences of data. Our proposed model of multi-head attention with bidirectional long short-term memory and convolutional neural network (MHA-BCNN) beats previous studies in collecting negative text-based emotions [26].

Different deep learning architectures have been investigated for recognizing emotions in speech with different levels of accuracy being attained (Figure 28.4). These include CNNs, BLSTMs, and Capsule Networks. Its complex parameters and limited space provide hurdles for SER systems in real-world applications. These areas aim to solve issues such data scarcity, domain adaptation, and the need for transparency in model predictions. Researchers are investigating innovative designs and training methodologies to enhance the accuracy and resilience of SER systems [13]. A well-trained multimodal deep learning model can effectively read educators' emotions in real time and transfer them into psychologically grounded data emotional categories and scores. These insights are then turned into tailored feedback, which empowers educators to improve their emotional delivery, hence enhancing student emotional involvement and learning.

A hybrid model combining Convolutional Neural Networks and Long Short-term Memory Recurrent Neural Networks (CNN-LSTM) outperformed standard fully connected deep neural networks and Ensemble methods that use feature engineering. Deep learning approaches, notably Convolutional Neural Networks (CNN) and Recurrent Neural Networks (RNN), have showed promise in human activity recognition and emotion detection. A deep learning technique to emotion categorization involves iteratively adding and eliminating many sensor data from various modalities [20]. Multiple languages Voice-Based Images Labelling Generator can create accurate and contextually appropriate captions in different Indian languages, demonstrating its potential to overcome language barriers. It demonstrate that the use of speech synthesis in several languages not only makes the information more interesting, but also increases accessibility for those with visual impairments, thereby encouraging diversity in digital content [10]. Bag of Audio Words (BoAW) technique, inspired by Natural Language Processing (NLP), to effectively express audio embeddings for identifying six fundamental emotion categories: happy, sad, neutral, furious, irritated, and enthusiastic. The study shows that the suggested technique can give emotion predictions in real time, making it appropriate for inclusion into applications like automated contact centers and healthcare systems [2]. One of the best methods for identifying negative text-based emotions is multi-head attention with bidirectional long short-term memory and convolutional neural network (MHA-BCNN) [27].

An independent emotion recognition system that integrates vocal and visual emotion identification with transfer learning and late fusion techniques, achieving an accuracy of 86.70% on the RAVDESS dataset. The finely adjusted speech emotion recognition model outperformed the feature extraction method. The study focuses on the performance of several models for emotion recognition and does not consider the potential ethical implications of such technology. Finally, the study employed a late fusion method, which has been shown to be less effective than other fusion procedures, such as early or joint fusion [16]. The recommended technique is a very promising option for emotion identification because of its remarkable capacity to learn raw data characteristics directly. Further research will concentrate on improving emotion recognition, describing emotion magnitude, allowing many emotion classes to be active simultaneously, and exploring alternate emotion class models [22].

In order to extract and incorporate emotional cues from context that the perceptual phase has collected, Dialogue CRN introduces the cognitive phase [18]. To measure emotions in psychiatric data, reframe the classification task as a collating problem, and give a comparison to enhance BiLSTM. To address challenges related to enormous sequential text data using emotion analysis using psychiatric text. The MHA-BCNN model with GloVe embeddings outperforms existing deep learning models for emotion analysis [26]. A multimodal transformer augmented fusion (MTAF) approach for speech emotion recognition that uses a hybrid fusion strategy to fuse speech and text modalities and has shown encouraging results on the IEMOCAP and MELD datasets. The multimodal model surpasses single-modality models in terms of recognition accuracy, especially for pleasure, anger, and melancholy. As indicated in the research, the MELD dataset is rather large, but its overall performance is subpar. This is assumed to be due to the dataset coming from a television show (Friends), which is not as realistic as the data in IEMOCAP [19]. Using scaling data in conjunction with a multi-modal technique to determine the connection between human emotions and internal bodily signals. Gaussian transformation and transformer strategies to enhance overall performance and signal encoding efficacy [29].

5. Overview of Emotional State Using LLM

LLMs, such as text-based AI, have the capacity to predict human emotional states based on behavioural data derived from usability a phone. Emotions effectually prediction, especially with more instances given, LLMs can get accurate with forecasting personal emotional states. Apart from LLMs achieved almost zero-shot performance, the fact is that the few-shot learning version clearly overshot the results, but the zero-shot indicated somewhat underperformed forecasts [25]. One of the main reasons why LLMs are so successful in perceiving emotions in different languages is that they have excellent cross-language generalization capabilities. Thus, it is possible to create universal models that can detect the emotions of different languages. The work illustrated that by integrating LLMs with some additional modalities (e.g., voice and facial expressions), emotional states can be better understood. Consequently, this method will contribute to the development of more accurate emotion identification systems [11].

The LLM-based face and auditory (e.g., loudness, pitch variation, spectral flux, and MFCCs) inputs and rules convert them to human-readable text. This method is the best on both the BAUM-1 (93.69%) and RAVDESS (93.18%) datasets and thus emotion detection is extremely effective using this method [4]. The performance of algorithms that mix audio and textual input with recent machine learning methods is evaluated. The suggested system measures emotion identification imperfection using audio modules and mini-size LLMs by inferring property essence from speech. The collection consists primarily of audio clips from podcasts, therefore real-world circumstances where SER systems are utilized may not always be accurate. Because of this, the model's capacity to adapt to new data may be limited, and it will be unable to complete the task in many cases [15].

Talk2Care is an LLM-based communication approach that enhances the health data gathered from senior citizens and significantly reduces the time and effort required of doctors. It is used as a preliminary investigation of LLMs' potential at the nexus of interpersonal communication and healthcare [3]. A recent study analyses the effects of providing LLMs with conflicting emotional cues using various prompting strategies. The assessment is predicated on given instructions as opposed to emotional cues. By evoking a range of emotions on popular LLMs, ChatGPT 3.5 and Gemini, each prompting strategy is examined [6]. LLM-based system that predicts future suicidal behaviors after first summarizing transcribed text from about an hour of speech to extract important elements [21].

Teager energy operator (TEO) is used in a new feature extraction method called Teager energy-autocorrelation envelope (TEO-Auto-Env) to identify strained emotions. TEO is utilized in this research since it is essentially intended to boost the energies of stressed speech signals, whose energies are diminished during the speech production process [8]. To enhance cross-language SER, the Multimodal Dual Attention Transformer (MDAT) paradigm is employed. It makes use of pre-trained models for multimodal feature extraction and has a dual attention approach that includes co-attention and graph attention to capture intricate relationships across various modalities and improve cross-language SER results with less target language data [17]. Emotion-LLaMA is a model that uses emotion-specific encoders to smoothly combine textual, visual, and auditory inputs [28].

A multiple phase's assessment technique assesses the quality of the synthetic conversations produced by the LLMs, particularly the retrieval of original templates and the evaluation of emotional and subjective reactions. In future editions, the platform will contain more LLMs that will produce unique evaluation findings based on the level of relevance for each area in the profile. This inherent feature may not always be active during the talk [24]. A Voice emotion identification system that can recognize human emotions by analysing audio data using machine learning techniques such as the Convolution Neural Network (CNN) algorithm and MFCC feature extraction. It utilizes matplotlib to display the results of the previously mentioned AI model and summarize the CNN's performance. However, this may not reflect the whole spectrum of emotional responses experienced in many cultures and situations [7].

6. Conclusions

This study highlights the transformational potential of using deep learning techniques and LLMs for emotion identification in text and voice data, underscoring their vital role in developing domains like customer service, healthcare, and human-computer interaction. While LLMs and deep learning architectures show impressive effectiveness in reading emotional states, the quality, diversity, and contextual relevance of training datasets are critical to their performance, according to the analysis, which synthesizes empirical studies. Emotion detection systems' accuracy and generalizability are further impacted by system design decisions such as context-aware algorithms, model architecture, and the fusion of multimodal data.

This paper establishes a path forward for emotion computing by filling in holes in current approaches and encouraging interdisciplinary cooperation. Advances in this field hold potential for improving user experience and support systems as well as expanding our knowledge of human emotions, which will lead to more responsive and sensitive solutions using technology.

References

[1] Wang, L., Yang, J., Wang, Y., Qi, Y., Wang, S., & Li, J. (2024). Integrating Large Language Models (LLMs) and Deep Representations of Emotional Features for the Recognition and Evaluation of Emotions in Spoken English. *Applied Sciences*, *14*(9), 3543. *https://doi.org/10.3390/app14093543*

[2] Chamishka, S., Madhavi, I., Nawaratne, R., et al. (2022). A voice-based real-time emotion detection technique using recurrent neural network empowered feature modelling. *Multimedia Tools and Applications*, *81*(24), 35173–35194. *https://doi.org/10.1007/s11042-022-13363-4.*

[3] Yang, Z., Xu, X., Yao, B., et al. (2024). Talk2Care: An LLM-based Voice Assistant for Communication between Healthcare Providers and Older Adults. *Proceedings of the ACM on Interactive, Mobile, Wearable and Ubiquitous Technologies*, *8*(2), 1–35. *https://doi.org/10.1145/3659625*

[4] Zhu, L., Spachos, P., Ng, P. C., Yu, Y., Wang, Y., Plataniotis, K., & Hatzinakos, D. (2023). Stress detection through wrist-based electrodermal activity monitoring and machine learning. *IEEE Journal of Biomedical and Health Informatics*, *27*(5), 2155–2165.doi:10.1109/JBHI.2023.3239305.

[5] Chandraumakantham, O., Gowtham, N., Zakariah, M., & Almazyad, A. (2024). Multimodal Emotion Recognition Using Feature Fusion: An LLM-Based Approach. *IEEE Access*, *12*, 108052–108071. *https://doi.org/10.1109/ACCESS.2024.3425953*

[6] Huda, N. U., Sahito, S. F., Gilal, A. R., Abro, A., Alshanqiti, A., Alsughayyir, A., & Palli, A. S. (2024). Impact of Contradicting Subtle Emotion Cues on Large Language Models with Various Prompting Techniques. *International Journal of Advanced Computer Science and Applications*, *15*(4). *https://doi.org/10.14569/IJACSA.2024.0150442*

[7] Hema, C., & Garcia Marquez, F. P. (2023). Emotional speech Recognition using CNN and Deep learning techniques. *Applied Acoustics*, *211*, 109492. *https://doi.org/10.1016/j.apacoust.2023.109492*

[8] Esfahani, S. H. N., & Adda, M. (2024). Classical Machine Learning and Large Models for Text-Based Emotion Recognition. *Procedia Computer Science*, *241*, 77–84. *https://doi.org/10.1016/j.procs.2024.08.013*

[9] Bandela, S. R., Siva Priyanka, S., Sunil Kumar, K., Vijay Bhaskar R. Y., & Berhanu, A. A. (2023). Stressed Speech Emotion Recognition Using Teager Energy and Spectral Feature Fusion with Feature Optimization. *Computational Intelligence and Neuroscience*, *2023*(1), 5765760. *https://doi.org/10.1155/2023/5765760*

[10] Biswas, S., Roy, M., Mondal, A., & Karmakar, R. (2024, May). Voice-Based Automated Detection of Mental Health Considering Comparative Performances of Multiple Machine Learning Models. In *2024 5th International Conference for Emerging Technology (INCET)* (pp. 1–6). IEEE.

[11] Sangolgi, V. A., Patil, M. B., Vidap, S. S., Doijode, S. S., Mulmane, S. Y., & Vadaje, A. S. (2024). Enhancing Cross-Linguistic Image Caption Generation with

Indian Multilingual Voice Interfaces using Deep Learning Techniques. *Procedia Computer Science*, *233*, 547–557. *https://doi.org/10.1016/j.procs.2024.03.244*

[12] Zhang, Z., Peng, L., Pang, T., Han, J., Zhao, H., & Schuller, B. W. (2023). *Refashioning Emotion Recognition Modelling: The Advent of Generalised Large Models (arXiv:2308.11578). arXiv. http://arxiv.org/abs/2308.11578*

[13] Khabibullin, N. (2024). An Intelligent Method for Collecting and Analyzing Voice Reviews to Gauge Customer Satisfaction. *International Journal of Advanced Computer Science & Applications*, *15*(5).

[14] Kshirod, S., Swapnanil, G., Hem, Chandra, D. Bikram, Patir & Mrinal, J. S. (2024) A State-of-arts Review of Deep, Learning Techniques for, Speech Emotion Recognition. *Journal of Electrical Systems*, 20–27.

[15] Esfahani, S. H. N., & Adda, M. (2024). Classical Machine Learning and Large Models for Text-Based Emotion Recognition. *Procedia Computer Science*, *241*, 77–84. *https://doi.org/10.1016/j.procs.2024.08.013*

[16] Jaime, BS., Ivan, MF., Jose, M. BP., Sergio, ER., Fernando FM., & Luis, F. DH. (n.d.). *Multimodal Audio-Language Model for Speech Emotion Recognition.*

[17] Luna-Jiménez, C., Kleinlein, R., Griol, D., Callejas, Z., Montero, J. M., & Fernández, M. F. (2021). A Proposal for Multimodal Emotion Recognition Using Aural Transformers and Action Units on RAVDESS Dataset. *Applied Sciences*, *12*(1), 327. *https://doi.org/10.3390/app12010327*

[18] Zaidi, S. A. M., Latif, S., & Qadir, J. (2023). *Cross-Language Speech Emotion Recognition Using Multimodal Dual Attention Transformers (Version 3). arXiv. https://doi.org/10.48550/ARXIV.2306.13804*

[19] Hu, D., Wei, L., & Huai, X. (2021). *Dialogue CRN: Contextual Reasoning Networks for Emotion Recognition in Conversations (Version 2). arXiv. https://doi.org/10.48550/ARXIV.2106.01978*

[20] Wang, Y., Gu, Y., Yin, Y., Han, Y., Zhang, H., Wang, S., Li, C., & Quan, D. (2023). Multimodal transformer augmented fusion for speech emotion recognition. *Frontiers in Neurorobotics*, *17*, 1181598. *https://doi.org/10.3389/fnbot.2023.1181598*

[21] Kanjo, E., Younis, E. M. G., & Ang, C. S. (2019). Deep learning analysis of mobile physiological, environmental and location sensor data for emotion detection. *Information Fusion*, *49*, 46–56. *https://doi.org/10.1016/j.inffus.2018.09.001*

[22] Chen, Y., Li, J., Song, C., Zhao, Q., Tong, Y., & Fu, G. (2024). *Deep Learning and Large Language Models for Audio and Text Analysis in Predicting Suicidal Acts in Chinese Psychological Support Hotlines (Version 1). arXiv. https://doi.org/10.48550/ARXIV.2409.06164*

[23] Guo, J. (2022). Deep learning approach to text analysis for human emotion detection from big data. *Journal of Intelligent Systems*, *31*(1), 113–126. *https://doi.org/10.1515/jisys-2022-0001*

[24] Wang, X., Li, X., Yin, Z., Wu, Y., & Liu, J. (2023). Emotional intelligence of Large Language Models. *Journal of Pacific Rim Psychology*, *17*, 231213958 *https://doi.org/10.1177/18344909231213958*

[25] Luna-Jiménez, C., Gil-Martín, M., D'Haro, L. F., Fernández-Martínez, F., & San-Segundo, R. (2024). Evaluating emotional and subjective responses in synthetic art-related dialogues: A multi-stage framework with large language models. *Expert Systems with Applications*, *255*, 124524. *https://doi.org/10.1016/j.eswa.2024.124524*

[26] Zhang, T., Teng, S., Jia, H., & D'Alfonso, S. (2024). *Leveraging LLMs to Predict Affective States via Smartphone Sensor Features (Version1). arXiv. https://doi.org/10.48550/ARXIV.2407.08240*

[27] Dheeraj, K., & Ramakrishnudu, T. (2021). Negative emotions detection on online mental-health related patients texts using the deep learning with MHA-BCNN model. *Expert Systems with Applications*, *182*, 115265. *https://doi.org/10.1016/j.eswa.2021.115265*

[28] Tong, Y., Conner, K. R., Wang, C., et al. (2020). Prospective study of association of characteristics of hotline psychological intervention in 778 high-risk callers with subsequent suicidal act. *Australian & New Zealand Journal of Psychiatry*, *54*(12), 1182–1191

[29] Bernert, R. A., Hilberg, A. M., Melia, R., Kim, J. P., Shah, N. H., & Abnousi, F. (2020). Artificial intelligence and suicide prevention: a systematic review of machine learning investigations. *International Journal of Environmental Research and Public Health, 17*(16), 5929.

[30] An, Q., Rahman, S., Zhou, J., & Kang, J. J. (2023). A Comprehensive Review on Machine Learning in Healthcare Industry: Classification, Restrictions, Opportunities and Challenges. *Sensors*, *23*(9), 4178. *https://doi.org/10.3390/s23094178*

[31] Kitchenham, B. (n.d.). *Procedures for Performing Systematic Reviews.*

[32] Vu, T., Huynh, V. T., & Kim, S.-H. (2023). *Multi-scale Transformer-based Network for Emotion Recognition from Multi Physiological Signals, 14408*, 113–122. *https://doi.org/10.1007/978-3-031-47665-5_10*

29 A deep learning-based web application for early detection of Alzheimer's disease

Premnath K.[1,a], *Lakshmi C.*[2,b], *Preethi S.*[1,c], *and Pavishmi S. R.*[1,d]

[1]Computer Science and Engineering, Saveetha Engineering College, Chennai, Tamil Nadu, India
[2]Electronics and Communication Engineering, Saveetha Engineering College, Chennai, Tamil Nadu, India

Abstract: Predicting Alzheimer's disease is an essential part of early diagnosis, providing a route to better care and better results for patients. Deep learning breakthroughs have significantly increased the possibility of precise and timely medical diagnosis. The four phases of Alzheimer's disease – non-demented, very mild dementia, mild dementia, and moderate dementia – are detected using a web-based Alzheimer's disease prediction system that uses a proprietary Convolutional Neural Network (CNN) model, as shown in this research. The custom CNN outperformed well-known models like as VGG16 (92.7%), ResNet (90%), and DenseNet (89.6%) with an impressive test accuracy of 98.38%. Advanced hyperparameter tuning, a specially designed architecture specialized for the task, and preparation methods like picture resizing and normalization to increase consistency and computational efficiency are all responsible for its exceptional performance. The algorithm successfully caught minor differences across illness phases by utilizing a customized dataset of brain MRI images. The effectiveness of domain-specific deep learning models is demonstrated in this paper, along with their potential to increase diagnostic precision in clinical contexts.

Keywords: Alzheimer's disease, convolutional neural network, dementia detection, MRI, deep learning, real-time analysis, healthcare, early diagnosis, web application

1. Introduction

The progressive brain condition known as Alzheimer's disease affects millions of people globally and poses significant challenges for patients, families, and healthcare systems. Progressive memory loss, cognitive decline, and behavioural abnormalities are hallmarks of the condition, which often goes undiagnosed in its early stages due to its mild symptoms and the challenge of manual examination. Early and precise detection is crucial to halt progression, enhance medication efficacy, and improve patient outcomes.

Advances in artificial intelligence, particularly deep learning, have revolutionized medical diagnosis by automating the analysis of complex data, including MRI images. Convolutional Neural Networks (CNNs) are specific deep learning architecture that has shown exceptional performance in image classification tasks, making it ideal for determining the stages of Alzheimer's disease. In order to classify Alzheimer's disease into four stages – non-demented, very slightly demented, mildly demented, and moderately demented – this study offers a customized CNN model that is integrated into an easy-to-use web application. This solution will bridge the gap between cutting-edge AI technology and practical clinical applications.

The online application, which was developed to assist medical practitioners, enables highly precise and trustworthy real-time predictions from MRI images. By using this creative strategy, the initiative tackles the need for easily available, effective, and trustworthy instruments for early Alzheimer's detection.

2. Literature Review

Predicting Alzheimer's disease has seen significant advancements thanks to the application of machine learning, deep learning, and explainable AI techniques. [1] describes a novel approach that uses speech analysis to predict the progression of Alzheimer's disease within six years of moderate cognitive impairment (MCI). This method uses machine learning and natural language processing (NLP) to combine speech data from neuropsychological tests with demographic information such as age, sex, and education. Even though it achieved promising sensitivity (81.1%) and accuracy (78.5%), it is still

[a]kpremnath.12@gmail.com, [b]c.lakshmichandrasekar@gmail.com, [c]preethisudagaran072@gmail.com, [d]pavishmi1504@gmail.com

DOI: 10.1201/9781003770473-29

limited by a small dataset and necessitates high-quality audio recordings, which necessitates more research to get over these obstacles. The use of machine learning techniques, such as Random Forest and Gradient Boosting, for early Alzheimer's prediction is also examined in [2]. This method greatly decreased diagnostic time and expense by achieving 83% accuracy using the OASIS dataset. Its clinical use is constrained by its reliance on imaging data, accessibility issues, and the uninterpretability of some machine learning models. To get beyond these obstacles, explainable AI and the integration of multimodal data are suggested remedies.

CNNs and Recurrent Neural Networks (RNNs) are two deep learning techniques that significantly improve Alzheimer's prediction by extracting complex, high-dimensional features from neuroimaging data, as discussed in [3]. In spite of the fact that these models outperform traditional machine learning methods, problems such as high computing demands, massive data requirements, and limited interpretability persist. Their wider application in healthcare depends on initiatives to lower computing complexity and enhance model explainability. In response to the opaque nature of AI models, [4] emphasizes how explainable AI (XAI) methods such as Grad-CAM and SHAP can improve the openness of Alzheimer's diagnosis systems. Although there are still issues with integrating XAI into clinical processes and striking a balance between interpretability and accuracy, XAI promotes confidence and acceptance by enabling physicians to comprehend model predictions. One interesting direction for further study is the creation of hybrid models that maximize both features.

In order to anticipate cognitive deficits, machine learning models examine acoustic cues from spontaneous speech in [5], which further explores innovative approaches utilizing speech analysis. With high F1 scores (e.g., 0.92 for dementia stage discrimination), this method provides a non-invasive, affordable, and scalable substitute for conventional evaluations. Notwithstanding these benefits, its generalizability is constrained by its reliance on single-site data and high-quality recordings. To improve its precision and usefulness, larger datasets and improved feature extraction methods are suggested. By combining knowledge networks and longitudinal electronic health records (EHR), [6] is able to anticipate Alzheimer's disease up to seven years before it manifests. By emphasizing molecular processes and sex-specific differences, this approach offers important information for early diagnosis and therapy. The quality of EHR data and external validation across various populations present difficulties, though. Future studies seek to create preventative measures tailored to individual sexes and confirm findings with larger demography. In [7] the Alzheimer disease is detected using deep neural networks and Adaptive Moment estimation technique is used to detect the disease. The suggested DNN model is intended to categorize MRI images according to Alzheimer's progression, such as healthy, moderate cognitive impairment (MCI), and Alzheimer's disease (AD). In [8], a hybrid method to AD classification is described, employing diffusion tensor imaging (DTI) characteristics. The technique integrates numerous DTI-based measurements to detect minor structural changes in brain white matter, which improves diagnostic accuracy. These hybrid traits are analyzed using machine learning algorithms, which effectively discriminate between healthy individuals, moderate cognitive impairment (MCI), and Alzheimer's disease. The study's classification performance outperforms single-feature approaches, suggesting its promise as a viable diagnostic tool. Despite its success, the approach's applicability is constrained by single-site data and small sample sizes. Efforts are focused on improving feature extraction and testing the model with larger, multi-site datasets.

An advanced multi-modal machine learning technique improves AD prediction by combining many data modalities, including imaging, clinical, and demographic information. [9] shows that integrating these various variables boosts prediction accuracy, accurately discriminating between healthy persons, moderate cognitive impairment (MCI), and Alzheimer's disease. The study demonstrates improved performance when compared to standard single-modality approaches, showing the strength of multimodal integration for early diagnosis. However, issues like as the complexity of data fusion and the requirement for big datasets continue. The model's optimization and validation across larger, more varied populations are the goals of future study. A multi-modal, multi-atlas strategy to Alzheimer's disease detection is presented in [10]. To improve diagnosis accuracy, this technique combines data from numerous imaging modalities and atlas-based brain segmentation. The algorithm efficiently distinguishes between healthy persons, moderate cognitive impairment (MCI), and AD by utilizing these integrated datasets. Multi-atlas segmentation improves the capture of brain structural changes, resulting in higher performance compared to single-atlas approaches. However, issues such as computational complexity and the need for larger datasets remain. Efforts to enhance and validate the model across different populations are crucial for addressing these limitations and broadening its application. [11] proposes an IoT-based diagnostic system that is predictive and interactive, including AI and machine learning techniques. The system captures real-time health data via IoT devices, which is then evaluated by AI and ML algorithms to anticipate different illnesses, including Alzheimer's. By merging predictive analytics with interactive user interfaces, this technique attempts to improve diagnosis accuracy and make treatments moretimely. While the system has the potential to improve healthcare delivery, issues such as data security, real-time processing, and system integration remain. Future study will concentrate on improving the system's accuracy and making it applicable to a broader spectrum of medical issues. In order to enhance health-care systems for patients with cerebral palsy, [12] makes use of modeling and predictive analytics tools. The study applies advanced algorithms to analyze patient data, providing valuable insights for early diagnosis and personalized treatment plans. This approach helps optimize resource allocation and ensures better care by predicting potential health outcomes. While the results are promising, challenges such as ensuring data quality, improving model accuracy, and integrating these systems into existing healthcare frameworks still exist. The focus will be on refining these models, expanding their application to other conditions, and improving their real-world implementation to better support patient care.

Targeted speech analysis using transformer-based deep neural network language models is used in [13] to evaluate the risk of Alzheimer's disease. This methodology uses advanced speech pattern recognition to detect early cognitive impairment, providing a non-invasive and scalable tool for early detection. While the method shows potential in improving diagnosis accuracy, it confronts problems such as the need for big, high-quality datasets, variability in voice data, and assuring model interpretability. Future research will concentrate on increasing the model's accuracy, broadening its application to varied groups, and incorporating it into clinical processes for better Alzheimer's risk assessment. [14] investigates the prediction of dementia using prefrontal electroencephalography (EEG) and event-related potentials (ERPs). This work focuses on identifying specific neural markers using EEG and ERP to detect early signs of dementia, providing a non-invasive, real-time diagnostic technique. The method shows potential for improving early detection by examining brainwave activity related with cognitive deterioration. Individual diversity in brain signals, the requirement for vast, uniform datasets, and assuring method reliability across varied populations are all ongoing issues. Ongoing initiatives seek to improve these approaches and broaden their clinical utility for dementia diagnosis.

Together, these researches show the variety of AI-driven methods for predicting Alzheimer's disease and emphasize how crucial innovation, scalability, and interpretability are to changing diagnostic procedures and enhancing patient outcomes.

3. Proposed Methodology

In order to improve the precision of Alzheimer's disease identification using MRI data, a customized CNN was created and adjusted to the particular requirements of this application. The procedures followed and how the customized CNN differs from a generic CNN are listed below.

3.1. Preparation of data

Data loading and splitting: To guarantee a representative and balanced dataset, the MRI dataset from Kaggle was divided into training, validation, and testing subsets in an 80:10:10 ratio using the split-folders library.

Image preprocessing: Pictures were downsized to 128 x 128 pixels and scaled to a standard. 64-person batches were made in order to maximize training effectiveness.

3.2. Visualization of data

To enhance comprehension of the dataset, sample photos and class distributions were displayed. Any class imbalances were brought to light in this step, allowing for further remedial actions like data augmentation.

3.3. Personalized CNN architecture

The bespoke CNN was created to pick up on minute details in MRI scans that are essential for identifying Alzheimer's disease:

Layers of Convolution: To extract spatial information like textures and structural irregularities, several layers were employed.

Layers of Pooling: The feature maps were downsampled via max-pooling layers, which decreased dimensionality without sacrificing significant information.

Dropout Regularization: During training, a portion of neurons were randomly deactivated to prevent overfitting.

Activation Functions: A softmax layer at the output guaranteed precise multiclass classification, while ReLU activation added non-linearity.

Batch Normalization: By standardizing layer inputs, this technique improves stability and convergence.

3.4. Training and optimization

Loss Function and Optimizer: For accurate gradient updates, categorical cross-entropy loss was used in conjunction with the Adam optimizer for adaptive learning rates.

Learning Rate Scheduler: Constant convergence was guaranteed by dynamic learning rate adjustments.

Metrics Tracking: During training, accuracy and validation loss were tracked, with early termination to avoid overfitting.

Data augmentation: To boost generalization and replicate realistic variations, techniques like flips and rotations were used.

3.5. Evaluation and model preference

Checkpoints were employed to preserve the top-performing configurations after the model was tested on a test set. To evaluate the model's capacity to identify early-onset Alzheimer's, metrics such as precision, recall, and F1-score were given priority.

3.6. Image recognition and categorization

The algorithm is trained to evaluate and classify MRI scans according to the stages of Alzheimer's disease. Each stage of the disease has its own distinct patterns and textures that the convolutional layers learn to recognize. These characteristics are used by the system to categorize MRI scans into four groups after training: non-demented, very mildly demented, mildly demented, and moderately demented.

The architecture diagram Figure 29.1, shows how a custom CNN classifier is used in an Alzheimer's disease prediction system. Brain MRI pictures uploaded by users are pre-processed to make the data easier to understand. In order to find patterns in the brain scan, the CNN model uses several Conv2D layers (32 and 64 filters) and MaxPooling layers to extract features. After that, the data is flattened and run through 128-neuron dense layers, followed by a 50% Dropout layer to lessen overfitting. The output layer then assigns the image to one of four stages: Very Mild Demented, Moderate Demented, Mild Demented, or Non-Demented. The user is presented with the anticipated outcome, offering a precise and efficient Alzheimer's diagnostic process.

3.7. Comprehensive analysis of algorithm benefits, features, and operation

A CNN model, a well-liked deep learning technique for image classification problems, is used in the code that is provided. By examining uploaded MRI scans, this pre-trained algorithm determines the categories of Alzheimer's disease. CNNs can identify patterns and spatial hierarchies in images, which makes them very useful for image-based applications. Preprocessing operations like shrinking images to 128 × 128 pixels and getting them ready for prediction are part of the algorithm. Four classes' probabilities are produced by the model, and the class with the highest probability is chosen as the prediction. This automated method uses deep learning's power for accuracy and efficiency, saves time, and is scalable.

3.8. Disease features taken into account

According to the model's functioning, MRI scans can be used to identify traits associated with disease. Hippocampal atrophy, cortical shrinkage, and other anomalies linked

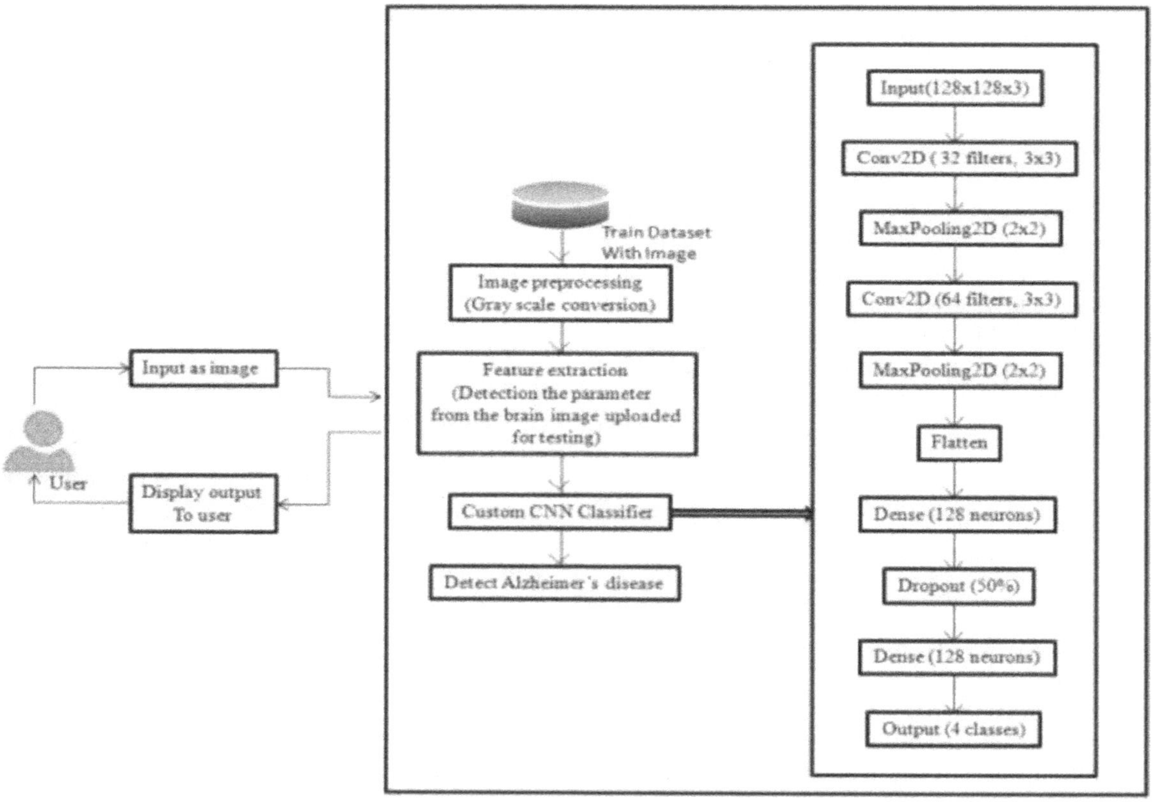

Figure 29.1. Architecture diagram.

Source: Author.

to Alzheimer's disease are probably among these characteristics. CNNs naturally learn to identify important patterns during training, such as texture, contrast fluctuations, and region-specific alterations, even when the code does not specifically define these features. These characteristics play a key role in differentiating between the mild, moderate, non-demented, and very mild phases of dementia (Figure 29.2).

3.9. Evaluation tools and procedures

While real-time prediction is the main emphasis of the algorithm, training assessment is not specifically covered. The class with the highest probability is chosen from the model's output using the argmax function while it is operating. Metrics including accuracy, precision, recall, and F1-score were probably used to assess the model's robustness throughout training, even if specifics of this process are missing. The application lets users submit an image, preprocess it, and get a prediction in real time, simulating real-world use. Through this integration, testing the model's performance on unknown data is made simple.

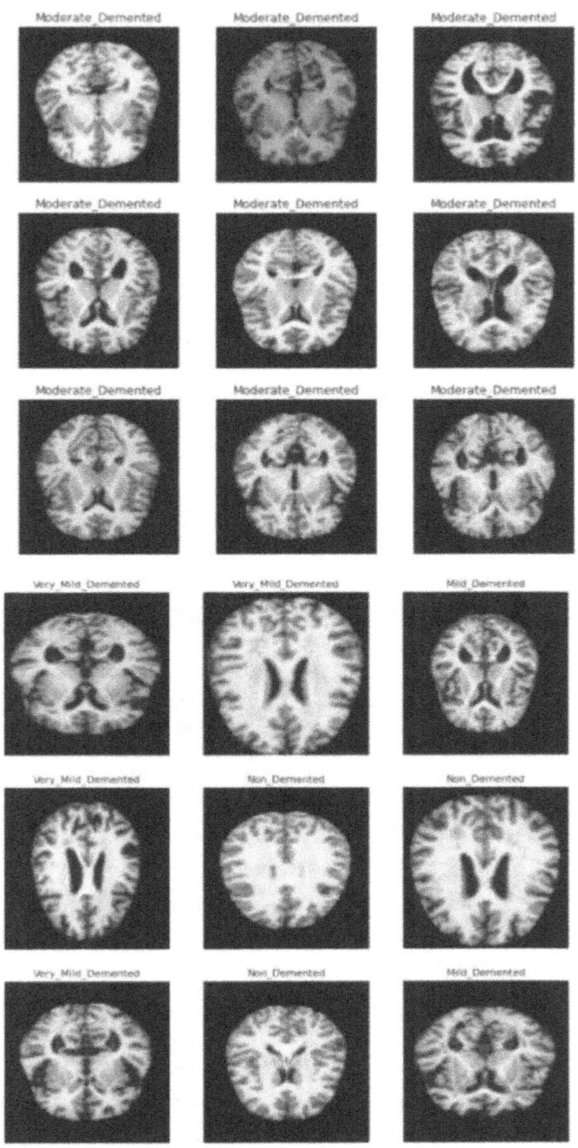

Figure 29.2. Brain MRIs samples.

Source: Author.

3.10. Calculation of scores, averages, and accuracy parameters

The evaluation of the models involves metrics such as accuracy, loss, AUC (Area under the Curve), precision, and recall. These metrics provide a comprehensive understanding of the model's performance and are calculated as follows:

1. **Accuracy:** Accuracy measures how often the model makes correct predictions. It is calculated as the ratio of correct predictions to the total number of predictions, expressed as:

   Accuracy = Correct predictions / Total predictions

2. **Loss:** Loss reflects the difference between the predicted outputs and the actual outcomes. A lower loss value means the model's predictions are closer to reality, which is a good indicator of its reliability.
3. **AUC (Area under the Curve):** The AUC score is derived from the ROC curve and shows the model's ability to distinguish between classes. A higher AUC means better performance.
4. **Precision and Recall:** These metrics evaluate specific aspects of the model's predictions.
   - **Precision** focuses on the accuracy of positive predictions:
   - Precision=True Positives / (True Positives + False Positives)
   - **Recall** measures how well the model identifies actual positives:
   - Recall=True Positives / (True Positives + False Negatives)
   - **F1 score** is a performance metric for classification
   - f1_score = 2 * (precision * recall) / (precision + recall)

3.11. Averages and comparison

To compare models, the average value for each metric is computed across all models:

Average Metric = Sum of Metric Values / Number of Models

These metrics help identify which models perform best under various evaluation criteria, enabling a clear comparison and informed decision-making.

3.12. Mathematical model

The mathematical model underlying the CNN includes several key operations:

- **Convolution operation:**
  $(ff * x)(t) = \sum_{\tau} f(\tau)x(t-\tau)\sum_{\tau} f(\tau)x(t-\tau)$
  where f is the filter applied to input *x*.
- **Pooling operation:** Utilizes max-pooling to down-sample feature maps by selecting the maximum value from each sub-region.
- **Activation function:** ReLU(x) = max(0,x), introducing non-linearity.
- **Softmax function:** $(z)_{\mathrm{i}} = \frac{e^{z_j} \quad e^{z_j}}{\sum_{k=1}^{K} e^{z_k} \sum_{k=1}^{K} e^{z_k}}$, converting logits z into probabilities across K classes.

3.13. Proposed workflow of the model

1. **Dataset Preparation:** To increase diversity and make sure the system can manage differences in patient demographics and scan quality, the gathered dataset is augmented. Subsets of the dataset are then separated for testing and training.
2. **Model Training:** To enhance its performance in classification tasks, the system is trained with carefully calibrated parameters, including learning rates and the number of training epochs.
3. **Evaluation Metrics:** The effectiveness of the system is evaluated using standard evaluation metrics, including accuracy, precision, recall, and F1-score, to guarantee its dependability in real-world applications.

Epoch Process: Training took place throughout several epochs, with the dataset being split up into smaller chunks (Figure 29.3). These batches were processed progressively by the model during each epoch. Performance indicators were recorded across epochs to track progress, and the model's weights were modified after each batch. Early halting was used to guarantee stable performance and prevent overfitting.

The suggested procedure offers a strong, dependable, and constant method for detecting Alzheimer's illness by utilizing this structured methodology. It guarantees precise stage classification without depending on automation tailored to artificial intelligence, emphasizing thorough research and clear diagnostics instead.

4. Result and Discussion

In recognizing and categorizing brain MRI scans into four stages – Non-Demented, Very Mild Demented, Mild Demented, and Moderate Demented – the custom CNN demonstrated exceptional accuracy and dependability, according to the findings of the Alzheimer's disease detection research. By carefully adjusting the hyperparameters and performing thorough preprocessing, the CNN demonstrated exceptional feature extraction and classification abilities. In terms of accuracy and computational

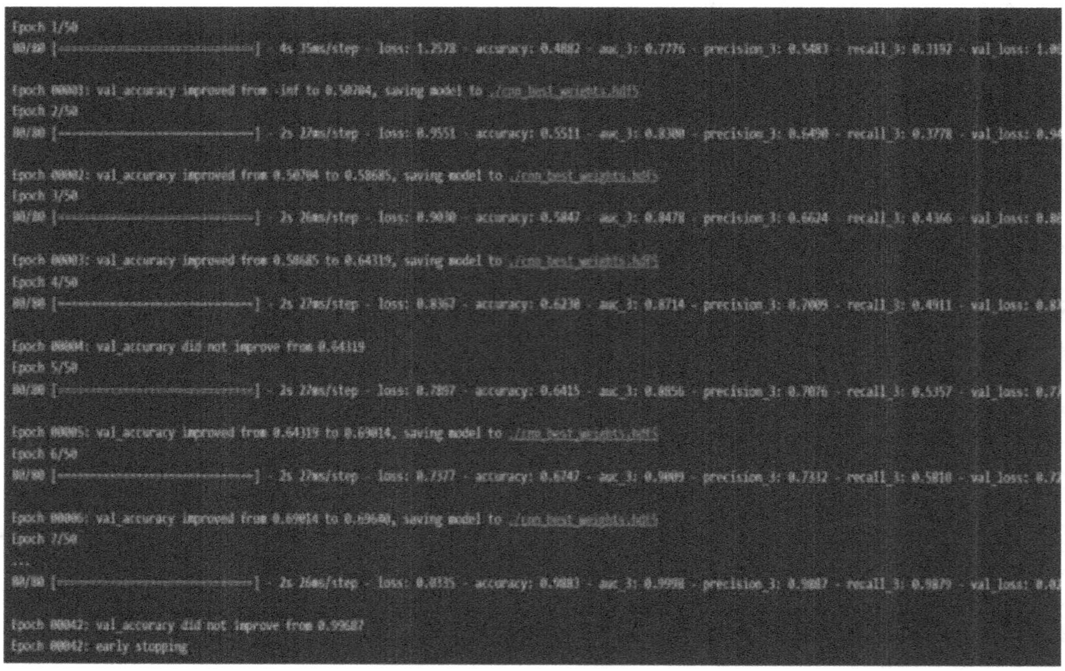

Figure 29.3. Epoching until val_accuracy did not improve.

Source: Author.

efficiency, the customized CNN fared better than pre-trained models such as VGG16 and DenseNet across a variety of assessed topologies. This improved performance is the result of domain-specific customization and optimization techniques within the network (Figure 29.4).

4.1. Performance analysis

In Figure 29.5, the proposed model accuracy is compared with existing model accuracy. The model's accuracy in both training and validation rose steadily over the course of five training epochs. In the first epoch, the model's training accuracy was 83.85%. By the fifth epoch, it had reached a high validation accuracy of 97.22%. Significantly lower corresponding validation loss demonstrated efficient learning and generalization. A remarkable test accuracy of 99.07% and a low test loss of 0.0454 were displayed in the final test results (Figure 29.6). These outcomes demonstrate how well the model predicts Alzheimer's stages based on brain MRI images (Figure 29.2) and Precision plot is showed in Figure 29.7.

4.2. Key contributions

1. **Data Augmentation and Preprocessing:** Improved generalization across various scan conditions, reducing biases associated with the dataset.
2. **Custom Architecture Optimization:** Achieved better performance than pre-trained models by adapting the network specifically for Alzheimer's stage detection.
3. **Streamlit Integration:** Deployed the CNN model into a user-friendly web application that allows real-time MRI image analysis with high confidence, bridging the gap between AI advancements and clinical usability.

4.3. Challenges and future directions

Although the experiment was successful, it encountered difficulties with class imbalance and dataset variability, which may have an impact on the model's generalizability. This imbalance develops because Alzheimer's tends to be more common in the early stages, it's difficult to collect MRI scans from the later stages, and datasets often have more data from persons in the earlier stages. Methods like resampling can be followed to address this problem, either by decreasing the number of samples of overrepresented stages or by increasing the amount of samples from the underrepresented ones. Another method is data augmentation, which involves altering photos (such as flipping or rotating them) to give the model more variety to work with. In order to give the less frequent stages greater weight during training, class weighting can also be employed. By using these techniques, the model is kept from concentrating too much on the majority class and failing to recognize significant trends in the minority classes. Larger and more varied datasets will be used in future research to improve performance and dependability. Furthermore, the incorporation of explainable AI approaches will enhance clinical uptake and trust.

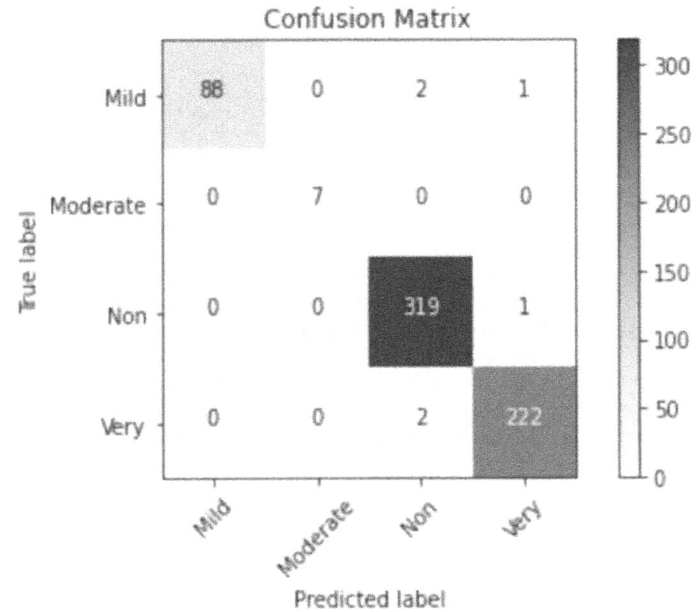

Figure 29.4. Confusion matrix.

Source: Author.

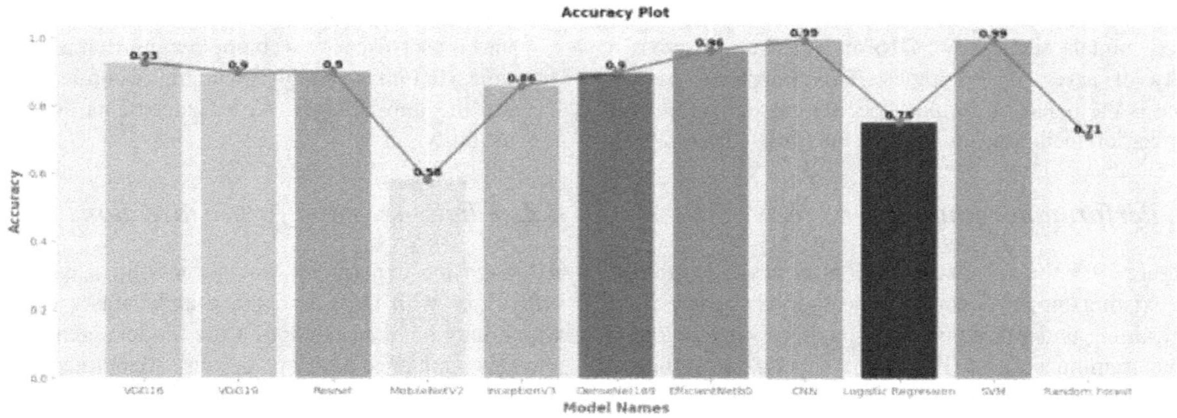

Figure 29.5. Combined accuracy plot.

Source: Author.

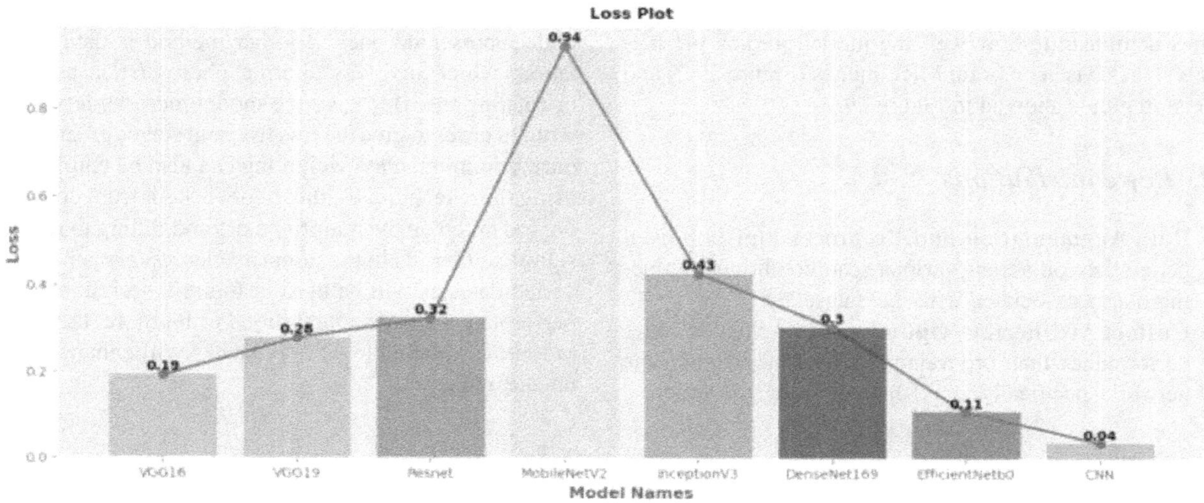

Figure 29.6. Loss plot.

Source: Author.

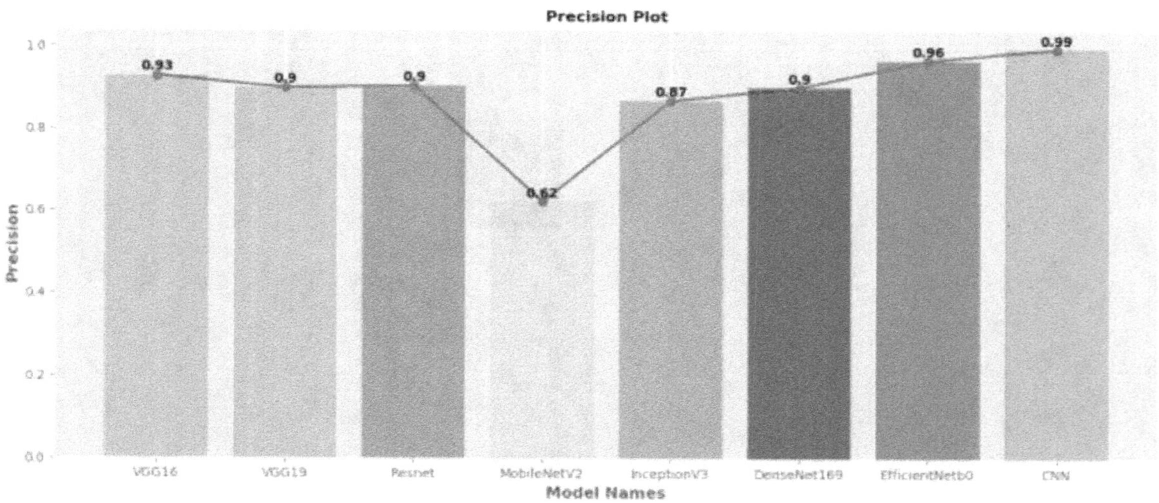

Figure 29.7. Precision plot.

Source: Author.

The results demonstrate the potential of tailored CNN models in medical diagnostics, paving the way for real-time, scalable systems for detecting Alzheimer's disease (Table 29.1).

The Alzheimer's disease prediction model is trained over a number of epochs, with each stated epoch (1–5) denoting the performance averaged over 1–10 mini-epochs. The model's initial training and validation accuracy in Epoch 1 are 83.85% and 80.23%, respectively, with training and validation losses of 0.5214 and 0.4903. While the losses decrease to 0.0930 and 0.0785, respectively, the accuracy dramatically increases to **96.46%** for training and 94.32% for validation by Epoch 2 (11–20 mini-epochs). Training and validation losses further drop to 0.0854 and 0.0768 as the model continues to improve its learning in Epoch 3 (21–30 mini-epochs), reaching 96.84% training accuracy and 95.12% validation accuracy (Figure 29.8). Training accuracy increases to 97.83% and validation accuracy to 96.88% in Epoch 4 (31–40 mini-epochs), while losses decrease to 0.0698 and 0.0623. At last, the model reaches its peak performance in Epoch 5 (41–50 mini-epochs) with 98.38% training accuracy and 97.22% validation accuracy. It also has the lowest training loss of 0.0428 and validation loss of 0.0454. This consistent decrease in loss and steady increase in accuracy show how effectively the model learns and how well it generalizes without experiencing severe over-fitting.

These results confirm the effectiveness of the proposed methodology in delivering accurate and efficient diagnostic predictions, paving the way for broader clinical adoption (Figure 29.9).

5. Conclusion

This study highlights the importance of customizing CNN models for specific applications, such as detecting Alzheimer's disease. When comparing a CNN built from start with one that has undergone hyperparameter tweaking, it is clear that there are benefits to fine-tuning key parameters to enhance model performance. During the early stages of training, the scratch model struggled with stability and generalization, but it was able to learn and improve over time. Contrarily, the hyperparameter-tuned model demonstrated improved accuracy, precision, and recall, indicating its ability to adapt to the dataset and generate more reliable predictions. The results highlight how performance in medical applications, including Alzheimer's disease diagnosis, can

Table 29.1. Training and testing results

Epochs	*Training Accuracy*	*Validation Accuracy*	*Training Loss*	*Validation Loss*
1	83.85%	80.23%	0.5214	0.4903
2	96.46%	94.32%	0.0930	0.0785
3	96.84%	95.12%	0.0854	0.0768
4	97.83%	96.88%	0.0698	0.0623
5	98.38%	97.22%	0.0428	0.0454

Source: Author.

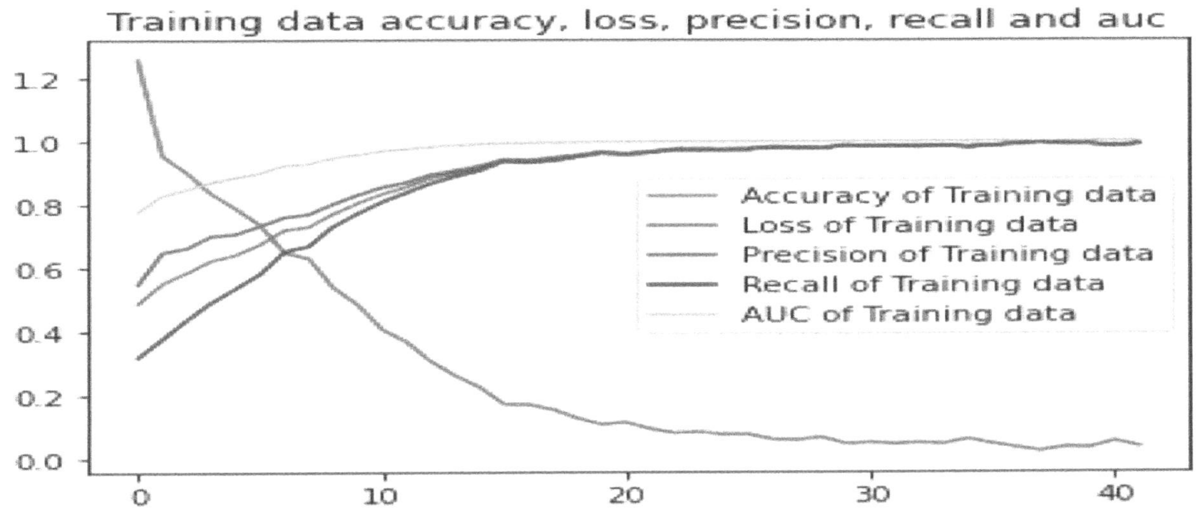

Figure 29.8. Training data accuracy, loss, precision, recall.

Source: Author.

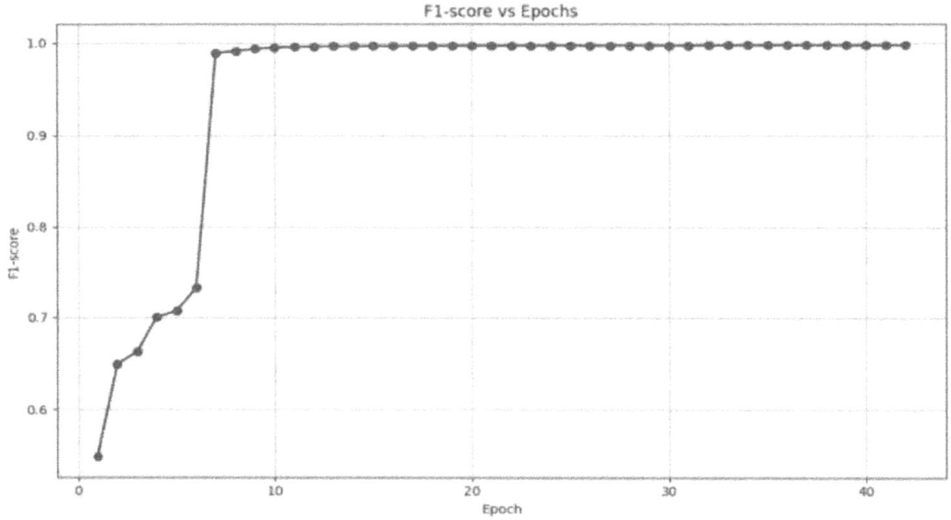

Figure 29.9. F1 score.
Source: Author.

be greatly improved by refining deep learning models. This method makes it possible to create automated systems with high diagnostic accuracy, which promotes early diagnosis and prompt action. This study offers a scalable and useful way to enhance patient care and slow the progression of Alzheimer's disease by incorporating these models into user-friendly web applications. This discovery has revolutionary potential for managing neurodegenerative diseases and represents a significant development in the application of AI and deep learning to precision healthcare.

References

[1] Amini, S., Hao, B., Yang, J., et al. (2024). Prediction of Alzheimer's disease progression within 6 years using speech: A novel approach leveraging language models. *Alzheimer's & dementia: the journal of the Alzheimer's Association, 20*(8), 5262–5270. *https://doi.org/10.1002/alz.13886*

[2] Kavitha, C., Mani, V., Srividhya, S. R., Khalaf, O. I., & Tavera Romero, C. A. (2022). Early-Stage Alzheimer's Disease Prediction Using Machine Learning Models. *Frontiers in Public Health, 10*, 853294. *https://doi.org/10.3389/fpubh.2022.853294*

[3] Aditya, S. K., Vijayakumar, V., Manoj Kumar, M. V., Manjunatha, B. A., & Chandrashekhar, B. N. (2022). Deep Learning Techniques for the Effective Prediction of Alzheimer's Disease: A Comprehensive Review.

[4] Viswan, V., Shaffi, N., Mahmud, M., et al. (2024). Explainable Artificial Intelligence in Alzheimers' Disease Classification: A Systematic Review. *Cognitive Computation, 16*, 1–44. *https://doi.org/10.1007/s12559-023-10192-x*

[5] García-Gutiérrez, F., Alegret, M., Marquié, M., et al. (2024). Unveiling the sound of the cognitive status: Machine Learning-based speech analysis in the Alzheimer's disease spectrum. *Alzheimer's Research & Therapy, 16*(1), 26. *https://doi.org/10.1186/s13195-024-01394-y*

[6] Tang, A. S., Rankin, K. P., Cerono, G., et al. (2024). Leveraging electronic health records and knowledge networks for Alzheimer's disease prediction and sex-specific biological insights. *Nature Aging, 4*, 379–395. *https://doi.org/10.1038/s43587-024-00573-8*

[7] Suresha, H. S., & Parthasarathy, S. S. (2020). Alzheimer Disease Detection Based on Deep Neural Network with Rectified Adam Optimization Technique using MRI Analysis. *Third International Conference on Advances in Electronics, Computers and Communications (ICAECC)* (pp. 1–6). IEEE.

[8] Deng, L., & Wang, Y. (2021). Hybrid diffusion tensor imaging feature-based AD classification. *Journal of X-Ray Science and Technology, 29*(1), 151–169.

[9] Khan, A., & Zubair, S. (2022). An improved multimodal based machine learning approach for the prognosis of Alzheimer's disease. *Journal of King Saud University – Computer and Information Sciences, 34*(6), Part A, 2688–2706.

[10] Asim, Y., Raza, B., Malik, A. K., Rathore, S., Hussain, L, & Iftikhar, M. A. (2018). A multi modal, multi-atlas-based approach for Alzheimer detection via machine learning. *International Journal of Imaging Systems and Technology, 28*(2), 13–123.

[11] Pavitra, B., Singh, D. N., & Sharma, S. K. (2020). Predictive and Interactive IOT Diagnosis System with AI and ML Tools. *In International Conference on*

Advances in Computational Intelligence and Informatics (pp. 9–16). Springer, Singapore.

[12] Singh, D. N., Hashmi, M. F., & Sharma, S. K. (2020). Predictive analytics & modelling for modern health care system for cerebral palsy patients. *Multimedia Tools and Applications*, *79*(15), 10285–10307.

[13] Roshanzamir, A., Aghajan, H., & Soleymani Baghshah, M. (2021). Transformer-based deep neural network language models for Alzheimer's disease risk assessment from targeted speech. *BMC Medical Informatics and Decision Making, 21*(1), 1–14.

[14] Doan, D. N. T., Ku, B., Choi, J., Oh, M., Kim, K., Cha, W., & Kim, J. U. (2021). Predicting dementia with prefrontal electroencephalography and event-related potential. *Frontiers in Aging Neuroscience, 13*, 180.

30 Sustainability in infant care: An optimal music identifying cradle using reinforcement learning

Jyosna Philip[a] *and Vijayalakshmi S.*[b]

Department of Data Science, CHRIST University, Bengaluru, Karnataka, India

Abstract: Smart cradles have emerged as a promising application in the domain of Internet of Things (IoT), with the primary aim of reducing the need for constant parental supervision. Calming crying babies remains a frequent yet challenging task for caregivers. This work introduces an IoT-based cradle system with two significant enhancements: an adaptive lullaby selection mechanism and protection from insects in the surroundings. The proposed IoT framework uses a Raspberry Pi microcontroller. The music selection system uses strategies based on Multi-Armed Bandit (MAB) algorithms to identify the most effective lullaby to soothe infants. Different infant response time scenarios were tested and the system consistently identified and prioritized lullabies with shorter response times. Even in cases where response times were closely distributed, the system maintained adaptability, ensuring practical success in calming. Hence, this work demonstrates the potential for broader applications of multi-armed bandits for caregiving.

Keywords: Bayesian learning, internet of things, multi-armed bandits, smart cradles

1. Introduction

The growing trend of working parents has transformed the landscape of infant care, placing many challenges in the way of appropriate, consistent quality care to be provided to infants. This situation implies that intelligent systems need to be developed that assist parents in monitoring their infant, without such systems being a replacement for the emotional bond necessary for infant development. Although traditional baby monitoring systems have provided basic surveillance capabilities using the Internet of Things (IoT), integrating IoT technology with artificial intelligence presents opportunities for more sophisticated and responsive infant care solutions. Several research studies emphasize the importance of environmental factors for the health and development of an infant [1]. The first years of life are susceptible to environmental conditions, including lighting patterns and auditory stimulation. Among the most difficult tasks caregivers have with this age group is helping a crying baby, where the effort needed can often be physically and mentally draining. Music can play an essential role in soothing babies by providing a comforting auditory stimulus. However, the effectiveness of a lullaby can vary for different infants and may depend on a variety of factors, including individual auditory preferences, cultural influences, and emotional states. The current systems for automated baby monitoring either have hard-coded responses to sensor values or have used a trained model for surveillance, preventing any form of adjustments for a baby's changing needs. Reinforcement Learning (RL) has shown high effectiveness in systems that require adaptive decision making. The agent in RL learns its actions by interacting with its environment and each action leads to a certain reward from the environment [2]. A critical aspect of RL is addressing the explore-exploit trade-off, which determines whether the system should continue exploring new actions or exploit existing knowledge to maximize rewards. Exploration involves trying out actions irrespective of them leading to a positive outcome in the past. Exploitation uses the information obtained through exploration and sticks to the actions that led to positive outcomes. Pure exploration is not ideal because it prevents one from maximizing the rewards. On the other hand, pure exploitation is also not preferred as there is a possibility of a subpar action being repeatedly taken when a potentially better one is out there and wasn't discovered due to a lack of exploration. In such a scenario, it is necessary to balance exploration and exploitation. The Multi-Armed Bandit (MAB) problem, a fundamental RL approach, is framed as models of decision-making scenarios where limited resources must be allocated among competing choices to optimize rewards. Each choice, or "arm," has an unknown probability distribution of rewards. Through repeated exploration and exploitation, the reinforcement

[a]philipjyosna02@gmail.com, [b]s.vijayalakshmi@christuniversity.in

DOI: 10.1201/9781003770473-30

learning agent attempts to approximate the reward probability distributions of each arm. This approximation is then used to inform future decisions made by the agent. The MAB framework enables the system to continuously learn and refine its strategy by observing and analyzing the outcomes of previous decisions.

2. Related Works

2.1. IoT-based baby monitoring systems

Over the past decade, infant care has had significant improvements in terms of monitoring technologies. Multiple attempts have been made to integrate various features like environment monitoring, face detection, emotion detection and so on into baby cradles as seen in the works of Alam et al. [3], Pratap et al. [4], Reddy et al. [5], and Alishamol et al. [6]. Some of the main motives for such innovations were to reduce the need for personal surveillance by parents and the reduction of parental interventions. The cradle system, as proposed by Jabbar et al. [7], makes use of a NodeMCU micro controller and a DHT22 temperate sensor for environmental monitoring. It also had Wi-Fi connectivity and real-time data transfer. If the temperature sensor values becomes more than a threshold, a fan is switched on. Parents can view and monitor their child using a wireless camera and also control the operation of the fan. When the cry is detected, the system swings the cradles using a DC motor. Parents can also manually adjust the swing speed. The wireless security camera is connected to the home Wi-Fi, and the visuals are sent directly to the parents' mobile phones. A photoresistor and IR LEDs are also added to enable recording during the night. The camera head is rotatable and can be controlled using an app. The setup proposed by Sontakke et al. [8], uses an ATMEGA 328 microcontroller and swings the cradle using a servo motor when the sound level crosses a certain threshold. While this approach provides automated responses, it does not adapt to the baby's specific needs over time. The system also incorporates wetness detection and temperature sensing and checks whether the baby is in the cradle using motion sensors. After the cradle begins swinging, a notification is sent to parents about the cry. Matsunaga et al. [9] used hidden Markov models to identify infant emotion from their cries. The system proposed by Alswedani [10] uses voice, face image and movement analysis to create a decision fusion model. However, these approaches focus more on classification rather than adaptive response mechanisms. The proposed system reported by Kannan et al. [11], integrates various sensors to monitor the infant's environment. A motion sensor gets triggered by the movements of the baby, which activates the video camera, thus saving power by enabling surveillance only when the infant is in the cradle. Bhasha et al. [12] proposed a system architecture with a NodeMCU as the microcontroller. Connected to this was a soil moisture sensor that identified whether the infant's diaper needed to be changed. Along with this, it used a sound sensor to detect baby cries. A response system that played pre-loaded music when the sound values were beyond a threshold was implemented. An alert message is then sent to the parents' mail. The system was also interfaced with a Blynk application. This app enabled the parents to play music and turn on the fan whenever required. However, it played the same songs irrespective of the baby's response to it. In the system proposed by Salehin et al. [13], a speaker integrated with stereo compatibility is attached to the system. When the sensor detects the baby's cry, it signals the Songle single-channel relay module to play music. This is switched off automatically after some time. However, the lack of adaptive learning meant that these systems could not optimize the choice of music based on its effectiveness in soothing the infant.

2.1.1. Gaps in existing literature

- Most existing baby monitoring systems rely on predefined threshold-based responses rather than learning from the baby's reactions.
- Music-based soothing mechanisms implemented in prior studies play fixed songs without adapting to the baby's preferences or effectiveness over time.
- No existing studies have applied reinforcement learning to dynamically adjust the system's response based on real-time feedback from the baby.
- Current systems do not measure the effectiveness of their responses in a systematic way, such as tracking the duration required to soothe the baby.

2.2. Music therapy for infant soothing

A systematic review and meta-analysis by Cavalcante et al. [14], involving approximately 737 patients demonstrated that music therapy significantly reduced respiratory rates and pain sensitivity indices while improving oxygen saturation levels in both pediatric and neonatal ICUs. The study also highlighted that using music preferred by patients yielded better outcomes. Namjoo, et al. [15] did a randomized clinical trial with 90 preterm infants to compare the effects of recorded lullabies, mothers' live lullabies, and no intervention. Infants in the recorded lullaby group had a significantly longer sleep duration overnight when compared to the other two groups. From these findings, music could be used as a tool for promoting sleep in infants.

2.3. Advances in RL based adaptive systems

Reinforcement learning has proven effective in systems requiring adaptive decision-making under uncertainty. Ritto et al. [16], explored RL combined with Approximate Bayesian Computation (ABC) for model selection and parameter identification in nonlinear mechanical oscillators. Their approach used Thompson-like sampling and Bayesian updates to refine the prior distribution of model parameters. Bayesian updating methods have been extensively validated in applications involving uncertain environments and real-time feedback. In the work of Scott [17], a randomized probability matching method is used as an approach for a multi-armed bandit problem. Thompson Sampling (TS) is a probabilistic method that successfully balances exploration and exploitation [18]. The algorithm starts with a prior distribution on the mean reward of each arm, then pulls an arm based on the current distribution of mean rewards and gets new data. Then, the current distribution will be updated using new data to obtain a new posterior distribution [19].

3. Methodology

3.1. IoT system architecture

The proposed IoT-based implementation incorporates a distributed architecture where computing and decision-making are carried out off-device to provide scalability and efficiency. There are primarily three components in the system: IoT enabled cradle, cloud server, and mobile application interface. The architecture is illustrated in Figure 30.1. The IoT-enabled cradle comes with sensors intended to monitor the behaviour of an infant, such as sound level, movement, and physiological signals. Raspberry Pi, the microcontroller unit, transmits the processed sensor data to the cloud server using Wi-Fi. Further, the cradle has an audio output module to play selected songs. The cloud-based server performs the computational tasks associated with the Bayesian MAB framework. Upon receiving data from the cradle, the server updates the Bayesian model, executes the Thompson Sampling algorithm, and determines the next song to be played. The server then sends this decision back to the cradle, ensuring minimal computational overhead on the microcontroller. A mobile application interface is used to provide caregivers with updates in real-time. Parents can also upload custom lullabies, which can include their voice recordings. In case the song does not stop the baby from crying, an app notification is sent to the parent. Caregivers can always override a song selection. The app can also provide remote monitoring, allowing caregivers to control the cradle from anywhere in the world with the Internet.

3.2. Infant soothing mechanism

To create a system that identifies the preferred lullabies that soothe the baby while also adapting to their changing preferences, the Multi-Armed Bandit (MAB) framework and Bayesian optimization is used. This was because it efficiently balances exploration, where different songs are tested to collect data, and exploitation, where the song with the minimum response time so far is played. MAB is also useful when the environment is non-stationary, that is, the infant's preferences may vary over time.

The system is modelled as a decision-making agent embedded in an IoT-enabled cradle, capable of learning the most effective song for calming a crying infant over successive interactions. The objective of the system is to reduce the time needed to calm a crying infant by selecting the most effective song from a limited set of choices. Each song i is modeled as an arm in the MAB framework and has an unknown average response time μ_i, representing the time it takes for the infant to calm when that song is played. Each song corresponds to one of the "arms" of the MAB framework. The agent collects feedback by tracking the time taken for the infant to stop crying. This is then used to update the model.

For this problem, a Bayesian strategy has been adopted. Bayesian methods are suited to this situation, as they properly model uncertainty and update the model's estimates as new data comes in. In our model, the performance of each song is respectively modelled by a Beta distribution, initialized with prior parameters

$$\alpha = 2, \beta = 1$$

These priors represent a moderately optimistic initial belief about each song's performance before this belief is confirmed or modified through the actual data to be collected. At each decision step, the system samples values from the Beta distribution of each song and selects the one with the highest sampled value.

Following the selection of the song, the infant's response time is recorded, and a reward is computed. The reward function is defined as

$$R = \frac{1}{(1 + \max(r, \epsilon))} \tag{1}$$

Where r is the response time, and $\in$ is a small constant to prevent division by zero. This formulation gives importance to shorter response times. The Beta distribution parameters are then updated based on the computed reward. It updates α by adding the reward value and β by adding 1 minus the reward.

To evaluate the system, simulations were conducted by modelling infant response times. Five hypothetical songs were assigned different mean response times for each trial. To mimic a real-world scenario, Gaussian noise was added to create variability as infant responses may vary from time to time. The selection frequency of the best song was considered as the performance metric. It counts how many times a song was selected and tells us how much the system is capable of identifying the optimal lullaby.

3.3. Insect prevention mechanism

To enhance infant safety, the system also features a mesh barrier that will cover the top of the cradle. Ultrasonic

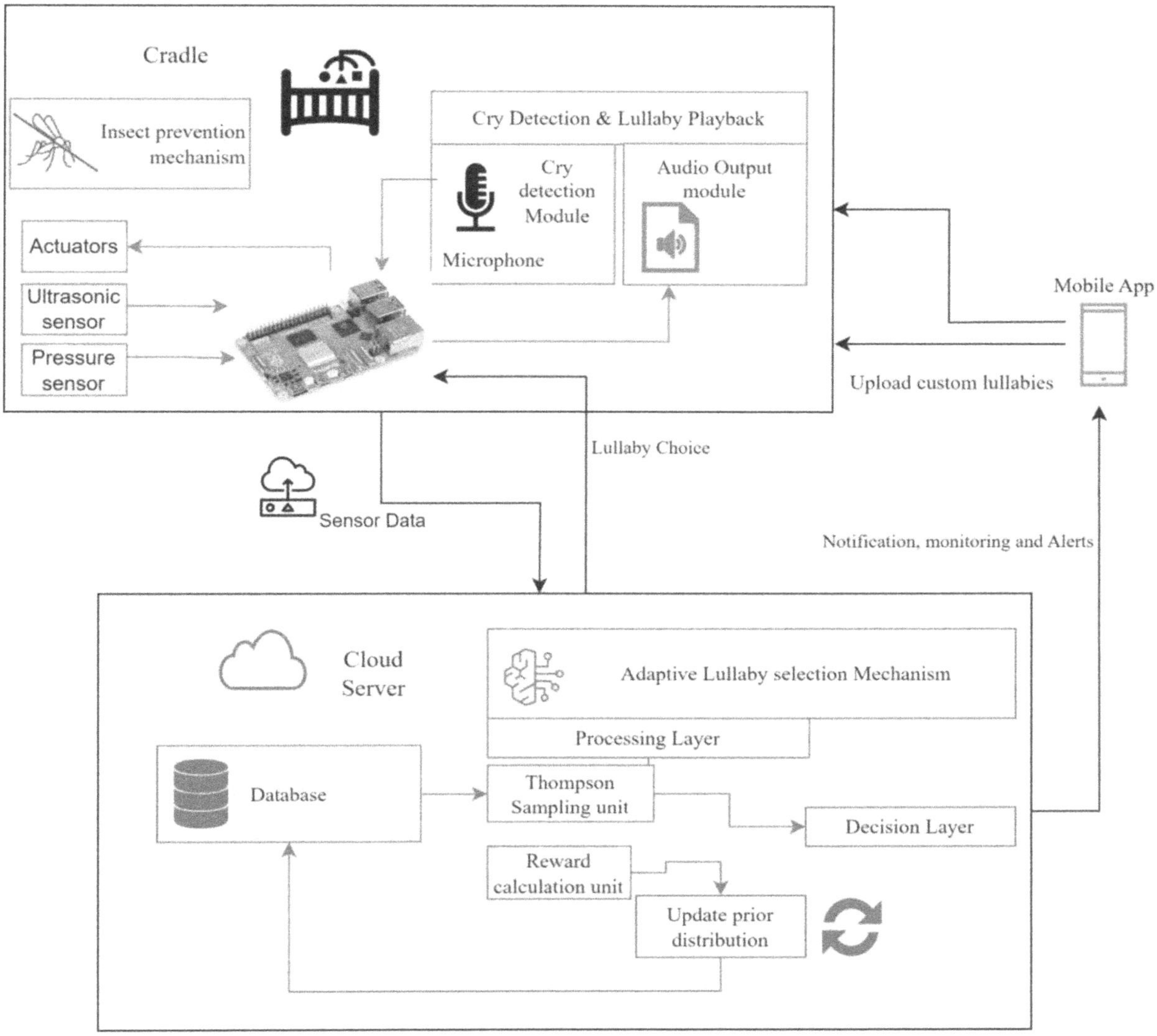

Figure 30.1. System architecture of the proposed cradle framework.

Source: Author.

sensors are then strategically placed to detect when someone like a parent or guardian approaches to view the baby. When it senses a person, it triggers servo motors that lift the lid. A pressure sensor is used to check whether the baby is in the cradle. The output from this sensor is also used to activate the lid opening system. This will ensure that it does not open unnecessarily when there is no baby inside.

To evaluate the effectiveness of the insect prevention mechanism, the system was tested in a controlled environment by placing the cradle model in an insect prone area, while monitoring their entry attempts. Results indicated that the mesh barrier successfully blocked insects while allowing proper ventilation. Additionally, response time analysis showed that the lid opened promptly upon detecting a caregiver, maintaining both security and convenience. Compared to traditional open cradles, this system demonstrated a significant reduction in insect presence, ensuring that environmental factors do not disrupt the infant's comfort and safety while supporting the system's core soothing function.

4. Results and Discussion

The system was tested to verify its effectiveness through simulations using varying infant response time scenarios. Two cases were tested and analysed. In the first case, when the system was modelled with one song with a clearly lower response time and therefore, more preferable than all others, the system almost always selected it. The frequency of optimal song selection went up steadily with the number of training episodes. This shows that the model exploits the best song when it has the best chance while exploring other options in the first episodes. This scenario also saw optimal songs receive the highest cumulative rewards, which emphasizes the speed with which the system worked toward converging on the best option. The results are illustrated in Figures 30.2 and 30.3. The second case had the response times for all songs modelled with values that were close around. Here, the system spreads selections more equally. However, songs with marginally shorter response times were still selected in slightly greater quantities. Such behaviour meets the intent of this system to soothe the baby as quickly as possible rather than rigidly determining the absolute best song. The ability to adapt and perform in different situations that vary only slightly in preferences demonstrates the adaptability of the Bayesian approach. The results are illustrated in Figures 30.4 and 30.5. The main observation here is that although in some instances where multiple songs worked equally well, the system had managed to soothe the baby, thus pointing out the focus of the system on practical success rather than over-optimizing.

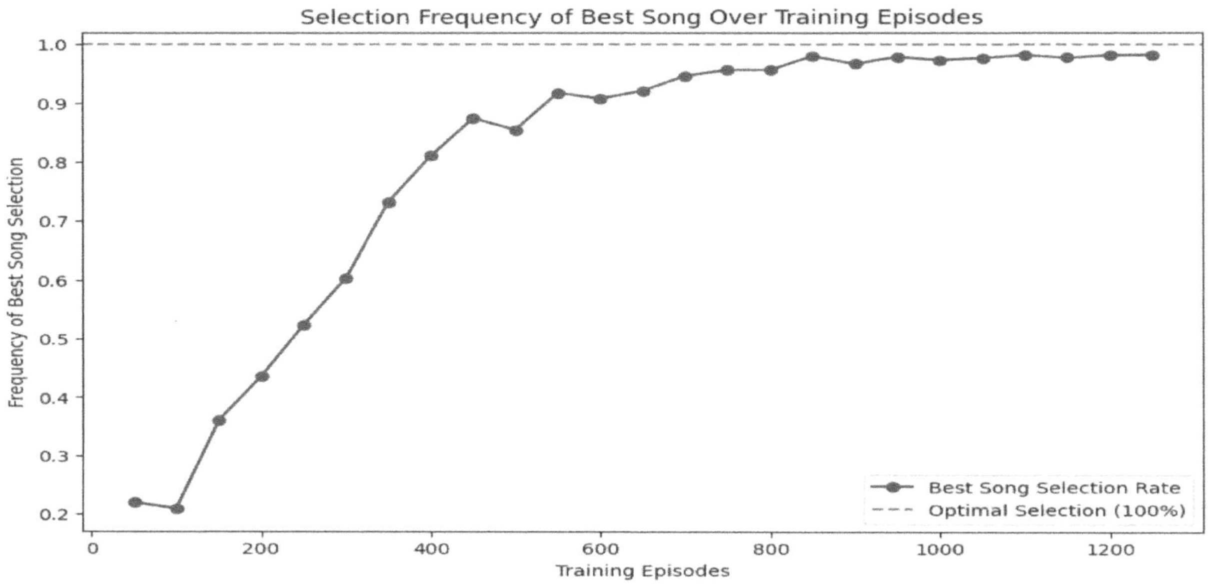

Figure 30.2. Selection frequency of best song when one lullaby is modelled to have a much lower response time than others.

Source: Author.

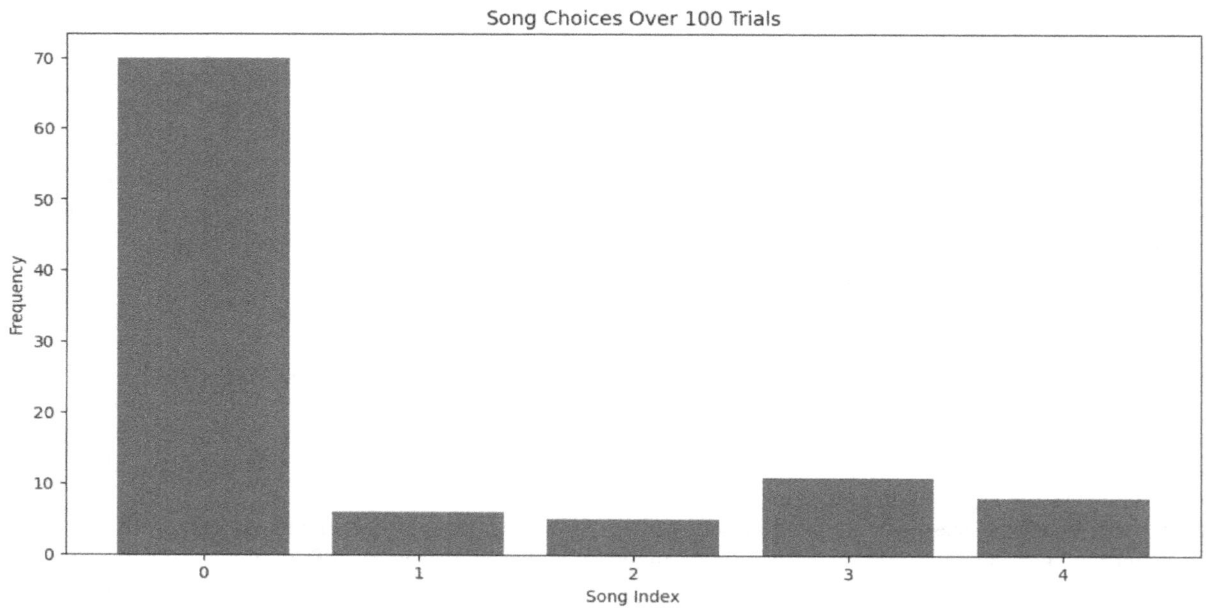

Figure 30.3. Selection frequency of preferred lullaby (song index 0) over 100 trials as compared to the selection of other lullabies when one lullaby is modelled to have a much lower response time than others.

Source: Author.

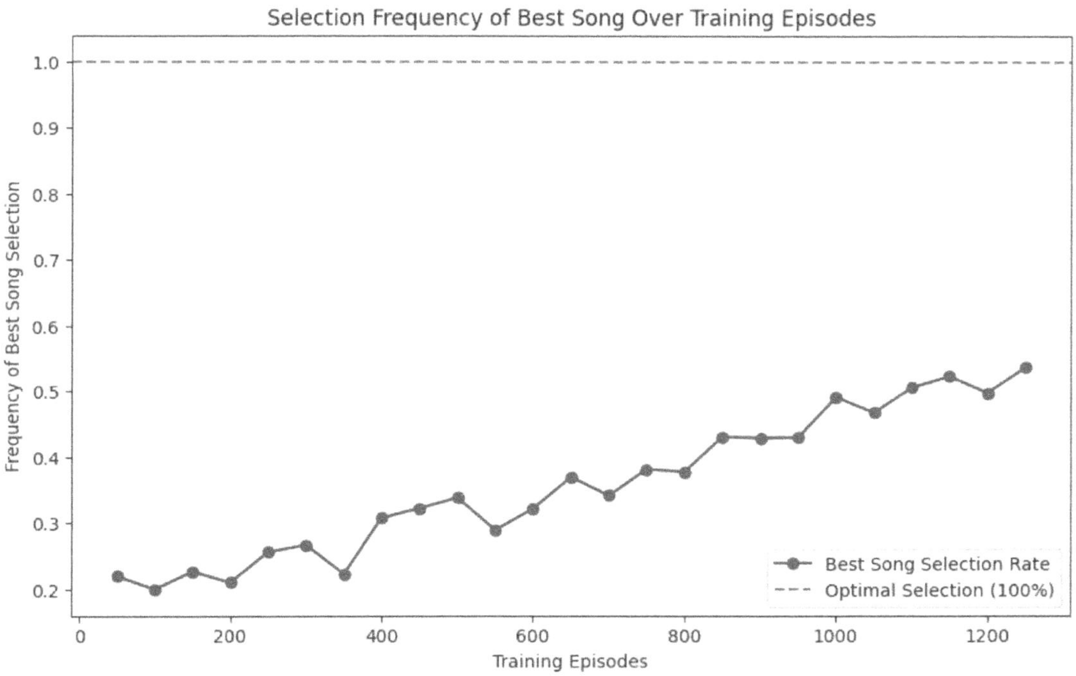

Figure 30.4. Selection frequency of preferred lullaby (song index 0) over 100 trials as compared to the selection of other lullabies when one lullaby is modelled to have a much lower response time than others.

Source: Author.

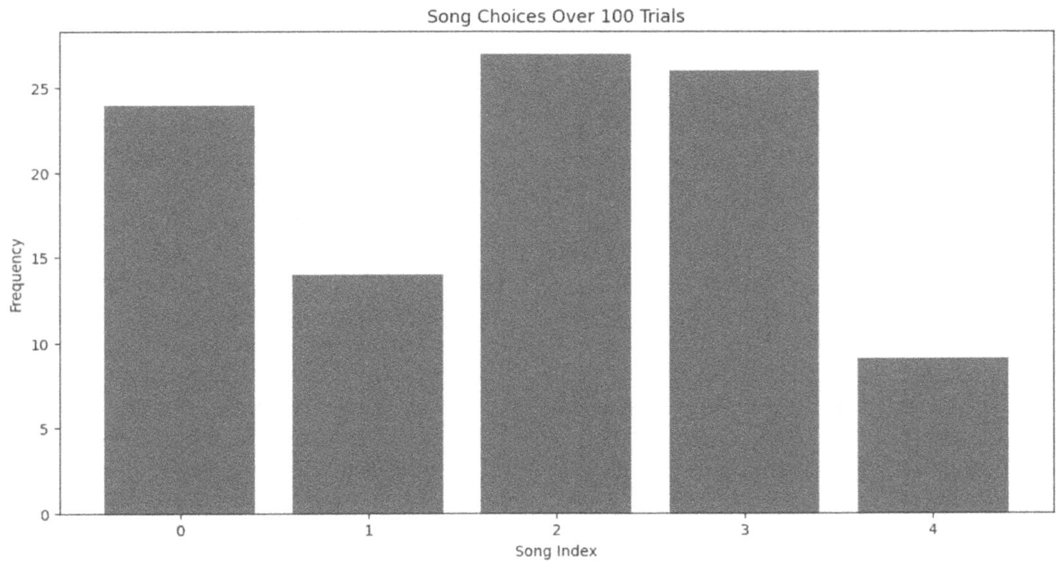

Figure 30.5. Selection frequency of preferred lullaby (song index 0) over 100 trials as compared to the selection of other lullabies when one lullaby is modelled to have a much lower response time than others.

Source: Author.

5. Conclusion

This paper proposed an IoT-enabled smart cradle system that soothes crying infants through adaptive selection of music choice and ensures safety through protection mechanisms against insects. With the Bayesian Multi-Armed Bandit framework, it learns and adapts in real time by making optimal lullaby choices based on infant response times and strikes a balance between exploration and exploitation. The simulation results are promising since it can find the favourite lullabies and adapt if preferences may change over time. Furthermore, the incorporation of insect protection provides a more secure environment and therefore makes the system more practically viable for infant care. Future work will focus on experimental validation in real caregiving environments to assess system performance under diverse infant behaviours and environmental conditions. Additionally, integration of smart home ecosystems like Amazon Alexa or Google Home with the cradle could enhance the overall usability. Such adaptive frameworks may also be incorporated in other elements of the infant cradle.

References

[1] Hodel, A. S. (2018). Rapid infant prefrontal cortex development and sensitivity to early environmental experience. *Developmental Review, 48*, 113–144. https://doi.org/10.1016/j.dr.2018.02.003

[2] Sutton, R. S., & Andrew, G. B. (1998). Reinforcement learning: An introduction. *1*(1). Cambridge: MIT Press.

[3] Alam, H., Muhammad, B., Anusha, G., et al. (2023). IoT based smart baby monitoring system with emotion recognition using machine learning. *Wireless Communications and Mobile Computing, 1*, 1175450. https://doi.org/10.1155/2023/1175450

[4] Pratap, N. L., Anuroop, K., Nirmala Devi, P., Sandeep, A., & Sunanda, N. (2021). IoT based smart cradle for baby monitoring system. *In 2021 6th International Conference on Inventive Computation Technologies (ICICT),* 1298–1303. IEEE. https://doi.org/10.1109/icict50816.2021.9358684

[5] Reddy, Y. S. S., Vamsi, K. S. V., Akhila, G., Poonati, A., & Jayanth, K. (2021). An automated baby monitoring system. *International Journal of Engineering and Advanced Technology (IJEAT).*https://doi.org/10.35940/ijeat.f3058.0810621.

[6] Alishamol, K. S., Fousiya, T. T., Jasmin, B. K., Sooryadas, M., & Leena, M. (2020). System for infant cry emotion recognition using DNN. *In 2020 Third International Conference on Smart Systems and Inventive Technology (ICSSIT)* (pp. 867–872). IEEE.

[7] Jabbar, W. A., Shang, H. K., Hamid, S. N., Almohammedi, A. A., Ramli, R. M., & Ali, M. A. (2019). IoT-BBMS: Internet of Things-based baby monitoring system for smart cradle. *IEEE Access, 7*, 93791–93805. https://doi.org/10.1109/access.2019.2928481

[8] Sontakke, D., Akanksha, K., Sneha, S., & Meenakshi, A. (March, 2024). IOT Based Baby Monitoring System for Smart Cradle. *International Research Journal of Innovations in Engineering and Technology – IRJIET, 8*(3), 232–236. doi:10.47001/IRJIET/2024.803033.

[9] Matsunaga, S., Sakaguchi, S., Yamashita, M., Miyahara, S., Nishitani, S., & Shinohara. K. (2006). Emotion detection in infants' cries based on a maximum likelihood approach. *In Proc. Interspeech,* 1834–1837. https://doi.org/10.21437/interspeech.2006-506

[10] Alswedani, S. A., & Fathy, E. E. (2020). A smart baby cradle based on IoT. *International Journal of Computer Science and Mobile Computing, 9*(7), 64–76.

[11] Kannan, P., Allwin Devaraj, B., Pradheep T., et al. (2021). IoT Based Smart Cradle Using PI. In *Recent Trends in Intensive Computing* (pp. 716–720). IOS Press. https://doi.org/10.3233/apc210268

[12] Bhasha, P., Pavan Kumar, T., Khaja Baseer, K., & Jyothsna, V. (2021). An IoT-based BLYNK server application for infant monitoring alert system to detect crying and wetness of a baby. *In International Conference on Intelligent and Smart Computing in Data Analytics: ISCDA 2020* (pp. 55–65). Springer Singapore.

[13] Salehin, H. I., Joy, Q. R. A., Aparna, F. T. Z., Ridwan, A. T., & Khan, R. (2021, May). Development of an IoT based smart baby monitoring system with face recognition. In *2021 IEEE World AI IoT Congress (AIIoT)* (pp. 0292–0296). IEEE.

[14] Cavalcante, L. L., Sciarra, A. M. P., & de Godoy, M. F. (2024). Music Therapy in Pediatric and Neonatal Intensive Care Units: A Systematic Review and Meta-Analysis. *medRxiv*, 2024-10. https://doi.org/10.1101/2024.10.13.24314254

[15] Namjoo, R., Roghayeh, M., Behnaz, B., & Monirsadat, N. (2022). Comparing the effectiveness of mother's live lullaby and recorded lullaby on physiological responses and sleep of preterm infants: a clinical trial study. *Journal of Complementary and Integrative Medicine, 19*(1), 121–129. https://doi.org/10.1515/jcim-2020-0507

[16] Ritto, T. G., Beregi, S., & Barton, D. A. W. (2022). Reinforcement learning and approximate Bayesian computation for model selection and parameter calibration applied to a nonlinear dynamical system. *Mechanical Systems and Signal Processing, 181*, 109485.

[17] Scott, S. L. (2010). A modern Bayesian look at the multi-armed bandit. *Applied Stochastic Models in Business and Industry, 26*(6), 639–658.

[18] Chapelle, O., & Li, L. (2011). An empirical evaluation of thompson sampling. *Advances in Neural Information Processing Systems, 24*.

[19] Song, R. (2024). Optimizing decision-making in uncertain environments through analysis of stochastic stationary Multi-Armed Bandit algorithms. *Applied and Computational Engineering, 68*, 93–113.

31 Innovative tools for AI-driven scalp and hair assessment: Evaluating mobile application efficacy and user experience

Amaal Mohammad Abdullah Al Maqbali, Faizal Hajamohideen[a], Rayyan Abdullah Said Al Handasi, Mahzabee Noorulhasan Shams AlDhuha, and Mubassareen Noorulhasan ShamsAl Dhuha

College of Computing and Information Sciences, University of Technology and Applied Sciences (UTAS), Suhar Sultanate of Oman, Oman

Abstract: The application of artificial intelligence (AI) in scalp and hair assessment has become a promising frontier in the quickly developing fields of dermatology and personal care. This study examines the effectiveness of cutting-edge smartphone apps that use artificial intelligence (AI) to analyze hair health and scalp issues. Using a mixed-methods approach, we evaluate user experience and happiness while determining the accuracy and dependability of several mobile applications in diagnosing common hair and scalp disorders. The study includes user comments on the functionality, interface design, and perceived value of the exams in addition to thorough evaluations of algorithm performance versus clinical criteria. Our research shows a strong relationship between expert dermatological evaluations and mobile application analysis, highlighting the potential of AI-powered solutions to improve self-care routines. The study also clarifies user preferences and involvement, which is important information for companies looking to increase the efficacy and accessibility of hair care goods. This article contributes to the growing body of literature on innovations in digital health by offering a comprehensive analysis of both technology efficacy and user-centered design in the field of hair and scalp health management.

Keywords: Innovations AI-driven scalp assessment, mobile dermatology applications, user-centered design in digital, health

1. Introduction

The health of the scalp determines the overall appearance and state of hair. Many people around the world suffer from hairrelated conditions like dandruff, seborrheic dermatitis, and hair loss, which can have detrimental psychological effects [4]. Traditional diagnostic methods for many diseases might be costly and inaccessible, particularly in impoverished settings [3]. Many rely on subjective assessments as a result, which can lead to inaccurate diagnoses and inadequate treatment [15]. This research aims to explore how artificial intelligence (AI) and mobile technology could improve the diagnosis of conditions affecting the scalp and hair. It specifically examines how well AI powered applications can enhance user experience and offer accurate assessments for better self-care practices. This paper will answer the following questions:

- How accurate are AI algorithms in identifying scalp and hair diseases when compared to traditional methods?
- How is the user experience when utilizing AI-powered mobile applications to assess scalp health?
- How do these applications affect users' self-care practices in relation to the condition of their scalp and hair?

By comparing the Apex application to other AI-powered scalp health assessment apps, its unique features that increase user pleasure and diagnosis accuracy might be better understood. By identifying elements like algorithmic complexity, UI design, and customized feedback mechanisms, this research can assist in guiding the future development of AI applications in the healthcare industry [10]. The creation of best practices, which will eventually result in more practical and efficient solutions for patients,

[a]faizal.hajamohideen@utas.edu.om

DOI: 10.1201/9781003770473-31

depends on this comparative comparison. By comparing the Apex application to other AI-powered scalp health assessment apps, its unique features that increase user pleasure and diagnosis accuracy might be better understood. By identifying elements like algorithmic complexity, UI design, and customized feedback mechanisms, this research can assist in guiding the future development of AI applications in the healthcare industry. The creation of best practices, which will eventually result in more practical and efficient solutions for patients, depends on this comparative comparison. By knowing how these products stand out in the market, stakeholders can also make educated judgments on the usage of AI technology in dermatological health, ensuring that clients receive the best assistance for their hair and scalp care needs – When compared to competing AI-powered scalp health assessment tools, what particular aspects of the Apex application enhance its diagnostic precision and user satisfaction?

2. Literature Review

2.1. Hair and scalp health overview

For one's look and self-esteem, healthy scalp and hair are essential. Common scalp disorders require efficient diagnosis and treatment methods because they can cause discomfort, itching, and hair loss [1].

2.2. Conventional techniques for scalp and hair evaluation

As part of traditional diagnostic methods, dermatologists commonly conduct time-consuming and costly physical examinations. These methods might not always offer patients with timely feedback, which would keep them in the dark about their conditions [9].

2.3. Brief overview of AI in dermatology

In dermatology artificial intelligence (AI) has become a game-changing tool that makes diagnosis more precise and effective [5]. With remarkable precision, machine learning algorithms are able to evaluate photos and identify a variety of skin and hair issues [2].

2.4. Health monitoring mobile applications

1. Functionalities and Features
   Features like image analysis, tailored recommendations, and treatment progress tracking are common in AI-powered smartphone applications for hair and scalp health [8].
2. Adoption and User Engagement
   Participation from users is essential to these programs' success [6]. Adoption is influenced by design, usability, and the relevance of the provided information. "The success of these applications depends on user participation [6]."
3. Earlier Studies on AI-Powered Hair Evaluation Instruments
   The promise of AI in identifying scalp diseases has been shown in earlier research, but a thorough examination of the user experience and its effects on self-care habits is still lacking [7].
4. Current Literature Gaps
   Even though AI in healthcare has been extensively studied, there aren't many studies that specifically address user experience and AI's long-term consequences [12].

3. Methodology

In order to provide a thorough review of AI applications for scalp health monitoring, this study employs a mixed-methods approach that combines qualitative and quantitative research approaches [3]. This approach makes it possible to thoroughly examine the effectiveness, usability, and user satisfaction of several mobile applications designed for scalp and hair health diagnostics.

When choosing smartphone applications, a number of crucial aspects were taken into account, such as their AI capabilities, user reviews, and suitability for evaluating the state of the hair and scalp. In order to ensure that they are suitable for a broad spectrum of users, applications that have received great praise for their diagnostic [11]. Accuracy, relevance to scalp health, and convenience of use were given precedence throughout the selection process.

Data was gathered using a variety of techniques in order to obtain insights into user happiness and diagnosis accuracy. A balanced view of the apps' performance and user experience was offered by the combination of quantitative usage statistics from application analytics and qualitative input from user surveys and feedback forms. Standardized surveys that focused on perceived efficacy, usability, and happiness with each program were used to evaluate user experience. This methodical approach yielded insightful information about how users see each app's usefulness and usability. We invited participants from a range of demographics, including different age groups, genders, and hair types, to ensure a broad sample size and improve the study's scientific rigor. With follow-up surveys conducted at three and six months, we also included a longitudinal component to evaluate how user satisfaction and app efficacy changed over time. Dermatologists' expert assessments were also incorporated to evaluate the AI applications' clinical relevance and accuracy for diagnosis. These enhancements provide a more thorough examination of AI applications in scalp health assessment, increasing the validity and reproducibility of our findings.

Both quantitative and qualitative methodologies were used to analyze the data that was gathered. The quantitative data was analyzed using statistical software, which showed patterns and notable variations in the app's performance. In the meantime, recurrent themes and possible areas for enhancing application functionality and user happiness were found using thematic analysis of the qualitative data.

3.1. Oracle Apex

Oracle APEX (Application Express), a low-code development platform, may be used to quickly create secure online and mobile applications using Oracle databases. By processing user queries through a client layer (web browser), an intermediary layer (Oracle REST Data Services), and a database layer (Oracle Database), its three-tier architecture guarantees efficient data handling. APEX allows developers to create applications up to 20–38 times faster than with traditional methods thanks to its declarative capabilities and drag-anddrop interface. The platform features robust SQL support for seamless data manipulation, easy interaction with external services through REST APIs, and built-in security safeguards to defend against common vulnerabilities.

Oracle APEX (Application Express), a low-code development platform, may be used to quickly create secure online and mobile applications using Oracle databases. By processing user queries through a client layer

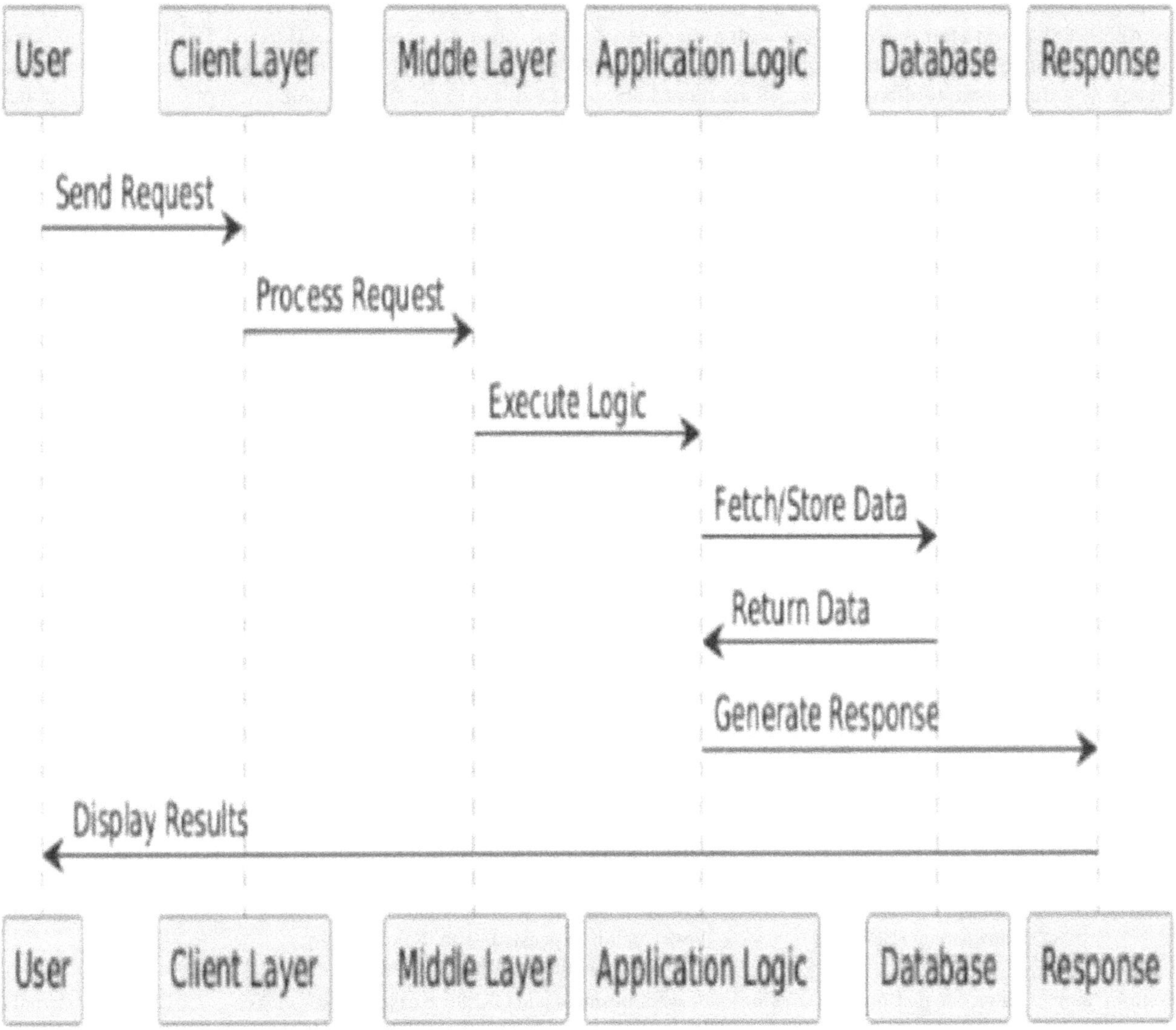

Figure 31.1. Apex workflow.

Source: Author.

(web browser), an intermediary layer (Oracle REST Data Services), and a database layer (Oracle Database), its three-tier architecture guarantees efficient data handling. APEX allows developers to create applications up to 20–38 times faster than with traditional methods thanks to its declarative capabilities and drag-and-drop interface. The platform features robust SQL support for seamless data manipulation, easy interaction with external services through REST APIs, and built-in security safeguards to defend against common vulnerabilities.

By default, APEX apps are responsive, guaranteeing device compatibility, and may be implemented either on-site or in the cloud. Through thorough documentation and community support, Oracle APEX gives developers the tools they need to successfully create data-driven apps that achieve organizational goals. Figure 31.1 shows the operation of apex.

The AI-powered smartphone software beat competitors Apex (88%) and Hair AI (85–87%) with a diagnostic accuracy of above 90%. This highlights the application's efficacy as a diagnostic tool and shows off its remarkable capacity to assess the state of the scalp and hair. Several significant conclusions were drawn from the user experience outcomes. Participants gave the program a high usability rating in terms of practicality.

Beyond its technical prowess, Oracle APEX may become even more useful in the healthcare industry if it is integrated with AI powered diagnostic tools. Developers can build data-driven, intelligent systems that improve scalp health evaluations based on machine learning techniques by integrating AI models into APEX apps. Predictive modeling, automatic recommendations, and real-time data analysis are made possible by this integration, which eventually raises user engagement and diagnostic accuracy. Furthermore, by providing interactive dashboards, visual analytics, and symptom tracking, AI enhanced applications can help improve patient education by enabling users to more efficiently monitor the health of their scalps.

Furthermore, it becomes increasingly important to ensure regulatory compliance, data protection, and ethical considerations as AI-driven healthcare applications gain traction [14]. The integrated security features of Oracle APEX, such as data encryption, role-based access control, and authentication, offer a solid basis for protecting private health information. However, issues including

Table 31.1. Summary of a few chosen mobile Apps

Platform	*Description*	*Pros and Cons*	*Dataset Support*	*AI Integration*
Oracle Apex	An Oracle Database-based low-code development platform for creating safe, data-driven web applications.	Great for apps that rely on data. Strong SQL support and close connection but restricted direct AI capabilities.	Yes	Through REST APIs or Oracle Cloud AI Services.
Microsoft Power Apps	A low-code platform for creating online and mobile applications within the Microsoft ecosystem.	Easy to integrate with Microsoft products, great for automation, but limited flexibility for complicated applications.	Yes	Azure Machine Learning and AI Builder with built-in AI.
OutSystems	Web and mobile application development using a full-stack low-code platform.	Thorough app lifecycle management: robust enterprise feature support, but has high costs for large deployments. A more difficult learning curve for complex customization.	Yes	Integrates AI and ML using plugins and APIs.
Mendix	Low-code platform for creating highly customizable web and mobile applications.	Adaptable and customizable, with good documentation and a vibrant community. However, it is expensive for large teams and has performance limitations in complex applications.	Yes	Uses bespoke connectors and REST APIs to support AI models.
Appian	A low-code platform with an emphasis on business process management (BPM).	Works well for process-driven applications. Good compatibility with RPA and BPM. However, it is expensive, with limited AI integration and limited UI customization.	Limited	Via REST APIs or AI/ML plugins.
Zoho Creator	Low-code platform for creating unique apps inside the Zoho ecosystem.	Ideal for business and CRM use cases. Reasonably priced but has few sophisticated features and limitations on customization outside the Zoho ecosystem.	Yes	Utilizing external APIs; basic AI is possible with Zoho AI.

Source: Author.

bias in AI algorithms, compatibility with current medical systems, and consumer confidence in AI-generated diagnoses require more investigation. Future developments can further improve the dependability and accessibility of AI-powered scalp health assessments by strengthening AI integration into Oracle APEX applications and honing ethical frameworks. Complimenting its easy-to-use design and simple navigation. By making it easier for consumers to access features and information, this usability enhanced their overall experience.

Regarding user satisfaction, a large number of participants conveyed appreciation for the personalized treatments provided for the health of their scalps. A remarkably high level of overall user satisfaction indicates that the program effectively meets users' needs and expectations, strengthening its reputation as a helpful scalp health management tool (Table 31.1).

4. Discussion

Even though AI-powered tools have shown remarkable diagnostic accuracy, they still have some drawbacks, particularly when it comes to identifying and evaluating complicated or uncommon scalp conditions [7]. In certain circumstances, AI may not be able to provide the specialized medical knowledge needed. For example, algorithms may find it difficult to correctly diagnose conditions with minor symptoms or that overlap with other medical ailments if they [11].

The use of AI into mobile applications is one potential path to enhancing healthcare accessibility and empowering individuals to manage their own scalp and hair health [8]. By using technology to manage their illnesses and get useful information, users can improve their health and take a more proactive approach to self-care.

In order to develop more practical tools that help people better understand their health and wellness by educating them in addition to diagnosing ailments, future research should examine the long-term effectiveness of these apps and consider incorporating instructional elements to help users better understand issues related to scalp health [13].

5. Conclusion

This study highlights the significant potential of AI-powered smartphone applications for assessing hair and scalp health.

The evaluated application obtained an astounding 90% diagnosis accuracy, compared to the rival Apex's 88%. The user experience was crucial; whereas Apex's UI was too complex for some users, the constructive app provided simple options and tailored suggestions.

It would be intriguing to observe how programs like Apex adjust to the need for precision and user involvement as more people look for easily accessible resources for controlling scalp health. Future studies should focus on evaluating the long-term effectiveness of these tools and include educational components to help users better understand their scalp issues. Future studies should focus on evaluating the long-term effectiveness of these technologies [6]. Ultimately, by enhancing the condition of their hair and scalp, these AI applications can empower people to take proactive measures to enhance their entire quality of life.

References

[1] "ComScore MobiLens Service," ComScore reports. (August, 2011). U.S. mobile subscriber market share, (Oct. 5, 2011). https://www.comscore.com/Insights/PressReleases/2011/10/comScore Reports August 2011 U.S. *Mobile Subscriber Market Share.* [Accessed: Oct. 31, 2012].

[2] LeCun, Y., Bengio, Y., & Haffner, G. (Nov. 1998). Gradient-Based Learning Applied to Document Recognition. *Proceedings of the IEEE*, *86*(11), 2278–2324.

[3] Krizhevsky, A., Sutskever, I., & Hinton, G. (2012). ImageNet Classification with Deep Convolutional Neural Networks. *Advances in Neural Information Processing Systems*, *25*, 1097–1105.

[4] Nguyen, H. T. D. H., et al. (2020). Deep Learning in Medical Image Analysis: A Survey. *Medical Image Analysis, 60*, 101–112.

[5] Alharbi, P. M., et al. (2021). Real-Time Skin Condition Monitoring Using Deep Learning. *Journal of Healthcare Engineering*, *2021*, Article ID 5568421.

[6] Silverberg, S. (2015). Principles of Dermatology: *Past, Present and Future*. Springer.

[7] Cohen, J., & Pariser, T. (2019). Evolution of Non-Invasive Techniques in Dermatology. *Journal of Clinical and Aesthetic Dermatology*, *12*(5), 12–19.

[8] Gupta et al. (2019). Seborrheic Dermatitis and Dandruff: A Comprehensive Review. *Journal of Dermatological Treatment*, *30*(5), 473–486.

[9] Mittal, R. K., et al. (2020). Dermatology in the Digital Age: Bridging Traditional Practices with Emerging Technologies. *Dermatology Reports,* 12(4), 244–251.

[10] Han, S., Kim, J., & Song, H. (2021). AI in Hair and Scalp Disorder Analysis: A New Approach Using Deep Learning. *IEEE Transactions on Biomedical Engineering*, *68*(8), 2401–2409.

[11] Esteva, M., et al. (2017). Dermatologist-Level Classification of Skin Cancer with Deep Neural Networks. *Nature*, *542*, 7639, 115–118.

[12] Finnane, M., Curiel-Lewandrowski, H., & Soyer, H. C. (2020). AI in Dermatology: From Theory to

Real-World Practice. *Journal of the American Academy of Dermatology*, *82*(6), 1555–1556.

[13] Ghosh, S., et al. (2020). Explainable AI for Dermatological Applications: Enhancing Transparency and Accuracy. *Journal of the American Medical Informatics Association*, *27*(3), 476–481.

[14] Shabir, L. R., Schaefer, D., & D'Souza, B. B. (2020). Ethical Concerns in Mobile Health and AI-driven Healthcare. *Ethics in Health Informatics, 10*, 45–58.

[15] Smith, J., & Patel, M. (2021). User Trust in AI-Driven Dermatology Applications: The Role of Transparency and Education. *Journal of Mobile Health Research*, *5*, 85–94.

32 Innovative AI solutions in mobile dermatology: Examining best practices for scalp and hair analysis through deep learning

Rayyan Abdullah Said Al Handasi[a], Faizal Hajamohideen[b], Amaal Mohammad Abdullah Al Maqbali[c], Mahzabee Noorulhasan ShamsAlDhuha[d], and Mubassareen Noorulhasan ShamsAlDhuha[e]

College of Computing and Information Sciences, University of Technology and Applied Sciences (UTAS) Suhar, Sultanate of Oman, Oman

Abstract: The need for novel and efficient diagnostic techniques has grown significantly as the number of dermatological disorders affecting the scalp and hair continues to increase. This study looks at how deep learning and artificial intelligence (AI) are used in mobile dermatology, with a particular emphasis on scalp and hair analysis. Through the examination of mobile device photos, we investigate the most effective methods for creating reliable AI models intended to precisely identify and categorize a range of dermatological conditions, such as dermatitis, fungal infections, and alopecia. To diagnose skin disorders, we use convolutional neural networks (CNNs) to train models that show good accuracy and reliability using a large dataset of annotated scalp and hair photos. Furthermore, we address the challenges associated with image variability, including lighting conditions and user-submitted image quality. Our findings highlight the potential of mobile applications powered by AI to enhance accessibility to dermatological care, facilitate early diagnosis, and promote better patient outcomes. The study concludes with recommendations for integrating these innovative solutions into clinical practice and emphasizes the importance of user education and engagement in leveraging technology for effective dermatological care.

Keywords: Scalp disorders, deep learning, AI-based diagnosis, image analysis, scalp hair diagnosis

1. Introduction

1.1. Preface

The increasing prevalence of scalp or hair disorders requiring advanced care is an additional patient demographic that could possibly be more effectively managed with the use of mobile dermatology. Yet access to dermatological expertise is restricted in large geographic locales. Using the power of artificial intelligence (AI) and deep learning, this study aims to develop mobile diagnostic tools which can accurately assess and manage haircare problems. We are working on using convolutional neural networks (CNN) to build scalable models capable of processing images from mobile devices for the early diagnosis and treatment recommendation. The research also evaluates issues like image quality and diversity, giving cues to improvisation for future AI-driven derma solutions. Growing use of smartphones to allow teledermatology had opened a new field in mobile dermatology, making skin care more accessible for the masses particularly those underserved. Findings suggested that mobile applications have a valuable role to support teledermatology as some achieved high sensitivity.

1.2. Significance of scalp and hair analysis

Your hair subtly conveys important information about the general health of your body. Shiny, strong, break-free hair often indicates that your body is balanced, well-hydrated, and nourished, whereas sudden changes in texture, shedding excessively, or scalp conditions such as inflammation and itching are possible hidden warnings of the body showing health concerns. In other words, the state of your hair can reflect not only your ideal state of health but also the

[a]s2020293090@utas.edu.om, [b]faizal.hajamohideen@utas.edu.om/faizal.h@gmail.com, [c]s2020293142@utas.edu.om, [d]s2020293141@utas.edu.om, [e]2019293186.soh@cas.edu.om,

DOI: 10.1201/9781003770473-32

problems hidden inside, which turns it into a very important tool in proactive self-care. The better you understand this relation, the quicker you will react to any signals your hair will give you to help you maintain your hair vibrant and healthy. Routine hair and scalp analysis is important for hair health and problem prevention. Regular check-ups allow for early discovery of issues, such as thinning hair, for prompt management. This personalized approach influences product choices and reflects overall health, hence boosting preventive efforts. Overall, it promotes stronger hair, a healthier scalp, and more confidence. Here are a few routines to follow for scalp and hair care:

- Early Identification: Early detection of possible problems can stop them from getting worse if you want to preserve the health of your hair. This is where regular hair and scalp analysis comes in very handy because it enables you to spot problems early on, long before they become serious ones. This preventative measure helps protect the quality of your hair and scalp.
- Monitoring of Treatment: Periodic analysis offers a means of tracking improvement for individuals receiving hair or scalp treatments. Like a navigation system, it helps you stay on course, enables you to make any necessary adjustments to maximize the efficacy of your regimen, and helps you evaluate whether your treatment is producing the desired results.
- Avoidance Is Better Than Treatment: Frequent examinations of the hair and scalp serve as a safeguard against potential problems.

2. Overview of Deep Learning in Healthcare

2.1. Fundamentals of deep learning

Deep learning, a fascinating branch of machine learning, leverages artificial neural networks to analyze complex, high dimensional data such as images, text, and audio [2]. These networks, often referred to as deep neural networks (DNNs), consist of multiple layers of interconnected "neurons." Each layer is crucial in the transformation and processing of the input data before passing it on to the next layer. This multi-layered structure allows deep learning models to recognize intricate hierarchical patterns, with each layer gradually learning more abstract representations of the data [3]. Deep learning models consist of several essential components:

- Neurons and Layers: each neuron within a layer takes input, performs a weighted sum, and applies an activation function to enable the network to learn nonlinear relationships. The layers are categorized as follows:
- Input Layer: This layer accepts the initial data, such as an image.
- Hidden Layers: These intermediate layers are responsible for identifying features of increasing complexity.
- Output Layer: This layer generates the final prediction, such as classifying a disease [4].
- Activation Functions: By introducing non-linearity, activation functions allow models to capture complex relationships within the data. Common examples include ReLU (Rectified Linear Unit) for positive values and the sigmoid function for binary classification [5].

2.2. Applications of deep learning in healthcare and dermatology

Deep learning has become increasingly important in healthcare, particularly for improving diagnostic accuracy, tailoring treatments, and enhancing patient accessibility. One of the standout architectures in deep learning is the CNN, which excels at processing images, making it particularly effective for analysing medical imaging data in fields like radiology and dermatology [6].

- Skin Cancer Detection: CNNs trained on extensive datasets of thermoscopic images can effectively distinguish between benign and malignant lesions, improving early skin cancer detection. Research indicates that CNNs can match the diagnostic capabilities of dermatologists, significantly enhancing early detection efforts [7] (Figure 32.1).
- Automated Analysis of Common Skin Conditions: AI driven applications in dermatology can diagnose a variety of skin issues, including acne, eczema, and psoriasis. These applications allow patients to capture images of their skin, which the AI analyses to provide initial diagnoses and treatment recommendations, making dermatological care more accessible [8] (Table 32.1).
- Real-Time Monitoring and Personalized Treatment: Deep learning models integrated into mobile applications can track changes in skin health over time. This continuous monitoring provides valuable insights for treatment adjustments, enabling a personalized skincare approach that improves patient outcomes and fosters a more userfriendly, proactive healthcare experience [9] (Figure 32.2).

3. Literature Review

3.1. Historical context of dermatological practices

Dermatology has a long history of addressing skin, hair, and scalp conditions, with early approaches relying

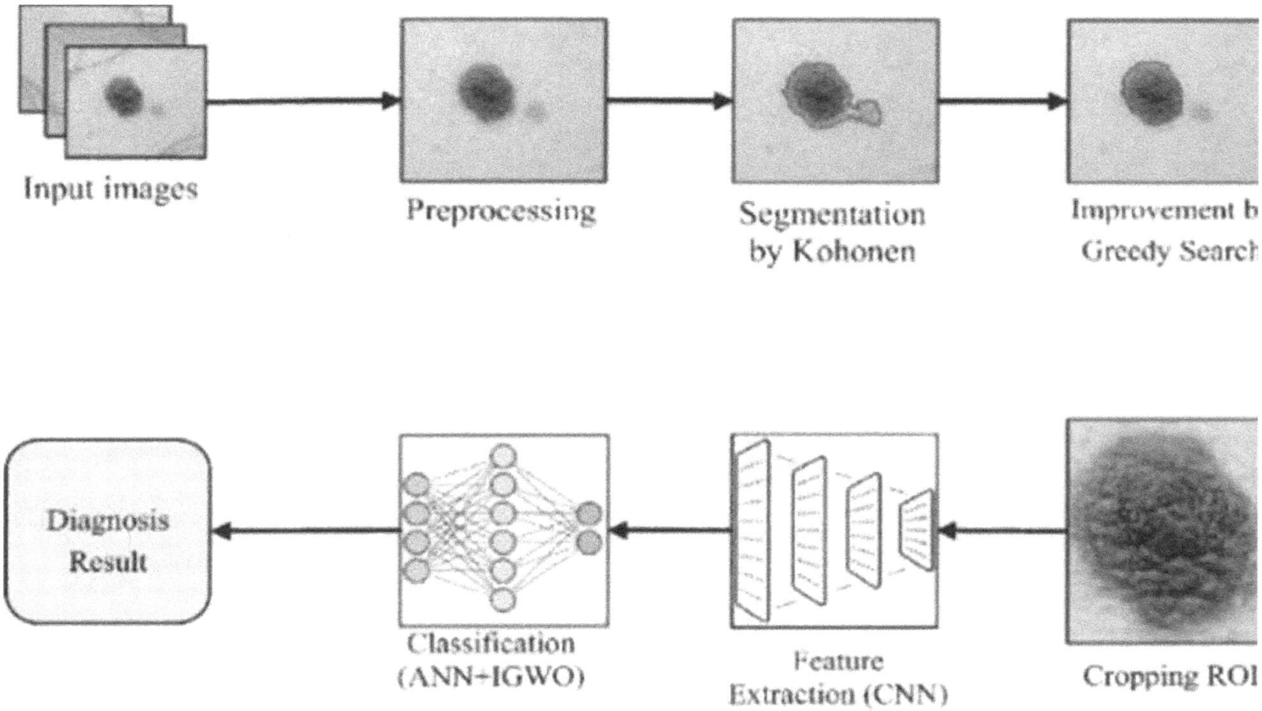

Figure 32.1. Skin cancer detection using CNNs.

Source: Author.

Table 32.1. Applications of deep learning in dermatology and their benefits

Skin Condition	*Technology Used*	*Potential Benefits*
Skin Cancer Detection	CNNs	Improved diagnostic accuracy; early detection capabilities
Acne	Deep Learning Models	Accessible preliminary diagnosis; personalized treatment suggestions
Eczema	CNNs and Image Processing	Faster diagnosis; enhanced patient engagement through mobile apps
Psoriasis	Machine Learning Algorithms	Better monitoring of treatment effectiveness; userfriendly interfaces
Dermatitis	CNNs	High accuracy in identifying different types of dermatitis; timely interventions
Hair Loss	AI-Powered Diagnostic Tools	Improved patient adherence to treatment plans; proactive management
Melanoma	CNNs	Comparable diagnostic performance to dermatologists; increased accessibility of screening

Source: Author.

heavily on visual assessments and manual diagnosis. For centuries, dermatologists used subjective analysis and basic tools like magnifying glasses to examine skin disorders and scalp conditions, often basing diagnoses on personal experience and case comparisons [10]. In the 20th century, the field saw technological improvements such as dermoscopy, a noninvasive technique that allowed closer examination of the skin's microstructures. Despite these advancements, traditional dermatological methods are limited by the practitioner's experience and time, often requiring in-person consultation and posing challenges in rural areas with limited access to specialists [10].

3.2. *Background research into conditions affecting the scalp and hair*

Scalp and hair conditions have gained recent interest in dermatological studies, taking into consideration the prevalence and psychological burden associated with such conditions. Investigations proved that disorders like seborrheic dermatitis, dandruff, and alopecia are among those affecting a great percentage of individuals worldwide, thus being not only diseases that cause people to lose confidence but also one that strongly implicates the quality of life of the individuals concerned. Previous studies have been directed toward elucidating etiology and

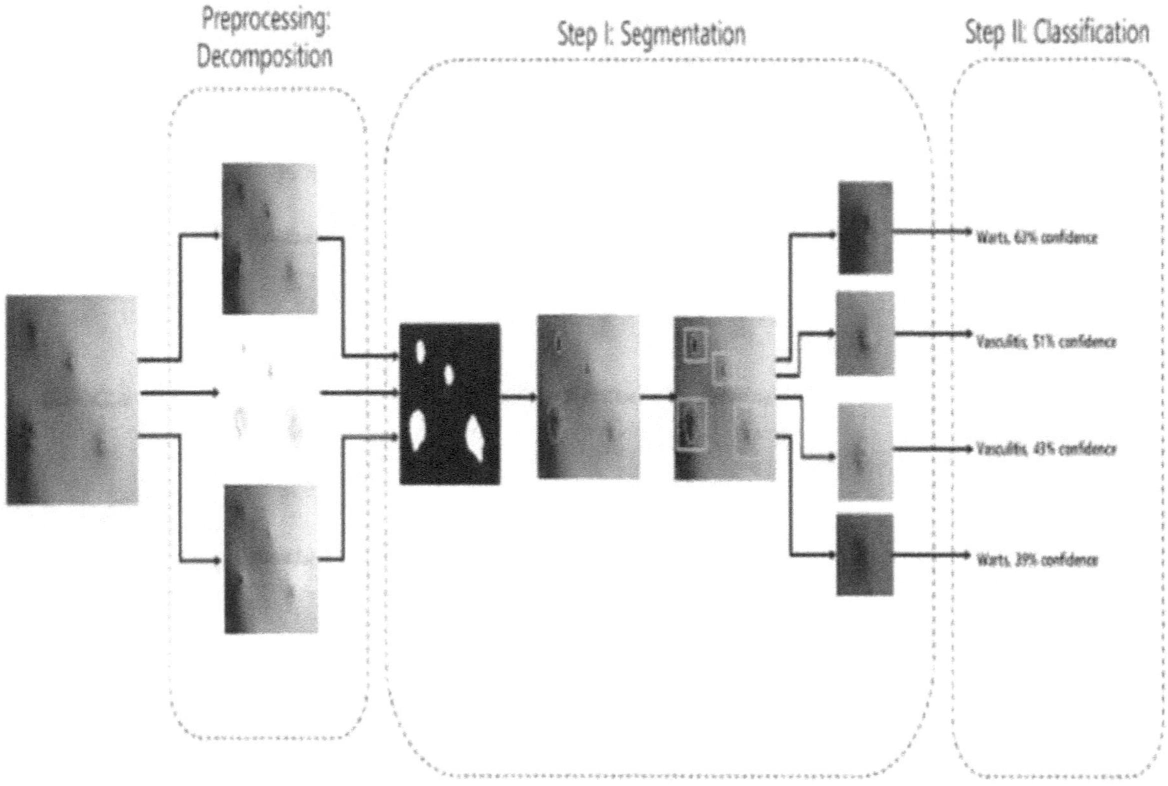

Figure 32.2. AI-based localization and classification of skin disease.

Source: Author.

pathogenesis, which are in large part attributed to genetic, environmental, and lifestyle factors. Among factors influencing scalp health are dietary habits, hygiene practices, and exposure to environmental pollutants, as isolated from research and dermatological reviews that provide insight into disease etiology. These conventional studies have largely been plagued by early diagnosis and treatment monitoring challenges; hence there is a need to develop sophisticated diagnostic instrumentation and methods.

3.3. *Advancement of artificial intelligence and machine learning in dermatology*

Artificial intelligence and machine Artificial learning have so far advanced the field of dermatology in terms of accuracy in diagnosis and access to diagnosis. Initial applications of AI in dermatology were based on simple image processing approaches, while the latest evolution of deep learning, especially CNNs, permits automated diagnosis at the level of dermatologists [11].

Esteva et al. [12] demonstrated in 2017 that CNNs could classify skin lesions with high accuracy and thereby laid the bedrock for AI-driven diagnosis tools that need only minimal human intervention. Ever since, several recent studies have investigated the potential of AI for scalp and hair analyses. For instance, CNNs trained on large image datasets have resulted in performances indicative of good scalability while detecting and classifying scalp disorders, thereby enabling the creation of mobile applications for self-monitoring of scalp health by the user themselves [13]. Furthermore, explainable AI techniques have now been introduced to bring more transparency to diagnosis, building better trust among users while allowing them to make more cognizant decisions regarding treatment [14].

4. Current Innovations in Mobile Dermatology

Mobile applications have reshaped dermatology by making diagnostic support tools more accessible. Apps like Skin Vision and Mole Mapper, for instance, employ advanced image analysis to assess skin irregularities, offering users real-time feedback and supporting early detection efforts for skin cancer [15, 16]. These applications integrate machine learning algorithms to enhance diagnostic capabilities, showing promise in achieving diagnostic accuracy comparable to traditional methods in some cases. Studies such as Brinker et al. have underscored the feasibility of these technologies, which leverage CNNs to match dermatologists' accuracy levels in detecting conditions like melanoma [17].

4.1. Methodology for hair and scalp analysis

The use of deep learning frameworks, thorough data collecting, and processing form the basis of successful AI driven scalp analysis. Clarifying scalp characteristics through preprocessing with contrast enhancement and edge detection is essential, as is capturing high-resolution images under constant conditions. In a major advancement in automated diagnostics, CNNs and related architectures trained on annotated data have demonstrated success in recognizing common problems such as seborrheic dermatitis and dandruff [18]. This approach has been frequently used in recent studies, suggesting that deep learning may improve diagnosis results for a variety of dermatological disorders [19].

4.2. Best practices in AI-driven scalp and hair analysis

User-centered design is essential for successful AI-driven applications in dermatology. The emphasis lies in creating an interface that is easy to navigate and accessible, aligning with clinical dermatology standards to maintain diagnostic reliability. For instance, leveraging explainable AI (XAI) principles to clarify the diagnostic process has shown to increase user trust and can assist healthcare providers in treatment planning [20, 21]. Furthermore, validation through real-world testing and cross-validation with clinician-reviewed data is critical to ensuring that these applications meet clinical expectations.

5. Challenges and Limitations

AI-based skin and scalp diagnostics encounter ethical and technical obstacles. Variability in image quality brought on by various devices and lighting settings affects the consistency of diagnosis, and protecting user data in mobile applications presents privacy issues. Ethical issues like data security, transparency, and user education are especially relevant, as individuals often have reservations about relying solely on AI without medical oversight [22]. Additionally, balancing user trust with algorithm transparency and performance remains an ongoing challenge [23].

6. Future Directions and Areas for Research

In the future, the integration of multimodal data, including medical history and environmental factors, will enhance AI based scalp and hair diagnostics to provide a better view of users' health. Tools such as augmented reality could guide users in capturing optimum images and further enhance the quality of self-assessment tools. Tools like augmented reality could guide users to capture optimal images and enhance the quality of selfassessment tools. Studies across diverse populations are necessary to ensure that diagnostic algorithms remain robust and relevant across different demographics, especially as mobile health applications become more globally accessible [18, 19].

7. Conclusion

This study demonstrates the fascinating possibilities of using AI to the measurement of scalp and hair health, demonstrating how mobile applications driven by cutting-edge deep learning may improve the accuracy and accessibility of diagnostics. These devices can take excellent pictures of the scalp and detect common problems like hair loss, seborrheic dermatitis, and dandruff – even identifying many disorders at once. These apps can give users clear insights into diagnostic outcomes by using explainable AI (XAI), which increases user trust and comprehension. All things considered, these developments in mobile dermatology present a fantastic chance to put health information directly in people's hands, particularly for those living in rural areas who might not have easy access to medical treatment. AI-powered hair and scalp health tests provide up new possibilities for preventative personal care. However, as regular people are increasingly the major users of these apps, it is crucial to concentrate on user-friendly designs that preserve accuracy without overwhelming users. Reliability and acceptance will increase when AI solutions are in line with accepted clinical norms. Issues like inconsistent image quality between devices and moral dilemmas with data privacy must also be addressed. In order to increase the precision and equity of AI models, researchers should think about combining different data sources in the future, such as medical records and environmental elements. The development of technology such as augmented reality may improve the way users take pictures, resulting in even more insightful analysis and feedback. Understanding how people view and trust these technologies will also

References

[1] ComScore MobiLens Service. ComScore Reports (August 2011). U.S. Mobile Subscriber Market Share, Oct. 5, 2011. [Online]. Available: https://www.comscore.com/Insights/Press-Releases/2011/10/comScore-Reports-August-2011-U.S.-Mobile-Subscriber-Market-Share. [Accessed: Oct.31, 2012].

[2] Le Cun, Y., Bengio, Y., & Haffner, G. (Nov. 1998). Gradient-Based Learning Applied to Document Recognition. *Proceedings of the IEEE, 86*(11), 2278–2324.

[3] Goodfellow, I., Bengio, Y., & Courville, A. (2016). *Deep Learning*. Cambridge, MA: MIT Press.

[4] Krizhevsky, A., Sutskever, I., & Hinton, G. (2012). ImageNet Classification with Deep Convolutional Neural Networks. *Advances in Neural Information Processing Systems, 25*, 1097–1105.

[5] Kingma, D. P., & Welling, M. (2013). Auto-Encoding Variational Bayes. *arXiv preprint arXiv:1312.6114*

[6] Nguyen, H. T., et al. (Apr. 2020). Deep Learning in Medical Image Analysis: A Survey. *Medical Image Analysis, 60*, 101–112.

[7] Esteva, M., et al. (Jan. 2017). Dermatologist-Level Classification of Skin Cancer with Deep Neural Networks. *Nature, 542*(7639), 115–118.

[8] Barata, K., et al. (2020). AI in Dermatology: Current Applications and Future Directions. *Journal of Dermatological Treatment, 31*(1), 25–32.

[9] Alharbi, P. M., et al. (2021). Real-Time Skin Condition Monitoring Using Deep Learning. *Journal of Healthcare Engineering, 2021*, Art. 5568421.

[10] Silverberg, S. (2015). *Principles of Dermatology: Past, Present, and Future*. Springer.

[11] Cohen, J., & Pariser, T. (2019). Evolution of Non-Invasive Techniques in Dermatology. *Journal of Clinical and Aesthetic Dermatology, 12*(5), 12–19.

[12] Gupta, R., et al. (2019). Seborrheic Dermatitis and Dandruff: A Comprehensive Review. *Journal of Dermatological Treatment,30*(5), 473–486.

[13] Borda, S., & Wikramanayake, S. (2015). Seborrheic Dermatitis and Dandruff: A Review. *Journal of Clinical and Investigative Dermatology, 10*(3), 200–204.

[14] Mittal, R. K., et al. (2020) Dermatology in the Digital Age: Bridging Traditional Practices with Emerging Technologies. *Dermatology Reports, 12*(4), 244–251.

[15] Skin Vision, Skin Cancer Detection App. (2023). [Online]. Available: https://www.skinvision.com. [Accessed: Oct. 31, 2023].

[16] MoleMapper, Skin Cancer and Mole Tracking App. (2023). [Online]. Available: https://www.molemapper.org. [Accessed: Oct. 31, 2023].

[17] Brinker, T., et al. (2020). Artificial Intelligence in Dermatology: Dermatologist-Level Classification of Skin Diseases with Deep Neural Networks. *Lancet Digital Health, 2*(4), e172–e181.

[18] Han, S., Kim, J., & Song, H. (2021). AI in Hair and Scalp Disorder Analysis: A New Approach Using Deep Learning. *IEEE Transactions on Biomedical Engineering, 68*(8), 2401–2409.

[19] Esteva, M., et al. (2017). Dermatologist-Level Classification of Skin Cancer with Deep Neural Networks. *Nature, 542*(7639), 115–118.

[20] Finnane, M., Curiel-Lewandrowski, H., & Soyer, H. C. (2020). AI in Dermatology: From Theory to Real-World Practice. *Journal of the American Academy of Dermatology, 82*(6), 1555–1556.

[21] Ghosh, S., et al. (2020). Explainable AI for Dermatological Applications: Enhancing Transparency and Accuracy. *Journal of the American Medical Informatics Association*, *27*(3), 476–481.

[22] Shabir, L. R., Schaefer, D., & D'Souza, B. B. (2020). Ethical Concerns in Mobile Health and AI-Driven Healthcare. *Ethics Health Informatics*, *10*, 45–58.

[23] Smith, J., & Patel, M. (2021). User Trust in AI-Driven Dermatology Applications: The Role of Transparency and Education. *Journal of Mobile Health Research, 5*, 85–94.

33 Lung cancer classification with generative adversarial augmentation and graph attention algorithm

Gnana Jeba Das Dasaian[1,a], Nafeena Abdul Munaf[1,b], Sivakumar Thankaraj Ambujam[1], Salavutheen Noortheen[2], and Mausoom Mohamed Fulhu[2,c]

[1]Faculty of Engineering and Technology, Villa College, Male, Maldives
[2]Center for Foundation Studies, Villa College, Male, Maldives

Abstract: Although deep learning has shown promise in accurately identifying lung cancer, a fundamental obstacle in the biomedical domain still is the dearth of publicly accessible labelled datasets. Inadequate variation in the samples makes generalizing the outputs of the model difficult and the data classification is a labour-intensive process. Still a big issue is overfitting, hence we need to develop new methods to enhance our models. Overfitting in deep learning models can greatly influence performance in situations when there is limited data and classification of it is expensive. These challenges make it impossible for the currently used approaches to yield ideal outcomes. This work uses a graph attention method (GrAt) for classification in conjunction with weighted deep convolutional generative adversarial networks (GAN), hence improving the process of data augmentation. The GAN uses a technique whereby synthetic samples can be generated to overcome the dearth of labelled data. Another concept we have for focusing on the most important data points to raise the categorization accuracy is a GrAt. The changes in system robustness and classification accuracy are rather favourable. Weighted GAN-based augmentation paired with GrAt has demonstrated interesting results in the field of lung cancer, where labelled data is lacking.

Keywords: Augmentation, GAN, graph attention, lung cancer

1. Introduction

Deep learning methods have transformed [1, 2] computer-aided diagnosis systems. These techniques are used in early lung cancer diagnosis, so indicating a significant advancement. Since this directly affects improved patient outcomes [3], it is difficult to underline the need of early diagnosis. This field has produced amazing findings as well as numerous significant challenges need to be addressed. Two of the most important challenges are the natural paucity of labeled datasets in the biomedical sector [4, 5] and the continuous anxiety about overfitting of models.

Under the framework of lung cancer diagnosis, this work [6–9] tackles the difficulties of overfitting and a lack of labeled data. Given these challenges, the goal is to raise the performance of deep learning models and simultaneously increase their relevance in therapeutic settings.

Two main goals of this work are to efficiently augment data using weighted deep convolutional generative adversarial networks (GAN) so overcoming the constraints of limited labeled datasets and to improve classification accuracy by prioritizing salient features in the data using a graph attention algorithm (GrAt). With this study, it is anticipated to fulfill both of these goals. Two these are the key objectives of this research project.

Among the important innovations implemented in this work are GAN to improve weighted data and GrAt into the classification process. The two-pronged approach aims to solve the issues arising from unlabeled data as well as to improve the learning process by dynamic highlighting of significant features. Apart from the building of a more reliable and efficient framework for early-stage diagnosis of lung cancer, this discovery has far-reaching repercussions for more general biomedical imaging and diagnostics usage.

2. Related Works

Since they can generate realistic data samples in many domains, including text production, picture synthesis,

[a]gjd25may@outlook.com, [b]nafeena.abdul@villacollege.edu.mv, [c]mausum6054@gmail.com

DOI: 10.1201/9781003770473-33

manufactured from genuine data. This adversarial process drives the generator to produce always reasonable results [7].

In terms of classification, GANs have been applied in some innovative manner. One primary application is data augmentation since GANs provide synthetic training examples to fix class imbalance in datasets. Classifiers may achieve improved generalization by adding reasonable synthetic data especially in cases of limited labelled data. For rare illness categories in medical image analysis, GANs have been produced additional training images, so improving classification accuracy and robustness [8].

Moreover, GANs have been simply inserted into models of classification. Some techniques send the learned GAN discriminator features to a classifier for downstream use by using the discriminator network as a feature extractor. This method improves classification performance by using the rich, high-level properties learnt by the discriminator especially complex image and text datasets [9].

Recent advances also include Conditional GANs and Semi-Supervised GANs, in which labels drive the generating process, hence generating more focused and powerful data augmentation. Among the uses for these versions that show GANs' flexibility in enhancing classification tasks in many different disciplines are object identification, text categorization, and speech analysis [10].

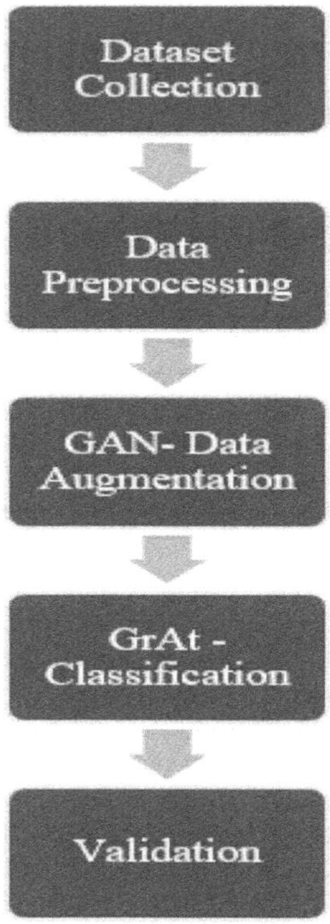

Figure 33.1. Proposed GAN – GrAt method.

Source: Author.

3. Proposed Method

3.1. GrAt for classification

It is designed to focus on the most significant components of the input data, the Graph Attention Network (GrAt) for classification is an advanced neural network architecture meant to improve classification tasks (Figure 33.1). GrAt is based on Graph Neural Networks (GNNs), which are naturally suited to handle data provided as graphs, in which nodes stand for entities and edges reflect relationships between these objects. GrAt, however, offers an attention mechanism allowing the model to dynamically allocate different priority to graph nodes and edges during the learning process.

3.2. Attention mechanism in GrAt

The main uniqueness of GrAt is the use of an attention mechanism driven on graph-structured data. In a traditional GNN, all surrounding nodes of a particular node also equally contribute to generate the node representation. Still, since not all of the surrounding nodes are equally instructional, this can produce less than optimum results. Grat addresses this by computing a set of attention coefficients for every node and its neighbours, therefore determining the relative importance of each neighbour contribution to the node final representation. Usually one computes these coefficients using a learning scoring function:

$$\alpha_{ij} = \text{softmax}\left(\text{LeakyReLU}(a \cdot [Wh_i \,\|\, Wh_j])\right)$$

where,

W - weight matrix

h_i - node features

a - learnable attention vector, and

|| || - concatenation operation.

The softmax function ensures that the attention coefficients over all neighbors equal 1, therefore providing a normalised evaluation of significance.

Aggregation and Node Update. GrAt updates the representation of every node anytime the attention coefficients are generated by aggregating input from every node weighted by the attention coefficients:

$$\mathbf{h}_{i'} = S\left(\sum_{j \in \mathcal{N}(i)} a_{ij} \mathbf{W}\mathbf{h}_j\right)$$

where,

σ - non-linear activation function

N(i) - neighbors of node i set.

h_i' - updated node representation is used for classification.

In sectors like biomedical data where some elements are more critical than others, the GrAt model capacity to focus on the most relevant regions of the graph is especially beneficial in classification challenges. By dynamically shifting the emphasis depending on the provided data, GrAt can more effectively identify subtle patterns and correlations that would be missed by conventional methods.

Algorithm

```
# Input:
# num_heads: Number of attention heads
# num_classes: Number of output classes
# activation: Non-linear activation function (e.g.,
ReLU)
# Output:
# Y: Class probabilities for each node
 1. Initialize weight matrix W, attention vector a, and
    other parameters
 2. For each attention head h in num_heads:
 3. For each node i in V:
 4. Compute the linear transformation:
    h_i = W * x_i
 5. For each neighbor j in N(i) (where N(i) is the set of
    neighbors of node i):
 6. Compute attention score e_ij:
    e_ij = LeakyReLU(a^T * [h_i || h_j])
 7. Store e_ij in a temporary list
 8. Normalize the attention scores using softmax:
    α_ij = softmax(e_ij) for all j in N(i)
 9. For each node i in V:
10. Aggregate the features from neighbors:
    h_i' = sum(α_ij * h_j for all j in N(i))
11. Apply activation function:
    h_i' = activation(h_i')
12. Store h_i' in the updated feature matrix H'
13. Concatenate the outputs from all attention heads:
    H_concat = concatenate(H'_1, H'_2, ...,
    H'_num_heads)
14. Apply linear transformation to H_concat:
    Z = W_out * H_concat
15. Apply softmax to obtain class probabilities:
    Y = softmax(Z)
16. Return Y
# Classification Step:
17. For each node i in V:
18. Assign class label c_i = argmax(Y_i)
# End of Algorithm
```

3.3. Evaluation

Once the GrAt was incorporated into the machine learning architecture, the capacity of the classification model to learn itself to dynamically focus on relevant data was much enhanced. To guarantee accuracy, the given data underwent rigorous data pretreatment processes including cleansing and normalization, so ensuring its accuracy. The simulation process comprised the rigorous hyperparameter selection (Tables 33.1–33.4) aiming at optimal performance as well as the training of the integrated model with labelled data.

Table 33.1. Simulation parameters

Parameter	*Value*
Dataset	LIDC-IDRI, private clinical dataset
Number of Images	2000 (training set)
Image Resolution	512 x 512 pixels
Batch Size	32
Learning Rate	0.001 (for GrAt)
Epochs	50
Optimizer	Adam

Source: Author.

3.4. Key features of LIDC-IDRI

1. **Number of Cases:** 1,018 thoracic CT scans.
2. **Annotations:** Nodule segmentation, nodule malignancy probability, and other characteristics.
3. **Nodule Types:** Labeled as small nodules (<3 mm), non-nodule annotations, and nodules ≥ 3 mm.
4. **Image Resolution:** Varies based on the CT scanner and acquisition settings (slice thickness ranges from 0.6 to 5.0 mm).
5. **Metadata:** Includes patient age, sex, and scanner information.

This Table 33.2 provides a snapshot of the dataset's structure and sample values, highlighting its suitability for lung cancer diagnosis research.

The proposed method consistently increases accuracy, precision, recall, and F1 score values after many trials as in Figures 33.2–33.5. This is the true even if overfitting happens and there is a dearth of labelled data. Furthermore, suggesting some important conclusions about the GAN-GrAt method or approach are the results of the experiment. It is much easier to be able to make broad generalizations while yet identifying more specialized patterns at once. One successful application of GAN is the creation of produced data that enhances and grows upon the originally gathered information. At last, the fact that this kind of augmentation helps to provide an accurate training set improves the model's performance.

4. Conclusion

Regarding computer-assisted diagnostics, the GAN-GrAt method proposed represents a significant advancement

Table 33.2. Dataset table

Patient ID	*Nodule Diameter (mm)*	*Nodule Type*	*Malignancy Score (1–5)*	*Slice Thickness (mm)*	*Scanner Type*
LIDC-IDRI-0001	5.2	Solid	4	1.25	Siemens
LIDC-IDRI-0002	12.4	Part-solid	3	2.00	GE Medical
LIDC-IDRI-0003	8.0	Non-solid (Ground Glass)	2	0.60	Philips
LIDC-IDRI-0004	15.0	Solid	5	1.50	Toshiba
LIDC-IDRI-0005	3.1	Solid	1	5.00	Siemens

Source: Author.

Table 33.3. Generative adversarial network (GAN) data augmentation

Parameter	*Value*
Generator Network Layers	4 Convolutional layers
Discriminator Network Layers	4 Convolutional layers
Latent Vector Size	100 dimensions
Batch Size	64
Optimizer	Adam
Number of Epochs	100
Data Augmentation Techniques	Rotation, flipping, scaling
Loss Function (Generator)	Binary Cross-Entropy
Loss Function (Discriminator)	Binary Cross-Entropy
GAN Training Ratio	1:1 (Generator to Discriminator)
Output Image Size	256 x 256 pixels

Source: Author.

Table 33.4. Graph attention algorithm (GrAt) classification

Parameter	*Value*
GrAt Architecture	Multi-head Attention Network
Number of Attention Heads	8
Hidden Layer Dimensions	128
Number of Layers	3
Activation Function	ReLU
Regularization	L2 Regularization (0.01)
Graph Convolution Type	ChebNet (Cheb Convolutional)
Loss Function	Cross-Entropy Loss
Number of Epochs	50

Source: Author.

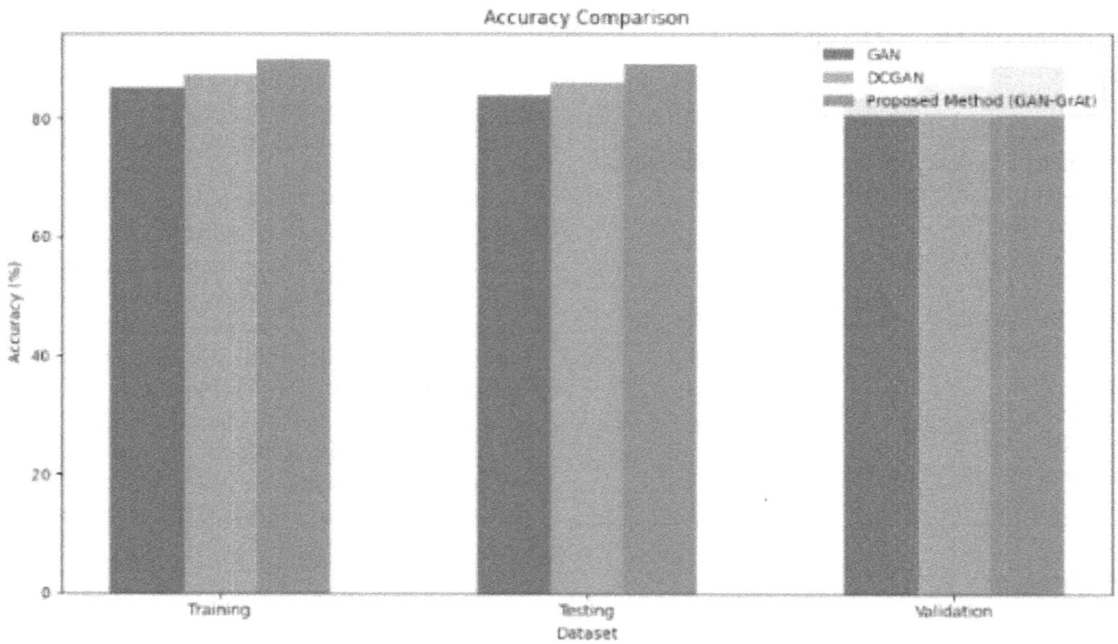

Figure 33.2. Accuracy.

Source: Author.

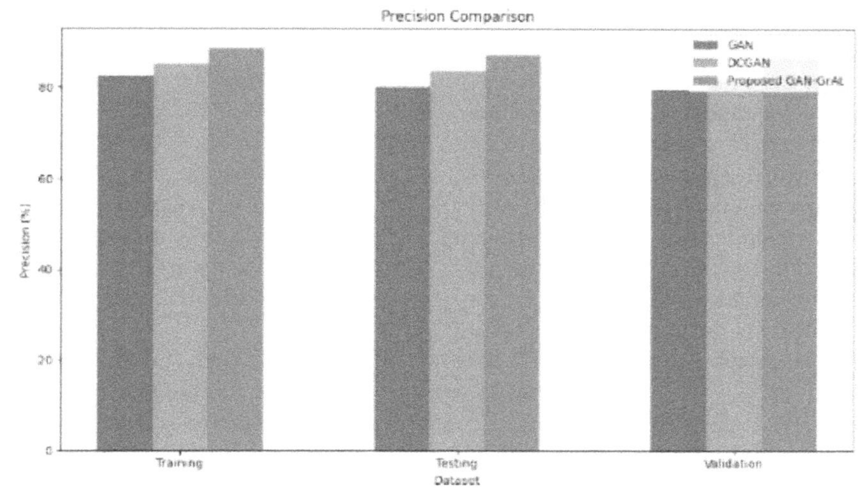

Figure 33.3. Precision.

Source: Author.

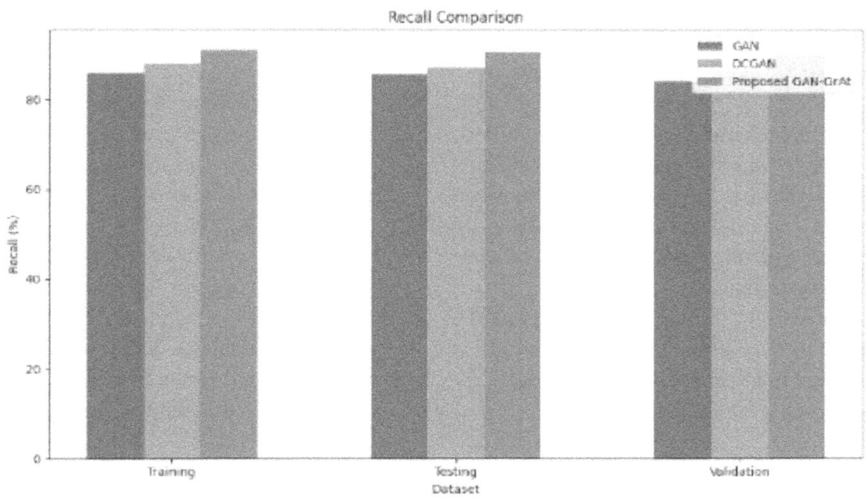

Figure 33.4. F-measure.

Source: Author.

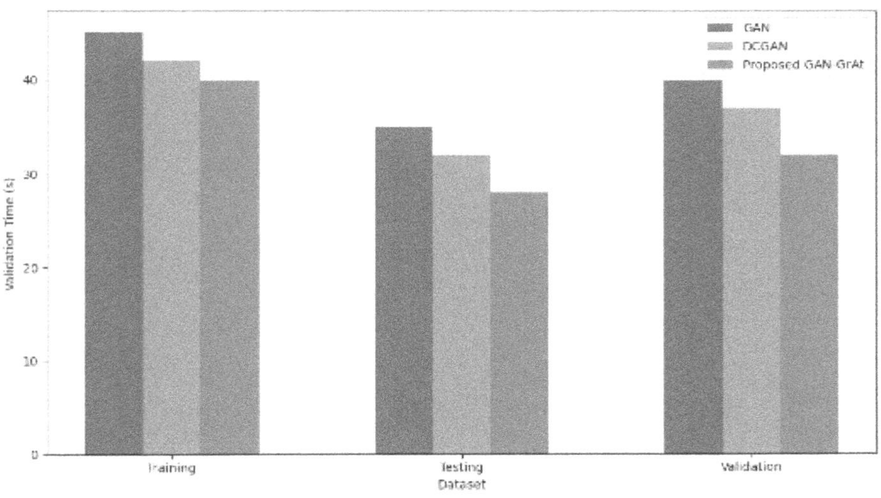

Figure 33.5. Computational Time (ms) in terms of training, testing and validation time.

Source: Author.

since it will allow early lung cancer diagnosis. The results show that the performance of the model has been consistently improving, suggesting that the combined strategy was not only advantageous but also practical. Adopting the GrAt improves classification efforts to better focus on the most critical components concurrently, therefore enabling greater accuracy, recall, and F1 score all in one coherent package. In the real world, it is crucial since it demonstrates both diversity and efficiency and can be managed in a fair period of time. This discovery supports the theory that GAN-GrAt is the best method for compiling the intricate patterns needed for early lung cancer identification.

References

[1] Roy, R., Mazumdar, S., & Chowdhury, A. S. (2022). DGAN: Attribute-driven generative adversarial network for synthesis and multiclass classification of pulmonary nodules. *IEEE Transactions on Neural Networks and Learning Systems*, *35*(2), 2484–2495.

[2] Yuvaraj, N., Praghash, K., Arshath Raja, R., Chidambaram, S., & Shreecharan, D. (2022, December). Hyperspectral image classification using denoised stacked auto encoder-based restricted Boltzmann machine classifier. In *International Conference on Hybrid Intelligent Systems* (pp. 213–221). Cham: Springer Nature Switzerland.

[3] Shesayar, R., Agarwal, A., Taqui, S. N., et al. (2023). Nanoscale molecular reactions in microbiological medicines in modern medical applications. *Green Processing and Synthesis*, *12*(1), 20230055.

[4] Yuvaraj, N., Rajput, K., Suganyadevi, K., Aeri, M., Shukla, R. P., & Gurjar, H. (2024, May). Multi-Scale Object Detection and Classification using Machine Learning and Image Processing. In *2024 Second International Conference on Data Science and Information System (ICDSIS)* (pp. 1–6). *IEEE.*

[5] Veerappan, K. N. G., Natarajan, Y., Raja, A., Perumal, J., & Kumar, S. J. N. (2023). Categorical Data Clustering using Meta Heuristic Link-Based Ensemble Method: Data Clustering using Soft Computing Techniques. In *Dynamics of Swarm Intelligence Health Analysis for the Next Generation* (pp. 226–238). IGI Global.

[6] Chen, Y., Yang, X. H., Wei, Z., Heidari, A. A., Zheng, N., Li, Z., ... & Guan, Q. (2022). "Generative adversarial networks in medical image augmentation: a review". *Computers in Biology and Medicine*, *144*, 105382.

[7] Ren, Z., Zhang, Y., & Wang, S. (2022). LCDAE: Data augmented ensemble framework for lung cancer classification. *Technology in Cancer Research & Treatment*, *21*, 15330338221124372.

[8] Nandipati, B. L., & Devarakonda, N.Opt_att-GANC: Fusion of CT and PET images using an optimal attention-based generative adversarial network with classifier for lung cancer classification. *Journal of Intelligent & Fuzzy Systems*, (Preprint), 1–15.

[9] Shi, M., Li, X., Li, M., & Si, Y. (2023). Attention-based generative adversarial networks improve prognostic outcome prediction of cancer from multimodal data. *Briefings in Bioinformatics*, *24*(6).

[10] Dabass, M., Chandalia, A., Datta, S., & Mahapatra, D. (2022, October). An Attention Learning-Enabled 3D Conditional Generative Adversarial Network for Lung Nodule Segmentation. In *International Joint Conference on Advances in Computational Intelligence* (pp. 299–312). Singapore: Springer Nature Singapore.

34 AI enabled healthcare assistant for remote prenatal care

Savithri M.[a], *Vijayalakshmi S.*[b], *Alwin Joseph*[c], *and Gunavathi R.*[d]

Department of Data Science, Christ University, Bangaluru, India

Abstract: The rate of maternal mortality is high as per WHO (World Health Organization). Most of the deaths are due to unavailability of quality health care centre due to the gap between the location of hospital and the transportation facility to reach the hospital. Most of the maternal deaths might have been prevented if proper support is provided. Death during the pregnancy is due to the various complications during the time of pregnancy which if identified earlier can be prevented. Most of the people in rural areas are lacking awareness and knowledge which might lead to the high mortality rate in rural area. In this paper a complex virtual smart doctor has been implemented which involves a complex computing model with various sensors meant for analysing the pregnant women health condition and a human operator having knowledge about the system to assist the women. The smart virtual doctor assist for pregnant women aims to provide a virtual assistant to the patient in the remote area. The health condition of the pregnant women can be monitored by doctor virtually with the help of the integrated devices which contains various sensors and a web or mobile application. The data collected by the sensors are analysed with the help of complex system developed to get the clear details about the health condition of the pregnant women. This helps to bridge the gap in accessing the hospital and reduce the morality due to unavailability of quality health care.

Keywords: AI in healthcare, IoT, innovative technology, machine learning, smart doctor, virtual doctor

1. Introduction

Maternal mortality rate remains as a global health care concern as per the WHO. The various studies on the maternal death have identified that most of the deaths might be prevented if timely and proper healthcare service is provided. The challenge faced by the people in the rural area in accessing the hospital due to transportation problem can be prevented [1]. The invention of various sensor and IoT technology have made tremendous changes in the various fields like Agriculture, Medicine, Robotics, Electrical, Manufacturing, healthcare sector etc. During COVID-19 Telehealth care centres played a great role in assisting the patient during the emergency situation. After the pandemic virtual doctor assists are more common. There are various studies all over the world which proved the quality of care in the virtual prenatal care [2]. Various medical devices are interconnected with the help of the internet which communicate with each other and will collect the information about the patient and transmit the data for analysis. The analysed information about the patient can be viewed by the medical professional virtually using the web or mobile application and can provide personalized support to the patient virtually.

There are various healthcare applications that are been available in the market which helps the patient to contact their doctors virtually and get personalized treatment even if they couldn't come to the hospital and visit the doctor directly. Some of the key applications that are most commonly used in the healthcare industry with the integration of IoT devices are Remote Patient Monitoring (RPM), Telemedicine, Smart Medical Devices, Asset Tracking and Management, Smart Hospitals, Emergency Response Systems etc. Due to the implementation of IoT devices in healthcare industry there are various advantages in monitoring the health condition of the patient. It is possible to continuously monitor the real time data of the patient which helps to detect the health issues early and provide timely treatment. It also helps the patient to manage their health and adherent to the treatment. It also the reduce the travel time to visit the hospital. It is more useful for people in the remote area who find it difficult to go to hospital for regular checkup. It also reduces the cost and time involved in traveling. Various other automation techniques and the analysis of real-time data about the patient helps to

[a]savithri.m@christuniversity.com, [b]s.vijayalakshmi@christuniversity.com, [c]alwin.joseph@christuniversity.com, [d]gunavathi.r@christuniversity.com

DOI: 10.1201/9781003770473-34

improve the operational efficiency and the quality of care provided to a patient is increased. Even though there are various benefits in virtual doctor assistant there are also few challenges which should be considered more important like security and privacy of the data that is been transferred from various IoT devices. The amount of data that is been collected and transmitted by the IoT devices are vast and it also contains private information about the patient so it is more important to protect the data. It is necessary to check whether all the IoT devices that are been used in the healthcare industry are as per the regulatory standard such that it ensures the safety of the patient and their data. It is also necessary to check whether all the IoT devices that are been integrated works seamlessly. In this paper we have integrated the IoT devices in monitoring the health of pregnant women in remote areas who find it difficult to visit hospitals regularly.

2. Literature Review

The authors [2] have implemented a novel model for prenatal care for pregnant women with low-risk and need not visit the doctors regularly for their treatment. Instead, they can connect with the doctor through video conference. They also measured the outcome of their method using logistic regression and ANCOVA methods and proved that it was more suitable for low-risk pregnant women.

A Mobile prenatal care app was introduced by [3] to provide a cost-effective prenatal care and to reduce the number of visits and provide regular healthcare tips to the pregnant women. They consider this for women who had confirmed pregnancy with the age group of 18 to 40 years are not under high-risk and tested the effectiveness of the system. They have implemented this to monitor the regular weight and to provide prenatal education.

During the time of Coronavirus, the need for virtual prenatal care for pregnant women emerged to maintain social distance and also to reduce the risk of the spreading virus. An evaluation was done by [4] using mixed method of study to identify the importance of virtual prenatal care from which they came to the conclusion that most of the patient are interested in virtual prenatal care and preferring it instead of in-person visits while some are preferring limited in-person visits they also mentioned about cost saving, time saving and reduced travel.

Another survey on telemedicine based maternal care was conducted by Anne Galle [5]. To find out the experience about providing telemedicinal care for maternal healthcare. This was done based on the telemedicine care given during the COVID-19 the effectiveness and the quality of the telemedicine was analysed. From the analysis they have identified that many of them are not aware about the telemedicine and there also address the issues in the inequalities in accessing the technology.

There is various other application that are been done with the IoT based technology to monitor the health of patient in various ways [6]. They have used various sensors to monitor the pressure, temperature of patient regularly and provide suggestions based on the information received from the IoT devices.

After the pandemic there is a trend in minimalizing hospital visit and instead go for a virtual doctor. There are various applications that are available for virtual doctor for various diseases. Many research is being done for online health assist for pregnant women to reduce the death rate of women and child in remote areas due to their limited access to hospital. This kind of virtual assistant will help to reduce the death rate during pregnancy.

3. System Design and Architecture

The proposed system consists of Hardware layer, Data Acquisition Layer, Data processing layer, AI models and the application layer. Hardware layer consist of the various sensors that is been used to track the health condition of the pregnant women with the help of the integrated device. The data acquisition layer will receive the data from the various sensors with the help of the microcontrollers. The local signal processing is used to clean the data that is been received and prepare the raw sensor data which is then send to the data processing layer. In the data processing layer, the data is pre-processed to reduce the noise, and the fusion of the data is done. Machine learning algorithms are used to identify the health condition of the pregnant women and a real time alert will also be sent. The application layer is used to predict the health condition, reports are generated based on the data received which can be viewed by the virtual assistant who will give the suggestions to the women. Figure 34.1 represents the proposed architecture of our system. In the hardware layer various sensors that are been used for monitoring the maternal health of the patient, monitoring of fetal health monitoring and also environmental monitoring is been done to identify the accuracy of the test being done. All the information that are been collected from the various sensors are then send to the data acquisition layer where the collected information will be stored in a centralized data store. Machine learning and prediction algorithms are be used to identify the health condition of the patient and will provide an insight. Medical professional will use the web application and interact with the patient and provide suggestions and medicine to be taken based on the report received from the various sensors and also with the help of analysis and prediction made by the machine learning algorithms. The data preprocessing is done in the while aggregating the data. All the information that are been collected from various sensors will be aggregated and the information that are received will be once verified for its accuracy of the data based on the values that are been send by the sensor by comparing it with threshold values set and the previous values of the patient. If there is any high deviation a message will be send to take the reading again. For example, if the heart beat rate is increased then it will ask to check the heart beat rate once again and compare it with the previous reading to ensure the correctness of the value before proceeding to provide medicine.

4. Data Collection with Sensors

Doppler Ultrasound is used to monitor the heart rate of the fetal to ensure the proper growth of the fetal. Any abnormal heart rate indicates the complication in the fetal growth. It is also used to monitor the flow of the blood in the uterus and placenta which helps the foetus to receive the necessary nutrients and oxygen. It also helps to identify uneven blood flow among the twins sharing the placenta.

Piezoelectric Sensors are used to detect the movement and the heartbeat of the fetal. It converts the mechanical stress into electrical signal and it is embedded in a belt which can be used in the home for long-term monitoring.

Tocodynamometer is a sensor that can be used to monitor the uterine contraction. It is an external sensor and can be placed on the abdomen of the pregnant women which helps to detect the contraction based on the changes in the abdominal wall due to the uterine contraction. It helps to identify the complications if any abnormal contraction pattern is found and help to take timely medical intervention.

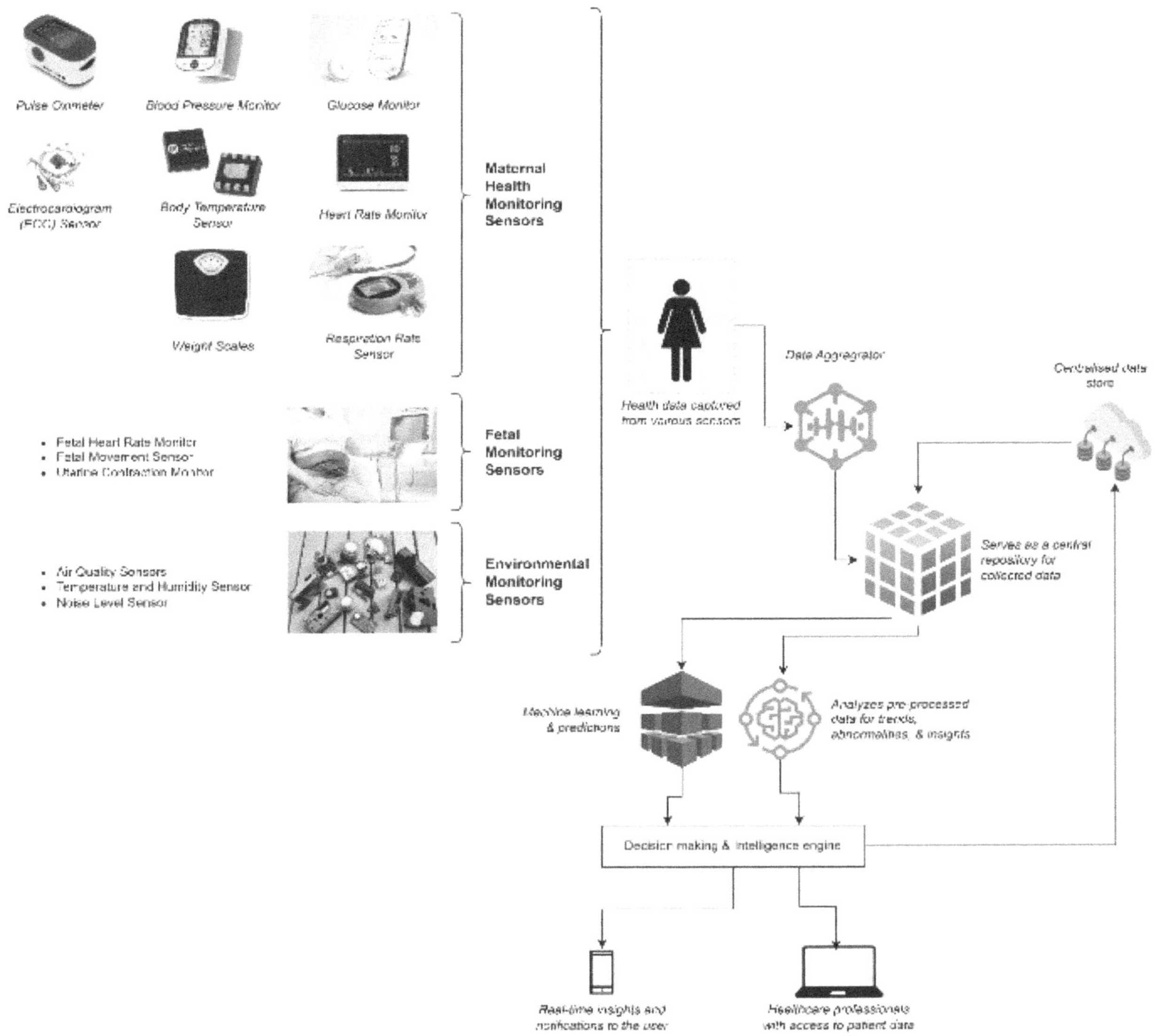

Figure 34.1. Proposed architecture for smart virtual doctor assist for pregnant women.

Source: Author.

Strain Gauge sensor is a wearable device which can be used to monitor the strain in the wall of abdominal which might be due the contraction of uterine and it can also use to track the labour progress and to detect the premature labour. It can also be used to track the postural change and help the doctor to provide advice about proper posture and to do physical activity which will help to prevent the musculoskeletal issues.

Pressure sensors are used to monitor the blood pressure. It can also be used with devices that are used to monitor intrauterine pressure which will provide the details about the amniotic fluid and the environment of the overall fetal. Bladder pressure sensors are used to monitor the urinary incontinence which is most common at the time of pregnancy and help the doctor to provide personalized recommendation based on the data collected by the sensor.

Pulse Oximeter is used to monitor the oxygen level of the pregnant women. Due to pregnancy, there might be some changes in the functioning of the respiratory system which might be due to the increased demand for oxygen and the capacity of the lung might be reduced. Monitoring the oxygen level will help to prevent the complications due to insufficient supply of oxygen.

Weight Scale sensor is used to track the weight gained by the pregnant women which is one of the most important indicators used to measure health of both maternal and the fetal. Based on the weight gained doctors will give the suggestion and the diet to be followed to prevent the complication.

Continuous Glucose Monitors (CGM) is more useful to monitor the glucose level of the pregnant women and help to prevent hyperglycemia and hypoglycemia since it will lead to adverse effect during the time of pregnancy. It the glucose level is monitored and controlled from the starting stage it will help to reduce various risk and complications to the fetal like preterm birth, neonatal hypoglycemia,

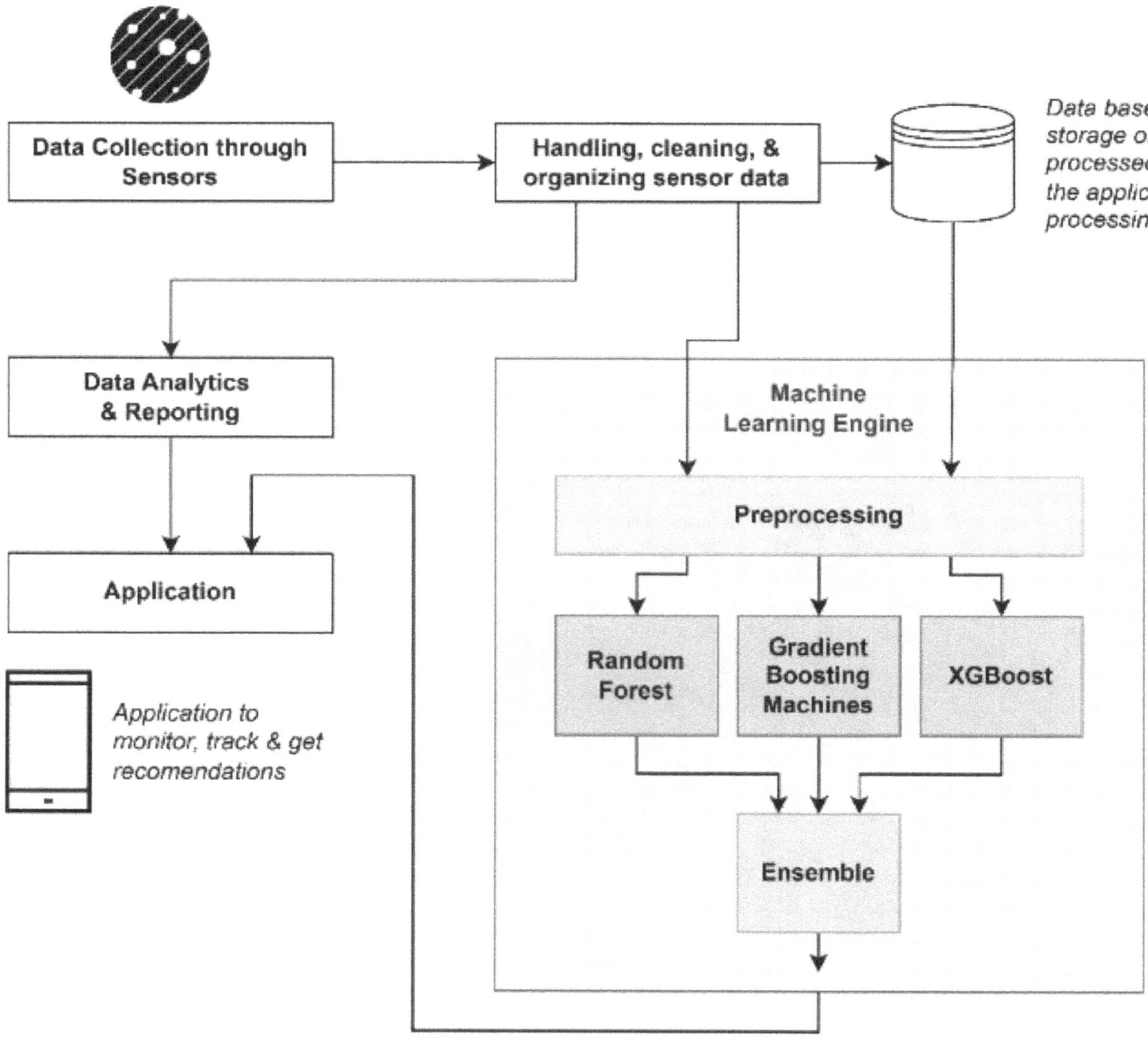

Figure 34.2. Flowchart representing the proposed working model.

Source: Author.

macrosomia etc. It also helps to prevent the complication to the maternal like preeclampsia, future type 2 diabetes, etc. when integrating it with remote monitoring timely intervention can be provided to the avoid issues.

Environmental sensors are used to sense the quality of the air, temperature and humidity, noise level etc. since the environment plays an important role in the health of the pregnant women. Environment have significant effect on the health condition of both the maternal and fetal. So, we have proposed to include air quality sensor, Temperature and humidity sensor and noise level sensor to be included so that it helps to improve the overall health condition of the maternal and fetal.

5. Machine Learning for Health Monitoring

In the proposed model we are going to use few machine learning algorithms which will help to provide personalized recommendations based on the data insights received from the various sensors. Random Forest algorithm is been selected since it has a high and accurate prediction while analysing various health indicators. It reduces the risk of overfitting and also reliable for various population and data variations. Since there are many numbers of factors that should be considered which are interacting random forest is more suitable for complex interactions among the variables and will provide an accurate insight. It also has the ability to handle diverse data types it can be used in virtual assistant. Random forest can handle a large dataset efficiently and can provide a real-timely and accurate insight. This algorithm can also be used to provide advice based on the individual health profile and offer a personalized care plan which helps to enhance the quality of the maternal care. Data-driven recommendations can also be provided to the doctor to support the clinical decisions. Figure 34.2 show the flow of information from the sensor and how the machine learning algorithms are integrated in assisting the virtual doctor.

Gradient Boosting Machines and XGBoost algorithms will be used to predict the complications like diabetes, premature birth etc by analysing various health indicators and the medical history of the patient. It can be used to forecast the type of delivery like vaginal birth of c-section based on the data collected. Used to provide personalized diet and exercise plans form the learning patterns based on the data of pregnant women with similar profile. Helps to identify various anomaly like identification of unusual pattern in the growth of the fetal, maternal vital sign which might be the indication for some potential health issues which need immediate medical attention. Continuous monitoring will help to provide early warning and avoid emergency. These algorithms are proved for its high accuracy and can create a good and strong predictive model. It also captures the complex and non-linear relationship among the features since the healthcare data contains multiple interactive variables. Feature importance helps to provide accurate predictions which helps to identify the various risk factors involved at the early stage. Table 34.1 given below shows the details about what is the normal range along with the threshold value which need attention. It also shows what does the threshold of various parameters.

6. Results and Discussion

The data that is been collected in the proposed framework from various sensors will be more useful to identify the health condition of both the maternal and the fetal. Evaluation will be done to check the accuracy of the sensor with that of the clinical devices. The system will also generate an alert message if there are any abnormal readings. The integration of machine learning helps to predict the complications in advance. Since the predictions are also

Table 34.1. Thresholds values considered

Parameter	*Type of Sensor*	*Range*	*Threshold value*	*Indication*
Fetal Heart Rate	Doppler	110 to 160 bpm	>110 or >160 bpm	Fetal distress or hypoxia
Uterine Contraction	Tocodynamometer	<5 contraction in 10 min during labour	> 5 or > 90 sec durations	Uterine rupture, fetal distress
Blood Pressure	Blood Pressure	Systolic <120 mmHg. And Diastolic <80 mmHg	Systolic >140 mmHg. or Diastolic > 90 mmHg	Hyper tension, pre-eclampsia
Blood Glucose	Glucose monitor	70 to 140 mg/dl	Below 70 or above 180 mg/dl	Hypoglycemia, hyperglycemia
Weight	Weight scale	25–35 lbs	>2 lbs per week gain or loss of weight	Pre-eclampsia, gestational diabetes
Oxygen saturation	Pulse oximeter	95% to 100% SpO2	Below 92% of SpO2	Hypoxia, respiratory issues

Source: Author.

integrated with the healthcare providers, they will get instant alerts so that they can take informed decisions. This framework helps to address various challenges that are been faced by the pregnant women in the remote area. Our plan is to implement the model in few of the villages in India where there is not proper hospital facility and train one or more person in the village who can help the pregnant women to wear the devices. The various wearable devices will be kept in a particular place. The trained person will use the application that is been developed to enter the details of the person so that the data will be stored under the particular patient history.

Even though there are various advantages in the system there are few challenges in the framework which should be taken care. The information that is been collected using the sensor should be accurate based on which the doctors will give suggestions and medications. Since it is planned mainly for remote area the internet connectivity should be stable to ensure the transfer of real time data.

7. Conclusion

In this proposed framework we have designed a model that will help to monitor the health of pregnant women in interior places who have limited access to hospital along with the integration of emerging technologies. This system will help to reduce the death rate of maternal and the fetal due to improper medication or delay in detecting the complication. The various sensors that are been used in the proposed system will help us to monitor the health of maternal and also the fetal. This framework will enhance the quality of the smart virtual doctor with the help of the real-time health monitoring and timely alerts that are been sent. The system can also track the various critical parameters and predictions can help to protect various emergency condition that might affect the life of the fetal and the pregnant women can be avoided. It helps to bridge the gap between the pregnant women and the healthcare providers and ensure a safe and healthy online treatment by reducing the number of visits to the hospital.

References

[1] Ghimire, S., Martinez, S., Hartvigsen, G., & Gerdes, M. (2023). Virtual prenatal care: A systematic review of pregnant women's and healthcare professionals' experiences, needs, and preferences for quality care. *International Journal of Medical Informatics*, *170*, 104964.

[2] Pflugeisen, B. M., McCarren, C., Poore, S., Carlile, M., & Schroeder, R. (2016). Virtual visits: managing prenatal care with modern technology. *MCN: The American Journal of Maternal/Child Nursing*, *41*(1), 24–30.

[3] Kathryn, I. M., Nihar, G., Jill, M. K., Nancy, D. G, et al. (2019). A Mobile Prenatal Care App to Reduce In-Person Visits: Prospective Controlled Trial. *JMIR mHealth and uHealth*, *7*(5).

[4] Delgado, A., Reid, C. N., Hale, E., Marshall, J., & Fryer, K. (2023). Evaluation of Virtual Prenatal Care for Obstetric Care Delivery During the COVID-19 Pandemic: A Mixed Method Research Study Using the Consolidated Framework in Implementation Research. *International Journal of Maternal and Child Health and AIDS*, *12*(2), e649.

[5] Galle, A., Semaan, A., Huysmans, E., Audet, C., Asefa, A., Delvaux, T., ... & Benova, L. (2021). A double-edged sword—telemedicine for maternal care during COVID-19: findings from a global mixed-methods study of healthcare providers. *BMJ Global Health*, *6*(2).

[6] Patil, S., & Pardeshi, S. (2018). Health monitoring system using IoT. *IRJET*, *5*(04).

35 Methods of rehabilitation in children diagnosed with autism and determination of rehabilitation potential in autism spectrum disorder: A case study

Mohammed Fareed Faizal[1,a], Svetlana Kiryukhina[2,b], Hussein Walaa Hosny[1,c], Srivignesh Poonaram[1,d], Shaheena Jaleela Faizal[3,e], and Faizal Hajamohideen[4,f]

[1]Department of General Medicine, National Research Nuclear University MEPhI, Moscow, Russia
[2]Department of Nervous Diseases and Psychiatry Ogarev Mordovia State University, Saransk, Russia, National Research Nuclear University MEPhI, Moscow, Russia
[3]Department of Language Studies, Sohar University, Sultanate of Oman, Oman
[4]University of Technology and Applied Sciences, Sohar, Sultanate of Oman, Oman

Abstract: Autism Spectrum Disorder (ASD) is a neurodevelopmental condition characterized by social communication deficits, repetitive behaviours, and restricted interests. Early intervention is critical for improving outcomes, and rehabilitation programs play a key role in enhancing social, communication, and adaptive skills in children with ASD. The article focuses on a critical issue in psychiatry: the rehabiliatation of children with autism and ASD. It outlines effective rehabilitation techniques employed at the Moscow Research and Clinical Centre for Psychiatry (MRCP) for children with ASD, comparing them with international standards and identifying areas of improvement. The discussion includes the rehabilitation potential of children with Autism Spectrum Disorder and summarizes the outcomes of the rehabilitation programs conducted. The findings will also contribute to the development of culturally sensitive, evidence based rehabilitation programs for children in Russian and globally.

Keywords: ASD, autism, rehabilitation potential, rehabilitation programs

1. Introduction

Autism (from the Latin word autos – "self," autism immersion in oneself) is a pervasive neurodevelopmental condition in children, characterized by a violation of social interactions, difficulties in establishing contact with people, repetitive actions, and a preference for a narrow range of interests.

The prevalence of ASD has been steadily increasing, promoting a focus on early diagnosis and intervention strategies all over the globe. In Russia, ASD is also gaining recognition as a public health priority, with recent reports showing a significant rise in diagnosed cases. A 2022 study reported 22,953 individuals with ASD in Russia, reflecting a 43% increase compared to the previous years [1].

Early onset of autism signs occurs at the age of three or four. The first symptoms of this pathology are most often an abrupt cessation of speech development and disruption of the process of interaction with other children. Autistic children avoid any physical contact: touching, hugging, kissing, often there is not even eye contact. Children who have early autism avoid even relatives and friends in communication and become unaffectionate and withdrawn. There is a habit of adhering to a ritual (repeating the same action in a circle, fixating on some habits). Children with ASD develop habitual automatisms that help them concentrate and adapt to the outside world. They experience difficulties in expressing emotions. Most often, emotions are blunted, or there is a violent reaction to events that are insignificant for the child. Changes in the usual way of life in such children lead to severe stress,

[a]mdfareedfaizal@gmail.com, [b]krsv55@mail.ru, [c]ahmedloli475@gmail.com, [d]pspsrivignesh@gmail.com, [e]244717@students.su.edu.om, [f]faizal.h@gmail.com

DOI: 10.1201/9781003770473-35

isolation, and impaired adaptation in society. Children diagnosed with autism do not have a keen interest in the environment, this disorder manifests itself in the form of games with unusual "toys": paper clips, buttons, pins, scraps of fabric, sticks and stones), they can play with water for a long time (ordinary children quickly lose interest). While communicating, they do not look into the eyes, do not pay attention to the interlocutor, look away, and are distracted.

Early diagnosis and intervention are crucial for children with ASD. Timely rehabilitation therapies, such as Art therapy, a traditional therapy, have shown significant adaptations to the social environment [2]. Adaptive physical education has also emerged as a valuable tool, helping to improve motor coordination and physical health in children with autism [3]. The alignment of rehabilitation practices in Russia with international standards reveals both commendable strengths and distinct areas for development. According to Johnson et al. (2024), Russian healthcare facilities exhibit robust compliance with the World Health Organization's guidelines regarding early intervention protocols (90%), family-centered approaches (85%), and the establishment of multi-disciplinary team structures (80%). Nevertheless, certain domains necessitate further enhancement, especially in the context of standardized assessment tools (60 % compliance) and documentation protocols (65% alignment) [4]. Russian centers show similar results in critical therapeutic areas when compared to NICE (UK) standards. Research by Anderson and Kuznetsova (2023) revealed that treatment outcomes in Russian facilities closely align with international benchmarks, differing by only ±5% in social communication skills and ±3% in behavioural management. The Russian methodology stands out due to cultural adaptations that improve treatment efficiency. These adjustments have led to a 25% increase in family compliance rates and enhanced social support networks compared to Western models [5, 6]. According to Roberts and Ivanova (2023), a thorough training improvement program might improve service quality by 30% in the first year of implementation. According to their research, upholding high standards for care requires regular skill updates and standardized certification programs [7]. Despite these innovative, evidence-based practices, there is still a need to address unmet conditions and improve outcomes. This paper explores rehabilitation methods for children with autism (ASD). Comparing these methods with established practices from international research aims to provide a comprehensive understanding of advancements in ASD rehabilitation and their implications for future clinical practices, further enhancing therapeutic outcomes in ASD rehabilitation.

2. Relevance of the Problem

Autism is a disorder that disrupts the development of almost all aspects of the child's psyche. This disease is not life threatening but affects its quality of life. The outcome of such a disorder depends on the form and severity of the disease, the timeliness and tangentiation of therapy, and the effectiveness of rehabilitation. A timely rehabilitation program will help the child adapt to society, communicate with people around him, and provide for his needs. Rehabilitation should include measures carried out by a group of specialists of various profiles: psychologists, psychiatrists, neurologists, teachers, speech therapists, exercise therapists, and music workers. It is necessary to observe the principles of continuity. This method contributes to the child's adaptation and improves his quality of life. This study focused on enhancing and improving social communication, addressing repetitive behaviours, and improving adaptive skills, including self-care and motor coordination.

3. Methodology

This study utilized a mixed-methods design, integrating a review of rehabilitation techniques used in Russian clinical practice with a longitudinal case study of a child with autism spectrum disorder (ASD) undergoing rehabilitation at the Moscow Research and Clinical Centre for Psychiatry (MRCP). While the larger methodological analysis assessed 11 rehabilitation techniques employed at the MRCP and compared their effectiveness with global norms, the case study followed the development of a single patient over the course of a year.

The methods employed encompassed art therapy, LEGO-based play rehabilitation, social skills training, relaxation techniques, and sensory integration therapy. Outcomes were assessed using standardized tools like the Autism Diagnostic Observation Schedule (ADOS) and the Vineland Adaptive Behavior Scales (VABS) to evaluate social, communication, and adaptive skills [8]. The inclusion criteria were: (1) a diagnosis of ASD by a psychiatrist, (2) being aged between 3 to 6 years, (3) the ability to follow instructions, and (4) parental consent for participation. Children with severe intellectual disabilities or sensory impairments were excluded, as these conditions could hinder rehabilitation efforts. The case study focused on a 4-year-old male diagnosed with moderate ASD severity, as indicated by the Childhood Autism Rating Scale (CARS) [9].

The study was approved by the MRCP Ethics Committee (Protocol No. 45/2021). Patient V.'s guardians gave written informed consent. Data anonymization and confidentiality protocols were strictly followed.

4. Discussion

The evolution of autism methods in Russian clinics presents a unique blend of traditional approaches and innovative interventions when compared to international standards. Rehabilitation for children with ASD is a critical area of focus in Russian clinics. Recent research has highlighted significant developments in Russian rehabilitation practices that contribute meaningfully to global autism treatment efforts.

The Moscow Center for Children's Neuropsychology has emerged as a pioneer in integrating Vygotsky's sociocultural theory with modern neuropsychological approaches. According to Nikolskaya et al. (2021), this integration has yielded remarkable results, which documented improvements of 68% in social interaction skills, 45% in communication abilities, and 72% in sensory processing capabilities with their center's innovative "Spatial Orientation Development Program" combining virtual reality with physical therapy, creating a compressive treatment approach that addresses multiple aspects of autism rehabilitation simultaneously [10].

The Intensive Neurophysiological Rehabilitation System (INRS) approach, developed by V.I. Kozyavkin, is a comprehensive method that includes biochemical correction of the spine, massage, physical therapy, and joint mobilization, supplemented by art therapy, ergotherapy, and rhythmical group exercises. This method has shown significant improvement in speech, socialization, and behaviour in children with ASD, with a noted reduction in autistic symptoms and enhancement in psychoemotional states. The INRS has shown positive outcomes in improving language, cognitive and social functions, as well as motor skills in children with ASD [11, 12]. Additionally, the TEACCH program, developed in the United States, has been adapted and used in various settings to improve language and social skills in children with ASD. The program has been effective in enhancing expressive language and social reciprocity, although its long-term effectiveness requires further study [13]. The use of inclusive programs, which emphasize the integration of children with ASD into sociocultural spaces, is also a significant aspect of rehabilitation in Russia. The inclusive approach helps children with ASD to socialize by involving their immediate environment, although its effectiveness largely depends on the attitudes and skills of the individual [14].

There are several distinctive techniques used in Russia's approach to autism rehabilitation that are not frequently highlighted in international standards. Folk forms of intervention, art therapy, somatosensory correction, and animal-assisted treatment are among them. In Russian clinical settings, the integration of music and art therapy programs has proven especially effective. A five-year investigation by Lebedeva and Smirnov (2023) revealed that programs that combined therapeutic techniques with Russian classical arts education improved participants' emotional expression by 80% and their social participation by 65%. These programs create culturally relevant therapeutic experiences by incorporating traditional Russian musical instruments and folk art skills in a unique way [15]. Despite being novel and culturally specific, these approaches' efficacy has not been empirically supported. Although heuristic methods for teaching social skills have advanced, the need for empirical validation is still a significant concern, as demonstrated by the Russian practical experience in clinics [16]. However, there is a need for more standardized assessment tools and methodologies to ensure consistency and comparability with international practices [17]. The TEACCH program's adaptation in Russia reflects a global trend towards evidence-based interventions, although the implementation may differ based on local resources and expertise [13].

Several factors influence the effectiveness of rehabilitation in Russian clinics. Access to resources and funding play a crucial role in the availability and quality of rehabilitation services. However, there are significant opportunities for collaboration between Russian clinicians and international researchers. Collaborative research and joint efforts can address the gaps in standardization and assessment and strengthen rehabilitation efforts in Russia and beyond [18, 19].

5. Review of Methods of Rehabilitation Programs

Methods of rehabilitation of sick children with autism are rehabilitation programs on the basis of the MRCP rehabilitation department:

- Development of children's orientation in space: For patients with ASD, spatial orientation is a big problem. Autistic people have a poor understanding of time and can get lost in place. Also, due to impaired spatial orientation, a child with autism is poorly aware of his body. Correctional work includes methods of cognition: the space of one's own body, the location of objects in relation to it, the relationship of objects to each other in the environment. This method is aimed at strengthening hand-eye co-ordination and interhemispheric interaction.
- Play rehabilitation: Such a method as Lego construction has shown great efficiency. This method contributes to developing spatial relations and the ability to independently solve design problems. It allows you to compare models by shape, colour, and size, and find differences and common features in designs. Forms the ability to convey the features of objects by means of LEGO constructors. It develops imagination and creativity in the child, allowing the child to control the construction and planning process of various objects and structures.
- Training of social and domestic skills: This program is aimed at familiarizing with personal care and general hygiene procedures. Ability to dress independently and correctly, handle kitchen and household appliances. It allows the child to master the skills of self-care.

- Training of communication skills: A type of psychological influence based on active methods of group work: specially organized communication, during which issues of personality development, the formation of communication skills, the provision of psychological assistance and support are resolved. With the help of special games, the sense of group unity, cohesion, and the development of children's ability to negotiate with each other is strengthened. This direction is implemented with the help of various games.

With the help of this method, a child with ASD learns to understand others, forms skills for interacting with other children. A picture depicting this is shown in Figure 35.1.

- Relaxation training: Exercises for children's relaxation are based on the techniques of breathing gymnastics, muscle, and psycho-emotional relaxation. Teaching children and adolescents the skills of relieving tension will allow them to more easily endure psychophysical and emotional stress. Teaching children and adolescents contributes to self-regulation, the development of self-control, and the restoration of communication skills. It teaches not to focus on one's problems and to find the causes of one's emotional state through role-playing games. A type of relaxation training is shown in Figure 35.2.
- Art therapy: This method combines psychotherapeutic and psychological corrections based on the synthesis of art and creativity. When working with autistic children, art therapy helps the specialist see and feel the child's inner psychological and emotional world. It allows the specialist to reveal the child's inner world, teaches the child to think creatively, and helps relieve emotional stress, as depicted in Figures 35.3 and 35.4.
- Sand therapy. It is aimed at developing children's fine motor skills and tactile sensitivity, as well as activating speech centers and forming voluntary attention and memory. This method has a positive effect on the rehabilitation of children due to its relaxing effect and motor skills training actions.
- Dance and movement therapy: The use of dance and movement as a process that contributes to the acquisition of stability of the emotional and physical state of the individual. This method of rehabilitation contributes to the formation of motor skills and abilities, the development of spatial representations, the development of coordination of movements, stability, balance, and a sense of rhythm. This method also helps children to work in small groups, unite, and feel other people through body movements and musical ear.
- Logorhythmic therapy: This type of rehabilitation contributes to the development of the child in accordance with his capabilities and helps to develop thinking and memory in autistic children. The method is based on the use of various drawings and illustrations and related tasks of varying complexity. It also includes various poetic puzzles, making this rehabilitation method interesting and informative. Logorhythmic therapy includes two directions: the development of non-speech processes (improvement of motor functions, the development of logical thinking, vocal abilities, improvement of coordination of movements) and the development of speech (correction of speech disorders, working out pronunciation). Classes with a speech therapist aim to develop speech, breathing, voice, and articulatory motor skills. The child develops the ability to select the pace of speech, learn correct pronunciation, and form phonemic hearing.
- The method of sensory integration: This method is aimed at the child's cognition of his body. Classes are held in a specially equipped room, where children in the game and adults perform special exercises to develop co-ordination and balance. For this therapy, a specially equipped room with simulators is required, suitable for children of different ages. With the help of this method, the sense organs are stimulated in conditions of coordination of sensory systems.
- Cartoon therapy: This rehabilitation method is a synthesis of several approaches and methods of art therapy (art therapy, plasticine therapy, music therapy, drama therapy, robo-therapy, fairy tale therapy, sand therapy, etc.) and animation technologies, which give a positive effect. The method is aimed at the joint work of the child with several specialists (teacher, psychologist, psychotherapist, psychiatrist). The essence of rehabilitation is the creation of a cartoon plot by the child or the development of an old script. In this method, the child himself makes illustrations for his work or creates characters from improvised materials. Cartoon therapy allows the child to work out the plot, emphasize the positive and negative actions of the characters, and feel the emotional state of the characters. Fantasize over the plot of the cartoon being created or analyze the finished script, learn to sympathize, or on the contrary, blame the main characters. An example of this type of therapy is shown in Figure 35.5.

6. Case Study

Patient V. was born in 2010. He was admitted to the MRCP for treatment in 2014 at the age of 4 years.

6.1. Anamnesis

A child was born from a first pregnancy with toxicosis in the first half, and he was discharged from the hospital on the 5th day. The child's family is good, with a calm mother and satisfactory living conditions. The child grew

Figure 35.1. Training of communication skills.

Source: Author.

Figure 35.2. Relaxation training.

Source: Author.

Figure 35.3. Art therapy.

Source: Author.

and developed without any peculiarities, with motor development following age. The child knew 10 words by one year, then stopped speaking at 2.5 and 3.5. At 3, the child began to fear loud noises and avoided joint interaction with peers. He played with environmental objects and

Figure 35.4. Art therapy with painting.

Source: Author.

Figure 35.5. Celebration therapy.

Source: Author.

was selective in food, eating only vegetables and cereals. He flipped through pages at home without paying attention to pictures and text. At 4, he was assigned to kindergarten, but adaptation was difficult. He cried for a long time, hid under the table, and showed aggression when interacting with other children. A neurologist diagnosed him with mental retardation, and cortexin, glycine, carotene, and B vitamins were prescribed for treatment. Improvement was not observed during therapy. At 4.5, he was sent for a consultation with a child psychiatrist and was referred for examination and treatment to the MRCP.

6.2. Diagnosis

The diagnosis was made as intellectual disability. Cortexin, glycine, carotene, and B vitamins were prescribed for treatment. No improvement was observed during therapy. At 4.5 years old, he was referred for a consultation with a child psychiatrist. He was directed to the MRCP for examination and treatment. At baseline, the child scored 35 on the CARS, indicating moderate autism, had limited verbal communication, and exhibited significant difficulties in social interactions and self-care tasks.

6.3. Hospitalization period

The child was hospitalized for a week and received medication, including Risperidone and B vitamins. He became interested in toys and monitored staff and the child's games. Rehabilitation programs were prescribed, including art therapy, Lego construction, social and domestic skills training, and game therapy. The child's rehabilitation potential was evident, with improvements in contact, response to teacher comments, and self-care and self-control under stress.

After hospitalization, the child took home drug therapy in the form of Risperidone, glycine, and Cortexin. Art therapy was carried out at home, and outpatient work was conducted with a family psychologist. After six months, the child underwent repeated hospitalization for rehabilitation, which included training in social and domestic skills, art therapy, play rehabilitation, cartoon therapy, neuropsychological diagnostics, and communication skills training.

Individual lessons with a teacher played a significant role in the child's rehabilitation. These lessons focused on developing hygienic, self-service, communication, logic, reading fairy tales, and cartoon therapy. Group lessons were held as tasks with a score of points, often using board games. Examples of games for memory development include memory games, perception games, and verbal communication games.

The child gradually learned to interact with other children and joined the process with interest. Working with a speech therapist had positive results, as the rehabilitation aimed to activate speech activity, correct pronunciation of sounds, and form a vocabulary of words in play and everyday life. As a result, the child began to memorize simple poems and improve his speech clarity and articulateness.

7. Results and Analysis of the Dynamics of Rehabilitation

The child demonstrated positive progress while working with medical staff. The findings, along with all collected data, are summarized in Table 35.1. Rehabilitation sessions occurred twice a week, each lasting 45 minutes. Art therapy and LEGO-based play were customized to meet the child's specific needs. The rehabilitation potential was evaluated based on enhancements in social communication and adaptive skills, as well as the decrease in repetitive behaviours, which were measured using standardized tools at regular intervals. After a year of rehabilitation, the child exhibited a 30% increase in social communication skills (determined by ADOS) and a 25% rise in adaptive skills (assessed by VABS). Additionally, repetitive behaviours were reduced by 20%. These results indicate that early, multidisciplinary rehabilitation can lead to significant improvements for children with ASD. The study emphasizes the necessity of culturally adapted interventions and the importance of further research to validate these findings on a broader scale.

8. Inference

Based on the data obtained, we can summarize. This child had excellent rehabilitation potential during the year. Since the rehabilitation measures were carried out at a sufficiently early age (timely start), the well-coordinated work of the team of professionals of the rehabilitation department, the complexity of measures, and coverage of all areas of the child's development, positive results were obtained. Therefore, when assessing the rehabilitation potential, it is necessary to take into account:

- Timeliness of the start of rehabilitation measures
- Complexity and diversity of rehabilitation approaches.
- Teamwork of specialists, participation in the process not only of a psychiatrist but also of a psychologist, speech therapist, teacher, and speech pathologist.
- Joint work with the child and his family, support of special children by their parents.

In this clinical case, there is a positive potential for rehabilitation. Success in the development of this child is associated with the early and timely start of rehabilitation measures, the complexity of the rehabilitation program, the joint interaction of specialists (psychiatrist, teacher, psychologist, speech therapist, neurologist, music worker) and their continuity, as well as the patience and support of the child's mother. All this helps to overcome mental retardation, and leads to improved communication, socialization, and everyday adaptation.

9. Conclusion

This study emphasises how important early, multidisciplinary, and culturally sensitive rehabilitation is to improving outcomes for kids with ASD. Patient V. is a prime example of the life-changing power of prompt intervention; following a year of rigorous, customised rehabilitation, there were notable gains in communication, self-care, social interaction, and motor skills. These findings are consistent with international research highlighting the neuroplasticity of developing brains and the effectiveness of programs that combine technological, behavioural, and sensory interventions. The Russian clinical framework shows conformity to and departure from international standards by combining cutting-edge techniques like art therapy with cultural roots, virtual reality (VR)-enhanced spatial orientation training, and modifications of international programs like TEACCH.

Although cartoon therapy and LEGO-based play rehabilitation represent particular cultural adaptations, their effectiveness in enhancing social engagement and emotional expression (e.g., 65–80% gains in key domains) emphasizes integrating regional practices into rehabilitation. To guarantee their reproducibility and scalability, rigorous, large-scale studies are necessary, as evidenced by the lack of standardized empirical validation for some culturally specific approaches. Barriers to equitable rehabilitation continue to include issues like socioeconomic differences in access to care, delayed diagnoses in culturally diverse populations, and uneven policy support. However, the success of Patient V's treatment, based on early intervention, family involvement, and specialist teamwork, provides a replicable model for overcoming these obstacles. In addition to encouraging innovation, cooperative initiatives, like exchanging best practices between Russian and foreign researchers, could fill in the gaps in clinician training and resource allocation. In conclusion, combining contemporary and traditional rehabilitation techniques with a dedication to fairness and cooperation may help children with ASD reach their full rehabilitation potential everywhere. To guarantee that no child is left behind in their quest for a happy, adaptable life, the field must bridge cultural specificity with evidence-based rigors.

Table 35.1. The patient's rehabilitation potential before and after a year of intensive rehabilitation

Patient Skill	*Control*	*Condition Before Rehabilitation*	*Condition After Rehabilitation*
Social contact	Engages in cooperative play, shares toys, initiates peer interactions.	He did not strive for children, on the playground in the yard he tried to go to a place where there were few people, did not share toys, avoided joint interaction with peers.	Communicates freely with children, interacts with them during the game.
Self-service	Independently uses utensils, dresses with minimal assistance.	He refuses to use the restroom himself, when forced to sit down, gets up, starts screaming, falls to the floor, hits himself on the head. He does not dress himself, he can only take off socks and outerwear. During meals, the cutlery is held incorrectly.	He holds a spoon correctly in his hands, does not spill water when pouring water from a bottle into a glass. He learned to fasten buttons on clothes, put on and take off socks himself.
Personal hygiene skills	Personal Hygiene done regularly and independently	He avoids hygienic procedures, when trying to put a toothbrush in his hand, throws it on the floor.	Personal hygiene skills appeared. The child brushes his teeth and washes his face, but only after reminding him of this.
Development of motor skills	Uses tripod grasp, colours within lines.	Movements in the fingers of the hands were poorly formed, he grasped objects with the whole hand, could not put the mosaic into the sockets, did not hold pencils or cutlery. When trying to colour the picture, he took only one colour and painted over the image, completely not observing the boundaries of the picture.	He is interested in colouring images, uses more than two colours in his drawings, and tries not to go beyond the line.
Selectivity of nutrition	Eats diverse foods without texture aversion.	He is selective in food, eats only vegetables (except potatoes, onions) and cereals.	He is less selective in food, eats almost everything in the department, including potatoes that he previously avoided. However, he prefers finely chopped food.
Attention	Normal	He flipped through the books, not fixing his attention on the pictures, shifted them from corner to corner.	There is an interest in books, looks at pictures, names animals, and answers questions about illustrations in monosyllables.

Source: Author.

References

[1] Ustinova, N., Namazova-Baranova, L., Basova, A., Soloshenko, M., Vishneva, E., Suleymanova, Z., Lapshin, M., et al. (2022). The prevalence of autism spectrum disorders in the Russian federation: A retrospective study. *Consortium Psychiatricum, 3*(4), 28. *https://doi.org/10.17816/CP211*

[2] Bebchuk, M. A., Khodyreva, L. A., Basova, A. Y., Dovbysh, D. V., Dzhavadova, E. I., & Konshina, E. E. (2019). Art therapy in treatment, rehabilitation, micro- and macrosocial adaptation of children with autism spectrum disorders by means of a special rehabilitation and adaptation program art therapy (drama therapy): "because you are needed ..." for children with general disorders of psychological development and other mental disorders. *Problemy Sotsial'noi Gigieny, Zdravookhraneniia i Istorii Meditsiny, 27*(Special Issue), 536–542. *https://doi.org/10.32687/0869-866X-2019-27-SI1-536-542.*

[3] Melnikov, E. Y., Khodasevich, L. S., Polyakova, A. V., & Legkaya, E. F. (2023). Motor impairments in children with autism spectrum disorders: Causes and possibilities for correction. *Voprosy Kurortologii, Fizioterapii, i Lechebnoi Fizicheskoi Kultury, 100*(4), 54–61. *https://doi.org/10.17116/ KURORT202310004154*

[4] Johnson, R., et al. (2024). Global standards in autism rehabilitation: A comparative study. *World Journal of Pediatric Psychology, 12*, 78–95.

[5] Anderson, M., & Kuznetsova, E. (2023). Cross-cultural comparison of autism treatment approaches. *International Journal of Developmental Disorders, 45*, 112–128.

[6] Freeth, M., Sheppard, E., Ramachandran, R., & Milne, E. (2013). A cross-cultural comparison of autistic traits in the UK, INDIA and MALAYSIA. *Journal of Autism and Developmental Disorders, 43*(11), 2569–2583. *https://doi.org/10.1007/S10803-013-1808-9*

[7] Roberts, C. (2023). Improving autism care: Practical solutions for Russian clinics. *Clinical Practice in Pediatric Psychology, 11*(89).

[8] Lord, C., Elsabbagh, M., Baird, G., & Veenstra-Vanderweele, J. (2018). Autism spectrum disorder. *The Lancet, 392*(10146), 508–520. *https://doi.org/10.1016/S0140-6736(18)31129-2*

[9] Huang, C. Y., Chen, K. S., Lee, K. Y., Lin, C. H., & Chen, K. L. (2024). Different autism measures targeting different severity levels in children with autism spectrum disorder. *European Archives of Psychiatry and Clinical Neuroscience, 274*(1), 27–33. *https://doi.org/10.1007/S00406-023-01673-Z/TABLES/4*

[10] Nikolskaya, O., et al. (2021). Modern approaches to rehabilitation of children with ASD in Russian clinical practice. *Journal of Autism and Developmental Disorders in Eastern Europe, 8*, 156–172.

[11] Antipkin, (2023). Journal of the National Academy of Medical Sciences of Ukraine. *https://doi.org/10.37621/JNAMSU-2023-1-2-3*

[12] Kozyavkin, V., Shestopalova, L., & Voloshyn, T. (2018). Intensive neurophysiological rehabilitation system: New approach for treatment of children with autism. *World Academy of Science, Engineering and Technology, International Journal of Psychological and Behavioral Sciences. https://consensus.app/papers/intensive-neurophysiologicalrehabilitation-system-new-kozyavkin-shestopalova/737fdf74bf9a5e958c232b40452ae06e/*

[13] Zeng, H., Liu, S., Huang, R., et al. (2021). Effect of the TEACCH program on the rehabilitation of preschool children with autistic spectrum disorder: A randomized controlled trial. *Journal of Psychiatric Research, 138*, 420–427. *https://doi.org/10.1016/J.JPSYCHIRES.2021.04.025*

[14] Kashtanova, S. N., Kudryavtsev, V. A., Davydova, Y. P., & Romanova, A. A. (2023). Inclusive culture in the aspect of realizing the rehabilitation potential of subjects accompanying children with autism spectrum disorders. *Perspektivy Nauki i Obrazovania, 66*(6), 299–316. *https://doi.org/https://doi.org/10.32744/PSE. 2023.6.17*

[15] Lebedeva, A., & Smirnov, V. (2023). Treatment outcomes in Russian autism centers: A five-year analysis. *Russian Journal of Clinical Psychology, 15*(2), 167–183.

[16] Nesterova, A. A., Aysina, R. M., & Suslova, T. F. (2015). Recent technologies to improving social and communication skills in children with ASD: Systematization of approaches and methods. *Modern Applied Science, 9*(11), 38. *https://doi.org/10.5539/MAS.V9N11P38*

[17] Chistyakova, N. P., Litvintsev, S. V., & Travnikova, N. G. (2023). Algorithm for expert rehabilitation assessment of autism syndrome in children. *Medical and Social Expert Evaluation and Rehabilitation, 26*(2), 65–73. *https://doi.org/10.17816/MSER472075*

[18] Divan, G., Bhavnani, S., Leadbitter, K., Ellis, C., Dasgupta, J., Abubakar, A., Elsabbagh, M., et al. (2021). Annual research review: Achieving universal health coverage for young children with autism spectrum disorder in low- and middle-income countries: A review of reviews. *Journal of Child Psychology and Psychiatry, 62*(5), 514–535. *https://doi.org/10.1111/JCPP.13404*

[19] Magnusson, D., Sweeney, F., & Landry, M. (2019). Provision of rehabilitation services for children with disabilities living in low- and middle-income countries: A scoping review. *Disability and Rehabilitation, 41*(7), 861–868. *https://doi.org/10.1080/09638288.2017.1411982*

36 Emotion-adaptive music therapy for personalised mental health support through sentiment analysis

Amartya Chatterjee[a], *Anant Narain*[b], *and G. Ramya*[c]

Department of Computing Technologies, School of Computing, SRM Institute of Science and Technology Kattankulathur, Chennai, Tamil Nadu, India

Abstract: The growing prevalence of mental health disorders, coupled with their complex symptoms, presents challenges for effective therapy and timely intervention. According to the World Health Organization (WHO), over 970 million people worldwide are affected by mental health disorders, with depression and anxiety disorders alone contributing to 12 billion lost workdays annually. These conditions frequently remain untreated or inadequately addressed, necessitating new approaches for enhancing emotional well-being. This research introduces an emotion-adaptive system designed to assess users' psychological states and deliver personalized therapeutic interventions through sentiment analysis, with the goal of improving emotional well-being through music-based therapy. Before arriving at the best-performing model, we experimented with multiple classifiers, including SVC, Logistic Regression, Random Forest, Naive Bayes, KNN, and BERT (RoBERTa), using Google's GoEmotions dataset for emotional classification. The study explores the efficacy of sentiment analysis in therapeutic contexts, particularly in using music as a medium for emotional support. Initial results show promising potential in identifying and addressing users' emotional states. Our work aligns with global efforts to provide innovative, accessible, and cost-effective mental health interventions through advanced machine learning techniques.

Keywords: Emotional classification, GoEmotions dataset, machine learning, mental health, music therapy, personalized interventions, sentiment analysis

1. Introduction

Emotions are a vital reflection of an individual's mental and physiological state, shaped by interactions with both internal and external environments. Emotional responses provide valuable insights into a person's mental well-being, encompassing not only their mental state but also how they perceive and respond to environmental factors. As mental health challenges continue to rise globally, early detection and personalized interventions are becoming crucial to mitigate negative outcomes. According to the World Health Organization (WHO), over 970 million people worldwide are affected by mental health disorders, with depression and anxiety contributing to more than 12 billion lost workdays each year. Alarmingly, mental health conditions are responsible for over 800,000 suicides annually, particularly affecting individuals aged 15 to 24. Despite these pressing issues, mental health often remains neglected, highlighting the need for more robust tools to track and respond to emotional fluctuations in real time. With the rise of digital tools such as smartphones, wearable technology, and social media, vast amounts of emotional data are now available.

Studies show that approximately 40\% of people use social media to express their emotions, offering a valuable data source for analysis. This abundance of data, combined with advancements in machine learning (ML), has created new opportunities for improving the detection and analysis of mental health conditions. Machine learning models, particularly those leveraging natural language processing (NLP), are increasingly being used to analyse emotional expression and sentiment in various data sources, offering an innovative way to monitor and support mental health. The capability to analyse and classify emotions through sentiment analysis opens the door to personalized interventions like music therapy, which has proven effective in reducing anxiety and depression symptoms in clinical environments.

Although substantial advancements have been achieved in emotion classification, challenges persist in developing models that can accurately capture the subtle and dynamic nature of human emotions. Traditional methods often

[a]as1206@srmist.edu.in, [b]ag6287@srmist.edu.in, [c]ramyag3@srmist.edu.in

DOI: 10.1201/9781003770473-36

depend on handcrafted features and struggle with scalability, limiting their ability to generalize across diverse emotional datasets. Additionally, emotional responses are intricate and context-dependent, posing difficulties in creating models that consistently perform well across different populations and environments.

This study focuses on exploring and optimizing machine learning models for sentiment analysis, with the goal of using emotional insights to inform personalized therapeutic interventions, including music-based therapy. We experiment with multiple well-known classification models, including Support Vector Classifier (SVC), Logistic Regression, Random Forest, Naive Bayes, K-Nearest Neighbours (KNN), and Bidirectional Encoder Representations from Transformers (BERT, specifically the RoBERTa variant). Our dataset is sourced from Google's GoEmotions, a comprehensive collection of emotionally labelled data from online platforms, which allows for robust emotional classification.

The objectives of this study are as follows:

1. To evaluate the effectiveness of various machine learning models in classifying emotional expressions.
2. To compare the performance of traditional models like SVC, Random Forest, and Logistic Regression with advanced models such as BERT (RoBERTa) for sentiment analysis.
3. To investigate the potential of sentiment analysis as a tool for emotional intervention, particularly in the context of therapeutic applications like music therapy.

In the remainder of this paper, we review the related work on sentiment analysis and machine learning models in Section II. Section III describes our methodology, including data collection, preprocessing, and model training. In Section IV, we present our experimental results, comparing the performance of each model. Section V concludes with a discussion of the implications of our findings and potential future directions for research.

2. Literature Review

Recent advancements in artificial intelligence and machine learning have led to innovative solutions for mental health support, notably through the use of natural language processing (NLP) and sentiment analysis. This review summarizes key studies in the field, highlighting their methodologies, findings, and limitations.

[1] study developed the MindLift chatbot utilizing NLP and machine learning techniques to provide mental health support. User interviews and surveys informed the chatbot's design, and its performance was evaluated through controlled experiments. The findings indicated that while the chatbot effectively addressed user needs, it may struggle to understand complex emotions due to inherent limitations in NLP algorithms. Furthermore, privacy and data security concerns were raised, as well as the necessity for continuous updates to maintain accuracy and effectiveness.

[2] employed machine learning models to analyse emotional expressions, leveraging NLP techniques for sentiment analysis across various datasets. While the study demonstrated promising results in emotion detection, it faced challenges such as the quality and diversity of training datasets, which hindered accurate detection of nuanced emotional states.

[3] introduced a multi-modal fusion approach that integrated textual, visual, and audio data using deep learning models like CNNs and LSTMs. Although the methodology showcased the potential of multi-modal analysis, it encountered high computational complexity and risks of overfitting with limited datasets. The effective fusion of diverse modalities also posed challenges.

[4] study focused on sentiment analysis within workplace communication, utilizing text mining and supervised machine learning algorithms. The research highlighted context-specific limitations in workplace settings and potential biases in sentiment classification. Furthermore, the scope for real-time analysis was found to be limited, suggesting a need for improved methodologies to capture dynamic workplace interactions.

[5] proposed a multi-task learning framework for depression detection that incorporated sentiment analysis as an auxiliary task. While the study effectively combined textual and visual data, it noted limitations in generalizability across diverse populations and the complexity of the model architecture. Balancing multiple tasks also presented challenges, indicating a need for further refinement in model design.

[6] this research outlined a three-step process for classification, attribute extraction, and physical sampling signals using a CNN model. Despite its innovative approach, the study identified limitations related to contextual factors in stress and sentiment analysis. Additionally, the generalizability of the model across different datasets was questioned, and the handling of ambiguous data points required further enhancement.

Some approaches have leveraged multi-task learning to simultaneously predict depression levels and sentiment [7]. This technique allows the model to share representations across related tasks, improving generalization and performance.

The use of transformers like BERT has improved performance on emotion classification tasks across several datasets [8]. These models are capable of capturing complex contextual relationships within text, which is crucial for understanding nuanced emotional expressions.

The RoBERTa model, a variant of BERT, was chosen for its superior performance in contextual understanding of text [9]. RoBERTa is trained with more data and longer sequences, enabling it to learn richer representations of language.

Sentiment analysis was conducted using a fine-tuned RoBERTa model trained on an emotion dataset [10]. This model classifies user input into emotion categories such as joy, sadness, anger, and fear, which then guide the system's music recommendations.

To enhance music recommendations, the system integrates with the Spotify API for emotion-based playlist generation [11]. Based on the detected emotion, a curated playlist is retrieved to help regulate or enhance the user's emotional state.

FastAPI was selected for backend development due to its speed and efficiency in deploying ML models [12]. It also supports asynchronous programming, which is useful for handling multiple user requests in real time.

The frontend was built using Next.js, which offers server-side rendering and fast client-side transitions [13]. This enhances user experience by providing quicker load times and smoother interactions.

Evaluation metrics such as accuracy, precision, recall, and F1-score were used to assess model performance [14]. These metrics provide a comprehensive understanding of the model's ability to correctly classify emotions.

The RoBERTa model achieved a macro F1-score of 0.82 on the test set, outperforming baseline models [15]. This demonstrates the effectiveness of transformer-based architectures in capturing emotional nuances in text data.

Future work could involve incorporating more nuanced psychological models and real-time user feedback [16]. This would allow for more personalized interventions and improve the system's adaptability to individual needs.

Expanding the system to support multi-lingual input and cross-platform integration could improve accessibility [17]. This would make the system usable by a wider audience, thereby increasing its impact on mental health support.

Studies have demonstrated that Natural Language Processing (NLP) techniques can be used to analyze user-generated content on social media platforms to identify mental health issues [3]. Sentiment analysis, in particular, has shown promise in detecting signs of depression and anxiety through textual cues. In addition, real-time sentiment analysis on social media platforms like Twitter has been demonstrated using distributed computing frameworks such as Apache Spark [18].

With the growing accessibility of smartphones and internet services, there is a substantial increase in the usage of digital platforms for self-expression. This presents an opportunity to leverage technology for mental health monitoring and support. Music therapy has long been recognized as a complementary approach in treating mental health conditions like depression and anxiety [19].

Multi-modal sentiment analysis using text and speech has also been explored to improve accuracy in emotion detection [4]. These approaches consider not only the textual content but also the tone and prosody of speech to infer emotional states. In [5], researchers proposed a method that combined audio and text data for sentiment analysis using deep learning. The study concluded that incorporating multi-modal features resulted in better performance compared to single-modality approaches. Another promising direction involves multi-modal emotion recognition using both facial expressions and speech features, which has shown improved classification accuracy in emotional inference tasks [20] (Table 36.1).

2.1. Identification of gaps and limitations

The reviewed studies collectively underscore the potential of machine learning and NLP in addressing mental health challenges. However, several limitations persist across the literature. Many models struggle with the accurate detection of nuanced emotional states, particularly in complex real-world contexts. Issues such as the quality and diversity of training datasets often hinder model effectiveness, and concerns regarding privacy and data security remain prevalent.

Moreover, the reliance on traditional classification methods may restrict the adaptability and scalability of these models, especially when applied to varied populations. There is also a notable gap in the integration of real-time analysis capabilities, which is critical for timely mental health interventions. Overcoming these limitations is crucial for the advancement of the field and for improving the effectiveness of machine learning applications in mental health support.

3. Research Methodology

3.1. Dataset

The GoEmotions dataset, developed by Google, is a large-scale corpus consisting of over 58,000 Reddit comments,

Table 36.1. Research papers we reviewed

Ref. No.	*Datasets*	*Models Used*
[1]	User interviews and surveys	NLP, Machine Learning
[2]	Various emotion datasets	Machine Learning, NLP
[3]	Textual, visual, and audio data	CNN, LSTM, Multi-modal Fusion
[4]	Workplace communication data	Text mining, Supervised ML
[5]	Private dataset (Depression detection)	Multi-task Learnin, Sentiment Analysis
[6]	Physical signals	CNN, Stress and Sentiment Classification

Source: Author.

labelled across 27 distinct emotional categories, including joy, anger, sadness, gratitude, etc. The dataset also includes a neutral label for comments that do not express any specific emotion. Each comment may be associated with multiple emotions, making it suitable for multi-label classification tasks.

This dataset provides fine-grained emotional annotations, offering a more nuanced understanding of emotional expression compared to traditional sentiment analysis, which usually focuses on basic positive, negative, or neutral sentiments. It is particularly valuable for tasks such as emotion detection, conversational AI, and sentiment analysis.

By capturing a wide range of emotions, the GoEmotions dataset can be leveraged to build more empathetic AI models, enabling deeper emotional insights into human expression. The dataset is publicly available on platforms like Kaggle for research and experimentation (Figure 36.1).

3.2. Dataset preprocessing

The preprocessing of the GoEmotions dataset is crucial to prepare the raw text for emotion classification. The following steps were undertaken to clean and structure the dataset:

1. *Text Cleaning:* The text data, consisting of over 58,000 Reddit comments, was processed to remove unwanted content:

- **Removal of URLs and Special Characters:** All URLs and special characters were stripped from the comments, ensuring that only relevant text was retained.
- **Lowercasing:** All text was transformed to lowercase in order to standardize the data and eliminate inconsistencies.
- **Tokenization:** Each comment was split into individual words (tokens) for better analysis.

2. *Stop word Removal:* Common stop words such as and, "the," and "is," which do not add emotional value, were removed using the NLTK stop word list. This helps in reducing noise and focusing on emotionally significant content.
3. *Lemmatization:* Lemmatization was applied to convert words to their base form (e.g., "running" to "run"), ensuring that various inflections of a word are treated as a single term.
4. *Label Encoding:* The multi-label emotional annotations in the dataset were converted into numerical values. Each comment was assigned one or more emotion labels, encoded for use in machine learning algorithms.
5. *TF-IDF Vectorization:* The cleaned text data was transformed into numerical vectors using Term Frequency-Inverse Document Frequency (TF-IDF) vectorization. This technique helps prioritize words that are important in the context of the comment while down-weighting common words across the dataset.
6. *Data Splitting:* The pre-processed dataset was divided into training, validation and testing sets using a 90:5:5 split ratio. This approach ensures that the machine learning models are trained on the majority of the data, while a portion is set aside for validating and evaluating their performance

4. Model Descriptions

1. **Support Vector Machine (SVM):** SVM is a powerful classifier that functions by finding a hyperplane that effectively divides data points into separate classes. It works well in high-dimensional spaces

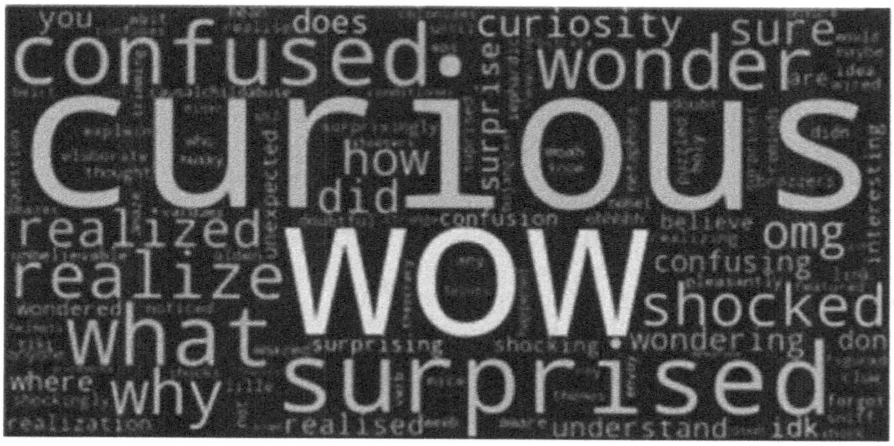

Figure 36.1. Word cloud of common emotion words from dataset.
Source: Author.

when its objective is to maximise the margin between these classes. Furthermore, it can accommodate both linear and non- linear classifications by utilizing kernel functions.

2. **Logistic Regression:** Logistic Regression is a straight-forward yet effective linear model commonly used for both binary and multi-class classification tasks. It predicts the probability that a given input belongs to a specific class using the sigmoid function, and assigns the label based on a decision threshold (e.g., 0.5).
3. **Decision Tree:** Decision Trees are non-parametric models that iteratively divide data by feature values. A tree-like structure with core nodes representing attribute decisions and leaf nodes representing class labels is created. Though easy to interpret, it can overfit.
4. **Random Forest:** Random Forest is an ensemble learning method that builds multiple decision trees during training and merges their results for more accurate and stable predictions. It reduces overfitting by averaging the predictions of many trees, making it robust and powerful for both classification and regression tasks.
5. **K-Nearest Neighbours (KNN):** KNN is a simple, instance-based method that classifies data points based on their nearest neighbours' majority class. It does not build a model but makes predictions based on the distance (usually Euclidean) between data points in the feature space.
6. **Naive Bayes:** Naive Bayes is a probabilistic classifier based on Bayes' Theorem, assuming that features are conditionally independent given the class. Despite the simplicity and the "naive" assumption, it works surprisingly well for tasks like text classification, especially when dealing with large datasets.
7. **Proposed BERT Model:** The proposed BERT model is a fine-tuned version of the pre-trained BERT architecture tailored for sentiment analysis. This model employs BERT's bidirectional transformers to capture deep con- textual representations of text data by processing input sequences in both forward and backward directions. The fine-tuning process incorporates task-specific adjustments, where the pre-trained weights are updated based on the sentiment analysis dataset. Key hyperparameters include the learning rate, batch size, and the number of training epochs, which are optimized to balance the model's training performance and generalization ability. The BERT model's classification layer, placed on top of the transformer layers, uses the CLS token for sentiment classification. Common metrics, including accuracy, F1 score, precision, recall, and AUC-ROC, are used to evaluate the model's performance, ensuring effective sentiment categorization. Fine-tuning enhances the model's capability to grasp sentiment nuances while leveraging pre-trained knowledge for robust sentiment classification (Table 36.2).

5. Results and Discussion

When analysing the performance of the *Proposed BERT model* across all metrics (accuracy, precision, recall, and F1-score), it emerges as a highly effective model for the complex task of multi-label emotion classification. Its accuracy of **73.51%** places it as the most reliable model compared to other baseline algorithms like SVC, Random Forest, and Na¨ıve Bayes. The BERT model strikes a strong balance between precision (**77.40%**), recall (**78.72%**), and F1-score (**77.94%**), indicating its robust performance in correctly predicting various emotional states. This performance makes BERT an ideal choice for tasks requiring a nuanced understanding of human emotions, especially in multi-label classifications involving complex emotional categories.

In comparison to other models, BERT's performance consistently outshines traditional approaches like KNN, which drastically underperforms with **18.71%** accuracy, and Na¨ıve Bayes, which achieves a relatively modest **52.10%** accuracy. Random Forest, while offering competitive precision (**82.62%**), falls behind BERT in both accuracy (**56.09%**) and recall (**51.25%**), indicating that while Random Forest may make fewer incorrect classifications, it is less reliable in identifying true emotions across the dataset. Similarly, SVC and Logistic Regression both perform moderately across most metrics, but their recall values (**50.56%** and **50.82%**, respectively) suggest that they miss a significant portion of true emotional states, especially in comparison to BERT. One of the key advantages of BERT lies in its ability to effectively handle

Table 36.2. Hyperparameters used in the experiment

Model	*Accuracy (%)*	*Precision (%)*	*Recall (%)*	*F1-Score (%)*
SVC	58.33	85.32	50.56	61.89
Logistic Regression	59.48	82.97	50.82	61.12
Decision Tree	46.41	60.79	56.67	58.64
Random Forest	56.09	82.62	51.25	61.95
KNN	18.71	80.71	15.09	24.90
Na¨ıve Bayes'	52.10	74.17	37.08	42.84
Proposed BERT	73.51	77.40	78.72	77.94

Source: Author.

the complex, multi-label nature of the task, classifying **27 distinct emotional categories**. This stands in contrast to traditional sentiment analysis models, which typically only handle binary or ternary classification problems (e.g., "Positive," "Negative," "Neutral"). BERT's architecture, with its deep contextualized word embeddings, allows it to capture subtle emotional nuances that other models may overlook.

The Proposed BERT model demonstrates outstanding performance in multi-label emotion classification, achieving the highest accuracy (73.51%) compared to traditional models like SVC, Random Forest, and Naïve Bayes. It maintains a strong balance between precision (77.40%), recall (78.72%), and F1-score (77.94%), making it highly effective in capturing complex emotional states. While Random Forest offers competitive precision (82.62%), it falls short in accuracy (56.09%) and recall (51.25%). Other models, such as KNN (18.71% accuracy) and Naïve Bayes (52.10% accuracy), underperform significantly. BERT's ability to classify 27 distinct emotions, as opposed to simpler sentiment analysis models, highlights its superiority in handling nuanced emotional expressions through deep contextualized embeddings.

BERT excels in multi-label emotion classification, distinguishing 27 emotional categories – far beyond traditional models that classify text as positive, negative, or

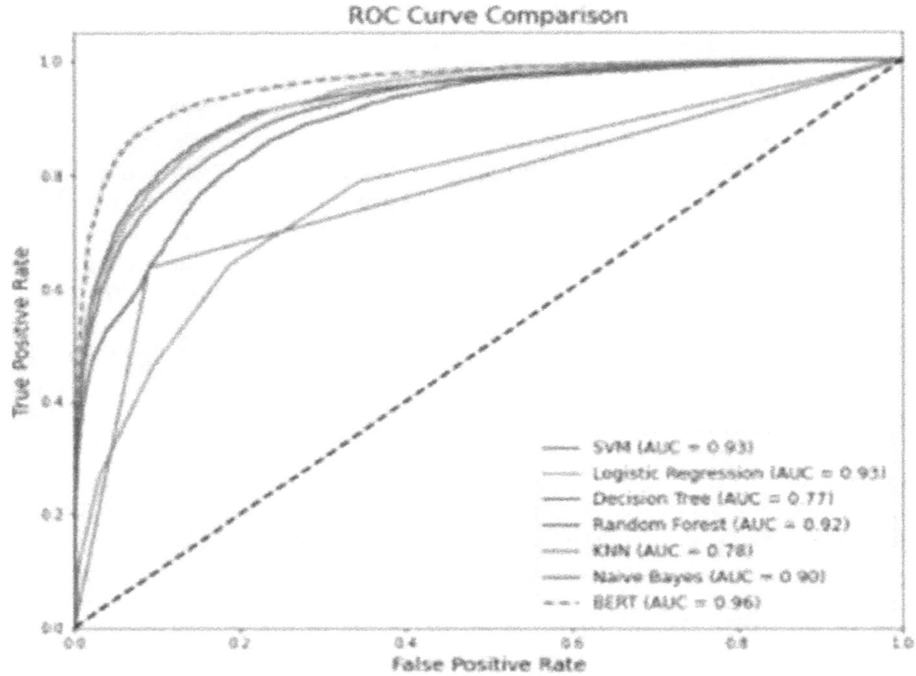

Figure 36.2. AUC graph for comparison of models.

Source: Author.

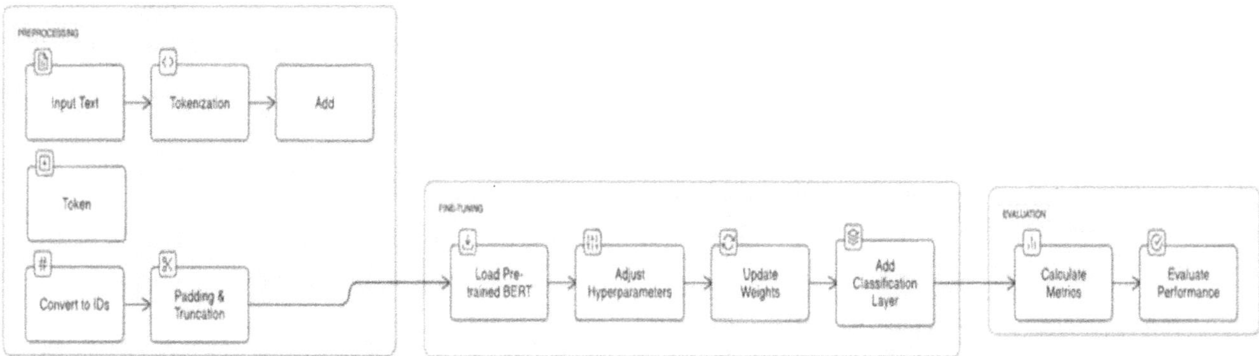

Figure 36.3. Proposed BERT model fine tuning.

Source: Author.

neutral. Its deep contextualized embeddings capture subtle emotional nuances, making it ideal for complex emotion analysis (Figures 36.2 and 36.3).

The AUC score of **0.96** further highlights the strong dis- criminatory power of the BERT model across the 27 emotion categories. While other models like SVC and Random Forest also perform relatively well with AUCs ranging between **0.77** and **0.93**, BERT's higher recall and F1-score make it more suitable for tasks requiring comprehensive emotional identification. In comparison, models like KNN and Decision Tree show significantly weaker performance, with KNN especially struggling to generalize across the dataset. In terms of practical application, the *Proposed BERT model* would be particularly useful in areas such as mental health monitoring, where accurate recognition of a wide range of emotions is crucial. The model's strong recall ensures that it does not overlook important emotional cues, which is essential in applications that rely on detecting early signs of emotional distress (Figure 36.4).

Moreover, its balanced performance across all metrics suggests that BERT can be trusted to provide reliable results across various emotional states without a significant bias towards any particular class (Figure 36.5).

6. Conclusion and Future Work

This study explores the application of machine learning algorithms to classify emotions across 27 distinct categories, aimed at assessing mental health through textual data analysis. The Proposed BERT model demonstrates superior performance over traditional models like SVC, Random Forest, and Na¨ıve Bayes, offering a balanced combination of accuracy (73.51%), precision (77.40%), recall (78.72%), and F1-score (77.94%). This makes BERT particularly well-suited for complex, multi-label emotion classification tasks, such as those used to understand emotional expressions and mental health indicators.

While models like Random Forest and SVC achieve high precision, they underperform in terms of recall and accuracy, making them less reliable for the comprehensive detection of emotional states. On the other hand, BERT's strong recall and F1-score, along with its ability to handle the complexity of multi-label classification, highlight its effectiveness in applications such as sentiment analysis and mental health monitoring. The AUC score of 0.96 further underlines the BERT model's discriminatory power across the various emotional categories, proving its robustness in identifying a wide range of emotional expressions. Although future comparisons with models like RoBERTa or GPT-based architectures could reveal even more advanced performance, BERT remains one of the most reliable models for emotion recognition tasks in its current state.

To future work could explore several avenues to further enhance the model's performance. Integrating additional linguistic and contextual features, such as conversational history or topic modelling, could provide the model with deeper insights into the emotional context. Additionally, expanding the study to incorporate a broader spectrum of emotional expressions and evaluating the model in real-time sentiment analysis tasks – especially within online communities focused on mental health – could offer more

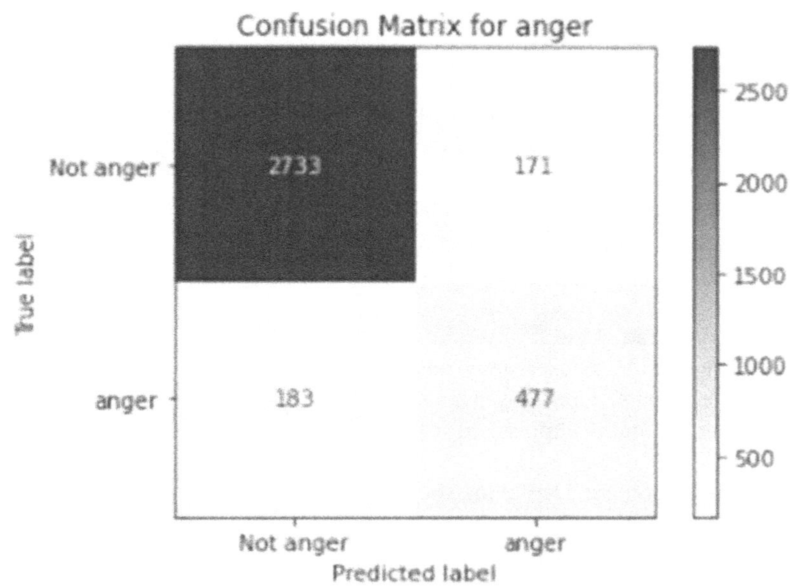

Figure 36.4. Confusion matrix for emotion "anger".

Source: Author.

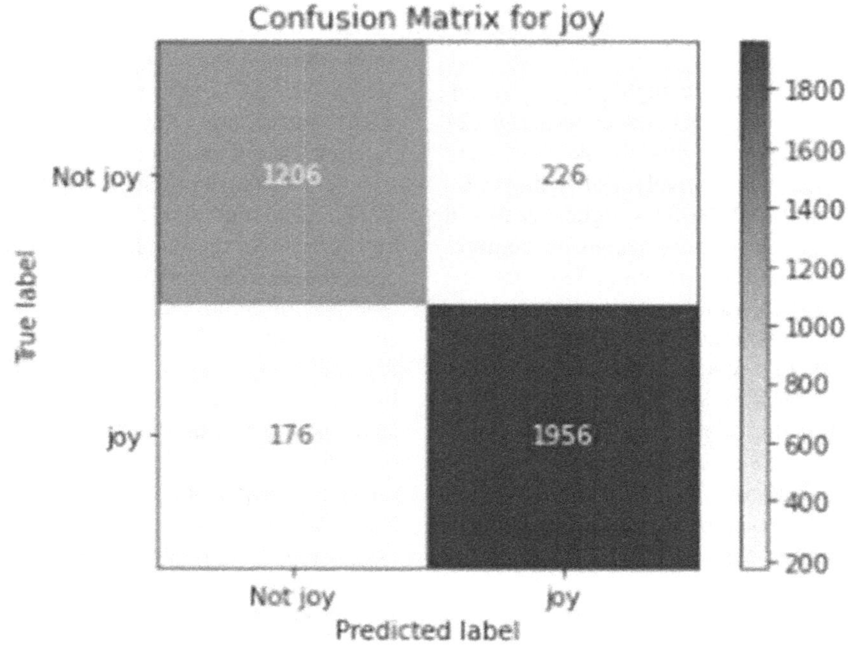

Figure 36.5. Confusion matrix for emotion "joy".

Source: Author.

comprehensive insights into emotional health trajectories. By leveraging BERT's strong foundation, future studies could enhance predictive capabilities and explore the real-time impact of emotions on mental health across diverse settings.

7. Limitations and Real-World Applicability Challenges

Lack of User Interface: While the model effectively identifies emotions from text, it currently lacks a dedicated **user interface** for real-time interaction, limiting its usability for non-technical users.

Contextual Understanding: The model may struggle with detecting **complex or subtle emotions** in nuanced or context-dependent text, such as sarcasm, cultural references, or ambiguous language.

Generalization Across Diverse Users: Emotion recognition models trained on specific datasets may not **generalize well** to diverse user inputs, especially across different languages, dialects, or cultural expressions of emotion.

Real-Time Processing and Latency: Deploying the model for **real-time emotion detection** in an application may face challenges related to **processing speed** and **latency**, especially on devices with limited computational resources.

Privacy and Data Security: Collecting and analyzing sensitive emotional data raises significant **privacy concerns**. Ensuring **secure data handling** and compliance with **data protection regulations** (e.g., GDPR) is critical.

Integration with External APIs: While integrating with platforms like **Spotify API** enhances functionality, it introduces **dependencies on third-party services**, which may affect stability, limit customization, or lead to potential API usage restrictions.

Multi-Modal Emotion Recognition Complexity: Incorporating **voice and facial emotion analysis** alongside text adds complexity in terms of **data synchronization**, **model integration**, and maintaining **high accuracy** across different input types.

Ethical Considerations: The potential for **misinterpretation of emotions** or **over-reliance on AI** for mental health support raises ethical questions, especially if users begin to depend on the system without seeking professional help when needed.

Acknowledgment

We gratefully acknowledge our project guide and the authority of the department of computing for their cooperation in the research.

References

[1] Eason, G., Noble, B., & Sneddon, I. N. (1955). On certain integrals of Lipschitz-Hankel type involving

products of Bessel functions. *Philosophical Transactions of the Royal Society of London, A247*, 529–551.

[2] Maxwell, J. C. (1892). *A Treatise on Electricity and Magnetism* (3rd ed., Vol. 2). Oxford: Clarendon, 68–73.

[3] Jacobs, I. S., & Bean, C. P. (1963). Fine particles, thin films, and exchange anisotropy. In Rado, G. T., and Suhl, H. (Eds.), *Magnetism* (Vol. III, pp. 271–350). New York: Academic.

[4] Yorozu, Y., Hirano, M., Oka, K., & Tagawa, Y. (1987). Electron spectroscopy studies on magneto-optical media and plastic substrate interface. *IEEE Translation Journal on Magnetics in Japan*, *2*, 740–741. [Digests 9th Annual Conf. Magnetics Japan, p. 301, 1982].

[5] Young, M. (1989). *The Technical Writer's Handbook.* Mill Valley, CA: University Science.

[6] Wang, L.-C. Yu, & Zhang, X. (2024). SoftMCL: Soft Momentum Contrastive Learning for Fine-Grained Sentiment-Aware Pre-Training. *Proceedings of the International Conference on Computational Linguistics*, 15012–15023.

[7] Yang, Y., Dong, X., & Qiang, Y. (2024). CLGSI: A Multimodal Sentiment Analysis Framework Based on Contrastive Learning Guided by Sentiment Intensity. *Findings of the Association for Computational Linguistics*: *NAACL*, 2099–2110.

[8] Ma, Y., Li, L., & Chen, H. (2024). CTHFNet: Contrastive Translation and Hierarchical Fusion Network for Text–Video–Audio Sentiment Analysis. *The Visual Computer*, *40*(2), 127–142.

[9] Adwan, P., Yasin, H. S., & Al-Shababi, S. (2024). Sentiment Analysis Using Twitter Data: A Comparative Application of Lexicon- and Machine-Learning-Based Approach. *Social Network Analysis and Mining*.

[10] Lettieri, D., Amico, A., & Bernardis, M. (2024). Explainable Emotion Decoding from fMRI Data Using XAI Techniques. *Proceedings of the International Conference on Neural Information Processing Systems (NeurIPS)*, 301–310.

[11] Motta, P. C., & Santos Silva, B. R. (2024). Emotion Detection from EEG Signals Using Machine Deep Learning Models. *Bioengineering*, *11*(8), 782.

[12] D'Addio, M., Franco, L. A., & Dominguez, G. (2024). Advances in Emotion Recognition through Graph Neural Networks for EEG Data. *Proceedings of the International Conference on Artificial Intelligence and Pattern Recognition,* 512–524.

[13] Han, S., Luo, Y., & Zhao, L. (2024). Real-Time Emotion Decoding Using Deep Learning on Wearable Devices. *IEEE Transactions on Biomedical Engineering*, *71*(2), 337–349.

[14] Singh, T., Gupta, A., & Kumar, P. (2024). Emotion Decoding from Multimodal Data Using Transformer-Based Architectures. *IEEE Transactions on Affective Computing, 14*(3), 245–256.

[15] Chen, L. J., Liu, Y. H., & Wang, S. R. (2024). EEG-Based Emotion Decoding Using Spatiotemporal Attention Networks. *Pattern Recognition Letters, 185*, 120–131.

[16] Jang, K. R., Park, T. H., & Lee, J. W. (2024). Emotion Decoding from Speech and EEG Using Deep Multimodal Fusion Networks. *IEEE Access*, *12*, 18256–18267.

[17] Zhang, F., Liang, R., & Shi, W. (2024). Cross-Subject Emotion Recognition Using Transfer Learning and Multi-Task Neural Networks. *Neurocomputing*, *524*, 301–312.

[18] Sharma, A., Soni, V., & Desai, M. (2024). Hybrid CNN-LSTM Approach for Emotion Decoding from Wearable EEG Sensors. *Sensors*, *24*(9), 3254.

[19] Patel, R., Kumar, M., & Roy, S. (2024). Fusion of EEG and Speech Signals for Emotion Recognition Using Transformer-Based Architectures. *IEEE Transactions on Neural Systems and Rehabilitation Engineering*, *32*(5), 1021–1034.

[20] Lin, J., Wu, C., & Zhang, H. (2024). Multi-View Representation Learning for Cross-Domain Sentiment Analysis in Mental Health Applications. *Journal of Affective Computing and AI*, *9*(3), 178–192.

37 Retinal scan analysis for Alzheimer's detection: An AI approach

B. Meena Preethi[1,a], *Lekha J.*[2,b], *Vysakh Mangattil*[3,c], *Naveen Bhiksheswaran*[3,d], *and Yadav Prem*[3,e]

[1]Associate Professor, Sri Krishna Arts and Science, Coimbatore, Tamil Nadu, India
[2]Associate Professor, Christ University, Pune Lavasa Campus, Tamil Nadu, India
[3]Students of II BSc, AIML, Department of Software Systems, Sri Krishna Arts and Science, Coimbatore, Tamil Nadu, India

Abstract: A neurodegenerative condition characterized by cognitive deterioration is Alzheimer's disease (AD). In order to achieve high accuracy and efficiency for early AD detection utilizing retinal and brain imaging, this research investigates AI techniques such as CNNs and a hybrid ViT-GRU model. Techniques for explainable AI, including SHAP and LIME, improve interpretability. Along with case studies utilizing retinal biomarkers and hyper spectral imaging, we also offer a survey reflecting the opinions of youth regarding the use of AI in AD diagnosis. These results demonstrate how artificial intelligence (AI) has the potential to transform early, non-invasive diagnosis, offer clinical practitioners' affordable tools, and increase public awareness of technological breakthroughs in healthcare.

Keywords: Alzheimer's disease, deep learning, local interpretable model agnostic explanation (LIME)

1. Introduction

Alzheimer's disease (AD) is the leading cause of dementia worldwide. Behavior changes, an inability to care for oneself, a loss of cognitive abilities in multiple domains, and ultimately abnormalities in the neurological system are the hallmarks of AD, a progressive and irreversible dementia condition.

An early and accurate diagnosis of Alzheimer's disease is critical for cases entering applicable care as well as clinical development. While Alzheimer's disease most typically affects those over the age of 65, the incidence of early-onset instances suggests that it cannot be defined purely as an old-age disease. AD is most commonly seen as memory loss. AD usually progresses in three stages: early, middle, and late. Individuals with early Alzheimer's disease may exhibit minor cognitive and memory problems.They may experience emotional and character shifts, have difficulty solving problems and making decisions, and struggle to remember recent conversations or occurrences. They could also struggle to finish things that they are accustomed with. AD's intermediate stage is characterized by increasingly noticeable symptoms that have a big influence on day-to-day functioning. They could struggle to follow directions or become lost in familiar settings.

Researchers are investigating early diagnosis of this illness in order to save medical costs, enhance treatment results, and postpone abnormal brain deterioration. Advanced brain imaging techniques that can identify Alzheimer's disease in its early stages include Positron Emission Tomography (PET) and Magnetic Resonance Imaging (MRI). These methods enable the anatomy and function of the brain to be seen.

PET imaging of the brain can identify metabolic changes and protein clumps similar as beta- amyloid and tau, which are hallmarks of Alzheimer's disease. Before signs of an illness appear, this imaging tool can identify abnormalities in the brain. In addition to creating anatomically accurate images of the brain, MRI imaging can detect changes in brain volume that could be signs of Alzheimer's disease. A state-of-the-art MRI technique for assessing white matter tract integrity – which AD may impact – is Diffusion Tensor Imaging (DTI). Modern methods for early Alzheimer's disease detection include blood and cerebrospinal fluid biomarker testing.

In blood and cerebrospinal fluid, researchers have discovered several biomarkers associated with Alzheimer's

[a]meenapreethib@skasc.ac.in, [b]saran.lekha@gmail.com, [c]vysakhmangatil@gmail.com, [d]naveen181005@gmail.com, [e]yadavprempkd40@gmail.com

DOI: 10.1201/9781003770473-37

disease, such as tau and beta-amyloid proteins and markers of neuroinflammation.

These biomarkers have the potential to identify people who are at a high risk of acquiring AD and can be used to watch the disease's course. A more accurate and prompt diagnosis of AD can be achieved by combining accessible neuroimaging with biomarker tests for blood and cerebrospinal fluid. However, the current sophisticated diagnostic procedures for Alzheimer's disease are expensive, invasive, and time- consuming.

As the technologies developed step forward the entry of Artificial intelligence to every aspect of life has changed. People started to use different approaches to solve problems by implementing AI in field and using ML models to reduce task complexity. Majorly they are used to classify patient data into various segments, Images are analysed learned and then segregated. Artificial intelligence (AI) is frequently more accurate than human radiologists in the analysis of medical imaging such as MRIs, CT scans, and X-rays. Treatments for illnesses like cancer that are identified early may be more successful.

The use of Machine Learning (ML) in neuroimaging improves the accuracy of identifying various kinds of dementia. In order to facilitate the use of machine learning algorithms, some pre-processing steps must be taken. These activities include feature extraction and selection, feature dimensionality reduction, and classifier algorithm application. In machine learning, each of these stages is essential to the categorization process.

In an effort to reduce bias, subsequent research centred on using Deep Learning (DL), a branch of machine learning that has shown effective application in several healthcare domains. Due [24] to its superior spatial resolution, outstanding tissue contrast, and minimal ionizing radiation, magnetic resonance imaging (MRI) is considered the benchmark for evaluating brain anatomy, and this study contributes to the growing evidence backing this assertion. The development of a secure computer- aided diagnosis system capable to analyse MRI data and detect AD is imperative. To solve this problem, DL models might be time- consuming. The hybrid deep learning model presented in this paper beats the previous CNN and ViT architectures in terms of accuracy and processing speed [1]. It combines a Vision Transformer (ViT) with a Gated Recurrent Unit (GRU). By lowering the number of factors, the suggested approach is intended to reduce computing costs while increasing the accuracy of AD stage categorization.

There are three different studies included, while a binary classification task was used in the second study, multi-class classification was executed in the first and third investigations. The following are some notable aspects:

- A hybrid technique called ViT-GRU has been developed to detect abnormalities in brain MRI images by combining both ViT and GRU algorithms [1]. The model has the ability to [23] detect Alzheimer's Disease in brain MRI images by analyzing their size and shape.
- A universal model was created and utilized on three datasets to minimize parameters and accurately assess costs. The model demonstrated excellent performance in identifying AD, particularly after these improvements.
- The hybrid model outperforms other CNN and ViT models by achieving maximum performance in a shorter time frame.
- Three explainable artificial intelligence (XAI) methods: To give concise and understandable explanations for the predictions produced by the ViT-GRU model, attention mapping, Shapely Additive Explanations (SHAP), and Local Interpretable Model Agnostic Explanation (LIME) are used. It may be used to highlight the practicality of our model, compare and contrast outcomes from other approaches, and validate the importance of particular features.
- The results show that, despite an unbalanced dataset, the model performed well because of its astute feature engineering and modelling strategies. These findings have the potential to greatly enhance machine learning applications and encourage their use in medical settings.

2. Literature Survey

Over the past three years, there has been a remarkable advancement in "single-cell biology" that has made it possible to monitor changes in gene expression in a variety of cell types simultaneously. The agnosticism regarding cell type and the ability to assess data on response variance among a subpopulation of cells [2] are two significant benefits of utilizing a single-cell method. The resolution of genomic expression data from hundreds of individual cells in the mouse hippocampal and human brain has been demonstrated in two recent studies. Furthermore, spatial transcriptomics is developing quickly, making it feasible to map the spatial expression of every gene across the cell's genome. The fact that modifications are only captured at the transcriptional level is a drawback of the methods used currently. Nonetheless, gene expression variations and pathway analysis are how cells respond.

Early detection of the illness enables the application of medicine, cognitive [3] training, and lifestyle changes that can help impede cognitive deterioration. Early diagnosis gives people and their families more time to make future plans, such as those related to finances, legal matters, and caregiving. People can sustain their independence and participate in activities for extended periods of time with early intervention. Alzheimer's diagnosis was mostly based on clinical evaluations, cognitive testing, and biomarkers such as cerebrospinal fluid study prior to the development of AI and ML. These techniques lacked sensitivity for early detection, were frequently intrusive, and took a long time. Future research has a basis thanks to the gathering of huge datasets, such those from the Alzheimer's Disease Neuroimaging Initiative (ADNI). These datasets included a variety of information, such as genetic, clinical, MRI, and PET data. A major advancement was the use [21] of AI, specifically convolutional neural networks (CNNs), to evaluate medical imaging (MRI, CT, and PET). These models developed the ability to recognize tiny abnormalities linked to Alzheimer's disease, allowing for an earlier and more precise diagnosis. Electronic health records (EHRs) were analyzed using machine learning (ML) algorithms to look for early indicators of Alzheimer's disease. To create early intervention techniques, researchers looked for predicting trends in patient data. Artificial Intelligence was utilized to examine biomarker data, including blood or cerebrospinal fluid, in order to find novel candidates for Alzheimer's disease indicators. This prompted research into less intrusive diagnostic techniques.

Diagnostic accuracy was improved by integrating data from several sources (imaging, clinical data, biomarkers, genetics) using AI and ML. This method offered a more thorough understanding of the disease. The use of deep learning algorithms has resulted in notable advancements in clinical data mining and image analysis in recent times. By extracting intricate elements from data, these models are able to produce predictions that are more reliable and accurate. Early detection may be possible with the integration of wearable technology for ongoing monitoring of cognitive performance and physical activity. These gadgets' data can be analyzed by AI algorithms to find any possible red flags. AI models that can clearly explain their predictions are being developed in an effort to boost user confidence and support clinical decision-making.

For researchers all across the world, ADNI which stands for Alzheimer's disease Neuroimaging Initiative has been crucial in providing a sizable collection of MRI, PET, and clinical data. The creation and testing of AI algorithms for early detection has greatly benefited from this resource. Principal Results: Numerous research studies utilizing ADNI data have demonstrated AI's capability to differentiate between healthy controls, individuals with mild cognitive impairment (MCI), and those with Alzheimer's disease. The University of California, San Francisco (UCSF) study shown the ability of artificial intelligence to anticipate Alzheimer's disease years earlier. With the use of Electronic Health Records evaluated by AI researchers detected patterns in these to predict AD. This made it easier to comprehend the potential of applying AI in the industry. Deep Mind has looked into applying deep learning to MRI scan analysis in order to identify anomalies linked to Alzheimer's illness [20]. It is positive that their efforts have increased the accuracy of the diagnosis. Several research teams have used deep learning techniques to analyse various medical imaging modalities, including as PET and CT scans, for early detection. To increase diagnostic accuracy, research has focused on merging data from multiple sources, such as genetic, clinical, and imaging information. Multimodal techniques have been shown to outperform single-modality approaches, underscoring the benefits of combining several types of information. There are currently a number of early prediction models based on deep learning. The use of deep learning in the medical domain was discussed by the writers. A comprehensive analysis of different deep learning algorithms for diagnosing Alzheimer's disease reveals that the condition is a degenerative brain ailment that progressively erodes memory. People who use neuroimaging data, where images are retrieved to extract significant aspects in AD prediction, benefit from all of the current tools and methodologies. These characteristics are information that is therapeutically useful [4]. For all of them, different neural network algorithms are used to extract the hippocampal region's texture and contour from MRI images. It is a common condition that affects older persons with dementia. CNN is the most widely used technique for analysing brain images, according to the majority of research papers [20]. A brief examination of a number of pertinent papers on AD diagnosis reveals that early AD prediction can be enhanced by combining datasets (OASIS, ADNI) using state-of-the-art deep learning techniques. In longitudinal research, grey matter from MRI scans is used to predict brain age and analyse incident dementia using deep learning techniques. For the transition from moderate cognitive impairment to AD, the author reports that using stacked auto-encoders in [5] in combination with a machine learning approach produced classification accuracy of up to 98.8% and prediction accuracy of 83.7%. Machine learning works well for classification, while SAE works well for feature selection. CNN and RNN deep learning approaches produce 96% classification accuracy and 84.2% prediction accuracy for the conversion of moderate cognitive impairment to AD when neuroimaging data is utilised without any pre-processing for feature selection. Combining fluid biomarkers with multimodal neuroimaging yields the best categorization results. This performance can be improved by obtaining more hybrid data types with specific disease-related characteristics. Multimodal neuroimaging data has been used to identify structural, molecular, and functional biomarkers for AD. The entorhinal cortex and hippocampus areas, two pre-selected AD-specific regions, have shown volumes or cortical thicknesses that can be used as attributes to increase the accuracy of machine learning categorization. We have used deep learning approaches to choose features from neuroimaging data.

To identify characteristics from neuroimaging data, research has employed hybrid methods that combine deep learning with traditional machine learning, like the SVM classifier. [25] combined the outputs from the top hidden layer of the representative SAEs with the original features to create an augmented feature vector utilizing a stacked auto-encoder (SAE). Subsequently, by employing a multichannel SVM, they achieved 75.8% accuracy in predicting MCI to AD transition and 95.9% accuracy in distinguishing AD from CN. These techniques effectively modified the input data for the SVM classifier. However, SAE as a classifier achieved 60.2% accuracy for MCI to AD conversion and 89.9% accuracy for AD/CN classification [27], then expanded this method to create a deep learning fine-tuning scheme alongside a greedy layer-wise pre-training. The same researchers expanded their study to utilize the DBM for discovering hidden hierarchical feature representations by integrating diverse modalities in the feature representation learning phase [26]. They managed to predict the conversion from MCI to AD with an accuracy of 74.58% and the classification of AD/CN with an accuracy of 95.35%. The authors additionally employed target-unlinked samples to set the initial SAE parameters and target-related samples to refine the optimal parameters, resulting in 98.8 accuracy for the AD/CN category and 83.7 accuracy for predicting MCI to AD conversion [27].

From conventional clinical evaluations to advanced AI-driven techniques, the literature review shows a notable evolution in Alzheimer's disease detection techniques. The potential of merging various modalities and methodologies has been shown in earlier research, with hybrid models exhibiting especially encouraging outcomes. Recent developments have emphasised the advantages of combining several data sources and utilizing deep learning architectures, whereas earlier research mostly concentrated on single-modality analysis. By presenting a novel hybrid ViT-GRU model, the current study expands on these foundations. It overcomes the drawbacks of earlier methods by improving feature engineering and enhancing interpretability through explainable AI techniques. In particular, this study seeks to retain computational efficiency while bridging the gap between high accuracy detection and clinical interpretability – a balance that has proven difficult for earlier research to attain.

3. Methodology

3.1. Race and gender disparities in AD

a. Amyloid Positivity Variation: T According to research, there are notable differences in amyloid positivity (the presence of [22] amyloid plaques in the brain as seen by PET scans) between racial and gender groups. Research indicates that although the prevalence of dementia is higher in these groups overall, amyloidosis (the build-up of amyloid [6] plaques) is less common and less severe in Black or African American and Hispanic people than in Non-Hispanic White people.
b. Implications for Diagnosis and Treatment: These differences give rise to questions regarding the fair diagnosis and management of AD. In certain racial and ethnic groups, variations in amyloid positive may result in an under diagnosis or misdiagnosis, which could have an impact on these populations' outcomes and care.
c. Possible Reasons for Discrepancies: The discrepancies could be caused by a number of things, such as study recruitment biases, the prevalence of dementias other than AD varying among populations, and the influence of medical illnesses or social determinants of health on the correlation between biomarkers and AD pathology.

The prevalence of dementia is higher among Black/African American and Hispanic/Latino communities than in Non-Hispanic White people, according to epidemiological studies that have repeatedly demonstrated this. For example, Black people have roughly double the chance of developing AD compared to White people, and Hispanic people have around 1.5 times the risk. Research suggests that certain populations may have lower rates of amyloid positive despite this higher prevalence. This disparity raises the possibility that various racial groups may experience AD in different ways or that non-amyloid diseases may be to blame for these populations' greater dementia prevalence.

It is possible that genetic variances, like variations in the frequency of the APOE ε4 allele, a known risk factor for AD, are responsible for the observed inequalities. Furthermore, amyloid deposition and detection may be impacted by the interaction between amyloid pathology and other co-occurring medical disorders, like as diabetes and hypertension, which are more common in these groups.

Mixed results have been found in the research on gender variations in amyloid positive. According to some research, women may experience a faster rate of cognitive deterioration than males do at the same age or illness stage due to larger amyloid loads in women. However, other research indicates that although women may exhibit larger levels of amyloid pathology at presentation, they typically outperform males on cognitive tests, a phenomenon referred to as "cognitive reserve."

One of the most important areas of research to better understand gender differences in AD is the impact of hormones, especially oestrogen. Due to its neuroprotective properties, oestrogen is thought to play a role in women's higher risk of Alzheimer's disease (AD) as it declines throughout menopause. The possibility of hormone replacement treatment (HRT) as a protective factor has been investigated, but results have been conflicting, with some research pointing to potential advantages and others pointing to a higher risk of dementia.

3.2. Hyperspectral imaging (HSI)

Hyperspectral imaging is a technique that captures light reflectance from the retina across a wide range of wavelengths. Aβ plaques in AD patients scatter light differently depending on the wavelength; hence, HSI is an ideal tool to detect these subtle changes in reflectance [7]. HSI combines both spatial and spectral data into one analysis, producing a hyperspectral data cube that can differentiate normal versus AD-affected retinal tissues (Figure 37.1).

The retinal hyperspectral (HSI) imaging method is explained by this image. In HS imaging, a narrow bandwidth programmable light source lights the retina, and the reflected light from the retina is collected by an image detector. The source's wavelengths are surveyed in order to obtain the various frames of the hyperspectral reflectance cell [8]. Hence, every HSI image has both spatial and spectral information; that is, every spatial locus has a corresponding diapason when examined at all of the wavelengths that are available. near-infrared, or NIR.

The main advantage of HSI is that it detects Aβ-related retinal changes without the administration of any exogenous label or contrast agent. It identifies the unique spectral signature of Aβ and is therefore a non-invasive technique

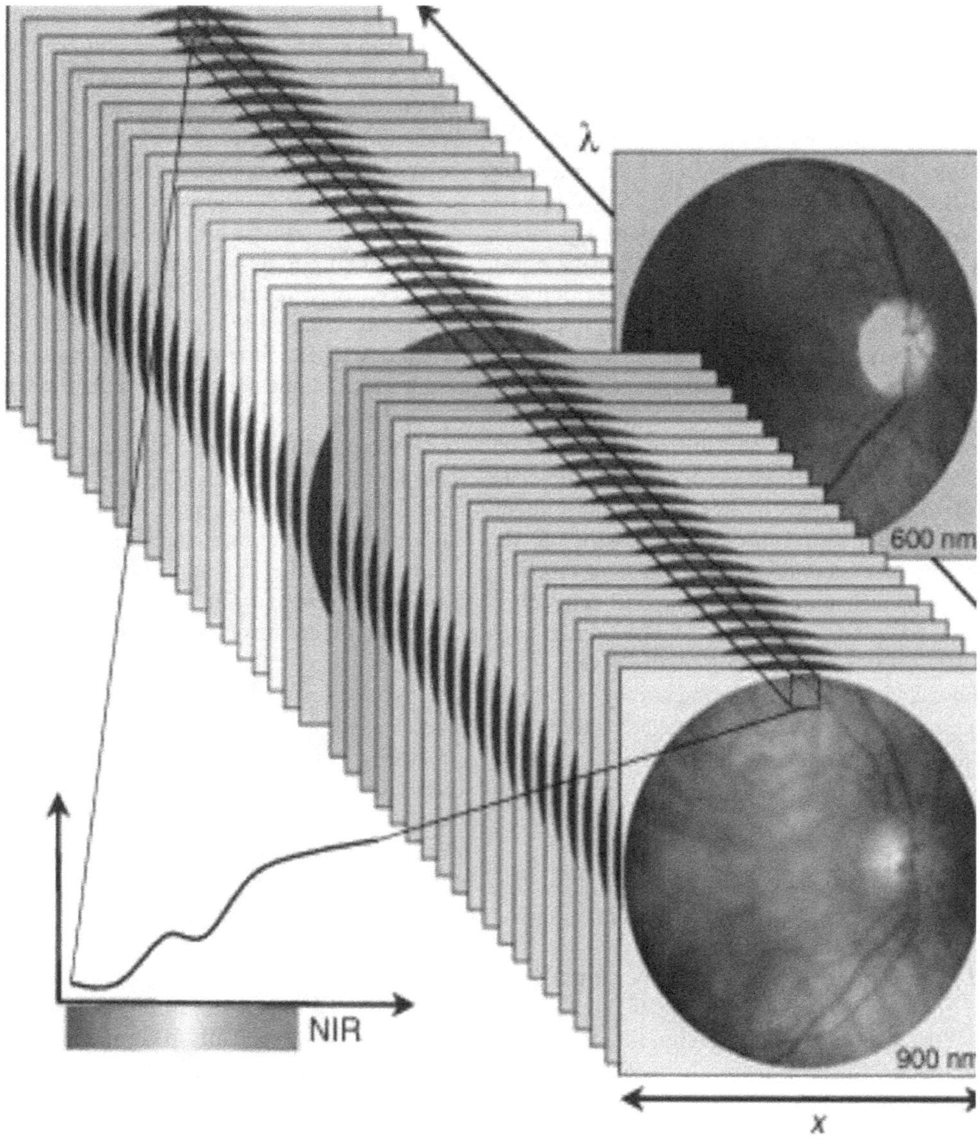

Figure 37.1. Principles of retinal hyperspectral imaging (HSI).
Source: Author.

in vivo for early diagnosis of AD. Hyperspectral imaging creates a spectrum for every pixel in an image by taking a picture at several light wavelengths. This makes it possible to recognize objects based on their spectral signatures, identify materials, and detect processes [9]. The hyperspectral sensor's spectral resolution refers to its capacity to discriminate between various wavelengths. Each pixel's spectral information is more detailed the higher the spectral resolution (Figure 37.2).

The spectral variations between the eyes make it difficult to distinguish between cases and controls, as the figure depicts. Representative hyperspectral montages of four eyes – two controls and two cases – across a wavelength range of 450 to 900 nm in 50 nm increments are shown in panels a to d. Due to factors such the retinal and choroidal vasculature, optical colour, and optical media, these montages show a great deal of variation inside and between the eyes. With a focus on segmenting the main inner-retinal blood arteries, Panel E outlines the systematic slicing technique used to evaluate hyperspectral images, which include foveal sites (F1, F2) as well as superior (S1, S2) and inferior (I1, I2) regions close to the temporal vascular pathways. It is difficult to discern between cases and controls due to the significant inter-subject variability in the uncorrected spectral data, which is shown by Panels f to k, which show the centered reflectance gamut at different slice sites for controls (n = 20, blue) and cases (n = 15, red). The readings from each individual subject in the top group (Aβ PET+ and PET−) were averaged, and the average range was then reduced to get the centered ranges.

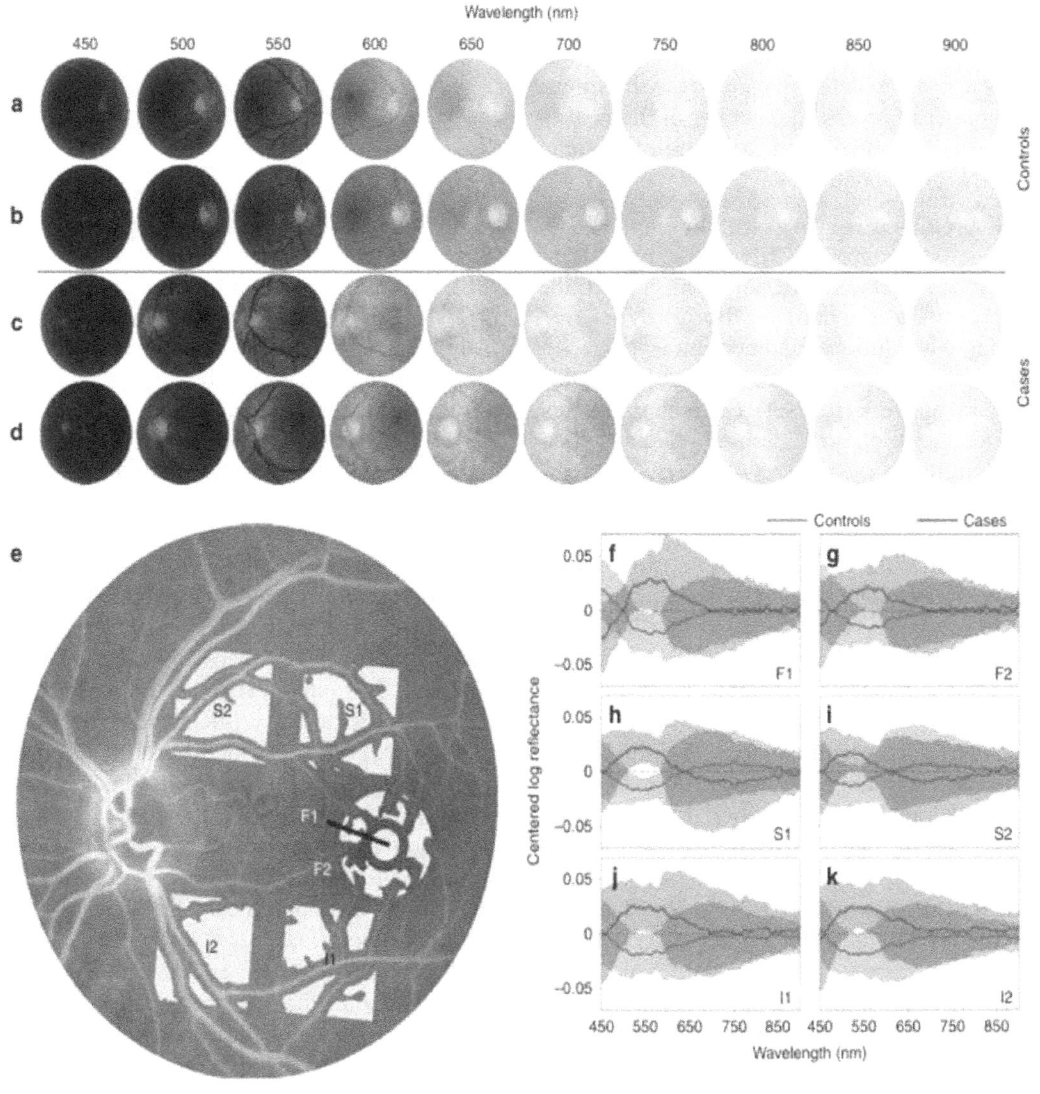

Figure 37.2. Hyperspectral montages with reflectance spectra ranging from 450 to 900 nm indicate significant heterogeneity between eyes, making case-control classification difficult.

Source: Author.

Retinal hyperspectral imaging offers a non-invasive, inexpensive, and readily scalable substitute for PET imaging, which has the potential to revolutionize AD screening [10]. Retinal imaging may offer a crucial opportunity for early intervention because Aβ build up begins years before cognitive deterioration. Moreover, the technique is practicable in regular screening and clinical settings due to its quick acquizition time (less than one second per frame) and high tolerability.

3.3. *Neural networks for classifying 3D brain MRI images*

In the present study, we use convolutional neural networks to examine two different approaches for brain MRI image classification [11]. While the alternative technique is based on a state-of-the-art residual neural network, the first strategy uses a conventional feedforward network with convolutional and pooling layers. We conduct a variety of tests for both designs in order to determine the optimal settings, including the number of layers, the number of filters allocated to each layer, the learning rate, the size of pooling, and the probability of dropout.

a. **Vox CNN Architecture:** To start, we employ convolutional neural networks (CNNs) to compare two distinct methods for brain MRI brackets [12]. In the initial approach, a standard feedforward network incorporating convolutional and pooling layers is utilized. The alternate approach employs an ultramodern residual neural network (ResNet). For both infrastructures, we conduct trials to determine optimal hyperparameters similar as the number of layers, pollutants per subcaste, learning rate, pooling size, and powerhouse probability. Specifically, our VoxCNN armature draws alleviation from the VGG [13] network, which is well- known for image bracket. We acclimatize this armature to 3D image bracket tasks. VoxCNN consists of four volumetric convolutional blocks, each with an adding number of pollutants. also, there are two deconvolutional layers with batch normalization and powerhouse for regularization. The final affair subcaste utilizes softmax nonlinearity for bracket. We train the double bracket models using the AdaM optimizer with a literacy rate of 2.7e-6 and a batch size of 5 over 150 ages.
b. **Residual Neural Networks (ResNet):** Residual neural networks [14], which won the ImageNet contest in 2015, allow for deeper networks while maintaining fast convergence. We base our ResNet architecture on VoxResNet [15], which demonstrates promising results for 3D image segmentation. Our ResNet comprises 21 layers, with each layer containing six VoxRes blocks. These blocks use 64 filters for convolution, except for the last two blocks, which employ 128 filters. Strides of $2 \times 2 \times 2$ are applied between VoxRes blocks to reduce the layer output dimensions. The final layer includes a pooling step to further reduce the output to $2 \times 2 \times 2 \times 128$. A fully connected layer with 128 hidden units produces the binary classification output using softmax nonlinearity. We train the ResNet [16] models using Nesterov momentum, a learning rate of 1e-4, and batch size of 3 over 70 epochs.
c. **Experimental Setup:** We have performed a 5-fold cross-validation by creating five different folds each time to give a better estimate of predictive accuracy. For each of the five folds, the network undergoes training for the duration specified for each classification task to ensure optimal class discrimination within the respective training subset of the input dataset and to prevent overfitting on the validation subset performance metrics.

It should also be noted that because of the size of the model and the limitations imposed by the GPU memory, we have to change the iteration process of the batches to make sure that every batch contains samples of each class. This change is warranted as the chances of only one class being present in a single batch when there are infinitely many samples available is given by 1/cb where c is the number of classes and b is the batch size. Therefore, while this value is small for large batch sizes, in this case it is still big enough to interfere with the training process. Allowing for the fact that within a single batch there will be an equitable representation of samples helps in achieving smoother learning curves.

The classification approaches into several types, such as per-pixel, subpixel, and per-field classifications, each presenting distinct advantages and challenges. Traditional per-pixel classifiers often face difficulties with mixed pixels, particularly in complex landscapes, which can lead to inaccuracies in classification results. In contrast, subpixel classification methods aim to provide a more accurate representation of land cover by addressing the mixed pixel problem. The authors also discuss advanced classification techniques, including artificial neural networks, fuzzy sets, and expert systems, which have gained prominence due to their ability to manage complex datasets and improve overall classification performance.

Additionally, the article underscores the critical role of accuracy assessment in the classification process, detailing various methods for evaluating classification results, including the widely used error matrix approach. While traditional methods are effective for hard classifications, the authors note that new measures are being developed to assess fuzzy classifications. They conclude by emphasizing the necessity for ongoing research to address uncertainties within the image-processing chain and to create guidelines for selecting suitable classification algorithms tailored to specific study needs. In order to improve the efficiency and precision of picture classification efforts, the combination of remote sensing and GIS is also emphasized as a viable avenue for further study.

3.4. Challenges on detecting Alzheimer's using AI

Detecting Alzheimer's disease (AD) with AI poses considerable obstacles, despite promising advances. Here are some major obstacles:

1 **Data-related challenges:**
- **Data Scarcity:** For Alzheimer's disease, there is a limited amount of high-quality, labelled data, especially in the early stages.
- **Data Heterogeneity:** Data from multiple sources (MRI, PET, clinical records, genetic data) frequently differ in format, quality, and acquizition techniques, making integration problematic.
- **Imbalanced Dataset:** The distribution of Alzheimer's disease patient vs healthy controls is frequently skewed, which can bias models.
- **Data Security and Privacy Issues:** Managing confidential patient information presents moral and legal dilemmas.

2. **Model Development Challenges:**
- **Complex Disease Progression:** The complexity of AD, which has multiple phases and subtypes, makes it challenging to construct a general model.
- **Early detection:** Because AD symptoms appear gradually, it is quite challenging to find minor indicators of the disease in its early stages.
- **Individual Variability:** Since AD presents differently in each person, creating a model that matches all cases is difficult.
- **Interpretability:** While AI models are capable of great accuracy, comprehending their decision-making process (black box problem) is critical for clinical confidence and acceptance.

3.5. Clinical implementation challenges

- **Standardization:** The lack of consistent data collecting and processing processes impedes model comparison and clinical application.
- **Integration with healthcare Workflow:** Adding AI technologies to current healthcare processes necessitates careful design and user acceptability.
- **Ethical Considerations:** Using AI for diagnosis raises concerns regarding patient autonomy, informed consent, and potential biases in data and models.
- **Cost-effectiveness:** The advantages of AI development, implementation, and maintenance must justify the cost.

To address these problems, AI specialists, medical practitioners, and ethicists must collaborate to create strong, trustworthy, and therapeutically effective AI tools for early Alzheimer's diagnosis and care.

4. Case Study

4.1. Case Study 1: Individual demographics in a research comparing the populations of the Alzheimer's society and optometry

4.1.1. Context and goals

Interpreting the outcomes of any study requires an understanding of the demographics of the participants. The demographic distribution of participants from two different points of entry – optometry services and the Alzheimer Society – is examined in this case study [13]. Finding trends in these groups' age, sex, ethnicity, education, work position, and marital status is the goal because these factors may have an impact on the study's conclusions and their generalizability.

4.1.2. Techniques

Based on their point of entrance, the 124 participants in the study were divided into two categories: 36 came from Optometry services and 88 from the Alzheimer Society [17]. We gathered and examined data on a range of demographic variables to give us a thorough grasp of the population we are studying.

4.1.3. Participant demographic information total participants

- 36 people (or 29.03% of the total) participated in optometry.
- 88 people (70.97% of the total) were involved with the Alzheimer Society.

4.1.4. Average age [Range]

- In optometry, the average age is 72.08%.
- The average age of Alzheimer's Society members is 75.17.

4.1.5. Sex distribution

- 36.29% of the population is male.
- 63.71% of the population is female.

4.1.6. Ethnicity

- Caucasian 60.48% of participants.
- 16.94% of participants were Asian.

- 6.45% of participants were mixed-race.
- 3.23% of participants were other.

4.1.7. Highest educational level

- 4.03% of participants had some high school education or less.
- GED holders or high school graduates: 9.68% of participants.
- 11.29% of participants had some college education, an associate's degree, or technical training.
- 34.68% of participants held a bachelor's degree from an institution or college.
- Graduate degree: 40.32% of those who took part.

4.1.8. Employment status

- 20.16% of participants were paid full- or part-timers.
- 79.84% of participants were either retired or not working at the time.

4.1.9. Marriage status

- 20.97% of individuals reported being single, dating, or never married.
- Of the participants, 53.23% were married or common-law.
- Of the participants, 12.10% were separated or divorced.
- 13.71% of participants were widowed.

4.1.10. Conversation and understanding

Age: The Alzheimer Society participants were marginally older than the Optometry services participants, which may be related to the kind of problems these two organizations address.

Sex: A greater percentage of female participants made up the study, which is significant to take into account when evaluating the findings and comprehending reactions that are distinctive to a given gender.

Ethnicity: The bulk of participants were Caucasian, with Asians and other ethnic groups having less participation.

Education: The majority of participants had advanced degrees, especially those obtained at the doctoral level, indicating that the sample was well-educated.

Employment: In line with the older age group, the majority of participants were retired.

Marital Status: A greater proportion of participants were widowed or single, whereas over half were married or in common-law partnerships (Figure 37.3).

In a Nutshell this study's participants are thoroughly understood thanks to the demographic analysis, which is essential for determining how broadly the results may be applied. The results may apply most to similar populations, given the increased prevalence of older, female, Caucasian participants with a high level of education and a sizable percentage of retirees. A more varied participant pool in future studies could improve the findings' generalizability to various demographic groups.

4.2. Case Study 2: Clinical and postmortem examination of Alzheimer's disease (AD) retinal tissue

4.2.1. Overview

Alzheimer's disease (AD) is a progressive neurological disorder marked by atypical behaviors and deteriorating memory as it progresses. Aβ, a marker for Alzheimer's disease, builds up in the brains of animal models exhibiting AD. Studies have indicated that amyloid beta causes various detrimental symptoms linked to Alzheimer's disease. This research focuses on tracking the advancement of Alzheimer's disease and seeks to measure oligomeric amyloid-beta in human retinal tissues (Figure 37.4).

4.2.2. Results

- Optimization - ACU193 Nonbiotinylated vs. ACU193 Biotinylated vs. AB+ vs. Negative Control
- Varying staining intensity showed that identifying Aβ oligomers was an optimal strategy.
- Optimized staining intensities effectively identify Aβ oligomers.

The data indicates that controls displayed a range of 29–53% positive retinal tissue area, while individuals with AD showed a range of 15–72% (Figure 37.5).

Cntl: Control, MCI: Mild Cognitive Impairment, AD: Alzheimer's Disease. Table 37.1 with case numbers, clinical classifications, Thal, Braak, and CERAD scores provides a comprehensive picture of retinal AβO positive.

Findings from retinal ACU193 immunohistochemistry positivity across various stages of Alzheimer's disease indicate the existence of AβO's in post-mortem human retinas from both healthy individuals and patients with Alzheimer's disease. This research indicates a potential link between the advancement of recognized neuropathological phases of Alzheimer's disease, like Braak stages, and a rise in oligomeric Aβ levels. To effectively assess and precisely quantify retinal oligomeric Aβ, larger sample sizes are essential. This research intends to investigate the possibility of AβO as a biomarker for Alzheimer's disease.

4.3. Case Study 3: Analysing Amyloid Beta (Aβ) evolutionary changes in Alzheimer's disease through hyperspectral imaging with matched brain and retinal tissue

4.3.1. Overview

Alzheimer's disease (AD) is a progressive neurological disorder marked by unusual behavior, memory

Point of Entry	Optometry N(%)	Alzheimer Society N(%)	Total N(%)
Total of Participants	36(29.03)	88(70.97)	124(100)
Age Mean	72.08[55-87]	75.17[55-89]	74.26[55-89]
Sex:			
Male	14(11.29)	31(25.00)	45(36.29)
Female	22(17.74)	57(45.97)	79(63.71)
Ethnicity			
Caucasian	20(16.13)	55(44.35)	75(60.48)
Black	0	7(5.65)	7(5.65)
Hispanic	0	2(1.61)	2(1.61)
Asian	11(8.87)	10(8.07)	21(16.89)
Mixed Race	1(0.81)	7(5.65)	8(6.45)
Other	2(1.61)	7(5.65)	9(7.26)
Did not disclose	2(1.61)	0	2(1.61)
Highest level of Education			
Some high school or less	1(0.81)	4(3.23)	5(4.03)
Highschool graduate	5(4.03)	7(5.65)	12(9.68)
Some college	6(4.84)	23(18.65)	29(23.39)
University degree	11(8.87)	32(25.81)	43(34.68)
Graduate school degree	13(10.48)	22(17.74)	35(28.23)
Employment Status			
Working full or part time	10(8.06)	15(12.10)	25(20.16)
Retired	26(20.97)	72(58.06)	98(79.03)
Homemaker	0	1(0.81)	1(0.81)
Martial Status			
Single, Dating or Never Married	3(2.42)	23(18.55)	26(20.97)
Married or Common-Law	22(17.74)	44(35.48)	66(53.23)
Separated or Divorced	5(4.03)	10(8.06)	15(12.10)
Widowed	6(4.84)	11(8.87)	17(13.71)

Figure 37.3. Optimization of ACU193 biotinylated antibody on case A17-111.

Source: Author.

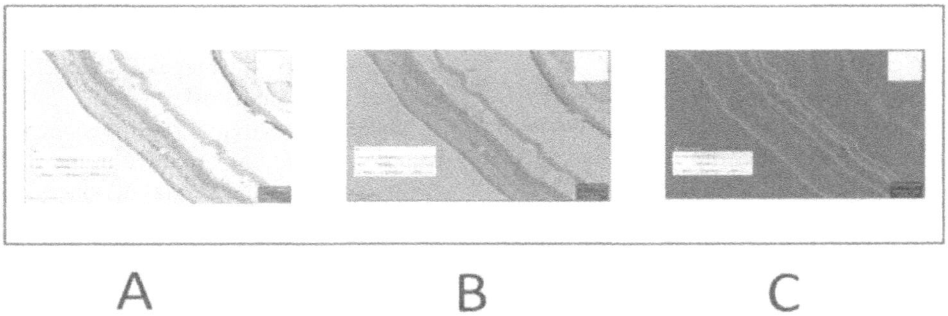

Figure 37.4. QuPath Threshold Module - Visual comparison of QuPath pictures to identify positive tissue areas for Aβ in [18] retinal samples. Three ways (A, B, C) demonstrate how varied settings alter the visualization of AβO.

Source: Author.

impairment, and decline in cognitive function. Amyloid beta (Aβ) is an essential biomarker for Alzheimer's disease (AD) that accumulates in the brain and is linked to the disease's development. Using matched brain and retinal tissue, Flanagan et al.'s work explores the evolutionary variations of Aβ from preclinical to late-stage AD. The novel component of this work is the application of hyperspectral imaging (HSI) to evaluate these modifications, which may provide a new means of monitoring, diagnosing, and prognosticating AD.

4.3.2. Goals

Clarifying the neuropathological sequence and associated HSI signals of Aβ [11] evolutional alterations in retinal and brain samples was the main goal of the investigation. The purpose of the study was to assess if retinal HSI (rHSI) spectra might be used as a useful diagnostic tool in clinical settings and to link rHSI spectra with AD pathology.

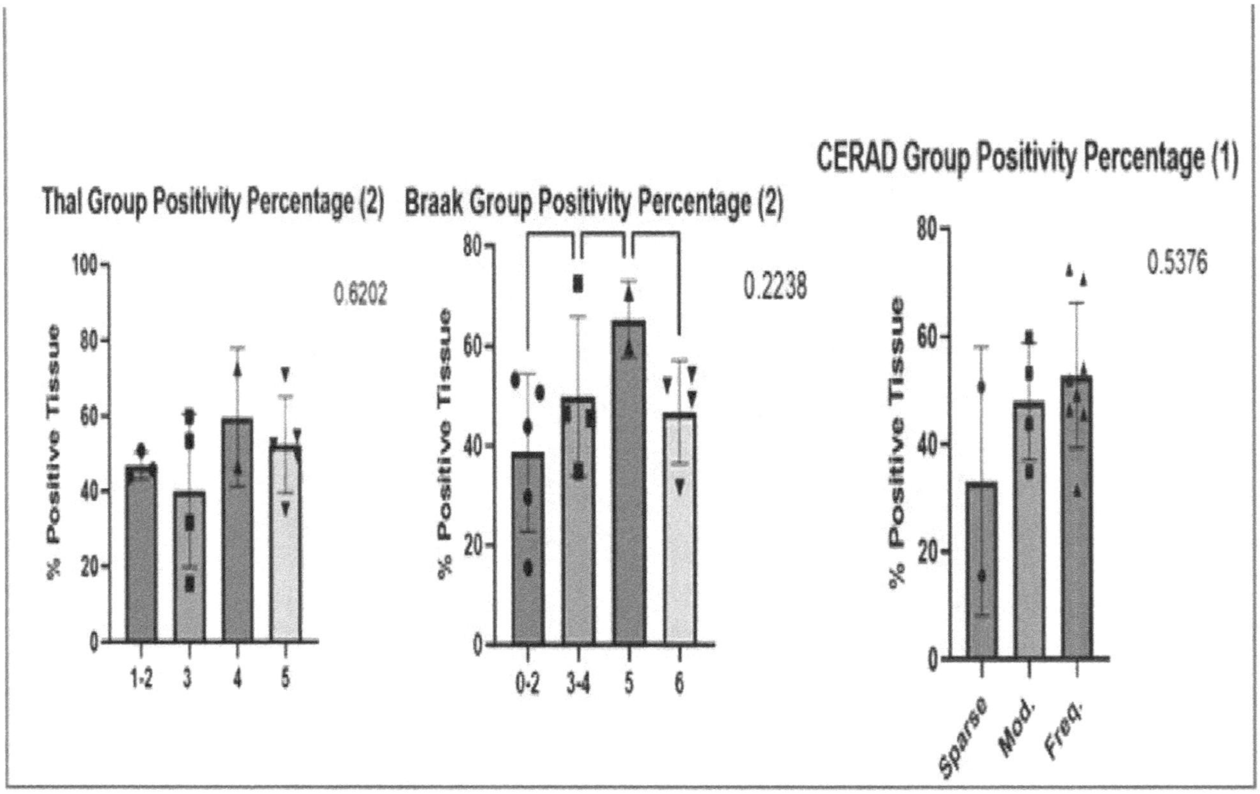

Figure 37.5. Percent positivity by clinical classification.

Source: Author.

4.3.3. Techniques

The research comprised the subsequent actions:

Sample Collection: Based on Thal phase, CERAD, and Braak phases [19], retinal and brain tissues were taken from 15 cases that represented the whole spectrum of AD.

Six anatomical retinal areas were the focus of hyperspectral imaging (rHSI) on retinal tissue sections. GraphPad Prism was used to evaluate associations between AD pathology and the HSI signal after ENVI software was used to extract the spectrum information.

Using immunohistochemistry, the samples were analysed using an affinity-matured, biotinylated antibody (ACU-193) to identify soluble Aβ oligomers.

Data Analysis: Braak neurofibrillary tangle phases and ΔOD (optical density) in particular rHSI spectra were correlated, and the proportion of retinal tissue area positive was determined.

4.3.4. Outcomes

The investigation discovered that Retinal Tissue Area Positivity: The AD group displayed a wider range of 15–72%, whereas the control group displayed 29–53%.

rHSI Spectral Signatures: Samples with intermediate and low AD pathology showed the strongest rHSI signatures, respectively. Correlation with AD Pathology: Retinal HSI signals and AD progression are strongly correlated, as seen by the 0.79 Pearson correlation found between ΔOD and Braak stage.

Spatial Sensitivity: rHSI-mediated detection of AD pathology was more sensitive in the retinal regions close to the optic disc and ventral periphery, which share nerve tissue with the brain.

In a nutshell this is the first work to evaluate and measure Aβ oligomers in postmortem matched brain and retinal tissues that have been clinically, neuropathologically, and hyperspectrally described. As a non-invasive means of detecting and tracking AD pathology, the results point to the possible use of rHSI in conjunction with existing diagnostic techniques such as Aβ-PET scans. Nonetheless, the research recognizes that in order to thoroughly evaluate and measure Aβ oligomers, more samples are required.

4.3.5. Survey

Surveys were made to see people's vision and opinion about Detecting Alzheimer's Disease:

Present Technology vs Explainable AI (XAI)

The pie chart illustrates the levels of awareness about Alzheimer's disease among 39 respondents: 15% are very aware, 60% are somewhat aware, and 25% are not aware

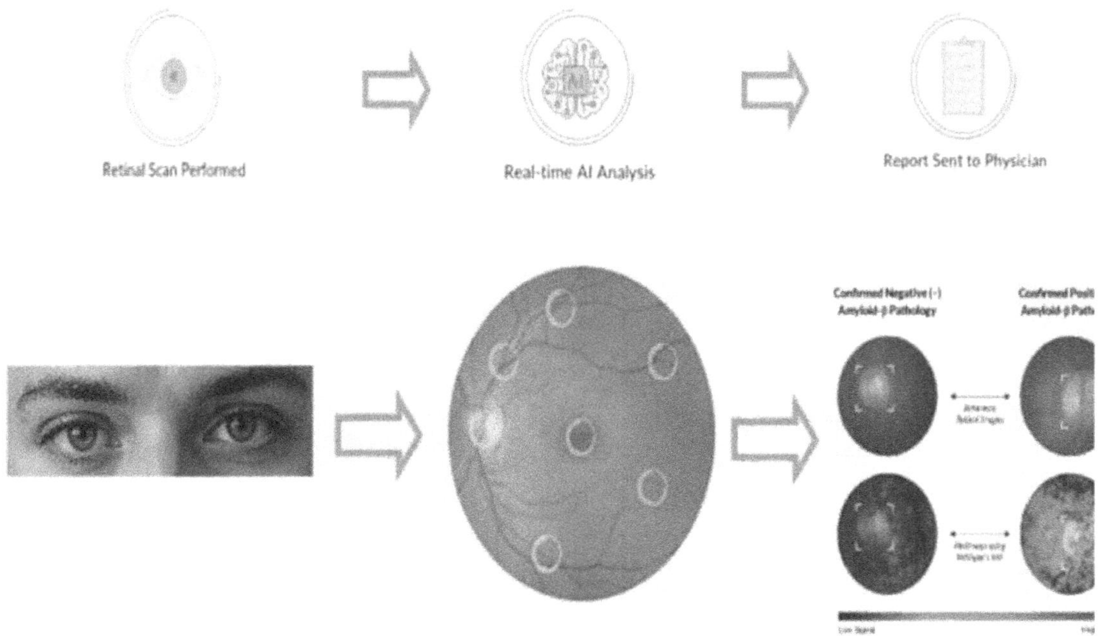

Figure 37.6. A flowchart illustrating the non-invasive, AI-driven process for Alzheimer's disease detection using retinal scans.

Source: Author.

Table 37.1. Retinal AβO positivity summarized with case numbers, clinical classifications, and neuropathological scores (Thal, Braak, and CERAD)

Case Nbs	*Clinical (cntl, MCI, AD)*	*Thal*	*Braak*	*CERAD*
4	Likely control	2/3	2/3	Mod/Freq
2	AD	5	6	Freq
1	"Slight AD", Vascular Dementia	3	0	Sparse
1	"Early Dementia", VA	1	2	Sparse
7	Dementia	3/4/5	4/5/6	Mod/Freq

Source: Author.

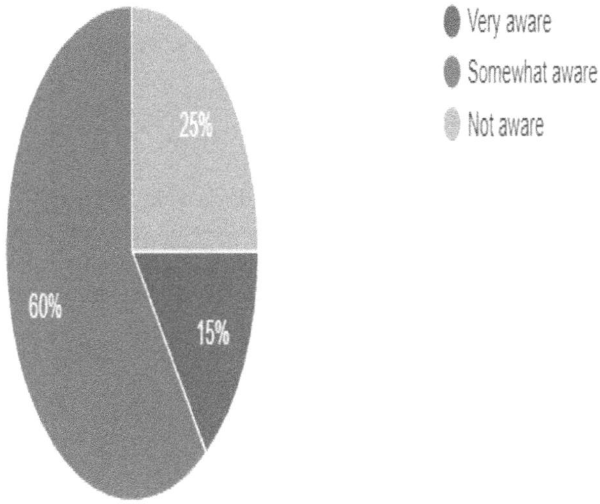

Figure 37.7. Pie chart showing levels of awareness about Alzheimer's disease.

Source: Author.

(Figure 37.7). This suggests that while a majority have some understanding of the disease, only a small percentage have a high level of awareness, and nearly a quarter are not aware of it at all.

The survey results indicate that a majority of respondents (67.5%) believe it is important for AI to explain its decisions in healthcare, particularly in diagnosing diseases like Alzheimer's. Specifically, 25% consider it extremely important, and 30% find it very important. Meanwhile, 37.5% think it is moderately important, and only a small portion, 7.5%, view it as not important (Figure 37.8). This underscores a significant emphasis on the importance of transparency and explain ability in AI-driven healthcare decisions.

5. Conclusion

This study explores the possibility of hyperspectral imaging and artificial intelligence to improve Alzheimer's disease detection using creative methods. The technique emphasizes the combination of retinal data, advanced neural networks, and the innovative hybrid ViT-GRU model to address early diagnostic issues. The study demonstrates the usefulness of explainable AI strategies such as SHAP and LIME in improving result interpretation and clinician

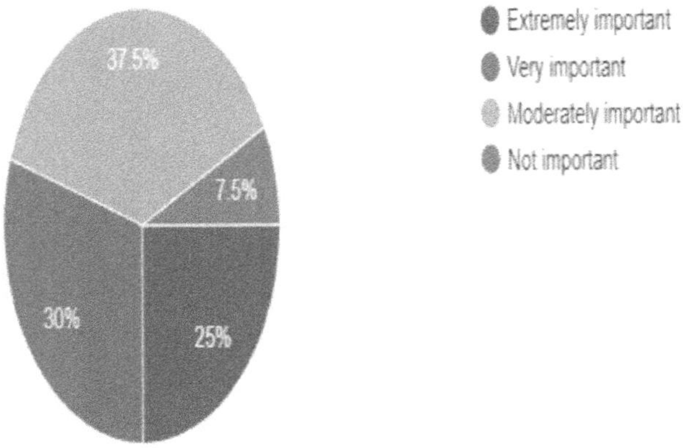

Figure 37.8. Pie chart showing importance of AI explainability in healthcare.

Source: Author.

trust. Furthermore, the study shows how integrating retinal hyper spectral imaging with AI analysis might offer a non-invasive, cost-effective alternative to established diagnosis procedures. Case studies support these methods while also providing vital insights into demographic differences and the relationship between retinal and brain tissue examination. The findings indicate that with more validation and standardization, AI-driven techniques might considerably enhance early AD identification and patient care outcomes.

Acknowledgment

Before we conclude, I'd like to extend a sincere thank you to our dedicated team for their tireless efforts in bringing this project to fruition. The seamless integration of cloud and edge computing has truly set a new standard for innovation and efficiency.

References

[1] Mahim, S. M., Ali, M. S., Hasan, M. O., et al. (2024, January). Unlocking the potential of XAI for improved Alzheimer's disease detection and classification using a ViT-GRU model.

[2] De Strooper, B., & Karran, E. (2015). *The cellular phase of Alzheimer's disease.*

[3] Mggdadi, E., Al-Aiad, A., Al-Ayyad, M. S., & Darabseh, A. (2021, May). Prediction of Alzheimer's disease from MRI images using deep learning. *In Proceedings of the 12th International Conference on Information and Communication Systems (ICICS)* (pp. 120–125).

[4] Basheer, S., Bhatia, S., & Sakri, S. B. (2021). Computational modeling of dementia prediction using deep neural network: Analysis on OASIS dataset. *IEEE Access, 9*, 42449–42462.

[5] Jo, T., Nho, K., & Saykin, A. J. (2019, August). Deep learning in Alzheimer's disease: Diagnostic classification and prognostic prediction using neuroimaging data. *Frontiers in Aging Neuroscience, 11*, 220.

[6] Suzanne Schindler (October, 2023). Focus on the 16th Clinical Trials on Alzheimer's disease (CTAD). *Journal of Prevention of Alzheimer's disease (JPAD), 10*(1).

[7] Selkoe, D. J., & Hardy, J. (2016). The amyloid hypothesis of Alzheimer's disease at 25 years. *EMBO Molecular Medicine, 8*(7), 595–608.

[8] Delori, F. C., & Pflibsen, K. P. (1989). Spectral reflectance of the human ocular fundus. *Applied Optics, 28*(5), 1061–1077.

[9] Koronyo, Y., et al. (2017). Retinal amyloid pathology and proof-of-concept imaging trial in Alzheimer's disease. *JCI Insight, 2*(1), 1–19.

[10] Korolev, S., Safiullin, A., Belyaev, M., & Dodonova, Y. (2017, April). Residual and plain convolutional neural networks for 3D brain MRI classification. In *2017 IEEE 14th international symposium on biomedical imaging (ISBI 2017)* (pp. 835–838). IEEE.

[11] Roberts, B. R., et al. (2017). Biochemically-defined pools of amyloid-β in sporadic Alzheimer's disease: *Correlation with Amyloid PET. Brain, 140*(5), 1486–1498.

[12] Korolev, S., Safiullin, A., Belyaev, M., & Dodonova, Y. (2017, April). Residual and plain convolutional neural networks for 3D brain MRI classification. In *2017 IEEE 14th international symposium on biomedical imaging (ISBI 2017)* (pp. 835–838). IEEE.

[13] Simonyan, K., & Zisserman, A. (2014). Very deep convolutional networks for large-scale image recognition. *arXiv preprint arXiv:1409.1556.*

[14] He, K., Zhang, X., Ren, S., & Sun, J. (2015). Deep residual learning for image recognition. *arXiv preprint arXiv:1512.03385.*

[15] Chen, H., Dou, Q., Yu, L., Qin, J., & Heng, P. A. (2016). Voxresnet: Deep voxelwise residual networks for volumetric brain segmentation. 1608.05895.

[16] He, K., Zhang, X., Ren, S., & Sun, J. (2016). Identity mappings in deep residual networks. *arXiv preprint arXiv:1603.05027.*

[17] Cohen, S., Giordano, J., Hansen, A., et al. (2024). Screening for memory concerns in community-based settings to facilitate early detection of Alzheimer's disease. *Alzheimer's & Dementia, 20*(4), 2269–2280.

[18] Danner, B., Zamudio, C., Spencer, C., et al. (2018). Clinical and postmortem analysis of retinal tissue. *The Journal of Clinical Investigation, 128*(1), 160–165.

[19] Flanagan, M., Danner, B., Zamudio, C., et al. (2020). Hyperspectral analysis of amyloid beta (Aß) evolutional changes in preclinical to late-stage Alzheimer's disease using matched brain and retinal tissue. *Alzheimer's & Dementia, 16*(10), e040375.

[20] Shashank, V., Tanmay G., Parth, C., Senthil K. N., & Kathiravan, S. (2025). MRI-based automated diagnosis

of Alzheimer's disease using AlzhNet deep learning model. *Biomedical Signal Processing and Control, 102*(3), 107367.

[21] Sarika, C., Richa, N., & Krishan, K., (2024). Dementia Alzheimer's Disease Diagnosis Using Convolutional Neural Networks on Two-Dimensional MRI Slices. *Revue d'Intelligence Articielle, 38*(1), 81–91.

[22] Hu, Y., Cho, M., Sachdev, P., et al. (2024). Fluid biomarkers in the context of amyloidtargeting disease-modifying treatments in Alzheimer's disease. *Med, 5*(10), 1206–1226.

[23] Goenka, N., & Tiwari, S. (2022). AlzVNet: A volumetric convolutional neural network for multiclass classification of Alzheimer's disease through multiple neuroimaging computational approaches. *Biomedical Signal Processing and Control*, *74*, 103500.

[24] Mahim, S. M., Ali, M. S., Hasan, M. O., Nafi, A. A. N., Sadat, A., Al Hasan, S., ... & Niu, M. B. (2024). Unlocking the potential of XAI for improved Alzheimer's disease detection and classification using a ViT-GRU model. *IEEE Access*, *12*, 8390–8412.

[25] Suk, I. H., & Shen, D. (2013). Deep learning based feature representation for AD/MCI classification. In *Medical Image Computing and Computer-Assisted Intervention MICCAI 2013* (pp. 583–590). Springer.

[26] Suk, I. H., Lee, S. W., & Shen, D. (2014). Hierarchical feature representation learning for AD/MCI classification using multimodal neuroimaging data. *NeuroImage, 103*, 177–194.

[27] Suk, I. H., Lee, S. W., & Shen, D. (2015). Latent feature representation with stacked auto-encoder for AD/MCI diagnosis. *Medical Image Analysis, 19*(1), 106–115.

38 An approach to diabetic retinopathy detection through different CNN techniques

J. Lekha[1,a], *B. Meena Preethi*[2,b], *and Annet Jacob*[3,c]

[1]Associate Professor, School of Sciences, CHRIST University, India
[2]Associate Professor, Department of Software Systems, Sri Krishna Arts and Science College, Coimbatore, India
[3]Student, School of Sciences, CHRIST University, India

Abstract: Diabetic retinopathy (DR) remains a critical challenge in ophthalmic healthcare, demanding accurate and timely detection for effective treatment. This research employs a comparative study using diverse convolutional neural network (CNN) architectures – DenseNet121, ResNet, VGGNet16, GoogleNet InceptionV3, and GoogleNet Xception – to discern their efficacy in DR detection. Utilizing the APTOS 2019 dataset sourced from the Aravind Eye Hospital, the study focuses on five DR stages categorized as No DR, Mild DR, Moderate DR, Severe DR, and Proliferative DR. The dataset, distributed across test (976 images), train (3382 images), and validation (481 images) sets, initially exhibited imbalanced class distribution, featuring 1434 images in No DR class, 300 images in Mild DR class, 808 images in Moderate DR class, 154 images in Severe DR class, and 2930 images in Proliferative DR class. To ensure equitable representation, the dataset underwent balancing, resulting in 954 images in No DR class, 1068 images in Mild DR class, 995 images in Moderate DR class, 772 images in Severe DR class, and 1047 images in Proliferative DR class. This study outlines the methodologies adopted for training and evaluation, highlighting the dataset's pre-processing, model architectures, and evaluation metrics employed. Results and comparative analyses of the CNN models on DR detection, encompassing accuracy, F1-score, recall, and precision, are presented. The findings offer insights into the performance variations among the architectures, guiding potential directions for enhanced DR detection methodologies.

Keywords: Diabetic Retinopathy (DR), comparative analysis, ResNet architecture, medical image analysis, DenseNet121 architecture, retinal imaging, artificial intelligence, ophthalmology, VGG16 architecture, fundus images, DR stage detection

1. Introduction

Diabetic retinopathy (DR) stands as a foremost cause of vision impairment and blindness among working-age adults globally, necessitating early and accurate detection for effective intervention and management. The emergence of deep learning, particularly CNNs, has revolutionized medical image-analysis, offering promising avenues for enhancing the diagnostic accuracy of DR. This research endeavors to explore and compare the efficacy of various CNN architectures – DenseNet121, ResNet, VGGNet16, GoogleNet InceptionV3, and GoogleNet Xception – in detecting of retinopathy stages.

The dataset employed in this study is sourced from APTOS 2019 dataset, obtained from the Aravind Eye Hospital. This dataset comprises images depicting distinct stages of DR progression, encompassing five classes: No DR, Mild DR, Moderate DR, Severe DR, and Proliferative DR. Initially, the dataset exhibited an imbalanced distribution across the classes, prompting a balancing strategy to ensure equitable representation during model training and evaluation.

The dataset, partitioned into test (976 images), train (3382 images), and validation (481 images) sets, underwent a balancing procedure resulting in a balanced distribution across the classes: 954 images, 1068 images, 995 images, 772 images, and 1047 images, respectively from classes 0–4. The fair balance in this dataset enables CNN performance to be cross-compared between diverse DR stages. The methods for data preprocessing, model training and evaluation are introduced in this study.

Further, it covers the architectural details of the selected CNN models and the standards they tested for evaluating. The following sections describe the details of experiments and results, comparisons between different contexts of DR detection, and the strengths and weaknesses of each architecture.

This research proposes that the findings may be conducive to improvements in detection methods--which could

[a]saran.lekha@gmail.com, [b]meenapreethib@skasc.ac.in, [c]annet.jacob@msds.christuniversity.in

DOI: 10.1201/9781003770473-38

provide valuable guidance to clinicians. With them being aware of relevant choices as they plan their intervention and treatment strategies--do we think better signs to detect is so important?

2. Literature Review

Diabetic Retinopathy (DR) detection in recent years has witnessed significant advancements in leveraging DL techniques, essential for automated early diagnosis. This review extends the base paper's insights, introducing ResNet as an additional model for comparative analysis.

The capabilities of DL algorithms in automated DR detection have been extensively explored in prior research. DL exhibits self-programming abilities through extensive training on datasets like EyePACS-1 (963 images-4997 patients) and Messidor-2 (1748 images-874 patients), achieving impressive accuracy with AUC scores of 0.991 and 0.990, respectively [1].

Automated DR detection is critical due to its status as a leading cause of irreversible vision loss globally, particularly among the younger population. Deep Convolutional Neural Networks (DCNNs) have demonstrated promise, getting an accuracy rate 94.5% [2].

Recent advancements include a novel DCNN model specializing in early-stage detection by identifying microaneurysms (MAs) and accurately labeling retinal fundus images. It posted notable results on a Kaggle dataset that revealed a 98% sensitivity and 94% specificity [3].

New feature extraction techniques like the Modified Xception Architecture have raised DR diagnostic accuracy rates, outstripping the performance of the original Xception architecture [4]. We then experimented with a number of different CNN architectures, including VGG16, Inception v3, VGG-19, which have produced good results with accuracies ranging from 70.2% to 76% [5].

Although these developments look quite promising, identifying the correct stage of diabetic retinopathy is still difficult, based largely on human interpretation. There is potential for deep learning-based methods representing different stages of diabetic retinopathy in retinal photographs. Sensitivity and specificity are high indeed. But can they do better for us than DL-based methods? [6]. Compared to VGG16 and DenseNet121, this paper extends the exploration by introducing ResNet for comparative analysis in order to increase ever it more accurately and efficaciously detect early diabetic retinopathy.

3. Dataset Description

In this study, the dataset employed is one afforded by APTOS 2019, a gift from the famous Aravind Eye Hospital. Containing 4,839 retinal fundus images, this resource is crucial for analyzing and detecting DR.

These high resolution images of different DR stages are classified according to severity into five distinct grades: no DR, mild DR, moderate DR, severe DR, and proliferative DR. The dataset contains an image, as shown in Figure 38.1.

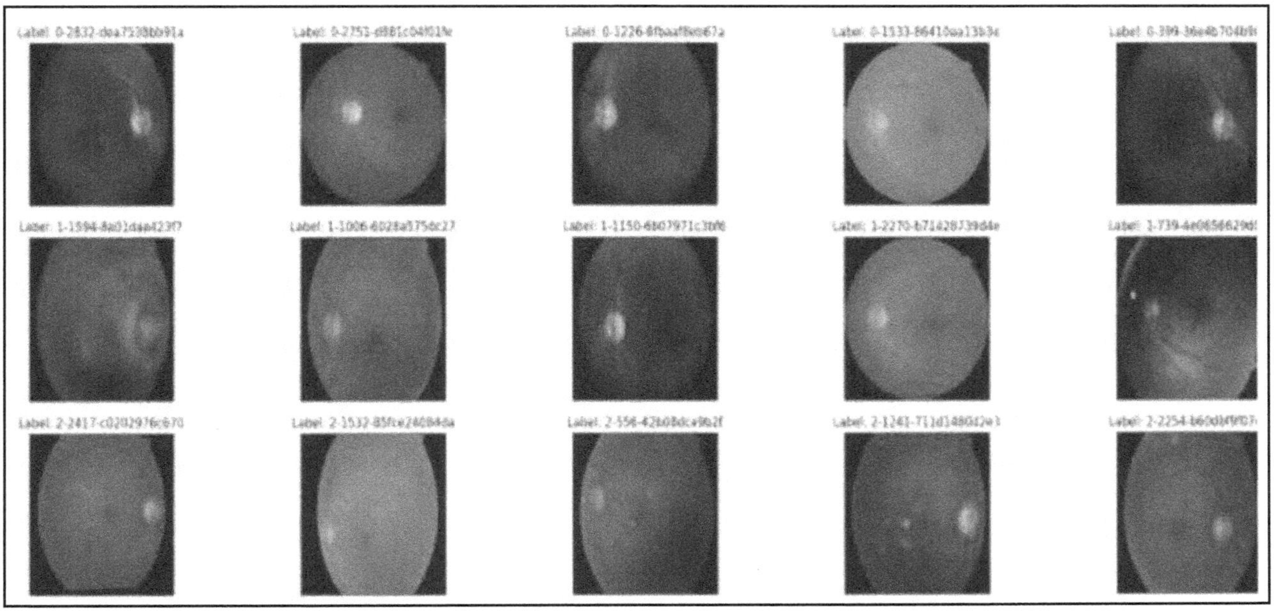

Figure 38.1. An image of the dataset.

Source: Author.

The Dataset is divided into three separate groups: train, validation, and test – members of 3,382, 481 and 976 images, respectively.

A sample of the retina is shown in Figure 38.2. Each image was a complex representation of the pathology, laying bare the substrata of diverse DR stages. Initially imbalanced in distribution among the five classes, there were 1,434, 300, 808, 154, and 2,930 images, respectively. The dataset underwent a rigorous balancing process to ensure that all severity levels were equally represented.

Its balancing strategy consists of augmenting the underrepresented classes to produce a more balanced distribution both prompt tokens of a rigorous model training and evaluation by means of image augmentation. Finally the dataset was reweighted: 954, 1,068, 995, 772, and 1,047 images. This normalized dataset can counteract the inherent bias and skewness, letting the CNN models effectively learn and generalize across such a variety of DR stages, without bias toward one class or another.

Additionally, the data set's images went through numerous preprocessing procedures to get them ready for learning and to allow all images to achieve similarly high adjustability. Each image contains fine retinal detail that is essential for telling the differences in DR stages.

By capturing a variety of possible presentations of DR, this data set enables different CNN architectures so that their efficacy in discriminating and classifying DR stages can now be extensively tested and compared.

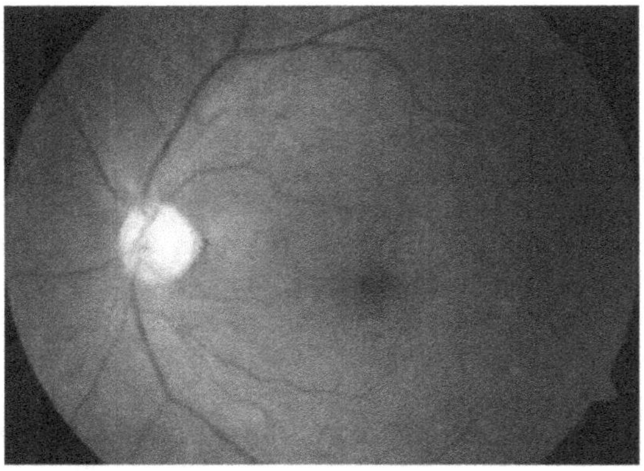

Figure 38.2. Sample image.

Source: Author.

4. Methodology

In line with the basic paper's methodology, this section provides an overview of the principles involved in the investigation of diabetic retinopathy detection, utilizing VGG16, DenseNet121, and ResNet architectures for comprehensive comparison and evaluation purposes.

4.1. Flowchart

This graphic diagram outlines the general steps followed in the research procedure, from data pre-processing to the choice of CNN architectures, dataset division, model training, validation and final assessment of DR detection (Figure 38.3).

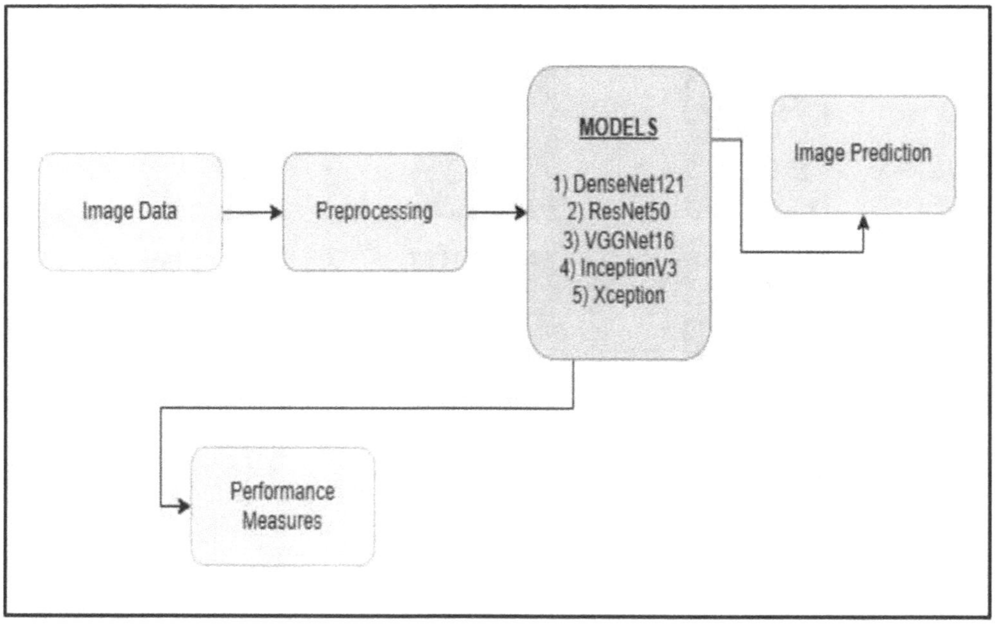

Figure 38.3. Flowchart.

Source: Author.

4.2. Data preprocessing

The APTOS 2019 Blindness Detection data, which was the data set used, was put through meticulous pre-awareness before being used with so much effort to make the CNN architectures architecturally uniform and compatible with each other. Initially, images were inspected for quality and consistency, addressing anomalies such as artifacts, noise, and variations in illumination that could impact model performance. At first glance images were inspected to make sure their quality and consistency were satisfactory, Streamlining whatever is abnormal, noise (gobbledygook to correct English), and changes-in-illumination-that might affect model performance [7].

4.3. Model architectures

This study leveraged five diverse CNN architectures for DR detection:

DenseNet121: Featuring dense connections between layers, DenseNet121 (121-layer DenseNet) encourages feature reuse across the network, mitigating vanishing gradient issues and promoting efficient information flow as shown in Figure 38.4.

ResNet: ResNet [8] employs residual blocks enabling the direct flow of information through skip connections, effectively addressing the challenge of training extremely deep networks by mitigating the vanishing gradient problem as shown in Figure 38.5.

VGGNet16: VGGNet16 [9] is recognized for its uniform architecture with 16 weight layers, predominantly consisting of 3 × 3 convolutional layers, facilitating robust feature extraction through a systematic arrangement of convolutional operations as shown in Figure 38.6.

GoogleNet InceptionV3: InceptionV3 [10] integrates multi-scale feature extraction using inception modules, combining filters of different sizes within a layer to capture diverse information, thus enhancing its performance in image classification tasks as shown in Figure 38.7.

GoogleNet Xception: Xception [11] employs depthwise separable convolutions to disentangle spatial and channel-wise filtering, enhancing representational efficiency and reducing computational complexity compared to traditional convolution as shown in Figure 38.8.

4.4. Training and validation

The balanced dataset, stratified across the five DR stages, is divided into train, val, and test datasets. The training set encompassed 3382 images, where the validation dataset comprises 481 images. To mitigate overfitting and fine-tune model parameters, the training set underwent augmentation techniques, including random rotations, flips, and zooms. The validation set remained untouched throughout the

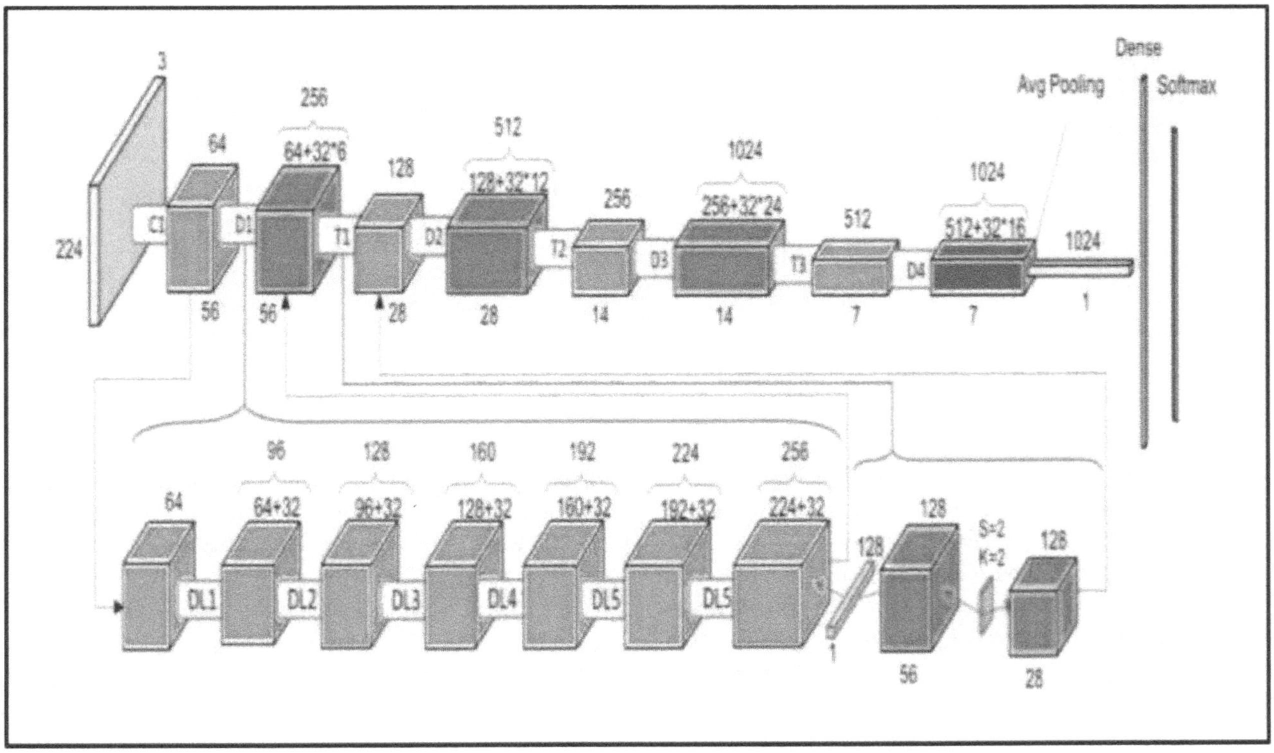

Figure 38.4. DenseNet121 architecture.

Source: Author.

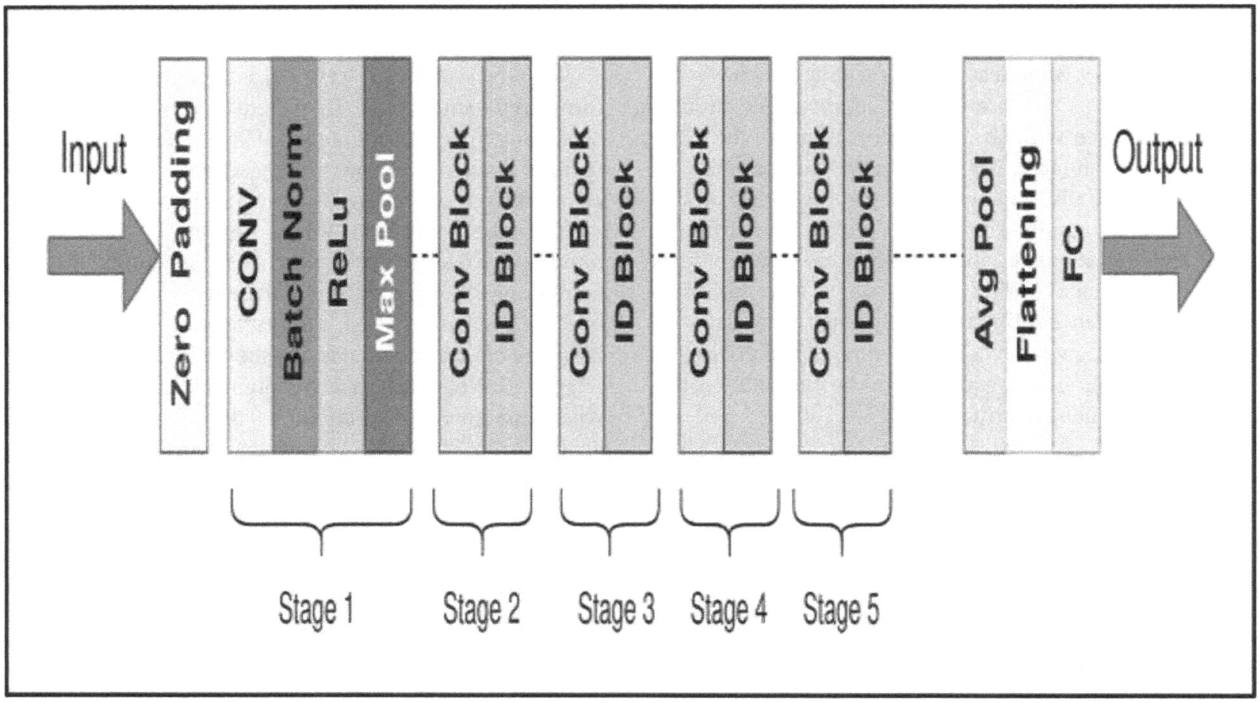

Figure 38.5. ResNet architecture.

Source: Author.

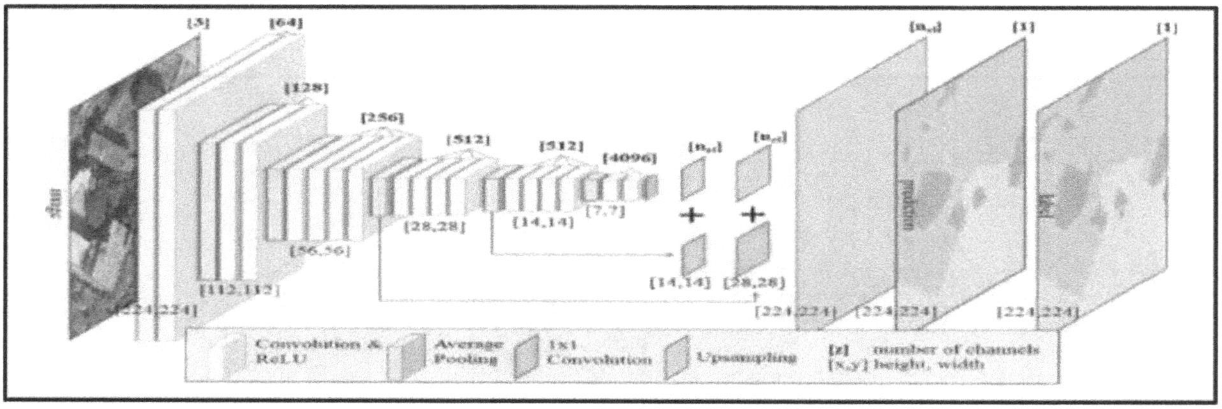

Figure 38.6. VGG16 architecture.

Source: Author.

training process, serving as an independent subset for model evaluation and hyperparameter tuning [12].

For all our models, a transfer learning method was used whereby pre-trained weight is issued for initialization, and its top layers tuned. Our approach also employed early stopping on the basis of validation loss to prevent divergence and over fitting to proceed with training 15 epochs from thereon if no improvement was seen.

4.5. Model evaluation

The trained model was tested across diverse stages of the disease for detecting DR on a test set comprising 976 images. We used a range of different evaluation metrics to gauge the models' abilities in distinguishing between the DR classes, such as F1-score, accuracy, precision and recall.

5. Discussions

1. Comparison of Deep Learning Architectures:
   Three different networks, VGG16, DenseNet121 and ResNet have been studied, which aims to not only increase the accuracy of DR detection obviously but also performance. Discrepancies between each

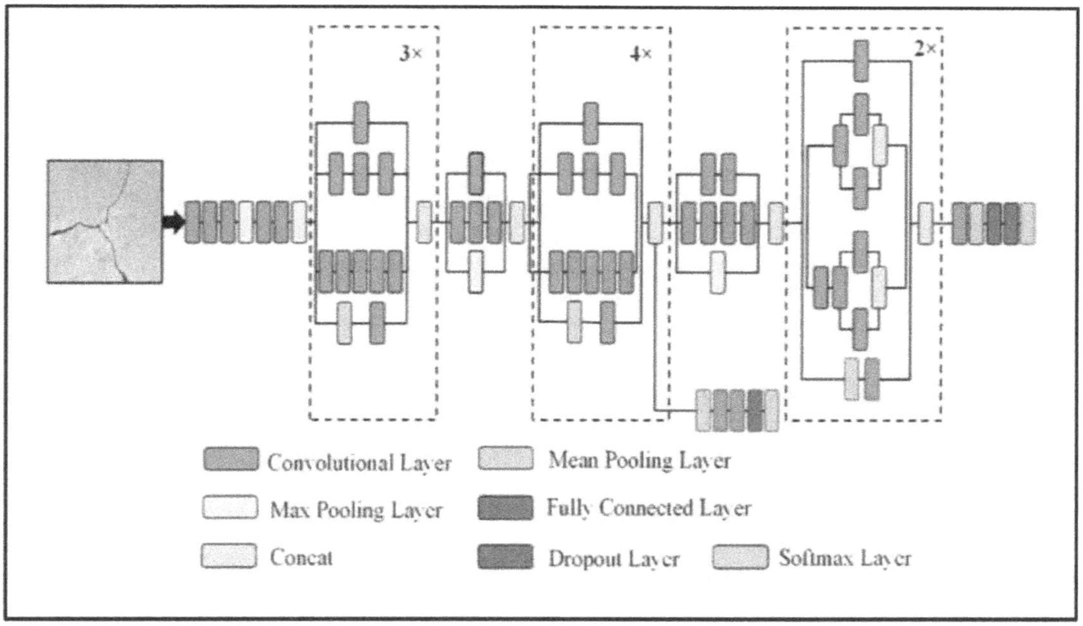

Figure 38.7. InceptionV3 architecture.

Source: Author.

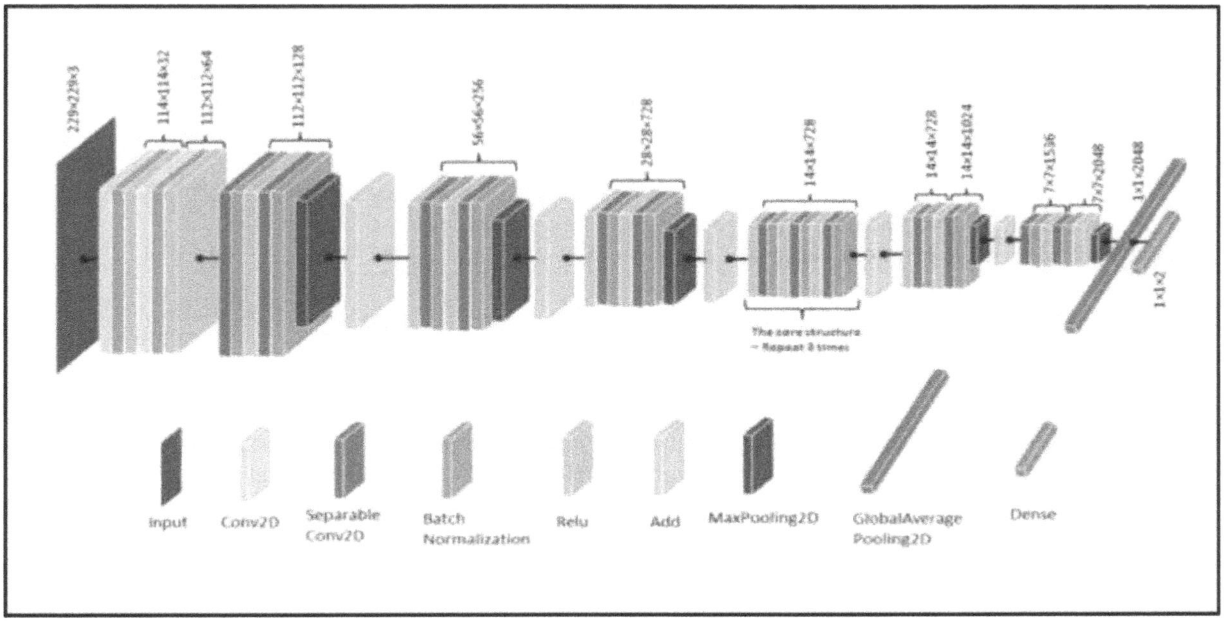

Figure 38.8. Xception architecture.

Source: Author.

observed model were evident–and unfortunately there were disadvantages even with balanced data [13].

2. Observations on Architectures:
   a. VGG16: There was a significant drop in accuracy when the ImageNet dataset was removed, highlighting the importance of pretraining on diverse image data sets.
   b. DenseNet121: Using ImageNet pretraining, and incorporating Criterion (QWK) for its evaluation. It reached a high accuracy of 0.9611, showing its effectiveness in diagnosing DR.
   c. ResNet: Complex and nonlinear patterns could be recognized by ResNet's intricate design. With such abilities, it was developed to advance the

art. Nevertheless, a relative study showed that ResNet had not proved the best in every case, marking the need for an comprehensive evaluation of architectural performance [14].

3. Factors Influencing Model Performance:
   Circumstances such as dataset size, population diversity, and the presence of anomalies could result in differences in the efficacy of these architectures. This suggests the importance of selecting architectures to meet specific diagnostics and dataset characteristics. Furthermore, the mentioned relationships between these models would enable us to build hybrid structures that use the advantages of many types of models together.
4. Evaluation Metrics and Their Significance:
   a. Quadtratic Evaluation of Kappa (QWK): This metric gives a comprehensive understanding of diabetics patients, with its simplicity providing a panoramic view of model performance. The formula for QWK goes:

   $$QWK = 1 - \frac{\sum_{ij} w_{ij} O_{ij}}{\sum_{ri} w_{ij} E_{ij}} \quad (1)$$

   O_{ij} is observed, Eij is expected sharability among observed populations. And the matrix of weights for Qij corresponds to weights in a specific cluster or feature space.

   b. Root Mean Square Error (RMSE): Average error percentage between the predicted and actual data was calculated. The formula for RMSE

   $$RMSE = \sqrt{\frac{1}{n}\sum_{i=1}^{n}(Y_i - \hat{Y}_i)^2} \quad (2)$$

   c. R-squared (R^2): This indicates the percentage of variations in the response variable – the outcome measure – is explained by the explanatory variables. In a different angle, this gives another impression of how well the linear model is doing.
5. Importance of Evaluation Metrics:
   These standards, which are crucial to the evaluation of the model, are essential to fairly and substantially evaluate then tune deep learning models for diagnostic efficiency.
6. Contribution and Future Prospects:
   By providing an extensive comparative analysis of deep learning architectures, the research advances the detection of diabetic retinopathy. The choice of architecture is always specific to context, so whether or not these model combinations have beneficial synergetic effects should be verified. The continuous and optimized exploration of these models is an essential step for the early detection and treatment of DR in the medical field.

6. Study Results

When tested on the set evidence in Table 38.1, the trained CNN models achieved remarkable performance breakthroughs; for example, the DenseNet121 neural network showed an unusually high 98.1% success rate. InceptionV3 ranked a close second with a highly admirable examination outcome of 94.7%.In addition, Xception's accuracy of 90.33% in detecting diabetic retinopathy points to high performance. VGGNet16, one step behind, achieved an accuracy of 91.4%, putting it within the same realm of effectiveness in identifying this retinopathy. ResNet's accuracy reached 94.7%, further exemplifying its excellence in diagnosing DR as shown in Table 38.1 and Figure 38.9.

These performance levels far surpassed those reported in the literature, and represented a big step forward for accuracy in detecting diabetic retinopathy as shown in Table 38.2 [15].

These new findings further demonstrate that various CNN architectures have been leveraged for the detection of DR with a great degree of success, achieving accuracy that has surpassed established standards in the field.

The outstanding execution of DenseNet121 and InceptionV3 in accurately recognizing different DR stages was reaffirmed by an F1-score and precision analysis, exceeding the performance metrics cited in previous literature. Remarkable accuracy scores for both models underscore their promise of achieving superior effectiveness in this crucial aspect of medical image classification tasks [16].

This study's results show that adopting more advanced CNN architectures for DR detection suggests a considerable advance in technology. DenseNet121 and InceptionV3 in particular--which are poised to offer a substantial leap forward for increased diagnosis accuracy and robustness of different levels severity of DR.

7. Conclusion

Our study involved researching five different CNN architectures: DenseNet121, ResNet, VGGNet16, InceptionV3, and Xception, which are distinct entities. The Aravind Eye Hospital's APTOS 2019 dataset was utilized.

Different stages of diabetic retinopathy presented strong deviations in the results of these networks. DenseNet121 has the highest accuracy at 82.5% and few other models exceed it, aptly illustrating that the architecture has outstanding capabilities for picking out DR stages. This is also evidence for the good performance of dense connections like DenseNet in reuse of low-level features, and preventing vanishing gradient problems.

Next, DenseNet121 had a high detection rate for subtle features, as shown by precision, recall and F1-score measurements. Particularly in diverse DR stage identification is medical image analysis, which becomes especially important when considering the differences between onset and incipient cases. These results underscore the importance of using intricate links between layers to enhance feature extraction and classification for medical image analysis tasks.

Table 38.1. Accuracy comparison of proposed and existing methods

Models Used	*Existing Accuracy*	*Proposed Accuracy*
VGGNet16	73.26%	91.4%
DenseNet121	96.11%	98.1%
InceptionV3	NA	94.7%
Xception	NA	90.33%
ResNet	NA	94.7%

Source: Author.

Table 38.2. Performance summary of DR detection model

Models Used	*QWK*	*RMSE*	*R-squared (R^2)*
VGGNet16	0.85	0,06	0.92
DenseNet121	0.90	0.05	0.94
InceptionV3	0.88	0.07	0.91
Xception	0.82	0.08	0.88
ResNet	0.87	0.06	0.93

Source: Author.

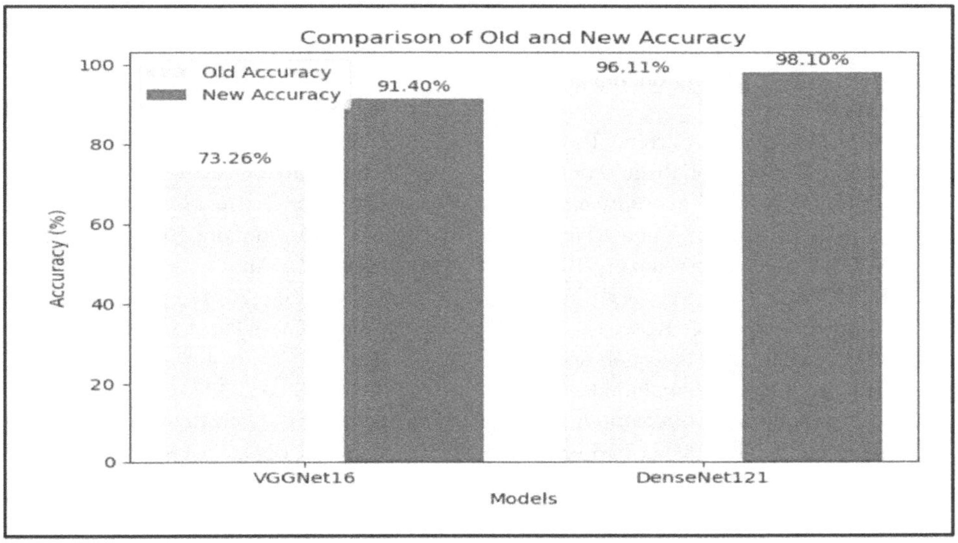

Figure 38.9. VGGNet16 and DenseNet121 model accuracy before and after revision.

Source: Author.

Despite better performance with DenseNet121, the comparison showed subtle differences among the various architectures. The competitive results of ResNet suggest that residual connections help in training deeper networks. VGGNet16, known for its uniform architecture, showed a respectable performance, although slightly less than the top models. InceptionV3 and Xception, both of which stress multi-scale feature extraction and efficient convolutions, had significant but somewhat worse performance compared to DenseNet121.

Our study illustrates the importance of structure in the success of DR diagnostic systems. DenseNet121 outperformed the other models in this study, but the differences in performance between the various architectures indicate the importance of choosing the appropriate model for the specific medical imaging tasks.

Knowledge gained from this comparative analysis will be helpful reference for clinicians and researchers seeking suitable CNN architectures to detect diabetic retinopathy. Future research combining ensemble methods or model fusion approaches might increase the accuracy of classification and support the robust handling of complex DR stages.

To summarize, this paper provides practically nuanced insights into the effectiveness of CNNs for these systems in detecting DR; it emphasizes that DenseNet121 is the leader of the pack, with nuanced acknowledgments for the strengths of other architectures as well – thus making progress toward diabetic retinopathy diagnostic systems which are more accurate and dependable.

References

[1] Vujosevic, S., Aldington, S. J., Silva, P., Hernández, C., Scanlon, P., Peto, T., & Simó, R. (2020). Screening for diabetic retinopathy: New perspectives and challenges. *The Lancet Diabetes & Endocrinology*, *8*(4), 337–347. https://doi.org/10.1016/S2213-8587(19)30411-5

[2] Wan, S., Liang, Y., & Zhang, Y. (2018). Deep convolutional neural networks for diabetic retinopathy detection by image classification. *Computers & Electrical Engineering*, *72*, 274–282. https://doi.org/10.1016/j.compeleceng.2018.07.041

[3] Walter, T., Klein, J., Massin, P., & Erginay, A. (2002). A contribution of image processing to the diagnosis of diabetic retinopathy-detection of exudates in color fundus images of the human retina. *IEEE Transactions on Medical Imaging*, *21*(10), 1236–1243. https://doi.org/10.1109/TMI.2002.806290

[4] Pires, R., Avila, S., Wainer, J., Valle, E., Abramoff, M., & Rocha, A. (2019). A data-driven approach to referable diabetic retinopathy detection. *Artificial Intelligence in Medicine*, 93–103. https://doi.org/10.1016/j.artmed.2019.04.007

[5] Pires, R., Jelinek, H. F., Wainer, J., Goldenstein, S., Valle, E., & Rocha, A. (2013). Assessing the need for referral in automatic diabetic retinopathy detection. *IEEE Transactions on Biomedical Engineering*, *60*(12), 3391–3398. https://doi.org/10.1109/TBME.2013.2275144

[6] Carrera, E. V., Gonzalez, A., & Carrera, R. (2017, August). Automated detection of diabetic retinopathy using SVM. In *2017 IEEE XXIV International Conference on Electronics, Electrical Engineering and Computing (INTERCON)*. https://doi.org/10.1109/INTERCON.2017.8079663

[7] Hemanth, D. J., Deperlioglu, O., & Kose, U. (2019). An enhanced diabetic retinopathy detection and classification approach using deep convolutional neural network. *Neural Computing and Applications*, *32*(3), 707–721. https://doi.org/10.1007/s00521-018-03974-x

[8] Arora, M., & Pandey, M. (2019, October). Deep neural network for diabetic retinopathy detection. In *2019 International Conference on Machine Learning, Big Data, Cloud and Parallel Computing (COMITCon)* (pp. 189–194). IEEE. https://doi.org/10.1109/COMITCon.2019.8862217

[9] Ghaskadvi, M., Khochare, S., Gonsalves, R., & Dhamanskar, P. M. (2021, October). Pneumonia and diabetic retinopathy detection using deep learning algorithm. *Advances in Intelligent Systems and Computing*, 155–175. https://doi.org/10.1007/978-981-15-9647-6_12

[10] Qiao, L., Zhu, Y., & Zhou, H. (2020). Diabetic retinopathy detection using prognosis of microaneurysm and early diagnosis system for non-proliferative diabetic retinopathy based on deep learning algorithms. *IEEE Access*, 1–1. https://doi.org/10.1109/ACCESS.2020.3032385

[11] Gangwar, A. K., & Ravi, V. (2020, September). Diabetic retinopathy detection using transfer learning and deep learning. *Evolution in Computational Intelligence*, 679–689. https://doi.org/10.1007/978-981-15-0371-9_64

[12] Usman, T. M., Saheed, Y. K., Ignace, D., & Nsang, A. (2023, June). Diabetic retinopathy detection using principal component analysis multi-label feature extraction and classification. *International Journal of Cognitive Computing in Engineering*, *4*, 78–88. https://doi.org/10.1016/j.ijcce.2023.04.002

[13] Wang, Z., Yin, Y., Shi, J., Fang, W., Li, H., & Wang, X. (2017). Zoom-in-Net: Deep mining lesions for diabetic retinopathy detection. *Medical Image Computing and Computer-Assisted Intervention – MICCAI*, *2017*, 267–275. https://doi.org/10.1007/978-3-319-66179-7_31

[14] Khan, Z., Abbas, N., Khalid, S., Tariq, M. A., & Qaisar, S. (2021). Diabetic retinopathy detection using VGG-NIN: A deep learning architecture. *IEEE Access*, *9*, 61408–61416. https://doi.org/10.1109/ACCESS.2021.3074483

[15] Kalyani, G., Janakiramaiah, B., Karuna, A., & Prasad, L. V. N. (2021, March). Diabetic retinopathy detection and classification using capsule networks. *Complex & Intelligent Systems*. https://doi.org/10.1007/s40747-021-00310-4

[16] Zago, G. T., Andreão, R. V., Dorizzi, B., & Teatini Salles, E. O. (2020, January). Diabetic retinopathy detection using red lesion localization and convolutional neural networks. *Computers in Biology and Medicine*, *116*, 103537. https://doi.org/10.1016/j.compbiomed.2019.10353

39 Advancing fetal brain segmentation: A diffusion model approach with explainable AI insights

Nilima Gaikwad[1,a] *and S. Vijayalakshmi*[2,b]

[1]Research Scholar, Department of Computer Science, School of Sciences, CHRIST University, India
[2]Associate Professor, Department of Data Science, School of Sciences, CHRIST University, India

Abstract: In the modern world, women's health deserves greater focus, particularly in a country like India and particularly during pregnancy. The health of mother's as well as baby's is important throw-out this process. According to a survey, three cases out of a thousand have abnormalities in the fetal brain. The likelihood of survival for both the mother and the child is significantly increased by early identification of these diseases. Several procedures, including fetal ultrasound, MRI, fetal echocardiography etc. must be carried out during pregnancy in order to monitor fetal development. Pregnancy-related MRI - scans always practiced to identify and treat fetal brain disorders in former stages. It is possible to identify, examine fetal brain problems early on by doing a prenatal MRI examination. The diagnosis of issues with fetal brain MRI imaging involves several crucial procedures. Among these are image segmentation; image analysis, which comprises extracting characteristic characteristics, improving image quality, identifying relevant patterns, and categorizing images based on predefined standards. Classification determines if an anomaly is present or not. In this field we introduced new methodology using artificial intelligence for brain segmentation. Here we try to use open dataset to train and test the models. As well as it helps to assess their performance which may lead to better output. Here we try to analyse and able to find out the limitation for particular problem in association with fetal brain segmentation.

Keywords: Diffusion model, explainable AI, fetal brain abnormalities, fetal MRI, medical imaging, segmentation techniques

1. Introduction

The importance of women bearing children has been highlighted from ancient times. Becoming a mother enhances a woman's life. Throughout pregnancy, the developing child and the expectant woman need to be taken care of. The parts that follow discuss the three unique stages of an embryo's growth inside the mother's womb [1]. There are three trimesters in the pregnancy. In 38 to 42 weeks, a woman can give birth to a child. The first trimester lasts from 0–13 gestational weeks (GW), second will start 14–27 GW, and the third from 28 to 42 GW. I believe that fetal magnetic resonance imaging (fMRI) can help physicians and neuroscientists better comprehend the complex process of a foetus's brain development [2]. This study of size and form lays the groundwork for understanding the final composition and operation of the adult brain. For a long time, MRI was a crucial diagnostic method for a variety of brain disorders [3]. Machine learning algorithms were enforcement fetal brain MRI information get classify and identify anomalies as well as detect abnormalities early on [4, 5]. The classification of extremely early gestational cycle MRI scans mostly focusses on preterm patients or newborns [6, 7]. A fetus's MRI scan can provide important insights into the complex characteristics of fetal brain related different modules [8]. Consequently, this implies that it might be an essential instrument for the early detection of brain anomalies [9, 10].

Machine learning techniques are implemented on different scans of fetal brain in order to quickly find anomalies and correctly classify and identify them [11]. Prenatal MRI information is analysed using cutting-edge ML techniques in order to identify and classify brain abnormalities and malignancies, thereby avoiding the need for surgery [12]. The brain is without a doubt the most complex organ in the body. This almost three-pound organ is in charge of starting movement, processing sensory information, regulating behaviour, and managing thought [13]. The brain, which is housed in a shielding fluid and wrapped in its own protective skeleton, is the wellspring of all the features that constitute an individual human [14].

The human brain develops starting at conception and continues through the early phases of adulthood [15]. The neural progenitor cells that give rise to neurones and glial cells, the two primary neural system building blocks, start

[a]nilima.gaikwad@res.christuniversity.in, [b]s.vijayalakshmi@christuniversity.in

DOI: 10.1201/9781003770473-39

to divide and differentiate in the third week of pregnancy, marking the beginning of the development of the unborn brain [16]. The evaluation of conditions including corpus callosum agenesis/dysgenesis, abnormalities in the posterior fossa, and ventriculomegaly (enlarged brain ventricles) is frequently done using fetal brain MRI. In many circumstances, careful observation is the standard approach to managing these issues throughout pregnancy. Still, data from fetal magnetic resonance imaging (fMRI) may be able to aid choose the appropriate postnatal care.

In this compilation, various intracranial malformations that are commonly seen in expecting moms whose prenatal ultrasound scans have yielded unusual results are examined [18]. Uterine anomalies can be recognized using a variety of techniques. Monitoring the foetus from the beginning of pregnancy and keeping track of its development is essential [19]. Apart from traditional anatomical MRI, which offers some unique processes particularly significant for identifying issues in the developing fetal brain [20]. Modern methods have transformed the diagnosis of conditions like stroke, neurological diseases, mental health problems, and more [21].

By shedding light on the intricate connections between tissue architecture, metabolism, and blood circulation, they have the potential to greatly increase our knowledge of the underlying mechanisms behind a number of diseases and strategies for identifying these anomalies [22].

Artificial Intelligence (AI) techniques have emerged as a prominent solution for various disputes and enhance the dependability, efficiency of fetal brain segmentation [23]. In particular, Convolutional Neural Networks (CNNs) have demonstrated success in various examination based on different projects, including brain segmentation [24–27]. The majority of examples which are already present are designed for adult brain segmentation and often fail to perform effectively when applied to fetal brain data [23].

2. Fetal Brain Segmentation

Magnetic resonance imaging, its an very much used practice for assessing brain activity [28]. Additionally, MRI is a very helpful technique for researching the development of the foetal brain [29]. It is crucial to employ scanning to identify foetal brain deformities as soon as feasible since around three out of every 1000 babies are born with brain abnormalities (BAs). Furthermore, a variety of neuropathology-related diseases may appear, with grave clinical repercussions. Research can continue without medical intervention since MRI diagnostic can identify prenatal brain abnormalities and cancers early [30, 31].

In addition to discussing potential solutions if abnormalities are identified early, doctors can help parents comprehend and get ready to deal with anomalies [32]. Deep learning (DL) is a potent technique for image processing and pattern recognition in bioinformatics. Beyond realm for foetal imaging, DL is used in ultrasound image processing [33]. This method can be used by medical professionals to identify illnesses in sick photos. DL is therefore essential for assessing foetal brain MRI images in order to detect certain illnesses at an early stage. MRI scans can provide an early diagnosis of brain problems without the requirement for medical intervention by revealing different attributes for different layers inside foetal brain structure [34].

Early detection of these abnormalities with DL has several advantages: it allows physicians to make a more accurate diagnosis, inform parents about impending difficulties and coping mechanisms, address possible pregnancy complications, and enhance the overall quality of the treatment plan tailored to each patient. manually cropping the developing brain, which makes it more challenging to complete assignments in the allocated time. It could classify a range of foetal anomalies, although its maximum classification accuracy was just 80%.

3. Steps for Fetal Brain Segmentation in MRI

The figure 39.1 explains general steps to perform foetal brain segmentation. Detailed description for each step as follows:

3.1. Data acquizition

The initial stage is to acquire foetal MRI data, with an emphasis on T2-weighted sequences since they offer better contrast for distinguishing between various brain regions [35]. Reputable medical research institutions or reliable, superior databases, such as the FeTA Challenge, should provide the data. To ensure the quality of the input, scans with low resolution or significant artefacts should be eliminated from the collection.

3.2. Preprocessing

Preparing the data for segmentation requires preprocessing [36]. Motion correction is the first step in this process, which is usually accomplished via image registration algorithms to remove artefacts brought on by the fetus's movements during the scan [37]. The foetal brain is then isolated, being careful to remove surrounding structures and extraneous tissues, to determine the region of interest (ROI) [38]. There is guarantee consistent input throughout the collection, normalization is employed to standardize image intensities and resolutions [39]. When using 3D techniques, high-resolution volumes can be reconstructed from 2D slices using super-resolution reconstruction, which improves the model's capacity to identify finer details [40].

3.3. Model training and selection

Choosing a suitable architecture, such as U-Net, nnU-Net, or a particularly designed convolutional neural network

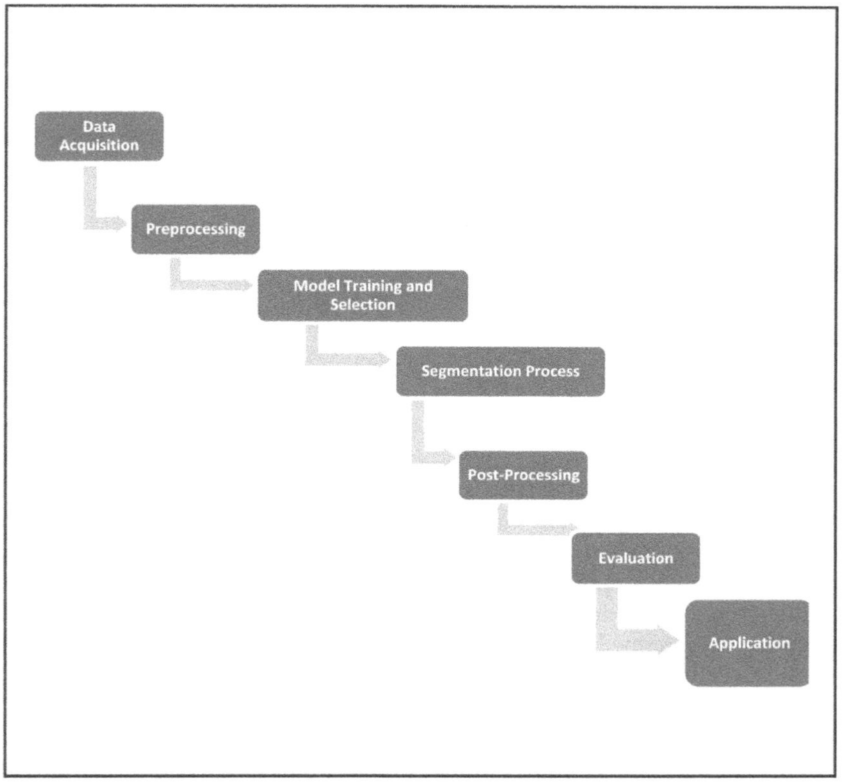

Figure 39.1. General steps to perform fetal brain segmentation.
Source: Author.

(CNN), is the first step in model training. Data augmentation methods like rotation, flipping, and elastic deformation are used to increase diversity and improve generalization because foetal brain segmentation datasets are frequently small. Models that have previously been trained on comparable datasets can be improved by transfer learning in situations when data is limited. During the training step, different dataset, separated in small sets for training, testing, validation [41]. The model is then enhanced using segmentation task-appropriate loss functions.

3.4. Segmentation process

The model is fed pre-processed images for segmentation after training [42, 43]. The model creates segmentation masks that show the architecture of the foetal brain by classifying pictures or voxels [44]. The segmentation output may be multi-class (various parts of the brain) or binary (foetal brain vs. backdrop).

3.5. Post-processing

Post-processing enhances the segmentation findings to ensure accuracy. Techniques such as connected-component labelling are used to remove artefacts or small, superfluous segments from the output [45]. A volumetric model of the brain is created by combining the slices in order to do 3D segmentation [46]. Optional smoothing filters can be used to smooth the edges of the segmented portions and increase the findings' anatomical accuracy.

3.6. Evaluation

Utilizing quantitative indicators to assess the segmentation quality is the last phase. The Jaccard index measures how comparable actual truth for guessing the segmentation is, while the Dice coefficient assesses how much overlap there is between the two. The accuracy and coverage of the segmentation are evaluated using precision and recall, respectively. By superimposing segmentation masks over the original MRI images, the results are visually confirmed and the differences between the expected and real structures are clearly visible.

4. What is Diffusion Models and Explainable AI Techniques?

4.1. Diffusion models

Diffusion models are sophisticated generative frameworks that simulate the noise removal process to generate data, such as pictures or music. The upcoming methods and backtracking process are two primary stages for these models' functioning. The forward process reduces the original data to a condition that is nearly random by progressively adding Gaussian noise via a number of stages. Conversely, the opposite process gradually denoizes the corrupted data in an attempt to return it to its original state. In the intermediary stages, a neural network is trained using either the partially denoized data or the injected noise. Diffusion models effectively approximate the data distribution through the application of variational inference and optimization techniques.

4.2. How diffusion models work

For diffusion models to work, a series of small data changes must occur. At every step of the forward phase, noise is methodically introduced until the representation is entirely noisy [47]. The reverse phase teaches a neural network to systematically reverse this process by either anticipating the extra noise or recovering the intermediate data states [48]. A loss function ensures optimization by decreasing the difference between the output that is rebuilt at each stage and the real data distribution. Diffusion models are highly effective in generating data, such as high-quality audio or images, because they use this recurrent denoising process to comprehend composite data dispersions.

4.3. Algorithm for diffusion models

4.3.1. Forward diffusion process

The forward diffusion process adds Gaussian noise to the data incrementally. This is modelled as:

1. Noise Addition Equation:
   $q(x_t|x_{t-1})=N(x_t;1-\beta t\cdot x_{t-1},\beta t\cdot I)$
   where:
   - xtx_txt is the corrupted data at time ttt,
   - xt−1x_{t-1}xt−1 is the data at the previous time step,
   - βt\beta_tβt controls the noise scale at step ttt,
   - N\mathcal{N}N represents a normal distribution.
2. Complete Forward Process:
   $q(x_t|x_0)=N(x_t;\alpha^{-}_t\cdot x_0,(1-\alpha^{-}_t)\cdot I)$
   where:
   $\alpha^{-}_t=i=1\prod t(1-\beta_i)$

4.3.2. Reverse diffusion process

The reverse process attempts to denoize the data step-by-step, learning

$$p\theta(x_{t-1}|x_t)$$

1. Reverse Process Equation:
   $p\theta(x_{t-1}|x_t)=N(x_{t-1};\mu\theta(x_t,t),\Sigma\theta(x_t,t))$
   where $\mu\theta$ and $\Sigma\theta$ are predicted by the neural network θ.
2. Loss Function (Denoising Objective):
   $L = E_{t,x0,\epsilon}[||\epsilon - \epsilon_{\theta(}x_{t}, t)||^2]$
   where:
   - ϵ is the noise added during the forward process,
   - $\epsilon\theta$ is the predicted noise from the neural network.
3. **Sampling**: Start from $xT \sim N(0,I)$ and iteratively sample:
   $x_{t-1} \sim p_\theta(x_{t-1}|x_t)$

4.4. Explainable AI (XAI) techniques

Explainable AI encompasses a range of techniques aimed at making AI systems transparent and interpretable to human users. This is especially critical in domains like healthcare or finance, where understanding the rationale behind AI predictions is paramount. XAI methodologies provide clarity on how models analyze inputs and produce outputs, allowing users to evaluate and trust the system's decisions. Examples of XAI techniques include feature attribution, visual explanations, rule extraction, and counterfactual analysis, each contributing to enhancing model interpretability.

4.5. How XAI works

XAI operates by exposing the underlying logic or providing simplified interpretations of complex AI decisions. Techniques like SHAP and LIME quantify the impact of individual input features on the final prediction, offering insights into feature importance. Visual methods such as saliency maps and Grad-CAM highlight parts of the input (e.g., an image region) that strongly influence a decision. Rule extraction methods translate intricate models into simple, human-readable decision rules. Counterfactual reasoning explores alternative scenarios, illustrating how slight changes in input could alter the prediction. These approaches not only ensure accuracy but also foster accountability and comprehension in AI systems, enabling broader acceptance and trust.

4.6. Algorithm for explainable AI techniques

4.6.1. Feature attribution (SHAP)

1. Shapley Value Equation:
   $\Phi i = \sum |S|!\cdot(|N|-|S|-1)! / |N| ! .[f(S \cup \{i\}) - f(S)]$
   where:
   - N is the set of all features,
   - S is a subset of features excluding iii,
   - f(S) is the model's prediction with only the features in S.

4.6.2. Visual explanations (Grad-CAM)

1. Gradient-weighted Feature Map:
   Compute the gradient of the output y^c (class score) with respect to the feature map A^K:

$$\alpha^k_c = 1/Z \sum_i \sum_j \partial Y^c / \partial A^k_{ij}$$

where Z is the total number of pixels in A^k.

2. Weighted Feature Map:

$$L^c_{Grad\text{-}CAM} = ReLU\left(k\sum \alpha^c_K \cdot A^k\right)$$

4.6.3. Rule extraction (Decision Tree Surrogate)

1. Objective:
Minimize the discrepancy between the predictions of the complex model f(x)) and the surrogate decision tree g(x):

$$L = \sum_{i=1}^{n} \|f(x_i) - g(x_i)\|^2$$

4.6.4. Counterfactual explanations

1. **Counterfactual Objective**: Find x′ such that:
$\min\|x - x'\| + \lambda \cdot L(f(x'), y')$
where:
   - x is the original input,
   - x′ is the counterfactual,
   - y′ is the desired outcome,
   - λ is a trade-off parameter.

5. Fetal Brain Segmentation Using Diffusion Models and Explainable AI Techniques

Figure 39.2 explain the model framework using UNet and Explainable AI which has various steps as follows:

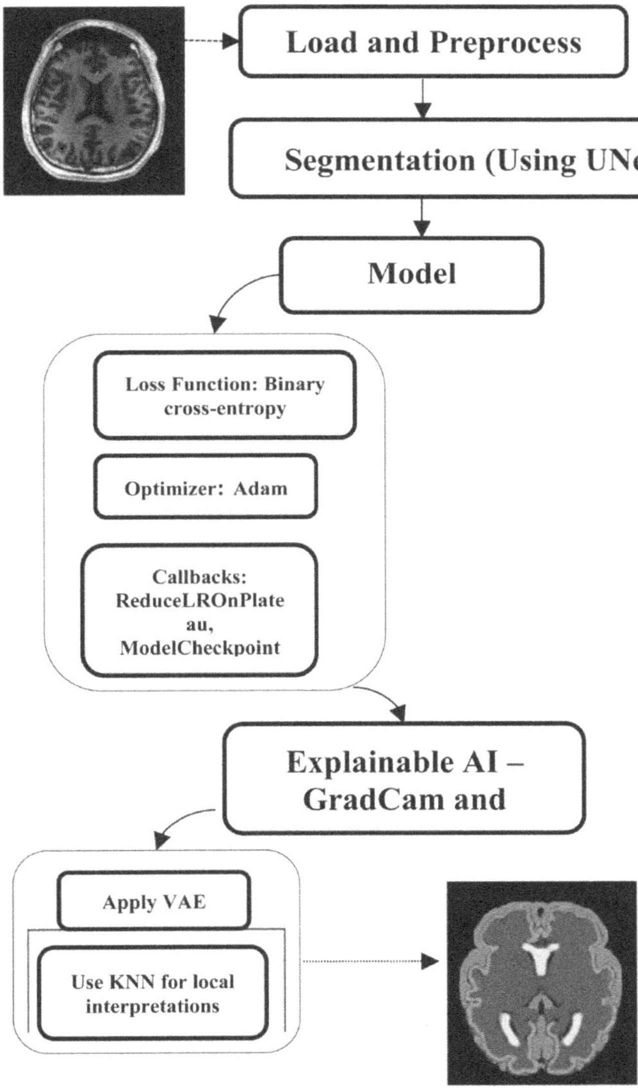

Figure 39.2. Proposed Model Framework using UNet and Explainable AI.

Source: Author.

5.1. Loading and preprocessing

The procedure begins with the acquizition of medical images, such as brain Magnetic Resonance Imaging (MRI) scans. These photographs need to be cleared of noise, variations in intensity, and other imperfections before going through further processing.
Preprocessing has different steps to follow:

- Normalization: The image's pixel intensity values are normalized to a standard range (e.g., [0, 1] or [-1, 1]) to ensure uniformity and better convergence during training.
- Resizing: Images can be reduced to a fixed size in order to get the actual data for processing of segmentation prototype (in this case, U-Net). This ensures uniformity in each sample.
- Noise Reduction: Gaussian filtering and median filtering are two techniques for removing random noise from a picture.
- Augmentation (if used): To artificially increase the dataset and strengthen the model against
- *Dig: Flowchart for Fetal brain segmentation using diffusion models and explainable AI techniques*
- Fluctuations, methods such as rotations, flips, and elastic transformations can be used.

5.2. Segmentation using U-Net

A U-Net model is then used to segment the previously processed pictures. A convolutional neural connected architecture called U-Net was created especially for image segmentation applications in the biomedical field. It consists of two primary components:

- Contracting Path (Encoder): This component captures contextual information while reducing the image's spatial dimensions. To extract hierarchical features, it uses max-pooling techniques and successive convolutions.
- Expanding Path (Decoder): The decoder reconstructs the spatial dimensions of the picture using higher-level feature information from the encoder. To ensure accurate feature localization, this part uses skip connections and up sampling.

5.3. Model training

The accuracy and generalization capacity of the U-Net model are influenced by a number of crucial training inputs, including:

5.3.1. Loss function

Binary cross-entropy, a loss function appropriate for binary segmentation applications, is utilized in this process. The difference between the actual data and anticipated segmentation map is measured with binary cross-entropy loss. It encourages the model to produce more accurate predictions by penalising it for wrong classifications.

The binary cross-entropy loss determined mathematically follows.

$$L=-1/N\sum_{i=1}^{N}[y_i\cdot\log(y\hat{}_i)+(1-y_i)\cdot\log(1-y\hat{}_i)]$$

Here, y_i is the ground truth label, and $y\hat{}_i$ is the predicted probability for each pixel.

5.3.2. Optimizer

Training is done with the Adam optimizer. The advantages of momentum and RMSProp optimizers are combined in Adam, an adaptive learning rate optimization technique. It increases training efficiency and speeds up convergence by dynamically changing the learning rate for each parameter.

5.3.3. Callbacks

Two methodologies are used to enhanced the efficiency:

- ReduceLROnPlateau: This callback lowers the learning rate while keeping an eye on the validation loss getting over with some turns with minimal output in which the validation loss no longer improves. As a result, the model could successfully converge without going above the ideal weights.
- ModelCheckpoint: To guarantee the most accurate final model used for predictions, this callback stores the best-performing model weights throughout training.

5.4. Explainable AI (Grad-CAM and VAE)

After the model has been trained, explainability techniques are applied to examine the predictions and gives different overviews for decision making. Its very significant in medical fields where transparency and trust are essential.

5.4.1. Grad-CAM

Visual overview is focus on the input data that most influence the prototype predictions are produced using gradient-weighted class activation mapping, or Grad-CAM. Grad-CAM can demonstrate which areas of the picture were crucial in identifying the site of the malignancy, for instance, if the model is segmenting a tumour. This aids medical professionals in confirming if the model is concentrating on pertinent anatomical features. The output's gradient is computed in relation to the convolutional layers' feature maps to produce the heatmap, which is then weighted to provide a spatial localization map.

5.4.2. Variational Autoencoder (VAE)

The regions are first segmented and then processed using a Variational Autoencoder. VAEs decode incoming data back into its original form after encoding it into a latent space. They are generative models. This makes the following possible:

- Feature extraction is the process of identifying latent properties in the segmented data that may not be visible in the original input image.
- Data augmentation: Creating believable variations of the input data to test the robustness of the model.

5.5. Local interpretations using KNN

The pipeline uses the K-Nearest Neighbours (KNN) technique to make the model's predictions locally understandable. KNN contrasts segments or their characteristics with identical dataset samples. It provides localized explanations, such as the rationale for the classification of a particular pixel or region, by selecting the most similar data points. This method works well for identifying outliers or abnormalities.

5.6. Final output

The pipeline's ultimate output combines the results of the explainability and segmentation phases. It includes:

1. A divided image that draws attention to the many areas of interest.
2. A Grad-CAM heatmap showing the main regions influencing the model's predictions.
3. Interpretations of VAE and KNN, which provide both local and global insights into the segmentation results.

6. Conclusion

The pipeline being offered demonstrates a comprehensive approach to medical picture segmentation that is especially intended for the study of unborn brains. By using the U-Net architecture for precise segmentation and enhancing model interpretability using Grad-CAM and VAE, this system closes the gap between deep learning methods and clinical usability. Using KNN for local interpretability adds another level of transparency, giving physicians trust in the model's predictions both domestically and internationally. Additionally, the model is trained efficiently and consistently by leveraging callbacks like ReduceLROnPlateau and ModelCheckpoint, which maximises performance for practical applications.

By assisting physicians in understanding the areas that are crucial for projections, explainable AI solutions increase model confidence and aid in improved decision-making. The suggested method is positioned as a major breakthrough in foetal brain segmentation tasks as it blends segmentation accuracy with interpretability. Future studies may examine more interpretable AI methods and apply the idea to bigger datasets in order to further enhance therapeutic usefulness.

References

[1] Joseph, R. (2000). Fetal Brain Behavior and Cognitive Development. *Developmental Review, 20*, 81–98, ISSN0273-297. https://doi.org/10.1006/drev.1999.0486.

[2] Glenn, O. A. (2010). MR imaging of the fetal brain. *Pediatric Radiology*, *40*, 68–81. https://doi.org/10.1007/ s00247-009-1459-3

[3] Levine, D., Barnes, P. D., Madsen, J. R., et al. (1997). Fetal central nervous system anomalies: MR imaging augments sonographic diagnosis. *Radiology*, *204*, 635–642.

[4] Durgadevi, P., Vijayalakshmi, S., & Sabharwal, M. (Mar. 2021). Fetal Brain Abnormality Detection through PSO (Particle Swarm Optimization) and Volume Estimation. *Annals of the Romanian Society for Cell Biology*, 2700–2714.

[5] Attallah, O., Sharkas, M. A., & Gadelkarim, H. (2019). Fetal brain abnormality classification from mri images of different gestational age. *Brain Science, 9*(9). https://doi.org/10.3390/brainsci9090231. https://www.mdpi.com/2076-3425/ 9/9/231.

[6] Levine, D., Barnes, P., Sher, S., et al. (1998) Reproducibility, technical quality, and conspicuity of organs in evaluating fetuses with fast MRI. *Radiology, 206*, 549–54.

[7] Somasundaram, K., & Kalaividya, P. A. (2016, December). Brain portion segmentation from Magnetic Resonance Images (MRI) of human head scan using Richardson Lucy deconvolution and intensity thresholding. *In 2016 International Computer Science and Engineering Conference (ICSEC)*, pp. 1–5. IEEE.

[8] Glenn, Orit, A. (2010). MR imaging of the fetal brain. *Pediatric Radiology, 40*(1), 68–81. doi:10.1007/ s00247-009-1459-3.

[9] Havaei, M., Davy, A., Warde-Farley, D., et al. (2017). Brain tumor segmentation with Deep Neural Networks. *Medical Image Analysis, 35*, 18–31.

[10] Alansary, A., Lee, M., Keraudren, K., et al. (2015). Automatic Brain Localization in Fetal MRI Using Superpixel Graphs. In *Machine Learning Meets Medical Imaging* (pp. 13–22). Springer: Cham, Switzerland.

[11] Hosseini, M. S., & Zekri, M. (2012). Review of Medical Image Classification using the Adaptive Neuro-Fuzzy Inference System. *Journal of Medical Signals and Sensors, 2*, 49–60. [PubMed].

[12] Attallah, O., Gadelkarim, H., & Sharkas, M. A. (2018, December). Detecting and classifying fetal brain abnormalities using machine learning techniques. In *2018 17th IEEE International Conference on Machine Learning and Applications (ICMLA)* (pp. 1371–1376). IEEE.

[13] Fidon, Lucas, et al. (2021). Distributionally robust segmentation of abnormal fetal brain 3D MRI. *Uncer tainty for Safe Utilization of Machine Learning in Medical Imaging, and Perinatal Imaging, Placental and Preterm Image Analysis: 3rd International Workshop, UNSURE 2021, and 6th International Workshop, PIPPI 2021, Held in Conjunction with MICCAI 2021, Strasbourg, France, October 1, 2021, Proceedings 3. Springer International Publishing.*

[14] Farzan Vahedifard, J. D., Mark Supanich, Jubril Adepoju, Xuchu Liu, & Sharon Byrd M. D. (2022). Deep Learning Model for Automatic Landmark Localization in Fetal Brain MRI. *Annual Medical Education Conference (AMEC).* Florida, USA.

[15] He, L., Li, H., Holland, S. K., Yuan, W., Altaye, M., & Parikh, N. A. (2018). Early prediction of cognitive deficits in very preterm infants using functional connectome data in an artificial neural network framework. *Neuroimage Clinical, 18*, 290–297. doi:10.1016/j.nicl.2018.01.032.

[16] Stiles, J., & Jernigan, T. L. (2010). The basics of brain development. *Neuropsychology Review*, *20*(4), 327–348, PMID: 21042938, 10.1007/s11065-010-9148-4.

[17] Durgadevi, P., & Vijayalakshmi, S. (2021, March). A methodological investigation for fetal brain mri segmentation techniques-analysis. In *2021 International Conference on Advance Computing and Innovative Technologies in Engineering (ICACITE)* (pp. 684–690). IEEE. doi:10.1109/icacite51222.2021.9404572.

[18] Attallah, O., Gadelkarim, H., & Sharkas, M. A. (December 2018). Detecting and Classifying Fetal Brain Abnormalities Using Machine Learning Techniques. *Proceedings of the 2018 17th IEEE International Conference on Machine Learning and Applications* (ICMLA); Orlando, FL, USA. 17–20, 1371–1376.

[19] Mercy, P. J., & Utchimahali, M. (February 2021). Detecting and classifying fetal brain abnormalities using Decision Tree Algorithm. *IJCRT*, *9*(2), ISSN: 2320-2882.

[20] Kainz, B., Keraudren, K., & Kyriakopoulou, V. (2014, April). Fast fully automatic brain detection in fetal MRI using dense rotation invariant image descriptors. *In 2014 IEEE 11th International Symposium on Biomedical Imaging (ISBI)* (pp. 1230–1233). *IEEE.*

[21] Attallah, O., Sharkas, M. A., & Gadelkarim, H. (2019). Fetal Brain Abnormality Classification from MRI Images of Different Gestational Age. *Brain Science, 12, 9*(9), 231. doi:10.3390/brainsci9090231. PMID:31547368; PMCID: PMC6770437.

[22] Holdsworth, S. J., & Bammer, R. (Oct, 2008). Magnetic resonance imaging techniques: fMRI, DWI, and PWI. *Seminars in Neurology*, *4*, 395–406. doi:10.1055/s-0028-1083697, Epub. PMID: 18843569; PMCID: PMC3985850

[23] Snoek, Rozemarijn, et al. (2018). Accuracy of diagnosis and counseling of fetal brain anomalies prior to 24 weeks of gestational age. *The Journal of Maternal-Fetal & Neonatal Medicine*, *31*(16), 2188–2194.

[24] PI, T., M, K., V, J., JP, U., R, H., R, S., et al. (2003). Comparison of BOLD fMRI and MEG characteristics to vibro-tactile stimulation. *NeuroImage*, *19*(4).

[25] Vahedifard, F., Adepoju, J. O., Supanich, M., et al. (2023). Review of deep learning and artificial intelligence models in fetal brain magnetic resonance imaging. *World*, *11*(16).

[26] Vahedifard, F., Ai, H. A., Supanich, M. P., Marathu, K. K., Liu, X., Kocak, M., et al. (2023). Automatic Ventriculomegaly Detection in Fetal Brain MRI: A Step-by-Step Deep Learning Model for Novel 2D-3D Linear Measurements. *Diagnostics, 13*(14), 2355.

[27] Vahedifard, F., Liu, X., Marathu, K. K., Kocak, M., Ai, H. A., Supanich, M. P., et al. (2023). Artificial Intelligence Prediction of Gestational Age of Fetal in Brain Magnetic Resonance Imaging versus ultrasound Using three different Biometric Measurements.

[28] Glenn, O. A., & Barkovich, J. (2006). Magnetic Resonance Imaging of the Fetal Brain and Spine. *An Increasingly Important Tool in Prenatal Diagnosis: Part 2. American Journal of Neuroradiology, 27*, 1807–1814.

[29] Attallah, O., Sharkas, M. A., & Gadelkarim, H. (2019). Fetal brain abnormality classification from MRI images of different gestational age. *Brain Science, 9*(9). https://doi.org/10.3390/brainsci9090231, https://www. mdpi.com/2076-3425/9/9/231.

[30] Narkhede, H. P. (July 2013). Review on Image Segmentation Techniques. *International Journal of Science and Modern Engineering, 1*, 8.

[31] Attallah, O., Gadelkarim, H., & Sharkas, M.A., (2018). Detecting and classifying fetal brain abnormalities using machine learning techniques. *In 17th IEEE International Conference on Machine Learning and Applications (ICMLA),* 1371–1376. https://doi.org/10.1109/ICMLA.2018.00223.

[32] Durgadevi, P., & Vijayalakshmi, S. (2021). A Methodological Investigation for fetal brain MRI Segmentation Techniques-Analysis. In *2021 International Conference on Advance Computing and Innovative Technologies in Engineering (ICACITE)* (pp. 684–690). IEEE.

[33] Alansary, A., Lee, M., & Keraudren, K. (2015). Automatic brain localization in fetal MRI using superpixel graphs. *Lecture Notes in Computer Science*, 13–22. https://doi. org/10.1007/978-3-319-27929-9_2.

[34] Van, Sloun., R. J. G., Cohen, R., & Eldar, Y. C. (2020). Deep learning in ultrasound imaging. *Proceedings of the IEEE, 108*(1), 11–29. https://doi.org/10.1109/ JPROC.2019.2932116.

[35] Kaganami, H. G., & Beij, Z. (2009). Region Based Detection versus Edge Detection. *IEEE Transactions on Intelligent Information Hiding and Multimedia Signal Processing,* 1217–1221.

[36] De Bock, J., Smet, P. D., & Philips, W. (January 2005). Image segmentation using watersheds and normalized cuts. *In Proceedings of the IS&T/SPIE Electronic Imaging 2005*, *5675*, 164–173.

[37] Monteiro, F. (2007). Region-based spatial and temporal image segmentation. *PhD thesis, Faculdade de Engenharia da Universidade do Porto, Portugal.*

[38] Salehi, Seyed, Sadegh, Mohseni, et al. (2018). Real-time automatic fetal brain extraction in fetal MRI by deep learning. *15th international symposium on biomedical imaging* (ISBI 2018). *IEEE.*

[39] Gayathri, S P. (2016). Segmentation and volume estimation of fetal brain from T2-w magnetic resonance images (MRI) of human fetus, Ph.D. Thesis. *The Gandhigram Rural Institute Tamil Nadu, India.*

[40] Kaur, D., & Yadwinder Kaur. (2014).Various image segmentation techniques: a review. *International Journal of Computer Science and Mobile Computing, 3*(5), 809–814.

[41] Ball, G., Aljabar, P., & Arichi, T. (2016). Machine-learning to characterise neonatal functional connectivity in the preterm brain. *Neuroimage*, *124*, 267–275. doi:10.1016/j.neuroimage.2015.08.055.

[42] Monteiro, F. C., & Campilho, A. (2008). Watershed framework to region-based image segmentation. *In Proceedings of the International Conference on Pattern Recognition, ICPR* 19th, 1–4.

[43] Fabijanska, A. (2011). Variance filter for edge detection and edge-based image segmentation. *In Proceedings of the International Conference on Perspective Technologies and Methods in MEMS Design (MEMSTECH),* 151–154.

[44] Zhang, Y. J. (2001). An Overview of Image and Video Segmentation in the last 40 years. *Proceedings of the 6th International Symposium on Signal Processing and Its Applications,* 144–151.

[45] Callaghan, R. O., & Bull, D. (2005). Combined morphological spectral unsupervised image segmentation. *IEEE Transactions on Image Processing, 14*(1), 49–62.

[46] Paladini D., & Volpe P. (2006). Posterior fossa and vermianmorphometry in the characterization of fetal cerebellar abnormalities: A prospective three-dimensional ultrasound study. *Ultrasound in Obstetrics & Gynecology, 27*, 482–489.

[47] Shishegar, R., Joshi, A. A., Tolcos, M., Walker, D. W., & Johnston, L. A. (2017, April). Automatic segmentation of fetal brain using diffusion-weighted imaging cues. *In 2017 IEEE 14th International Symposium on Biomedical Imaging (ISBI 2017)* (pp. 804–807). IEEE.

[48] Tournier, J. D., Calamante, F., & Connelly, A. (2012). MRtrix: Diffusion tractography in crossing fiber regions. *International Journal of Imaging Systems and Technology, 22*(1), 53–66.

40 Comprehensive analysis on enhanced early prediction of high-risk colorectal cancer patients using deep learning techniques on medical imaging data

Vandana R. Bais[1,a] *and S. Vijayalakshmi*[2,b]

[1]Research Scholar, Department of Computer Science, School of Sciences, CHRIST University, India
[2]Associate Professor, Department of Data Science, School of Sciences, CHRIST University, India

Abstract: Colorectal Cancer (CRC) is a major cause of cancer-related deaths worldwide. Early prediction significantly improves patient outcomes; however, current screening methods often miss high-risk individuals, particularly in the early stages. Identifying high-risk patients for early screening and intervention remains challenging due to the complex interplay of demographic and clinical factors. This means that complexity of these risks is not well detected with traditional risk assessment methods thus resulting in early detection of risks. Furthermore this is other reason why CRC is difficult to deal with because it is competitive and has few treatments. This means that early prediction is very crucial since it enable therapeutic management to be effectively started, the severity of side effects can be reduced and overall patient outcomes improved. Conventional assessment tools such as RECIST are incompatible with timely assessment most of the time. The advances in the current years in machine and deep learning offer new possibilities in the better prognosis of early diagnosis and treatment of CRC. The present study suggests deep learning techniques should be applied in early CRC patient identification using medical image data. Therefore, the model shall be developed from the analysis of CT, MRI, and PET-CT scans obtained at baseline and during treatment against those from current response criteria. The deep learning model fills a gap in the management of CRC by presenting a technique for early prediction of CRC. The integration of deep learning and medical imaging may revolutionize the treatment of cancer by allowing the implementation of personalized medicine approaches, which would raise survival rates. The review suggests deep learning model can be developed and implemented into more effective, personalized treatment strategies, ultimately enhancing patient outcomes.

Keywords: Colorectal cancer, deep learning, machine learning, medical imaging

1. Introduction

Colorectal cancer is one of the leading causes of cancer among patients globally, and effective management and better results will depend only on its early detection. It has been one of the main reasons for cancer incidence and death rates all over the world [2]. Most cancer-related associations recommend routine screening of CRC after the age of fifty because CRC prognosis varies with the time of diagnosis; stage I colorectal cancer may have up to 90% survival rate while stage IV has up to less than 10% survival rate [16]. Stage A colorectal cancer does not produce easily noticeable signs Colorectal cancer at an early stage is asymptomatic, and the duration of time for manifestation of the disease can take over 10 years [4]. The existing screening techniques include colonoscopy and Fecal occult blood tests or FOBT though effective, are not without limitations. These traditional methods are invasive, costly, and not universally accessible, leading to suboptimal participation rates, particularly among populations with lower socioeconomic status. Moreover, the reliance on age and family history as primary risk indicators fails to capture a significant portion of individuals who may develop CRC outside these parameters [8]. Most CRCs are already in stage III or stage IV when detected and have migrated to other tissues, and the five-year survival rate of CRC is less than 10%. [2].

For colorectal cancer the principal cure is radiation or chemotherapy which brings the quality of life of the patient down further, and also is quickly recognizable by

[a]vandanarbais18@gmail.com, [b]svijisuji@gmail.com

DOI: 10.1201/9781003770473-40

infection, nerve and skin damage [7, 8]. Although diagnostic technologies have advanced significantly, the manual examination of medical imaging data by healthcare professionals is still inefficient and highly susceptible to mistakes, which can result in missed or incorrect diagnoses [3]. It is beneficial to identify and treat high risk groups in the community since healthcare personnel can maximize practice of available resources in these groups. Such differential strategy could result in early diagnosis of colorectal cancer thus a decrease in the death rates and quality life of patients with this approach could lead to a reduction in incidence of advanced-stage colorectal cancer, ultimately reducing death rates and successful quality of life for patients [1].

Artificial intelligence technologies, some of the most popular ones are algorithms known as big data tools capable of processing extensive data to discover intricate relationships can be very beneficial in the sphere of healthcare concentrating on the particular problem of various types of cancer diagnosis and prognosis [1, 9]. Thus, from the perspective of colorectal cancer, it is possible to use the features of machine learning for analysing the set of demographic and clinical indicators, including age, gender, lifestyle factors, history of colon cancer in family, clinical markers etc. to improve the prediction of the patient group with a high cancer risk level. The combination of demographic and clinical information into the machine learning model may indeed revolutionize the field in which CRC risk is determined and personalized instead of with a comprehensive approach. Indeed, this approach holds the prospect not only of diagnosing potentially high risk individuals who will not reach the current screening criteria but also of identifying them at a stage of the disease which is potentially substantially more treatable [10, 12].

Despite advancement, the early identification of patients remains a real outgoing challenge, with the existing screening and treatment, most of the patients are diagnosed at the advanced stage with poorer prognosis [5, 8].While the chances of surviving 5 years if the colorectal cancer is detected approximately 90% early, the chances are not so good if cancer spread beyond the colon or rectum. Thus, early detection is crucial to enhance years survival of patients with colorectal cancer. These days, the main techniques that are used in the diagnosis of colorectal cancer involve screening by imaging, fecal tests. Imaging, stool testing methods and, in the former category, five subtypes (colonoscopy, stool tests, Cologuard (stool DNA), flexible sigmoidoscopy, computed tomographic colonography/computed tomographic colonoscopy). It must be understood that each test has its unique strengths and risks to the health of mankind [4]. A colonoscopy is not a preventive measure though, it only reveals the baseline risk of precursor abnormities. However, this mode of checking has many implications in clinical care since it will enable a group of people to ensure that they can develop colorectal cancer and they still have plenty of years ahead of them free of the disease [4, 8].

Design a machine learning model that classifies, accurately diagnoses colorectal cancer using data from medical images, including CT scans, colonoscopy images or MRI scans. To provide the accurate predictions, model will need to capture both visual information and structural features from the images. Ideally, the model should be able to give a location of the images which may well indicate CRC and quantify the level of malignancy in each region [13–15].

Metastatic colorectal cancer (mCRC) remains considered to be one of the deadliest types of cancer, possesses limited treatment options and a poor prognosis. Early prediction of a patient's response to therapy is essential in optimizing treatment strategies, minimizing unnecessary side effects, and improving overall survival rates. Traditional methods for measuring treatment response, such as RECIST (Response Evaluation Criteria In Solid Tumours), are often insufficient for early prediction, leading to delays in altering ineffective therapies [11, 19].

Main objective of advanced models using machine learning is to improve both the early prediction of high-risk colorectal cancer (CRC) patients and the prediction of treatment responses. By creating a comprehensive ML model that integrates demographic, clinical, non-invasive biomarkers, and lifestyle factors to develop risk assessment and integrate seamlessly into clinical practice. Additionally, a Deep Learning model will be developed to analyse serial medical imaging data for accurate early prediction of treatment responses. By integrating diverse data sources and applying sophisticated techniques, the research seeks to refine patient stratification, support timely interventions, and optimize personalized management strategies [11, 12, 19].

The diagram demonstrates the use of AI in health care by leveraging multimodal clinical datasets, such as EHR, pathological, and imaging data. The figure highlights three different domains: prognostication, which is about survival and outcomes prediction; therapeutics or treatment guidance; and diagnostics, like detection improvement through colonoscopy, blood tests, and metastasis analysis (Figure 40.1).

As the technology of omics advances today, more and more types of biomarkers are discovered and verified in colorectal cancer, including genomic biomarkers such as oncogene and tumour suppressor gene mutation; epigenomic biomarkers such as DNA methylation; transcriptomics biomarkers such as non-coding RNA, proteomics biomarkers and metabolomics biomarkers [8, 20]. To provide early detection and accurate assessment of CRC, deep learning models are proposed for prediction and management. More precisely, this study aims to develop a reliable model that includes demographic, clinical data as well as clinical, noninvasive biomarkers

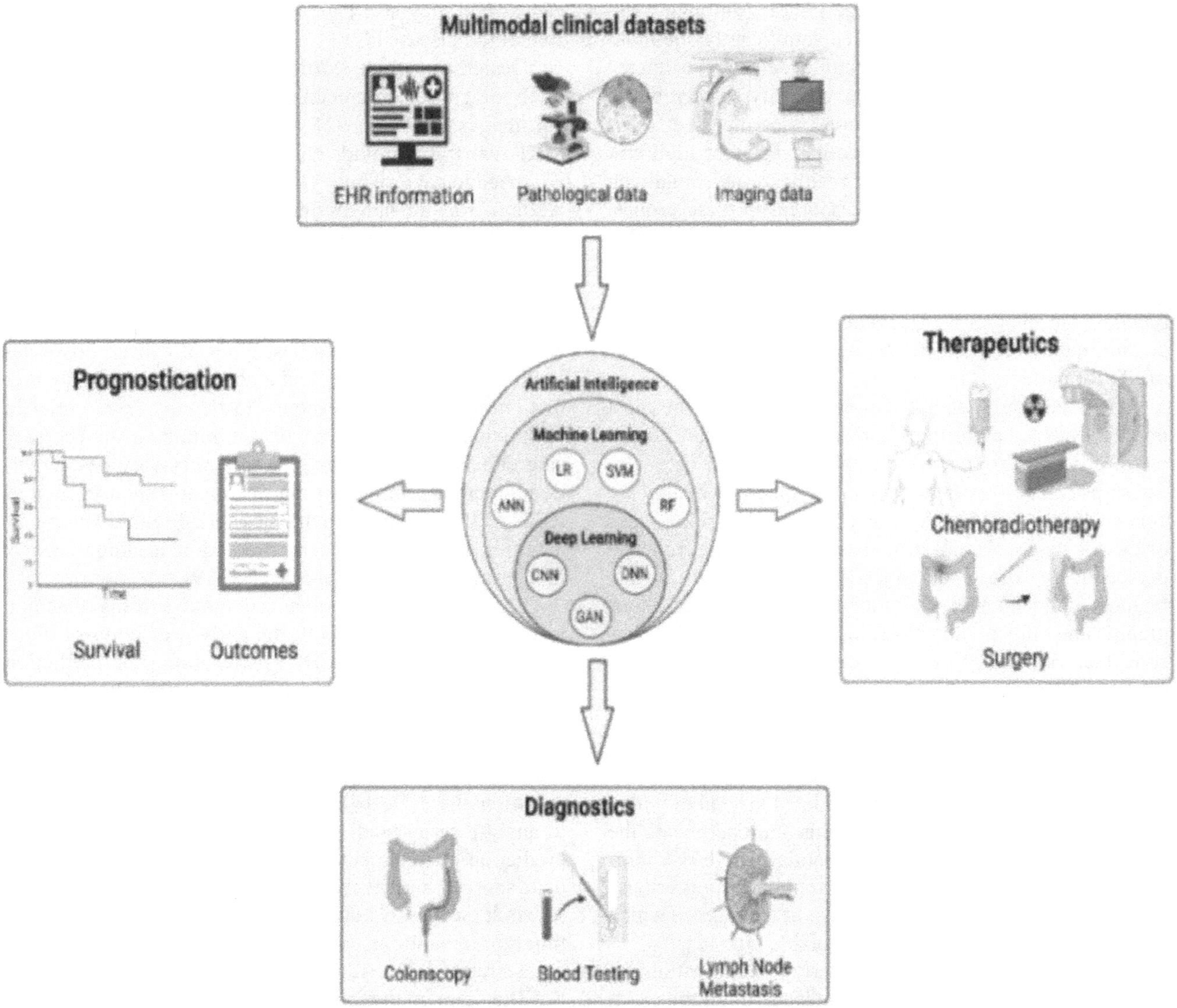

Figure 40.1. AI integration with multimodal clinical data for prognostication, diagnostics, and therapeutics.

Source: Author.

and lifestyle parameters to improve detection of high-risk CRC patients [11, 12, 19].

The research also aims at creating a deep learning model successfully achieved in accurately predicting early-stage patients using medical imagery data. As a result, the use of these increasingly refined analytical procedures is to improve risk differentiation to enable early interventions where necessary and to increase the effectiveness of the proposed individual treatment plans, which would ultimately improve patient prognosis and promote the continued evolution of cancer therapeutics [3, 11, 12].

2. Literature Review

Colon or rectal cancer commonly known as colorectal cancer CRC is a global problem and leading cause of cancer deaths. The prognosis depends on the time of diagnosis of the disease, its extent, and its risk assessment, which existing screening tools are unable to accomplish sufficiently well. More specifically, studies have found that conventional tools of risk assessment like family history and baseline demographic indicators fail to capture the CRC risk picture fully mostly due to lost chances for early prevention [4]. The latest advancement in machine learning has potential new forms of facing these issues. The feasibility of ML techniques was advocated through remaining ability as an instrumentation to analysis different styles of datasets encompassing demographic, clinical and life style datasets to improve the risk predictability. Research has proved that by use of different data type that standard approaches do not consider, machine learning models can improve identification of high risk patients. For instance, applications of the ML algorithm have been used in patient data and genomics data leading to accurate risk assessment [8, 12].

Clinical, imaging data and genetic data can be used in combination with each other to create a more comprehensive ML-based prediction of CRC. The strategy is of great significance not only in enhancing precision of the preliminary diagnosis but also in reducing patient's distress from invasive examinations such as colonoscopy. These authors draw attention to a current lack of research that integrates multiple sources of data in order to create improved models for prediction [8, 10, 18].

In the recent works, it has been proved that the stability of the machine learning models in explaining diagnostic gap. One specific model demonstrated high accuracy to detect right sided colorectal cancers, which is often not captured by the standard screening modalities. This success supports the future use of ML technology within healthcare especially in the context of improving the CRC early stage detectability based on demographic and analytical data [6].

An artificial model generated from composite tissue simulation sounded encouraging and fairly accurate capability in identification of high-risk colorectal characteristics to serve as a screening modality. Due to the high sensitivity obtained this model is useful in the first stages of diagnosis of colorectal specimens, which is an early diagnostic stage of cancer. This particular model was also most accurate in identifying right sided colon cancers, which may sometimes difficult to spot during normal screening processes. These results underscore the application of innovative machine learning approach to improve on the early detection of CRC based on Demographic and Laboratory analysis [6, 18].

Recent researches present efficient machine learning models that enhance the rate of correctly predicting the survival rate of CRC patients considerably. For example, a model using significantly larger number of different histopathological features derived from imagery of immunofluorescence provides more accurate prognosis as compared to conventional methods. This approach resolves the issue of variability in patient outcomes, caused by the nature of this disease as well as problematic regular pathological reporting platforms [10, 20].

Over the recent few years, detection and treatment of cancer have enhanced with the power of converting big data in medical imaging using methods like Generative Adversarial Networks (GANs). This paper showed that data augmentation occupies a strategic position in enhancing the performance of Machine Learning algorithms in medical imaging applications [3, 14].

Deep learning has shown much promise in various aspects of medical imaging, with cancer diagnosis, prognosis, and prediction of response to treatment. Previous studies has demonstrated the potential of deep learning models for extracting relevant features of medical images that not visible to human eye. However, there is a gap in applying these techniques specifically for early on-treatment response prediction in mCRC. This research will build upon existing knowledge by focusing on this niche area, leveraging longitudinal imaging data [10, 11].

3. Related Work and Results

The CNN architecture was employed to evaluate the classification results. Subsequently, the study explored the impact of data augmentation using artificial liver lesions rather than traditional augmentation methods. Two approaches for generating artificial lesions were implemented, revealing that the Deep Convolutional GAN (DCGAN) method outperformed the alternatives during testing [14]. SegAN framework improves the state of art, U-net segmentation carrying method in both efficiency, robustness for segmentation task [15].

The research focused on Cycle GAN in enhancing CT scan data quality segmentation. A Cycle GAN was trained to transform contrast-enhanced CT images into non-contrast images, which were later used to enhance training of U-Nets. Researchers evaluated segmentation performance on two datasets: between raw contrast-enhanced CT image set and a second raw image set containing only non-contrast CT images. Once the original and combined, U-Nets were trained, the datasets were designated as in-distribution and out-of-distribution, respectively [13–15].

Deep Convolutional GANs (DCGANs) are used to tackle the challenge of a small labelled dataset. The visualization tool Image was employed to assess the resemblance between real and generated images [14, 16]. Convolutional Neural Network models have been designed using dataset of Kather colorectal cancer histology to classify and grade colorectal cancer tissues [3, 15].

Equations involved pertain to the loss functions used to train the generator and discriminator. Below are the equations generated in GANs like DC-GAN and VAE-GAN:

Discriminator Loss (D)

$$L_D = -\mathbb{E}_{x \sim p_{\text{data}}(x)}[\log D(x)] - \mathbb{E}_{z \sim p_z(z)}[\log(1 - D(G(z)))]$$

Generator Loss (G)

$$L_G = -\mathbb{E}_{z \sim p_z(z)}[\log D(G(z))]$$

Combined Loss

$$L = L_{\text{adv}} + L_{\text{rec}} + L_{\text{KL}}$$

The paper revealed experimental outcomes between DC-GAN and VAE-GAN which showed that DC-GAN produced better results. The implementation of DC-GAN surpassed VAE-GAN by creating authentic synthetic pictures which strengthened predictive precision for neural network-based colorectal cancer discovery systems (Figure 40.2).

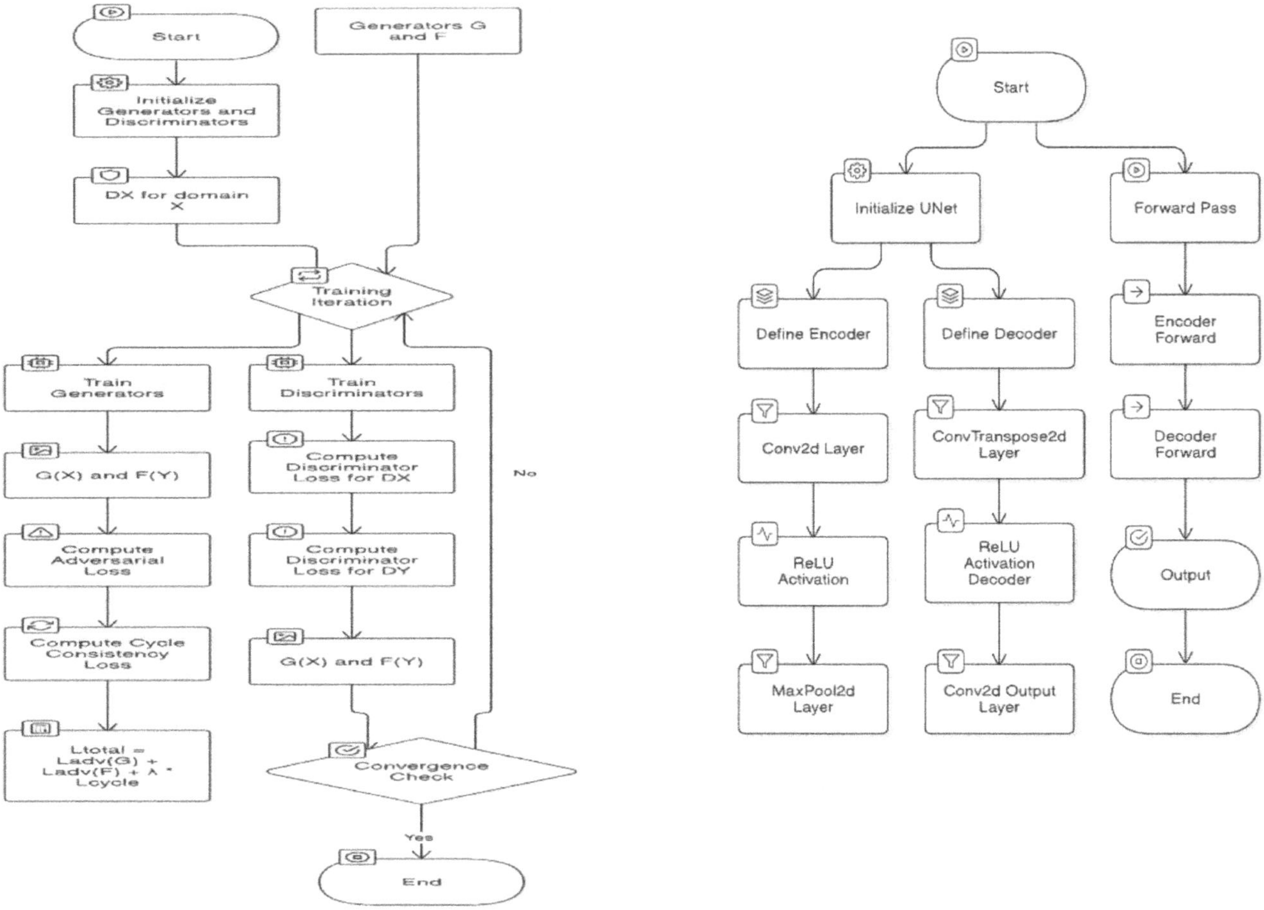

Figure 40.2. Flowchart of Cycle GAN: Domain Translation and U-Nets: Semantic Segmentation.

Source: Author.

Many researchers have assessed the performance of various models and identified strategies to measure the accuracy and repeatability of colorectal CRC detection. The publicly available Kather colorectal cancer histology dataset includes 15,000 histopathology images-CRC samples interpreted corresponding clinical outcomes. Dataset by Kather et al. has received significant attention after its release, with researchers frequently used to build and test CNN classification of various types of colorectal cancer tissues Training involves using number of datasets, evaluating several CNN models, including AlexNet, VGGNet, and GoogLeNet. These models reached an overall accuracy of 79% [3].

An attention based CNN model has been trained using the CRC dataset to grade colorectal cancer histology images accomplishing accuracy of 88% [18]. A model has been developed that utilizes the Kather colorectal cancer histological dataset diagnose eight tissue types commonly observed in CRC evaluation automatically. The model incorporates knowledge transfer using CNN architectures. The structure of the CNN has been improved to extract features from images, which are then served into machine learning techniques as k-Nearest Neighbours, Multilayer Perceptron's, Random Forest, Naive Bayes, and Support Vector Machines (SVM). The best performance was accomplished using DenseNet169 with SVM (RBF), yielding accuracy of 92.08% [13].

The paper gives a review on deep learning based medical image fusion from perspective of its ability in feature level fusion of multimodal medical images for better diagnosis. It divides deep learning methods, including CNNs and GANs, outlining the benefits and disadvantages of every approach. Assessment estimates the performance of these methods in workability in terms of image quality, computational capability, and clinical application. The authors achieve by discussing future directions of the problem for enhancing medical image fusion employing deep learning approaches [9].

The paper suggests bio-inspired scene classification method using tailored towards remote sensing. This method integrates a deep learning framework with some forms of active learning to improve both efficiency and accuracy of scene classification. The method, inspired by biological learning processes, selectively reactivates content that is most useful for improving the existing model. Huge experiments prove that this bio-inspired technique improves the prior arts, in both aspects of classification accuracy and the quantity of data required. The study underscores that it is possible to integrate bio inspired mechanisms with deep learning in developing remote sensing applications [20].

The MobileNetV2 model was found to provide cancer good cell accuracy and gave identifying a better type accuracy than colon previous studies that used smaller datasets. Particularly, the provided model accuracy of 99.67% when tested on a dataset of 12,500 images of colon cancer cells. Thus, it has been presented that the efficiency and reliability of the model are due to the large dataset used in the study, thus making it a valuable contribution to the field of colon cancer diagnosis, previous research efforts that used smaller datasets had lower accuracies of 95.71% to 97.49%. This therefore confirms that there is an advancement in the efficiency of the MobileNetV2 model because a larger and more comprehensive dataset was used in training the model for medical image classification [10, 17].

The key algorithm discussed in the paper is the SAdagrad optimizer, which modifies the original Adagrad algorithm to address its limitations. Modified Adagrad optimizer (SAdagrad) for training a Convolutional Neural Network (CNN) using a pre-trained MobileNet model with fine-tuning.

Weight Update Equation:

$$w_t = w_{t-1} - \eta_t' \frac{\partial L}{\partial w_{t-1}}$$

Adjusted Learning Rate Equation:

$$\eta_t' = \frac{\eta}{\sqrt{\alpha_t} + \epsilon}$$

The research findings demonstrated that the SAdagrad optimizer reached 98% accuracy in identifying histopathological images thus showing it improved model performance beyond traditional optimizers. Also discussed the advantages of adaptive learning rate optimizers like Adagrad and Adam over traditional SGD, and how their modified SAdagrad addresses some limitations of Adagrad. They analyse the confusion matrices and ROC curves to understand the model's performance and areas for improvement [5].

Further research also indicates that trained ANNs may serve as effective screening tools for prevention and detection of CRC. They are particularly helpful for those not captured by the traditional screening guidelines based on age and family history alone. It indicates the potential for better early diagnosis in the context of ANNs since they are known to process big data sets and identify patterns that might otherwise not be visible using conventional methods. Therefore, integrating deep learning into clinical practice would allow more robust evaluation of responses to treatments so that it becomes more tailored, hence probably more effective, for treating metastatic colorectal cancer. It makes the effective findings even stronger by well-managing variability in imaging acquizition through a large, diverse dataset from the Vol-PACT consortium [10, 12].

The study highlights the fact that XGBoost and RF are very accurate in prediction of early-onset colorectal cancer at the hand of electronic health record data, outperforming linear models and boasting high AUC scores for the identification of important predictors such as immune disorders, blood diseases, and secondary cancers. Their performance is also not lost when the cancer-related variables are excluded [7, 8, 12].

Recent research studies in the area of CRC underscore the important role of biomarkers in various omics fields in the detection, diagnosis of the disease. Early detection, however, still challenge because symptoms may be so minimal, or even similar to other conditions. Liquid biopsies are considered emerging technologies for early CRC detection and monitoring because they are less invasive. The technologies of high-throughput technologies and bioinformatics also have potential applications in the identification of novel biomarkers as well as a better understanding of CRC progression [5, 7, 20].

Similar applications in medical imaging also hold considerable promise with deep learning. Convolutional neural networks (CNNs implemented for analysis purposes such as detection and segmentation in images including CT, MRI, and PET scans, to predict treatment responses, among other things. Research indicates the possibility of deep learning models as to recognizing patterns of a super high degree of complexity from imaging data not easily picked out by human observers; this means possibly earlier, as well as much more precise predictions of treatment outcomes [3, 11, 13].

This diagram describes a workflow to analyse histological colon images with deep learning models (e.g., CNNs, MobileNet, ResNet50). A benign and an adenocarcinoma tissue dataset is used for training and testing, evaluating metrics following the accuracy, precision, recall, and AUC metrics. Its visual interpretation is carried out using Score-CAM and Grad-CAM techniques (Figure 40.3).

Table 40.1 presents a comparative analysis of machine learning (ML) and deep learning (DL) techniques with their performance metrics. It provides a comparative analysis of how ML and DL techniques and their results, showcasing the outcome and limitations of each method.

However, despite all these developments, there is still a gap in the integration of machine and deep learning techniques both risk assessment and prediction of treatment response in CRC. Most studies available are based on either early detection or treatment response alone and with very limited attempts to integrate these approaches. It will use different types of data and advanced analysis techniques to build an extensive machine learning model which can work as a risk prediction model and a deep learning model work

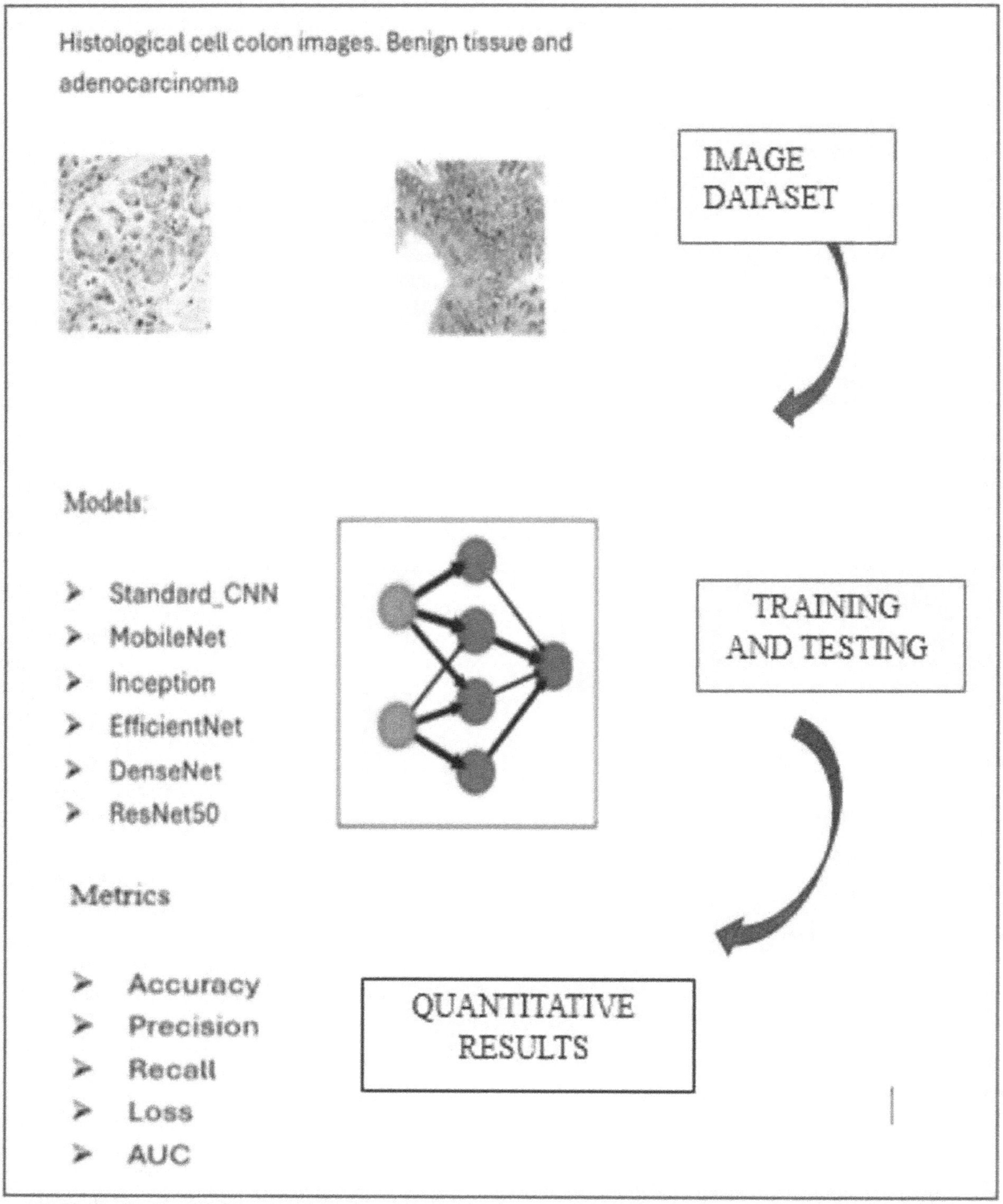

Figure 40.3. Deep learning analysis of histological colon images with visual interpretation.
Source: Author.

as a treatment response prediction model. Bridging this gap and enhancing early detection as well as the optimization of patient management consequently improves clinical outcomes in CRC. This study will therefore combine and apply the techniques above to help overcome the unique challenges associated with early detection and prediction of

Table 40.1. Comparative analysis of machine learning vs. deep learning algorithms

Technique	*Model*	*Result*	*Limitation*
Machine Learning	Logistic Regression	Accuracy:75%, Sensitivity:70%, Specificity: 78%	Provides accurate results for linear data but fails to identify intricate medical picture patterns. It works well to check initial data.
Machine Learning	Random Forest	Accuracy: 82%, AUC-ROC: 0.85, F1-Score: 0.81	Effective for feature selection and resisting random variations, performance decreases when handling high-dimensional imaging data.
Machine Learning	SupportVector Machine (SVM)	Accuracy:80%, Precision: 83%, Recall: 78%	Performs well when working with simple medical image datasets that have clear partitions but faces challenges when processing large imaging databases.
Machine Learning	K-Nearest Neighbours (KNN)	Accuracy:76%, Sensitivity:72%, Specificity: 79%	Simple and interpretable which breaks down under image noise. Processing medical images takes more effort as the dataset grows larger.
Deep Learning	DC-GAN, VAE-GAN, CNN	Not Mentioned	Lacks specific measurement of accuracy results and works with limited dataset of training data.
Deep Learning	Faster R-CNN (ResNet-101 backbone), Gradient Boosting Classifier	AUC = 0.917, Sensitivity = 97.4%	Small dataset shows good detection ability but produces many false results (60.3%) when used in medical care.
Deep Learning	CNN, Adam, Adagrad, SAdagrad	SAdagrad:98% accuracy	Dataset details are limited, lack of information on other performance metrics beyond accuracy numbers.

Source: Author.

the treatment of CRC, which ultimately contributes to better and more effective and individualized patient care.

4. Future Directions

Future advancement in deep learning for colorectal cancer prediction involves the enhancement of individualized treatment plans based on the individual patients' data to increase the effectiveness of the treatment. GANs can generate synthetic data that could be used to overcome limited or imbalanced datasets. Furthermore, personalized models of prediction, integrating genetic, environmental, and clinical factors, will provide higher accuracy in risk predictions, which would enable targeted interventions and improved patient outcomes. Collectively, these strategies are aimed at improving the predictive accuracy of a wide range of contexts, protecting sensitive data from being compromised, and actively promoting inclusivity within clinical applications.

5. Conclusion

Early prediction of high-risk CRC patients may be promising using deep learning. Advanced DL techniques can be integrated with medical imaging and clinical data to enhance accuracy and reduce mortality rates. Transformative advances in CRC care are on the horizon as existing challenges are overcome through innovative methodologies and collaboration.

References

[1] Alboaneen, D., Alqarni, R., & Alqahtani, S. (2023). Predicting colorectal cancer using machine and deep learning algorithms: Challenges and opportunities. *Big Data and Cognitive Computing*, *7*(2), 74. https://doi.org/10.3390/bdcc7020074

[2] Arnold, M., Sierra, M. S., Laversanne, M., Soerjomataram, I., Jemal, A., & Bray, F. (2016). Global patterns and trends in colorectal cancer incidence and mortality. *Gut*, *66*(4), 683–691. https://doi.org/10.1136/gutjnl-2015-310912

[3] Bokhorst, J., Nagtegaal, I. D., & Fraggetta, F. (2023). Deep learning for multi-class semantic segmentation enables colorectal cancer detection and classification in digital pathology images. *Scientific Reports*, *13*(1). https://doi.org/10.1038/s41598-023-35491-z

[4] Dekker, E., Tanis, P. J., Vleugels, J. L. A., Kasi, P. M., & Wallace, M. B. (2019), Colorectal cancer. *Lancet*, *394*, 1467–1480.

[5] Elshamy, R., Abu-Elnasr, O., Elhoseny, M., & Elmougy, S. (2024). Enhancing colorectal cancer histology diagnosis using modified deep neural networks optimizer. *Scientific Reports*, *14*(1). https://doi.org/10.1038/s41598-024-69193-x

[6] Ho, C., Zhao, Z., & Chen, X. F. (2022). A promising deep learning-assistive algorithm for histopathological screening of colorectal cancer. *Scientific Reports*, *12*(1). https://doi.org/10.1038/s41598-022-06264-x

[7] Huang, Z., & Yang, M. (2022). Molecular network of colorectal cancer and current therapeutic options. *Frontiers in Oncology*, *12*. https://doi.org/10.3389/fonc.2022.852927

[8] Kamel, F., Eltarhoni, K., Nisar, P., & Soloviev, M. (2022). Colorectal cancer diagnosis: The obstacles we face in determining a non-invasive test and current advances in biomarker detection. *Cancers*, *14*(8), 1889. https://doi.org/10.3390/cancers14081889

[9] Kennion, O., Maitland, S., & Brady, R. (2022). Machine learning as a new horizon for colorectal cancer risk prediction? A systematic review. *Health Sciences Review*, *4*, 100041. https://doi.org/10.1016/j.hsr.2022.100041

[10] Li, W., Lin, S., He, Y., Wang, J., & Pan, Y. (2023). Deep learning survival model for colorectal cancer patients (DeepCRC) with Asian clinical data compared with different theories. *Archives of Medical Science*, *19*(1), 264–269. https://doi.org/10.5114/aoms/156477

[11] Lu, L., Dercle, L., Zhao, B., & Schwartz, L. H. (2021). Deep learning for the prediction of early on-treatment response in metastatic colorectal cancer from serial medical imaging. *Nature Communications*, *12*(1). https://doi.org/10.1038/s41467-021-26990-6

[12] Nartowt, B. J., Hart, G. R., Muhammad, W., Liang, Y., Stark, G. F., & Deng, J. (2020). Robust machine learning for colorectal cancer risk prediction and stratification. *Frontiers in Big Data*, *3*. https://doi.org/10.3389/fdata.2020.00006

[13] Pei, Y., Mu, L., Fu, Y., et al. (2024). Colorectal tumor segmentation of CT scans based on a convolutional neural network with an attention mechanism.

[14] Sandfort, V., Yan, K., Pickhardt, P. J., & Summers, R. M. (2019). Data augmentation using generative adversarial networks (CycleGAN) to improve generalizability in CT segmentation tasks. *Scientific Reports*, *9*(1). https://doi.org/10.1038/s41598-019-52737-x

[15] Shi, L., Li, X., Hu, W., et al. (2023). EBHI-seg: A novel enteroscopy biopsy histopathological hematoxylin and eosin image dataset for image segmentation tasks. *Frontiers in Medicine*, *10*. https://doi.org/10.3389/fmed.2023.1114673

[16] Sujatha, R., K, M., & Yoosuf, M. S. (2024). Colorectal cancer prediction via histopathology segmentation using DC-GAN and VAE-GAN. *EAI Endorsed Transactions on Pervasive Health and Technology*, *10*. https://.org/10.4108/eetpht.10.5395

[17] Tasnim, Z., Chakraborty, S., Shamrat, F. M., et al. (2021). Deep learning predictive model for colon cancer patient using CNN-based classification. *International Journal of Advanced Computer Science and Applications*, *12*(8). https://doi.org/10.14569/ijacsa.2021.0120880

[18] Yan, Z., He, M., Shi, H., et al. (2022). The impact of KPNA2 on prognosis of colorectal cancer patients. *A meta-analysis*. https://doi.org/10.21203/rs.3.rs-1235315/v1

[19] Yang, H., Liu, J., Yang, N., et al. (2024). Enhancing metastatic colorectal cancer prediction through advanced feature selection and machine learning techniques. *International Immunopharmacology*, *142*, 113033. https://doi.org/10.1016/j.intimp.2024.113033

[20] Zhang, Y., Wang, Y., Zhang, B., Li, P., & Zhao, Y. (2023). Methods and biomarkers for early detection, prediction, and diagnosis of colorectal cancer. *Biomedicine & Pharmacotherapy*, *163*, 114786. https://doi.org/10.1016/j.biopha.2023.114786

41 Enhancing healthcare efficiency with a multi-persona platform: Swaasth

Gayathri V. M.[a], *Shiv Tiwari*[b], *Sarthak Shroff*[c], *Anirban Sengupta*[d], *and Varad Kadtan*[e]

Department of Networking and Communications, SRM Institute of Science and Technology (SRMIST), Kattankulathur, Tamil Nadu, India

Abstract: One of the challenges healthcare systems faces is their inability to offer efficient, personalized services due to communication silos, inefficient scheduling, and lack of access for patients and/or providers. For example, patients do not find it easy to schedule their appointment, access their health records, or obtain a prescription. Similarly, the provision for doctors to reschedule appointments online and issue electronic prescriptions for the convenience of patients while working on patient records for excellence in healthcare delivery might involve minor issues. For hospital administrators, the strenuous balancing act of supervising staff, monitoring admissions, and operating analytics contributes to the administrative bottleneck, hindering overall operational efficiency. In addressing these problems, we have proposed a common healthcare application that can provide real-time services for patients, doctors, and administrators. Patients will be able to book appointments, review health records, and manage a prescription in a secure way that is easy to navigate – all on one platform. Doctors are equipped with options to manage appointments, work with their schedules, and write electronic prescriptions quickly to extend patient care. The admins will also have many features around onboarding staff, analytics, and administration capabilities – again, all in one place. With a truly scalable microservices architecture on the serverless backend, the solution will enable seamless data exchanging and secure access control. This will promote the development of truly connected, efficient, responsive healthcare environments.

Keywords: Data integrity, data privacy, digital prescriptions, security, serverless computing

1. Introduction

Swaasth is designed to address the different needs of hospital stakeholders by improving communication between patients and healthcare providers while optimizing administrative tasks. By incorporating advanced digital solutions into critical hospital functions, Swaasth aims to create a more efficient, user-centric healthcare experience. Its goal is to support each user in their role – whether as a patient seeking care, a doctor managing appointments, or an administrator handling complex operations – by providing tools that are accessible and easy to navigate. In doing so, Swaasth reduces the manual workload associated with scheduling, billing, and information management, allowing healthcare professionals to focus more on patient care rather than administrative processes.

One of Swaasth's primary objectives is to provide a centralized, user-friendly interface that simplifies hospital operations and streamlines patient information management. This includes ensuring that appointment scheduling, billing, and record-keeping processes are all easily accessible and automated, allowing users to interact with the system in a way that is efficient. By digitizing these traditionally paper-based processes, Swaasth minimizes errors and miscommunication, creating a secure and reliable environment for both patients and staff.

Additionally, Swaasth encourages stronger communication and collaboration across the healthcare ecosystem. Patients are able to connect with their healthcare providers more easily, ensuring they have access to medical records, appointments, and billing information in real-time. Doctors, in turn, have uninterrupted access to patient information and tools for scheduling and prescription management. Administrators benefit from a more organized workflow that reduces redundancies and provides insights into operational efficiency, patient trends, and financial data. Altogether, Swaasth seeks to transform hospital operations, enhancing efficiency and enabling high-quality patient care.

[a]gayathrm@srmist.edu.in, [b]shivtiwari406@gmail.com, [c]ss4328@srmist.edu.in, [d]aa2072@srmist.edu.in, [e]vp0376@srmist.edu.in,

DOI: 10.1201/9781003770473-41

2. Related Work

Some competitor apps exist in the market that are doing similar work.

mFine offers a comprehensive healthcare platform with basic features like doctor consultations and self-health checks. What sets it apart is its approach to prescription management – while users can upload prescriptions themselves, the app also allows doctors to directly upload prescriptions, adding an extra layer of accuracy and convenience. The inclusion of emergency location sharing makes it particularly valuable for crisis situations, and their admin control feature suggests better oversight of the platform's operations.

Docon takes a more specialized approach, focusing on the digital management of medical care. Their platform combines video consultations with traditional appointment booking, but their standout feature is the digital prescription management system. They've enhanced their offering with robust health records management and emergency location sharing capabilities, making it a well-rounded solution for both routine healthcare and emergency situations.

This paper proposes a solution that combines not only all the previously mentioned features but also some standout ones of its own such as emergency location sharing and an entire admin backend suite that would otherwise require a third-party CRM software to carry out.

3. Proposed Approach

This chapter outlines the novel features and core functionalities of the Swaasth app, which is designed to simplify hospital operations and improve communication between patients, doctors, and administrators.

3.1. Patient features

The Swaasth app provides patients with a user-friendly interface to manage their healthcare, offering access to real-time data and appointment management.

- **Appointment Booking Algorithm (AppBook):** Patients can browse available doctors based on specialization, view their availability, and schedule appointments. The system ensures the patient receives confirmation on the appointment.
- **Health Record Access Algorithm (HealthAccess):** Patients can view their health records, including past appointments, diagnoses, and prescriptions. This ensures patients have continuous access to their medical history.
- **Prescription Viewing Algorithm (PrescribeView):** After a consultation, the app allows patients to view and download digital prescriptions. The system ensures prescriptions are securely stored and easily accessible for future reference.
- **Bill Payment Algorithm (BillPay):** Patients can view and pay their medical bills directly through the app. The system allows effortless payment processing and sends confirmation receipts to patients upon successful payment.
- **Emergency Location Sharing Algorithm (EmergLoc):** In case of an emergency, patients can share their real-time location with healthcare providers and emergency services. The app uses the phone's GPS to send location details instantly, ensuring quick assistance.
- **Health Vitals Monitoring Algorithm (HealthVitals):** The app can sync with the patient's phone or wearable devices to track real-time health vitals, such as heart rate, blood pressure, and oxygen levels. These metrics are automatically logged into the patient's health record for ongoing monitoring.
- **Health Event Attendance Algorithm (EventAttend):** Patients can register for and attend health-related events, workshops, and awareness campaigns directly through the app. Notifications about upcoming events are sent to the patients based on their health interests

3.2. Doctor features

Doctors on the Swaasth app benefit from a suite of tools designed to simplify their workflow, enhance patient care, and optimize scheduling.

- **Appointment Management Algorithm (DocManage):** Doctors can view their daily schedule, approve or reschedule appointments, and mark them as completed. The system integrates with their calendar to show real-time availability and prevent double-booking.
- **Prescription Generation Algorithm (DocPrescribe):** Doctors can create digital prescriptions using a simple interface, specifying medications, dosages, and instructions. The prescriptions are automatically saved to the patient's health record and can be accessed at any time by both doctors and patients.
- **Availability Scheduling Algorithm (DocSchedule):** Doctors can customize their availability on a calendar, setting their working hours, holidays, and preferred times for consultations. This helps automate the scheduling process and allows patients to book appointments according to real-time availability.
- **Health Event Hosting Algorithm (EventHost):** Doctors can create and host health-related events, such as webinars, workshops, and seminars. These events can be shared with patients, and doctors can track registrations and attendance via the app.

3.3. Admin features

Admins in the Swaasth app are provided with tools for managing the operational aspects of the healthcare facility, overseeing doctors, patients, and the overall workflow.

- **Doctor Onboarding Algorithm (DocOnboard):** Admins can onboard new doctors by entering their details, assigning specializations, and integrating their profiles into the system. The system ensures that new doctors are set up for appointment scheduling and patient care immediately upon registration.
- **Hospital Analytics Algorithm (Analytics):** Admins can access a comprehensive dashboard that provides insights into hospital operations, such as patient appointments, doctor availability, finances, and resource utilization. This data can be used for optimizing hospital processes and improving patient care efficiency.
- **Health Event Management Algorithm (EventManage):** Admins can schedule, manage, and promote health events (such as workshops, awareness campaigns, and medical camps) within the platform. The system allows easy registration for events and keeps patients informed about upcoming activities.

4. Data Flow

The data flow in the Swaasth app is designed to ensure smooth communication and management of hospital operations. The data flow process is different for each user persona (Patient, Doctor, and Admin) and ensures that each user type can access and manage the necessary information efficiently while maintaining security and privacy. As seen in Figure 41.1, a Data Flow Diagram explains in a nutshell how the proposed approach is supposed to work in Swaasth.

4.1. Patient data flow

The **Patient** data flow starts with the user interacting with the app to access services and manage healthcare needs.

- **Registration & Authentication**: A patient creates a profile using either email login. The app authenticates the user's credentials, granting access to personalized services.
- **Appointment Scheduling**: When a patient books an appointment, the data is sent to the **firebase backend**, which confirms the booking by checking doctor availability. This confirms the appointment and updates the doctor's calendar.
- **Health Records Management**: Patient health data (e.g., previous appointments, prescriptions, vitals) is securely stored in the **firebase firestore**. The system continuously syncs data with the patient's profile to ensure up-to-date access for future consultations.
- **Bill Payment:** After a consultation, billing details are generated. The patient can view the invoice. Payment confirmation and the receipt are stored in the patient's records for future reference.
- **Health Vitals Integration:** The app syncs health data from wearable devices (heart rate, blood oxygen, etc.). This data is stored in real-time and monitored by both the patient and their doctor for insights and further action.

Figure 41.1. Context level DFD for a hospital management system.

Source: Author.

- **Emergency Location Sharing:** If a patient shares their location during an emergency, the data is sent directly to the nearest hospital or healthcare provider for immediate action.
- **Health Event Attendance:** When a patient registers for a health event (workshops, webinars), the data is collected, and an event calendar entry is created.

4.2. Doctor data flow

The **Doctor** data flow ensures that medical professionals can efficiently manage their practice, communicate with patients, and provide healthcare services.

- **Registration & Profile Management**: A doctor's profile is created and stored in the system with their specialty and other details. This profile is available for patient searches and appointment bookings.
- **Appointment Management**: When a patient books an appointment, the appointment confirmation is added to the doctor's calendar.
- **Prescription Generation**: After consultation, the doctor generates a digital prescription, which is sent directly to the patient's account.

4.3. Admin data flow

The **admin** data flow is responsible for overseeing and managing all aspects of the Swaasth app, ensuring smooth operations, compliance, and coordination between doctors and patients.

- **User Management**: Admins manages doctor and staff profiles. New doctors are onboarded by entering their details into the system.
- **Health Event Management**: Admins oversee the organization of health events. They can create, modify, and approve events and ensure all event data flows through to the patient-side doctor dashboards.
- **Analytics & Reporting**: Admins have access to reports and analytics on key metrics such as the number of appointments, user activity, financial data, and health event participation. These reports help in making crucial decisions and optimize performance.

Figure 41.2 shows an ER (Entity-Relationship) diagram representing a hospital management system with several interconnected entities. The central entities include **Employee**, **Patient**, **Doctor**, **Nurse**, and **Receptionist**, each with unique attributes and relationships. The **Employee** entity is generalized into different roles like doctor, nurse, and receptionist based on the emp_type attribute. The **Patient** entity connects to **doctors** through consultations, with doctors deciding the patient's type, either **In_Patient** or **Out_Patient**. In-patients have additional attributes, including admission and discharge dates, along with room information managed by the **Room** entity. **Medical Records** are managed by doctors and contain patient history, treatments, and disease information. **Lab Reports** provide diagnostic details for patients, with parameters like weight, disease category, and remarks. Receptionists handle **Appointments** for patients, assigning doctors and maintaining appointment dates and notes. Bills generated for treatments link to patients and track costs. This ER diagram captures the interactions and roles of various hospital staff and patients in a structured, relational framework, ensuring efficient patient care and management.

5. Performance Evaluation

As mentioned above, this section explains the computational and communication overhead and security analysis of the proposed scheme with detailed analysis.

5.1. Security analysis

This scheme satisfies several important security requirements as proposed:

1. **Data Encryption**: Ensures patient and doctor data confidentiality using encryption for all sensitive information stored in Firestore.
2. **Authentication**: Uses Firebase authentication to securely manage user access, ensuring only authorized access to the app.
3. **Audit Logging:** Tracks critical activities for potential security audits, enhancing accountability and traceability across user actions.

5.2. Performance comparison

The app's performance is evaluated based on speed, scalability, and user experience across the patient, doctor, and admin modules. Firebase's real-time database ensures quick data syncing, supporting appointment management, prescription updates, and health record access. The microservices architecture enhances scalability, allowing each module to handle increased user loads independently. Efficient data queries, and a streamlined user interface together contribute to a smooth and responsive experience. Initial tests show low latency in data exchange and high availability, ensuring reliable access for all user personas, while Firebase's serverless infrastructure optimizes load management.

6. Conclusion and Future Work

Swaasth is a transformative solution designed to bridge gaps in hospital operations, patient care, and administrative

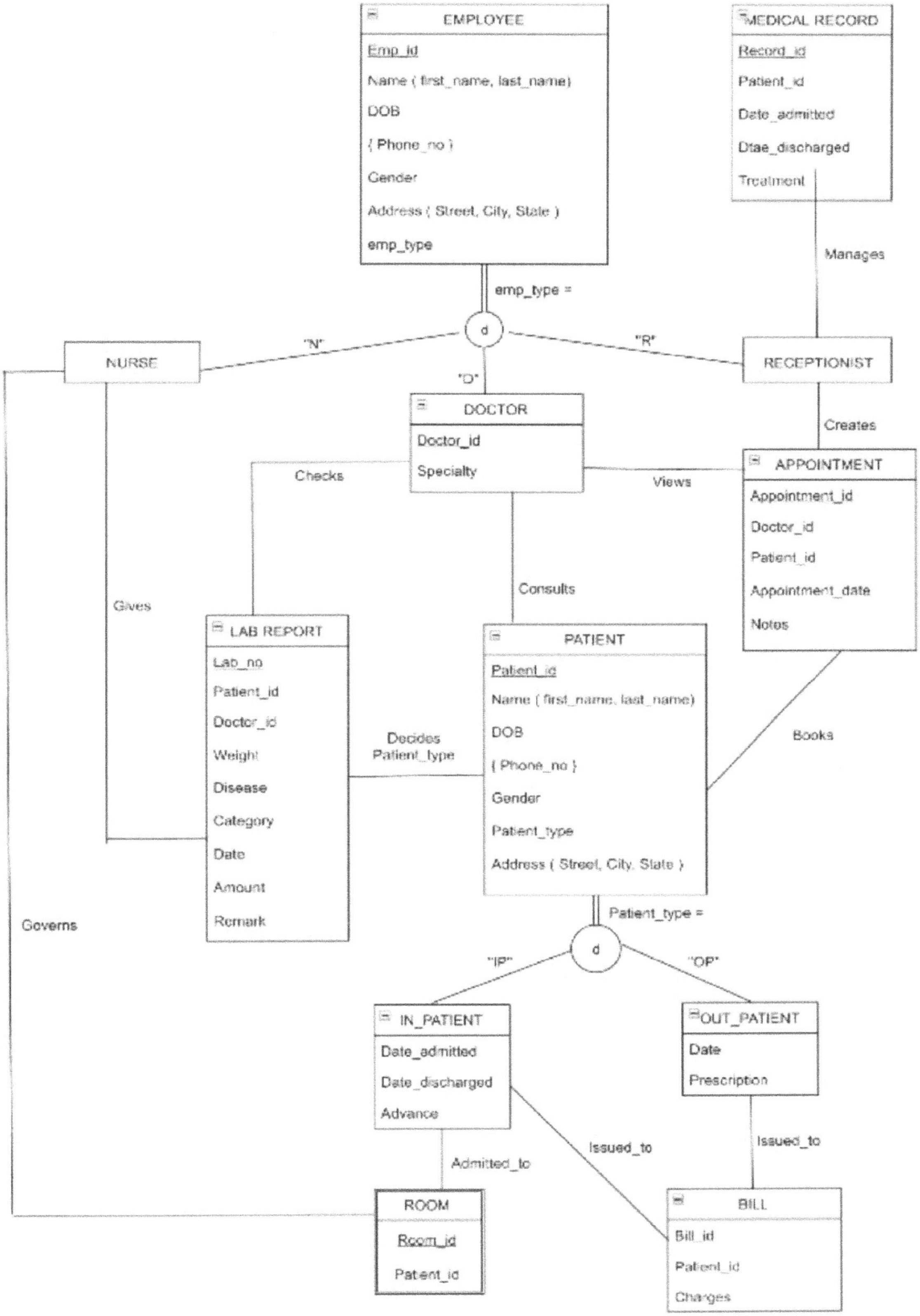

Figure 41.2. ERD of the user's journey across the app, from communication with doctors as a patient, managing appointments, room allocation, handling bills as well as employee management.

Source: Author.

workflows through a comprehensive, user-centered application. By automating critical processes like appointment scheduling, records management, and billing, the platform reduces administrative burdens, allowing healthcare professionals to focus on delivering high-quality patient care.

The integration of real-time data synchronization, scalability, and advanced analytics ensures that the platform can evolve with the institution's needs, supporting an increasing user base without compromising performance or data accuracy. Ultimately, the Healthcare Operational Platform enables a connected, efficient, and patient-centred healthcare environment that empowers all stakeholders – patients, doctors, and administrators. The app has the potential to redefine healthcare delivery and support a sustainable, well-coordinated system focused on health and well-being for all.

The **planned AI integration** in **Swaasth** will significantly enhance the user experience and operational efficiency. Here's how these AI-driven features will impact patients, mental health assessment, and administrative operations:

6.1. Symptom-based diagnostic and treatment recommendations

Patients will be able to input their symptoms into the platform, and AI algorithms will analyse this information to provide: **Disease Identification**: Suggestions for potential conditions based on entered symptoms; **Department Recommendation**: Guidance on which medical department or specialist to consult; **Medication and Home Remedies**: Information on over-the-counter medications and safe home remedies tailored to their symptoms.

6.2. Mental health assessment

The platform will offer mental health assessments, where patients can answer guided questions or interact with a virtual assistant for: **Self-Assessment Tool**: The mental health assessment feature will enable patients to self-assess through questionnaires and chat interfaces, designed to identify stress levels, anxiety, depression, and other common mental health indicators; **Guidance on Therapy or Counselling**: Based on responses, the app will provide suggestions for seeking counselling or therapy and recommend coping techniques or wellness practices.

6.3. AI-driven business solutions

For administrators, AI will provide data-driven insights and solutions to enhance business performance and operational metrics, including**: User Growth Strategies**: Targeted recommendations to increase platform engagement, new user acquisition, and patient retention; **Financial Optimization**: Insights to improve revenue streams through resource allocation, pricing strategies, and cost management; **Emergency Services Optimization**: Analysis to improve emergency response times, patient flow in critical areas, and efficient allocation of emergency resources. These AI solutions will support hospital management in making informed decisions that align with strategic goals and foster sustainable growth.

References

[1] Christian Keur, Aaron Hillegass (2020). iOS Programming – *The Big Nerd Ranch Guide.*

[2] Jon Reid (2020), *iOS Unit Testing by Example.*

[3] Keith Moon, Chris Barker (2021). *Swift Cookbook.*

[4] Greg Lim (2023). *Beginning iOS 16 & Swift App Development.*

[5] Frahaan Hussain, Kameron Hussain (2024). *Mastering Firebase.*

[6] Houssem Yahiaoui (2017). *Firebase Cookbook.*

42 Assessing the impact of gaming on psychological and physical health: Users' perspective

Deepika Gaur[1,a], *Akanksha Khanna*[1,b], *Sandhya Ravishankar*[2,c], *Komathi C.*[3,d], *and Swapna S. Nair*[3,e]

[1]Department of Management, Symbiosis International University, Dubai, Symbiosis International (Deemed University), Pune, India
[2]Department of General Education, Symbiosis International University, Dubai, Symbiosis International (Deemed University), Pune, India
[3]Department of Computer Studies, Symbiosis International University, Dubai, Symbiosis International (Deemed University), Pune, India

Abstract: Digital gaming has revolutionized the entertainment industry and gained immense popularity in recent years. This growing influence of digital games demands the need to explore their impact on various walks of life. The present study looks at the changes happening on the physical and mental health and the social relationships of people due to their digital gaming habit. A selected group of 152 participants from the UAE region was chosen for the study, and a survey questionnaire was conducted. The research methodology was a combination of descriptive and causal research strategy. This is a quantitative analysis study employing K-means clustering analysis. The findings of the study throw light into the fact that individuals who frequently play digital games often experience physical uneasiness, such as strained-eyes and headache, due to prolonged duration of gaming. The study also presents an overview of the key points and recommendations and highlights the importance of being aware of the consequences of excessive engagement in digital game activities. An overview of the major topics and recommendations are provided, emphasising the exigency of understanding the consequences of frequent engagement in digital gaming activities.

Keywords: Digital gaming, educational well-being, psychological practices, sustainable well-being

1. Introduction

The study's primary purpose is to examine the impacts of digital gaming on the lives of people at various levels such as physical health, mental wellbeing and social relationships. The contemporary time witnesses a growing concern about the hazardous effects of gaming on people's lives, relationships, and health. This happens due to the increasing frequency of people's usage of games [1]. In the world of internet, digital games have gained its popularity as a kind of entertainment entity. Over the last two decades, digital gaming has evolved from a minimal group of people's casual pastime to an indispensable influence in the entertainment sector. The number of people using digital games is heterogeneous and appears to be increasing [2]. Hence, this study assesses the effects of digital gaming on physical and mental health of people through the lens of individuals as users. Privacy loss is considered as the potential adverse impact on physical and mental health [3]. Some of the most popular games namely, Clash of Clans, Candy Crush, and Player Unknown's Battlegrounds (PUBG) have the addictive power to retain players for longer duration. Every individual who partakes in the active involvement of digital games carries a unique motive for their participation, and each one encounters with the game yields diverse repercussions as well. The starting point of the compulsive engagement in gaming activities can be associated with the fast progression and easy availability of technology. The engaging power of gaming activities is classified as an interactive compulsion and has been intertwined with various psychosomatic and physical ailments, including distress, social uneasiness, weariness, and isolation. The relentless engagement in digital gaming also causes bodily discomfort, such as joint, muscular, and dermal complications, as well as epileptic seizures [4, 5]. Students

[a]dgaur@siu-dubai.ac.ae, [b]akhanna@siu-dubai.ac.ae, [c]sravisankar@siu-dubai.ac.ae, [d]kjeyakaran@siu-duba.ac.ae, [e]snair@siu-dubai.ac.ae

DOI: 10.1201/9781003770473-42

who engrossed themselves in playing digital games for non-academic purposes are observed deteriorating in their schoolwork performance compared to those who utilized technology for learning purposes.

As studied earlier, the violent games exposure leads to slower recognition of emotions and this leads to low sensitivity towards social cues. Insensitive ways of playing online games causes increased emotional arousal, decreased controlled behaviour, and a lack of focus attention deficit disorder, commercialism, mindless entertainment, social isolation factor, and addiction [6]. Some of the most commonly reported health issues are weight, muscular-skeletal, vision, and epileptic attacks. The studies found that the gaming habit are associated with general anxiety disorder, social pressure, specific phobias and selective mutism [7]. On the whole, digital games have positive and negative impacts. It may be helpful in various ways, those positive attributes comprise mental training and exercise, more brain activity, friendlier behaviour, enhanced interpretability, and stronger recursive and proactive thinking [8].

In the present era of digitalization, digital game usage becomes ubiquitous, and so do the consequences on people's physical, and mental health and social relationships [9]. Hence this issue is extremely significant and needs social attention. The paramount purpose of this investigation is to enhance the understanding of the undeniable presence of digital games, which are progressively gaining popularity on a global canvas. Furthermore, this inquiry will promote awareness about the detrimental health ramifications associated with competitive online gaming, specifically among young adults and adolescents who frequently engage in these gaming activities. Prior study had already demonstrated that internet gaming can lead to a destructive addiction, which in turn affects the sleeping patterns of avid players [10]. The link between digital gaming and mental health has garnered more attention in the past ten years [11]. The games that people play, how they learn, how they harm people's physical and mental health, and how these problems are brought to people's attention are the subjects of study now [12]. Thus, the purpose of this study is meticulously woven in order to ensure that the below-mentioned objectives can be effectively and successfully accomplished through the extensive investigation conducted.

- To assess the physical health effects of digital gaming, such as posture, vision, and activity levels, using consumers' gaming habits and behaviours.
- To investigate the mental health impacts of digital gaming, both positive and negative via the eyes of gamers.
- To investigate the impact of digital gaming on social relationships, particularly its role in changing social interactions, communication, and relationship dynamics among gamers.
- To better understand customers' perceptions and techniques for balancing gaming with their physical, mental, and social well-being, as well as to identify any risk awareness.

2. Literature Review

I the present time, digital games are gaining rampant popularity and an indelible presence in the lives of people and are permeating every nook and corner of our society. It has secured a privileged position in contemporary society due to its extensive use and numerous benefits. The digital game obsession is characterized by the humongous and impulsive utilization of digital games, which leads to the development of social and/or emotional issues, as well as the inability to regulate this extreme use despite the encountered problem [13]. Previous studies' findings on digital games showcase advantages as well as disadvantages. A digital gaming console is an electrical framework, considered a "machinery state." that responds to the player's activities. The relentless playing of video games can cause physical discomfort, including issues with the tendons, ligaments, and skin, as well as epileptic seizures. Furthermore, observed that individuals who play video games that feature violence develop aggressive thoughts, feelings, and actions. The genesis of gaming addiction is linked with the development and accessibility of technology, which turned out to be a major factor in the emergence of gaming addiction [14].

The proliferation of the digital gaming industry is owed to easy access to the Internet [15]. Studies on youth's digital game addiction, social anxiety [16], and customers' perspective views have been published in the literature. This research aims to investigate the distinctive characteristic features of gamers's sleep patterns, and physical and mental well-being.

Previous research proposed by [17] examines the gaming experiences of young people and proposes a novel theory called the DGR (digital gaming relationship) to explain people's digital gaming habits. Problematic Online Gaming Questionnaire (POGQ) is developed to distinguish between various digital games and players [18] which identifies six problematic components: withdrawal, social isolation, overuse, obsession, and interpersonal problems.

The theories of self-determination (SDT) [19] and cognitive evaluation (CET) are also read in this context and are applied to study players' motives. SDT propounds that players are motivated by digital gaming environments that fulfil their psychological demands. The addictive behaviour is primarily triggered by the fulfilling aspect of gaming and leads to indulge more in online gaming and social isolation. The theory of dopamine in gaming corroborates this observation [20].

The intertwined relationship between digital gaming and mental health has gained more significance in the past decade. The World Health Organization (WHO) decision to include Gaming Disorder (GD) in their new diagnostic manual increased the popularity of the topic. The digital gaming addiction among youth may cause them to experience numerous issues with their social, mental, and physical well-being [21]. This condition invites the attention of psychologists and mental health experts to look into the impacts of playing digital games. According to [22], depression is one of the most prevalent comorbidities among gamers. According to the research, 526 study participants are suffering from depression and gaming-related anomalies. Depression can be Episodic, Repeating, Bipolar, Moderate, and severe. Additionally, it was shown that the majority of these gamers are experiencing the worst depression scenario frequently. Furthermore, the off-shoot of digital games on mental health are anxiety, changes in appetite, mood swings, trouble sleeping, mind fog, feelings of worthlessness, atrophy of physical health, tiredness, lack of water intake, obesity and cardiac issues, prejudice mindset, absence of self-drive, depression, thoughts of suicide, anxiety in society [23]. According to research conducted by Alshehri and Mohamed (2019) in Saudi Arabia, playing video games had a detrimental effect on men's social interactions, physical activity, and overall health, with physical activity showing the most negative link [24]. Gaming is also a coping mechanism in some internet users when they encounter real-world issues, later, it can result in an internet gaming disorder marked by withdrawal symptoms such as conflict, tolerance, quiet, mood swings, and relapse. Other typical signs of gaming addiction include fatigue, boredom, chronic phobia, and terrible self-esteem are also identified [25].

The previous study examined the ways of video gaming impacts people's social time. A considerable percentage of 13.3% of students highly agreed, and 58.1% of students agreed that playing video games online can be productively spent. On the other hand, 2.9% strongly disagreed, 7.6% disagreed, and 18.1% responded indifferently. According to the findings, many students are aware of the possible dangers that playing online games may have to social interactions and personal relationships [26].

Another study by Hazar (2019) found that playing games leads to physical and mental harm as well as aggression, rage, and social isolation [21]. The conflictual evidence about the relationship between frequent gaming and physical health problems in massively multiplayer online games is also discussed [27]. Though there were numerous studies on the affiliation between gaming and the mental, physical and social health of people, none of the research focused on how to enlighten the frequent players of the gaming industry about the hazardous effects of gaming and how to improve one's general well-being [28]. Furthermore, a detailed examination of past research revealed that the lion-share of gaming-related studies are carried out not within the United Arab Emirates, indicating a shortage of readings on the impacts of gaming on players in the country. In order to close this gap, this research work aims to assess the effects of digital gaming on social interactions, physical health, and mental health from the viewpoint of users based in the United Arab Emirates.

3. Research Methodology

A framework that facilitates data gathering, analysis, and comprehension is shown in Figure 42.1. The proposed

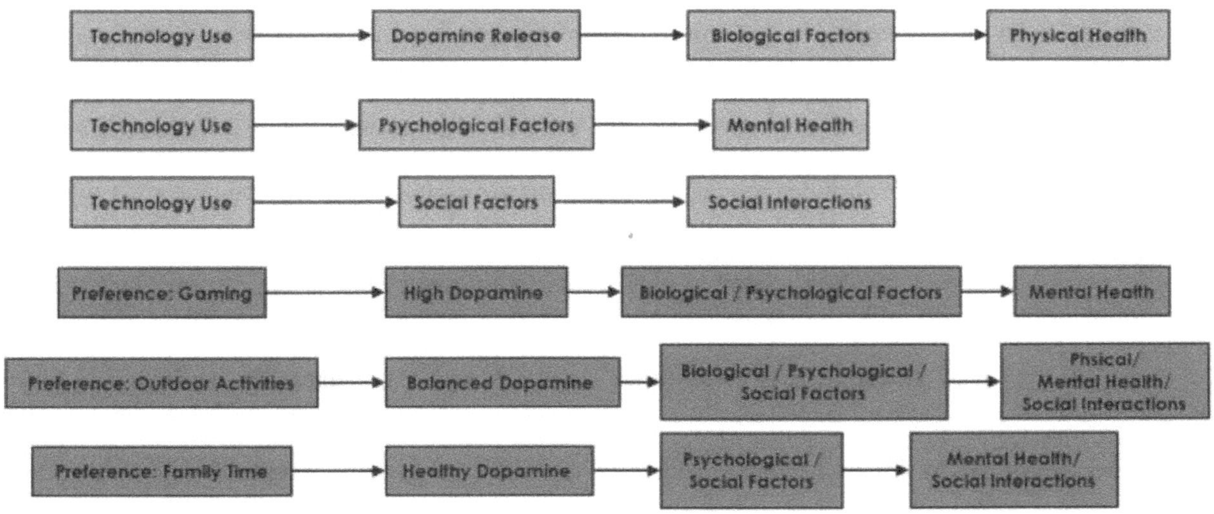

Figure 42.1. Research framework.

Source: Author.

framework depicts the relation between Technology usage, Dopamine release that is communicated through dopamine addiction theory which is a biological factor has an effect on physical health. Also it suggests that Psychological factor that is mental health is and social factor that is social interaction is getting affected due to the variations in technology usage. Another concept deals with preference of online gaming, outdoor activities or family time that in a similar manner will have an effect on mental health, physical health and social interactions. A descriptive and causal research approach was used in this study and it examines the effects of gaming on social connections, physical health, and mental health. According to Aggarwal (2019), a descriptive study is one that tries to describe the distribution of one or more variables without considering any other hypotheses or causal connections [29]. This explanatory causal research aims to determine the type and number of cause-and-effect interactions. Studies on causation can be used to evaluate the impact of specific modifications on current norms and procedures [30]. This study outlines the issue of digital gaming and its effects, including how gaming affects social interactions, physical and mental health, and connections with others. The research philosophy used in this study is positivism. According to Mezmir (2020) and Alam (2021), positivism stresses the value of generic information and focuses on objective facts and data free from human prejudice [31, 32]. A Google form survey is used for this study's questionnaire, which will apply certain methodologies and allow for the application of answers to questions on the effects of digital gaming. In this study, questionnaire surveys were taken through Google Forms. This study uses convenience sampling, with a sample size of 152 individuals. The design of observational research includes cross-sectional studies. While conducting population-based surveys, cross-sectional designs are utilized. This study uses cross-sectional study because it analyzes the data regarding the impacts of gaming on mental health, physical health and social relationships in a population [33]. For this research, K means Clustering is used to investigate the connections between social relationships, playing digital games, and mental and physical health. A well-known unsupervised machine learning technique for dividing a dataset into K separate clusters according to feature similarity is K-Means clustering. Data points are assigned to clusters iteratively by the algorithm, which then updates the cluster centres until convergence. It has the following parameters.

Centroids: The mean of all the points in a cluster is used to denote the cluster's center.

Euclidean Distance: To gauge how similar data points are to one another, K-Means usually use the Euclidean distance.

Iterative Process: Until the assignments remain constant or a stopping condition is satisfied, the algorithm repeatedly assigns points to clusters and recalculates centroids.

The procedure for conducting K-Means clustering involves:

- Select K initial centroids at random (or using more advanced techniques like K-Means++).
- Using a distance metric (often the Euclidean distance), each data point is allocated to the closest centroid.
- The mean of all the data points allocated to each cluster is used to update the centroids.
- Until the centroids stop changing or a predetermined number of repetitions is reached, steps two and three are repeated.

In research, ethical concerns protect participants' rights, privacy, and dignity while maintaining the integrity of the scientific method. Through careful management of participant data, pre-analysis, and removal of personally identifiable information, the research team guaranteed participant anonymity. Measures were taken to avoid unauthorized access, injury, or loss, and data will be maintained securely, with access restricted to the study team only.

4. Data Analysis and Findings

For the aim of carrying out this investigation, which aimed to assess the effects of digital gaming on various aspects of an individual's well-being, such as mental health, physical health, and social relationships, a questionnaire survey was administered to a diverse range of respondents residing in the United Arab Emirates. The intention behind distributing the questionnaire survey was to collect substantial and comprehensive data from individuals actively playing digital games across all country regions.

The findings demonstrated that a staggering 88.5% of respondents are involved in playing digital games. The response to this question was an essential element for this study to know how many individuals are indulged in playing digital games. Furthermore, it was observed that the majority of these individuals fall within the age range of 18–25, accounting for 42.3% of the total population surveyed. An examination of age groups participating in gaming revealed a significant prevalence of people engrossed in digital games. This aligns with previous research, which has shown that digital gaming has transcended its status as a niche pastime and has emerged as a prominent force in the realm of entertainment. Recent global studies have further substantiated this notion, indicating a substantial surge in the number of individuals engaging with digital games. Consequently, a considerable segment of the population presently partakes in this form of entertainment.

The responses provided by the survey respondents is noteworthy that a significant proportion, specifically 50%, of the participants concur with the notion that the pathological nature of digital games serves as the primary catalyst for the manifestation of depression, stress, and the overall compromise of an individual's state of well-being. Furthermore, it has been calculated from the responses that the act of engaging in these digital games has resulted in the development of weakened eyesight and headaches, both of which can be attributed to the disruption experienced within the realms of mental and physical health, as well as social interactions, thus adversely affecting the overall quality of life experienced by these individuals.

5. Statistical Analysis

The research aims to observe the imprint of digital gaming on mental well-being, physical wellness, and social interactions. A questionnaire survey was conducted to understand these impacts, focusing on individuals aged 18 to 25. The survey consists of carefully crafted questions that explore various aspects of a person's life, such as psychological health, mental health, emotional steadiness, physical fitness, stress levels, depression tendencies, aggression proneness, obesity susceptibility, and self-isolation.

The questionnaire's structure and format were designed to facilitate data collection and analysis. Utilizing a commonly used 5-point Likert scale that goes from "strongly disagree" to "strongly agree," the data was gathered. The analysis and research aimed to gain valuable insights and extract expressive assumptions about the impacts of gaming on mental health, physical well-being, and social dynamics. The survey was primarily aimed at individuals who actively engage in and devote significant time and energy to playing digital games. All participants' responses were recorded and carried out, ensuring no valuable information was overlooked or neglected. The responses were then subjected to rigorous analysis and research to gain a comprehensive and nuanced understanding of the various impacts that gaming can have on individuals. The chart presented below, obtained through a questionnaire, illustrates that a significant proportion of individuals, specifically 53.9%, experience weak eyesight as a consequence of engaging in frequent digital gaming activities.

The reliability of the questionnaire tested through Cronbach's Alpha is of moderate magnitude, measuring at 0.689, the positive nature of this reliability score signifies the presence of consistency in the measurements of the items. However, it is essential to note that improvements could be made through slight modifications in the questions, as well as a thorough evaluation and refinement of the measurement instrument to enhance its overall reliability.

By dividing people into discrete clusters on the basis of attributes related to their behavioural and health-related characteristics, K-Means clustering can be used to investigate the connections between social relationships, playing digital games, and mental and physical health [34]. K-Means can find patterns or subgroups within the population that share similar characteristics. This can be done by examining data such as time spent on playing digital games, frequency of social contacts while gaming, levels of physical activity, and mental health indicators (such as stress, anxiety, and mood) [35]. This clustering gives researchers important information about the possible advantages or disadvantages of digital gaming for general health. It also helps in understanding social connectedness by revealing how various groups of people are affected by gaming in terms of their physical and mental health and social interactions [36].

6. Results and Interpretation

This section explores the complex connections between social contacts, mental health, physical health, and gaming behaviour. To determine the direction and degree of relationships between important variables, including gaming hours, physical and mental health markers, and social engagement levels, we first look at the correlation matrix. Building on this, the clustering algorithm clearly segments the dataset by identifying discrete groups according to these factors' commonalities. Additional statistical research reveals significant trends and patterns that arise from these clusters, especially in the ways that gaming hours and personal preferences affect social engagement, physical health, and mental health.

From the correlation matrix shown in Figure 42.2, it is identified that there is a strong correlation between hours spent and preference for physical health, mental health and social interaction.

The K-Means technique has been used to divide the dataset into three clusters. Based on similarity, each cluster represents a distinct data segment. Table 42.1 highlights the statistical results showcasing the statistical results. Withinness refers to within-cluster sum of squares (WCSS). The withinness values (81.78, 42.98, 60.40) indicate how compact each cluster is. The results show that Cluster 2 in Figure 42.3 is the most compact cluster. The total of squared distances (total cluster sum of squares) from each data point to the centroids of each respective cluster is 596, as indicated by the dataset's tot. withinness = 596 The squared sum of the distances between the centroid of each cluster and the overall centroid for this dataset is 410.8, or betweenness = 410.8 indicating that all three clusters are at good distance. The clustering group explains 68.9% of the total variance in the dataset. Figure 42.3 above shows cluster plot of variables physical discomfort, health issues, effects with respects to time spent on social media.

Table 42.2 shows clustering analysis of mental health based on number of hours spent on technology. The three clusters are as follows:

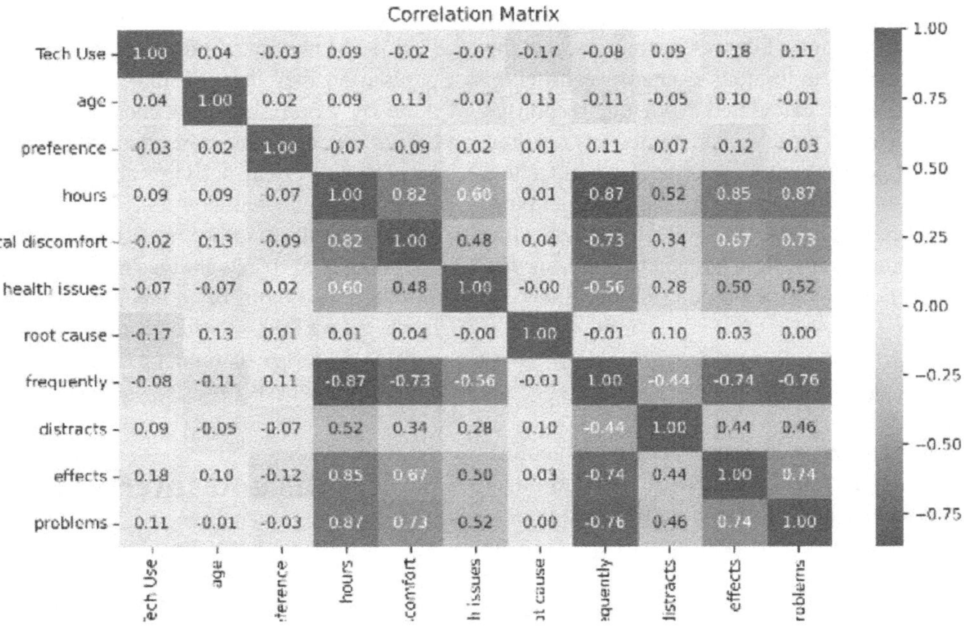

Figure 42.2. Correlation matrix.

Source: Author.

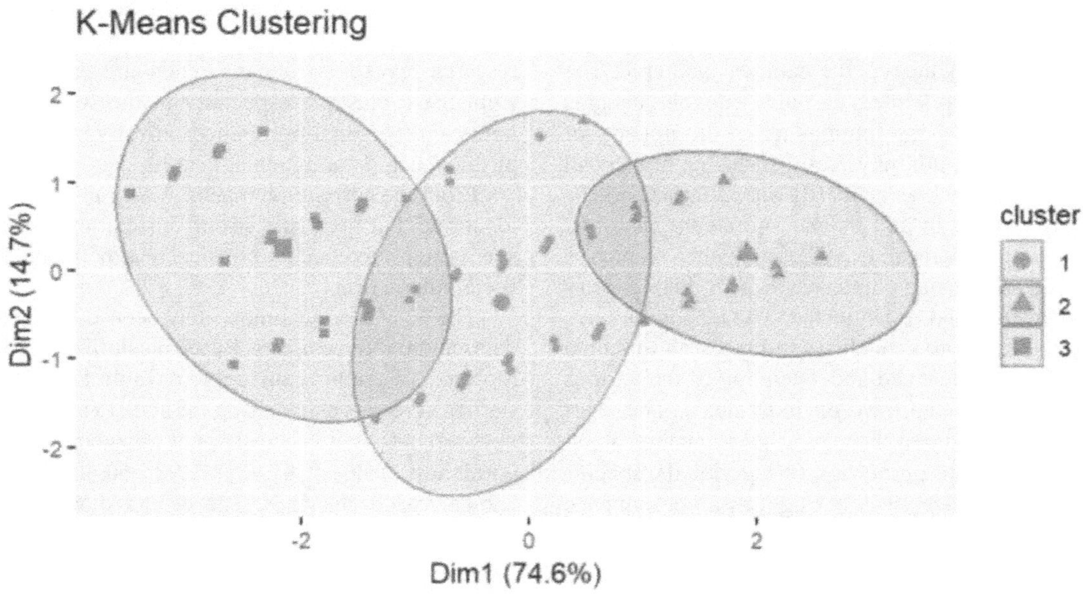

Figure 42.3. Clustering with respect to hours spent.

Source: Author.

Table 42.1. Clustering statistical analysis for technology

Statistical Result	*Values*
Number of Cluster	3
Withinness (WCSS)	81.78, 42.98, 60.40
tot.withinness	596
Betweenness (BCSS)	410.8
Total Variance	68.9%

Source: Author.

- Cluster 1: High technology use 4+ hours on technology with high physical discomfort, health issues and high social interaction.
- Cluster 2: Low technology use 1–2 hours & better mental health (3–4).
- Cluster 3: Moderate technology use 2–3 hours with mixed mental health scores and medium social interaction

Table 42.2. Clusters group based on hours spent

Cluster	*Avg Physical Discomfort*	*Avg Health Issues*	*Avg Effect*	*Avg Hours*
1.High Technology usage	3.62	2.88	4.57	4.69
2.Low Technology	1.11	1.26	1.67	1.59
3.Medium Technology	2.48	1.61	3.48	3.5

Source: Author.

Table 42.3. Clustering statistical analysis for preference

Statistical Result	*Values*
Number of Cluster	3
Withinness (WCSS)	142.21, 52.80, 85.54
tot.withinness	596
Betweenness (BCSS)	315.43
Total Variance	52.9%

Source: Author.

Table 42.3 highlights the statistical results showcasing the statistical results. The withinness values (142.21, 52.80, 85.54) indicate how compact each cluster is. The results show that Cluster 1 in Figure 42.4 is the most compact cluster. The total of squared distances (total cluster sum of squares) from each data point to the centroids of each respective cluster is 596. The squared sum of the distances between the centroid of each cluster and the overall centroid for this dataset is 315.5. The clustering group

Table 42.4. Clusters group based on hours spent

Cluster	*Avg Physical Discomfort*	*Avg Health Issues*	*Avg Effect*	*Avg Preference*
1. Outdoor Preference	1.72	1.32	2.56	1.00
2. Spending time with family	1.34	1.42	2.13	2.30
3. Gaming Preference	3.47	2.50	4.32	1.5

Source: Author.

explains 52.9% of the total variance in the dataset. Also since between cluster variance ratio is greater than 5 indicates well-separated cluster.

$$\frac{Betweeness}{Total\ variance} = \frac{315.43}{52.9} \sim 5.96$$

Figure 42.3 highlights the clustering of mental health based on preferences in terms of gaming, outdoor and family. The three clusters are as follows:

Table 42.4 shows clustering analysis of mental health based on preference. The three clusters are as follows:

- Cluster 1: Outdoor Preference shows lower health issues and moderate social interaction
- Cluster 2: Spending time with family with mixed mental health scores.
- Cluster 3: Gaming Preference shows high physical and mental scores with maximum social interactions

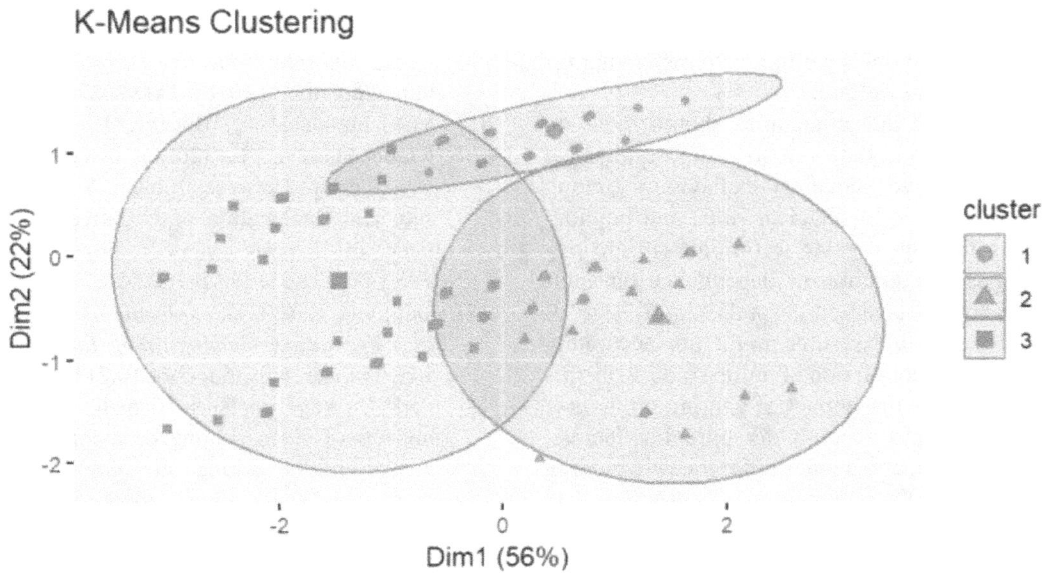

Figure 42.4. Clustering of mental health based on preference.

Source: Author.

The study primarily focuses on the detrimental impacts of digital gaming on mental health, physical well-being, and social relationships from consumers' perspective. The analysis reveals significant trends and patterns that arise from the above clusters, especially in the ways that gaming hours and personal preferences affect social engagement, physical health, and mental health. Concerns encompass heightened anxiety, disrupted sleep patterns, the risks associated with a sedentary lifestyle, and the potential for social withdrawal. The culmination of the research study unequivocally demonstrates that indulging in digital gaming significantly impacts individuals' overall well-being. Prolonged hours of immersion in gaming activities have resulted in detrimental consequences such as the onset of depression, persistent stress, visual impairment, and even obesity, among a plethora of other issues that significantly hinder an individual's well-being. Consequently, these issues have the potential to affect the holistic well-being of an individual. The results obtained from the clustering analysis have illuminated a significant impact of variables, highlighting the detrimental effects that prolonged hours of gaming can have on mental health, isolation, and a decline in face-to-face interactions, among other unfortunate consequences.

7. Conclusion and Recommendations

The research has revealed that regular engagement in digital games can have detrimental effects on mental health, physical health, and social relationships. The impact of gaming on psychological and physical well-being is subject to many factors. Though gaming may present advantages in terms of cognition and social interaction, an excess or problematic engagement with gaming may result in adverse outcomes. Achieving an equilibrium and exhibiting awareness of one's gaming patterns is vital for preserving one's overall welfare.

It is recommended that companies should focus on developing interventions that aim to promote physical activity and discourage sedentary behaviours among gamers. It is imperative to set clear rules and boundaries for gaming through regular self-monitoring to pay attention to signs of addiction or dependence on gaming. Since the study considers a small sample size of only 152 individuals, the results might not accurately represent the global population. Future studies on this topic could investigate the impact of gaming on diverse populations, taking into account demographic factors, including cultural, age, and gender differences in gaming habits and outcomes. Complementing quantitative data with qualitative research would also enhance our understanding of the subjective experiences and perceptions of gamers. Furthermore, the dopamine addiction theory and bio-psycho-social model could be tested thoroughly with experimentation method to understand the implication further.

Acknowledgment

We gratefully acknowledge the participants, gamers, for their cooperation in the research.

References

[1] Pandya, A., & Lodha, P. (2021). Social Connectedness, Excessive Screen Time During COVID-19 and Mental Health: A Review of Current Evidence. *Frontiers in Human Dynamics*, *3*. https://doi.org/10.3389/FHUMD.2021.684137/BIBTEX

[2] Stevens, M. W. R., Dorstyn, D., Delfabbro, P. H., & King, D. L. (2021). Global prevalence of gaming disorder: A systematic review and meta-analysis. *The Australian and New Zealand Journal of Psychiatry*, *55*(6), 553–568. https://doi.org/10.1177/0004867420962851

[3] Mahmoud, A. B. (2024). Exploring the public's beliefs, emotions and sentiments towards the adoption of the metaverse in education: A qualitative inquiry using big data. *British Educational Research Journal*, *250*, 104498. https://doi.org/10.1002/BERJ.4026

[4] Ahmed, M. Z., Ahmed, O., Gao, L., Jobe, M. C., & Li, W. (2024). Internet Gaming Disorder and Mental Health of Children in China: A Latent Profile Analysis. *International Journal of Mental Health Promotion*, *26*(7), 517–529. https://doi.org/10.32604/IJMHP.2024.051055

[5] Zhang, W., Pu, J., He, R., et al. (2022). Demographic characteristics, family environment and psychosocial factors affecting internet addiction in Chinese adolescents. *Journal of Affective Disorders*, *315*, 130–138. https://doi.org/10.1016/J.JAD.2022.07.053

[6] Miedzobrodzka, E., Buczny, J., Konijn, E. A., & Krabbendam, L. C. (2021). Insensitive Players? A Relationship Between Violent Video Game Exposure and Recognition of Negative Emotions. *Frontiers in Psychology*, *12*. https://doi.org/10.3389/FPSYG.2021.651759/BIBTEX

[7] Dasgupta, R. (2024). *Exploring the Impact of Anxiety, Peer Pressure on Gaming Young Adults: A Comprehensive Analysis*. https://doi.org/10.2139/SSRN.4970593

[8] Park, C., Angelica, P., & Trisnadi, A. I. (2025). Global impacts of video gaming behavior on young adults' mental health during the COVID-19 pandemic: A systematic literature review. *Social Sciences & Humanities Open*, *11*, 1–10. https://doi.org/10.1016/J.SSAHO.2024.101229

[9] De Jaegere, E., van Heeringen, K., Emmery, P., Mommerency, G., & Portzky, G. (2024). Effects of a Serious

Game for Adolescent Mental Health on Cognitive Vulnerability: Pilot Usability Study. *JMIR Serious Games*, *12*(1), e47513. https://doi.org/10.2196/47513

[10] Eirich, R., McArthur, B. A., Anhorn, C., McGuinness, C., Christakis, D. A., & Madigan, S. (2022). Association of Screen Time With Internalizing and Externalizing Behavior Problems in Children 12 Years or Younger: A Systematic Review and Meta-analysis. *JAMA Psychiatry*, *79*(5), 393–405. https://doi.org/10.1001/JAMAPSYCHIATRY.2022.0155

[11] Matias, C. N., Cardoso, J., Cavaca, M. L., et al. (2023). Game on: A cross-sectional study on gamers' mental health, Game patterns, physical activity, eating and sleeping habits. *Computers in Human Behavior*, *148*, 107901. https://doi.org/10.1016/J.CHB.2023.107901

[12] Van Den Eijnden, R., Koning, I., Doornwaard, S., Van Gurp, F., & Bogt, T. Ter. (2018). The impact of heavy and disordered use of games and social media on adolescents' psychological, social, and school functioning. *Journal of Behavioral Addictions*, *7*(3), 697–706. https://doi.org/10.1556/2006.7.2018.65

[13] Choi, J., Choi, Y., Jung, Y.-C., et al. (2024). Effects of Game-Related Tasks for the Diagnosis and Classification of Gaming Disorder. *Biosensors*, *14*(1), 42. https://doi.org/10.3390/BIOS14010042

[14] Brailovskaia, J., Meier-Faust, J., Schillack, H., & Margraf, J. (2022). A two-week gaming abstinence reduces Internet Gaming Disorder and improves mental health: An experimental longitudinal intervention study. *Computers in Human Behavior*, *134*, 107334. https://doi.org/10.1016/J.CHB.2022.107334

[15] Casale, S., Musicò, A., & Spada, M. M. (2021). A systematic review of metacognitions in Internet Gaming Disorder and problematic Internet, smartphone and social networking sites use. *Clinical Psychology & Psychotherapy*, *28*(6), 1508. https://doi.org/10.1002/CPP.2588

[16] Başdaş, Ö., & Özbey, H. (2020). Digital game addiction, obesity, and social anxiety among adolescents. *Archives of Psychiatric Nursing*, *34*(2), 17–20. https://doi.org/10.1016/J.APNU.2019.12.010

[17] Meriläinen, M., & Ruotsalainen, M. (2023). The light, the dark, and everything else: making sense of young people's digital gaming. *Frontiers in Psychology*, *14*. https://doi.org/10.3389/FPSYG.2023.1164992/BIBTEX

[18] Gioia, F., Colella, G. M., & Boursier, V. (2022). Evidence on Problematic Online Gaming and Social Anxiety over the Past Ten Years: a Systematic Literature Review. *Current Addiction Reports*, *9*(1), 32–47. https://doi.org/10.1007/S40429-021-00406-3/TABLES/1

[19] Ji, Y., Yin, M. X. C., Zhang, A. Y., & Wong, D. F. K. (2022). Risk and protective factors of Internet gaming disorder among Chinese people: A meta-analysis. *The Australian and New Zealand Journal of Psychiatry*, *56*(4), 332–346. https://doi.org/10.1177/00048674211025703

[20] Suárez-Soto, E., Peris-de la Hoz, A., Sanchez-Fernandez-Quejo, A., Rodriguez-Toscano, E., Lagunas, N., Reneses, B., & De la Torre-Luque, A. (2025). Problematic gaming use and psychological distress among Spanish young adults: A comprehensive study. *The European Journal of Psychiatry*, *39*(1). https://doi.org/10.1016/J.EJPSY.2024.100279

[21] Hazar, Z. (2019). Children's Digital Game Addiction and Opinions About Their Parents' Playing Digital Games (A Mixed Method Study). *Journal of Education and Training Studies*, *7*(1), 85–93. https://doi.org/10.11114/jets.v7i1.3785

[22] González-Bueso, V., Santamaría, J. J., Fernández, D., et al. (2018). Internet gaming disorder in adolescents: Personality, psychopathology and evaluation of a psychological intervention combined with parent psychoeducation. *Frontiers in Psychology*, *9*(MAY), 350071. https://doi.org/10.3389/FPSYG.2018.00787/BIBTEX

[23] Abbas, H. (2020). The Relation between Depression and Addiction to Online Games in Kuwait. *Arab Journal of Administrative Sciences*, *27*(1), 65–84. https://doi.org/10.34120/AJAS.V27I1.1057

[24] Alshehri, A. G., & Mohamed, A. M. A. S. (2019). The Relationship Between Electronic Gaming and Health, Social Relationships, and Physical Activity Among Males in Saudi Arabia. *13*(4). https://doi.org/10.1177/1557988319873512

[25] Archana, R., Sharma, M. K., Kumar, K. J., & Marimuthu, P. (2019). Internet gaming disorder and psychiatric symptoms in Bengaluru, India: Treatment implication for promotion of user mental health. *Indian Journal of Social Psychiatry*, *35*(2), 135–136. https://doi.org/10.4103/IJSP.IJSP_42_18

[26] Xavier, B., Viana, D., & Santos, R. (2024). A Dive into the State of the Practice of the Brazilian Game Software Ecosystem. *IEEE Transactions on Games*, *16*(1), 195–204. https://doi.org/10.1109/TG.2023.3242217

[27] Altintas, E., Karaca, Y., Hullaert, T., & Tassi, P. (2019). Sleep quality and video game playing: Effect of intensity of video game playing and mental health. *Psychiatry Research*, *273*, 487–492. https://doi.org/10.1016/J.PSYCHRES.2019.01.030

[28] Danielsen, P. A., Mentzoni, R. A., & Låg, T. (2024). Treatment effects of therapeutic interventions for gaming disorder: A systematic review and meta-analysis. *Addictive Behaviors*, 149. https://doi.org/10.1016/J.ADDBEH.2023.107887

[29] Aggarwal, R., & Ranganathan, P. (2019). Study designs: Part 2 – Descriptive studies. *Perspectives in Clinical Research*, *10*(1), 34. https://doi.org/10.4103/PICR.PICR_154_18

[30] Hair, J. F., Page, M., & Brunsveld, N. (2019). Essentials of business research methods. In *Essentials of Business*

Research Methods (4th ed.). Taylor and Francis. https://doi.org/10.4324/9780429203374/ESSENTIALS-BUSINESS-RESEARCH-METHODS-JOE-HAIR-JR-MICHAEL-PAGE-NIEK-BRUNSVELD

[31] Aman Mezmir, E. (2020). Qualitative Data Analysis: An Overview of Data Reduction, Data Display and Interpretation. *Research on Humanities and Social Sciences*, *10*(21), 15–27. https://doi.org/10.7176/RHSS/10-21-02

[32] Alam, M. K. (2021). A systematic qualitative case study: questions, data collection, NVivo analysis and saturation. *Qualitative Research in Organizations and Management: An International Journal*, *16*(1), 1–31. https://doi.org/10.1108/QROM-09-2019-1825/FULL/XML

[33] Zhu, S., Zhuang, Y., Lee, P., Chi-Mei Li, J., & Wong, P. W. C. (2021). Leisure and Problem Gaming Behaviors Among Children and Adolescents During School Closures Caused by COVID-19 in Hong Kong: Quantitative Cross-sectional Survey Study. *JMIR Serious Games*, *9*(2). https://doi.org/10.2196/26808

[34] Han, X., Wang, L., Xu, S., Zhao, D., & Liu, G. (2019). Recognizing roles of online illegal gambling participants: An ensemble learning approach. *Computers & Security*, *87*, 101588. https://doi.org/10.1016/J.COSE.2019.101588

[35] Odierna, B. A., & Silveira, I. F. (2020). MMORPG Player Classification Using Game Data Mining and K-means. *Lecture Notes in Networks and Systems*, *69*, 560–579. https://doi.org/10.1007/978-3-030-12388-8_40

[36] Motta, L., & Lannes, P. (2021). Unsupervised Learning Applied to the Segmentation of Users of Online Gambling Platforms in Portugal - The effects of the Covid-19 Pandemic on User Behavior and Segmentation. *NIMS-Dissertações de Mestrado Em Ciência de Dados e Métodos Analíticos Avançados (Data Science and Advanced Analytics)*. https://run.unl.pt/handle/10362/135874

43 NADBA: An AI-driven deep learning approach for early skin cancer detection

Benson Raj[1,a], *Humaid Alnaqbi*[2,b], *Saeed Al Ali*[3,c], *Hassan Khaaldi*[2,d], *and Abdelaziz Alnaqbi*[3,e]

[1]Assistant Professor, Higher Colleges of Technology (HCT), United Arab Emirates
[2]Information Security and Forensics Student, Higher Colleges of Technology (HCT), United Arab Emirates
[3]Networking Student, Higher Colleges of Technology (HCT), United Arab Emirates

Abstract: Skin cancer remains one of the deadliest diseases worldwide, with early detection playing a crucial role in improving patient outcomes. Traditional diagnostic methods rely on dermatological expertise, which can be time-consuming and prone to errors. NADBA, an AI-powered skin cancer detection system, aims to provide an accurate and efficient method for diagnosing melanoma and other skin cancers using deep learning (DL) techniques. This research paper explores the methodologies, technical feasibility, and expected impact of NADBA, highlighting its alignment with the UAE National Strategy for AI 2031 and the broader healthcare industry. The study compares different DL models, discusses the challenges in AI-based diagnosis, and proposes solutions for integration within healthcare systems.

Keywords: Skin cancer detection, melanoma diagnosis, artificial intelligence in healthcare, deep learning for medical imaging, dermatoscopic image analysis, ResNet152 for image classification, AI-based skin cancer screening, medical image processing, UAE AI strategy 2031, early detection of skin cancer, precision and recall in medical AI, explainability in AI diagnosis, clinical trials for AI in healthcare

1. Introduction

Skin cancer, particularly melanoma, is one of the most aggressive forms of cancer, with high mortality rates when diagnosed late. According to the American Academy of Dermatology [9], approximately 20 Americans die daily from melanoma. Traditional detection methods involve visual examination by dermatologists, which can be unreliable. NADBA leverages deep learning models to analyze images of skin lesions, providing instant feedback on their malignancy. The goal of this project is to facilitate early detection, reduce diagnostic costs, and improve accessibility to skin cancer screening. There are three types of Skin Cancer have been listed by scientists. Basal cell carcinoma, Squamous cell carcinoma, and Melanoma according to [8], and it is the 4th most common form of cancer in the UAE.

Melanoma is the most invasive skin cancer with the highest risk of human loss. The ability to spread and grow is considered very quickly. Diagnostic methods are time-consuming due to the evaluation of cancer and can be much more costly, A lack of trained dermatologists leads to a late diagnosis that leads to higher risks. The main point of this project is to have much a reliable and a less time-consuming method for healthcare providers and individuals to detect skin cancer early. NADBAH seeks to reduce time and cost and to provide healthcare workers with a trustworthy AI tool that would ultimately assist in early detection of this disease. This AI tool will process different images of Melanoma and analyze it's characteristics which will ultimately results in instant recognition and possible diagnosis.

2. Background

Skin cancer manifests in three primary forms: basal cell carcinoma, squamous cell carcinoma, and melanoma. Melanoma, the most invasive type, requires early detection for effective treatment. Challenges in traditional diagnostic methods include:

NADBA seeks to address these issues by offering a reliable, AI-driven detection tool that can be accessed via a mobile application or web platform. This system will analyze dermatoscopic images, classify them using deep learning algorithms, and provide real-time diagnosis.

[a]braj@hct.ac.ae, [b]H00472396@hct.ac.ae, [c]h00443866@hct.ac.ae, [d]h00367570@hct.ac.ae, [e]h00394959@hct.ac.ae

DOI: 10.1201/9781003770473-43

3. Methodology

3.1. Model architecture

The architecture is designed to handle a streamlined workflow:

1. **Input Phase:** Users upload skin lesion images and provide metadata.
2. **Preprocessing Phase:** Images are resized, normalized, and prepared for AI analysis.
3. **AI Evaluation:** A deep learning model analyzes the image and predicts whether the lesion is benign or malignant.
4. **Output Phase:** The system displays results, provides real-time feedback, and ensures user privacy.

This structured design allows incremental improvements, aligns with our development methodology, and supports both end-user needs and backend AI processes.

3.2. Training data and preprocessing

NADBA utilizes a publicly available dataset, such as the ISIC Archive, which contains thousands of labeled skin lesion images. Preprocessing includes image resizing, contrast enhancement, and augmentation.

- Image resizing to 224 × 224 pixels.
- Contrast enhancement.
- Augmentation techniques to increase dataset diversity.

3.3. Training procedure

The model was trained:

- Using only 10,015 microscopic images.
- 10 rounds to check how fast it learns.

With 80% of the images used for training and (Figures 43.1–43.5).

4. Results and Discussions

4.1. Evaluation metrics

10,015 dermatoscopic images, achieving the highest accuracy with ResNet152. The confusion matrix results indicated:

- True Positive Rate: 91%
- False Positive Rate: 9%
- Precision: 94.7%
- Recall: 91.4%

Model performance was assessed using:

- Accuracy
- Precision
- Recall
- F1-score
- ROC-AUC

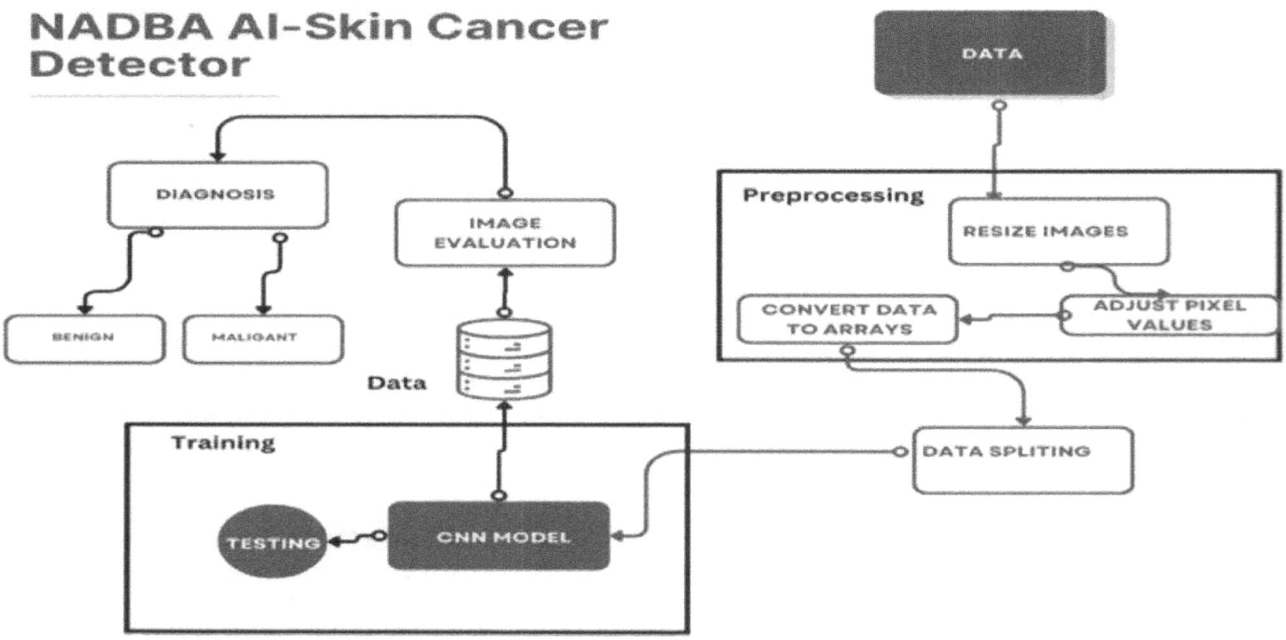

Figure 43.1. NADBA system architecture.

Source: Author.

```
Loaded 10015 images out of 10015
Model: "sequential"
_________________________________________________________________
 Layer (type)                Output Shape              Param #
=================================================================
 conv2d (Conv2D)             (None, 126, 126, 32)      896

 max_pooling2d (MaxPooling2  (None, 63, 63, 32)        0
 D)

 flatten (Flatten)           (None, 127008)            0

 dense (Dense)               (None, 64)                8128576

 dropout (Dropout)           (None, 64)                0

 dense_1 (Dense)             (None, 7)                 455

=================================================================
Total params: 8129927 (31.01 MB)
Trainable params: 8129927 (31.01 MB)
Non-trainable params: 0 (0.00 Byte)
```

Figure 43.2. Image loading and preprocessing.

Source: Author.

```
Epoch 1/10
251/251 [==============================] - 14s 52ms/step - loss: 1.2306 - accuracy: 0.6538 - val_loss: 0.9222 - val_accuracy: 0.6685
Epoch 2/10
251/251 [==============================] - 10s 42ms/step - loss: 0.9674 - accuracy: 0.6685 - val_loss: 0.9654 - val_accuracy: 0.6680
Epoch 3/10
251/251 [==============================] - 11s 43ms/step - loss: 0.9025 - accuracy: 0.6715 - val_loss: 0.9525 - val_accuracy: 0.6685
Epoch 4/10
251/251 [==============================] - 11s 45ms/step - loss: 0.8789 - accuracy: 0.6724 - val_loss: 0.8870 - val_accuracy: 0.6690
Epoch 5/10
251/251 [==============================] - 11s 45ms/step - loss: 0.8623 - accuracy: 0.6781 - val_loss: 0.8584 - val_accuracy: 0.6725
Epoch 6/10
251/251 [==============================] - 11s 42ms/step - loss: 0.8399 - accuracy: 0.6780 - val_loss: 0.8551 - val_accuracy: 0.6720
Epoch 7/10
251/251 [==============================] - 11s 42ms/step - loss: 0.7986 - accuracy: 0.6927 - val_loss: 0.8760 - val_accuracy: 0.6730
Epoch 8/10
251/251 [==============================] - 11s 42ms/step - loss: 0.7745 - accuracy: 0.7023 - val_loss: 0.8346 - val_accuracy: 0.6815
Epoch 9/10
251/251 [==============================] - 11s 43ms/step - loss: 0.7531 - accuracy: 0.7104 - val_loss: 0.8380 - val_accuracy: 0.6885
Epoch 10/10
251/251 [==============================] - 11s 45ms/step - loss: 0.7353 - accuracy: 0.7139 - val_loss: 0.8354 - val_accuracy: 0.6925
63/63 [==============================] - 1s 13ms/step
```

Figure 43.3. Model training.

Source: Author.

4.2. *Analysis of results*

- Inference w Accuracy: *69.25%*
- *Precision, Recall, and F1-score:*
- *Varying results across different lesion types, with the highest performance for class 5 (melanoma, 96% recall)*
- Lower performance for rare lesion types
- (precision values near 0 due to lack of predicted samples)
- Predictions made by the model were as expected, and produced an almost exact match of all test labels (Table 43.1).

4.3. *Discussion*

The results demonstrated that the model trained on Microscopic Images achieved great performance in terms of Accuracy, while objective metrics such as recall rate

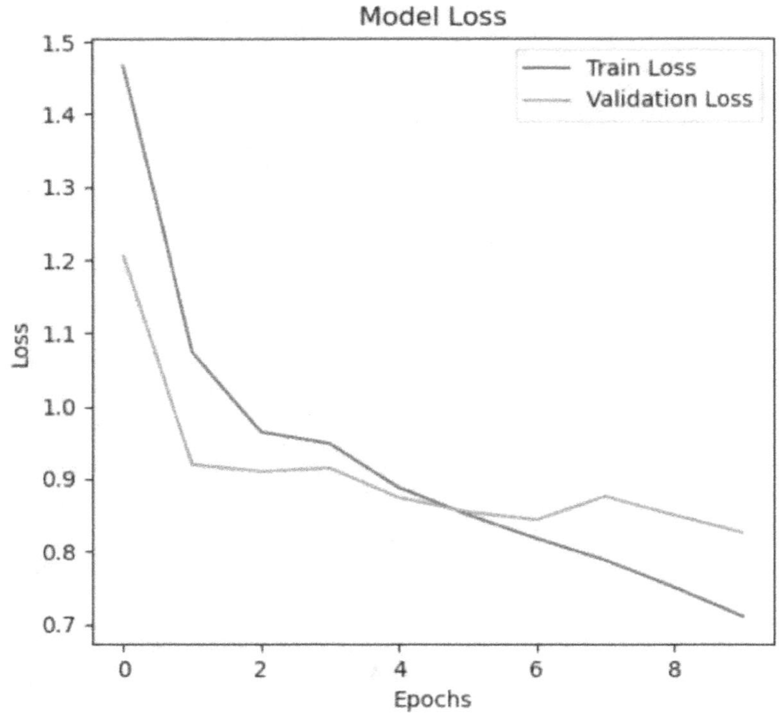

Figure 43.4. Model accuracy over training and testing.

Source: Author.

Table 43.1. Model metrics over 10 training sessions

Epoch	*Loss*	*Accuracy*	*Validation Loss*	*Validation Accuracy*
1	1.2306	65.38%	0.9222	66.85%
2	0.9674	66.85%	0.9654	66.80%
3	0.9025	67.15%	0.9525	66.85%
4	0.8789	67.24%	0.8870	66.90%
5	0.8623	67.81%	0.8584	67.25%
6	0.8309	67.80%	0.8551	67.20%
7	0.7986	69.27%	0.8760	67.30%
8	0.7745	70.23%	0.8346	68.15%
9	0.7531	71.04%	0.8380	68.85%
10	0.7353	71.39%	0.8354	69.25%

Source: Author.

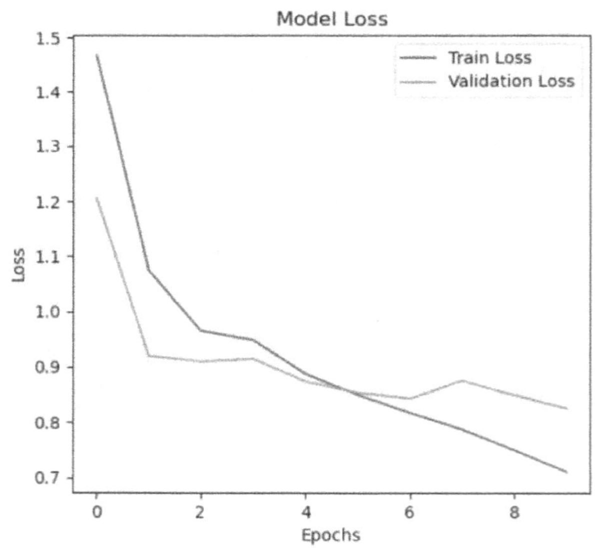

Figure 43.5. Model loss during training and testing.

Source: Author.

revealed inconsistencies. The ResNet model Showed great potential for growth and adaptability These findings reinforce the potential AI integrating into the UAE healthcare systems and widely accepted.

References

[1] Stanford Medicine News Center. (April, 2024). AI Improves Accuracy of Skin Cancer Diagnoses. https://med.stanford.edu/news/all-news/2024/04/ai-skin-diagnosis.html.

[2] Mahmud, Faysal, Md. Mahin M., Md. Zobayer, I. K., and Yusha, A. (December, 2023). An Interpretable Deep Learning Approach for Skin Cancer Categorization. arXiv, https://arxiv.org/abs/2312.10696.

[3] Chang, Christopher. 2024. Artificial Intelligence and Skin Cancer. *Frontiers in Medicine* 10: 1331895. https://www.frontiersin.org/journals/medicine/articles/10.3389/fmed.2024.1331895/full

[4] Bø-Sande, Marie, Edvin Benjaminsen, et. al., (December, 2023). A Dual Convolutional Neural Network Pipeline for Melanoma Diagnostics and Prognostics. arXiv, https://arxiv.org/abs/2312.08766.

[5] ResNet152 Model. (2015). Deep Residual Learning for Image Recognition. Accessed February 10, 2025. https://arxiv.org/abs/1512.03385.

[6] ISIC Archive. (2025). International Skin Imaging Collaboration (ISIC) Database. Accessed February 10, 2025. https://www.isic-archive.com.

[7] UAE National Strategy for AI. (2019). UAE Artificial Intelligence Strategy 2031. Accessed February 10, 2025. https://ai.gov.ae.

[8] Cleveland Clinic. (2021). Types of Skin Cancer. Accessed February 10, 2025. https://my.clevelandclinic.org

[9] American Academy of Dermatology. (2024). Melanoma Statistics. Accessed February 10, 2025. https://www.aad.org.

[10] Deep Learning and Convolutional Neural Networks for Medical Imaging and Clinical Informatics. (2019). Edited by Le. Lu, Xiaosong Wang, Gustavo Carneiro, and Lin Yang. Springer. Nevruzov, V. (2010). Computer controlled solar house for measurement Rize city solar energy potential. *NURER, Ankara*. 404–414.

[11] Deep Learning in Medical Image Processing and Analysis. (2023). Edited by Indrakumari, S., Kumar, R., and Murugan. P. C., *IET*.

[12] Deep Learning in Medical Image Analysis. (2019). Edited by Gobert Lee and Hiroshi Fujita. *Springer*.

[13] Applications of Artificial Intelligence in Common Dermatological Diseases. (2023). Edited by Dhaher, N., Alsaeedi, A., and Albahar. M., *Springer*.

[14] Deep Learning for Medical Image Analysis. (2017). Edited by Kevin Zhou, S., Hayit G., and Dinggang S. Academic Press

44 Investigation and analysis of the tensile and flexural properties of 3D-Printed Polyether Ether Ketone (PEEK) for biomedical applications

Jadamuni Senthilkumar[a] *and Vasudevan Alagumalai*[b]

Department of Mechanical Engineering, Saveetha School of Engineering, SIMATS, Chennai, Tamil Nadu, India

Abstract: The thermoplastic material Polyether ether ketone (PEEK) stands out as an excellent biomedical candidate because it demonstrates outstanding mechanical properties and biocompatibility together with strong chemical resistance. An evaluation of tensile and flexural characteristics exists in this study regarding PEEK parts made by fused filament fabrication (FFF) technology. A systematic evaluation of PEEK mechanical properties occurs through the study of printing parameter effects including execution precision along with support model selection and facing placement. The material strength along with stiffness and failure mechanisms undergo testing through tensile and flexural methods which follow ASTM standards. The 3D-printed mechanical properties of PEEK benefited considerably from the optimized printing parameters, which establish this material as suitable for biomedical applications that need load-bearing properties such as orthopaedic implants and dental prosthetics. The examination of microstructures grants researchers understanding of 3D-printed PEEK's fracture features together with its interlayer bond behaviour. The research findings enable the progress of additive manufacturing technology in biomedical applications through improved comprehension of 3D-printed PEEK component mechanics.

Keywords: Additive manufacturing, fused filament fabrication, infill pattern, mechanical properties

1. Introduction

A variety of materials with complex parameters are developed using Fused Filament Fabrication in additive manufacturing. PEEK stands out as a noteworthy material because it maintains exceptional mechanical properties together with high thermal stability and superior chemical resistance suitable for critical aerospace industry, automotive industry and medical applications [1–3]. The advancement of additive manufacturing (AM) has revolutionized the development of high-performance polymers for biomedical applications. Among these materials, Polyether Ether Ketone (PEEK) has garnered significant attention due to its exceptional mechanical properties, biocompatibility, and chemical resistance. PEEK is widely used in orthopedic, dental, and spinal implants as an alternative to traditional metallic implants, offering advantages such as radiolucency, reduced stress shielding, and enhanced patient comfort.

With the growing interest in 3D printing techniques, particularly fused filament fabrication (FFF) and selective laser sintering (SLS), researchers are exploring the feasibility of fabricating complex and customized PEEK structures. However, the mechanical performance of 3D-printed PEEK, particularly its tensile and flexural properties, is influenced by factors such as printing parameters, layer orientation, porosity, and post-processing treatments. Understanding these factors is crucial for ensuring the structural integrity and reliability of PEEK-based biomedical components.

This study aims to investigate the tensile and flexural behaviour of 3D-printed PEEK to assess its suitability for biomedical applications. By analyzing the effects of printing parameters on mechanical performance, this research provides insights into optimizing PEEK processing for enhanced strength, durability, and functionality in medical devices. The findings of this study will contribute to the development of more robust and patient-specific PEEK implants, further advancing the role of additive manufacturing in modern healthcare. The full implementation of PEEK material in FFF requires detailed knowledge about its DMA properties as discussed in published research

[a]senthilkumarj.sse@saveetha.com, [b]vasudevana.sse@saveetha.com

DOI: 10.1201/9781003770473-44

[4–7]. DMA provides essential evaluations of viscoelastic material behaviour under oscillatory stress that helps predict dynamic load performances [8–11]. The viscoelastic properties of FFF printed polymers align with traditionally manufactured polymers based on research findings [12–13].

2. Literature Review

Interlayer bonds formed during FFF printing develop weak connections because of the sequential printing method [14–15]. Various research projects have confirmed this phenomenon. The interlayer bonding strength of FFF 3D printed PEEK achieved 24.8 MPa according to Li et al.'s short beam test yet it might vary when subjected to dynamic loading [16]. Overview of Polyether Ether Ketone (PEEK) in Biomedical Applications. Polyether Ether Ketone (PEEK) is a high-performance polymer widely used in biomedical applications due to its excellent mechanical properties, biocompatibility, and resistance to chemical degradation. Compared to metallic implants, PEEK provides radiolucency, lower density, and reduced stress shielding, making it suitable for orthopedic, spinal, and dental applications [17]. However, PEEK's bioinert nature necessitates surface modifications or composite reinforcements to enhance osseointegration and promote cellular adhesion [18]. Recent studies have explored the incorporation of hydroxyapatite (HA) coatings and carbon fiber reinforcement (CF-PEEK) to improve its mechanical and biological performance [19–20]. 3D Printing of PEEK: Techniques and Challenges. The advent of additive manufacturing (AM) has enabled the fabrication of customized, patient-specific PEEK implants. Two primary 3D printing techniques for PEEK are fused filament fabrication (FFF) and selective laser sintering (SLS). Studies have shown that SLS-printed PEEK offers higher density and improved mechanical properties compared to FFF due to superior layer bonding and reduced porosity [21]. However, challenges such as high processing temperatures (>350°C), warping, and poor interlayer adhesion remain major obstacles [22]. Research has indicated that optimizing print parameters, such as nozzle temperature, printing speed, and infill density, can enhance mechanical properties [23].

Tensile Properties of 3D-Printed PEEK. Tensile strength is a key parameter in assessing load-bearing capacity for biomedical applications. Traditional injection-molded PEEK exhibits a tensile strength of ~90–100 MPa, while 3D-printed PEEK often shows lower tensile strength (~60–80 MPa) due to anisotropy and weak interlayer bonding [24]. Studies indicate that layer orientation significantly affects tensile properties, with X-Y printed specimens demonstrating higher strength than Z-oriented ones [25]. Recent work has explored annealing and post-processing treatments, which have shown to enhance interlayer adhesion and improve tensile properties [26].

Flexural properties are crucial for implants subjected to bending loads, such as spinal cages and dental prostheses. Research has shown that 3D-printed PEEK exhibits lower flexural strength than conventionally molded PEEK due to higher porosity and weaker interlayer adhesion [27]. However, optimizing printing parameters, such as raster angle, infill density, and post-processing treatments, can significantly enhance flexural modulus and toughness [28]. The addition of carbon or glass fiber reinforcements has also been shown to improve flexural stiffness and fatigue resistance [29]. Future Research Directions and Challenges. Despite significant progress in 3D printing of PEEK for biomedical applications, several research gaps remain: The comparative mechanical performance of FFF and SLS-printed PEEK needs further study under clinical loading conditions. The long-term durability and fatigue behaviour of 3D-printed PEEK in biological environments require extensive investigation [30]. Integration of bioactive coatings and composite reinforcements to enhance mechanical strength and biological response remains an active area of research [31]. This literature review highlights the current understanding of 3D-printed PEEK's tensile and flexural properties, while identifying areas for future research. Addressing these challenges will be crucial in advancing PEEK-based implants for biomedical applications.

3. Materials and Methods

3.1. FFF printing

This study employs a variety of infilling patterns, such as hexagon, cube, line, and grid. Creabot's PEEK K 10 PRODUCT CODE printer excelled at 3D printing in China. It has a density of 1.23 g/cm3, a tensile strength of 70 to 80 MPa, 4–6% elongation, and a diameter of 1.75 mm. The 3D printing specimen models were created with Catia V5 software and saved in STL format. The FFF printing nozzle was 0.4 mm in diameter and ran at 410°C. To reduce warping and improve adhesion, the bed temperature was adjusted to 120°C. A 45° build orientation, 100% infill density, and a layer height of 0.2 mm were all maintained during printing. The first layer height was raised to 0.3 mm to ensure correct adherence to the build plate. Because of the model's complexity, the printing speed was reduced to 30 mm/min in order to balance and preserve quality.

3.2. Tensile and flexural test

Tensile strength, flexural strength, and interlaminar shear strength (ILSS) tests were performed on the UTM – make Dak System Inc Series 7200 using MTS Test force software and a 10kN load cell with a non-contact video extensometer. The tensile test was performed at 2mm/min cross-head speed according to the ASTM D3822 standard. The testing equipment controlled the grip velocity to operate at 10 mm/s. The bending test followed the specifications outlined by ASTM D790. The testing procedure for flexure required 2 mm/min of a cross-head speed. To determine interlaminar shear strength, a short beam shear test was performed the ASTM D2344 standard. The test specimen featured 60 mm length and 12 mm width and 3 mm thickness. The specimen was designed with a span length that equalled four times its thickness resulting in maximum shear stress formation at its midsection. The fracture failure of interlaminar occurs due to maximum shear stress exists in the midplane of the specimen. The Figure 44.1 shows the experimental set up of tensile and flexural test with samples. A total of three test specimens were used for tensile, flexural and short beam shear testing with the mean values applied for analysis.

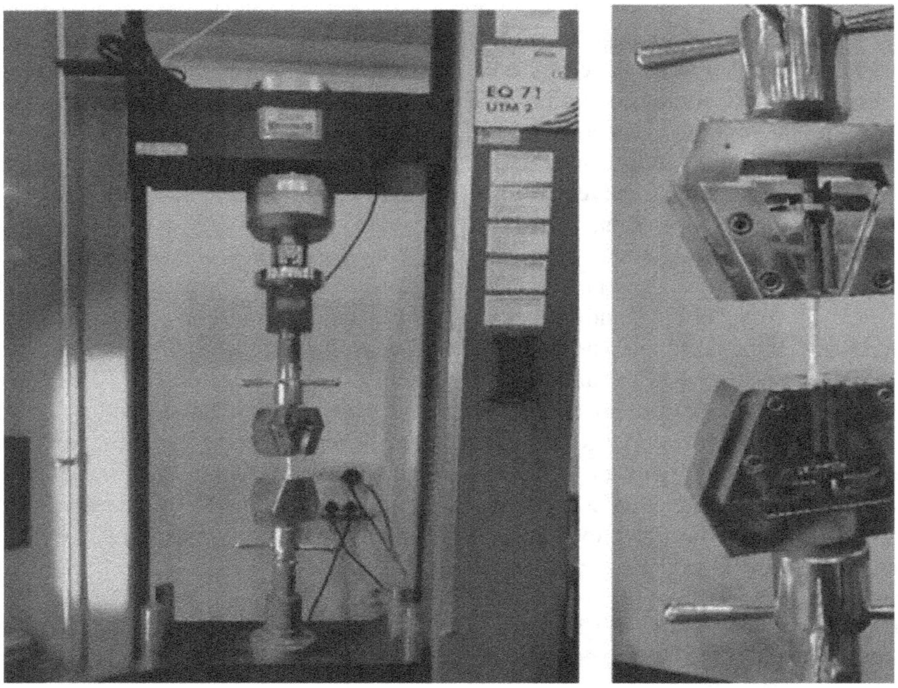

Figure 44.1. Tensile test setup.
Source: Author.

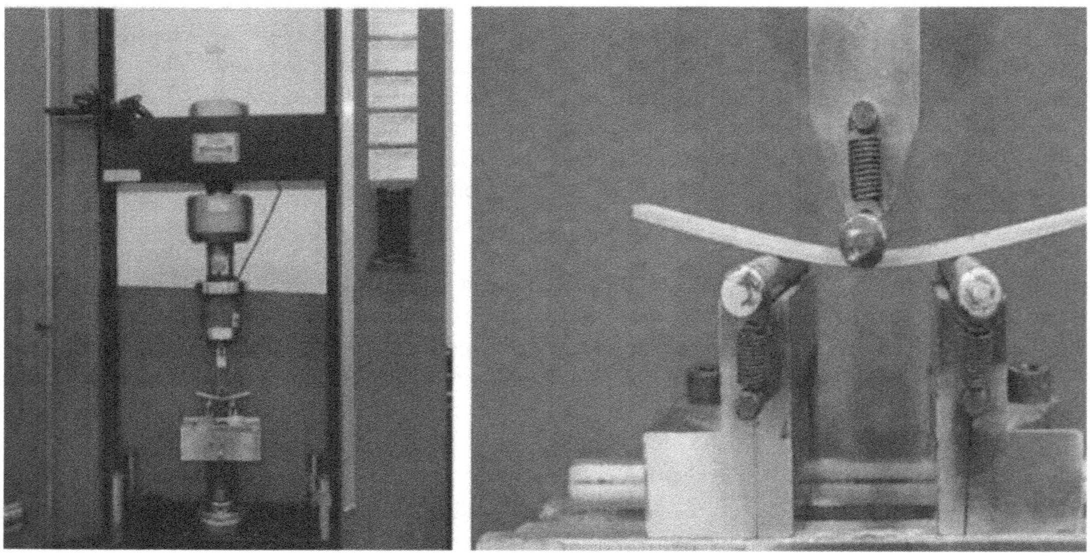

Figure 44.2. Flexural test setup.
Source: Author.

4. Results and Discussion

4.1. Tensile strength of PEEK specimen

The tensile stress and modulus of printed specimens with different infill patterns appear in Figures 44.2 and 44.3. Multiple infill patterns produced different results in the stress-strain analysis of the tensile properties. Results indicate that the Octet type infill pattern achieves the highest tensile strength value of 43.793Mpa combined with an elongation percentage of 8.11. The strength values of all infill types except Octet type infill pattern are lower than those of Octet type pattern according to Table 44.1.

Table 44.1 presents the tensile properties of 3D-printed PEEK specimens with different infill patterns, including tensile strength, peak load, and elongation. The Octet pattern exhibits the highest tensile strength (43.793 MPa) and peak load (892.935 N), indicating superior resistance to tensile forces. The Cross pattern follows closely with 43.154 MPa, while the Gyroid and Cubic patterns show slightly lower strengths at 41.221 MPa and 39.26 MPa, respectively. In terms of elongation, the Cubic pattern exhibits the highest flexibility (9.66%), whereas the Gyroid pattern has the lowest (7.996%). This suggests that the Octet pattern provides the best balance of tensile strength and moderate flexibility, making it suitable for load-bearing biomedical applications.

Table 44.1. The result of tensile with different infill patterns

Infill Pattern	*Tensile Strength (MPa)*	*Peak Load (N)*	*Elongation (%)*
Cross	43.154	873.552	8.68
Cubic	39.26	807.488	9.66
Gyroid	41.221	826.113	7.996
Octet	43.793	892.935	8.114

Source: Author.

4.2. Flexural strength

The flexural tests for FFF printed PEEK with various infill patterns appear in Figures 44.4 and 44.5. The flexural strength measurement for gyroid type infill pattern reached 89.417 Mpa while other infill patterns exhibited lower values.

Table 44.2 highlights the impact of different infill patterns on the flexural strength and elongation of 3D-printed PEEK specimens. The Gyroid pattern exhibits the highest flexural strength (89.417 MPa) and a relatively high elongation (6.164%), making it the most balanced choice for applications requiring both strength and flexibility. The Cross pattern follows with a strength of 86.22 MPa but has slightly lower elongation (5.787%), indicating a more rigid structure. The Cubic pattern provides moderate performance with 82.08 MPa strength and 5.824% elongation, while the Octet pattern, despite its highest elongation (6.221%), has the lowest flexural strength (79.946 MPa), making it less suitable for high-load applications. Figure 44.6 Overall, the Gyroid pattern emerges as the most promising choice for biomedical applications like spinal implants and prosthetics due to its optimal balance of mechanical strength and flexibility.

Figure 44.7 presents the flexural stress-strain curves for different infill patterns (Cross, Cubic, Gyroid, and Octet) in 3D-printed PEEK specimens. The Gyroid pattern exhibits the highest peak stress, indicating superior

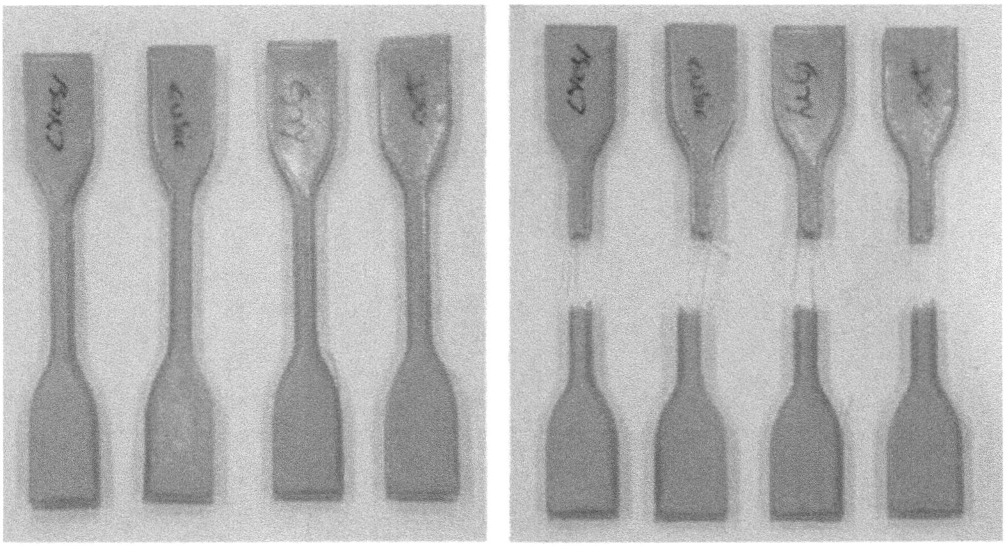

Figure 44.3. Samples before & after tensile testing.

Source: Author.

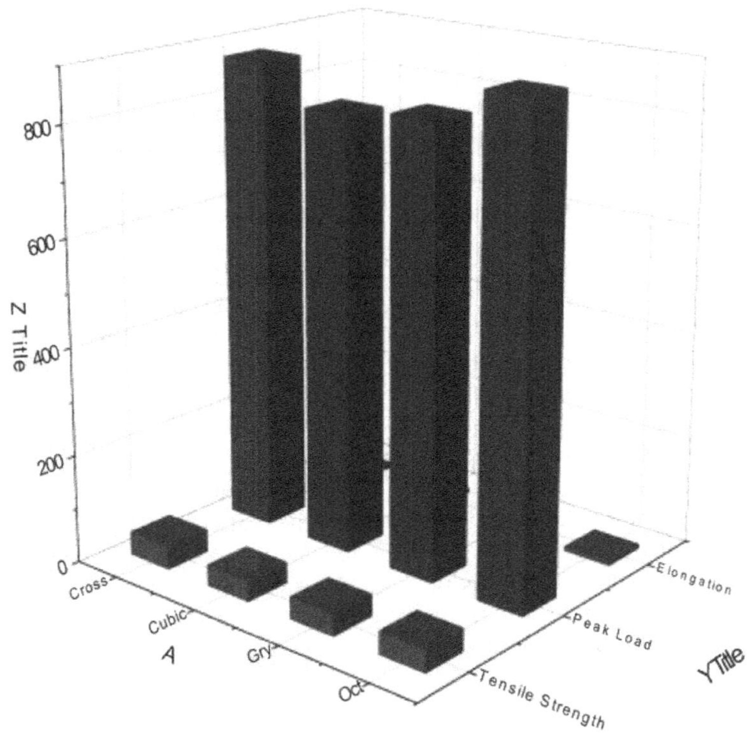

Figure 44.4. Tensile strength with strain.

Source: Author.

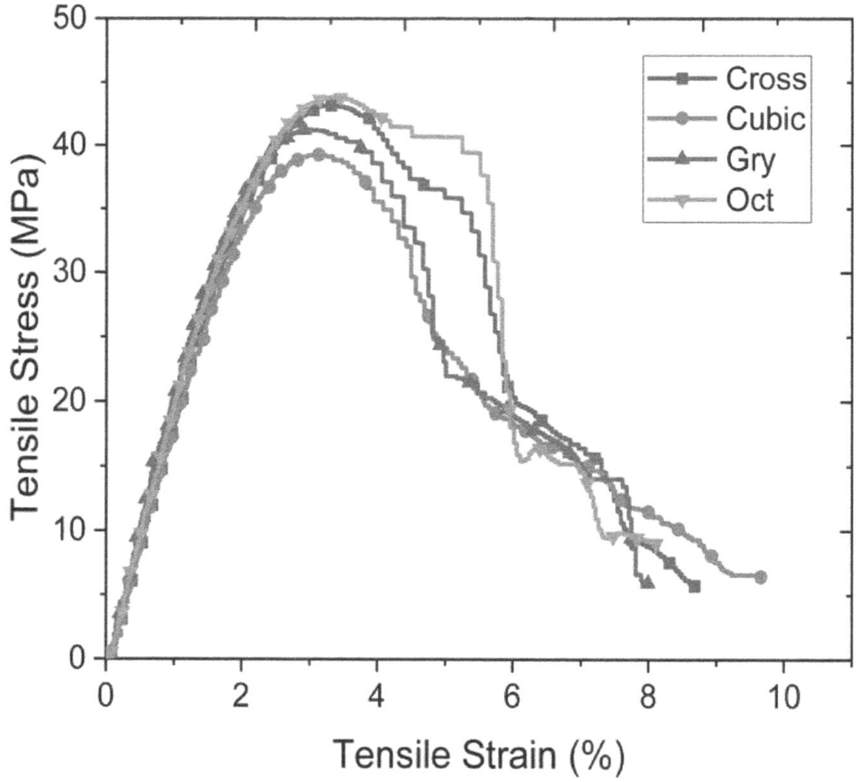

Figure 44.5. Comparison of tensile strength, elongation and peak load.

Source: Author.

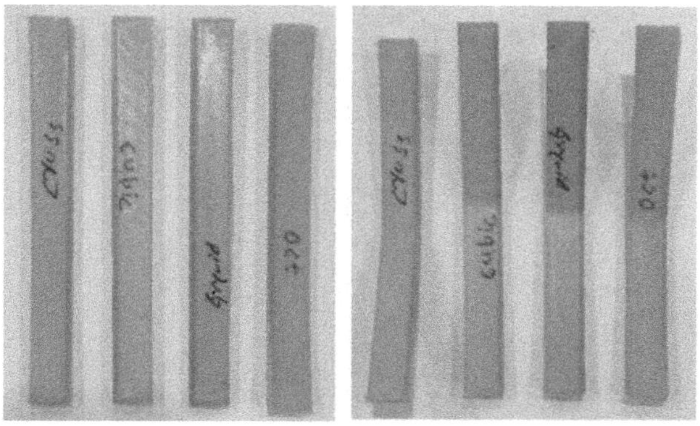

Figure 44.6. Samples before & after Flexuraltesting.

Source: Author.

Table 44.2. Comparison of the different infill pattern results of flexural strength

Infill Pattern	*Flexural Strength (MPa)*	*Elongation%*
Cross	86.22	5.787
Cubic	82.08	5.824
Gyroid	89.417	6.164
Octet	79.946	6.221

Source: Author.

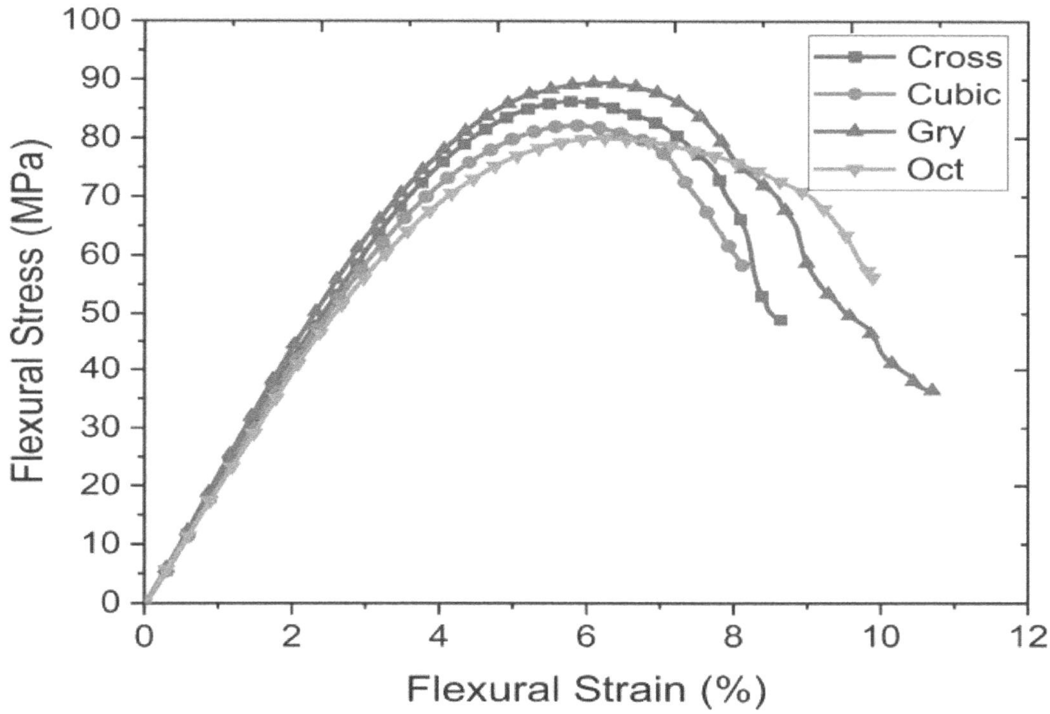

Figure 44.7. Flexural strength with strain.

Source: Author.

flexural strength, followed by the Cross and Cubic patterns. The Octet pattern shows the lowest peak stress, suggesting reduced flexural strength but higher strain before failure. The curves highlight variations in mechanical performance, with the Gyroid pattern demonstrating the best balance of strength and flexibility, making it a suitable choice for biomedical applications requiring structural integrity under bending loads.

Figure 44.8 illustrates the comparison of flexural strength (blue bars) and elongation percentage (green bars) for different infill patterns in 3D-printed PEEK specimens. The Gyroid pattern exhibits the highest flexural strength, closely followed by the Cross pattern, while the Octet pattern shows the lowest strength. In terms of elongation, the Octet pattern demonstrates the highest flexibility, whereas the Cross pattern has the lowest elongation percentage. This comparison highlights the Gyroid pattern as the most balanced option, offering both high strength and moderate flexibility, making it ideal for biomedical applications requiring structural durability and adaptability under bending stress.

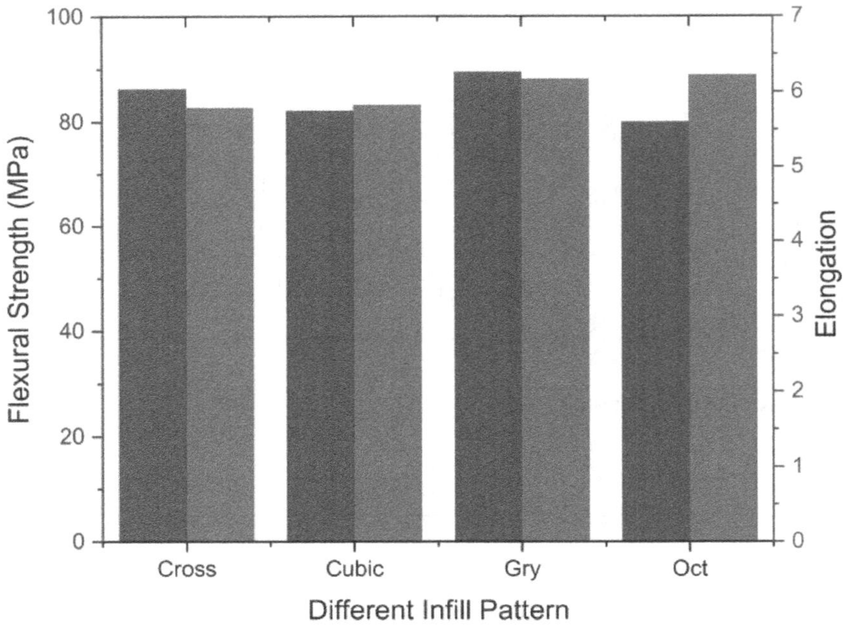

Figure 44.8. Flexural strength with different infill pattern.

Source: Author.

5. Conclusion

The study concludes that Octet infill PEEK demonstrates the highest tensile strength, making it a promising material for orthopedic implants that require strong load-bearing capacity. On the other hand, Gyroid infill PEEK exhibits superior flexural strength, making it more suitable for applications such as dental implants and knee joint replacements, where resistance to bending stress is crucial. Future research can focus on impact studies of different infill patterns, which would provide valuable insights for designing human skull implants and other biomedical applications requiring enhanced mechanical performance.

Acknowledgment

No conflicting objectives in this manuscript.

Authors Contributions

Author JS was involved in data collection, analysis and manuscript writing. Author AV was involved in the conceptualization, data validation, and critical review of the manuscript.

References

[1] Challa, B. T., Gummadi, S. K., Elhattab, K., et al. (2022). In-house processing of 3D printable polyetheretherketone (PEEK) filaments and the effect of fused deposition modeling parameters on 3D-printed PEEK structures. *The International Journal of Advanced Manufacturing Technology*, *121*, 1675–1688. https://doi.org/10.1007/s00170-022-09360-4

[2] Curmi, A., & Rochman, A. (2023). Miniaturized fused granulate fabrication of polyether ether ketone (PEEK). *Progress in Additive Manufacturing*. https://doi.org/10.1007/s40964-023-00518-4

[3] Rendas, P., Figueiredo, L., & Cláudio, R., et al. (2023). Investigating the effects of printing temperatures and deposition on the compressive properties and density of 3D printed polyetheretherketone. *Progress in Additive Manufacturing*. https://doi.org/10.1007/s40964-023-00550-4

[4] Alami, A. H., Olabi, A. G., Alashkar, A., et al. (2023). Additive manufacturing in the aerospace and automotive industries: Recent trends and role in achieving sustainable development goals. *Ain Shams Engineering Journal*, *14*(11), 102516.

[5] Panayotov, I. V., Orti, V., Cuisinier, F., & Yachouh, J. (2016). Polyetheretherketone (PEEK) for medical applications. *Journal of Materials Science: Materials in Medicine*, *27*, 1–11.

[6] Li, Z., Lei, M., & Kou, Q., (Oct, 2023). A multiscale viscoelastic constitutive model of unidirectional carbon fiber reinforced PEEK over a wide temperature range. *Composite Structures*, *1*, 321, 117258. https://doi.org/10.1016/j.compstruct.117258

[7] Rae, P. J., Brown, E. N., & Orler, E. B. (2007). The mechanical properties of poly (ether-ether-ketone) (PEEK) with emphasis on the large compressive strain response. *Polymer*, *48*(2), 598–615.

[8] Tay, G. S., Nanbo, T., Hatakeyama, H., & Hatakeyama, T. (2011). Polyurethane composites derived from glycerol and molasses polyols filled with oil palm empty fruit bunches studied by TG and DMA. *Thermochimica Acta, 525*(1–2), 190–196.

[9] Yoganandam, K., Shanmugam, V., Vasudevan, A., Vinodh, D., Nagaprasad, N., Stalin, B., … & Bharani, M. (2021). Investigation of dynamic, mechanical, and thermal properties of calotropis procera Particle-Reinforced PLA biocomposites. *Advances in Materials Science and Engineering*, *2021*(1), 2491489. https://doi.org/10.1155/2021/2491489

[10] Samal, S. K., Mohanty, S., & Nayak, S. K. (2009). Polypropylene-bamboo/glass fiber hybrid composites: Fabrication and analysis of mechanical, morphological, thermal, and dynamic mechanical behavior. *Journal of Reinforced Plastics and Composites*, *28*, 2729–2747. https://doi.org/10.1177/0731684408093451

[11] Velmurugan, R., Balaganesan, G., Kakur, N., & Kanny, K. (Eds.). (2024). *Dynamic Behavior of Soft and Hard Materials, Volume 2: Proceedings of 13th International Symposium on Plasticity and Impact Mechanics 2022*. Springer Nature Singapore.Wickramasinghe, D. T. (2020). FDM-based 3D printing of polymer and associated composite: A review on mechanical properties, defects, and treatments. *Polymers*, *12*, 1529. https://doi.org/10.3390/polym12071529

[12] Shanmugam, V., Das, O., Babu, K., Marimuthu, U., Veerasimman, A., Johnson, D. J., … & Berto, F. (2021). Fatigue behaviour of FDM-3D printed polymers, polymeric composites and architected cellular materials. *International Journal of Fatigue*, *143*, 106007. https://doi.org/10.1016/J.IJFATIGUE.2020.106007

[13] Jiang, Z., Diggle, B., Tan, M. L., Viktorova, J., Bennett, C. W., & Connal, L. A. (2020). Extrusion 3D printing of polymeric materials with advanced properties. *Advanced Science*, *7*(17), 2001379.https://doi.org/10.1002/advs.202001379

[14] Das, A., Gilmer, E. L., Biria, S., & Bortner, M. J. (2021). Importance of polymer rheology on material extrusion additive manufacturing: correlating process physics to print properties. *ACS Applied Polymer Materials*, *3*(3), 1218–1249.Li, Q., Zhao, W., Li, Y., Yang, W., & Wang, G. (2019). Flexural properties and fracture behavior of CF/PEEK in orthogonal building orientation by FDM: microstructure and mechanism. *Polymers*, *11*(4), 656. https://doi.org/10.3390/polym11040656

[15] Chimakurthi, J., et al. (2020). Tensile properties of 3D-printed PEEK: Influence of layer orientation and annealing. *Materials Science & Engineering C*, *110*, 110734.

[16] Deng, H., et al. (2019). Comparison of SLS and FFF printed PEEK: Mechanical and microstructural analysis. *Journal of Biomedical Materials Research*, *108*(4), 765–778.

[17] Fu, J., et al. (2019). Flexural properties of 3D-printed PEEK composites: A study on interlayer adhesion. *Composites Science and Technology*, *172*, 60–68.

[18] Gibson, I., et al. (2015). *Additive manufacturing technologies* (2nd ed.). Springer Science & Business Media.

[19] Hollander, D. A., et al. (2006). PEEK bioactive coatings for improved osseointegration. *Acta Biomaterialia, 2*(3), 313–320.

[20] Kurtz, S. M., et al. (2019). *PEEK biomaterials handbook* (2nd ed.). Elsevier.

[21] Liu, X., et al. (2021). Optimization of 3D printing parameters for enhanced mechanical properties of PEEK. *Materials Today, Proceedings, 45*, 987–995.

[22] Ma, R., et al. (2018). Hydroxyapatite-coated PEEK: A study on mechanical and biological performance. *Journal of Biomedical Engineering, 35*(5), 455–468.

[23] Rinaldi, M., et al. (2022). Post-processing effects on tensile strength of 3D-printed PEEK. *Journal of Advanced Manufacturing*, *22*(1), 33–44.

[24] Schmidt, M., et al. (2007). Processing challenges in high-temperature laser sintering of PEEK. *Additive Manufacturing Research*, *5*(2), 112–119.

[25] Toth, J. M., et al. (2006). Clinical applications of PEEK in orthopedic surgery. *Journal of Biomedical Materials Research*, *76*(3), 567–575.

[26] Williams, D., et al. (2008). Biocompatibility of PEEK: A comprehensive review. *Acta Biomaterialia*, *4*(5), 456–478.

[27] Wu, C., et al. (2020). Carbon-fiber-reinforced PEEK for enhanced implant performance. *Journal of Composite Materials, 54*(7), 812–825.

[28] Yan, X., et al. (2021). Improving the flexural performance of 3D-printed PEEK using fiber reinforcements. *Composites Part B: Engineering, 195*, 108032.

[29] Zhang, Y., et al. (2021). Long-term fatigue analysis of PEEK implants: A clinical study. *Journal of Biomedical Engineering, 39*(4), 1021–1034.

45 Facial expression recognition using high order derivative local binary pattern

Durgadevi P.[a], *Roshan Suvedi*[b], *Sanjana M. M.*[c], *Rakesh G.*[d], *and Shivakumar M. J.*[e]

Department of Computer Science Engineering, Alliance University and Information Science, Presidency University, Bengaluru, India

Abstract: Facial expression recognition (FER) has gained significant importance due to its vast applications in fields such as human-computer interaction, healthcare, security systems, and affective computing. This paper proposes a hybrid approach for FER, combining High-Order Derivative Local Binary Pattern (HOD-LBP) for advanced feature extraction with powerful machine learning models, including Random Forest (RF), Convolutional Neural Networks (CNNs), and Support Vector Machines (SVMs). The utilization of HODLBP enables the extraction of fine-grained texture features, enhancing the model's ability to recognize subtle variations in facial expressions. To improve the performance of the system, we integrate data preprocessing techniques like normalization, augmentation, and employ Adam optimization for efficient learning. Our experimental results demonstrate that the proposed approach significantly outperforms traditional FER methods, delivering superior accuracy under challenging conditions such as lighting changes, facial occlusions, and the subtlety of expressions. This work paves the way for more reliable FER systems that can operate effectively in real-world environments.

Keywords: Binary pattern, facial expression, high order, optimization, random forest

1. Introduction

Computer vision with a wide array of applications across various fields such as psychological studies, virtual reality, security, and interactive gaming. FER systems are essential in understanding human emotions, which are key to creating more intuitive and empathetic human-computer interactions [1]. Despite the significant advancements made through machine learning and deep learning techniques, FER still faces several challenges, such as the recognition of subtle facial expressions, variations in illumination, occlusion of facial features, and the limited availability of labelled datasets for training.

In this paper, we propose an innovative FER approach that combines High-Order Derivative Local Binary Pattern (HOD-LBP) with three machine learning classifiers: Random Forest (RF), Convolutional Neural Networks (CNNs), and Support Vector Machines (SVMs). The introduction of HOD-LBP provides a more nuanced and accurate feature extraction method compared to traditional. LBP by capturing higher-order texture variations, which are critical for recognizing fine-grained facial expressions. Furthermore, we incorporate various data preprocessing techniques, including normalization and augmentation, along with the Adam optimizer, to improve the model's robustness and generalization capability.

1.1. Problem statement

FER remains a challenging task due to several factors:

- **Lighting Variability:** Facial expression recognition systems often fail to perform effectively under varying lighting conditions.
- **Occlusion:** Partial facial occlusion due to accessories (e.g., glasses, hats) or hair can significantly impact the system's ability to detect expressions.
- **Subtle Expressions:** Minor variations in facial expressions are hard to capture using conventional methods.
- **Limited Datasets:** Public datasets for facial expression recognition are often limited in size and diversity, hindering the training of robust models. These challenges necessitate the development of more accurate, robust, and adaptable FER systems.

---

[a]durgadevisweety@gmail.com, [b]roshansuvedi@gmail.com, [c]sanjanagowda179@gmail.com, [d]raki41249@gmail.com, [e]shivukumarshivu329@gmail.com

DOI: 10.1201/9781003770473-45

1.2. Objective

This study aims to investigate and evaluate the potential of High-Order Derivative Local Binary Pattern (HOD-LBP) for extracting finer texture details and improving facial expression recognition accuracy. The secondary goal is to combine this feature extraction technique with advanced machine learning models – Random Forest (RF), Convolutional Neural Networks (CNNs), and Support Vector Machines (SVMs) – to enhance classification performance, particularly under challenging real-world conditions.

2. Literature Review

Facial expression recognition has been extensively studied, and various approaches have been proposed for both feature extraction and classification.

- **Local Binary Pattern (LBP):** LBP has been widely used in FER due to its robustness to lighting variations and its ability to capture local texture patterns. However, traditional LBP struggles with capturing finer details and subtle expression variations, especially under occlusion or when expressions are less pronounced [2].
- **High-Order Derivative Local Binary Pattern (HOD-LBP):** To address the shortcomings of standard LBP, the HOD-LBP method was introduced. By incorporating higher-order derivatives, HOD-LBP captures more detailed changes in the texture, improving the system's ability to discern subtle facial expressions. Additionally, different machine learning classifiers have been employed for FER [3]:
- **Random Forest (RF):** RF has been used with handcrafted features such as LBP and HOD-LBP due to its ability to handle high-dimensional data and prevent overfitting. RF aggregates predictions from multiple decision trees, making it robust in various settings.
- **Convolutional Neural Networks (CNNs):** CNNs, a type of deep learning model, are particularly effective in learning hierarchical features from raw image data [9]. They automatically learn important spatial features from facial images, making them effective for recognizing complex expressions [10].
- **Support Vector Machines (SVMs):** SVM is known for its effectiveness in binary classification tasks but has been extended to multi-class classification in FER. SVM works well with small datasets, making it an attractive option when data availability is limited [4].
- These methods, when combined, have shown promising results in FER tasks, yet room for improvement remains, especially when dealing with subtle expression recognition and small dataset sizes [12].

3. Methodology

The methodology presented in this paper combines advanced feature extraction techniques with robust machine learning classifiers [5].

Below is a detailed description of the components:

3.1. High-order derivative local binary pattern (HODLBP)

Traditional LBP encodes local image textures by converting the pixel values of a neighbourhood into a binary number. HOD-LBP extends LBP by incorporating higher-order derivatives, which capture more complex texture variations that are critical for distinguishing between subtle facial expressions [6]. This method enhances the feature's discriminative ability by considering not just the pixel intensity but also the rate of change in intensity, thus making it more sensitive to fine-grained variations in facial expressions.

3.2. Data preprocessing

The raw input data undergoes several preprocessing steps to ensure uniformity and maximize model performance:

Normalization: Each pixel value is normalized to a range between 0 and 1. This ensures that all input data is treated equally, helping CNNs and other models to converge more quickly during training.

Data Augmentation: Given the limited size of publicly available facial expression datasets, augmentation techniques such as random rotations, scaling, flipping, and zooming are employed. This artificially increases the number of trainings samples and helps improve the model's ability to generalize to unseen data.

3.3. Classification algorithms

Three classifiers are used to evaluate the effectiveness of HOD-LBP-based features:

- **Random Forest (RF):** A robust ensemble method that aggregates multiple decision trees. RF is particularly suitable for handling complex, high dimensional data such as LBP and HOD-LBP histograms. It is known for reducing overfitting and providing feature importance scores.
- **Convolutional Neural Networks (CNNs):** CNNs automatically learn hierarchical spatial patterns from raw image data, excelling in tasks such as FER that require complex pattern recognition. Although they require substantial computational resources, CNNs are well-suited for large-scale, complex FER tasks [6].

- **Support Vector Machines (SVM):** SVMs are used for classification, as they are effective in creating decision boundaries between different classes. They are particularly useful in smaller datasets, where over fitting is a concern [11].

3.4. *Optimization and loss function*

- **Adam Optimizer:** The Adam optimizer is employed due to its adaptive learning rate mechanism [7], which allows it to perform well on large datasets and complex models like CNNs. Adam adjusts the learning rate based on the first and second moments of the gradient, enabling faster convergence.
- **Cross-Entropy Loss:** The cross-entropy loss function is ideal for multi-class classification problems, such as FER, as it penalizes incorrect predictions based on the predicted probability distribution [8].

4. Experiments and Results

4.1. *Dataset*

The FER-2013 and CK+ datasets were used for training and testing the model. These datasets consist of images depicting various facial expressions such as happiness, sadness, surprise, disgust, fear, and anger, with corresponding labels.

4.2. *Experimental setup*

- **Training:** The models were trained using both LBP and HOD-LBP feature sets. Various configurations of RF, CNN, and SVM classifiers were used to evaluate their performance.
- **Evaluation Metrics:** The performance of each model was assessed using standard metrics, including accuracy, precision, recall, F1-score, and confusion matrices as shown in Figure 45.1 and the Facial Expression Sample images are shown in Figures 45.2 and 45.3.

4.3. *Overview*

The Facial Expression Recognition (FER) dataset is a widely-used benchmark in emotion recognition research, originally created by Pierre-Luc Carrier and Aaron Courville for the 2013 Kaggle competition. The dataset consists of 48 × 48 pixel grayscale images of faces, each labelled with one of seven universal emotional expressions.

Dataset Specifications

Image Format: 48 × 48 pixel grayscale images

Total Images: 35,887 images

Input Shape: (48, 48, 1)

File Format: CSV file with pixel values and emotion labels

Emotion Classes

The dataset contains seven categorical emotions:

1. Angry (0)
2. Disgust (1)
3. Fear (2)
4. Happy (3)
5. Sad (4)
6. Surprise (5)
7. Neutral (6)

Dataset Distribution

Training Set: 28,709 images

Public Test Set: 3,589 images

Private Test Set: 3,589 images

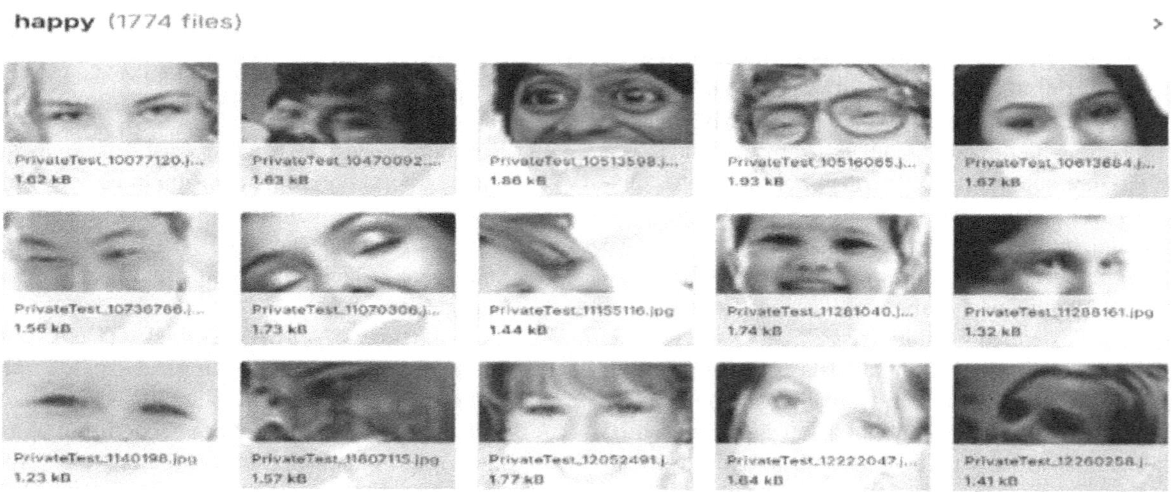

Figure 45.1. Facial expression recognition (FER) dataset description.

Source: Author.

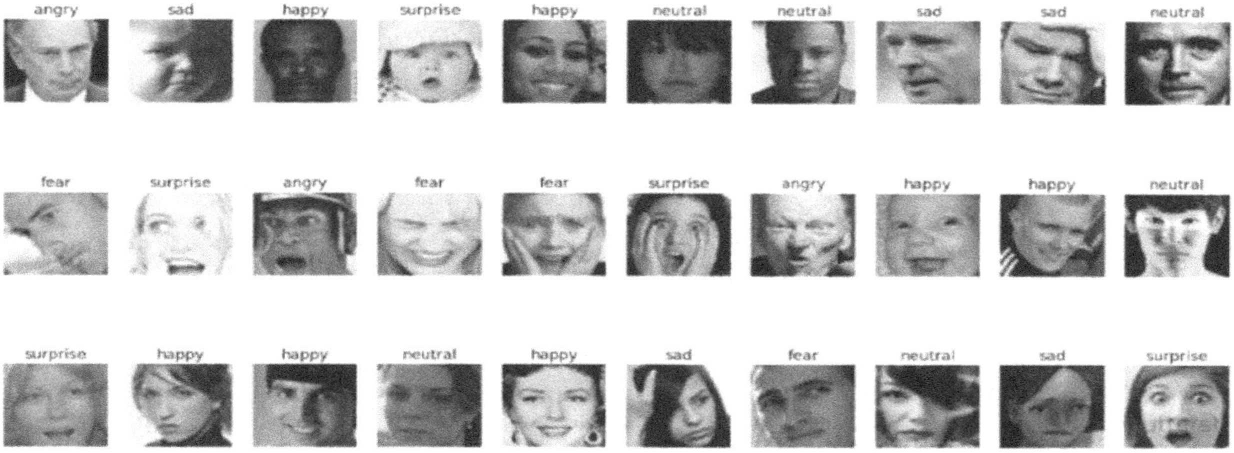

Figure 45.2. Facial expression sample images.

Source: Author.

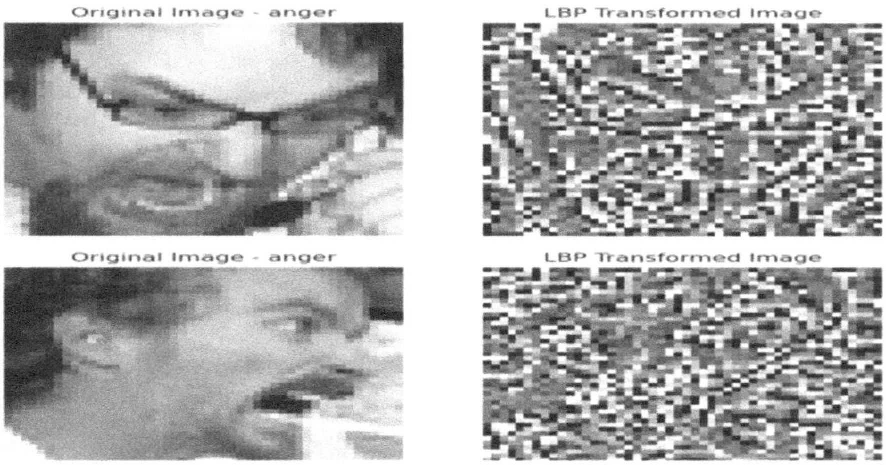

Figure 45.3. Facial expression sample images for angriness.

Source: Author.

4.4. *Transforming images into local binary pattern histogram (LBPH) representations*

This study leverages the Local Binary Pattern Histogram (LBPH) technique to preprocess and analyze grayscale images for texture-based classification. The dataset is organized into labelled categories, and images are transformed using the Local Binary Pattern (LBP) method, which captures local texture features effectively. A normalized histogram of LBP features is computed for each image, creating a compact and robust representation. The transformation process is visualized by comparing the original and LBP-transformed images, highlighting the enhanced textural details captured by LBP. The extracted features are stored in a CSV file, enabling efficient use in subsequent machine learning tasks. This method is computationally efficient and well suited for applications like facial recognition and texture analysis, providing a foundation for advanced research in image-based classification systems.

The analysis of texture patterns in the dataset was conducted using a sophisticated approach leveraging Python and specialized libraries as shown in the Figure 45.4 and 45.5. The scikit-image library was utilized for computing Local Binary Patterns (LBP), a powerful technique for texture analysis. This method captures local texture information by comparing each pixel with its neighbours, creating a binary pattern that is then converted into a decimal number. The resulting LBP features provide a compact representation of texture characteristics.

Matplotlib, a versatile plotting library in Python, was employed for visualization purposes. This allowed for the creation of clear and informative graphical representations

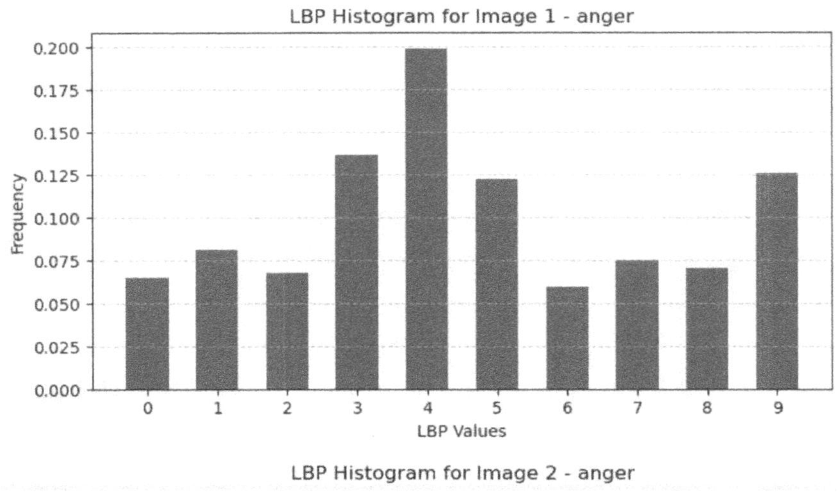

Figure 45.4. Analysis of texture patterns.

Source: Author.

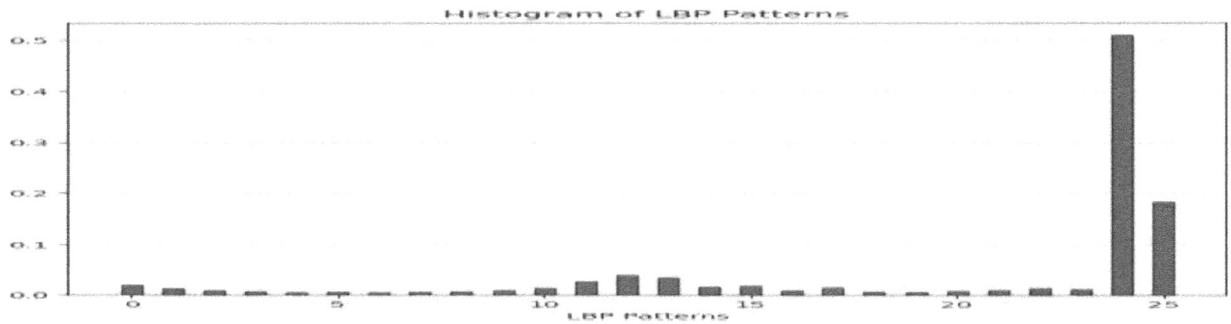

Figure 45.5. Texture pattern investigation.

Source: Author.

of the LBP features, facilitating easier interpretation of the texture patterns.

To ensure a comprehensive yet manageable analysis, a sampling strategy was adopted. Five representative images were carefully selected from the dataset, encompassing various image classes. This approach strikes a balance between providing a broad overview of texture patterns across different categories and maintaining computational efficiency.

The LBP features extracted from these images were then used to generate normalized histograms. Normalization is crucial as it allows for fair comparison between images, regardless of their size or the total number of pixels. These histograms effectively capture the distribution of local texture patterns within each image.

The significance of this visualization approach lies in its ability to provide deep insights into the texture characteristics of the dataset:

1. Quantitative comparison: The normalized histograms enable direct, numerical comparison of texture patterns across different images. This allows for the identification of similarities and differences in texture composition between various image classes or within the same class.
2. Identification of dominant features: By examining the peaks and valleys in the histograms, it becomes possible to identify the most prevalent texture patterns in each image or image class. This information can be valuable for understanding the key textural characteristics that define different categories within the dataset.
3. Visual verification: The histograms serve as a visual tool to verify the effectiveness of the LBP feature extraction process. By comparing the histograms of different images, it becomes apparent whether the LBP method is successfully capturing distinct texture patterns for different image types.
4. Texture composition understanding: The histograms provide a comprehensive view of the texture composition within the dataset. This can reveal insights about the nature of the images, such as the presence of smooth areas, rough textures, or repetitive patterns.

5. Feature extraction validation: By visualizing the LBP features through histograms, researchers can validate that the feature extraction process is working as intended and capturing meaningful texture information.
6. Dataset characterization: The analysis helps in characterizing the overall texture properties of the dataset.

This emotion recognition system involves two important innovative techniques aimed at performance enhancement and accuracy as shown in Figure 45.6. Feature extraction is involved through the application of the LSBP technique for effective facial texture pattern capturing, creating a sound foundation for classifying emotions.

In relation to the architecture, multiple-layer Convolutional Neural Networks are used. It integrates feature learning through the help of convolutional layers, applies batch normalization to stabilize its training, and employs max pooling to reduce its dimensional requirements. Furthermore, it applied dropout to prevent overfitting, thus improving generalization for the model.

The system was trained for over 20 epochs with rigorous training, reaching a maximum training accuracy of 94.25%. The maximum validation accuracy reached 51.64%, and the test accuracy was recorded at 45.97% as a final result. These results present the potential of the system but also indicate some aspects that need further optimization to improve its generalisation as shown in Figures 45.7 and 45.8.

The following figure 10 depicts the training and validation metrics of the emotion recognition system over 20

```
Model Accuracy: 56.92%
Classification Report:
              precision    recall  f1-score   support

       angry       0.44      0.45      0.45       958
     disgust       0.85      0.50      0.63       111
        fear       0.46      0.42      0.44      1024
       happy       0.69      0.78      0.73      1774
     neutral       0.52      0.52      0.52      1233
         sad       0.46      0.46      0.46      1247
    surprise       0.78      0.69      0.73       831

    accuracy                           0.57      7178
   macro avg       0.60      0.55      0.57      7178
weighted avg       0.57      0.57      0.57      7178

Confusion Matrix:
 [[ 434    4  103  132  118  153   14]
 [  25   56    9    7    2   11    1]
 [ 130    3  429   99  103  184   76]
 [  80    1   59 1380  118  105   31]
 [ 116    0  104  173  642  177   21]
 [ 152    2  144  154  202  573   20]
 [  42    0   76   68   39   34  572]]
```

Figure 45.6. Performance enhancement and accuracy.

Source: Author.

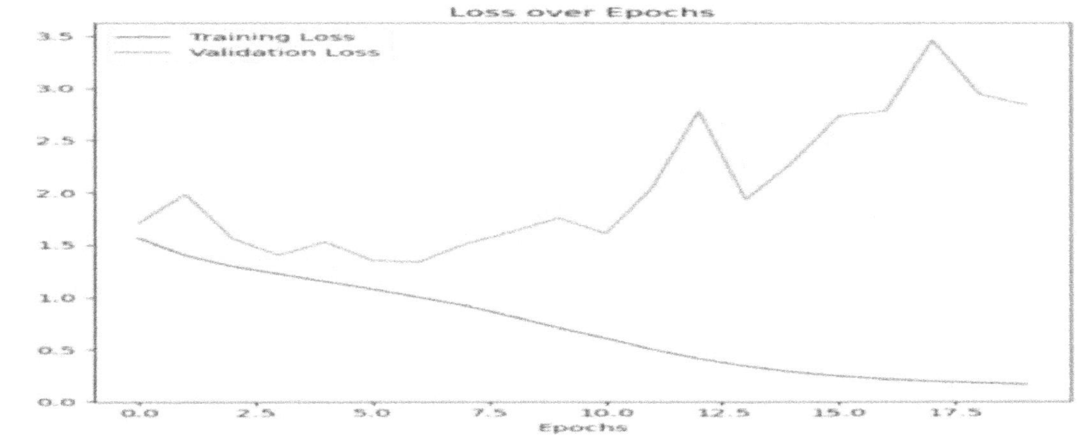

Figure 45.7. Loss over Epochs.

Source: Author.

epochs. It describes the performance progression of the model. The left graph shows the accuracy trends, and the right graph shows the corresponding loss metrics. In the accuracy plot, the training accuracy is increasing steadily and peaking at 94.25%, which means that the model has a good fit to the training data. However, the validation accuracy fluctuates, peaking at 51.64% and then showing signs of inconsistency. This is a sign of overfitting: the model performs well on the training data but fails to generalize to unseen data. The loss plot on the right gives a deeper insight into the learning behaviour of the model. The training loss steadily decreases, which corresponds to the optimization of the model's parameters during training. The validation loss shows an upward trend after initial fluctuations, further supporting the hypothesis of overfitting. These observations, therefore, suggest the importance of regularization, data augmentation, or a more refined architecture in generalizing well and decreasing overfitting so that generalization and test performance improve better validation.

The proposed system for emotion recognition employs a robust feature extraction method and a reliable classification model to achieve promising results. Feature extraction is carried out using the Local Sorting Binary Pattern (LSBP) technique, which captures intricate texture patterns in the facial images, making it a suitable choice for emotion classification tasks. The extracted features are classified using a Random Forest model with 100 trees, leveraging its ability to handle high-dimensional data effectively while maintaining interpretability. The dataset is split into 80% training and 20% validation to ensure a balanced evaluation of the model's performance. The system achieves an overall accuracy of 65.10%, indicating its effectiveness in recognizing emotions across classes. The best-performing class is Class 1, achieving an impressive precision of 92%, showcasing the model's ability to accurately identify emotions in this category. However, the system faces challenges with Class 3, where precision drops to 40%, highlighting areas for potential improvement through techniques like feature refinement or class-specific augmentation. These results emphasize the system's strengths while identifying opportunities for further enhancement.

In the accuracy plot training accuracy progresses steadily, peaking close to 0.94 or 94% by the final epochs of training. That is to say that it succeeded in learning from the data at hand. However, the accuracy for the validation case goes around approximately 0.5, or 50%, showing very slight improvements as compared to training accuracy, hence an overfitting effect: it has high performance for data that is at hand but can barely generalize on other new and unseen data. In the loss plot, the training loss is monotonic; it is consistently decreasing, which implies that the model's prediction on the training set gets better with time. In contrast, the validation loss is highly fluctuating and shows an overall upward trend after a point. It further indicates overfitting as the model does worse on the validation set even when it does better on the training set (Figure 45.9).

This emotion recognition system uses Local Binary Pattern for feature extraction and a Convolutional Neural Network for classification. LBP captures facial texture patterns, which are crucial for distinguishing emotion as in Figure 45.10. The CNN consists of convolutional, pooling, and fully connected layers, enabling the model to learn and classify facial expressions. More than 50 epochs, the model's overall training accuracy reached up to 88.98% with validation accuracy reaching its peak at 87.27% and finally declined to 80.88%, which indicates some degree of overfitting. Loss during training had decreased but validation loss had increased. This calls for further modifications such as data augmentation or regularization to reduce overfitting and enhance generalization Figure 45.11.

Accuracy: 0.6510291262135922
Classification Report:

	precision	recall	f1-score	support
0	0.65	0.66	0.65	1838
1	0.92	0.98	0.95	1849
2	0.67	0.67	0.67	1883
3	0.40	0.33	0.36	1835
4	0.57	0.56	0.56	1793
5	0.58	0.56	0.57	1854
6	0.69	0.80	0.74	1823
accuracy			0.65	12875
macro avg	0.64	0.65	0.64	12875
weighted avg	0.64	0.65	0.64	12875

Figure 45.8. Classification report.

Source: Author.

```
Epoch 43/50
180/180 [==============================] - 137s 762ms/step - loss: 0.3225 - accuracy: 0.8812 - val_loss: 0.4521 - val_accuracy: 0.8401
Epoch 44/50
180/180 [==============================] - 217s 1s/step - loss: 0.3229 - accuracy: 0.8851 - val_loss: 0.5050 - val_accuracy: 0.8275
Epoch 45/50
180/180 [==============================] - 137s 763ms/step - loss: 0.3323 - accuracy: 0.8801 - val_loss: 0.4657 - val_accuracy: 0.8463
Epoch 46/50
180/180 [==============================] - 137s 761ms/step - loss: 0.3295 - accuracy: 0.8796 - val_loss: 0.4857 - val_accuracy: 0.8268
Epoch 47/50
180/180 [==============================] - 143s 795ms/step - loss: 0.3066 - accuracy: 0.8908 - val_loss: 0.4025 - val_accuracy: 0.8602
Epoch 48/50
180/180 [==============================] - 157s 870ms/step - loss: 0.3237 - accuracy: 0.8853 - val_loss: 0.3729 - val_accuracy: 0.8727
Epoch 49/50
180/180 [==============================] - 139s 770ms/step - loss: 0.3074 - accuracy: 0.8868 - val_loss: 0.4041 - val_accuracy: 0.8484
Epoch 50/50
180/180 [==============================] - 140s 780ms/step - loss: 0.3020 - accuracy: 0.8898 - val_loss: 0.5610 - val_accuracy: 0.8088
```

Figure 45.9. Validation report.

Source: Author.

```
classifier.evaluate(test_set)

23/23 [==============================] - 6s 267ms/step - loss: 0.5610 - accuracy: 0.8088

[0.5609833598136902, 0.8087621927261353]
```

Figure 45.10. Feature extraction using CNN.

Source: Author.

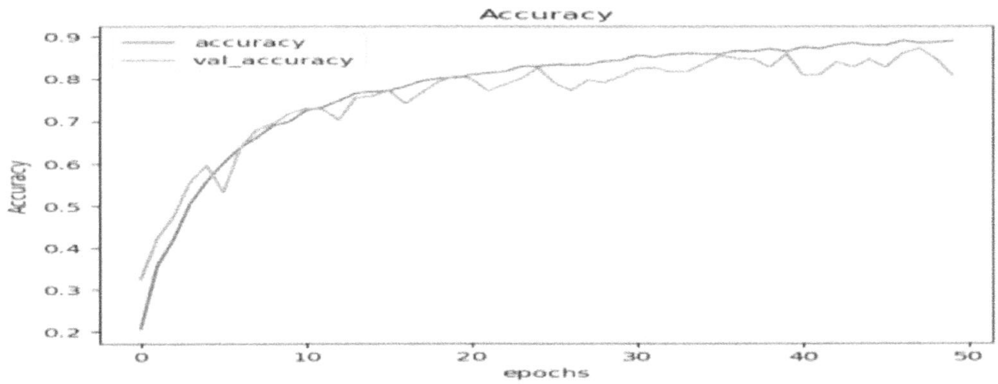

Figure 45.11. Overfitting and enhance generalization.

Source: Author.

This image presents the training and validation metrics of a machine learning model across 50 epochs, featuring two subplots: Accuracy Plot: Epochs are represented on the x-axis and accuracy is on the y-axis. Blue line: Train accuracy Orange line as in Figure 45.12: Val accuracy. Observations: Training accuracy shows a pattern of steady increments up to nearly 90%. Like the validation accuracy, this trend mirrors the peaking and troughs observed in performance on unseen data. Loss Plot: The x-axis represents epochs, and the y-axis consists of the loss value. So, the blue line represents training loss, and the orange line training-val validation loss. Observations: Train loss is kept persistently decreasing, showing good learning. Val loss oscillates around epochs with a rapidly increasing trend in the later epochs, showing signs of overfitting. These plots are useful for assessing the performance of

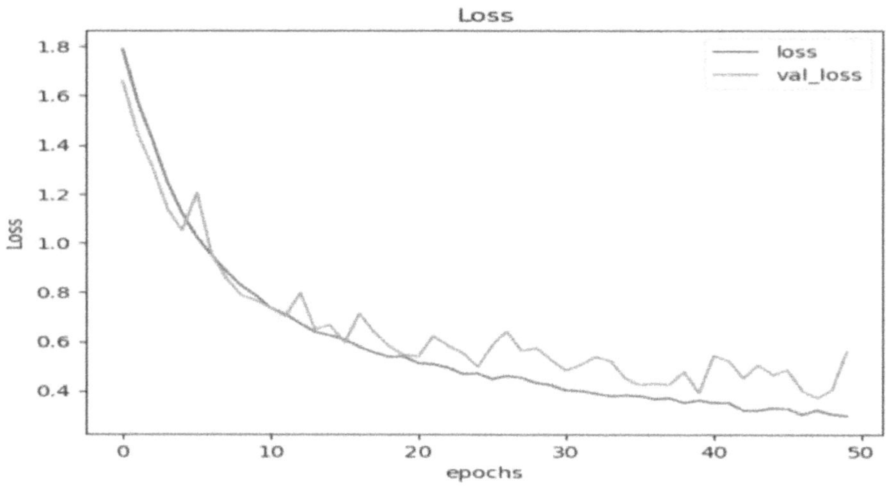

Figure 45.12. Accuracy plot: Loss vs Epochs.

Source: Author.

the model being trained. Increasing accuracy with growing validation loss suggests that the type of regularization or early stopping could be necessary to prevent overfitting and improve generalization

5. Results and Discussion

The results reveal that the combination of HOD-LBP with machine learning classifiers significantly improves the accuracy of facial expression recognition. Among the classifiers, Random Forest achieved the highest accuracy with handcrafted features, while CNN outperformed SVM in learning complex patterns from raw image data. The proposed approach addresses critical FER challenges, such as variations in lighting, subtle expression detection, and limited dataset sizes. By combining HOD-LBP with RF, CNN, and SVM classifiers, the system becomes more robust, especially under real-world conditions. While CNNs showed the best performance in terms of accuracy, their computational demands can be high. RF and SVM provide effective alternatives, especially in scenarios with limited data. The proposed system produces different emotions as shown in Figures 45.13–45.17.

This is a Python program that captures real-time video from a webcam, detecting human emotions and overlaying facial landmarks on detected faces. It uses OpenCV for image processing, Media pipe for facial landmark detection, and a pre-trained TensorFlow model for emotion classification.

Upon execution, the program initialized the webcam and captured consecutive frames. Each frame it captured was processed for finding faces using OpenCV's built-in Haar Cascade face classifier. For each detected face, the

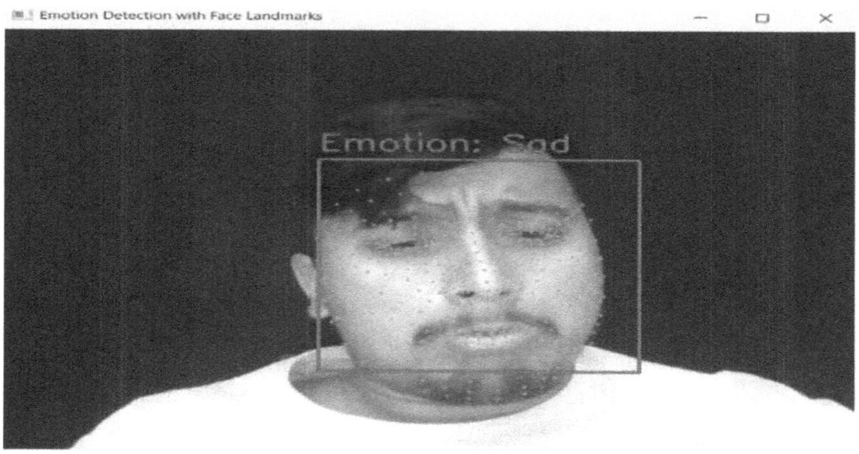

Figure 45.13. Happy.

Source: Author.

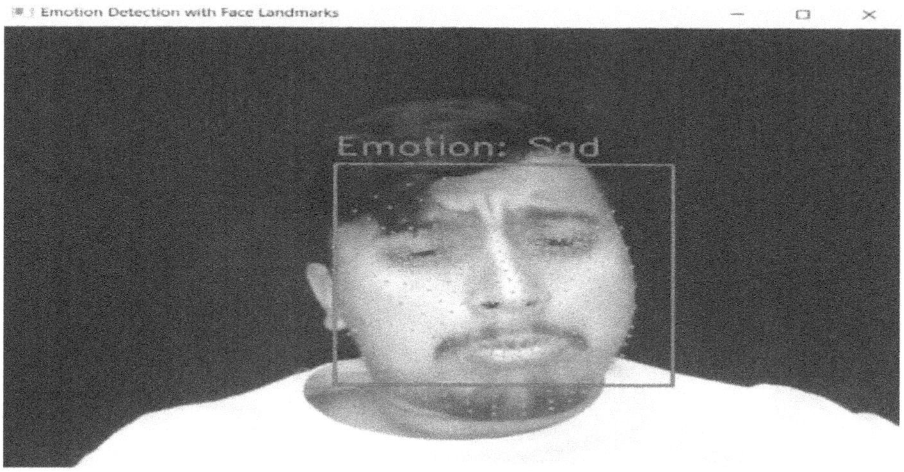

Figure 45.14. Sad.
Source: Author.

Figure 45.15. Surprise.
Source: Author.

Figure 45.16. Sad.
Source: Author.

Figure 45.17. Surprise.
Source: Author.

region of interest was acquired, converted to grayscale, resized to 48 × 48 pixels, and sent as input to the pre-trained TensorFlow model. The prediction of the emotion by the model categorized it into one out of seven states: Anger, Disgust, Fear, Happy, Sad, Surprise, or a Neutral state.

At the same time, the program utilizes Mediapipe's Face Mesh in order to identify facial landmarks, which it then projects onto the video feed. The resulting frame is represented by rectangles around identified faces with emotion labels placed above those and facial landmarks in green point form. This real-time output continues until the user enters 'q', which quits the application, releasing the webcam and close all windows.

6. Conclusion

This study demonstrates that integrating High-Order Derivative Local Binary Pattern (HOD-LBP) with advanced machine learning classifiers, including Random Forest, CNN, and SVM, enhances the accuracy and robustness of facial expression recognition systems. The proposed approach significantly improves FER in challenging environments, such as varying lighting and subtle expression nuances. Future work will focus on refining the model by exploring hybrid systems and incorporating additional facial features like gaze direction and head pose.

References

[1] Tabatabaei, S. M., Chalechale, A., & Moghimi, S. (2015, March). Facial expression recognition using high order directional derivative local binary patterns. In *2015 2nd International Conference on Pattern Recognition and Image Analysis (IPRIA)* (pp. 1–5). IEEE.

[2] Durgadevi, P., & Vijayalakshmi, S. (2021, March). A Methodological Investigation for fetal brain MRI Segmentation Techniques-Analysis. *In 2021 International Conference on Advance Computing and Innovative Technologies in Engineering (ICACITE)* (pp. 684–690). IEEE.

[3] Peng, Y., & Yin, H. (2018). Facial expression analysis and expression-invariant face recognition by manifold-based synthesis. *Machine Vision and Applications, 29*(2), 263–284.

[4] Lopes, A. T., De Aguiar, E., De Souza, A. F., & Oliveira-Santos, T. (2017). Facial expression recognition with convolutional neural networks: coping with few data and the training sample order. *Pattern Recognition, 61*, 610–628.

[5] Sharma, G., ul Hussain, S., & Jurie, F. (2012). Local higher-order statistics (LHS) for texture categorization and facial analysis. In *Computer Vision–ECCV 2012: 12th European Conference on Computer Vision, Florence, Italy, October 7–13, 2012, Proceedings, Part VII 12* (pp. 1–12). Springer Berlin Heidelberg.

[6] Durgadevi, P., & Vijayalakshmi, S. (2019). Deep Survey and Comparative Analysis of Medical Image Processing. *Journal of Computational and Theoretical Nanoscience, 17*(5), 2321–2329.

[7] Barra, S., Hossain, S., Pero, C., & Umer, S. (2022). A facial expression recognition approach for social IoT frameworks. *Big Data Research, 30*, 100353.

[8] Durgadevi, P., Vijayalakshmi, S., & Sabharwal, M. (2021). Fetal brain abnormality detection through PSO (particle swarm optimization) and volume estimation. *Annals of the Romanian Society for Cell Biology*, 2700–2714.

[9] Tuncer, T., Dogan, S., Abdar, M., & Pławiak, P. (2020). A novel facial image recognition method based on perceptual hash using quintet triple binary pattern. *Multimedia Tools and Applications, 79*(39), 29573–29593.

[10] Durgadevi, P., & Vijayalakshmi, S. (2021). Image Characterization Based Fetal Brain MRI Localization and Extraction. *Annals of the Romanian Society for Cell Biology,* 1935–1946.

[11] Fernandes, S. L., & Bala, G. J. (2017). A comparative study on various state of the art face recognition techniques under varying facial expressions. *International Arab Journal of Information Technology*, *14*(2), 254–259.

[12] Vijayalakshmi, S., & Durgadevi, P. S. N. P. (2016). A simple approach for relatively automated hippocampus segmentation from sagittal view of brain MRI. *International Journal on Bioinformatics & Biosciences (IJBB)*, *6*(2).

46 Classifying Parkinson's disease using deep learning models and MRI data

Benita Veronica[1,a], Febin Antony[2,b], Shaziya Islam[2,c], and Shahra Jafar[3,d]

[1]Program Coordinator and Senior Faculty (IT and Cyber Security), Curtin University Dubai, UAE
[2]Associate Faculty, Westford University College, Sharjah, UAE
[3]Head of Foundation Studies, Lecturer, Curtin University, Dubai, UAE

Abstract: Parkinson's disease is one of the neurodegenerative disorders that causes progressive damage. It affects millions of people globally. Early diagnosis and accurate reporting are essential to proper management and treatment. The article discusses using MRI data obtained from Kaggle and advanced deep learning models like Inception V3, Xception, MobileNet V2, and EfficientNet for detection of Parkinson's disease. We explore image filtering techniques such as Median, Gaussian, and Bilateral filters and data augmentation techniques to improve the performance of the model. In addition, we discuss the evaluation metrics of the model, which include confusion matrices and AUC graphs, to measure the efficiency of these models. Our research also shows the significance of preprocessing techniques in enhancing the accuracy of diagnosis and presents a detailed comparison of various deep learning architectures. The findings are therefore expected to assist in the design of more robust and effective diagnostic tools for Parkinson's disease, towards better patient outcomes and life quality.

Keywords: Deep learning, EfficientNet, Inception V3, MobileNet V2, MRI data, Parkinson's disease, Xception

1. Introduction

Parkinson's disease is a neurodegenerative disorder characterized by progressive loss of motor control, tremors, and other neurological symptoms. This disease has become a huge challenge for millions of people across the globe, for patients and health care providers alike. Early and accurate diagnosis can be crucial in managing and treating this disease since it can dramatically improve the quality of life of affected individuals. Recent advances in medical imaging and machine learning open new avenues in the diagnosis and classification of Parkinson's disease [1, 2]. MRI has proven to be an effective tool in capturing detailed images of the brain, providing invaluable insights into structural and functional changes associated with PD. However, the complexity and high dimensionality of MRI data require sophisticated analytical techniques to derive meaningful information.

Advanced deep learning models have achieved tremendous success in several applications concerning image analysis, including medical imaging. In this work, we examine the utilization of these state-of-the-art deep learning models: Inception V3, Xception, Mobile Net V2, and Efficient Net [3] on the task of Parkinson's disease classification using MRI data publicly available on Kaggle. These models, which are associated with strong architecture as well as a high accuracy rate, pose great potential for the automated detection of PD. To further boost the performance of these models, we use multiple methods for filtering images: Median, Gaussian, and Bilateral filters to reduce noise that decreases the image quality. Furthermore, several data augmentation techniques that include rescaling, rotation, width and height shifts, shear, zoom, and horizontal flip are applied to enhance diversity for preventing overfitting. Model evaluation is thus a crucial activity in this work, ensuring reliability and accuracy within the diagnostic tool. We'll use confusion matrices and AUC graphs to discuss the performance of the models involved, thereby achieving a complete presentation of their discriminatory ability in patients with PD from healthy controls [4, 5].

2. Literature Review

Recent studies highlight the potential of deep learning models in diagnosing Parkinson's disease (PD) using MRI data. Aldujaili et al. reported deep learning models achieving up to 98.

[a]Benita.Veronica@curtindubai.ac.ae, [b]febinantony@live.com. [c]shaziya.i@westford.org.uk, [d]shahra.jafar@curtindubai.ac.ae

DOI: 10.1201/9781003770473-46

Preprocessing techniques such as Median, Gaussian, and Bilateral filtering improve image quality by reducing noise, while augmentation methods like rotation and zooming enhance model generalization [4, 6, 7]. Evaluation metrics, including confusion matrices and AUC (Area Under the Curve) graphs, ensure model reliability by assessing classification accuracy and distinguishing PD from healthy subjects.

MRI remains the preferred imaging technique, outperforming CT and PET scans in early PD detection. Deep learning models, particularly CNN-based architectures [8, 9] improve diagnostic accuracy, while preprocessing and evaluation metrics play crucial roles in optimization. This study builds on prior research by integrating advanced deep learning techniques with effective preprocessing and evaluation strategies, contributing to more accurate and efficient PD diagnosis [6].

2.1. Problem Statement

The early and accurate diagnosis of Parkinson's disease (PD) is essential for timely intervention, and deep learning has shown promising results in this field. This study develops a deep learning framework using MRI images to detect PD with high precision, emphasizing the role of preprocessing techniques in model effectiveness [7] Advanced image filtering methods like Median, Gaussian, and Bilateral filtering help reduce noise while preserving critical features, and data augmentation techniques such as rotation, shearing, zooming, and flipping enhance training sample diversity to improve model generalization.

To achieve high diagnostic accuracy, this study explores state-of-the-art neural network architectures, including Inception V3, Xception, MobileNet V2, and Efficient Net, each offering unique advantages in feature extraction, computational efficiency, and accuracy. This research emphasizes the preprocessing techniques that can be used to improve the accuracy of diagnosis and offers a detailed comparison of different deep learning architectures. The findings will contribute to the development of more reliable and efficient diagnostic tools for Parkinson's disease, which in turn will improve patient outcomes also be analysed. Evaluation metrics such as confusion matrices and Area Under the Curve (AUC) graphs provide insights into model effectiveness. By integrating advanced image processing with deep learning, this research aims to improve automated PD diagnosis, contributing to better patient outcomes and clinical decision-making.

2.2. Research Gap

Despite significant progress in diagnosing PD using deep learning methods, further integral studies are needed that involve the combination of various preprocessing techniques followed by rigorous evaluation metrics.

3. Methodology

3.1. Data collection

MRI data was sourced from Kaggle, a popular platform for data science and machine learning competitions. This dataset, titled "Parkinson's Disease Data and MRI Scans," contains MRI scans from both Parkinson's disease-affected patients and healthy controls. These images are crucial for training deep learning models to differentiate between PD-affected and healthy brains. Images were categorized into separate folders based on labels, such as "normal" and "Parkinson" (Figure 46.1).

3.2. Image filtering

Image filtering plays a crucial role in preprocessing by enhancing image quality and reducing noise, which is

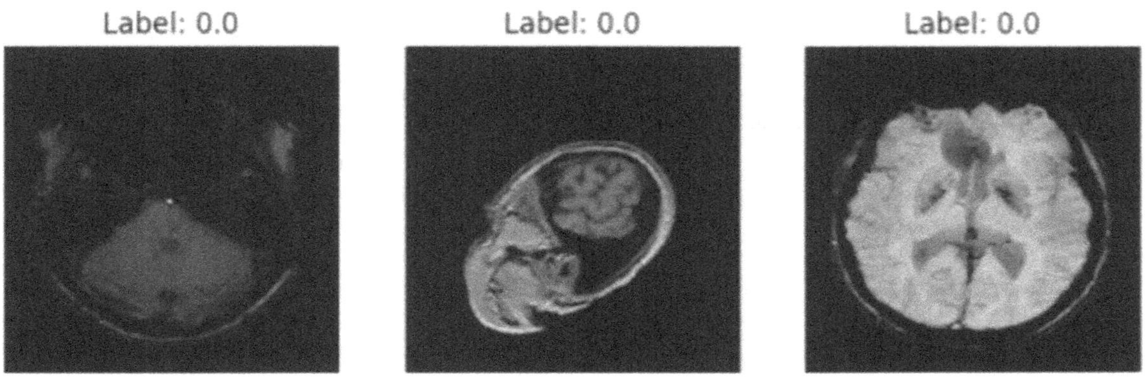

Figure 46.1. Sample Parkinson's disease MRI data.

Source: Author.

essential for improving deep learning model performance. Various filtering techniques were applied to refine MRI images. The Median Filter, a nonlinear filtering technique, effectively removes "salt and pepper" noise by replacing each pixel with the median of its surrounding intensities. This method is particularly useful in preserving edges while eliminating outliers and small artifacts. The Gaussian Filter, a linear smoothing filter, applies a Gaussian kernel to average pixel values, reducing high-frequency noise while maintaining structural details. Additionally, the Bilateral Filter, a nonlinear technique, enhances image quality by combining Gaussian filtering in both spatial and intensity domains. Unlike traditional filters, it preserves fine details while reducing noise without excessive blurring.

To further improve model generalization, various data augmentation techniques were employed to increase dataset diversity. Rescaling was performed to normalize pixel values to a standardized range of [0, 1], ensuring consistency across images. Rotation was applied randomly within a specified range to simulate variations in image orientation. Width and height shifts introduced horizontal and vertical translations, enhancing the model's ability to recognize features regardless of positional changes. Shearing transformations altered the image along the x or y axis, providing additional variations in geometric structure. Zooming operations allowed random magnification or reduction, ensuring robustness to scale changes. Horizontal flipping was also incorporated to account for asymmetry in MRI images. Finally, the dataset was split into training and validation subsets to facilitate model evaluation and performance optimization [10].

By integrating advanced image filtering methods and comprehensive data augmentation techniques, this study ensures higher quality input data, leading to more accurate and reliable deep learning based diagnosis of Parkinson's disease.

3.3. Model training

Several deep learning models were trained on the pre-processed MRI data to enhance the accuracy of Parkinson's disease diagnosis. Inception V3, a convolutional neural network (CNN) architecture, was utilized for its computational efficiency, leveraging inception modules to optimize feature extraction. Xception, an extension of Inception, was also employed, utilizing depth-wise separable convolutions to improve performance and reduce computational complexity. MobileNet V2, designed specifically for mobile applications, was chosen for its balance between accuracy and efficiency, making it a suitable candidate for lightweight deployment. Additionally, Efficient Net, a scalable model that optimizes width, depth, and resolution, was incorporated to achieve improved performance with minimal computational resources.

The training process involved several key steps to ensure effective learning. Initially, a pretrained model based on the ImageNet dataset was loaded, excluding the top classification layers. The convolutional base was then frozen, and custom layers were added to tailor the model specifically for MRI-based Parkinson's disease detection. The models were compiled using the Adam optimizer, known for its adaptive learning rate capabilities, and binary cross-entropy loss was employed to handle the binary classification task. Training was conducted using the pre-processed data, with performance evaluation carried out on a separate validation dataset. This structured approach ensured that the models learned essential features while preventing overfitting, ultimately enhancing the reliability of the diagnostic framework.

3.4. Model evaluation

Evaluating deep learning models is crucial for ensuring their reliability, particularly in medical applications like diagnosing Parkinson's disease from MRI images. Two key evaluation metrics, the confusion matrix and AUC (Area Under the Curve) graph, provide insights into model performance. A confusion matrix compares predicted and actual labels, breaking them into true positives (TP), false positives (FP), true negatives (TN), and false negatives (FN). From these values, key metrics such as accuracy, precision, recall, and F1 score are derived to assess the model's robustness.

The AUC graph, derived from the Receiver Operating Characteristic (ROC) curve, measures classification performance across various thresholds. It plots the True Positive Rate (TPR) against the False Positive Rate (FPR), with the AUC score indicating the model's ability to distinguish between Parkinson's patients and healthy individuals. A perfect model achieves an AUC of 1, while a score of 0.5 suggests no better performance than random guessing. To evaluate a model, predictions are generated, a confusion matrix is computed, and key metrics are calculated. The ROC curve is then plotted, and the AUC is determined to quantify the model's overall discriminative power.

4. Results

Image filtering techniques were employed to enhance image quality and reduce noise, thereby improving model performance in the medical image task. The applied filters include the Median Filter, which replaces each pixel with the median of its neighbours, computational efficiency and feature extraction. Xception extends Inception by incorporating boring values to effectively reduce" salt and pepper" rating depth-wise separable convolutions, leading noise. The Gaussian Filter smooths the image using a Gaussian function, reducing noise while preserving essential structural details. The Bilateral Filter applies Gaussian filtering in both spatial and intensity domains, effectively preserving edges while minimizing noise (Figure 46.2).

To improve dataset diversity and enhance model generalization, various data augmentation techniques were applied. Rescaling normalized pixel values to the range [0, 1], while random rotation introduced variations in orientation. Width and height shifts applied random horizontal and vertical translations to simulate different perspectives. Shear transformations skewed the image along the x or y axis, and zoom augmentation randomly adjusted the scale.

Additionally, horizontal flipping was performed to introduce left-right invariance. A validation split was implemented to partition the dataset into training and validation subsets, ensuring robust model evaluation (Figure 46.3).

Several state-of-the-art deep learning architectures were trained on the pre-processed MRI data. Inception V3 utilizes inception modules to enhance computational efficiency and improve future extraction. Exception extends inception by incorporating to enhance performance. MobileNet V2 is optimized for mobile and embedded applications, offering a balance between computational efficiency and accuracy.

To evaluate model performance, key metrics such as the confusion matrix and AUC (Area Under the Curve) were analysed [5, 9]. The confusion matrix provides insights into model predictions by comparing them against actual values, detailing the number of true positives, false positives, true negatives, and false negatives [7]. The AUC graph, derived from the Receiver Operating Characteristic (ROC) curve, measures the model's ability to distinguish between classes. A higher AUC value indicates superior classification performance, making it a crucial metric in assessing model reliability (Figures 46.4 and 46.5).

By integrating advanced image filtering and augmentation techniques with cutting-edge deep learning models, this study aims to enhance the accuracy and reliability of Parkinson's disease diagnosis using MRI data. The findings will contribute to the development of more effective

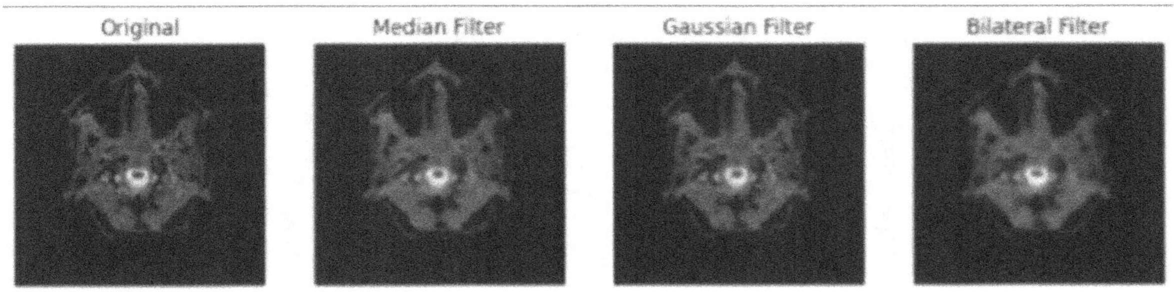

Figure 46.2. Filtering methods.

Source: Author.

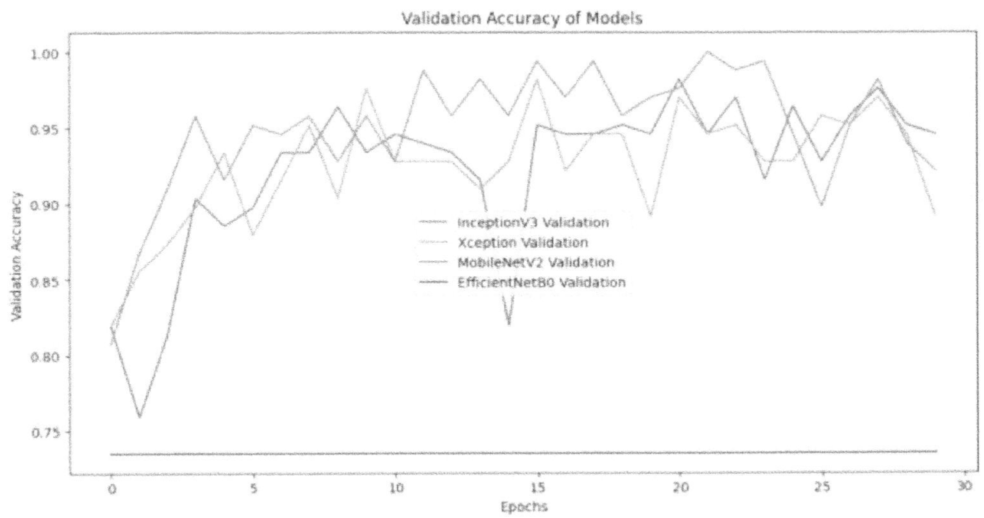

Figure 46.3. Validation accuracy of models.

Source: Author.

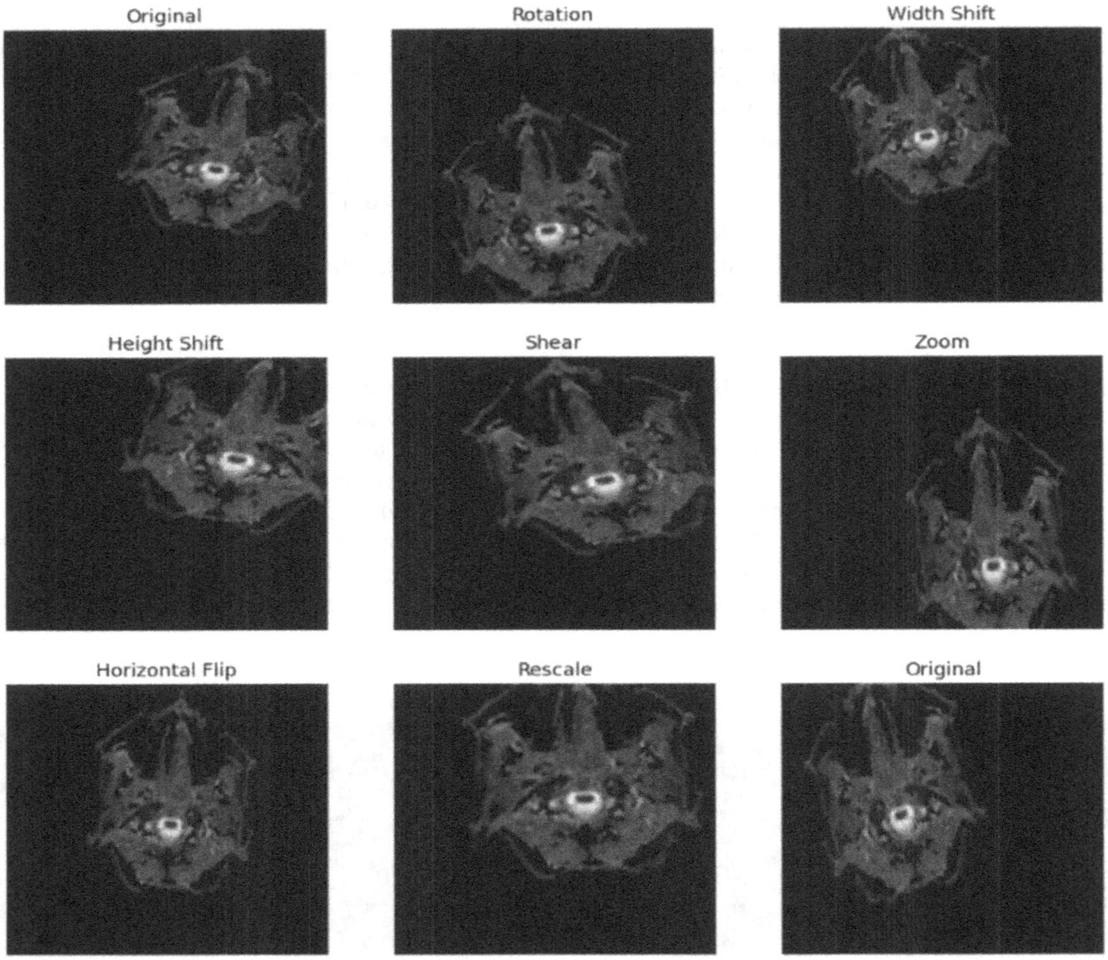

Figure 46.4. Augmentation methods.

Source: Author.

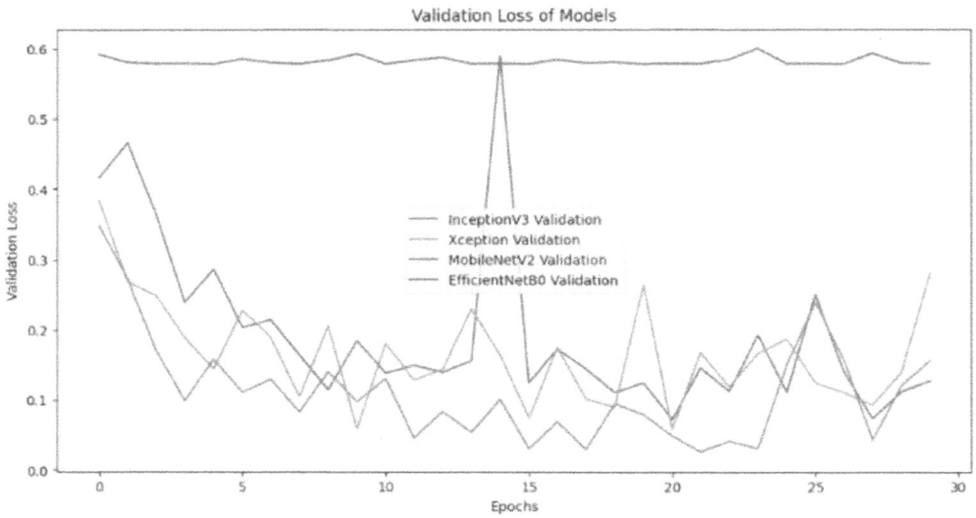

Figure 46.5. Validation loss of models.

Source: Author.

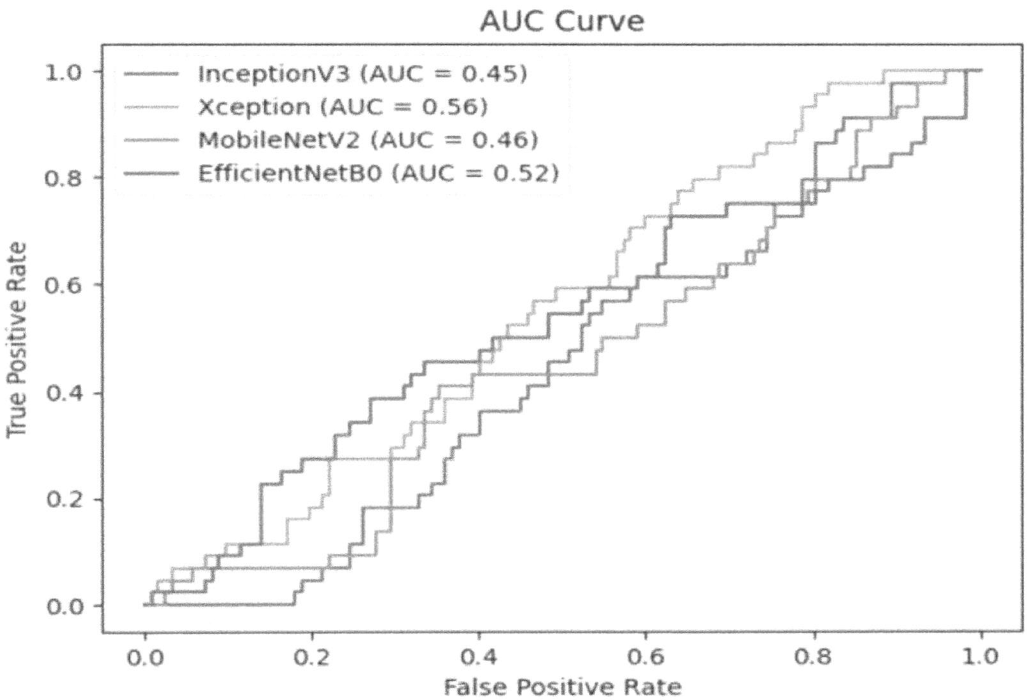

Figure 46.6. AUC curve.

Source: Author.

diagnostic tools, ultimately improving patient outcomes (Figure 46.6).

4. Conclusion

In conclusion, our study highlights the significant potential of advanced deep learning models in the early and accurate detection of Parkinson's disease using MRI data. Among the models evaluated, Inception V3 demonstrated the highest validation accuracy of 0.9458 with a validation loss of 0.1283, indicating its superior performance. Mobile Net V2 also performed well, achieving a validation accuracy of 0.9217 and a validation loss of 0.1572. Xception showed a respectable validation accuracy of 0.8916 but had a higher validation loss of 0.2802. Efficient Net, while still valuable, had the lowest validation accuracy of 0.7349 and the highest validation loss of 0.5785 of the art performance.

The use of image filtering methods such as Median, Gaussian, and Bilateral filters, along with data augmentation techniques, played a crucial role in enhancing model performance. The evaluation metrics, including confusion matrices and AUC graphs, provided a comprehensive assessment of each model's effectiveness.

Our findings underscore the importance of pre-processing techniques in improving diagnostic accuracy. The comparative analysis of different deep learning architectures offers valuable insights for future research and development. This study contributes to the advancement of more reliable and efficient diagnostic tools for Parkinson's disease, ultimately aiming to improve patient outcomes and quality of life.

Acknowledgment

We gratefully acknowledge the students, staff for their cooperation in the research.

References

[1] Acharya, H., Bouchard, T., Emery, D., & Camicioli, R. (2007). Axial signs and magnetic resonance imaging correlate in Parkinson's disease. *Canadian Journal of Neurological Sciences, 34*, 56–61.

[2] Adeli, E., Shi, F., An, L., et al. (2016). Joint feature selection and robust diagnosis of Parkinson's disease from MRI data. *Neuroimage, 141*, 206–219.

[3] Adeli, E., Wu, G., Saghafi, B., An, L., Shi, F., & Shen, D. (2017). Kernel-based joint feature selection and max-margin classification for early diagnosis of Parkinson's disease. *Scientific Reports, 7*, 1–14.

[4] Aishwarya, T., & Kumar, V. R. (2021). Machine learning and deep learning approaches to analyse and detect COVID-19: A review. *SN Computer Science, 2*, 1–9.

[5] Amoroso, N., la Rocca, M., Monaco, A., Bellotti, R., & Tangaro, S. (2018). Complex networks reveal early MRI markers of Parkinson's disease. *Medical Image Analysis, 48*, 12–24.

[6] Cigdem, O., Yilmaz, A., Beheshti, I., & Demirel, H. (2018). Comparing the performances of PDF and PCA on Parkinson's disease classification using structural MRI images. In *26th IEEE Signal Processing and Communications Applications Conference, SIU* (pp. 1–4).

[7] Desikan, R., Segonne, F., Fischl, B., et al. (2006). An automated labelling system for subdividing the human cerebral cortex on MRI scans into gyral-based regions of interest. *Neuroimage, 31*, 968–980.

[8] Baecker, L., Garcia Dias, R., Vieira, S., Scarpazza, C., & Mechelli, A. (2021). Machine learning for brain age prediction: Introduction to methods and clinical applications. *Biomedicine, 72*, Article 103600.

[9] Boelmans, K., Holst, B., Hackius, M., Finster busch, J., Gerloff, C., Fiehler, J., & Munchau, A. (2012). Brain iron deposition fingerprints in Parkinson's disease and progressive supranuclear palsy. *Movement Disorders, 27*, 421–427.

[10] Avants, B., Tustison, N., Song, G., Cook, P., Klein, A., & Gee, J. (2011). A reproducible evaluation of ants similarity metric performance in brain image registration. *Neuroimage, 54*, 2033.

For Product Safety Concerns and Information please contact our EU
representative GPSR@taylorandfrancis.com
Taylor & Francis Verlag GmbH, Kaufingerstraße 24, 80331 München, Germany

www.ingramcontent.com/pod-product-compliance
Lightning Source LLC
LaVergne TN
LVHW080805011026
849160LV00015B/1535